The Collected Works of George E. P. Box

Volume II

The Wadsworth Statistics/Probability Series

Series Editors

Peter J. Bickel, University of California, Berkeley
William S. Cleveland, AT&T Bell Laboratories
Richard M. Dudley, Massachusetts Institute of Technology

Richard A. Becker and John M. Chambers
S: An Interactive Environment for Data Analysis and Graphics

Peter J. Bickel, Kjell Doksum, and John L. Hodges, Jr.
Festschrift for Erich L. Lehmann

George E. P. Box
The Collected Works of George E. P. Box
Volumes I and II, edited by George C. Tiao

Leo Breiman, Jerome H. Friedman, Richard A. Olshen, and Charles J. Stone
Classification and Regression Trees

John M. Chambers, William S. Cleveland, Beat Kleiner, and Paul A. Tukey
Graphical Methods for Data Analysis

Franklin A. Graybill
Matrices with Applications in Statistics, Second Edition

John W. Tukey
The Collected Works of John W. Tukey
Volume I: *Time Series, 1949–1964,* edited by David R. Brillinger
Volume II: *Time Series, 1965–1984,* edited by David R. Brillinger

The Collected Works of George E. P. Box

Volume II

Editor in Chief

George C. Tiao
University of Chicago

Editors

C. W. J. Granger
University of California, San Diego

Irwin Guttman
University of Toronto

Barry H. Margolin
National Institute of Environmental Health Sciences

Ronald D. Snee
E. I. DuPont de Nemours & Co.

Stephen M. Stigler
University of Chicago

Wadsworth Advanced Books & Software
Belmont, California
A Division of Wadsworth, Inc.

Acquisitions editor: John Kimmel
Production editor: Mary Roybal
Cover and jacket: Lois Stanfield

Printed in the United States of America

1 2 3 4 5 6 7 8 9 10 — 89 88 87 86 85

Acknowledgments

Article 5.7 is reprinted with permission from ACADEMIC PRESS. Articles 5.9, 5.10 are reprinted with the permission of the AIR POLLUTION CONTROL ASSOCIATION. Article 4.12 is reprinted with the permission of ALMQVIST AND WIKSELL INTERNATIONAL. Articles 3.5, 3.9, 3.10, 3.15, 4.13 from the *Journal of the American Statistical Association* and articles 3.7, 4.7, 4.8, 4.10, 4.11 from *Technometrics* are reprinted with permission from the AMERICAN STATISTICAL ASSOCIATION. Article 5.3 is reprinted with permission from the BIOMETRIC SOCIETY. Articles 3.2, 3.12, 3.13, 3.14, 4.1 are reprinted with permission from the BIOMETRIKA TRUST. Articles 4.2, 4.3 are reprinted with the permission of the INSTITUTE OF MATHEMATICAL STATISTICS. Article 4.4 is reprinted with the permission of the INSTITUTION OF ELECTRICAL ENGINEERS. Article 4.5 is reprinted with the permission of the INTERNATIONAL INSTITUTE OF

(continued on page 710)

Library of Congress Cataloging in Publication Data

Box, George E. P.
The collected works of George E. P. Box.

(The Wadsworth statistics/probability series)
Includes bibliographies.
1. Mathematical statistics—Collected works. I. Tiao, George C., 1933- . II. Stigler, Stephen M. III. Title. IV. Series.
QA276.A12B6825 1984 519.5 84-17310
ISBN 0-534-03307-5 (v. 1)
ISBN 0-534-03308-3 (v. 2)

ISBN 0-534-03308-3

Contents

Volume II

Preface

The writings of George E. P. Box over the last thirty-five years have had a major impact on statistical theory and practice. His contributions cover a wide spectrum of topics, including statistical inference, robustness, time series analysis, design of experiments, response surface methodology, and evolutionary operation. He is a true believer in statistical methods as necessary tools in scientific investigation. To him, development of sound statistical theory and methods must go hand in hand with practice, and a statistician should be a genuine partner in scientific investigation, in the course of which his tools are judiciously applied and are modified and expanded as needs arise.

George Box's philosophy on statistics is amply reflected in the varied contributions he has made to the profession. These contributions include his books and articles on several major areas of statistics, some of which he himself initiated; the founding of a statistics department noted for its healthy blending of theory and practice in teaching and research; and his work as a consultant to industry, which provides a source, as well as an outlet, for his tools.

The collected works of Box presented in these volumes cover a major portion of his articles. The papers are divided into the following five main parts:

1. Statistical Inference, Robustness, and Modeling Strategy
2. Experimental Design and Response Surface Methodology
3. Time Series Analysis and Forecasting
4. Distribution Theory, Transformation of Variables, and Nonlinear Estimation
5. Application of Statistics

For each area, a member of the board of editors (Granger, Guttman, Margolin, Snee, Stigler, and Tiao) has selected articles as well as written an introduction to the articles highlighting their basic ideas, major methodological developments, and impact. A complete bibliography of the books and articles written by Box as of 1983 is given at the end of each volume.

On a personal note, as one of Box's students some twenty years ago and a long-time colleague, I am more pleased than I can say to see the publication of this collection. It represents a token tribute to his accomplishments, but, more important, it provides a perspective on the origins and development of his ideas. I wish to thank the other five editors for the time they have generously contributed to make publication possible, and David M. Steinberg and Chung Chen for their assistance in the preparation of the work. Finally, thanks are owed to many journals for allowing these articles to be reprinted.

George C. Tiao
University of Chicago

January 1984

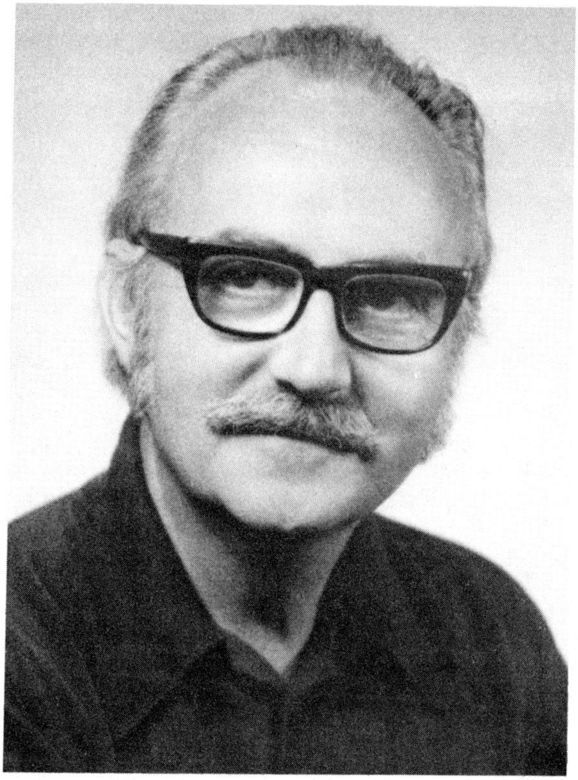

George E. P. Box

Biography

George E.P. Box was born October 18, 1919, in Gravesend, England, and did undergraduate work in chemistry at the University of London. While serving in the British Army during World War II, he was involved in experimental work to combat the effects of chemical weapons that required the use of statistics and the design of experiments. His formal education in statistics resulted in a Bachelor of Science degree in Mathematical Statistics in 1947 and a Doctor of Philosophy in Mathematical Statistics in 1952, both from the University of London. He also received a Doctor of Science degree from the University of London in 1961 and a honorary Doctor of Science degree from the University of Rochester in 1975.

Box began his career in statistics as Statistician and Head of the Statistical Techniques Research Section at Imperial Chemical Industries in England in the early- and mid-1950s. He took a leave of absence from Imperial Chemical Industries from 1953 to 1954 to accept a position as visiting Research Professor at the University of North Carolina and later returned to the United States as Director of the Statistical Techniques Research Group at Princeton University in 1957. He moved to the University of Wisconsin, Madison, in 1960, where he was founder, Professor, and Chairman of the Statistics Department until 1965. In 1965–1966, Box was a visiting Ford Foundation Professor at Harvard Business School. He returned to the University of Wisconsin as Professor and Chairman of the Statistics Department in 1966, leaving again in 1970 to become visiting Professor at the University of Essex in Colchester, England. In 1971, he was appointed to the newly created Ronald Aylmer Fisher Chair of Statistics at the University of Wisconsin, Madison, where he was appointed Vilas Research Professor of Mathematics and Statistics in 1980.

Dr. Box has published more than 110 research papers and is a coauthor of six books. The importance of his work has been recognized by the bestowal of many major awards, among them the British Empire Medal, the Royal Statistical Society Guy Medal in Silver, the Smith Reynolds Teaching Award from the University of Wisconsin, the Wilks Memorial Medal (ASA and U.S. Army), the American Society for Quality Control Shewhart Medal, and the American Institute of Chemical Engineers Professional Progress Award. He is a Fellow of the Institute of Mathematical Statistics, the American Academy of Arts and Sciences, the American Statistical Association, the Royal Statistical Society, and the American Society for Quality Control, as well as a member of the International Statistical Institute and the Biometrics Society. He has also served as president of the American Statistical Association and the Institute of Mathematical Statistics.

Few persons have had a greater impact on the use of statistics in science and engineering. Dr. Box is the principal architect of response surface methods and evolutionary operation and has made major contributions to experimental design, time series analysis, inference, nonlinear estimation, and robustness. In each instance, his contributions were stimulated by practical problems. He has served as a consultant to American Cyanamid, Monsanto Company, the World Bank, Pillsbury Company, and the Federal Reserve Board.

Dr. Box is also known worldwide for his pioneering work with the late Dr. Gwilym Jenkins on time series analysis. The Box-Jenkins approach, as it is fondly referred to, has been found useful in the modeling and control of industrial processes, economic forecasting, inventory management, and the analysis of environmental data. In addition to his professional activities, Dr. Box has many other interests. He is also known for his story telling, song writing, acting, poetry reading, and his interest in politics.

3

Time Series Analysis and Forecasting

Contents

3.0
Introduction

C. W. J. GRANGER
University of California, San Diego

The appearance in 1970 of the book *Time Series Analysis, Forecasting and Control* by George Box and Gwilym Jenkins had a widespread, immediate, and dramatic effect on the modeling and forecasting of time series. Before this publication, the forecasting field had been dominated by low-cost, adaptive models with no claim to optimality, autoregressive models of guessed order, or simple regression models whose specification depended more on the assumed usefulness of some underlying theory than on consideration of the temporal properties of the data. This landmark book emphasized ARIMA models and was based on a clear-cut and fairly easily used modeling strategy with the now-familiar three stages of identification, estimation, and diagnostic tests. The origins of this approach can be found in the 1962 paper with Jenkins (3.1), "Some Statistical Aspects of Adaptive Optimization and Control." In this paper, the forecasting models were developed merely to allow improved control, whereas later they had an important existence in their own right. This importance was already apparent in the 1968 paper, also with Jenkins (3.4), "Some Recent Advances in Forecasting and Control."

The basic model considered here is the ARIMA (p, d, q) for a series w_t of the form

$$\phi_p(B)(1-B)^d w_t = \theta_q(B)a_t \qquad (1)$$

where $\phi_p(B)$ and $\theta_q(B)$ are polynomials in the lag operator B of orders p and q, respectively, and a_t is white noise; d is the number of times w_t needs to be differenced before an ARIMA model is possible, so that w_t is said to be integrated of order d. Emphasis is on identification, that is, the choice of p, q, and d, and on parsimony, which is a preference for models with as few parameters as necessary to get an acceptable fit. This 1968 paper is an easy-to-read summary of the modeling strategy available at that time, including the seasonal models.

One of the more controversial aspects of the "Box-Jenkins procedures" was the frequent use of the differencing operator. Although this idea had been discussed for many years, it had rarely been explained within a convincing framework. The dangers of data analysis without differencing when necessary were explained in a most convincing manner in 3.6, "Some Comments on a Paper of Coen, Gomme and Kendall," coauthored with P. Newbold in 1971. Here it is shown that some apparently useful models for forecasting the stock market in fact had little empirical, or theoretical, support.

Since most economic data contains an important seasonal component, the basic model (1) was generalized to deal with this aspect in 3.3, "Models for Forecasting Seasonal and Non-Seasonal Time Series," in 1967 with Jenkins and Bacon. The model becomes, for a series z_t,

$$\phi_p(B^s)\,(1-B^s)\,\phi_p\,(B)\,(1-B)\,z_t \;=\; \theta_q\,(B^s)\,\theta_q\,(B)\,e_t \tag{2}$$

where e_t is white noise, $\Phi_p(B^s)$ is a polynomial of order P in the seasonal lag B^s, and so on. This paper is a particularly successful summary of the theory underlying the previously mentioned book. The book also considered a transfer function model in which one has one-way causation. This approach was generalized to the two-way causal, or feedback, cases in 3.7, "The Analysis of Closed-Loop Dynamic-Stochastic Systems," in 1974 with MacGregor, and 3.10, "Identification of Dynamic Regression (Distributed Lag) Models Connecting Two Time Series" with Haugh in 1977. Some aspects of this work are placed within various realistic applications in 3.8, "Some Recent Advances in Forecasting and Control, Part II," with Jenkins and MacGregor in 1974. In these papers, the three stages of analysis — i.e., identification, fitting, and diagnostic checking — are now applied to the more general class of bivariate ARIMA models.

The diagnostic tests used for the modeling procedures often look at the autocorrelations of residuals from estimated models. The 1970 paper with Pierce (3.5), "Distribution of Residual Autocorrelations in ARIMA Time Series Models," considered the statistical properties of such residuals and proposed the average of the first k-squared autocorrelations as a useful diagnostic statistic. In 1978 Ljung and Box produced a better approximation for the distribution of this widely used statistic in 3.13, "On a Measure of Lack of Fit in Time Series Models."

A contribution to the estimation question was made in 3.14, "The Likelihood Function of Stationary ARIMA Models," also with Ljung in 1979. The function is examined in some detail, and a method for its evaluation is preserved together with illustrations of its computational efficiency.

The most obvious direction for development is to multivariate models. The 1977 paper with Tiao (3.12), "A Canonical Analysis of Multiple Time Series," considers linear transformations of a vector of series into an equal number of series, ordered in level of predictability. The technique is applied to a group of agricultural series. A further major advance is presented in 3.15, "Modeling Multiple Time Series with Applications," in 1981 with Tiao. Here a tentative identification method is proposed using both auto- and cross-correlations, followed by the usual estimation and diagnostic checks. By now, the full-scale vector ARIMA models are in use. A more modern approach to modeling seasonal data, including multivariate cases and a discussion of seasonal adjustment, is 3.11, "Analysis and Modeling of Seasonal Time Series," published in 1976 with Hillmer and Tiao.

With the papers considered so far, there is a fairly clear sequence of development starting from a specific class of models. The ARIMA models that are good enough to provide a close approximation to the generating processes for univariate series are still capable of analysis. A number of generalizations are then considered, introducing a single causal series, bivariate feedback models, and then multivariate situations. As problems arise, such as with diagnostic tests or estimations, solutions are proposed. the final two technical papers included in this section do not quite fit into this sequence. ''A Change in Level of a Non-Stationary Time Series,'' published in 1965 with Tiao (3.2), considers a particular IMA (1,1) model and, though but an early contribution to the adaptive forecasting literature, is considerably more sophisticated statistically than the other papers in this field appearing at the same time. The paper ''Intervention Analysis with Applications to Economic and Environmental Problems,'' published in 1975 with Tiao (3.9), introduced an important simple and very realistic type of non-stationarity. A generating process is disturbed by a change in level, in the form of a single shock, that has both an immediate and a delayed effect. The model arose from consideration of the effects of the introduction of a new automobile pollution law in Los Angeles. A dynamic intervention component has been added to the usual ARIMA model. The situation is one that previously introduced models would have difficulty handling; in the real world, such rule changes frequently occur.

The final paper (3.16), ''Gwilym Jenkins, Experimental Design and the Time Series'' (1983), honors an excellent research worker and very pleasant colleague, Gwilym Jenkins. It summarizes much of the work done by Jenkins and Box and provides some very interesting historical insights into the development of their work. It is also a moving and pointed memorial to a most courageous man, whose early death saddened so many of us in the time series field.

This group of papers represents a substantial and highly influential body of research conducted over two decades. The Box-Jenkins modeling strategy and the models proposed are standard methods in many disciplines, but particularly in econometrics. Forecasts based on these methods, which are data-analysis intensive, are usually very difficult to beat by alternative methods, even when time-varying parameters, nonlinearities, or structural constraints are introduced. What makes this work most remarkable is that it represents only one of George Box's several interests.

3.1
Some Statistical Aspects of Adaptive Optimization and Control

G. E. P. BOX *University of Wisconsin*

G. M. JENKINS *Imperial College, London*

[Read at a RESEARCH METHODS MEETING of the SOCIETY, April 4th, 1962, Professor D. R. Cox in the Chair]

SUMMARY

It is often necessary to adjust some variable X, such as the concentration of consecutive batches of a product, to keep X close to a specified target value. A second more complicated problem occurs when the independent variables X in a response function $\eta(X)$ are to be adjusted so that the derivatives $\partial\eta/\partial X$ are kept close to a target value zero, thus maximizing or minimizing the response. These are shown to be problems of prediction, essentially, and the paper is devoted mainly to the estimation from past data of the "best" adjustments to be applied in the first problem.

1. INTRODUCTION

1.1. *Experimentation as an Adaptive Process*

OUR interest in adaptive systems arises principally because we believe that experimentation itself is an adaptive system. This iterative or adaptive nature of experimentation, as characterized by the closed loop, *conjecture–design–experiment–analysis* has been discussed previously by Box (1957).

In some circumstances, the probabilities associated with particular occurrences which are part of this iterative process may be of little direct interest, as may be the rightness or wrongness of particular "decisions" which are made along the way. For instance, on one admittedly idealized model of an evolutionary operation programme applied during the finite life of a chemical process, Box has shown in an unpublished report that if a virtually limitless number of modifications are available for trial, then to obtain maximum profit, each modification should be tested once and once only no matter what the standard deviation of the test procedure is.

Results of this kind are important but for the most part are not generally applicable to the experimental process itself. To see why this is so, we must distinguish between empirical and technical feedback.

1.2. *Empirical and Technical Feedback*

Empirical feedback occurs when we can give a simple rule which describes unequivocally what action should be taken and what new experiments should be done in every conceivable situation. In the above example the rule is: If after testing a modification once, it appears to give an improvement, then include it in the process and start to test another modification. If it fails to do this, leave it out and test another modification.

Technical feedback occurs when the information coming from the experiment interacts with technical knowledge contained in the experimenter's mind to lead to some form of action. The reasoning here is of a more complex kind not predictable

The Collected Works of George E. P. Box, 1984, Wadsworth, Inc., Belmont, CA 94002.
Originally published in *J. Roy. Stat. Soc.*, Series B, vol. 24, no. 2 (1962), pp. 297–343.

by a simple rule. Inspection and analysis of data will often suggest to the trained scientist explanations which are far from direct. These will provide leads which he follows up in further work. To allow technical feedback to occur efficiently, the scientist must have a reasonably sure foundation on which to build his conjectures. To provide this, the effects about which he has to reason will need to be known with fairly good and stated precision and this may require considerable replication, hidden or direct. This is one reason why the $n = 1$ rule is not recommended for most applications of evolutionary operation, or for any other procedure which benefits heavily from technical as well as from empirical feedback.

However, when the feedback is essentially empirical, the use of conservative rules appropriate for technical feedback can lead to considerable inefficiency. Two problems which we feel should be considered from the empirical feedback point of view are those of adaptive optimization and adaptive quality control. It is hoped that the discussion of these problems here may provide a starting point for a study of the much wider adaptive aspects of the experimental method itself.

1.3. *Adaptive Quality Control*

Quality control undoubtedly has many different motivations and aims. Quality control charts are used to spotlight abnormal variations, to indicate when something went wrong and, perhaps most important of all, to give a pictorial and readily understandable representation of the behaviour of the process in a form which is continually brought to the notice of those responsible for running it. These are all examples of the technical feedback aspects of quality control, the importance of which cannot be easily overrated but which will not be considered here.

Quality control charts are, however, also used as part of an adaptive loop. With Shewhart charts, for example, some rule will be instituted that when a charted point goes outside a certain limit, or a succession of points fall between some less extreme limits, then a specified action is taken to adjust the mean value. In this circumstance we are clearly dealing with empirical rather than with technical feedback and the problem should be so treated.

A paper which pointed out the advantages of a new type of quality control chart using cumulative sums was recently read before this Society by Barnard (1959). In it the important feedback aspect of the quality control problem was clearly brought out and we owe much to this work.

1.4. *Adaptive Optimization*

Our interest in adaptive optimization was stimulated in the first place by experiences with evolutionary operation. By evolutionary operation we mean a simple technique used by plant operators by means of which over the life of the plant many modifications are tried and, where they prove to be useful, are included in the process. In practice, evolutionary operation, like quality control, is used for a number of rather different purposes. Its main purpose seems to be to act as a permanent incentive to ideas and as a device to screen these ideas as they are born, either spontaneously or out of previous work. As with quality control, many important advances are made here by the use of technical rather than empirical feedback.

However, evolutionary operation has occasionally also been used as a kind of manual process of adaptive optimization. It sometimes happens that there exist some unmeasurable and largely unpredictable variables, ξ, such as catalyst activity, whose levels change with time. When this happens, evolutionary operation is sometimes used to indicate continuously how the levels of some other controllable variables X should be moved so as to ensure, as nearly as possible, that the current maximum value in some *objective function*, or *response*, η is achieved at any given time t.

Other examples of variables which affect efficiency and, like catalyst activity, are not easily measured when the chemical process is running, and for which adaptive compensation in other variables may be desirable, are: degree of fouling of heat exchangers, permeability of packed columns, abrasion of particles in fluidized beds and, in some cases, quality of feed stock. When such uncontrollable and unmeasurable variables ξ occur as a permanent and inevitable feature of chemical processes, it may become worth while to install special equipment to provide automatic adaptive compensation of the controllable and measurable variables X (Box, 1960). It also becomes worth while to consider the theory of such systems so that the adaptive optimization may be made as efficient as possible.

1.5. *Scope of the Present Investigation*

In section 2 we consider a simplified discrete-time model for the adaptive optimization problem. We find that we are led to problems of adaptive control which lead in turn to problems of prediction and smoothing of a time-series. Our results are then applicable to the problem of adaptive statistical quality control, which is given special attention in sections 2 and 3. These sections incorporate ideas well known to control engineers. In section 4, the methods are illustrated by means of examples and in section 5 the results are expressed in terms of power spectra.

The problem of maximizing or minimizing a response will be considered in a later paper.

2. A DISCRETE-TIME MODEL

2.1. *Discrete Adaptive Optimization Model*

We shall first consider a simplified model in which data are available only at discrete and equal intervals of time each of which we call a *phase*. We suppose further that the situation remains constant during a phase but may change from phase to phase. The model is thus directly appropriate to a batch-type chemical process and provides a discrete approximation to the operation of a continuous process from which discrete data are taken at equal intervals of time.

Suppose the uncontrollable and immeasurable variables have levels ξ_p during the pth phase and these levels change from one phase to the next. Then the conditional response function $\eta(X|\xi_p)$ will change correspondingly as in Fig. 1. Suppose that in the pth phase it may be approximated by the quadratic equation

$$\eta_p = \eta(X|\xi_p) = \eta(\theta_p) - \tfrac{1}{2}\beta_{11}(X-\theta_p)^2, \tag{2.1.1}$$

where $\theta_p = \{X_{\max}|\xi_p\}$ is the conditional maximal setting during the pth phase and β_{11} is known from prior calibration and does not change appreciably with ξ_p. We suppose finally that because of fluctuation in ξ_p, the conditional maximal value θ_p follows some stochastic process usually of a non-stationary kind. Let X_p be the *set-point* at which the controllable variable X is held in the pth phase (Fig. 1). Then the standardized slope of the response function at $X = X_p$,

$$\frac{1}{\beta_{11}}\left(\frac{d\eta_p}{dX}\right)_{X=X_p} = \theta_p - X_p = \epsilon_p,$$

measures the extent to which X_p deviates from θ_p. If experiments are performed at $X_p+\delta$ and at $X_p-\delta$ and the average response observed at these levels is $\bar{y}(X_p+\delta)$ and $\bar{y}(X_p-\delta)$, then an estimate e_p of $\epsilon_p = \theta_p - X_p$ is given by

$$e_p = \tfrac{1}{2}\{\bar{y}(X_p+\delta)-\bar{y}(X_p-\delta)\}/(\delta\beta_{11}) = \epsilon_p + u_p, \tag{2.1.2}$$

where u_p is called the measurement error, although it will be produced partly by uncontrolled fluctuations in the process. Now

$$e_p = \epsilon_p + u_p = \theta_p - X_p + u_p = z_p - X_p, \tag{2.1.3}$$

where $z_p = \theta_p + u_p$ is an estimate of the position of the optimal setting θ_p during the pth phase.

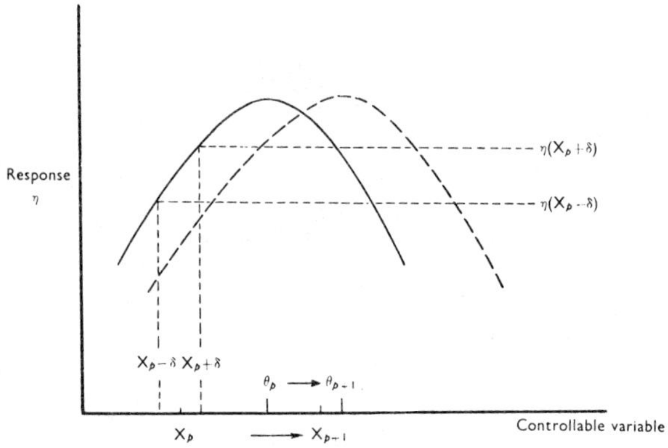

FIG. 1. Simple adaptive optimization.

If a series of adjustments has actually been made on some basis or other, we have a record of a sequence of set points $X_p, X_{p-1}, X_{p-2}, \ldots$ and of a sequence of deviations $e_p, e_{p-1}, e_{p-2}, \ldots$. From these it will be possible to reconstruct the sequence $z_p, z_{p-1}, z_{p-2}, \ldots$, of estimated positions of the maxima in phases $p, p-1, p-2, \ldots$. From such data we wish to make an adjustment x_{p+1} to the set point X_p so that the adjusted set point $X_{p+1} = X_p + x_{p+1}$ which will be maintained during the *coming* $(p+1)$th phase will be in some sense "best" in relation to the coming and unknown value of θ_{p+1}.

Now we suppose that the objective function η is such that the loss sustained during the $(p+1)$th phase is realistically measured by

$$\eta(\theta_{p+1}) - \eta(X_{p+1}\,|\,\xi_{p+1}) = \tfrac{1}{2}\beta_{11}(X_{p+1}-\theta_{p+1})^2, \tag{2.1.4}$$

the amount by which η falls short of the value theoretically attainable. The adjustment x_{p+1} will then minimize the expected loss if it is chosen so that $E(\theta_{p+1}-X_{p+1})^2$ is minimized. Thus our objective is attained if we set X_{p+1} equal to

$$\hat{\theta}_{p+1} = f(z_p, z_{p-1}, z_{p-2}, \ldots),$$

where $f(z_p, z_{p-1}, \ldots)$ is the minimum mean square estimate or *predictor* of θ_{p+1}, based on the observations $z_p, z_{p-1}, \ldots$.

For simplicity we take $\hat{\theta}_{p+1}$ to be a linear function of the z's,

$$\hat{\theta}_{p+1} = \sum_{j=0}^{\infty} \mu_j z_{p-j}, \tag{2.1.5}$$

and refer to the μ_j's as the *predictor* weights. If the model for the stochastic process $\{\theta_p\}$ is specified precisely, then it may well be that the best predictor is a non-linear function of the previous z's, as in the model used by Barnard (1959). In general, however, there is no empirical evidence for making such precise assumptions for the model so that the simplicity of the linear predictor is well worth maintaining.

At the beginning of the $(p+1)$th phase we should therefore apply to the previous setting X_p an adjustment $x_{p+1} = \hat{\theta}_{p+1} - \hat{\theta}_p$. Now in practice we do not observe the z's directly but only the e's. This means that we must calculate the adjustment from the estimated slopes $e_p, e_{p-1}, \ldots$, in the form

$$x_{p+1} = X_{p+1} - X_p = \hat{\theta}_{p+1} - \hat{\theta}_p = \sum_{j=0}^{\infty} w_j e_{p-j}, \tag{2.1.6}$$

where we call the w_j the *controller weights*. In section 2.3 we show that (2.1.5) and (2.1.6) imply that a definite relationship exists between the predictor weights μ_j and the controller weights w_j. The optimal adjustment is then obtained by choosing the w's, or equivalently the μ's, to minimize $E(\epsilon_{p+1}^2) = E(\theta_{p+1} - \hat{\theta}_{p+1})^2$.

Now $\hat{\theta}_{p+1} = f(z_p, z_{p-1}, \ldots)$ and $z_{p+1} = \theta_{p+1} + u_{p+1}$, so that if the "measurement error" u_{p+1} is distributed about zero with variance σ_u^2 *independently* of $u_p, u_{p-1}, \ldots$, and of $\theta_p, \theta_{p-1}, \ldots$, then

$$E(e_{p+1}^2) = E(z_{p+1} - \hat{\theta}_{p+1})^2 = E(\epsilon_{p+1}^2) + \sigma_u^2. \tag{2.1.7}$$

The loss $E(\epsilon_{p+1}^2)$ is then minimal when $E(z_{p+1} - \hat{\theta}_{p+1})^2 = E(e_{p+1})^2$ is minimized and $\hat{z}_{p+1}$, the best predictor of z_{p+1}, supplies in this case the best predictor $\hat{\theta}_{p+1}$ of θ_{p+1}. In general, σ_u^2 will depend on the number n of experiments performed in estimating the slope and also on δ, the magnitude of the perturbations. For any fixed values of n and δ, however, the best way of tracking θ_{p+1} is always to make an adjustment

$$x_{p+1} = \hat{\theta}_{p+1} - \hat{\theta}_p = \sum_{j=0}^{\infty} w_j e_{p-j}.$$

In those cases where the measurement errors u are independent of the θ's and of each other then $\hat{\theta}_{p+1} = \hat{z}_{p+1}$.

2.2. *Dynamics of Adjustment*

On our assumptions if, using data acquired during the pth and previous phases of operation, we wish to change the set point X_p of a controlled variable, such as temperature, to a level X_{p+1} so as to minimize the loss during the $(p+1)$th phase, we must set X_{p+1} equal to $\hat{\theta}_{p+1}$, the optimal value to be expected in the $(p+1)$th phase from the predictions of past data. In practice it would often not be possible to change a controlled variable X such as temperature directly. This would have to be done indirectly by adjusting some other variable X^* such as steam pressure. In general, we refer to the variable X^* which we can directly adjust as the manipulated variable. Now frequently, a change in the manipulated variable X^* (steam pressure) would not be immediately felt in the controlled variable X (temperature). The effect would build up over a considerable period of time.

It is convenient to suppose that the manipulated variable has been calibrated in units of its ultimate effect on the controlled variable. Then if we turn the steam valve forwards by "5 units", an increase of 5 units in temperature will ultimately result. Suppose that from data available in the pth phase an adjustment x^*_{p+1} can be made in the manipulated variable at a time such that proportions $v_0, v_1, v_2, \ldots$, of this change are experienced in the controlled variable in phases $p+1, p+2, p+3, \ldots$, where $v_0+v_1+\ldots = 1$. Then we call $v_0, v_1, v_2, \ldots$ the *dynamic weights*.

It is shown in section 5.4 that we can induce the controlled variable X to undergo the optimal adjustment

$$x_{p+1} = \sum_{j=0}^{\infty} w_j e_{p-j}$$

if we adjust the manipulated variable in accordance with

$$x^*_{p+1} = \sum_{j=0}^{\infty} w^*_j e_{p-j}, \tag{2.2.1}$$

where the w^*_j's are such that the w_j's are the convolution of the controller and dynamic weights,

$$w_j = v_0 w^*_j + v_1 w^*_{j-1} + v_2 w^*_{j-2} + \ldots. \tag{2.2.2}$$

It follows that if we know the optimal w_j's and the dynamic weights v_i, then provided $v_0 \neq 0$ we can compute the w^*_j's which tell us how to change the manipulated variable X^*.

2.3. *Discrete Adaptive Quality Control*

In this section we retain the previous symbols but change the context in which they are used. Suppose we are now concerned with some "quality characteristic", such as the concentration of consecutive batches, and that if no steps were taken to control its behaviour, its observed value at the pth phase would be

$$z_p = \theta_p + u_p. \tag{2.3.1}$$

As before, we assume that θ_p follows some stochastic process, in general non-stationary, and u_p is a measurement error. Suppose that our object is to hold θ as closely as possible to some *target value* which, without loss of generality, we can take to be zero. To achieve this, suppose a method is available whereby the mean value of the stochastic process z can be adjusted up or down at will. Suppose that at the pth stage a *total correction* of $-X_p$ is being applied to the mean. Then the quantity actually observed is the apparent deviation from target

$$z_p - X_p = \theta_p - X_p + u_p = \epsilon_p + u_p = e_p.$$

We wish now to calculate a further adjustment x_{p+1} computed from data in the pth and previous phases so that when the total correction $-X_{p+1} = -(X_p + x_{p+1})$ is applied during the pth phase the actual deviation from target $\epsilon_{p+1} = \theta_{p+1} - X_{p+1}$ will be small.

We now assume a quadratic loss function, that is we suppose that the loss involved through θ being off-target by an amount ϵ_p is proportional to ϵ_p^2. On this assumption we must choose x_{p+1} as before so that $E(\epsilon_{p+1})^2 = E(\theta_{p+1} - X_{p+1})^2$ is minimized.

Once more then, this requires that

$$x_{p+1} = \hat{\theta}_{p+1} - \hat{\theta}_p = \sum_{j=0}^{\infty} w_j e_{p-j},$$

where the w_j's are chosen so that $\hat{\theta}_{p+1}$ is the minimum mean square error estimate of θ_{p+1}. As before, if the measurement errors are uncorrelated with each other and with the θ's, then $\hat{z}_{p+1} = \hat{\theta}_{p+1}$ and we can then take

$$x_{p+1} = \hat{z}_{p+1} - \hat{z}_p = \sum_{j=0}^{\infty} w_j e_{p-j},$$

where now we choose the w's so that $\hat{z}_{p+1} = \hat{\theta}_{p+1}$ is the minimum mean square error estimate of z_{p+1}.

Finally, suppose the mean value could only be controlled indirectly via a manipulated variable. Then, as before, with

$$x_{p+1} = \sum_{j=0}^{\infty} w_j e_{p-j}$$

denoting the adjustment induced into the controlled variable by a correction

$$x^*_{p+1} = \sum_{j=0}^{\infty} w^*_j e_{p-j}$$

directly applied to the manipulated variable, we should obtain optimal control by choosing the w^*'s so that the weights

$$w_j = \sum_{i=0}^{\infty} v_i w^*_{j-1}$$

were those required for optimal prediction of θ_{p+1}.

2.4. *Relation Between Optimization, Control and Prediction Problems*

We have seen that with the assumptions and simplifications introduced, the adaptive optimization and adaptive control problems are identical. Furthermore, in both cases optimal action on data accumulated during the pth and previous phases is taken when X_p is adjusted either directly or indirectly to a value

$$X_{p+1} = X_p + x_{p+1} = \hat{\theta}_{p+1},$$

where $\hat{\theta}_{p+1}$ is the mean square error predictor of θ_{p+1}.

In Table 1, the roles which the various elements play in the three problems are summarized and in Fig. 2 the simultaneous behaviour of the three series z_p, X_p, ϵ_p is displayed for a typical situation. By reference to Table 1 we can interpret Fig. 2 for each problem.

In the prediction problem the z's are directly observed and the predictor $\hat{\theta}_{p+1}$ can be conveniently calculated from

$$\hat{\theta}_{p+1} = \sum_{j=0}^{\infty} \mu_j z_{p-j}, \qquad (2.4.1)$$

where the μ's are constants suitably chosen to minimize $E(\epsilon^2_{p+1}) = E(\theta_{p+1} - \hat{\theta}_{p+1})^2$.

In the optimization and control problems where optimal action is taken by setting X_{p+1} equal to $\hat{\theta}_{p+1}$ we do not observe the z's directly but only the e's. Now since $z_p = \hat{\theta}_p + e_p$ we could, as each new observation came to hand, reconstruct the z's and so use (2.4.1) to calculate the value $\hat{\theta}_{p+1}$. In practice, it is simpler to calculate

TABLE 1

Roles played by symbols in optimization, control and prediction problems

	e_p	$\epsilon_p = e_p - u_p$	z_p	$\theta_p = z_p - u_p$	X_p	x_{p+1}
Adaptive Optimization	Observed slope $\frac{\bar{y}(X_p+\delta)-\bar{y}(X_p-\delta)}{2\delta\beta_{11}}$ measures $\theta_p - X_p + u_p$, distance of X_p from max.; subject to error u_p	True slope, uncontaminated by measurement error u_p, measures $\theta_p - X_p$ distance of X_p from max.	Inferred position of max. can be calculated from $X_p + e_p$	Hypothetical true position of max. uncontaminated by error u_p	Set point of controlled variable	Calculated adjustment to be applied to set point X_p throughout pth phase
Adaptive control	Observed deviation of quality characteristic from its target	True deviation from target uncontaminated by measurement error u_p	Inferred value of quality characteristic if no control had been applied; can be calculated from $X_p + e_p$.	True value of quality characteristic if no control had been applied	Total correction applied negatively to mean	Calculated correction to be applied negatively to mean throughout the $(p+1)$th phase
Prediction	Observed deviation of z_p from prediction $\hat{\theta}_p = X_p$	True deviation from prediction uncontaminated by measurement error u_p	Directly observed characteristic	True value of characteristic uncontaminated by measurement error	Predicted value $\hat{\theta}_p$	Difference in predicted values $\hat{\theta}_{p+1} - \hat{\theta}_p$

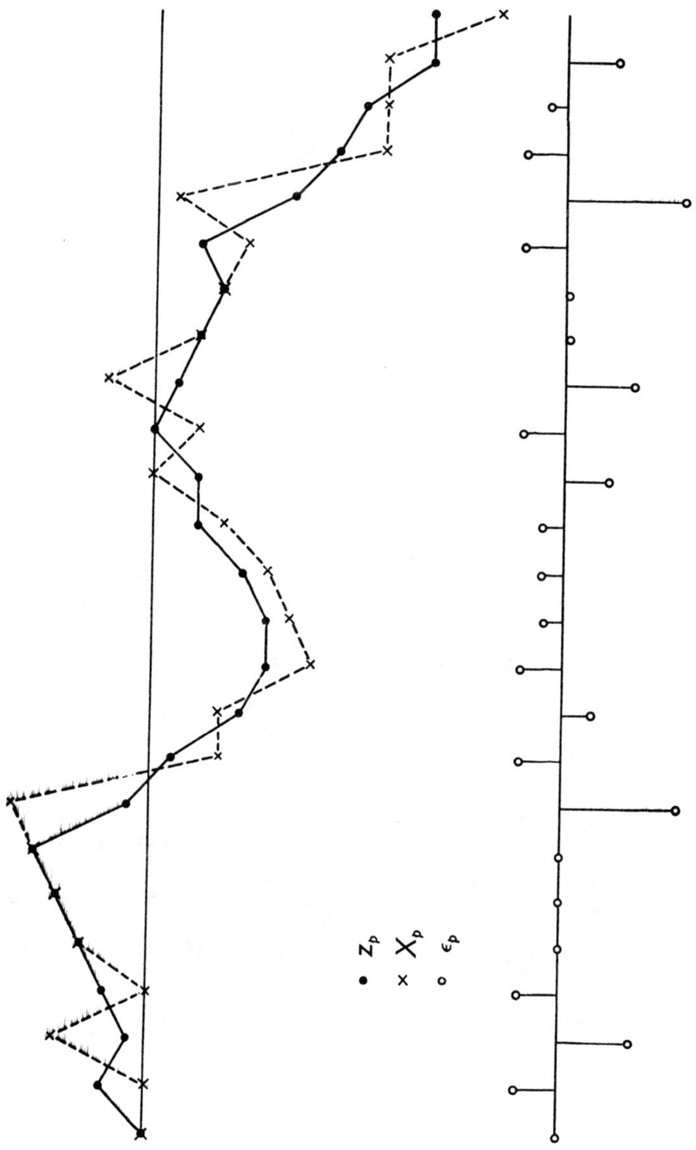

FIG. 2. Typical control and associated prediction problem.

the optimal adjustments $x_{p+1} = X_{p+1} - X_p = \hat{\theta}_{p+1} - \hat{\theta}_p$ from the directly observed e's according to

$$\hat{\theta}_{p+1} - \hat{\theta}_p = \sum_{j=0}^{\infty} w_j e_{p-j}, \tag{2.4.2}$$

where the weights w_j are functions of the μ's given by the recurrence relations

$$w_j = \mu_j - \mu_{j-1} + \sum_{k=0}^{j-1} \mu_k w_{j-1-k}. \tag{2.4.3}$$

To derive these relations we introduce the backward shift operator B such that $B^j z_p = z_{p-j}$. Then (2.4.1) may be written

$$\hat{\theta}_{p+1} = \sum_{j=0}^{\infty} \mu_j B^j z_p = \mathscr{L}_\mu(z_p), \tag{2.4.4}$$

where $\mathscr{L}_\mu$ is the linear operator on the z_p which produces $\hat{\theta}_{p+1}$. Similarly we may rewrite (2.4.2) in the form

$$(1-B)\hat{\theta}_{p+1} = \sum_{j=0}^{\infty} w_j B^j (z_p - \hat{\theta}_p),$$

so that

$$\left(1 - B + B\sum_{j=0}^{\infty} w_j B^j\right)\hat{\theta}_{p+1} = \sum_{j=0}^{\infty} w_j B^j z_p. \tag{2.4.5}$$

Denoting the linear operator

$$\sum_{j=0}^{\infty} w_j B^j$$

by $\mathscr{L}_w$, we obtain from (2.4.4) and (2.4.5) that

$$\mathscr{L}_w = (1 - B + B\mathscr{L}_w)\mathscr{L}_\mu. \tag{2.4.6}$$

Since the linear operators are infinite series in powers of B, it follows that this equality implies equality of every term in these series. If we identify coefficients of B^j on both

TABLE 2

Some examples of relationships between predictor and controller weights

Nature of weights	j	0	1	2	...	j
Exponentially weighted mean (proportional control)	μ_j	λ	$\lambda(1-\lambda)$	$\lambda(1-\lambda)^2$		$\lambda(1-\lambda)^j$
	w_j	λ	0	0		0
First order autoregressive	μ_j	a_1	0	0		0
	w_j	a_1	$-a_1(1-a_1)$	$-a_1^2(1-a_1)$		$-a_1^j(1-a_1)$
First order moving average (first difference control)	μ_j	β_1	$-\beta_1^2$	β_1^3		$(-1)^{j+1}\beta_1$
	w_j	β_1	$-\beta_1$	0		0

sides of (2.4.5) we are led to the recurrence relationship (2.4.3) between the controller weights w_j and the predictor weights μ_j. Some examples of this relationship are given in Table 2.

Example 1 corresponds to applying a weight λ to the current deviation and ignoring all the others and may be seen to correspond to the use of predictor weights which are exponentially weighted moving averages. These have been used with considerable success for the prediction of sales and production figures by Holt *et al.* (1960), Brown (1959), Muth (1960) and Cox (1961). The second and third examples take their names from the fact that they are best in a sense to be described in section 3 when the z_p process is of the type specified.

3. CHOICE OF WEIGHTS

3.1. *Choice of Predictor*

In this section we shall consider various specifications of the series z_p and how to choose the corresponding predictor weights with desirable properties. The associated controller weights obtained from the recurrence relation (2.4.3) will, of course, have the same desirable properties. The controller weights may be obtained directly but we adopt the present development via prediction partly because the prediction problem is of considerable interest in its own right.

As before, we suppose that the $(p-j)$th observation may be represented by

$$z_{p-j} = \theta_{p-j} + u_{p-j}, \tag{3.1.1}$$

where u_{p-j} is a measurement error and our object is to predict θ_{p+1} from previous values of z.

If our theory is to have practical value it would be unrealistic to assume that θ_{p-j} followed a stationary process. Instead we suppose that θ_{p-j} has a mean $E(\theta_{p-j}) = m_{p-j}$ which may vary. Our model can then be written

$$z_{p-j} = m_{p-j} + (\theta_{p-j} - m_{p-j}) + u_{p-j}. \tag{3.1.2}$$

The series $(\theta_{p-j} - m_{p-j})$ is assumed to be stationary and the $h \times h$ covariance matrix $\sigma_\theta^2 \mathbf{R_h}$ for h successive observations from this series is a Toeplitz matrix with elements $\{\rho_{|i-j|}\}\sigma_\theta^2$, where ρ_s is the autocorrelation of the θ's of lag s. The corresponding $h \times h$ covariance matrix for the u's, which we also assume stationary with mean zero, is denoted by $\sigma_u^2 \mathbf{S_h}$. The covariance matrix for the z's is thus $\sigma_\theta^2 \mathbf{T_h}$, where

$$\mathbf{T_h} = \mathbf{R_h} + \frac{\sigma_u^2}{\sigma_\theta^2} \mathbf{S_h}.$$

If the mean value $E(\theta_{p-j}) = m_{p-j}$ were known for every j then we should employ for our predictor of θ_{p+1} the quantity

$$m_{p+1} + \sum_{j=0}^{\infty} \mu_j \{z_{p-j} - m_{p-j}\}$$

and our prediction error would be

$$\epsilon_{p+1} = (\theta_{p+1} - m_{p+1}) - \sum_{j=0}^{\infty} \mu_j (z_{p-j} - m_{p-j}). \tag{3.1.3}$$

Now in most practical circumstances the predictor will place negligible weight on values of the series remote from the present time. Suppose all values of μ_j can be supposed zero for all $j \geqslant h$; then the predictor becomes

$$m_{p+1} + \sum_{j=0}^{h-1} \mu_j \{z_{p-j} - m_{p-j}\}.$$

The mean square error $E(\epsilon_{p+1})^2 = \text{var}(\epsilon_{p+1})$ may then be written in matrix form as

$$\text{var}(\epsilon_{p+1}) = \sigma_\theta^2 \{1 - 2\boldsymbol{\mu}'\boldsymbol{\rho} + \boldsymbol{\mu}'\mathbf{T}\boldsymbol{\mu}\}, \tag{3.1.4}$$

where $$\boldsymbol{\mu}'_h = (\mu_0, \mu_1, \ldots, \mu_{h-1})$$

and $$\boldsymbol{\rho}'_h = (\rho_1, \rho_2, \ldots, \rho_h).$$

Differentiating with respect to the vector $\boldsymbol{\mu}$, we find that the mean square error is minimized when

$$\boldsymbol{\mu} = \mathbf{T}^{-1}\boldsymbol{\rho}. \tag{3.1.5}$$

The optimal predictor of θ_{p+1} is thus

$$\hat{\theta}_{p+1} = \mathbf{z}'_p \mathbf{T}^{-1} \boldsymbol{\rho}, \tag{3.1.6}$$

where $$\mathbf{z}'_p = (z_p, z_{p-1}, \ldots, z_{p-h+1})$$

and this has variance $\boldsymbol{\rho}'\mathbf{T}^{-1}\boldsymbol{\rho}\sigma_\theta^2$. If in particular there were no measurement error and the z process, and hence the θ process, followed a kth order autoregressive scheme,

$$z_{p+1} = a_1 z_p + a_2 z_{p-1} + \ldots + a_{k+1} z_{p-k} \quad (k \leqslant h)$$

then we obtain the well-known result that $\boldsymbol{\mu} = \mathbf{a}$ where $\mathbf{a}' = (a_0, a_1, \ldots, a_{k+1})$.

3.2. *Constrained Predictors*

In most practical situations the mean m_{p-j} would not be known. If we employed the quantity $\mu_0 z_p + \mu_1 z_{p-1} + \ldots$ as a predictor of θ_{p+1} then the error of prediction would be

$$\epsilon_{p+1} = (\theta_{p+1} - m_{p+1}) - \sum_{j=0}^{h-1} \mu_j (z_{p-j} - m_{p-j}) + \left(m_{p+1} - \sum_{j=0}^{h-1} \mu_j m_{p-j}\right) \tag{3.2.1}$$

and the mean square error,

$$E(\epsilon_{p+1}^2) = \left(m_{p+1} - \sum_{j=0}^{h-1} \mu_j m_{p-j}\right)^2 + \text{var}(\epsilon_{p+1}), \tag{3.2.2}$$

would in general be inflated by an amount which depended on the behaviour of the mean m_{p-j}. One way in which this difficulty might be overcome would be to constrain the predictor weights μ_j so that the bias term

$$m_{p+1} - \sum_{j=0}^{h-1} \mu_j m_{p-j}$$

was eliminated or at least rendered small.

For example, if at the pth observation it was assumed that the mean, although unknown, was constant over the last h observations and would remain constant for one further observation, so that $m_{p+1} = m_p = m_{p-1} = \ldots = m_{p-h+1} = m$, then the bias term would vanish if the μ's were chosen subject to the restriction

$$\sum_{j=0}^{h-1} \mu_j = 1. \tag{3.2.3}$$

As observed by Cox (1961), by introducing this constraint we endow the predictor with the property of adjusting to the mean of the series. This is a property which is, of course, not possessed in general by the unrestricted Wiener predictors. For example, for predicting an autoregressive series the optimal weights $\boldsymbol{\mu} = \mathbf{a}$ will not in general sum to unity. This is because the existence of a known fixed mean is implicit in the assumption of an autoregressive series. For reasons which will be clear later, we would call the predictor constrained by (3.2.3) a *mean-projecting* predictor. More generally, if we can represent the behaviour of the $h+1$ values

$$m_{p+1}, m_p, m_{p-1}, \ldots, m_{p-h+1}$$

exactly by a polynomial of degree $0 \leqslant d \leqslant h-1$ then the bias term would vanish if

$$\sum_{j=0}^{h-1} j^k \mu_j = (-1)^k \quad (k = 0, 1, 2, \ldots, d). \tag{3.2.4}$$

Constrained predictors so obtained may be referred to as linear trend projecting, quadratic trend projecting, and so on.

Suppose then that the $h+1$ values $m_{p+1}, \ldots, m_{p-h+1}$ are exactly represented by a polynomial of degree $d < h-1$. Then since the bias term is now zero, the mean square error will be minimized by minimizing $\operatorname{var}(\epsilon_{p+1})$ subject to the constraints (3.2.4).

These constraints may be written

$$\mathbf{A}'\boldsymbol{\mu} = \mathbf{c}, \tag{3.2.5}$$

where $\mathbf{A}$ is an $h \times (d+1)$ matrix $\{a_{jk}\} = \{j^k\}$ with $j = 0, 1, \ldots, h-1$; $k = 0, 1, \ldots, d$, and $\mathbf{c}$ is a $(d+1) \times 1$ vector $\{(-1)^k\}$, $k = 0, 1, \ldots, d$. Introducing a vector of Lagrange multipliers $\boldsymbol{\lambda}' = (\lambda_0, \lambda_1, \ldots, \lambda_d)$, we have that the minimum mean square error is given by the unconditional minimum of

$$\{1 + 2\boldsymbol{\mu}'\boldsymbol{\rho} + \boldsymbol{\mu}'\mathbf{T}\boldsymbol{\mu}\} - 2(\boldsymbol{\mu}'\mathbf{A} - \mathbf{c})\boldsymbol{\lambda}. \tag{3.2.6}$$

On equating derivatives to zero, we obtain the simultaneous equations

$$\boldsymbol{\mu} = \mathbf{T}^{-1}\boldsymbol{\rho} + \mathbf{T}^{-1}\mathbf{A}\boldsymbol{\lambda}, \quad \mathbf{A}'\boldsymbol{\mu} = \mathbf{c}. \tag{3.2.7}$$

On multiplying the first equation by $\mathbf{A}'$ and substituting the result in the second, we obtain

$$\boldsymbol{\lambda} = (\mathbf{A}'\mathbf{T}\mathbf{A})^{-1}\{\mathbf{c} - \mathbf{A}'\mathbf{T}^{-1}\boldsymbol{\rho}\}; \tag{3.2.8}$$

hence finally

$$\boldsymbol{\mu}_c = \mathbf{T}^{-1}\boldsymbol{\rho} + \mathbf{T}^{-1}\mathbf{A}(\mathbf{A}'\mathbf{T}^{-1}\mathbf{A})^{-1}(\mathbf{c} - \mathbf{A}'\mathbf{T}^{-1}\boldsymbol{\rho}) \tag{3.2.9}$$

is the vector of optimal constrained predictor weights.

The corresponding predictor $\boldsymbol{\mu}_c'\mathbf{z}$ has variance

$$\{\boldsymbol{\rho}'\mathbf{T}^{-1}\boldsymbol{\rho} + (\mathbf{c} - \mathbf{A}'\mathbf{T}^{-1}\boldsymbol{\rho})'(\mathbf{A}'\mathbf{T}\mathbf{A})^{-1}(\mathbf{c} - \mathbf{A}'\mathbf{T}^{-1}\boldsymbol{\rho})\}\,\sigma_\theta^2, \tag{3.2.10}$$

where the second term in the curly brackets represents the inflation of the variance which occurs due to the constraints.

Given that the mean m_{p-j} follows a polynomial of the type assumed, this predictor $\boldsymbol{\mu}_c$ provides in fact the minimum variance linear unbiased predictor of θ_{p+1} as has been shown in an unpublished report by E. Parzen. We refer also to some earlier work on constrained predictors by Zadeh and Ragazzini (1950).

The predictor $\boldsymbol{\mu}_c$ has an alternative interpretation. Using it, the predicted value of θ_{p+1} may be written

$$\boldsymbol{\mu}_c'\mathbf{z} = \mathbf{c}'(\mathbf{A}'\mathbf{T}^{-1}\mathbf{A})^{-1}\mathbf{A}'\mathbf{T}^{-1}\mathbf{z}+\boldsymbol{\rho}\mathbf{T}^{-1}\{\mathbf{z}-\mathbf{A}(\mathbf{A}'\mathbf{T}^{-1}\mathbf{A})^{-1}\mathbf{A}'\mathbf{T}^{-1}\mathbf{z}\}. \qquad (3.2.11)$$

Now we have seen that if the mean was known we should use as the predictor for θ_{p+1} the quantity

$$m_{p+1}+\boldsymbol{\mu}'(\mathbf{z}-\mathbf{m}) \quad \text{with } \boldsymbol{\mu}' = \boldsymbol{\rho}'\mathbf{T}^{-1}, \qquad (3.2.12)$$

where $\mathbf{m}' = (m_p, m_{p-1}, \ldots, m_{p-h+1})$.

If in (3.2.12) we replace $m_{p+1} = \mathbf{c}'\boldsymbol{\beta}$ and $\mathbf{m} = \mathbf{A}'\boldsymbol{\beta}$ by the corresponding quantities where $\hat{\beta}$ is replaced by its best linear unbiased estimate

$$\hat{\boldsymbol{\beta}} = (\mathbf{A}'\mathbf{T}^{-1}\mathbf{A})^{-1}\mathbf{A}'\mathbf{T}^{-1}\mathbf{z},$$

then this is precisely the constrained predictor $\hat{\theta}_{p+1}$ given by (3.2.11).

Suppose, for example, we assume that over the last h observations z can be represented by a mean m subject to a linear trend, plus a stochastic variable $\theta-m$ for which the variance–covariance matrix $\mathbf{R}\sigma_\theta^2$ is known, plus a measurement error for which the variance–covariance matrix $\mathbf{S}\sigma_u^2$ is known.

Then the prediction of θ_{p+1} using the constrained predictor $\boldsymbol{\mu}_c'\mathbf{z}$ would be equivalent to the following:

(i) estimate m_{p+1} by projecting the line which best fits the last h z's through one further observation;

(ii) add to this an estimate of $\theta_{p+1}-m_{p+1}$, the deviation from the projected line at the next observation.

The latter estimate would be a weighted sum of the residuals from the line using the *unconstrained* predictor $\boldsymbol{\mu} = \mathbf{T}^{-1}\boldsymbol{\rho}$.

3.3. *A More Practical Predictor*

The above approach would be extremely difficult to apply in practice. To use it we would have to choose d and h, the degree of the polynomial and the length of series to which it was to be fitted. We should also need to know the covariance matrices $\mathbf{R}\sigma_\theta^2$ and $\mathbf{S}\sigma_u^2$. We are therefore led to seek an alternative approach which can be applied more readily whilst retaining so far as possible the advantages of the previous system.

We shall suppose in what follows either that the measurement errors u are uncorrelated one with another and with the θ's, so that $\hat{z}_{p+1} = \hat{\theta}_{p+1}$ or that the u's are negligible, in which case $z_{p+1} = \theta_{p+1}$ and again $\hat{z}_{p+1} = \hat{\theta}_{p+1}$.

Table 3 shows the optimal μ and w weights obtained for various series of interest. In each case the series is assumed to be represented over the last h observations by a polynomial of degree d, and the variation about the polynomial is assumed to follow the process indicated. In some instances the errors are supposed independent and in others that they follow a first order autoregressive series with coefficient a.

The first three series in the table are all examples of what may be called fully saturated predictors. We fit a polynomial of degree d to the last $h = d+1$ observations. The weight μ_j is then given by minus the coefficient of t^{j+1} in $(1-t)^{d+1}$. The form of the weights w_j is of special interest. In general, the change in the predictor $\hat{z}$ at the $(p+1)$th stage may be written as in (2.4.2) in the form

$$\Delta\hat{z}_{p+1} = \sum_{j=0}^{\infty} w_j e_{p-j}$$

TABLE 3

Some constrained predictors and their associated controller weights

Description	j	0	1	2	3	4	5	6	7	8	9
1 $d=0$ $h=1$	μ_j	1	.	.	.	.	.	.	.	.	.
Errors independent	w_j	1	.	.	.	.	.	.	.	.	.
2 $d=1$ $h=2$	μ_j	2	−1	.	.	.	.	.	.	.	.
Errors independent	w_j	2	1	1	1	1	1	1	1	1	1
3 $d=2$ $h=3$	μ_j	3	−3	1	.	.	.	.	.	.	.
Errors independent	w_j	3	3	4	5	6	7	8	9	10	11
4 $d=0$ $h=\infty$	μ_j	γ_0	$\gamma_0(1-\gamma_0)$	$\gamma_0(1-\gamma_0)^2$	$\gamma_0(1-\gamma_0)^3$	.	.	.	.	.	.
Errors independent	w_j	γ_0	0	0	0	0	0	0	0	0	0
5 $d=1$ $h=4$	μ_j	1·00	·50	·00	−·50	·00	·00	·00	·00	·00	·00
Errors independent	w_j	1·00	·50	·50	·25	·38	·38	·44	·38	·38	·38
6 $d=1$ $h=4$ *First order A. R.*	μ_j	1·14	·27	·04	−·45	·00	·00	·00	·00	·00	·00
$(a_1=0{\cdot}3)$	w_j	1·14	·43	·57	·32	·47	·45	·40	·45	—	—
7 $d=1$ $h=4$ *First order A. R.*	μ_j	1·22	·15	·04	−·41	·00	·00	·00	·00	·00	·00
$(a_1=0{\cdot}5)$	w_j	1·22	·42	·58	·37	·47	·48	·43	·47	—	—
8 $d=1$ $h=4$ *First order A. R.*	μ_j	1·33	·01	·00	−·34	·00	·00	·00	·00	·00	·00
$(a_1=0{\cdot}9)$	w_j	1·33	·45	·60	·46	·51	·53	·51	·52	—	—
9 $d=1$ $h=6$	μ_j	0·67	·47	·27	·07	−·13	−·33	·00	·00	·00	·00
Errors independent	w_j	0·67	·24	·27	·27	·22	·08	·32	·21	·19	·20

and for the first three series in Table 3 this becomes

series 1, $\Delta\hat{z}_{p+1} = e_p$;

series 2, $\Delta\hat{z}_{p+1} = e_p + S^1 e_p$;

series 3, $\Delta\hat{z}_{p+1} = e_p + S^1 e_p + S^2 e_p$,

where $$S^1 e_p = \sum_{j=0}^{\infty} e_{p-j}, \quad S^2 e_p = \sum_{j=0}^{\infty}\sum_{k=0}^{\infty} e_{p-j-k},$$

and in general $S^j e_p$ denotes the jth multiple sum over the past history of the e's. For a dth degree polynomial fitted to the last $h = d+1$ observations we would have

$$\Delta\hat{z}_{p+1} = e_p + S^1 e_p + S^2 e_p + \ldots + S^d e_p.$$

We now recall that a mean projecting predictor which has proved to be of great practical value is the exponentially weighted mean, shown here as series 4. Technically, $h = \infty$ for this predictor, but in practice the exponential weights become negligible after a moderate number of observations so that local changes in mean are satisfactorily followed. For this series

$$\Delta\hat{z}_{p+1} = \gamma_0 e_p.$$

This important mean projecting series does not fit naturally into our previous formulation but it is now seen to occur as a simple generalization of series 1.

In series 5, 6, 7, 8 and 9 we have tabulated sets of weights for various linear trend projectors. Bearing in mind that the properties of these series are not very sensitive to moderate changes in the weights, we see that all of them could be approximated by an obvious modification of series 2, namely

$$\Delta\hat{z}_{p+1} = \gamma_0 e_p + \gamma_1 S^1 e_p.$$

Finally, therefore, as a natural generalization, we consider the model where $\Delta\hat{z}_{p+1}$, the change to be applied at the $(p+1)$th stage to the previous predictor $\hat{z}_p$, is given by

$$\Delta\hat{z}_{p+1} = (\gamma_{-l} S^{-l} + \ldots + \gamma_{-2} S^{-2} + \gamma_{-1} S^{-1} + \gamma_0 + \gamma_1 S^1 + \gamma_2 S^2 + \ldots + \gamma_m S^m)\, e_p, \tag{3.3.1}$$

where $S^{-j} e_p = \Delta^j e_p$.

3.4. *The Nature of the Stochastic Process for which the Predictor is Optimal*

We now ask the question: what stochastic process would z need to follow for such a predictor to be optimal? Consider a process generated by

$$z_{p+1} = \sum_{j=0}^{\infty} \eta_j z_{p-j} + \alpha_{p+1} \tag{3.4.1}$$

which is, in general, non-stationary and where $\alpha_{p+1}, \alpha_p, \alpha_{p-1}, \ldots$ are uncorrelated identically distributed random variables, with mean zero. Suppose we predict the series using

$$\hat{z}_{p+1} = \sum_{j=0}^{\infty} \mu_j z_{p-j}. \tag{3.4.2}$$

Then since

$$E(e_{p+1})^2 = E(z_{p+1} - \hat{z}_{p+1})^2 = E\left\{\sum_{j=0}^{\infty} (\eta_j - \mu_j)\, z_{p-j}\right\}^2 + E(\alpha_{p+1})^2, \tag{3.4.3}$$

prediction is optimal if $\mu_j = \eta_j$ ($j = 0, 1, 2, \ldots$). It follows that the equivalent predictor

$$\Delta \hat{z}_{p+1} = \sum_{j=0}^{\infty} w_j \, e_{p-j}$$

is optimal for the equivalent stochastic process

$$\Delta z_{p+1} = \Delta \alpha_{p+1} + \sum_{j=0}^{\infty} w_j \, \alpha_{p-j}$$

and that when this optimal predictor is used the e's become α's and are uncorrelated.

In the present instance then our predictor is optimal for a series generated by

$$\Delta z_{p+1} = \Delta \alpha_{p+1} + \sum_{j=-l}^{m} \gamma_j \, S^j \, \alpha_p. \tag{3.4.4}$$

Differencing m times we have

$$\Delta^{m+1} z_{p+1} = \Delta^{m+1} \alpha_{p+1} + \sum_{j=0}^{l+m} \gamma_j \, \Delta^{l+m-j} \, \alpha_p.$$

On rearranging this expression, we find that our predictor would be optimal for any stochastic variable z whose $(m+1)$st difference may be represented by a moving average process of order $l+m+1$,

$$\Delta^{m+1} z_{p+1} = \alpha_{p+1} + \sum_{j=0}^{l+m} \delta_j \, \alpha_{p-j}, \tag{3.4.5}$$

so that all serial covariances after that of lag $l+m+1$ are zero. Thus we have a result which is of considerable practical value. If, after differencing our series z, which in general will be non-stationary, m times, we could render it stationary and if the population serial covariances of lag greater than some value $l+m+1$ were then zero, a predictor of the type (3.3.1) would be optimal.

The widely applied predictor obtained by taking an exponentially weighted mean

$$\hat{z}_{p+1} = \gamma_0 \sum_{j=0}^{\infty} (1-\gamma_0)^j \, z_{p-j}$$

corresponds simply to taking the single central term in our general series namely $\Delta \hat{z}_{p+1} = \gamma_0 \, e_p$. It is optimal for the stochastic process

$$\Delta z_{p+1} = \Delta \alpha_{p+1} + \gamma_0 \, \alpha_p = \alpha_{p+1} - (1-\gamma_0) \, \alpha_p,$$

$$\text{i.e.} \quad z_{p+1} = m + \alpha_{p+1} + \gamma_0 \, S^1 \, \alpha_p,$$

for which the first difference is a first order moving average. The addition of further terms can be thought of, therefore, as an appropriate generalization of this exponential predictor. In particular, for series which are highly non-stationary and exhibit marked trends, the additional term in $S^1 e_p$ will be of particular value since this will allow the predictor to adjust to changes in linear trend as well as to changes in mean.

Bearing in mind the great success of the exponential predictor, it might be expected that the simple generalization

$$\Delta \hat{z}_{p+1} = \gamma_{-1} \Delta e_p + \gamma_0 \, e_p + \gamma_1 \, S e_p \tag{3.4.6}$$

might be adequate for many practical purposes. Experience of two kinds indicates that this is so. Our own somewhat limited experience in applying this theory to industrial series has shown that for those series so far tried, this generalization has been adequate. In fact, so far as *prediction* is concerned, the term in Δe_p has not so far been needed. A further vast fund of experience in this area is possessed by control engineers. We have seen already that if there were no dynamics then the adjustment x_{p+1} of the control set point should be made equal to $\Delta \hat{z}_{p+1}$ the predicted change. A form of automatic control commonly used in industrial plants in continuous time makes a correction proportional to a linear combination of (i) the first derivative of the current deviation, (ii) the deviation itself, (iii) the integral of the deviations over all past history. If, therefore, we were using our predictor (3.4.6) for control purposes, we would employ a discrete time analogue of what control engineers have been using on automatic equipment for many years. The success of their efforts suggests that stochastic processes for which (3.4.6) is optimal adequately describe many industrial series.

The types of continuous control mentioned above are called *derivative*, *proportional* and *integral* respectively. For the discrete process we shall refer to the corresponding terms in (3.4.6) as *first difference*, *proportional* and *cumulative* terms and when we use the prediction equation for control purposes we shall talk of first difference control, proportional control and cumulative control. We shall adopt the three-term model in what follows bearing in mind that it is readily elaborated if need be.

3.5. *Fitting the Three-Term Model*

The stochastic process for which (3.4.6) is optimal is

$$z_{p+1} = m + \alpha_{p+1} + \gamma_{-1}\alpha_p + \gamma_0 S^1 \alpha_p + \gamma_1 S^2 \alpha_p. \tag{3.5.1}$$

For this process the second difference is the moving average of order 3,

$$\Delta^2 z_{p+1} = \alpha_{p+1} + (\gamma_1 + \gamma_0 + \gamma_{-1} - 2)\alpha_p + (1 - 2\gamma_{-1} - \gamma_0)\alpha_{p-1} + \gamma_{-1}\alpha_{p-2}. \tag{3.5.2}$$

We notice

(i) that in those common cases where no difference term is needed, γ_{-1} would be zero and only the first and second serial correlations would be non-zero;

(ii) if $\gamma_1 = \gamma_0 = 1$ and $\gamma_{-1} = 0$ the terms in α_p, α_{p-1} and α_{p-2} are all zero, $\Delta^2 z_{p+1} = \alpha_{p+1}$, and predictor number 2 in Table 3 is optimal. For this choice of the γ's the second differences are uncorrelated. The nature of the serial correlation among the second differences indicates in what way and how far we must move away from this "pivotal" predictor.

To gain appreciation of the adequacy of a model of the kind assumed, a fairly long length of series would first be differenced $m+1$ times until it appeared stationary. In those cases studied, second differencing has always proved adequate. In two of the series discussed in section 4 first differencing would, in fact, have been adequate but nothing is lost by taking a higher order difference than is strictly necessary. In our calculations we have taken a "typical" run of two or three hundred observations and begun by calculating the first twelve serial covariances of the second differences. If, as a result of inspection, we find that these are small after the third we have proceeded with the estimation of the parameters $\gamma_{-1}, \gamma_0, \gamma_1$ in the three-term model. Inadequate differencing is indicated when positive serial correlation persists for higher lag s. If serial covariances of fourth or higher order were appreciable, but higher covariances were small, we would introduce further parameters corresponding to high derivative control as described above.

A spectral method of estimation of the parameters is described in section 5.5. An alternative method which we have used extensively is to evaluate the sum of squares $S(\gamma_1, \gamma_0, \gamma_{-1}) = e_1^2 + \ldots + e_n^2$ for a grid of values of $\gamma_1, \gamma_0, \gamma_{-1}$ and hence pick out the best values of $\gamma_1, \gamma_0, \gamma_{-1}$ by inspection. In practice we have found that the minimum is fairly flat. Typically, a rather large near-optimal region is found in the space of $\gamma_1, \gamma_0, \gamma_{-1}$ from which we can choose a convenient solution.

Suppose we have a set of data $z_1, z_2, \ldots, z_n$. To make the calculation of the sum of squares $e_1^2 + \ldots + e_n^2$ corresponding to a particular choice of γ_{-1}, γ_0 and γ_1 we note that if z_p, e_p, Δe_p and $S^1 e_p$ are known we may calculate

$$\hat{z}_{p+1} = z_p + \gamma_{-1} \Delta e_p + \gamma_0 e_p + \gamma_1 S^1 e_p$$

and then knowing z_{p+1}, the quantities e_{p+1}, Δe_{p+1} and $S^1 e_{p+1}$ can be easily obtained. To start the process we set $\hat{z}_1 = z_1$ so that $e_1 = \Delta e_1 = S^1 e_1 = 0$. This simple repetitive calculation is ideally suited for the electronic computer which we feel it safe to assume is now universally available.

Under the assumption that the e's are normally distributed, the sum of squares surface $S(\gamma_1, \gamma_0, \gamma_{-1})$ is, in fact, equivalent to the log-likelihood function conditional on $e_1 = \Delta e_1 = S^1 e_1 = 0$. The actual log-likelihood function could be obtained by integrating over the joint distribution of e_1, Δe_1 and $S^1 e_1$ as described by Barnard *et al.* (1962). We have not used this refinement since it would only be important for short series and further, if the choice of the best control parameters were very dependent on e_1, Δe_1 and $S^1 e_1$, we would be disinclined to use them for controlling the future behaviour of the series anyway.

In choosing our grid of values, it is necessary to restrict the $(\gamma_1, \gamma_0, \gamma_{-1})$ point to a certain region or else the control procedure will become *unstable*. It will be shown in section 5.4 that this occurs because the optimum predictors corresponding to the moving average process (3.2.5) lead to infinite prediction variances if the control parameters lie outside a certain region. For the general form of control typified by (3.2.2), the stability condition is that the roots of the characteristic equation

$$1 + \sum_{j=0}^{l+m} \delta_j x^{j+1} = 0$$

should not lie on or inside the unit circle with the possible exception of $m+1$ roots which could all be equal to unity.

The computer programme for the Control Data Corporation's 1604 machine at Madison, Wisconsin, produces the first twelve serial covariances of the second differences and obtains a sum of squares grid for $\gamma_{-1} = -0{\cdot}5\,(0{\cdot}25)\,0{\cdot}5$, $\gamma_0 = 0{\cdot}0\,(0{\cdot}1)\,2{\cdot}0$ and $\gamma_1 = 0{\cdot}0\,(0{\cdot}1)\,2{\cdot}0$ for a series of 300 terms in about 60 seconds. Some examples of these calculations will be discussed in the next section.

3.6. *Representation of Dynamics by Exponential Stages*

Suppose the control dynamics can be represented by an exponential transfer function so that $v_i = (1-\phi)\,\phi^i$ $(i = 0, 1, 2, \ldots)$.

From (2.2.2) it may be shown that to induce action in the controlled variable according to

$$x_{p+1} = \gamma_{-1} \Delta e_p + \gamma_0 e_p + \gamma_1 S^1 e_p \tag{3.6.1}$$

an adjustment

$$x^*_{p+1} = \gamma^*_{-2} \Delta^2 e_p + \gamma^*_{-1} \Delta e_p + \gamma^*_0 e_p + \gamma^*_1 S^1 e_p \tag{3.6.2}$$

must be applied to the manipulated variable, where $\gamma^*_{-2} = k\gamma_{-1}$, $\gamma^*_{-1} = \gamma_{-1}+k\gamma_0$, $\gamma^*_0 = \gamma_0+k\gamma_1$, $\gamma^*_1 = \gamma_1$ and $k = \phi/(1-\phi)$. In other words the matrix **T**, which transforms the γ's into γ^*'s when exponential dynamics apply, is given by the matrix relation

$$\boldsymbol{\gamma}^* = \mathbf{T}\boldsymbol{\gamma} \tag{3.6.3}$$

with

$$\mathbf{T} = \begin{bmatrix} k & . & . \\ 1 & k & . \\ . & 1 & k \\ . & . & 1 \end{bmatrix}. \tag{3.6.4}$$

The corresponding matrix **T** appropriate for the more general control model (3.3.1) is of the same form. We see that to compensate for exponential dynamics we must add difference control to the manipulated variable of one order higher than we require to be applied to the controlled variable. Furthermore, if we know

(i) the γ's required for optimal prediction,

(ii) the constant k of the exponential transfer,

then the γ^*'s, which determine the appropriate combination of the various kinds of control to be applied to the manipulated variable, are very easily obtained.

It frequently happens in practice that only the two terms involving γ_0 and γ_1 which produce proportional and cumulative action are needed for near-optimal prediction (and hence, in the absence of dynamics, for near-optimal control). Yet as soon as we must compensate for simple exponential dynamics our third term γ_{-1} will be equal to $k\gamma_0$ and so difference action is needed.

In some cases the dynamics are such that more than one exponential stage is involved. If the action x_{p+1} which it was desired to induce at the controlled variable was represented by an expression involving c terms and the dynamic behaviour was as if s exponential stages were involved with constants $k_1, k_2, \ldots, k_s$, then control involving differences of s orders higher would be required at the manipulated variable. If $\mathbf{T}_i$ is the $(c+i)\times(c+i-1)$ transfer matrix for the ith stage of the same form as (3.6.2) but containing the constant k_i, then the $(c+s)\times c$ matrix **T** representing the overall transfer matrix is given by $\mathbf{T} = \mathbf{T}_1\mathbf{T}_2\ldots\mathbf{T}_s$. For example, if an exponentially weighted mean with constant γ_0 adequately predicts the next observation, so that simple proportional control according to $x_{p+1} = \gamma_0 e_p$ is required to be applied to the controlled variable, and if the dynamics can be represented by two exponential stages with constants k_1 and k_2, then $\mathbf{T}' = [k_1 k_2, k_1+k_2, 1]$ and we control the manipulated variable in acccordance with

$$x^*_{p+1} = k_1 k_2 \gamma_0 \Delta^2 e_p + (k_1+k_2)\gamma_0 \Delta e_p + \gamma_0 e_p.$$

4. EXAMPLES

4.1. *A Study of Three Series*

In Fig. 3 are shown two or three hundred observations from each of three series associated with typical chemical operations. The series show values respectively of concentration, temperature and viscosity when no control was applied. Each series consists of discrete observations although the plotted points are joined in this figure by lines. Series II was obtained by temporarily disconnecting an automatic temperature controller, series I and III were obtained by reconstructing the uncontrolled series. In each of these latter cases some manual control had been employed. From (i) the values for the controlled series, (ii) the control action taken, and, where appropriate, (iii) the approximate system dynamics, the course which the series would have taken had no control been exerted could be calculated to a sufficient degree of approximation. The concentration of active ingredient in successive

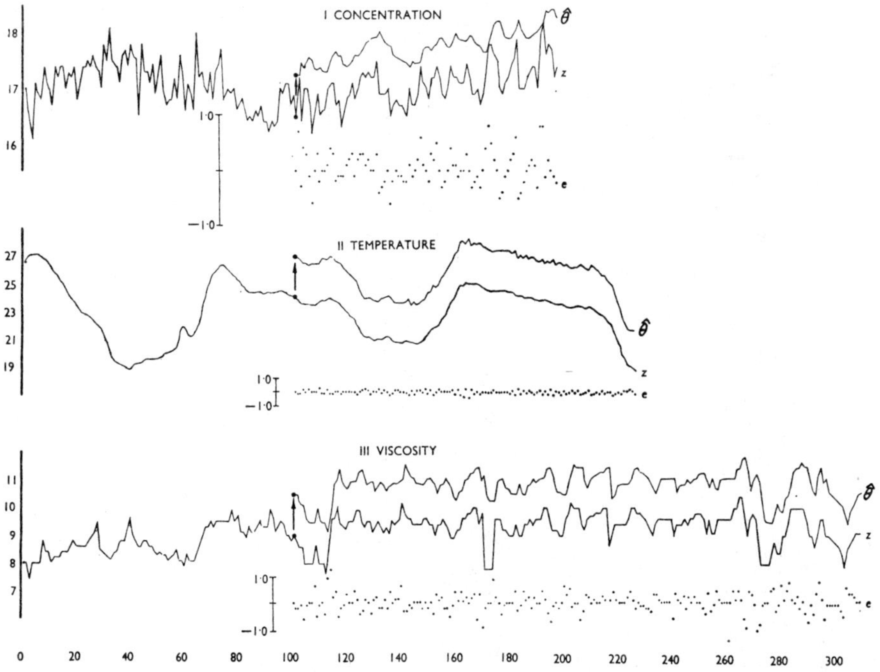

Fig. 3. Three uncontrolled industrial series, predicted series θ_p, and error e_p shown for latter sections.

batches shown in series I was controlled by the addition of the appropriate amount of pure chemical on a batch-to-batch basis. In this example, therefore, no dynamics had to be considered in reconstructing the series. The dynamics appropriate to series III are discussed later. It was found in each case that the serial covariances after the third are small, so that we feel justified in proceeding with our analysis based on the use of difference, proportional and cumulative control.

In Fig. 4 contour plots of the sums of squares surfaces are shown for different values of γ_0 and γ_1 when $\gamma_{-1} = 0$.

These diagrams show that:

(i) The choice of the coefficients is not very critical. A sum of squares very little greater than the smallest value can be obtained over a fairly wide area of the $(\gamma_{-1}, \gamma_0, \gamma_1)$ space.

(ii) For none of the series is there any strong indication that a difference term is necessary. Formal interpolation among the calculated sums of squares gives a minimum in each case in which γ_{-1} is slightly different from zero. However, the increase produced by setting $\gamma_{-1} = 0$ is entirely negligible.

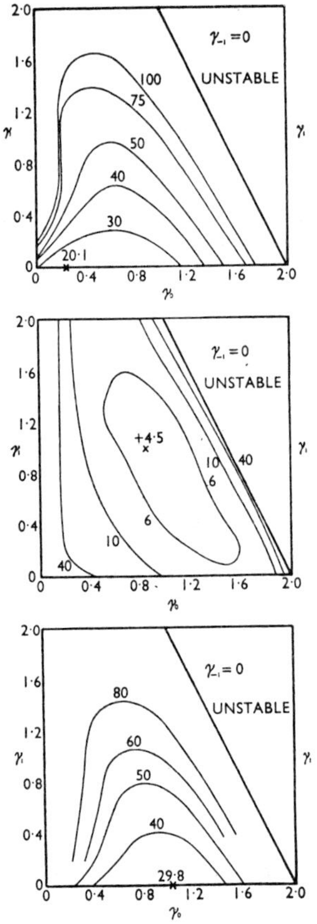

FIG. 4. Sums of squares contours for series I–III.

(iii) For series I the simple proportional predictor

$$x_{p+1} = 0{\cdot}3\, e_p$$

is near optimal. This corresponds to the prediction of the next value in the series by an exponentially weighted mean of previous observations. The small value $\gamma_0 = 0{\cdot}30$ ensures that the exponential μ weights (0·30, 0·21, 0·15, 0·10, 0·07, etc.) die out rather slowly. This is to be expected because of the rather noisy nature of the series. In general, the choice of the constant in a proportional (i.e. exponentially weighted) predictor is a compromise between the desirability of emphasis on "up-to-date values", obtained by choosing γ_0 close to unity and the desirability of "smoothing out the noise" obtained by taking a smaller value of γ_0.

(iv) For series II the necessity for cumulative control in addition to proportional control is clearly seen, the near-optimal predictor being

$$x_{p+1} = 1{\cdot}1\, e_p + 0{\cdot}8 \sum_{j=0}^{\infty} e_{p-j}.$$

Almost equally good prediction would be obtained by setting both γ_0 and γ_1 equal to unity. The corresponding μ weights would then be $\mu_0 = 2$, $\mu_1 = -1$ with all other values zero. The predictor $X_{p+1} = 2z_p - z_{p-1}$ is that obtained by projecting a straight line drawn through the last two observations and would be optimal when the second differences were independent random deviates. This type of prediction would be expected to be good for the present series (Fig. 3) in which linear trend extrapolation is clearly desirable and where the superimposed noise is small. We notice as expected that for this series the serial covariances of the second differences are *all* small in magnitude.

(v) For series III a near optimal predictor is $x_{p+1} = e_p$. This means that $X_{p+1} = z_p$ so that the previous observation is used to predict the next. Since an important characteristic of series III is the predominance of step changes, this is clearly sensible.

4.2. *Predicting the Series*

To illustrate the extent to which series of the various kinds can be predicted, we have applied predictors with optimal weights obtained from the first hundred values

TABLE 4

Serial correlations ρ_s and optimal control weights for series I–III

	Series I			*Series II*			*Series III*		
	1*st* 100 *obs.*	2*nd* 97 *obs.*	*Total* 197 *obs.*	1*st* 100 *obs.*	2*nd* 126 *obs.*	*Total* 226 *obs.*	1*st* 100 *obs.*	2*nd* 210 *obs.*	*Total* 310 *obs.*
Var (Δz^2)	·57	·22	·39	·032	·012	·020	17·7	22·2	20·7
ρ_1	−·69	−·53	−·67	·09	−·39	−·09	−·64	−·46	−·50
ρ_2	·22	·07	·22	−·18	·11	−·07	·25	−·07	·00
ρ_3	−·03	−·13	−·09	−·15	−·07	−·13	−·26	·08	·00
ρ_4	−·01	·22	·05	−·05	−·08	−·05	·30	−·13	−·01
ρ_5	·01	−·17	−·03	−·03	·10	·02	−·29	·06	−·02
ρ_6	−·06	−·02	−·06	·02	−·05	−·01	·25	−·01	·04
ρ_7	·12	·21	·15	·01	·17	·06	−·26	·06	−·02
ρ_8	−·05	−·24	−·12	−·07	−·05	−·06	·18	−·06	·01
ρ_9	·00	·14	·05	−·09	−·17	−·13	−·02	·03	·01
ρ_{10}	·03	·00	·03	·10	·14	·12	−·05	−·04	−·04
γ_0†	·30	·50	·30	1·20	1·00	1·10	·80	1·00	1·00
γ_1†	·00	·00	·00	·80	·60	·80	·00	·00	·00
Variance‡ before control	·139			2·313			29·0		
Variance‡ after control	·089			·013			11·0		

† Optimal values of γ_0 and γ_1, when $\gamma_{-1} = 0$.
‡ Variances refer to second part of series.

of each series to forecast values in the remainder. The predicted values are plotted above the second part of each series in Fig. 3. To avoid confusion between the original values and the predicted values, the latter have been moved upwards by the amount indicated and plotted above the original series. The errors of prediction are indicated by the series of dots plotted below each series. Table 4 shows for the first and the second part of each series, (i) the serial correlations of second differences, (ii) the optimal weights, (iii) the reduction in variance achieved by control of the second part of each series using weights calculated from the first part. Although no formal tests have been made there seems reason to doubt strict homogeneity particularly in the first series. The optimal weights for the first and second parts of each series agree remarkably well, however.

4.3. *An Example of an Adaptive Quality Control Scheme*

We now describe the institution of a practical adaptive quality control scheme based on the above discussion. For this we consider in more detail the system from which series III was obtained.

The values recorded in series III are hourly readings of viscosity. In a continuous reactor a gas was injected into a liquid to form a mixture of products which should have a viscosity between 86 and 98. The composition of the inlet gas was not constant and the reaction was controlled by feeding in more or less gas so as to achieve the required viscosity. Experiments with a gas of reasonably constant composition had shown that within the range considered a change in gas rate of 50 lb. per min. eventually produced a change in viscosity of about 10 units. After 82 min., 63 per cent. of this effect was produced and after 165 min. 86 per cent. If we assume a simple exponential response then this implies that approximately half the effect is achieved in one hour. Thus we may employ equation (3.5.3) with $\phi = \frac{1}{2}$, $\gamma_{-1} = 0$, $\gamma_0 = 1$, $\gamma_1 = 0$ to find the adjustment to be applied to the gas rate. We obtain simply

$$x^*_{p+1} = \Delta e_p + e_p = 2e_p - e_{p-1},$$

where x^*_{p+1} is the adjustment to be applied negatively to the gas rate measured in units of its *ultimate* effect on the viscosity. The logic of this control system is easily seen in this particularly simple example. Since only one half of the ultimate effect of any change realized will be produced at the next observation we must double the weight associated with e_p. However, the effect of having done this in the previous phase must also be undone; hence we must associate a weight of -1 with e_{p-1}.

The charts used to institute the scheme are shown in Fig. 5. The proportional (P) chart is simply a plot of hourly readings of viscosity. The origin is at the target value 92 midway between the limits 86 and 98 between which it is desired to control the response. A scale showing the "deviation from target value" is also shown. On the difference (D) chart is shown the difference between the last two readings plotted about an origin of zero. On the action (A) chart the sum of the deviations in charts P and D is plotted. This is referred to a scale showing directly the change in gas rate required. Each hour the operator obtains a new reading, he enters this on chart P, enters the difference from the previous reading on chart D, adds the two deviations and plots this on chart A. He then makes the appropriate adjustment to the gas rate which is shown on this last chart.

The above example happens to yield a particularly simple type of control. However, very little extra complication arises in other examples. When cumulative control is

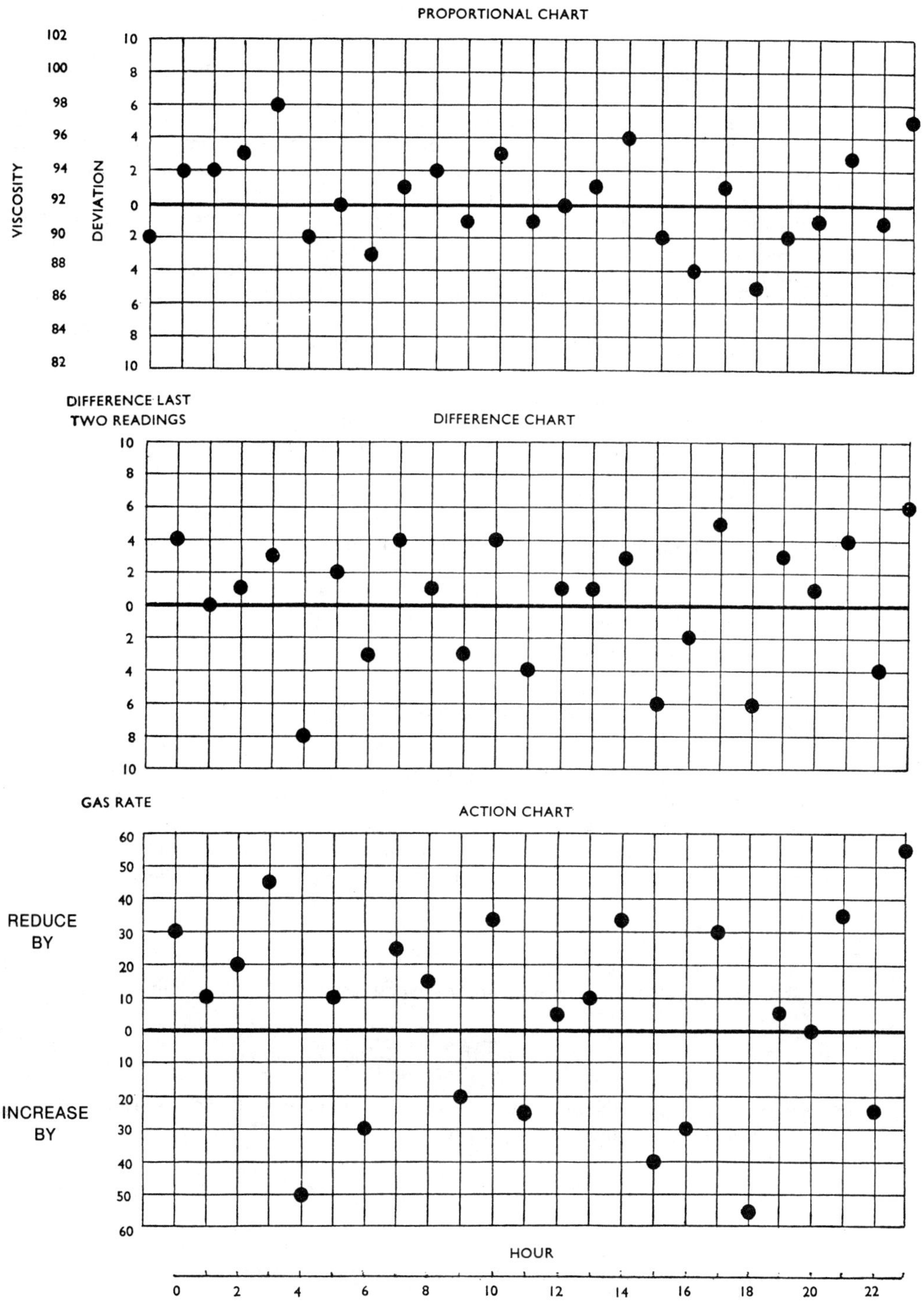

FIG. 5. Adaptive quality control chart for series III.

necessary a third (C) chart is added in which the cumulative sum of the deviations is recorded.

A suitably weighted combination of the readings from these charts is most easily obtained in practice by suitable adjustment of the scales so that simple addition of the deviations gives the required value for the action chart. For instance, if the values computed for the optimal scheme are $\gamma_{-1}^* = 0{\cdot}2$, $\gamma_0^* = 0{\cdot}6$, $\gamma_1^* = 0{\cdot}4$, the difference chart may be plotted on a scale in which one unit is 0·2 cm., the proportional chart on a scale in which one unit is 0·6 cm. and the cumulative chart on a scale in which one unit is 0·4 cm. The process operator then totals the deviations on a centimetre rule, plots the total on the action chart and reads off the appropriate adjustment in the scale of the manipulated variable.

Our experience so far, and the fact that fairly large changes in the weights can usually be tolerated without much effect on the efficiency of the scheme, leads us to believe that a fair amount of approximation can be tolerated. In particular, the dynamics need not be exactly represented and in adaptive optimization a moderate error in the estimate of β_{11}, the curvature of the response curve, would not greatly upset the scheme.

A small point which is perhaps worth noting is that the cumulative sums of errors which must be plotted when cumulative control is needed do not have the tendency to wander off the chart which is found when cumulative sums of the *original* deviations are plotted.

It is believed that many situations exist where a set of simple charts like the above could greatly improve the manual adjustments which are necessary in an infinite variety of industrial operations.

5. SPECTRAL APPROACH IN DISCRETE TIME

5.1. *Basic Spectral Formulae*

In this section we shall present some of the previous results in spectral form and thus prepare the way for a more realistic approach to the adaptive optimization problem to be considered in a later paper.

We have shown that the error ϵ_p could be expressed as a linear function over past history of the random variables θ_p and u_p. We represent such a linear function by

$$T_l = \sum_{j=0}^{\infty} \phi_j^{(l)}\, \theta_{p-j}, \tag{5.1.1}$$

and a simple calculation shows that the covariance between two such linear functions is

$$\operatorname{cov}(T_l, T_m) = \sum_{p=0}^{\infty} \sum_{q=0}^{\infty} \phi_p^{(l)}\, \phi_q^{(m)}\, \zeta_{|p-q|}, \tag{5.1.2}$$

where $\zeta_s = \sigma_\theta^2 \rho_s$ is the autocovariance of the θ's of lag s. If the θ's are observed at unit time intervals, then we may write

$$\zeta_s = \sigma_\theta^2 \int_{-\pi}^{+\pi} g_\theta^*(\omega)\, e^{i\omega s}\, d\omega, \tag{5.1.3}$$

where $g_\theta^*(\omega)$ is the spectral density function, defined in the frequency range $[-\pi, \pi]$ for operational convenience, and is equal to $\frac{1}{2} g_\theta(\omega)$, where $g_\theta(\omega)$ is defined in $[0, \pi]$.

If we now substitute for (5.1.3) in (5.1.2) we obtain

$$\operatorname{cov}(T_l, T_m) = \sigma_\theta^2 \int_{-\pi}^{+\pi} g_\theta^*(\omega)\, H_l(\omega)\, \bar{H}_m(\omega)\, d\omega, \tag{5.1.4}$$

where

$$H_l(\omega) = \sum_{j=0}^{\infty} \phi_j^{(l)}\, e^{-ij\omega}$$

is the discrete Fourier transform of the weights $\phi_j^{(l)}$ and the bar denotes a complex conjugate. In the special case where $l = m$, (5.1.4) reduces to

$$\operatorname{var}(T_l) = \sigma_\theta^2 \int_{-\pi}^{+\pi} g_\theta^*(\omega) |H_l(\omega)|^2\, d\omega. \tag{5.1.5}$$

More generally, we may write the variance–covariance matrix of several such linear combinations in the form

$$\sigma_\theta^2 \mathbf{R} = \sigma_\theta^2 \int_{-\pi}^{+\pi} \mathbf{H}(\boldsymbol{\omega})\, \overline{\mathbf{H}}(\boldsymbol{\omega})'\, g_\theta^*(\omega)\, d\omega, \tag{5.1.6}$$

where $\mathbf{H}(\boldsymbol{\omega})' = \{H_1(\omega), H_2(\omega), \ldots, H_k(\omega)\}$ and the matrix $\mathbf{H}(\boldsymbol{\omega})\overline{\mathbf{H}}(\boldsymbol{\omega})'$ is Hermitian.

Equation (5.1.5) says that the variance of any linear combination may be expressed as a *spectral average* with weight function or kernel $|H_l(\omega)|^2$. The generating function $H_l(\omega)$ for the weights $\phi_j^{(l)}$ is important and is called the *frequency response function* of the linear operation performed by (5.1.1), or else the *transfer function* when expressed as a function of $p = i\omega$. In engineering language, (5.1.1) is called a *linear filter* and is characterized by the important property that if the input θ_p is periodic of the form $Ae^{ip\omega}$, then the output is given by $AH(\omega)\, e^{i\omega p}$, i.e. a periodic disturbance with the same frequency. Since $H(\omega) = G(\omega)e^{i\phi(\omega)}$, it follows that the output is multiplied by a factor $G(\omega)$, the *gain*, and phase-shifted by an amount $\phi(\omega)$. We may now interpret (5.1.5) heuristically as follows; the random variable θ_p may be decomposed into contributions $\theta_p(\omega)$ at each frequency with variance $\sigma_\theta^2 g_\theta^*(\omega)$ which when passed through the linear filter (5.1.1) will be multiplied by the factor $|H(\omega)|$, so that its variance is $\sigma_\theta^2 |H(\omega)|^2 g_\theta(\omega)$. The total variance is then obtained by adding up the contributions at each frequency.

The importance of these ideas to statistical estimation becomes apparent when it is realized that any linear estimate may be regarded as a filtering operation performed on the observations. We illustrate this by considering the estimation of a mean

$$\bar{\theta} = \frac{1}{2n+1} \sum_{p=-n}^{n} \theta_p \tag{5.1.7}$$

and a harmonic regression coefficient in the form

$$b = \frac{2}{2n+1} \sum_{p=-n}^{+n} \theta_p \cos(\omega_0 p). \tag{5.1.8}$$

For convenience we have arranged that the number of observations in the sample $N = 2n+1$ is odd and that the time origin occurs at the $(n+1)$st observation. Regarding (5.1.7) and (5.1.8) as filtering operations, it is readily seen that their frequency response functions $H_1(\omega)$ and $H_2(\omega)$ are respectively given by

$$H_1(\omega) = \sin(\tfrac{1}{2}N\omega)/\sin(\tfrac{1}{2}\omega) \tag{5.1.9}$$

and

$$H_2(\omega) = \frac{\sin\{\frac{1}{2}N(\omega-\omega_0)\}}{\sin\{\frac{1}{2}(\omega-\omega_0)\}} + \frac{\sin\{\frac{1}{2}N(\omega+\omega_0)\}}{\sin\{\frac{1}{2}(\omega+\omega_0)\}}. \tag{5.1.10}$$

By choice of time origin we have arranged that the phase-shift is zero and hence the frequency response function is a gain only. Also, these functions differ only in that (5.1.9) is centred about zero frequency and (5.1.10) about the frequency ω_0. The function (5.1.10) is exactly the same as the spectral window associated with the truncated periodogram estimate of the spectral density and has been plotted along with other forms of $H(\omega)$ in a paper recently published by one of us (Jenkins, 1961). It has the property that it is highly peaked about the frequency ω_0 and in fact behaves like a Dirac δ function as $n, N \to \infty$. It is not surprising, therefore, that it may be shown rigorously that

$$\lim_{N\to\infty} N \operatorname{var}(\bar{\theta}) = \pi\sigma_\theta^2 g_\theta(0) \tag{5.1.11}$$

and

$$\lim_{N\to\infty} N \operatorname{var}(b) = 2\pi\sigma_\theta^2 g_\theta(\omega_0), \tag{5.1.12}$$

so that the variances for large samples may be approximated respectively by $\pi\sigma_\theta^2 g_\theta(0)/N$ and $2\pi\sigma_\theta^2 g_\theta(\omega_0)/N$. However, this approximation may be seen to be a good one even for low values of N by observing that the width of the spectral window (5.1.10) will be narrow in general. Thus if the spectral density remains fairly constant over the width of the windows associated with the two estimates T_l and T_m, then

$$\operatorname{cov}(T_l, T_m) \sim \sigma_\theta^2 g_\theta(\omega_0) \int_{-\pi}^{+\pi} H_l(\omega)\bar{H}_m(\omega)\,d\omega = 2\pi\sigma_\theta^2 g_\theta(\omega_0) \sum_{j=0}^{\infty} \phi_j^{(l)} \phi_j^{(m)}. \tag{5.1.13}$$

In the case of a variance, if $g_\theta(\omega_0)$ remains fairly constant over the bandwidth of the window, then

$$\operatorname{var}(T_l) \sim 2\pi\sigma_\theta^2 g_\theta(\omega_0) \sum_{j=0}^{\infty} (\phi_j^{(l)})^2 \tag{5.1.14}$$

from which (5.1.11) and (5.1.12) are readily obtained. From (5.1.13) it may be seen that T_l and T_m are approximately uncorrelated provided the weights $\phi_j^{(l)}$ and $\phi_j^{(m)}$ are orthogonal, which is true for the estimators $\bar{\theta}$ and b.

5.2. *Prediction in Discrete Time*

Suppose that we can observe the mean process θ_p exactly without any measurement error u_p, so that $z_p = \theta_p$. Then the problem of pure prediction one step ahead as defined by (2.4.1) is to minimize the prediction error ϵ_p given by

$$\epsilon_{p+1} = \theta_{p+1} - \sum_{j=0}^{\infty} \mu_j\,\theta_{p-j}. \tag{5.2.1}$$

With

$$\mu(\omega) = \sum_{j=0}^{\infty} \mu_j e^{i\omega j},$$

the frequency response function associated with the predictor weights, it is readily seen that the $H(\omega)$ associated with the prediction error (5.2.1) is $1 - e^{-i\omega}\mu(\omega)$ and hence that

$$\operatorname{var}(\epsilon_{p+1}) = \sigma_\theta^2 \int_{-\pi}^{+\pi} |1 - e^{-i\omega}\mu(\omega)|^2 g_\theta^*(\omega)\,d\omega. \tag{5.2.2}$$

If it is now assumed that the spectral density may be written in the form

$$g_\theta(\omega) = \frac{\sigma_\epsilon^2}{2\pi\sigma_\theta^2} |\psi(\omega)|^{-2} \quad (-\pi \leqslant \omega \leqslant \pi), \tag{5.2.3}$$

then, using fairly standard techniques in the calculus of variations, it may be shown that the loss (5.2.2) is minimized with respect to variations in $\mu(\omega)$ when

$$1-e^{-i\omega}\mu(\omega) = \psi(\omega). \tag{5.2.4}$$

Two special cases are of interest:

(i) If the θ_p series represents an autoregressive process, then $\psi(\omega)$ is a polynomial of the form

$$\psi(\omega) = 1-\sum_{j=1}^{k} a_j e^{-i\omega j}.$$

It follows that $\mu(\omega) = a_1 + a_2 e^{-i\omega} + \ldots + a_k e^{-i(k-1)\omega}$ so that the predictor weights μ_j are equal to a_j, as was shown in section 3.

(ii) If the θ_p series constitutes a moving average process

$$\theta_p = \epsilon_p + \beta_1 \epsilon_{p-1} + \ldots + \beta_m \epsilon_{p-m}, \tag{5.2.5}$$

then
$$\psi^{-1}(\omega) = 1+\beta_1 e^{-i\omega} + \ldots + \beta_m e^{-im\omega} \tag{5.2.6}$$

and the solution for the optimum weights as given by (5.2.4) becomes extremely complicated. An example when $m = 1$ has been given previously in Table 2.

5.3. *Control of the Mean when the Added Noise is Absent*

In this situation, $\epsilon_p = e_p$ and the *loop equation* (2.4.2) may be written

$$\Delta\epsilon_{p+1} = \Delta\theta_{p+1} - \sum_{j=0}^{\infty} w_j \epsilon_{p-j} \tag{5.3.1}$$

which may be expressed symbolically as

$$\epsilon_{p+1} = \left(1-\frac{B\mathscr{L}_w}{1-B+B\mathscr{L}_w}\right)\theta_{p+1} = (1-\mathscr{L}_\mu)\,\theta_{p+1}, \tag{5.3.2}$$

where $\mathscr{L}_w = \Sigma w_j B^j$ and B is the backward shift or delay operator. If the operator $\mathscr{L}_\mu$ is expanded in an infinite series in powers of B, then the terms of this series represent the weights given to previous θ's in obtaining a linear prediction of θ_{p+1}. The transform of these weights will be the frequency response function of the filter associated with (5.3.2) and this is obtained simply by replacing B by $e^{-i\omega}$, i.e.

$$H(\omega) = 1-\frac{e^{-i\omega}W(\omega)}{1-e^{-i\omega}+e^{-i\omega}W(\omega)}, \tag{5.3.3}$$

where $W(\omega) = \Sigma w_j e^{-i\omega j}$ is the frequency response function of the controller weights. Since $H(\omega) = 1-\mu(\omega)e^{-i\omega}$, it follows that the relation between the predictor and controller weights may be expressed in transforms by

$$\mu(\omega) = \frac{W(\omega)}{1-e^{-i\omega}+e^{-i\omega}W(\omega)} \tag{5.3.4}$$

yielding the inverse relation

$$W(\omega) = \frac{\mu(\omega)(1-e^{-i\omega})}{1-e^{-i\omega}\mu(\omega)}. \tag{5.3.5}$$

It follows from (5.2.4) that the optimum controller weights for the spectrum (5.2.3) are given by

$$W(\omega) = (e^{i\omega}-1)\{\psi^{-1}(\omega)-1\}. \tag{5.3.6}$$

In the case of an autoregressive scheme, $\psi^{-1}(\omega)$ is rather complicated, but for the moving average process with $\psi^{-1}(\omega)$ given by (5.2.6), the best controller weights are

$$W(\omega) = (e^{i\omega}-1)\sum_{j=1}^{m}\beta_j e^{-i\omega j}, \tag{5.3.7}$$

leading to various forms of difference control depending on the value of m. When $m = 1$, the weights correspond to first difference control. It is interesting to note the duality which exists between the autoregressive and moving average processes in so far as the optimum choice of predictor and controller weights is concerned.

If we use (5.3.4) and (5.3.5), it is possible to draw some general conclusions about the conditions which have to be satisfied in order for the predictor and controller weights to satisfy some of the constraints discussed in section 3, for example that they sum to unity. Since the sum of any set of weights is given by the value of the frequency response function at zero frequency, it follows from (5.3.5) that if the predictor weights do not sum to unity, the controller weights will always sum to zero. The disadvantages of the predictors which are not mean correcting are thus seen to carry over to the controller weights. If the latter sum to zero then in a situation where, for example, the last few deviations ϵ_p were all equal to some quantity δ, the controller would recommend no control action whereas a correction $-\delta$ is required for effective control.

When the predictor weights do add to unity, it is necessary to apply L'Hôpital's rule to evaluate the frequency response functions at zero frequency since both numerator and denominator vanish in (5.3.4) and (5.3.5). A simple calculation gives

$$\sum_{j=0}^{\infty}\mu_j = \sum_{j=0}^{\infty} j w_j \Big/ \left\{1+\sum_{j=0}^{\infty} j w_j\right\}, \tag{5.3.8}$$

$$\sum_{j=0}^{\infty} w_j = \sum_{j=0}^{\infty}\mu_j \Big/ \left\{\sum_{j=0}^{\infty}\mu_j+\sum_{j=0}^{\infty} j\mu_j\right\}. \tag{5.3.9}$$

It may be seen, therefore, that for a mean and trend correcting predictor, the controller weights must have an infinite sum and so we are led to some form of integral control.

5.4. *Control with Dynamics and Added Noise*

In discrete time there is a delay of one time unit before correction is made, so that even when all the correction becomes effective at the next instant, there are dynamic considerations present. For the moment we consider the dynamic aspects as referring to the extent to which the current correction becomes fully effective or not at the next instant, i.e. we ignore the initial delay of one time unit.

The loop equation may then be written, as before, in the form

$$\Delta\theta_{p+1} = \Delta\epsilon_{p+1}+x_{p+1}, \tag{5.4.1}$$

but now x_{p+1} is made up of two operations acting independently. The controller produces, from the past deviations ϵ_p, a correction whose negative value is $x^*_{p+1} = \mathscr{L}^*_w\,\epsilon_p$, where $\mathscr{L}^*_w$ is the linear operator associated with the controller. This is acted upon in turn by the dynamic characteristics of the process, characterized by a linear operator $\mathscr{L}_v$, finally producing $x_{p+1} = \mathscr{L}_v\mathscr{L}^*_w\,\epsilon_p$. From (5.4.1), it is seen that the linear operator connecting ϵ_{p+1} and the previous θ's is given by

$$\epsilon_{p+1} = \left(1 - \frac{B\mathscr{L}_v \mathscr{L}_w^*}{1-B+B\mathscr{L}_v \mathscr{L}_w^*}\right)\theta_{p+1}, \tag{5.4.2}$$

so that the loss may be written as

$$\operatorname{var}(\epsilon_{p+1}) = \int_{-\pi}^{+\pi} \left|1 - \frac{e^{-i\omega} V(\omega) W^*(\omega)}{1-e^{-i\omega}+e^{-i\omega} V(\omega) W^*(\omega)}\right|^2 g_\theta^*(\omega)\, d\omega. \tag{5.4.3}$$

Hence the effect of the dynamics is to replace the controller frequency response function $W(\omega)$ in (5.3.4) by the product of the controller transfer function and the process or valve transfer function in (5.4.3).

By writing down explicitly what is meant by the operation $x_{p+1} = \mathscr{L}_v \mathscr{L}_w^* \epsilon_p$, it is possible to determine the weights which are actually applied to the errors ϵ_p. We may write this operation in the form

$$x_{p+1} = \sum_{r=0}^{\infty} v_r x_{p-r}^*$$

which in turn may be written as

$$x_{p+1} = \sum_{r=0}^{\infty} v_r \sum_{s=0}^{\infty} w_s^* \epsilon_{p-r-s} = \sum_{l=0}^{\infty} w_l \epsilon_{p-l},$$

where

$$w_l = \sum_{r=0}^{l} v_r w_{l-r}^*. \tag{5.4.4}$$

It follows that the effective weights which are applied to the errors ϵ_p are the convolution of the controller weights w_j^* and the weights v_j characterizing the dynamic response of the process.

In order to consider the effect of the added noise, it is necessary to derive the modified loop equation in this case. We must now distinguish between ϵ_{p+1} and e_{p+1} because if the u_p process is not white noise, then it is no longer true that $E(\epsilon_{p+1}^2)$ is minimized when $E(e_{p+1}^2)$ is minimized. Ignoring dynamics, the loop equation becomes

$$\epsilon_{p+1} = \epsilon_p + (\theta_{p+1} - \theta_p) - x_{p+1} + u_{p+1}. \tag{5.4.5}$$

This differs from the previous loop equation in that the added error u_{p+1} now inflates the loss. The linear operators connecting ϵ_{p+1} and the past history of the series and the u series may be obtained by rearranging (5.4.5) in the form

$$(1-B+B\mathscr{L}_w)\,\epsilon_{p+1} = (1-B)\,\theta_{p+1} + u_{p+1},$$

so that the loss may be written as

$$\operatorname{var}(\epsilon_{p+1}) = \sigma_\theta^2 \int_{-\pi}^{+\pi} \left|1 - \frac{e^{-i\omega} W(\omega)}{1-e^{-i\omega}+e^{-i\omega} W(\omega)}\right|^2 g_\theta^*(\omega)\, d\omega$$
$$+\sigma_u^2 \int_{-\pi}^{+\pi} \left|\frac{W(\omega)}{1-e^{-i\omega}+e^{-i\omega} W(\omega)}\right|^2 g_u^*(\omega)\, d\omega. \tag{5.4.6}$$

This formula shows that a compromise must now be set up between direct control of the θ_p series and smoothing out the effect of the u_p series, since a choice of $W(\omega)$ which makes the first integral small would tend to make the second integral large and vice versa.

5.5. *Estimation of the Control Parameters from Past Records*

In previous sections, we have shown that if the z series is stationary, then the optimum control parameters may be determined from a knowledge of the spectrum of the process. However, since most practical series display non-stationarity of various types, we were led to consider controller weights corresponding to constrained predictors designed to deal with short-term non-stationarity. In section 3 this led to the consideration of a mixture of cumulative proportional and first difference control. A series for which this form of control was optimal, whilst non-stationary itself, is such that its second differences are stationary and follow a third order moving average process. We now proceed to exhibit these results in spectral form.

In Table 5 we give the frequency response functions corresponding to these three forms of control in discrete time.

TABLE 5

Frequency response functions for various types of control

Control	*Controller weights*	*Frequency response function*
Cumulative	$w_j = \gamma_1$ (all j)	$\gamma_1/(1-e^{-i\omega})$
Proportional	$w_0 = \gamma_0$ $w_j = 0\ (j \geqslant 1)$	γ_0
First difference	$w_0 = \gamma_{-1}$ $w_1 = -\gamma_{-1}$ $w_j = 0\ (j \geqslant 2)$	$\gamma_{-1}(1-e^{-i\omega})$

It follows that a mixture of these three forms of control has a frequency response function

$$W(\omega) = \gamma_{-1}(1-e^{-i\omega})+\gamma_0+\gamma_1/(1-e^{-i\omega}). \tag{5.5.1}$$

Substituting for this in (5.4.6) and assuming that the added noise is white, we may express the loss in the form

$$\operatorname{var}(e_{p+1}) = \sigma_z^2 \int_{-\pi}^{+\pi} \frac{|1-e^{-i\omega}|^4 g_z^*(\omega)}{|H(\omega)|^2}\, d\omega, \tag{5.5.2}$$

where $H(\omega) = 1+e^{-i\omega}(\gamma_{-1}+\gamma_0+\gamma_1-2)+e^{-2i\omega}(1-2\gamma_{-1}-\gamma_0)+\gamma_{-1}e^{-3i\omega}.$ (5.5.3)

From (5.5.2) it follows that $\operatorname{var}(e_{p+1})$ is finite only if the roots of $H(\omega)/(1-e^{-i\omega})^2$ lie outside the unit circle and this is the condition for stability mentioned earlier. We denote the second difference of the z_p series by d_p, and note that a second difference is equivalent to a filtering operation with frequency response function $(1-e^{-i\omega})^2$, which is a high pass filter removing low frequency trend. We may thus relate the spectrum of the second differences to that of the original series by a special case of a well-known formula relating the output and input spectral densities of a linear filter, viz.

$$\sigma_d^2 g_d(\omega) = \sigma_z^2 g_z(\omega)\,|1-e^{-i\omega}|^4. \tag{5.5.4}$$

It follows that the loss may be written as

$$\operatorname{var}(e_{p+1}) = \sigma_d^2 \int_{-\pi}^{+\pi} \frac{g_d^*(\omega)}{|H(\omega)|^2}\, d\omega. \tag{5.5.5}$$

This is similar to (5.2.2) and the converse of the result (5.2.4) shows that for fixed $H(\omega)$, this is minimized for variations in $g_d(\omega)$ of the form (5.2.3) when $\psi(\omega) = H(\omega)$, i.e. the second differences follow the third order moving average process whose generating function is given by (5.5.3).

At this stage we could proceed and find theoretically the values of $\gamma_{-1}, \gamma_0, \gamma_1$ which minimize (5.5.5) for a given $g_d(\omega)$. The variational argument no longer applies in this case and it would then be necessary to evaluate the integral (5.5.5) and find the values of $\gamma_{-1}, \gamma_0, \gamma_1$ which lead to minimum loss analytically or numerically.

From a practical point of view this is an academic exercise and needs to be replaced by a more empirical approach based on the following three stages:

(a) Given the reconstructed z series, to produce estimates of the parameters $\gamma_{-1}, \gamma_0, \gamma_1$.

(b) To keep a check on whether this form of control is adequate, i.e. whether the estimated deviations e_p after fitting the third order moving average are uncorrelated, or equivalently whether the spectrum of the e_p is effectively uniform or "white".

(c) To provide for updating or revision of the parameters as more information becomes available.

For (a) we may proceed as described in section 3.4 by evaluating the residual sum of squares $S(\gamma_{-1}, \gamma_0, \gamma_1)$ after fitting the third order moving average process for a grid of values of $\gamma_{-1}, \gamma_0, \gamma_1$ and then locate the best values of these parameters numerically. An alternative method is to write down the likelihood function of the second differences d_p. If the deviations e_p are independent, then the joint distribution P of these quantities in a sample of size N is given by

$$\log P = c - N \log \sigma_e - \frac{1}{2\sigma_e^2} \sum_{p=1}^{N} e_p^2.$$

Denoting by $c_s^{(e)}$ the serial covariance of lag s of the e's, then it is readily shown that, if N is even and equals $2n$,

$$c_s^{(e)} = \frac{\pi}{N} \sum_{j=-n}^{n-1} I_N(\omega_j)\, e^{i\omega_j s}, \tag{5.5.6}$$

where $$I_N(\omega_j) = \frac{1}{\pi N} \left| \sum_{p=1}^{N} d_p\, e^{-i\omega_j p} \right|^2 \quad (\omega_j = 2\pi j/N)$$

is the sample periodogram of the second differences. Since the sum of squares of the e_p in (5.5.5) is simply $Nc_0^{(e)}$, it follows that the latter quantity may be expressed as a weighted sum of the periodogram ordinates. Using the following approximate relation between the periodograms of the e_p and d_p series,

$$I_N^{(d)}(\omega_j) = I_N^{(e)}(\omega_j)\, |H(\omega_j)|^2, \tag{5.5.7}$$

where $H(\omega)$ has been defined by (5.5.3), we obtain the following approximate form for the log likelihood of the d_p's:

$$\log L = c - N \log \sigma_e - \frac{\pi}{2\sigma_e^2} \sum_{j=0}^{n} \frac{\gamma_j I_N(\omega_j)}{|H(\omega_j)|^2}, \tag{5.5.8}$$

where $$\gamma_j = \begin{cases} 2 & (\omega_j \neq 0, \pi), \\ 1 & (\omega_j = 0, \pi). \end{cases}$$

When N is odd, the expression is the same except that the ordinate $\omega_j = \pi$ does not appear. A similar form for the likelihood function has been derived previously by

Whittle (1950). Since

$$I_N(\omega_j) = \frac{1}{\pi}\left\{c_0 + 2\sum_{s=1}^{N-1} c_s \cos(\omega_j s)\right\}, \tag{5.5.9}$$

it is necessary, in order to evaluate the likelihood function, to calculate all the serial correlations in the sample. The expression (5.5.9) may be replaced to a high degree of accuracy by an estimate of the power spectral density, viz.

$$p_N(\omega_j) = \frac{1}{\pi}\left\{c_0 + 2\sum_{s=1}^{m} \lambda_s c_s \cos(\omega_j s)\right\}, \tag{5.5.10}$$

where the λ_s's are one of the sets of weights used in spectral analysis.

If we now substitute in (5.5.8) we have the following form for the log-likelihood function, accurate to a fairly high degree of approximation:

$$\log_e L \sim c - N \log \sigma_e - \frac{\pi N}{4k(m)\sigma_e^2} \sum_{j=0}^{k(m)} \frac{\gamma_j p_N(\omega_j)}{|H(\omega_j)|^2}, \tag{5.5.11}$$

where $\omega_j = \pi j/k(m)$ and $k(m)$ is the number of effectively independent spectral estimates which depends on λ_s and m and has been tabulated for the various weight functions used in spectral analysis by Jenkins (1961).

The advantage of (5.5.11) is that if the power spectrum $p_N(\omega)$ is known, the likelihood function may be calculated very easily for a grid of values of $\gamma_{-1}, \gamma_0, \gamma_1$ without recourse to any further calculations from the sample whilst the sum of squares method given in section 3.4 involves the fitting of the model to the data for each set of values of $\gamma_{-1}, \gamma_0, \gamma_1$. However, the latter method is easily programmed for an electronic computer and also gives information about the individual estimates e_p which would not be obtained directly in the likelihood-function approach.

The maximum likelihood estimator of σ_e^2 is readily seen from (5.5.11) to be

$$\hat{\sigma}_e^2 = \frac{\pi}{2k(m)} \sum_{j=0}^{k(m)} \frac{\gamma_j p_N(\omega_j)}{|H(\omega_j)|^2}, \tag{5.5.12}$$

and hence can be obtained easily when the optimum values of $\gamma_{-1}, \gamma_0, \gamma_1$ are available. An approximate test for the whiteness of the residual spectrum $g_e(\omega)$ may then be obtained by observing that

$$p_N(\omega) \sim \hat{\sigma}_e^2 g_N^{(e)}(\omega) |H(\omega)|^2 \tag{5.5.13}$$

and hence $g_N^{(e)}(\omega)$ may be estimated directly, since all the other quantities will have been evaluated in the calculation of the likelihood function.

Given the spectra of the second differences, it is also possible to draw some conclusions about the form of control which will be required. We illustrate this remark by referring to the spectra of series I–III which are plotted in Fig. 6. If no control is possible, i.e. $g_d(\omega)$ is a constant, then the second differences will have a spectrum of the form

$$\frac{2}{3\pi}(1-\cos\omega)^2$$

which increases rapidly at higher frequencies and has been plotted as a base in Fig. 6. The need for proportional control is indicated by a boosting up of the lower and intermediate frequencies as in series I but more strongly in series III; however, the zero frequency spectral density will, in this case, be zero, subject to sampling fluctuations. The desirability of some form of integral control is indicated by the presence of non-zero spectral density at $\omega = 0$ and in higher density at the lower frequencies generally. For series II, this effect is quite marked since the spectrum is effectively

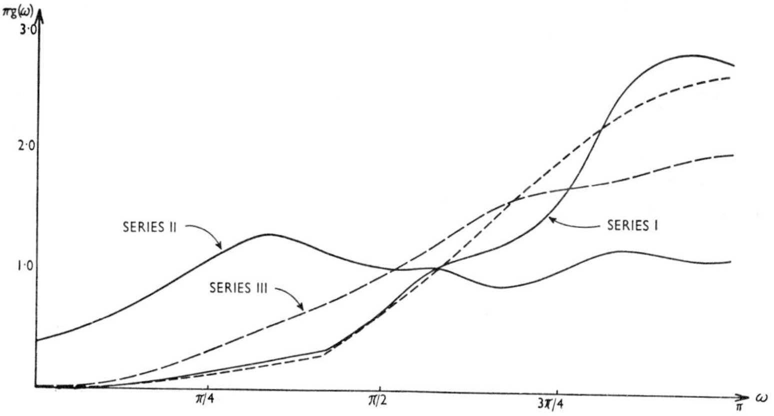

FIG. 6. Spectra of second differences for series I–III.

white and it is even possible for the spectrum to have a maximum at zero frequency if $\gamma_1 > 1$. This is due to the fact that the zero in the frequency response function of the second difference filter is compensated for by an effectively infinite spectral density at zero frequency in the original series due to the presence of trends.

ACKNOWLEDGEMENT

This research was supported by the National Science Foundation.

REFERENCES

BARNARD, G. A. (1959), "Control charts and stochastic processes", *J. R. statist. Soc.* B, **21**, 239–272.
—— JENKINS, G. M. and WINSTEN, C. B. (1962), "Likelihood inference and time-series", *J. R. statist. Soc.* A, **125**, 321–372.
BOX, G. E. P. (1957), "Integration of techniques in process development", *Trans. of the 11th Annual Convention of the Amer. Soc. Qual. Control*, 687.
—— (1960), "Some general considerations in process optimisation", *Trans. Am. Soc. mech. Engrs*, Ser. D (*Basic Eng.*), **82**, 113–119.
—— and CHANMUGAM, J. R. (1962), "Adaptive optimisation of continuous processes", *Industrial and Engineering Chemistry* (*Fundamentals*), **1**, 2–16.
BROWN, R. G. (1959), *Statistical Forecasting for Inventory Control.* New York: McGraw-Hill.
COX, D. R. (1961), "Prediction by exponentially weighted moving averages and related methods", *J. R. statist. Soc.* B, **23**, 414–422.
HOLT, C. C., MODIGLIANI, F., MUTH, T. F. and SIMON, H. A. (1960), *Planning Production, Inventories and Work Force.* London: Prentice-Hall.
JENKINS, G. M. (1961), "General considerations in the analysis of spectra", *Technometrics*, **3**, 133–166.
MUTH, J. F. (1960), "Optimal properties of exponentially weighted forecasts", *J. Amer. statist. Soc.*, **55**, 299–306.
WHITTLE, P. (1950), "The simultaneous estimation of a time-series harmonic components and covariance structure", *Trabajos de Estadistica*, **3**, 43–57.
ZADEH, L. A. and RAGAZZINI, J. R. (1950), "An extension of Wiener's theory of prediction", *J. appl. Phys.*, **21**, 645–655.

Discussion on Paper by Professor Box and Dr Jenkins

Dr K. D. Tocher: Those of you who have had time to read the paper carefully will have noticed the subtle comment tucked away at the beginning of the section on Spectral Methods which makes criticism of the paper difficult. There the authors prepare the way for a more realistic approach to a theory of optimization and so any criticisms may be answered with the retort "it will all be put right in the next paper".

I think this paper is important because it is one of the first entries of the statistician into the field of control. There are already two well-established groups concerned with system control: first, there are the engineers with a long history of mathematical analysis, and second, the more recent entrants, the cyberneticians.

I think it will be rewarding for us to discuss the two very different points of view of these two groups and see how the authors' problems and solutions fit into the pattern.

Both the control engineer and the cybernetician start by drawing the same box with lines on the left called the inputs, lines on the right called the outputs and other lines on the bottom which represent the control. They both discuss the problem of what to do with these control lines, so that in the face of certain behaviour of the input lines, they will obtain the required behaviour on the output lines.

The engineer assumes he understands the mechanism of the system and in the absence of control inputs, he can describe the behaviour of the system by means of a set of equations, usually differential ones, which he solves by using Laplace transforms. He can choose his control inputs and regards these as the outputs of another system whose inputs are the outputs of the original system and possibly some or all of the inputs. He now *designs* this new system so that the resulting behaviour of the dual system is that required.

This raises the specification of the objectives of the control and if this is many-fold they may be conflicting. Often, the problem then ceases to be mathematically soluble and certain *ad hoc* rules are introduced; there is a great art in finding adequate solutions. But this is typical of all applications of a scientific theory and does not nullify the position of the engineer who gives a precise answer to a precise question about a precise system.

Originally, the specification of the whole system was deterministic, but in the last decade or so, systems with inputs only described statistically have been studied. The outputs, whether under control or not, are then only describable statistically and for analysis of systems under these conditions spectral analysis has been developed into a powerful tool. That engineers should use this mode of description is not unexpected, since the frequency and phase concepts are the stock-in-trade of the electrical engineer; it is in the field of electrical systems that the most important applications have been made.

Nevertheless, the mechanism of the system has continued to be assumed deterministic and known. The cybernetician who wants to deal with much larger systems found this assumption unrealistic, since he did not know the mechanism of his system, much less the value of the parameters describing it. What can be done in these circumstances? The principal work in this field is due to Ashby who replaces the general inputs and outputs by special Boolean ones taking only one of two values. In principle, this can always be done for any system although it is not very practical in most cases. The description of the system is then an operational one, determining for each input state the outputs. The whole mechanism can then be described by a set of truth tables. This apparently unpromising abstract approach has yielded in Ashby's hands some remarkable results. By recognizing that *any* description is incomplete, he has simplified the description of many situations and given a general analysis of their behaviour. Perhaps the most remarkable of these is his explanation of habitation which he shows is an almost inevitable feature of any system. Habitation is the often-observed phenomenon whereby response to a stimulus becomes more rapid and reproducible after repetition.

Now the more naïve cyberneticians, objecting to the precise analysis of the engineer and rightly pointing out that his techniques only control the system against a specified set of disturbances, wanted to discuss the control of any system against *any* kind of disturbance. A little reflection will show that this is not possible without making some kind of restraining assumptions about the system, knowing something about it, if not its exact mechanism.

Ashby's technique enables a general theorem to be established about what can be done in this direction. His Law of Requisite Variety states, roughly speaking, that to control a system which can take on a large number of states requires a control mechanism with a similar number of states. It is not possible to control a complicated system with a simple control.

The exact formulation of the law involves the idea of the entropy of the system which is related to Shannon's measure of information in a communication channel.

Now the behaviour of the system is defined completely by exhaustive experiments; then a model can be set up which reproduces the exact behaviour of the system and so, in a sense, its mechanism can be described. Since exhaustive experiments cannot be conducted on a real system, its behaviour in untried circumstances can only be predicted by a model induced from the practical experiments. Now the inconsistency of the naïve view is revealed. For what system too complicated to understand will we be able to specify exactly the relevant input variables? Clearly the complications which deny us one thing deny us the other. These unknown but relevant inputs give the system a statistical behaviour and there is no way of differentiating between systems with statistical behaviour induced by unknown inputs and those with a statistical mechanism.

The randomness of the behaviour of the system is a reflection of the large variety of the system which must be matched by a corresponding variety in its control mechanism. This is loosely interpreted as a requirement that the control must have random behaviour and so the idea has grown up that the control mechanism must have a random character.

The idea seems to have achieved such popularity, due no doubt to its presentation as a panacea, that I decided to build a machine to demonstrate that randomness alone would never give adequate control of a system.

The artificial system studied consisted of a large electrical network, each limb of which contained one of the following: a resistance, a diode, an inverting amplifier, a quantifier (Schmidt trigger for the technically interested). Certain of the nodes were chosen arbitrarily as inputs and certain others as outputs. The system would have defied an Atlas to calculate its behaviour and so, in a sense, its mechanism was unknown.

Ten of the inputs were chosen as control inputs and the problem was to determine their settings to achieve some objective of the outputs. The simplest was that one output should take a specified value. The control inputs were connected to condensers which were either charging or discharging, and so the whole mechanism was in a dynamic state. The direction of movement is determined by a relay-like device which can be set at will. A detection mechanism is attached to the chosen output and gives a signal if the discrepancy between the output and the desired value starts to increase. This signal then changes the direction of movement of the inputs by resetting the relays to a new random configuration, the input then darting off in a new direction. The whole mechanism can work up to 20,000 changes a second and yet the "settling" time for difficult values of the required output level can be 10–15 minutes.

If a random element is used in the control of a system it simply expresses our inability to specify a method of selecting a point in a phase space and so we search the space "at random". The searching process itself gives evidence about the system which is not used and can always be replaced by some procedure which uses our partial knowledge of the system. The concept of a system about which we know nothing is as meaningless as the idea of "complete ignorance" so beloved of the exercise-solvers who confuse the whole field of inductive reasoning.

Thus we can summarize the two approaches of the engineer and cybernetician as the precise solution to a precise problem and an attempt to solve an imprecise problem.

Now, in case the idea arises that this vote of thanks is taking the second classical approach—talking about something other than the paper—we return to the approach of this paper. How does this fit in to the framework? Where between these two extremes do Box and Jenkins lie?

First, the inputs to the system have vanished after a brief introduction in the first page or two. Their place is taken by the quantity θ which acts as a summary of the whole mechanism. The statistical behaviour of θ can either be generated by the statistical behaviour of the suppressed inputs or by a stochastic mechanism; either is possible and a decision is irrelevant. Certainly, then, the system studied has an unknown mechanism.

However, now the operational definition of the system is also denied to us since we cannot experiment with or observe the effects of changes of input pattern. What alternative method of describing the system is used? The analysis is restrained to those systems whose θ behaviour can be described in terms of the past behaviour of θ and nothing else about the system. In the case of discrete time θ_{n+1} is a function of θ_n, θ_{n-1}, ... But it cannot be a function of these alone, as this would make the system deterministic and therefore of no interest to a statistician.

The stochastic element is restored in a very curious way. The quantities z, the observed or estimated values of θ, are used instead of the θ's in the functional relationship. The characteristics of the observational or estimating procedures are built into the stochastic nature of the system—this is certainly a radical extension of the Uncertainty Principle!

However, I cannot complain if the authors do not choose to analyse the systems I am interested in, and their approach will certainly arouse interest in the field of control by a third group, the statisticians.

At the beginning of the paper there is an interesting section about the difference between technical and empirical feedback. I think that this is a rather important idea. It really is a reflection of the difference in the kind of model of the system which is used. One never can deal with any system without making a model of it. Perhaps the authors would like to comment on the idea that technical feedback corresponds to a fairly detailed model, a model rather like the control engineers use, while empirical feedback corresponds to a rather vaguer type of model, the kind that cyberneticians use, and the model which is used for this particular paper is really somewhere halfway between. It is an operational model, since it describes what happens in terms of certain equations but does not postulate any mechanisms as to why the equations describe the behaviour.

I am sure the authors would agree that this is a very abstract model. The purpose of it was, I suspect, to gain experience. Although they give evidence that it does represent quite realistically some of the time series that one meets, it does not necessarily give a representation of everything that we meet. A model of this kind involves a series of assumptions. Nobody minds about this. Everybody knows that you have to make assumptions in a theoretical analysis, but I think it is useful to bear in mind clearly these assumptions: I think there is a danger that one may fall into a trap if we read this paper quickly (which, however, is not an easy thing to do).

First of all, of course, there is the assumption that θ_{n+1} is a linear function of the previous θ values; then there is the assumption of a quadratic loss function. Operational research workers are, I believe, coming to query this now that they are dealing increasingly with economic problems. Then there is an assumption that the trend relation between the effector variable and the control variable is linear. Finally, of course, there are assumptions about independence of the observational errors. Now, as a first step, this set of assumptions seems sensibly and carefully chosen so that some mathematics is possible. But several times in the paper, there are hints dropped that really the optimum choice of control is very robust. Even if the weights are chosen wrongly, control is not substantially degraded. But it would be dangerous to draw from this perfectly valid result the conclusion that this technique will work in a wide range of cases. In fact, all that is proved is if the real situation does satisfy all the assumptions that I outlined, i.e. is represented by the model analysed in the paper, then variations within that model are not important. But whether or not the model is applicable in any particular real case, and whether departures from these assumptions are important, is something that would need a lot more discussion.

In the last part of the paper on the spectral approach to discrete time, it is interesting to see the introduction of Fourier transforms. When the control engineer has to deal with problems in continuous time he uses the Laplace transform. In the last decade or so, mainly due to the impact of computers on the industrial world, his interest has moved to sampled data systems, in which a continuous function is sampled only at discrete intervals of time. For the analysis of systems of that kind, the Laplace transform is just about tolerable, but for the design of systems it was found that the complications caused by discreteness of the observations made the old techniques quite unusable. Consequently, the Z transform was invented; this is merely the familiar generating function for time series. It is called the Z transform because Z is used as the carrier variable.

There are many points of similarity between the treatment in the last part of this paper and that of Ragazzini and his co-workers in this field, but no reference is made to their work. Perhaps the authors would like to comment on the connections between the two pieces of work.

I have much pleasure in proposing a vote of thanks.

Mr G. A. Coutie: It is often recognized among statisticians that in some branches of their subject the rate of progress in theoretical thinking far outstrips the rate at which useful applications are made of the theory. The subject matter of the paper by Professor Box and Dr Jenkins concerns one such branch, and they are to be congratulated on the excellent way in which they have demonstrated the inter-relationships of adaptive optimization, adaptive quality control and prediction. I am sure that the authors will be keen to put the methods to the test in a research project either at Madison or in London. In the chemical industry in particular, in which the techniques of this paper are most likely to be of advantage, many of the methods that have been proposed by control engineers for the optimal control of chemical processes, particularly those involving feed-forward control or the use of an on-line digital computer, have not proved to be as successful in practice as most people hoped they would. Often the reason for this is nothing to do with the techniques themselves, but is due to a lack of understanding of the basic mechanism of a chemical process, poor plant design, inadequate methods for chemical analysis, lack of proper instrumentation, and so on. In their place a number of simpler, more empirical methods such as evolutionary operation or regression analysis have often been tried, and, although these methods tend to throw new light on a process, one feels that they are only being used for lack of more reliable, sophisticated tools. The present paper goes some way towards bridging the gap between the theory and practice of control, but the true value of the paper will only be realized when a plant chemist is able to claim that he has improved the efficiency of his process by being able to predict θ_{p+1} more accurately than hitherto.

I should like to confine the rest of my remarks to the problem of prediction. In recent years there has been considerable interest in many industrial concerns in the use of prediction methods for the short-term forecasting of demand. Usually the problem is to derive a simple formula, suitable possibly for routine use on an electronic computer, which will predict the customer demand of a sales product for a short period ahead in order that the resultant forecast may be used as a basis for stock control and production scheduling purposes. One of the most suitable forms of prediction in these circumstances has been shown, initially by Professor C. C. Holt, to be an exponentially weighted moving average, possibly corrected for trend and seasonal effects. Professor Box and Dr Jenkins point out in section 2 of their paper that what they describe as proportional control, which amounts to the use of an exponentially weighted moving average, has been used for the prediction of sales (or, preferably, demand) and they go on in section 3 to describe an extended system involving proportional, integral and first difference control.

At first sight it seems that this system may also be used for the prediction of demand, in addition to the control of production processes, but a difficulty may well arise because of a shortage of data on which to estimate γ_{-1}, γ_0 and γ_1. My experience of applying exponentially weighted moving averages to monthly demand data of sales products, where often 6 or 7 years' data are all that is available, particularly when a parameter is included to allow for trends in demand, is that although the prediction error is relatively insensitive to small changes in the estimated parameters, this feature is more than offset by the fact that the best values of the parameters may themselves be highly dependent on the particular stretch of data from which they are estimated. I wonder whether the same is true for the estimation of γ_{-1}, γ_0 and γ_1. If the parameters are estimated only once and then used for future predictions, one is sometimes doubtful about their validity if the nature of the series is liable to change.

In these circumstances we have been led to an alternative approach in which we make estimates of the parameters which will be as robust as possible to any changes in the nature of the time series, and in which we operate a control scheme on the cumulative prediction errors to allow for corrections to be made if the predictions are found to be lagging behind the time series. For many series which are expected to consist of runs at a constant mean, with occasional jumps in the mean, we have found it preferable to use only proportional control, with a fixed value of $\gamma_0 = 0{\cdot}2$, which seems to give a reasonable compromise between discounting random error and allowing for real shifts in mean. For series which have long-term trends as well, it would be necessary to include a trend parameter in Holt's method of prediction together with the one for proportional control.

Dr O. L. Davies has recently suggested that a backwards sequential test on the prediction errors may be used to detect lags in the predictions, and we have investigated empirically the estimation of suitable control limits from past data. To be explicit, if the demand in month p is θ_p we may predict the following month's demand by applying the correction $x_{p+1} = X_{p+1} - X_p = 0{\cdot}20\epsilon_p$, say, to the predicting series X_p, where $\epsilon_p = \theta_p - X_p$. Any lags in prediction are detected by comparing the cumulative sums of errors

$$S_{p,i} = \sum_{j=0}^{i-1} \epsilon_{p-j}$$

with a set of control limits by means of a backwards sequential test. In practice we find that the estimation of suitable control limits from past data is usually less sensitive than that of the best parameters for proportional control and allowance for trend in Holt's method. The only justification for such a relatively simple procedure is that we have applied it to a wide range of sales products, and that it works well. The three-term prediction method described by the authors appears to allow for changes in trend in a much more systematic way than previous methods, and I look forward to evaluating it in the best possible way—by trying it.

I have much pleasure in seconding the vote of thanks to the authors for their excellent paper.

The vote of thanks was put to the meeting and carried unanimously.

Professor E. Parzen: I would like to compliment Professor Box and Dr Jenkins on their most original paper. In a recent paper (*Ann. math. Statist.*, **32** (1961), 951–989), I wrote that, in my view, modern time-series analysis is a subject which embraces three fields which, while closely related, have tended to develop somewhat independently. These fields are (i) statistical communication and control theory, (ii) the probabilistic (and Hilbert space) theory of stochastic processes possessing finite second moments, and (iii) the statistical theory of regression analysis, correlation analysis, and spectral (or harmonic) analysis of time series. The paper presented here this evening is a brilliant synthesis of ideas from these three fields, and I regard it as a truly important contribution to modern time-series analysis. I hasten to add that there are a number of theoretical points in the paper requiring clarification which I feel sure will come in time. At present I have two comments to offer.

I believe that the distinction made by the authors between technical feedback and empirical feedback is a distinction which other researchers have made under other names.

Thus in my paper I write on p. 952:

> "It has been pointed out by various writers (see, for example, Neyman) that there are two broad categories of statistical problems: problems of stochastic model building for natural phenomena and problems of statistical decision making. These two categories of problems are well illustrated in the analysis of economic time series: some study time series in order to understand the mechanism of the economic system while others study time series with the simple aim of being able to forecast, for example, stock market prices."

I wonder if Professor Box and Dr Jenkins would agree that technical feedback corresponds to model building and empirical feedback to decision making?

Next, I would like to comment on the sentence following equation (3.4.3), with the aim of showing that the paper does in fact make use of the Hilbert space theory of time series, a fact which I am sure the authors will find horrifying. Given a time series $\{z_p, p = 0, 1, \ldots\}$, let $\hat{z}_p$ denote the minimum mean square error linear predictor of z_p, given $z_{p-1}, z_{p-2}, \ldots$. The prediction error $e_p = z_p - \hat{z}_p$ may be called the *innovation* at time p; this terminology is due to Wiener. The innovations $\{e_p\}$ form an uncorrelated sequence. By definition

$$\Delta z_{p+1} = \Delta e_{p+1} + \Delta \hat{z}_{p+1}, \tag{1}$$

so that

$$\Delta z_{p+1} = \Delta e_{p+1} + \sum_{j=0}^{\infty} w_j\, e_{p-j} \tag{2}$$

if it holds that

$$\Delta \hat{z}_{p+1} = \sum_{j=0}^{\infty} w_j\, e_{p-j}. \tag{3}$$

When in fact does (3) hold? If the process $\{z_p\}$ is purely non-deterministic (in the sense that it has an empty remote past), then its values z_p may be reconstructed from the innovations in the finite past by

$$z_p = \sum_{j=0}^{\infty} c_j\, e_{p-j}, \tag{4}$$

where the coefficients c_j may depend on p and satisfy

$$E(z_p\, e_{p-k}) = c_k\, E(e_{p-k}^2) \quad (k = 0, 1, \ldots). \tag{5}$$

Similarly

$$\Delta z_{p+1} = \sum_{j=-1}^{\infty} w_j\, e_{p-j}, \tag{6}$$

where

$$E(\Delta z_{p+1}\, e_{p-k}) = w_k\, E(e_{p-k}^2). \tag{7}$$

From the last equation, one sees the probabilistic meaning of the controller weights w_j, since

$$\Delta \hat{z}_{p+1} = \sum_{j=0}^{\infty} w_j\, e_{p-j}. \tag{8}$$

In particular, in order that the optimal predictor be an exponentially weighted mean and thus satisfy

$$\Delta \hat{z}_{p+1} = \gamma_0\, e_p, \tag{9}$$

it is necessary and sufficient that $w_k > 0$ for $k \geqslant 1$ which means that the increment Δz_{p+1} is orthogonal to all innovations at times $p-1, p-2, \ldots$. These results are of course contained in the paper. But I hope the formulation I have given here helps to clarify them.

In conclusion, let me mention one of the theoretical points which I believe needs clarifying. In the foregoing derivation, I assumed that $\{z_p\}$ is purely non-deterministic. How does one reconcile this with the assumption that $\{z_p\}$ may contain deterministic components?

Professor G. A. BARNARD: I wish to remark on the relationship between the methods which Professor Box and Dr Jenkins propose and the method of cumulative sum charts which was given in my paper in 1959. It seems to me that for practical purposes the methods of estimating the relative proportions of proportional, derivative, and integral control more or less empirically as, they suggest, will in practice largely supersede the use of cusum charts. I say *largely*, for one thing, because not all of us have computers, so that the actual evaluation may not be an easy matter and graphical methods may have to be used. But also there may be types of disturbance process for which the method based on cumulative sums is more uniquely appropriate. It is clear from some work which we have done at Imperial College, which Mr Clark will mention later on, that the cases where this can happen are extremely rare.

As has already been pointed out, the degree of control is fairly robust with respect to variations in the weights; and even in situations where one can say theoretically that the "V-mask" should be better, it appears that one rarely loses much by using Professor Box and Dr Jenkins's method. Thus, for practical purposes, cusum charts can be regarded as being superseded for all those who are able to carry through the computations.

The other point that I would like to make is really addressed to those of us in the audience who are not primarily control engineers. One should ask oneself, in connection with the design of any experimental programme, whether this may not best be regarded as a problem in control engineering. Because nowadays, to recall a phrase of Trotsky, the era of permanent experimentation is upon us, and the idea of an experiment as being something which is started and finished and then the results applied is no longer appropriate. Generally, whenever we are dealing with a system which can be thought of as changing as we are investigating it, we may do well to look at it as presenting a control problem.

This applies in particular to another type of experimental problem which has been receiving a lot of attention lately, namely, screening. We had a case just recently of someone who was concerned with testing a series of compounds as additives to animal feeding stuffs to see whether they did any good or not. He had formulated his problem as one of testing a given compound to know whether it did or did not do any good to the animals being fed with it. Instead it should have been recognized that when they stopped testing one compound they would simply start testing another. The question was thus not properly formulated as one of testing one compound and then making a decision; it was rather of testing a compound until it appeared more appropriate to start testing another. This general formulation of the screening problem represents a way of looking at continuous experimentation in general. One asks the question, not how large a sample should we take, or how long should we go on with this experiment, but, rather, at what point does it become sensible to start asking another question than the one that is being asked in this experiment. If one looks at continuous experimentation in that way, then one can see that the design and the formulation of problems to be answered are different. If one is concerned with processes in which one might think of controlling a large number of variables or controlling the process in many different ways, one can get mixtures of continuous experimentation with the screening model, and continuous experimentation with the control model. One would experiment with one variable and go on experimenting with it so long as it appeared reasonable to hope that by controlling that variable one would do some good, and then perhaps stopping controlling that variable and looking at another, and so on. In the modern world we have people who never stop doing experiments, and for these people the question of how long to go on and when to stop is not a meaningful question.

Dr F. R. HIMSWORTH: Like Dr Tocher, I am glad to see quality control and control engineering being considered together. It is surprising that this has not happened before; it has presumably come about only since digital computers focused attention on sampled data.

I am glad to see analysis of actual time series, and the important matter of checking the prediction against fresh data, not against the data used for analysis. I suspect that the disturbances which seriously affect a process are usually step changes in the input variables. This does not necessarily mean a step change in the output; for example, if the process contains one exponential transfer stage the output changes exponentially. This may be why the exponentially weighted moving average is so often a good predictor. Trends in input variables, such as those mentioned in the paper, are not likely to be fast enough to make control difficult.

I was surprised to find no mention of the extensive work of Ragazzini and Franklin, Goodman and Reswick, and others on sampled data control theory. Also, there is little or no mention of stability, though this is the first concern of the control engineer. Is an optimal control system necessarily stable? A remark in section 3.4 suggests it is not. It is usual to work at some comfortable distance from the point of instability.

Evolutionary operation is difficult to recognize in section 1.4. The original concept at least was concerned purely with "empirical feedback"; if "technical feedback" is required a research campaign rather than evolutionary operation seems appropriate. I am interested in the observation (section 1.2) that in the "empirical feedback" situation each modification should be tested once only; we reached the same conclusion in developing the regular simplex design for evolutionary operation.

Professor P. WHITTLE: Firstly, I should like to add my compliments to the authors on a very fine paper. There is a great deal that one could say, but I shall restrict my comments to one point.

In section 3.4 the authors show that the linear least-square predictor, when constrained to be exact for a polynomial in time, is the predictor that would have been optimal for a series obtained by n-fold partial summation of a certain stationary series. However, the converse proposition is a more natural one; if the series x_t is generated by the relation

$$\Delta^n x_t = y_t, \qquad (1)$$

where y_t is stationary, then, although x_t is not itself stationary, one can calculate a linear least-square predictor for it, and this predictor turns out to be automatically exact for x_t sequences which are polynomials of order $n-1$ or less. Prediction problems of this type have been treated by Yaglom (*Mat. Sbornik*, **37** (1955), 141–196).

This result is an extreme case of a general result: if a stationary series has covariance generating function $h(z)$, and if $h(z)$ has a pole of order n at $z = \alpha^{-1}$ ($|\alpha|<1$), then the Wiener predictor for this series is exact for sequences $t^j \alpha^t$ ($j = 0, 1, \ldots, n-1$). If the series x_t is generated by a relation

$$(1-\alpha T)^n x_t = y_t, \qquad (2)$$

where T is the backward translation operator, then the covariance generating function of x_t will in fact have an n-fold pole at $z = \alpha^{-1}$, and the predictor for the stationary series x will be exact for the sequences $t^j \alpha^t$. If we let α tend to 1 from below, then relation (2) will tend to relation (1), and the limiting predictor obtained in this fashion will be exact for polynomials.

I feel that model (1) is often a much more natural one for processes with drift than the conventional one of a polynomial trend with a superimposed stationary component, as the drift is likely to be built into the process, and to be stochastic in character if the process is itself stochastic.

Mr C. J. CLARK: Following on Professor Barnard's remarks, I would like to give a brief account of some work carried out recently at Imperial College in which the control procedure of Professor Box and Dr Jenkins, and Professor Barnard's "V-mask" technique were tried on some artificial series. The series of 100 terms each were of the type suggested by Professor Barnard (1959), i.e. the process mean is subject to jumps occurring at Poisson intervals, the size of the jumps being random normal deviates; the mean at any stage is the sum of all previous jumps. Superimposed on this is a random normal error term, independent of the process mean. Series with average numbers of observations per jump of 2, 5 and 10 and with different jump sizes were examined.

In the absence of the error term we would expect Professor Box and Dr Jenkins's method to give $\gamma_0 = 1$ as optimum control. For all the series examined it was found that proportional control was all that was required and the error term just caused the value of γ_0 to be modified from 1. On applying the "V-mask" method to the cumulative sums of the same series, it was found that in order to control the series to have a mean square error of a similar size to that using the previous method, it was necessary to reduce the angle of the mask until it indicated "out of control" at practically each step. We thus lose the idea of a lag time and in effect reduce to the method of Professor Box and Dr Jenkins. However, for series where the number of observations per jump are large the "V-mask" method may well have advantages over this new method.

Dr J. J. FLORENTIN: In the first part of the paper the authors quote several engineering problems requiring similar statistical techniques for their solution. There are, indeed, many others, problems in radar detection for example. Perhaps it would be simpler to say that many engineering problems can be formulated in decision theory form. From this general viewpoint a decision is made which is a linear operation on a time series; the outcome is an error e having a probability density $P(e)$. A cost function $f(e)$ is given. Now suppose that $P(e)$ is unimodal and symmetric, and that $f(e)$ is symmetric and convex, then obviously the best linear operation will put the mean of $P(e)$ at the centre of $f(e)$. This is the authors' solution when $f(e)$ is quadratic, but obviously their methods will work under a much wider range of practical circumstances, and are robust.

I want to query the use of the term "adaptive". Do the authors intend that the coefficients γ, and the corresponding controller weights w_j, are to be recalculated frequently from observations made during normal operation of the plant? If not, then most engineers would say that operation with a fixed set of w_j was conventional control. Perhaps the authors have considered the enormous amount of computation that would be necessary if the scheme were to be made truly adaptive. Not the least of the difficulties is that as the control is improved, the fluctuations become much smaller, then the fitting of an improved prediction scheme is hampered by the smallness of the readings with the consequent greater effect of other errors.

To the control engineer an outstanding contribution of this paper is the brilliant interpretation of the significance of the practical effectiveness of three-term controllers. The authors modestly say that this is common experience amongst control engineers, but it has remained for the authors to realize that this experience gives direct insight into the structure of the time series commonly met in industrial plants.

In discussing the dynamics of plants it has not been made clear enough that measurement of these exponential factors involves great experimental and theoretical difficulties. There is a disadvantage in the differencing method they propose for cancelling the effects of plant dynamics. The differencing is likely to produce large, abrupt changes in control settings; this is often undesirable. To some extent modifications in the theory could be introduced to avoid this; for example, the control effort could be costed.

The methods of this paper would be effective in many plants, but in others the response must be considered as a function of two, or more, interacting control variables. Have the authors considered the problems involved here?

Professor J. H. WESTCOTT: Speaking as a control engineer, I wish to reinforce what other speakers have said and to welcome this flirtation between control engineering and statistics. I doubt, however, whether they can yet be said to be "going steady".

I think I would have understood the paper more rapidly if the first sentence had read: "We propose to use Wiener's prediction method as modified by Zadeh and Ragazzini and our notation is as follows." Then I could have got on to the second part of the paper which, I suspect, may have been cut, and this is the part that I would like to have seen more of.

There is a large amount of statistics in the paper compared with the amount of control theory, but it is the latter part which puzzles me. The least squares method has been used for a long time, but it surprises me that the method works so very well for Professor Box and Dr Jenkins, and not so well for me. I wonder whether the reason is that only very simple forms of plant characteristic are used in the paper; in fact, the single time constant is probably the most complicated form of dynamic characteristic considered. Also, only single variable systems are analysed and most interesting practical plants are multivariate systems, which are much more difficult.

I want also to say that the techniques for adjusting the control, referred to in the paper as rather haphazard, have in fact been very highly systematized. There are very definite rules which operators are expected to work to.

The following written contribution was received after the meeting.

Professor D. R. Cox: I wish to join in the general congratulations to Professor Box and Dr Jenkins on a very interesting and important paper. My remarks are phrased in terms of prediction. In section 3.2 the authors obtain an optimum predictor subject to the condition of exact balance against a polynomial trend. They then remark that the resulting predictor would be difficult to apply in practice. It is worth noting also that except for the interesting, but perhaps rather artificial, series of section 3.4 such a procedure is even theoretically not the best thing to do. For in general complete removal of bias due to trend will increase variance, and some compromise will be preferable. Of course, this sorts itself out automatically in the authors' more empirical approach of section 3.5.

The authors have described a method for examining empirical series based on successive differencing. I should like to sketch briefly an alternative method based on Jowett's plot of mean square successive difference (*Appl. Statist.* **4** (1955), 32). An advantage of the mean square successive difference, $v_k = \frac{1}{2}\,\text{Ave}\,(x_i - x_{i+k})^2$, considered as a function of k, is, as Jowett stressed, that the behaviour for small k is relatively unaffected by irregular long-term components. Suppose that by examining v_k for small k, a predictor is derived making no allowance for trends (Cox, 1961). Thus if the local behaviour is consistent with a Markov process of parameter ρ, the Wiener predictor for predicting h steps ahead from observations $\{\ldots, x(t-1), x(t)\}$ is

$$\mu + \rho^h\{x(t) - \mu\},$$

where μ is the mean. This suggests using

$$\hat{x}_e(t;\lambda) + \rho^h\{x(t) - \hat{x}_e(t;\lambda)\},$$

where $\hat{x}_e(t;\lambda)$ is an exponentially weighted moving average with weights proportional to $1, \lambda, \lambda^2, \ldots$ and $\lambda = 0{\cdot}8$–$0{\cdot}9$. Alternatively, if the local serial correlations are all non-negative, a simple exponentially weighted moving average is likely to give good results.

Suppose now that we wish to make some allowance for linear trends. One simple estimate of local slope is

$$\hat{y}(t) = (1-\lambda')\{x(t) - \hat{x}_e(t;\lambda')\};$$

λ' would probably be taken rather large, say $0{\cdot}8$. If the basic predictor is $\hat{x}_e(t;\lambda)$, we can introduce the partially trend-adjusting predictor

$$\hat{x}_e(t;\lambda) + \kappa\left(h + \frac{\lambda}{1-\lambda}\right)\hat{y}(t),$$

where $0 < \kappa < 1$, with a corresponding form for the other basic predictors. If $\kappa = 1$, this is completely trend-adjusting. If $\kappa = 0$, there is no adjustment for trend. The quantity κ can be fixed in advance, depending on the expected ratio of trend to error, or alternatively estimated locally giving a more truly adaptive predictor.

A possible advantage of this sort of approach over the authors' method of fitting is that we can set up a predictor that works reasonably well if the trend-like behaviour of the series is quite different from that of the series used for initial analysis.

Dr JENKINS replied briefly at the meeting and the two authors subsequently replied in writing as follows:

We would like to thank the contributors to the discussion for their encouragement and constructive comments which we feel sure will influence the future course of this work.

We are grateful to Dr Tocher for his bird's-eye view of the field of control and especially the outlook of the cybernetician. However, we feel that he recasts our procedures and assumptions in such an elaborate mould that they emerge in a scarcely recognizable and somewhat terrifying state. Since we are undoubtedly responsible for some of the confusion, we may perhaps recapitulate: (1) We believe that a reasonable way to control a process is to act so as to cancel out our best estimate of future departure from target; (2) if, having been differenced p times, a time series can be represented by a moving average of order q then the incremental change in the optimal predictor will be a linear aggregate of the last error and of sums and differences of errors in past time. The first statement seems fairly innocuous and the second is a mathematical fact. The remaining but crucial question is whether this system of time series, with conveniently small values of p and q, can frequently represent happenings in the real world. (3) We believe it can, (a) because of the success of the exponentially weighted mean predictor ($p = 1$, $q = 1$), (b) because of the success of the three-term controller ($p \leqslant 2, q \leqslant 3$), (c) because all of the seven haphazardly chosen individual series studied up to the time of writing have been well represented by this system with $p \leqslant 2$, $q \leqslant 2$.

We do not imagine, of course, that this stochastic model necessarily represents mechanistically what is occurring, but only that it can provide an apt and flexible graduating function rather in the same way that a polynomial can represent a deterministic function. In this we envisage the proportional term playing a similar central role to that of the linear term in the polynomial. It is perhaps clearer from this summary that, as has been pointed out in the discussion by Dr Florentin, neither the quadratic nature of the cost function nor the use of least squares is an essential element in our development. This may help to resolve to some extent Professor Westcott's perplexity.

We would have thought that Mr Coutie's 72 or so observations would have provided him with reasonably good estimates and that each month a minute or so might be spent updating these. However, if insufficient past data are available we see no objection to using for the estimation of weights a constructed series representing best current opinion. In this connection one can easily discover by simple trial calculation how a given control system would react to some peculiar or catastrophic disturbance which is feared but which has not occurred in the past data employed for fitting. Such artificial disturbances may also be intermixed with a series of genuine past data from which weights are to be estimated. It seems to us that a good deal of flexibility is permissible. In one series of demand figures excellent results were obtained by using a simple moving average to eliminate seasonal effects and *then* applying three-term prediction to the residual series. Here very little change in the parameters occurred over a 12-year period. Concerning Mr Coutie's last point, we feel that rather than use $\Sigma\epsilon$ to tell when something has gone wrong, a device also used by Brown, it is better to include the cumulative term in the model as we do and so obtain a predictor which accommodates itself to the situation.

A number of speakers have suggested other formulations partially associated with various types of feedback. The particular distinction which we would emphasize is that between empirical or "idiot" feedback which could be based on a set of preformulated rules, but may equally employ a crude or a sophisticated model, in contrast to technical or "intelligent" feedback resulting from the interaction of the human mind with the data leading to new ideas and models.

We find ourselves entirely in agreement with Professor Barnard's remarks concerning continuous experimentation. However, we feel that cumulative sum charts have a long and useful future ahead particularly in answering the all-important question "when did it happen?" This is often a key question making technical feedback possible.

We do not follow Dr Himsworth's linking up of exponential transfer stages and exponentially weighted means. Our methods work well with step changes in the inputs as is for example attested by Mr Clark's interesting contribution. The conditions for stability are given in our paper and the unstable regions are shown in our log-likelihood plots. In practice, as would be expected, the sum of squares rapidly rises as the instability regions are approached and no difficulty arises. Dr Himsworth will find that he is mistaken concerning the original concept of evolutionary operation. In the original paper on this subject (Box, *Applied Statistics*, **6** (1957), 1), on p. 13, will be found a description of a special "Evolutionary Operation Committee" necessary to make the method "really effective". The paper says:

> "The major task of such a group is to discuss the implications of current results and make suggestions for future phases of operation. Their deliberations will frequently lead to the formulation of theories which in turn suggest new modifications that can be tried with profit."

Our experience, which we had thought partially overlapped his, is that many of the most worth-while advances using evolutionary operation have been made as a result of a progressively improved understanding of the process as it operated on the full scale.

Our emotions when we first learned that we had made use of Hilbert space theory are difficult to set down in words but certainly our admiration for both the perception and clarity of Professor Parzen was still further increased. We agree that some further theoretical qualifications of some aspects of our work are necessary.

Dr Florentin's doubts about the use of the word "adaptive" might be resolved if we agreed to employ "parameter adaptive" for a scheme in which, for example, the w_j were progressively modified. His remark that differencing is likely to produce abrupt changes in controller settings is a valid one and we will have to give some further thought to the question of placing a bound on the amount of control which we are allowed to make. The problem of multi-variable control raised by Dr Florentin and Professor Westcott is important and interesting. We have not yet tackled problems of this kind in the adaptive quality control context, but they play an important part in the adaptive optimization problem which we shall be discussing later.

Professor Cox's method for modifying exponentially weighted moving average predictors for polynomial trends is interesting and we look forward to seeing this method applied to data such as series I–III.

3.2
A Change in Level of a Non-Stationary Time Series*

G. E. P. BOX and GEORGE C. TIAO
University of Wisconsin

1. Introduction

Suppose that observations z_t of a time series are available at equally spaced time intervals. We consider the problem of making inferences about a possible shift in level of the series associated with the occurrence of an event E at some particular time. For example, the observations might be of some economic indicator and we might suspect a change in level to occur in a particular interval because of a change in fiscal policy. Alternatively, z_t could be the daily output of a chemical process and the event E might be a change in the supplier of raw material.

Suppose we have available n_1 observations before the event E and further n_2 observations afterwards. It is well known that if the observations $z_1, \ldots, z_{n_1}$ are a random sample from the normal distribution $N(\mu_1, \sigma^2)$ and $z_{n_1+1}, \ldots, z_{n_1+n_2}$ a random sample from $N(\mu_2, \sigma^2)$, then inferences concerning the change in level $\delta = \mu_2 - \mu_1$ can be made using the Student t-distribution. The criterion

$$\frac{(\bar{y}_2 - \bar{y}_1) - \delta}{s\sqrt{\left(\frac{1}{n_1} + \frac{1}{n_2}\right)}}$$

would follow a t-distribution and could be used for testing hypotheses and for obtaining confidence intervals for δ. The fiducial distribution of δ would be a scaled t-distribution. With appropriate prior assumptions this would also be the posterior distribution of δ. If the observations were not independent but the physical situations 'event E' and 'no event E' could be applied in random order, then the t-distribution would again be appropriate as a close approximation to the null randomization distribution and so could supply a valid basis for inference.

In examples like those quoted above where the observations are almost certainly dependent and no possibility for randomization exists, the usual procedure based on the t-distribution would of course be invalid and could be extremely misleading. Even moderate dependence between observations within a sample can seriously invalidate such a procedure.

2. The mathematical model

In practice not only would successive observations usually be dependent, but frequently the time series would be non-stationary. A type of stochastic model (Box & Jenkins, 1962) which has been used with considerable success in representing non-stationary time series is the integrated moving average process

$$z_p = M + (\gamma_{-l}\Delta^{l-1} + \ldots + \gamma_{-1} + \gamma_0 S + \ldots + \gamma_m S^{m+1})\,\alpha_{p-1} + \alpha_p, \qquad (2\cdot1)$$

where
$$\Delta\alpha_p = \alpha_p - \alpha_{p-1}, \quad S\alpha_p = \Delta^{-1}\alpha_p = \sum_{j=0}^{\infty} \alpha_{p-j},$$

* This research was supported by the Office of Naval Research.

The Collected Works of George E. P. Box, 1984, Wadsworth, Inc., Belmont, CA 94002.
Originally published in *Biometrika,* vol. 52, parts 1 and 2 (1965), pp. 181–192.

M reflects the initial location of the series at the remote past and the α's are independent random normal deviates having variance σ^2. In this paper the symbol ∞ will represent a large but finite positive integer, so that the start of the series is imagined to be in some remote past. We shall consider only the simple special case

$$\begin{aligned} z_p &= M + \gamma_0 S\alpha_{p-1} + \alpha_p \\ &= M + \gamma_0 \sum_{j=1}^{\infty} \alpha_{p-j} + \alpha_p \quad (0 \leqslant \gamma_0 < 2). \end{aligned} \tag{2·2}$$

Except when $\gamma_0 = 0$, the time series in equation (2·2) is non-stationary and in particular, it does not possess a mean. However, a measure of location of the series at any given time p is provided by the sum

$$L_p = M + \gamma_0 \sum_{j=1}^{\infty} \alpha_{p-j}. \tag{2·3}$$

The quantity L_p we shall define as the 'level' of the process at time p. If we start to observe the series at time $t = 1$, then the random normal deviates $\alpha_0, \alpha_{-1}, \ldots$ will have been realized, although unobserved. Thus, we can treat

$$L = L_1 = M + \gamma_0 \sum_{j=0}^{\infty} \alpha_j \tag{2·4}$$

as a fixed but unknown location parameter. The series in (2·2) may now be written

$$\left.\begin{aligned} z_1 &= L + \alpha_1, \\ z_p &= L + \gamma_0 \sum_{j=1}^{p-1} \alpha_{p-j} + \alpha_p \quad (p = 2, 3, \ldots). \end{aligned}\right\} \tag{2·5}$$

As has been pointed out by Muth (1960), the model implies that the system is subjected to periodic random shocks, a proportion γ_0 of each shock being absorbed into the 'level' of the series. The optimal estimate at time p of any future observation z_{p+t} is $\underset{p}{E}(z_{p+t})$, where $\underset{p}{E}$ denotes the conditional expectation at p. It is easy to show that

$$\underset{p}{E}(z_{p+t}) = \gamma_0 \sum_{j=0}^{\infty} (1-\gamma_0)^j z_{p-j}, \tag{2·6}$$

an exponentially weighted average of all previous observations. From (2·6), it follows that the stochastic process in (2·2) and (2·5) can be written as

$$\begin{aligned} z_p &= \gamma_0 \sum_{j=0}^{\infty} (1-\gamma_0)^j z_{p-1-j} + \alpha_p \\ &= \gamma_0 \sum_{j=0}^{p-2} (1-\gamma_0)^j z_{p-1-j} + (1-\gamma_0)^{p-1} L + \alpha_p, \end{aligned} \tag{2·7}$$

with

$$L = M + \gamma_0 \sum_{j=0}^{-\infty} \alpha_j = \gamma_0 \sum_{j=0}^{-\infty} (1-\gamma_0)^j z_j. \tag{2·8}$$

This simple process in (2·5) and (2·7) seems to provide adequate representation of a surprisingly large number of time series arising in economic and industrial applications (Brown, 1959; Holt *et al.* 1960; Box & Jenkins, 1962, 1963). Values of the constant γ_0 between zero and one are most frequently found in practice. When necessary, both the series and the predictor can be suitably generalized as in (2·1) to allow for more complicated situations and the procedure we shall discuss can be correspondingly modified.

3. The situation when γ_0 is known

For convenience in presentation we suppose at first that the parameter γ_0 in the stochastic process is known to a sufficient approximation. This is often the case in industrial applications where data from operating processes are available over long periods of time.

Suppose the parameter δ measures the shift in level of the series associated with the event E. Then, for the n_1 available observations before E,

$$\left.\begin{aligned} z_1 &= L+\alpha_1, \\ z_p &= L+\gamma_0 \sum_{j=1}^{p-1} \alpha_{p-j}+\alpha_p \\ &= \gamma_0 \sum_{j=0}^{p-2} (1-\gamma_0)^j z_{p-1-j}+(1-\gamma_0)^{p-1} L+\alpha_p \quad (p=2, \ldots, n_1). \end{aligned}\right\} \tag{3·1a}$$

For the n_2 available observations after E,

$$\begin{aligned} z_p &= L+\delta+\gamma_0 \sum_{j=1}^{p-1} \alpha_{p-j}+\alpha_p \\ &= \gamma_0 \sum_{j=1}^{p-2} (1-\gamma_0)^j z_{p-1-j}+(1-\gamma_0)^{p-1} L+(1-\gamma_0)^{p-(n_1+1)} \delta+\alpha_p \quad (p=n_1+1, \ldots, n_1+n_2). \end{aligned} \tag{3·1b}$$

Since γ_0 is assumed known, we can make the transformation

$$\left.\begin{aligned} y_1 &= z_1, \\ y_p &= z_p-\gamma_0 \sum_{j=0}^{p-2} (1-\gamma_0)^j z_{p-1-j} \quad (p=2, \ldots, n_1+n_2) \end{aligned}\right\} \tag{3·2}$$

and write (3·1a, b) as the familiar linear model,

$$\mathbf{y} = X\boldsymbol{\theta}+\boldsymbol{\epsilon},$$

where

$$\mathbf{y} = \begin{bmatrix} y_1 \\ \cdot \\ \cdot \\ y_{n_1} \\ \hdashline y_{n_1+1} \\ \cdot \\ \cdot \\ \cdot \\ y_{n_1+n_2} \end{bmatrix}; \quad X = \begin{bmatrix} 1 & 0 \\ (1-\gamma_0) & \cdot \\ \cdot & \cdot \\ (1-\gamma_0)^{n_1-1} & 0 \\ \hdashline (1-\gamma_0)^{n_1} & 1 \\ \cdot & (1-\gamma_0) \\ \cdot & \cdot \\ \cdot & \cdot \\ (1-\gamma_0)^{n_1+n_2-1} & (1-\gamma_0)^{n_2-1} \end{bmatrix}; \quad \boldsymbol{\theta} = \begin{pmatrix} L \\ \delta \end{pmatrix} \tag{3·3}$$

and $\boldsymbol{\epsilon}' = (\alpha_1, \ldots, \alpha_{n_1+n_2})$ is the $1 \times (n_1+n_2)$ vector of random normal deviates with common variance σ^2. Following Gauss, the least squares estimator of $\boldsymbol{\theta}$ is

$$\hat{\boldsymbol{\theta}} = \begin{pmatrix} \hat{L} \\ \hat{\delta} \end{pmatrix} = (X'X)^{-1} X'\mathbf{y}, \tag{3·4}$$

with

$$(X'X)^{-1} = \frac{\gamma_0(2-\gamma_0)}{[1-(1-\gamma_0)^{2n_1}][1-(1-\gamma_0)^{2n_2}]} \begin{bmatrix} 1-(1-\gamma_0)^{2n_2} & -(1-\gamma_0)^{n_1}[1-(1-\gamma_0)^{2n_2}] \\ -(1-\gamma_0)^{n_1}[1-(1-\gamma_0)^{2n_2}] & 1-(1-\gamma_0)^{2N} \end{bmatrix}, \tag{3·5}$$

$$N = n_1+n_2$$

and

$$X'\mathbf{y} = \begin{pmatrix} \sum_{s=1}^{N} (1-\gamma_0)^{s-1} y_s \\ \sum_{s=1}^{n_2} (1-\gamma_0)^{s-1} y_{n_1+s} \end{pmatrix}. \tag{3·6}$$

Using the definition of the y's in (3·2), it is straightforward to verify that, in terms of the observations z_i, we can write $(\hat{L}, \hat{\delta})$ as

$$\hat{L} = \frac{\gamma_0}{1-(1-\gamma_0)^{2n_1}} \left[\sum_{s=1}^{n_1} (1-\gamma_0)^{s-1} z_s + (1-\gamma_0)^{n_1} \sum_{s=1}^{n_1} (1-\gamma_0)^{n_1-s} z_s \right] \tag{3·7}$$

and

$$\hat{\delta} = f_1(z_{n_1+1}, \ldots, z_{n_1+n_2}; \gamma_0) - f_2(z_1, \ldots, z_{n_1}; \gamma_0), \tag{3·8}$$

with

$$f_1(z_{n_1+1}, \ldots, z_{n_1+n_2}; \gamma_0) = \frac{\gamma_0}{1-(1-\gamma_0)^{2n_2}} \left[\sum_{s=1}^{n_2} (1-\gamma_0)^{s-1} z_{n_1+s} + (1-\gamma_0)^{n_2} \sum_{s=1}^{n_2} (1-\gamma_0)^{n_2-s} z_{n_1+s} \right], \tag{3·8a}$$

$$f_2(z_1, \ldots, z_{n_1}; \gamma_0) = \frac{\gamma_0}{1-(1-\gamma_0)^{2n_1}} \left[\sum_{s=1}^{n_1} (1-\gamma_0)^{n_1-s} z_s + (1-\gamma_0)^{n_1} \sum_{s=1}^{n_1} (1-\gamma_0)^{s-1} z_s \right]. \tag{3·8b}$$

It is of course well known that, in general, the sampling distribution of the quantity

$$\frac{(\hat{\theta}_i - \theta_i)}{\left\{ \frac{(\mathbf{y}-\hat{\mathbf{y}})'(\mathbf{y}-\hat{\mathbf{y}})\, c^{ii}}{N-k} \right\}^{\frac{1}{2}}} \tag{3·9}$$

is a Student t-distribution where k is the number of regression coefficients, c^{ii} the ith diagonal element of $(X'X)^{-1}$ and $\hat{\mathbf{y}} = X\hat{\boldsymbol{\theta}}$.

In our case we have that

$$(\hat{\delta} - \delta) \left\{ \frac{[1-(1-\gamma_0)^{2n_1}][1-(1-\gamma_0)^{2n_2}]}{[1-(1-\gamma_0)^{2N}]\gamma_0(2-\gamma_0)\, s^2} \right\}^{\frac{1}{2}} \tag{3·10}$$

exactly follows a Student t-distribution with (n_1+n_2-2) degrees of freedom, with

$$s^2 = \frac{1}{n_1+n_2-2} \left\{ \sum_{t=1}^{n_1} [y_t - \hat{L}(1-\gamma_0)^{t-1}]^2 + \sum_{t=n_1+1}^{N} [y_t - \hat{L}(1-\gamma_0)^{t-1} - \hat{\delta}(1-\gamma_0)^{t-n_1-1}]^2 \right\}. \tag{3·11}$$

Alternatively, adopting a Bayesian approach with the assumption that L, δ, and $\log \sigma$ have locally independent uniform distributions *a priori*, one can easily show (Jeffreys, 1961; Savage, 1961; Tiao & Zellner, 1964) that the quantity in (3·10) again follows the same t-distribution. In the Bayesian interpretation, δ is of course the random variable and $\hat{\delta}$ and s^2 are known sample quantities.

Interpretation

At first sight, expression (3·10) appears somewhat complicated but in fact it has a remarkably simple interpretation as is seen by considering an example. The 50 observations z_t listed below and plotted in Fig. 1 were generated from a table of standard random normal deviates using (3·1 a, b) and setting $L = 5$, $\gamma_0 = 0·3$. A change in level, $\delta = 1·5$, was introduced after the 25th observation.

Suppose at first we know that $\gamma_0 = 0·3$ and we desire to make inferences about a possible change in level associated with the event E which occurs immediately after the 25th observation. The weight function applied to the observations z_t in calculating $\hat{\delta}$ in (3·8), which is our estimate of the change in level δ, is shown in Fig. 1 immediately above the series. It is seen that $\hat{\delta}$ is simply the difference between two exponentially weighted averages, one having maximum weight immediately prior to the event E and the other having maximum weight immediately after.

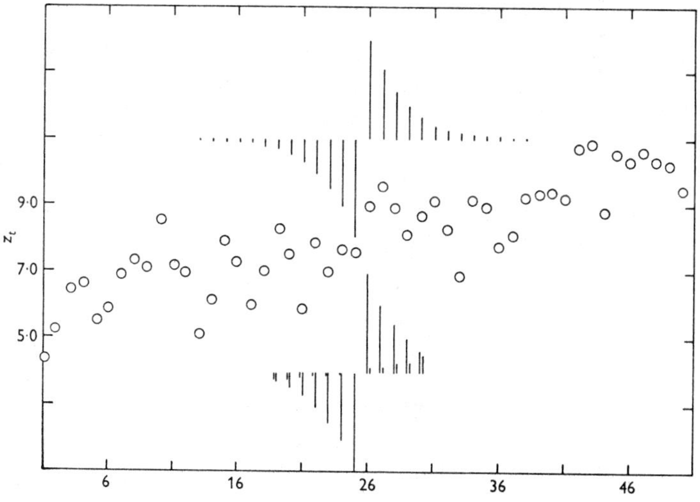

Figure 1

Table 1. *Data generated from the model* (3·1 *a*, *b*)

($L = 5$, $\delta = 1{\cdot}5$, $\gamma_0 = 0{\cdot}3$)

Sample no.	z_t	Sample no.	z_t	Sample no.	z_t	Sample no.	z_t
1	4·3931	13	5·1758	25	7·6149	38	9·2516
2	5·2961	14	6·1483	26	9·0002	39	9·4000
3	6·5456	15	8·0000	27	9·6242	40	9·4488
4	6·6551	16	7·2730	28	8·9715	41	9·2379
5	5·5611	17	6·0724	29	8·1689	42	10·8423
6	5·8530	18	7·0264	30	8·6201	43	10·8965
7	6·9774	19	8·2541	31	9·0731	44	8·8422
8	7·3000	20	7·5851	32	8·3121	45	10·6176
9	7·1386	21	5·8167	33	6·8790	46	10·4013
10	8·5815	22	7·8799	34	9·1923	47	10·6789
11	7·2307	23	7·0213	35	9·0164	48	10·4622
12	7·0464	24	7·6572	36	7·7885	49	10·3484
				37	8·0701	50	9·5311

Truncation in the weight function

Most of the complications which occur in the formulae are to take care of the possibility that the weight functions may not have 'died out' before the series begins and ends. The effect of truncation occurring in the weight function is illustrated by showing what happens in the example when only $n_1 = 7$ observations are available immediately prior to the event E and only $n_2 = 5$ after it. The weight function appropriate to this situation is shown in Fig. 1 immediately below the series. To allow for truncation, the weight function is seen to 'double back' on itself. It is the weight added in this doubling back, which is taken care of by the second terms in the square brackets of (3·8a) and (3·8b). When the truncation is negligible, the second term can be ignored and we then have a simple exponentially weighted average.

To see how many observations will be needed for truncation to be ignored, suppose we say that the weight is negligible when it has been reduced to 1/20 of its initial value. The number of observations which achieves this is the solution of the equation

$$\gamma_0(1-\gamma_0)^{n-1} = \tfrac{1}{20}\gamma_0. \tag{3·12}$$

For example, with $\gamma_0 = 0·25$, $\gamma_0 = 0·50$ and $\gamma_0 = 0·75$, the weight will be negligible after 14, 6, and 3 observations, respectively.

Simplification when truncation can be ignored

When the truncation effect can be ignored, considerable simplifications occur in the formulae. We have

$$(X'X)^{-1} \cong \gamma_0(2-\gamma_0)\, I_2, \tag{3·13}$$

where I_2 is a 2×2 identity matrix, and

$$\hat{L} \cong \gamma_0 \sum_{s=1}^{n_1} (1-\gamma_0)^{s-1} z_s, \tag{3·14}$$

$$\hat{\delta} \cong \gamma_0 \left[\sum_{s=1}^{n_2} (1-\gamma_0)^{s-1} z_{n_1+s} - \sum_{s=1}^{n_1} (1-\gamma_0)^{n_1-s} z_s \right]. \tag{3·15}$$

Thus, the quantity

$$\frac{(\hat{\delta}-\delta)}{s\sqrt{[\gamma_0(2-\gamma_0)]}} \tag{3·16}$$

is distributed approximately as Student t with (n_1+n_2-2) degrees of freedom.

For the present example, using the complete set of 50 observations, the above simplifications are clearly applicable. We obtain

$$\hat{\delta} = 1·5467, \quad s\sqrt{\{\gamma_0(2-\gamma_0)\}} = 0·6687.$$

Thus, the test of the hypothesis $\gamma_0 = 0$ against the alternative $\gamma_0 \neq 0$ is supplied by calculating the quantity

$$t = \frac{1·5467}{0·6687} = 2·313$$

which is significant at the 5 % level. Alternatively, an approximate 95 % confidence interval for δ is (0·236, 2·8574). This may correspondingly be interpreted as a 95 % Bayesian interval for δ.

Truncation not negligible

To illustrate the more general formulae where significant truncation occurs, we suppose that $n_1 = 7$ and $n_2 = 5$ so that the weight function shown in Fig. 1 below the series is appropriate. We then obtain

$$\hat{\delta} = 1·5297,$$

$$\left\{\frac{s^2\gamma_0(2-\gamma_0)}{[1-(1-\gamma_0)^{2n_1}][1-(1-\gamma_0)^{2n_2}]}\right\}^{\frac{1}{2}} = 0·5786,$$

and

$$t = 2·6438$$

which again is significant at the 5 % level. The corresponding 95 % confidence interval for δ is (0·2406, 2·8188).

Technique of calculation

The procedure described is one which should have many practical applications and the computations can readily be made on a desk calculator. In calculating the y's in (3·2), considerable time may be saved by employing the recursive relation

$$y_t = (z_t - z_{t-1}) + (1-\gamma_0)\, y_{t-1} \quad (t = 2, 3, \ldots). \tag{3·17}$$

Also, in calculating the residual sum of squares $(N-2)\,s^2$ in (3·11), we can make use of the identity

$$(\mathbf{y}-\hat{\mathbf{y}})'\,(\mathbf{y}-\hat{\mathbf{y}}) = \mathbf{y}'\mathbf{y} - \hat{\boldsymbol{\theta}}'X'X\hat{\boldsymbol{\theta}}. \tag{3·18}$$

Thus, $(N-2)\,s^2$ can alternatively be written

$$(N-2)\,s^2 = \sum_{j=1}^{N} y_i^2 - \frac{1}{\gamma_0(2-\gamma_0)}\{\hat{L}^2[1-(1-\gamma_0)^{2N}] + \hat{\delta}^2[1-(1-\gamma_0)^{2n_2}] + 2\hat{L}\hat{\delta}(1-\gamma_0)^{n_1}[1-(1-\gamma_0)^{2n_2}]\}. \tag{3·19}$$

4. An alternative model

The deduction of necessary consequences from a specific mathematical model has two uses. First, an adequate model may provide interesting conclusions. Secondly, unacceptable conclusions may announce unexpected inadequacies in the model. This point is illustrated when we consider the conclusions which one would be led to if an autoregressive model were adopted. Such models have been widely used—(Davis, 1941; Kendall, 1944*a*, *b*; Cochrane & Orcutt, 1949; Anderson, 1949, etc.)—to analyse time series data.

Suppose we consider the same problem but assume that the series is generated by a stationary first order autoregressive process. To simplify our analysis, we further suppose that the series starts at $t = 1$—for a discussion of the initial conditions, see e.g. Zellner & Tiao (1964). Then, instead of (3·1), we have

$$\left.\begin{aligned} z_1 &= L + \alpha_1, \\ z_p &= L + \rho\sum_{j=1}^{p-1}\rho^{j-1}\alpha_{p-j} + \alpha_p \quad (p = 2, \ldots, n_1) \end{aligned}\right\} \tag{4·1}$$

and

$$z_p = L + \delta + \rho\sum_{j=1}^{p-1}\rho^{j-1}\alpha_{p-j} + \alpha_p \quad (p = n_1+1, \ldots, n_1+n_2),$$

with $|\rho| < 1$. Thus, we may regard L as the mean and δ the shift in mean associated with the event E. For any specific value of the autoregressive coefficient ρ, say $\rho = \rho_0$, we can make the transformation

$$\left.\begin{aligned} y_1 &= z_1, \\ y_p &= z_p - \rho_0 z_{p-1} \end{aligned}\right\} \tag{4·2}$$

and write (4·1) as the linear model.

$$\mathbf{y} = X\boldsymbol{\theta} + \boldsymbol{\epsilon}, \tag{4·3}$$

where

$$\mathbf{y} = \begin{bmatrix} y_1 \\ \cdot \\ \cdot \\ \cdot \\ y_{n_1} \\ \hdashline y_{n_1+1} \\ \cdot \\ \cdot \\ \cdot \\ y_{n_1+n_2} \end{bmatrix}; \quad X = \begin{bmatrix} 1 & 0 \\ (1-\rho_0) & \cdot \\ \cdot & \cdot \\ \cdot & \cdot \\ (1-\rho_0) & 0 \\ \hdashline (1-\rho_0) & 1 \\ \cdot & (1-\rho_0) \\ \cdot & \cdot \\ \cdot & \cdot \\ 1-\rho_0) & (1-\rho_0) \end{bmatrix}; \quad \boldsymbol{\theta} = \begin{pmatrix} L \\ \delta \end{pmatrix}$$

and $\boldsymbol{\epsilon}$ is again the $N\times 1$ vector of random normal deviates with common variance σ^2. It can then be verified that

$$(X'X)^{-1} = c\begin{bmatrix} 1+(n_2-1)(1-\rho_0)^2 & -(1-\rho_0)-(n_2-1)(1-\rho_0)^2 \\ -(1-\rho_0)-(n_2-1)(1-\rho_0)^2 & 1+(N-1)(1-\rho_0)^2 \end{bmatrix}, \tag{4·4}$$

$$c^{-1} = n_1(n_2-1)(1-\rho_0)^4-2(n_2-1)(1-\rho_0)^3+(n_1+2n_2-3)(1-\rho_0)^2+1. \tag{4·5}$$

In terms of the z's, the estimates $(\hat{L},\hat{\delta})$ are given by

$$\hat{L} = \frac{1}{1+(N-1)(1-\rho_0)^2}\left\{(1-\rho_0)^2\sum_{i=1}^{N-1} z_i+\rho_0 z_1+(1-\rho_0)z_N-[(1-\rho_0)+(n_2-1)(1-\rho_0)^2]\hat{\delta}\right\}, \tag{4·6}$$

$$\hat{\delta} = g_1(z_{n_1+1},\ldots,z_{n_1+n_2};\rho_0)-g_2(z_1,\ldots,z_{n_1};\rho_0), \tag{4·7}$$

where
$$g_1(z_{n_1+1},\ldots,z_{n_1+n_2};\rho_0) = c\left[a_1 z_{n_1+1}+a_2\sum_{i=n_1+2}^{N} z_i+a_3 z_{n_1+n_2}\right],$$

$$g_2(z_1,\ldots,z_{n_1};\rho_0) = c\left[b_1 z_1+b_2\sum_{i=2}^{n_1-1} z_i+b_3 z_{n_1}\right]$$

with
$$a_1 = n_1(1-\rho_0)^4-N(1-\rho_0)^3+N(1-\rho_0)^2+\rho_0,$$
$$a_2 = (1-\rho_0)^2[n_1(1-\rho_0)^2+\rho_0],$$
$$a_3 = (1-\rho_0)[n_1(1-\rho_0)^2+\rho_0],$$
$$b_1 = (1-\rho_0)[(1-\rho_0)^2+\rho_0][n_2(1-\rho_0)+\rho_0],$$
$$b_2 = (1-\rho_0)^3[n_2(1-\rho_0)+\rho_0]$$
and
$$b_3 = (1-\rho_0)^2[n_2(1-\rho_0)+\rho_0]+\rho_0[1+(N-2)(1-\rho_0)^2-(n_2-1)(1-\rho_0)^3].$$

It follows that for a specific value, $\rho=\rho_0$, the sampling distribution of the quantity

$$\frac{(\hat{\delta}-\delta)}{\{s^2 c[1+(N-1)(1-\rho_0)^2]\}^{\frac{1}{2}}} \tag{4·8}$$

is the Student t-distribution with (n_1+n_2-2) degrees of freedom, with

$$s^2 = \frac{1}{n_1+n_2-2}\left\{(z_1-\hat{L})^2+\sum_{i=2}^{n_1}[z_i-\rho_0 z_{i-1}-\hat{L}(1-\rho_0)]^2 + [z_{n_1+1}-\rho_0 z_{n_1}-\hat{L}(1-\rho_0)-\hat{\delta}]^2+\sum_{i=n_1+2}^{N}[z_i-\rho_0 z_{i-1}-(1-\rho_0)(\hat{L}+\hat{\delta})]^2\right\}. \tag{4·9}$$

Once again, we obtain a t-distribution which relates the deviation $\hat{\delta}-\delta$ to its standard error. As before, $\hat{\delta}$ is the difference between two weighted averages as given in (4·7). However, the fact that apart from end-effects the weight function for these averages in (4·7) is *uniform* emphasizes the restrictiveness of the auto-regressive model. Specifically, our results imply that this model is only acceptable if observations near the beginning and the end of the series should have as much weight in the estimation of δ as those close to the event E. In many industrial and economic applications, it seems much more reasonable to suppose that as we move away from E, the observations should become less and less informative about δ. This is precisely what we find with the integrated moving average model. After a certain point, extra observations add nothing except to provide a better estimate of σ^2.

5. Inferences concerning δ when γ_0 is unknown

Returning now to the integrated moving average model, we consider the situation when the parameter γ_0 is unknown. The Bayesian approach may still be used to make inferences about δ, using sample information about the unknown γ_0.

Suppose as in § 3 that the prior distributions of $(L, \delta, \log \sigma)$ are locally uniform and independent and suppose in addition that the prior distribution of γ_0 is $p_0(\gamma_0)$, which is independent of $(L, \delta, \log \sigma)$. Then the joint posterior distribution of $(L, \delta, \sigma, \gamma_0)$ is

$$p(L,\delta,\sigma,\gamma_0|\mathbf{z}) \propto p_0(\gamma_0)\,\sigma^{-(N+1)}\exp\left\{-\frac{1}{2\sigma^2}Q(\gamma_0,L,\delta)\right\}, \tag{5·1}$$

where

$$Q(\gamma_0,L,\delta) = \sum_{t=1}^{n_1}[y_t - L(1-\gamma_0)^{t-1}]^2 + \sum_{t=n_1+1}^{N}[y_t - L(1-\gamma_0)^{t-1} - \delta(1-\gamma_0)^{t-n_1-1}]^2$$

$$(-\infty < L < \infty,\ -\infty < \delta < \infty,\ 0 < \gamma_0 < 2,\ 0 < \sigma < \infty)$$

and the y_t's are as defined in (3·2).

Upon integrating out L and σ from (5·1) we obtain the posterior distribution of (δ, γ_0) as

$$p(\delta,\gamma_0|\mathbf{z}) \propto p_0(\gamma_0)\left\{\frac{\gamma_0(2-\gamma_0)}{1-(1-\gamma_0)^{2N}}\right\}^{\frac{1}{2}}(s^2)^{-\frac{1}{2}(N-1)}\left(1+\frac{t^2}{N-2}\right)^{-\frac{1}{2}(N-1)}, \tag{5·2}$$

where

$$t = (\delta-\hat{\delta})\left\{\frac{[1-(1-\gamma_0)^{2n_1}][1-(1-\gamma_0)^{2n_2}]}{[1-(1-\gamma_0)^{2N}]\gamma_0(2-\gamma_0)s^2}\right\}^{\frac{1}{2}} \tag{5·3}$$

and $\hat{\delta}$ and s^2 are defined in (3·8) and (3·11), respectively. It is instructive to write (5·2) as the product

$$p(\delta,\gamma_0|\mathbf{z}) = p(\delta|\gamma_0,\mathbf{z})\,p(\gamma_0|\mathbf{z}). \tag{5·4}$$

As mentioned earlier in § 3, the conditional distribution $p(\delta|\gamma_0, \mathbf{z})$ is related to the Student t-distribution with $(N-2)$ degrees of freedom by the transformation (5·3). It follows that the conditional mean and standard error of δ are

$$E(\delta|\gamma_0) = \hat{\delta}, \tag{5·5}$$

$$\sigma_{(\delta|\gamma_0)} = \left\{\frac{s^2\gamma_0(2-\gamma_0)[1-(1-\gamma_0)^{2N}](N-2)}{[1-(1-\gamma_0)^{2n_1}][1-(1-\gamma_0)^{2n_2}](N-4)}\right\}^{\frac{1}{2}}. \tag{5·6}$$

Now the second factor on the right-hand side of (5·3) can be written

$$p(\gamma_0|\mathbf{z}) \propto p_0(\gamma_0)\,h(\gamma_0|\mathbf{z}) \tag{5·7}$$

with

$$h(\gamma_0|\mathbf{z}) \propto \frac{\gamma_0(2-\gamma_0)}{\{[1-(1-\gamma_0)^{2n_1}][1-(1-\gamma_0)^{2n_2}]\}^{\frac{1}{2}}}(s^2)^{-\frac{1}{2}(N-2)}. \tag{5·8}$$

Expression (5·7) is then the marginal posterior distribution of γ_0 which in (5·3) behaves like a weight function multiplying the conditional distribution $p(\delta|\gamma_0)$.

To illustrate how our inferences about δ may be affected by changes in the value of γ_0, the values of $\hat{\delta}$ and the conditional standard error $\sigma_{(\delta|\gamma_0)}$ calculated using the set of data in Table 1 are shown in Fig. 2 for various values of γ_0 together with the corresponding $t(\gamma_0)$-ratio where

$$t(\gamma_0) = \hat{\delta}\left\{\frac{[1-(1-\gamma_0)^{2n_1}][1-(1-\gamma_0)^{2n_2}]}{[1-(1-\gamma_0)^{2N}]\gamma_0(2-\gamma_0)s^2}\right\}^{\frac{1}{2}}. \tag{5·9}$$

It is seen that as γ_0 decreases from one to zero, $\hat{\delta}$ changes as the weight function stretches further and further away from E. The changes depend upon the particular behaviour of the series. Changes in the value of the standard error $\sigma_{(\delta|\gamma_0)}$ are partially accounted for by the fact that as γ_0 becomes smaller, the averaging process extends to more and more values of z_t.

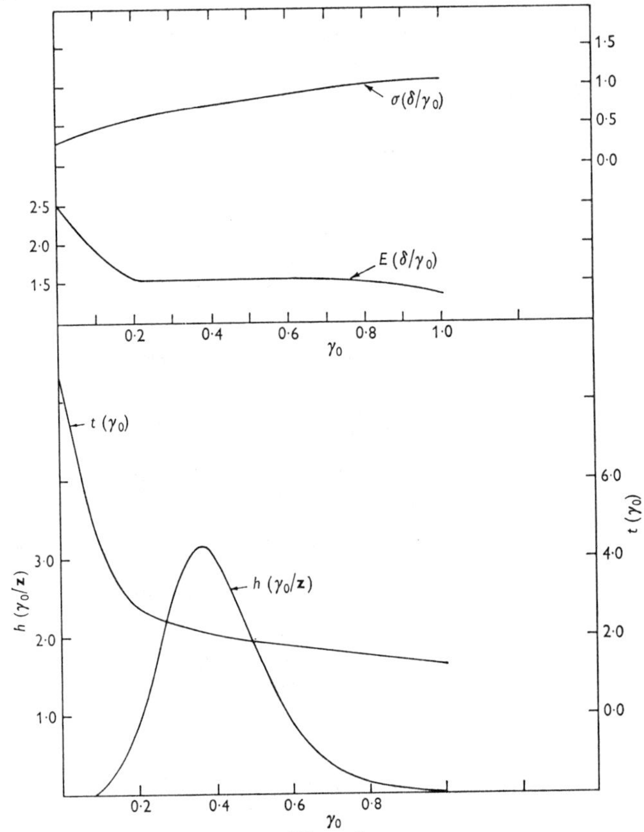

Figure 2

Also shown in Fig. 2 is the function $h(\gamma_0|\mathbf{z})$. This would be the posterior distribution of γ_0 if the prior distribution $p_0(\gamma_0)$ were taken locally uniform. It can thus be used to calculate the posterior distribution of γ_0 by combining it with any desired $p_0(\gamma_0)$. If as in many situations 'no information' about γ_0 were available *a priori*, then it could be regarded as indicating what we know about γ_0 from the sample itself. It may be remarked here that one can verify that the Fisherian measure of information for γ_0 is

$$I(\gamma_0, N) \simeq \frac{N}{1-(1-\gamma_0)^2} \tag{5·10}$$

which, using Jeffreys' invariance principle (1961), would lead us to the assumption that

$$p_0(\gamma_0) \propto \{1-(1-\gamma_0)^2\}^{-\frac{1}{2}}. \tag{5·11}$$

However, except in small samples, this would not alter our conclusions very much.

In many cases, an adequate guide to inferences to be drawn would be supplied by the graph of $t(\gamma_0)$ and $h(\gamma_0|\mathbf{z})$ as shown in Fig. 2. In the present example, the conditional distribution of δ is related to a t-distribution with 48 degrees of freedom and therefore a normal approximation is appropriate. We see that over the range in which $h(\gamma_0|\mathbf{z})$ is appreciable the value of $t(\gamma_0)$ is close to 2 and consequently for plausible values of γ_0, 95 % of the conditional posterior probability mass of δ is over the positive range.

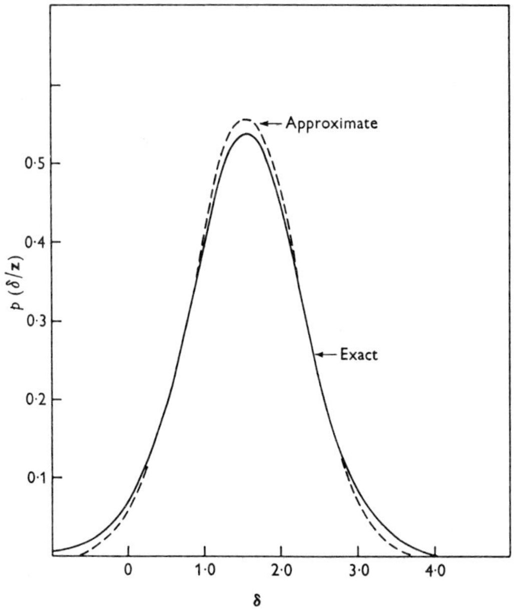

Figure 3

If desired, the overall inference concerning δ can be deduced from the marginal posterior distribution of δ,

$$p(\delta|\mathbf{z}) = \int p(\delta, \gamma_0|\mathbf{z})\, d\gamma_0. \tag{5·12}$$

Unfortunately, it does not seem possible to express this distribution in terms of simple functions. For our set of data, we have calculated the distribution using numerical methods and the result is shown by the solid curve in Fig. 3. In obtaining this result, the prior distribution $p_0(\gamma_0)$ was taken to be uniform. In practice, a fairly close approximation to the marginal distribution of δ would usually be provided by substituting the maximum-likelihood estimate of γ_0 in the conditional distribution $p(\delta|\gamma_0, \mathbf{z})$. The approximation for this example is shown by the dotted curve in Fig. 3.

References

Anderson, R. L. (1949). The problem of autocorrelation in regression analysis. *J. Amer. Statist. Ass.* **44**, 113–29.

Box, G. E. P. & Jenkins, G. M. (1962). Some statistical aspects of adaptive optimization and control. *J. R. Statist. Soc.* B, **24**, 297–343.

Box, G. E. P. & Jenkins, G. M. (1963). Further contributions to adaptive quality control: Simultaneous estimation of dynamics: Non-zero costs. *Bull. Int. Statist. Inst.* **40**, 943–74.

Brown, R. G. (1962). *Smoothing, Forecasting and Prediction of Discrete Time Series.* London: Prentice Hall.

Cochrane, D. & Orcutt, G. H. (1949). Application of least squares regressions to relationships containing autocorrelated error terms. *J. Amer. Statist. Ass.* **44**, 32–61.

Davis, H. T. (1941). *The Analysis of Economic Time Series.* Indiana: Bloomington Press.

Holt, C. C. *et al.* (1960). *Planning Production, Inventory, and Work Force.* London: Prentice-Hall.

Jeffreys, H. (1961). *Theory of Probability,* 3rd Edition. Oxford: Clarendon Press.

Kendall, M. G. (1944*a*). Oscillatory movement in English agriculture. *J. R. Statist. Soc.* **106**, 91.

Kendall, M. G. (1944*b*). On autoregressive time series. *Biometrika,* **33**, 105–22.

Muth, J. F. (1960). Optimal properties of exponentially weighted forecasts. *J. Amer Statist. Ass.* **55**, 299–306.

Savage, L. J. (1961). *The Subjective Basis of Statistical Practice.* Manuscript, University of Michigan.

Tiao, G. C. & Zellner, A. (1964). Bayes theorem and the use of prior knowledge in regression analysis. *Biometrika,* **51**, 219–30.

Zellner, A. & Tiao, G. C. (1964). Bayesian analysis of the regression model with autocorrelated errors. *J. Amer. Statist. Ass.* **59**, 763–78.

3.3
Models for Forecasting Seasonal and Non-Seasonal Time Series

G. E. P. BOX, G. M. JENKINS, and D. W. BACON

We hope we may be forgiven for introducing into an advanced seminar on Spectral Analysis of Time Series a paper which contains no mention of such analysis. Our excuse is that, important though spectral analysis is, there are many problems in time series analysis which are perhaps better approached in terms of a parametric model and here is perhaps a good place to remind ourselves of this fact.

The optimal forecasts of future values of a time series are determined by the nature of the stochastic model which describes that series. The main effort then in statistical analysis directed to forecasting must be in obtaining a suitable stochastic model for the series. The following paper outlines the approach which has been taken in a forthcoming book [1].

The stochastic models we employ are empirico-mechanistic in the sense that while they can be interpreted as descriptions of physical phenomena having the right general character they do not claim to represent exact physical reality and are fitted to data empirically.

An important principle in the choice of such models is that they should, whilst adequately representing the data, contain as few parameters as possible. Following John Tukey we call this the principle of parsimony.

1. NON-SEASONAL STOCHASTIC MODELS

1. 1. A linear stochastic model.

We denote values of a time series at equispaced times t, $t-1$, $t-2, \ldots$ by $z_t, z_{t-1}, z_{t-2}, \ldots$. Let B be the backward shift operator and let ∇ be the backward difference operator so that

$$Bz_t = z_{t-1} \quad \text{and} \quad \nabla z_t = z_t - z_{t-1} = (1-B)\, z_t \, .$$

Also let $a_t, a_{t-1}, a_{t-2}, \ldots$ be a sequence of uncorrelated random

The Collected Works of George E. P. Box, 1984, Wadsworth, Inc., Belmont, CA 94002.
Originally published in *Spectral Analysis of Time Series,* ed. B. Harris (New York: John Wiley & Sons, 1967), pp. 271-311.

normal variables having mean zero and variance σ_a^2.

For the representation of a time series which may be non-stationary such as occurs in economic and control applications we find that a useful model is

$$\varphi_p(B)\nabla^d z_t = \theta_0 + \theta_q(B)\,a_t \tag{1}$$

where

$$\varphi_p(B) = 1 - \varphi_1 B - \varphi_2 B^2 \ldots - \varphi_p B^p$$

$$\theta_q(B) = 1 - \theta_1 B - \theta_2 B^2 \ldots - \theta_q B^q$$

and we assume that the roots of $\varphi_p(B) = 0$, $\theta_q(B) = 0$ lie outside the unit circle. We shall say that the model (1) is of <u>order</u> (p, d, q) where in practice the positive integers p, d, and q will usually be 0, 1, or 2. We shall refer to $\varphi_p(B)$ as the autoregressive operator, $\Phi_{p+d}(B) = \varphi_p(B)(1-B)^d = \varphi_p(B)\nabla^d$ as the generalized autoregressive operator, and to $\theta_q(B)$ as the moving average operator.

When $d > 0$, a non-zero value for $\theta_0 = E(\nabla^d z_t)$ implies the existence of an underlying <u>deterministic</u> polynomial trend of degree d. To obtain an entirely stochastic representation for the examples we discuss in the present paper we omit θ_0 from the model when $d > 0$. For certain applications, for example, in estimating a possible linear trend in the presence of non-stationary noise, it is useful to retain a possibly non-zero θ_0 in the model.

The model is essentially a device for transforming the highly dependent series z_t to a sequence of random deviates, that is to "white noise".

1. 2. Motivation for the model.

The reason for introducing both a finite moving average operator $\theta(B)$ as well as a finite autoregressive operator $\varphi(B)$ is that a finite moving average is equivalent to an infinite autoregression and vice-versa, so that the inclusion of both types of terms makes for parsimony. The requirement that $\varphi(B) = 0$ has roots outside the unit circle, ensures stationarity of the series $\nabla^d z_t = w_t$ since the coefficients in $w_t = \varphi^{-1}(B)\theta(B)a_t$ then form a convergent series. The corresponding requirement that $\theta(B) = 0$ has roots outside the unit circle ensures that the coefficients of z_t in $a_t = \theta^{-1}(B)\varphi(B)\nabla^d z_t$ form a convergent series. We shall call this the requirement of invertibility.

The object of introducing the difference operator ∇^d is to allow for what may be called homogeneous non-stationarity.

Consider the situation where $d = 0$ in equation (1) so that $\varphi(B) z_t = \theta_0 + \theta(B) a_t$. The requirement that the roots of $\varphi(B)$ lie outside the unit circle would ensure not only that the series z_t was stationary with mean θ_0 but also that $\nabla z_t, \nabla^2 z_t, \nabla^3 z_t, \ldots$ etc. were each stationary with mean zero. Figure 1(a) shows one kind of non-stationary series we would like to be able to represent. This series is homogeneous except in level in the sense that except for a vertical translation one part of it looks very much the same as another. We can represent such behavior by retaining the requirement that each of the differences be stationary with zero mean but letting the level "go free". This we do by using the model

$$\varphi(B) \nabla z_t = \theta(B) a_t \; .$$

Figure 1(b) shows a second kind of non-stationarity of fairly common occurrence.

The series has neither a fixed level nor a fixed slope but its behavior is homogeneous if we allow for differences in these characteristics. We can represent such behavior by the model

$$\varphi(B) \nabla^2 z_t = \theta(B) a_t$$

which ensures stationarity and zero mean for all differences after the first and second but allows the level and the slope to "go free".

1. 3. Non-linear transformation of z.

A considerable widening of the range of useful application of the model (1) is achieved if we allow the possibility of transformation. Thus we may substitute $z_t^{(\lambda)}$ for z_t in (1) where $z_t^{(\lambda)}$ is some non-linear transformation of z_t involving one or more transformation parameters λ. A suitable transformation may be suggested by the situation or in some cases be estimated from the data. For example if we were interested in the sales of a recently introduced comodity, we might well find that sales volume was increasing at a rapid rate and that it was percentage fluctuation which showed stability rather than the absolute fluctuation.

In such a case it would clearly be sensible to analyze the logarithm of sales. When the transformation is to be estimated from the data one way to procede is to follow the route indicated by Box and Cox [2].

1. 4. Examples.

Rather simple representation in terms of the model (1) does seem to be possible for a large number of empirical time series.

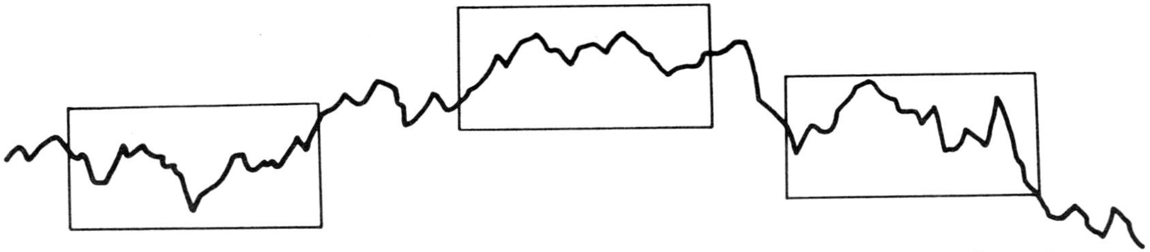

FIGURE 1(a). A series showing non-stationarity in level such as can be represented by the model $\varphi(B)\nabla z_t = \theta(B)a_t$.

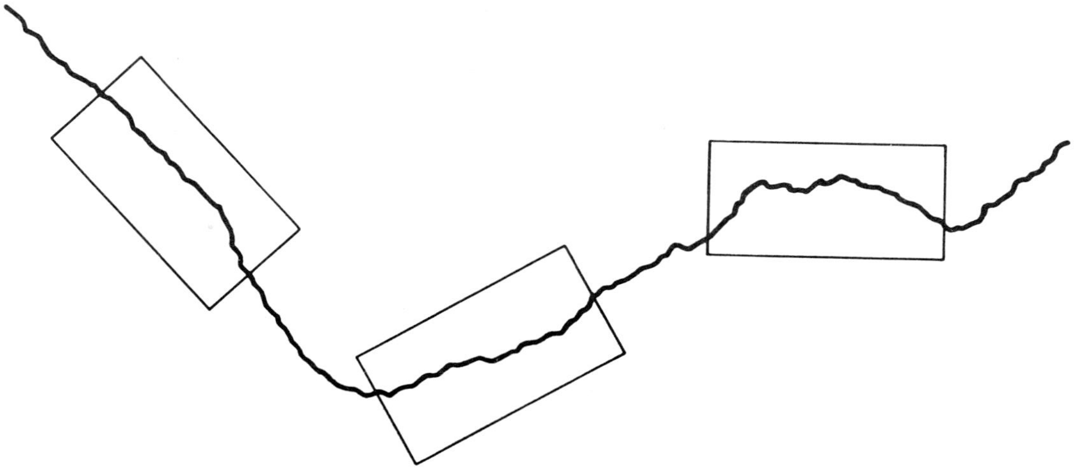

FIGURE 1(b). A series showing non-stationarity in level and in slope such as can be represented by the model $\varphi(B)\nabla^2 z_t = \theta(B)a_t$.

Figure 2 shows four different non-seasonal series arising from economic and control applications together with the models by which they are closely represented [1].

2. AN INTERATIVE APPROACH TO MODEL BUILDING

2. 1. Stages in the iterative process.

We now suppose that n consecutive observations $z_1, z_2, \ldots, z_n$ from a series are available and we wish to determine a suitable model. If possible n should be at least 50 and preferably more than 100. Although it will usually not be possible to obtain a very precise estimate of the model when fewer than 50 observations are available, yet even here such an estimate, suitably updated as new data become available will supply better forecasts than ad hoc methods. In practice such model determination must be done iteratively using processes of identification, fitting, diagnostic checking, refitting, and rechecking until a satisfactory representation is found. We can define these processes as follows:

By identification we mean the use of the data in the light of the information on how the series was generated to suggest a subclass of parsimonious models worthy to be entertained.

By estimation we mean efficient use of the data to make inferences about parameters conditional on the adequacy of the entertained model.

By diagnostic checking we mean checking the fitted model in its relation to the data with the hope of revealing model inadequacies and so proceding to model improvement.

2. 2. Identification.

Our primary tool of identification has been the inspection of the sample autocorrelation functions for z_t, ∇z_t, and $\nabla^2 z_t$. We first select the degree of required differencing d by noticing for what values of d the sample autocorrelation function dies out fairly quickly. For example Table 1a shows the first 24 sample autocorrelations for the series B and C from which it is tentatively concluded that stationarity may be induced by setting $d = 1$ in each case.

Having chosen a value for d, values to be entertained for p and q may usually be deduced by inspecting the pattern of the sample autocorrelations for $\nabla^d z_t$. In this inspection we use knowledge of the behavior of the theoretical autocorrelation function ρ_k for various types of models. The characteristics of $\rho_k(\nabla^d z)$ for models of order (1, d, 0), (2, d, 0), (0, d, 1), (0, d, 2) and (1, d, 1) are shown in Table 2. The boundaries of the admissible parameter space are indicated by the inequalities. By substituting sample estimates for ρ_k in Table 2, preliminary values for the parameters(which, however, are not in general efficient estimates) may be obtained.

FIGURE 2. Some series arising in forecasting and control with appropriate stochastic models.

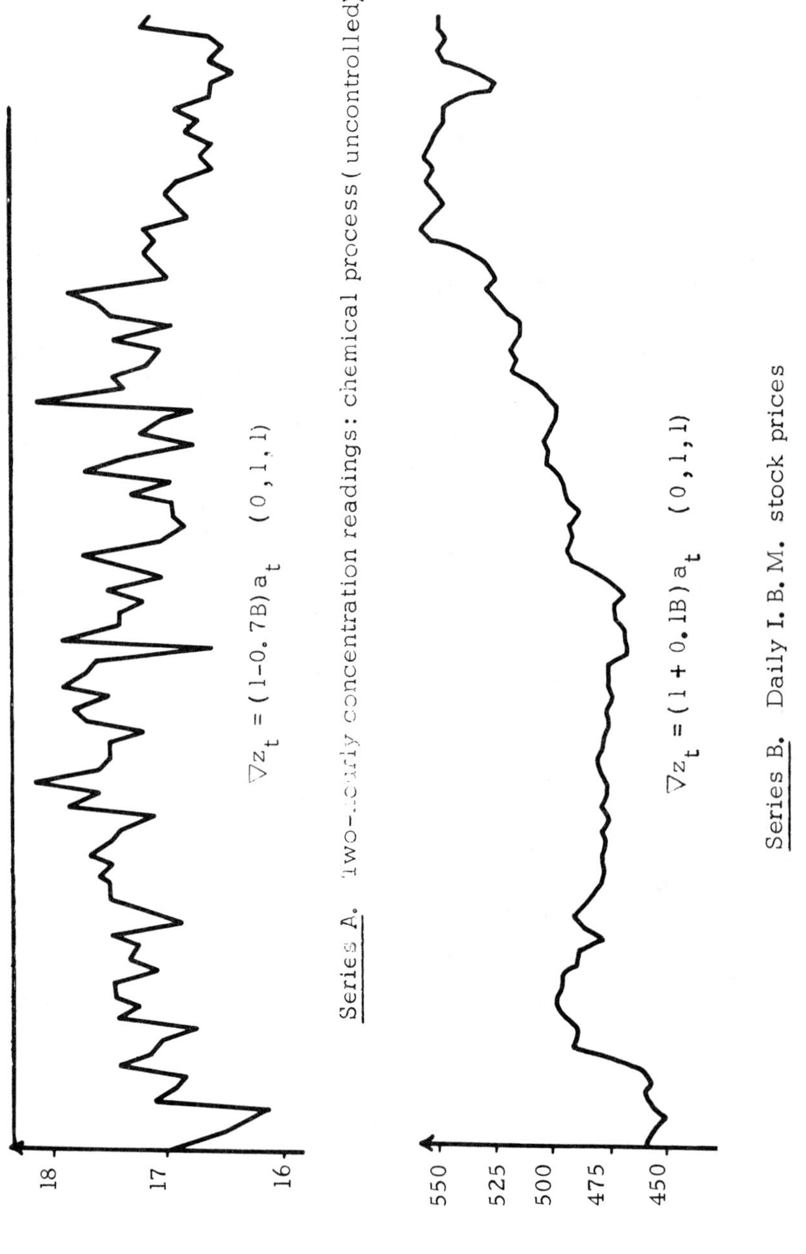
10
20
30
40
50
60
70
80
90
100
18
17
16
$\nabla z_t = (1-0.7B)a_t$ (0, 1, 1)
Series A. Two-hourly concentration readings: chemical process (uncontrolled)
550
525
500
475
450
$\nabla z_t = (1+0.1B)a_t$ (0, 1, 1)
Series B. Daily I. B. M. stock prices

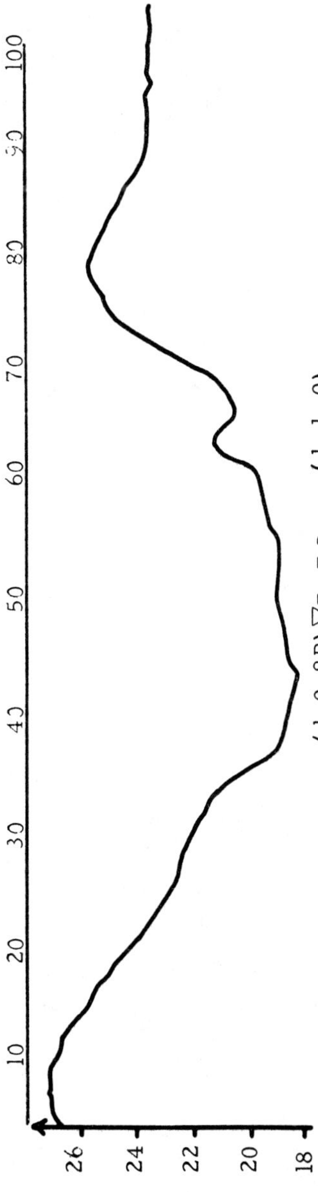

$(1-0.8B)\nabla z_t = a_t \quad (1,1,0)$

Series C. Temperature readings every minute: chemical process (uncontrolled)

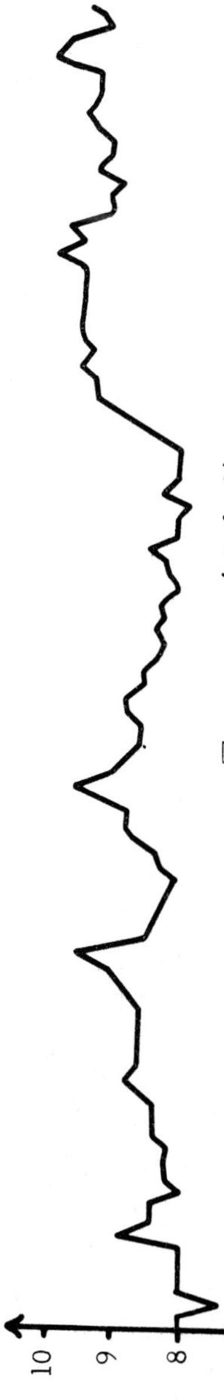

$\nabla z_t = a_t \quad (0,1,0)$

Series D. Hourly viscosity readings: chemical process (uncontrolled)

AUTOCORRELATIONS

SERIES B DAILY I. B. M. STOCK PRICES (369 OBSERVATIONS)

Lag	1	2	3	4	5	6	7	8	9	10	11	12	13	14	15	16	17	18	19	20	21	22	23	24
z	.99	.99	.99	.98	.98	.97	.97	.96	.95	.95	.94	.94	.93	.92	.92	.91	.90	.90	.89	.88	.87	.87	.85	.84
∇z	.08	.00	-.05	-.04	-.02	.13	.07	.03	-.07	.02	.08	.06	-.05	.07	-.07	.12	.13	.05	.05	.07	-.09	-.03	.07	.03
$\nabla^2 z$	-.45	-.02	-.04	.00	-.07	.11	-.01	.04	-.11	.02	.04	.04	-.12	.14	-.18	.10	.05	-.04	-.01	.10	-.12	-.02	.07	-.01

SERIES C TEMPERATURE READINGS EVERY MINUTE (226 OBSERVATIONS)

Lag	1	2	3	4	5	6	7	8	9	10	11	12	13	14	15	16	17	18	19	20	21	22	23	24
z	.99	.95	.91	.86	.81	.76	.71	.65	.62	.54	.48	.43	.37	.32	.27	.21	.16	.10	.05	.00	-.06	-.12	-.17	-.23
∇z	.82	.67	.54	.46	.42	.34	.27	.21	.16	.16	.11	.12	.08	.09	.09	.09	.11	.06	.06	.05	.05	.06	.00	-.05
$\nabla^2 z$	-.08	-.07	-.12	-.06	.01	-.02	.05	-.05	-.13	.13	-.13	.08	-.08	.03	-.01	-.06	.19	-.11	-.01	-.02	.00	.15	.02	-.02

TABLE la. Sample autocorrelations for series B and C.

PARTIAL AUTOCORRELATIONS

SERIES B DAILY I. B. M. STOCK PRICES (369 OBSERVATIONS)

Lag	1	2	3	4	5	6	7	8	9	10	11	12	13	14	15	16	17	18	19	20
z	.996	-.09	.01	.05	.02	.02	-.12	-.05	-.02	.06	-.05	-.09	-.03	.07	-.08	.06	-.14	-.10	-.01	-.08
∇z	.09	-.01	-.05	-.03	-.02	.13	.05	.02	-.06	.05	.09	.03	-.08	.08	-.06	.14	.10	.00	.07	.08
$\nabla^2 z$	-.45	-.28	-.24	-.20	-.29	-.17	-.13	-.03	-.14	-.16	-.09	.02	-.13	.01	-.19	-.13	-.03	-.10	-.10	-.06

SERIES C TEMPERATURE READINGS EVERY MINUTE (226 OBSERVATIONS)

Lag	1	2	3	4	5	6	7	8	9	10	11	12	13	14	15	16	17	18	19	20
z	.99	-.81	-.03	-.02	-.10	-.07	-.01	-.03	.04	-.04	-.15	.10	-.14	.01	-.10	-.02	-.07	-.11	.11	-.13
∇z	.81	-.01	-.01	.06	.03	-.03	-.01	-.08	.00	.10	-.14	.10	-.05	.05	.02	.06	.06	-.17	.09	.00
$\nabla^2 z$	-.08	-.08	-.14	-.10	-.03	-.05	.02	-.06	-.16	.09	-.14	.01	-.09	-.02	-.05	-.09	.13	-.13	-.03	-.05

TABLE 1b. Sample partial autocorrelations for series B and C .

Order (1, d, 0) ρ_k decays exponentially $\varphi_1 = \rho_1$ $-1 < \varphi_1 < 1$	Order (0, d, 1) Only ρ_0 and ρ_1 are non-zero $\rho_1 = \frac{-\theta_1}{1+\theta_1^2}$ $-1 < \theta_1 < 1$
Order (2, d, 0) ρ_k is mixture of exponentials or a damped sine wave $\varphi_1 = \frac{\rho_1(1-\rho_2)}{1-\rho_1^2}$ $\varphi_2 = \frac{\rho_2-\rho_1^2}{1-\rho_1^2}$ $\begin{cases} \varphi_2 > -1 \\ \varphi_2 + \varphi_1 < 1 \\ \varphi_2 - \varphi_1 < 1 \end{cases}$	Order (0, d, 2) Only ρ_0, ρ_1 and ρ_2 are non-zero $\rho_1 = \frac{-\theta_1(1-\theta_2)}{1+\theta_1^2+\theta_2^2}$ $\rho_2 = \frac{-\theta_2}{1+\theta_1^2+\theta_2^2}$ $\begin{cases} \theta_2 > -1 \\ \theta_2 + \theta_1 < 1 \\ \theta_2 - \theta_1 < 1 \end{cases}$
Order (1, d, 1) ρ_k decays exponentially after 1st lag correlation $\rho_1 = \frac{(1-\theta_1\varphi_1)(\varphi_1-\theta_1)}{1+\theta_1^2-2\varphi_1\theta_1}$ $\rho_2 = \rho_1\varphi_1$ $-1 < \varphi_1 < 1$ $-1 < \theta_1 < 1$	

TABLE 2. Behavior of theoretical autocorrelation function of dth difference of series for various sample models.

The partial autocorrelation function.

A useful tool which supplements the autocorrelation function in identification is the partial autocorrelation function. This can be calculated by fitting autoregressive processes

$$w_t = \phi_1 w_{t-1} + \dots + \phi_j w_{t-j} + a_t$$

of order $j = 1, 2, \dots$ to the series. If we denote by $\hat{\phi}_{jj}$ the last or jth coefficient in a process of order j, the plot of $\hat{\phi}_{jj}$ versus j is called the sample partial autocorrelation function. This has the property that if an autoregressive process of order p is appropriate, the partial autocorrelations beyond the pth will be small.

Examples.

Table 1(a) shows that whereas the autocorrelation function of the first difference of series B dies out very slowly, the autocorrelations of ∇z are very small. This suggests a model of the form $\nabla z_t = a_t - \theta a_t$ with $\theta \approx -0.1$. This is confirmed by the partial autocorrelations of z in Table 1(b) which are effectively zero beyond the first. The large value of the first partial autocorrelations shows that $\hat{\varphi}_1 \approx 1$ and thus that the autoregressive operator is equivalent to a first difference.

Table 1(a) shows that whereas the autocorrelations of series C die out very slowly, those of ∇z die more quickly and correspond to an autoregressive process with $\varphi \approx 0.8$. This is confirmed by the partial autocorrelations of z which are effectively zero beyond the second.

Uniqueness of identification.

While a given linear model has a unique convariance structure the converse is not true. This would at first sight exclude the use of the sample autocorrelation function to identify the model.

Now with $w_t = \nabla^d z_t$ we can write the model

$$\varphi(B) w_t = \theta(B) a_t$$

as

$$\prod_{i=1}^{p} (1 - G_i B) w_t = \prod_{j=1}^{q} (1 - H_j B) a_t$$

with the G's and H's inside the unit circle. The autocovariance generating function for w_t is then

$$C(B) = \sigma_0^2 \prod_{i=1}^{p} (1 - G_i B)(1 - G_i B^{-1}) \prod_{j=1}^{q} (1 - H_j B)(1 - H_j B^{-1})$$

whence it follows that there are 2^{p+q} representations

$$\prod_{i=1}^{p} (1 - G_i B^{\pm 1}) w_t = \prod_{j=1}^{q} (1 - H_j B^{\pm 1}) a_t$$

having the same covariance structure. It will be noticed however that only one of these expresses w_t in terms of past values of w and of a. Also although we can write for example

$$(1 - G_i B^{-1}) = -G_i B^{-1} (1 - G_i^{-1} B)$$

to produce a representation in terms of past history G_i^{-1} will now lie outside the unit circle. Thus there will only be one linear representation of a stationary invertible process in which w_t is expressible exclusively in terms of past values of w and of a.

Reversed Model.

In what follows we shall need to use the fact that if a sequence of observations $z_1, z_2, \ldots, z_t, \ldots, z_n$ can be described by the model

$$\varphi(B)(1 - B)^d z_t = \theta(B) a_t ,$$

they can equally well be described by the reversed process

$$\varphi(F)(1 - F)^d z_t = \theta(F) a_t \tag{2}$$

where F is the forward shift operator B^{-1}.

2.3. Estimation.

Suppose the model has been tentatively identified as of some specific form within the family

$$\varphi(B) \nabla^d z_t = \theta(B) a_t .$$

Then if for a sequence of observations z_t and for given $\underset{\sim}{\varphi}$ and $\underset{\sim}{\theta}$ we can compute

$$a_t = \theta^{-1}(B)\varphi(B)(1-B)^d z_t \qquad t = 1, 2, \ldots, n,$$

then the log likelihood for $\underset{\sim}{\varphi}$ and $\underset{\sim}{\theta}$ is a linear function of the sum of the squares $S(\underset{\sim}{\varphi}, \underset{\sim}{\theta}) = \Sigma_{t=1}^n a_t^2$.

Recursive calculation of the a's.

In practice the a_t's are conveniently calculated recursively. For example, if we were fitting the model

$$(1 - \varphi B)(1 - B) z_t = (1 - \theta B) a_t ,$$

we could write

$$a_t = \theta a_{t-1} + z_t - (1+\varphi) z_{t-1} + \varphi z_{t-2} ,$$

and knowing a_{t-1} and the z's we could calculate a_t and hence a_{t+1} and so on. A difficulty occurs at the beginning of the series since for example, a_1 depends on a_0, z_0, and z_{-1} all of which are unknown. In most cases if n were large, little would be lost by starting with a_3 and setting a_2 equal to its expected value of zero. More exactly we can proceed as follows. For any chosen values of the parameters φ and θ we can compute a_1 by substituting maximum likelihood estimates $\hat{a}_0(\varphi,\theta)$, $\hat{z}_0(\varphi,\theta)$, $\hat{z}_{-1}(\varphi,\theta)$ to obtain

$$a_1 = \theta\hat{a}_0(\varphi,\theta) + z_1 - (1+\varphi)\hat{z}_0(\varphi,\theta) + \varphi\hat{z}_{-1}(\varphi,\theta) .$$

Because the series satisfies the reversed model (2) the maximum likelihood estimates $\hat{z}_0(\varphi,\theta)$, $\hat{z}_{-1}(\varphi,\theta)$ will turn out to be the minimum mean square error forecasts calculated from the reversed series and $\hat{a}_0(\varphi,\theta)$ will be zero.

Although theoretically this same difficulty of unknown starting values occurs in the recursive calculation of the forecasts, in practice this calculation can be begun so far back in the reversed series that transients due to the initial approximation will have died out by the time the beginning of the series is reached.

As has been emphasized by G. A. Barnard [3] there is much to be gained by inspecting the likelihood function, or equivalently for the present Normal assumption, the sum of squares function. Such inspection allows us to obtain an impression of the overall estimation situation and is particularly important when we are studying an unfamiliar estimation problem or one that could give rise to anomalies. The present situation is of this kind, for while a_t is a linear function of the autoregressive parameters φ, it is non-linear in the moving average parameters θ. Furthermore it happens not infrequently in practice that the minimum sum of squares is on a boundary of the admissible parameter square.

Confidence Regions.

When we have a sum of squares surface which is roughly quadratic near the minimum, an approximate $1-\epsilon$ confidence region is included within the contour

$$(3) \qquad S_{1-\epsilon}(\underset{\sim}{\varphi}, \underset{\sim}{\theta}) = S(\hat{\underset{\sim}{\varphi}}, \hat{\underset{\sim}{\theta}})\left\{1 + \frac{\chi^2_\epsilon(p+q)}{n}\right\}.$$

For example Figure 3 shows the sum of squares function $S(\theta)$ for the model $\nabla z_t = (1-\theta B)a_t$ fitted to the IBM data (Series B). The minimum sum of squares is $S(\hat{\theta}) = 19{,}216$ at $\hat{\theta} \simeq -0.09$.

The critical sum of squares

$$S_{0.95}(\theta) = 19{,}416 = 19{,}216\left\{1 + \frac{3.84}{369}\right\}$$

defines an approximate 95% confidence interval $(-0.19, 0.03)$. Other sums of squares plots for two and three parameter systems with approximate confidence regions have been given in [1], [4], [5], [6].

Bayesian Viewpoint.

Alternatively if we adopt a Bayesian rather than a sampling theory viewpoint then with a prior distribution

$$p(\underset{\sim}{\varphi}, \underset{\sim}{\theta}, \sigma) = \frac{1}{\sigma} p(\underset{\sim}{\varphi}, \underset{\sim}{\theta})$$

we find the posterior distribution* of $\underset{\sim}{\varphi}$ and $\underset{\sim}{\theta}$ to be simply

$$p(\underset{\sim}{\varphi}, \underset{\sim}{\theta} \mid y) = \text{constant} \times S^{-\frac{n}{2}}(\underset{\sim}{\varphi}, \underset{\sim}{\theta})\, p(\underset{\sim}{\varphi}, \underset{\sim}{\theta}).$$

For a locally uniform reference prior the posterior distribution may be obtained simply by plotting $S^{-n/2}$ as is done for the I. B. M. example in Figure 3. The contours of S are with this assumption posterior probability contours, and the approximate confidence region (3) can

---

* The substitution of maximum likelihood estimates for values of the series z_0, z_{-1} etc. which are unavailable will be approximately equivalent to integrating out the unknown values over a locally uniform prior.

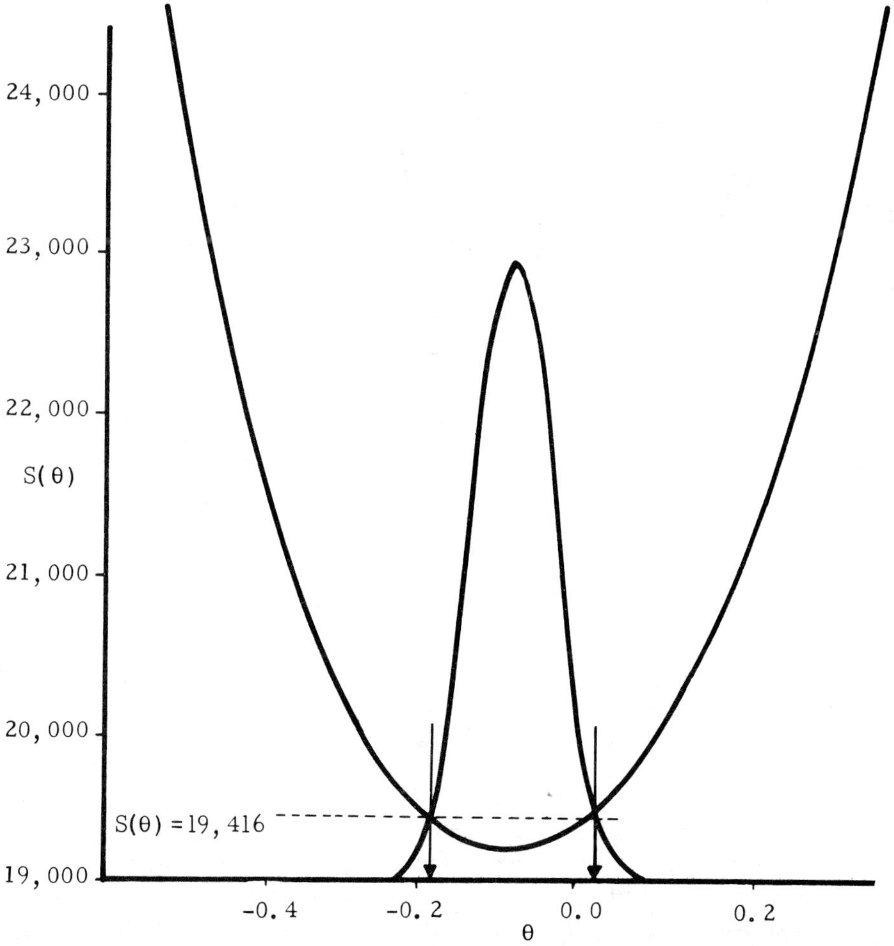

FIGURE 3. Sum of squares function for I. B. M. data (series B) with approximate 95% confidence region for θ and posterior distribution for locally uniform reference prior.

alternatively be interpreted as an approximate Bayesian region containing that proportion $1-\epsilon$ of this posterior probability distribution having highest density.

Multi-parameter Situations.

When there are many parameters we cannot so easily appreciate the whole likelihood surface. However it will often be desirable to look at particular features of this surface. For example, Figure 4 was constructed in the process of fitting and checking the model for series C. For this series the identification had suggested the model $(1 - 0.8B)\nabla z_t = a_t$ of order $(1,1,0)$. In the fitting process the opportunity was taken of entertaining the somewhat more elaborate model

$$(1 - \varphi B)\nabla z_t = (1 - \theta_1 B - \theta_2 B^2) a_t \quad \text{of order} \quad (1, 1, 2) .$$

In fitting such a model we can take advantage of the linearity of the autoregressive part by writing it in the form

$$(4) \qquad (1 - \varphi B) e_t = a_t$$

with

$$(5) \qquad (1 - \theta_1 B - \theta_2 B^2) e_t = \nabla z_t .$$

For each pair of fixed values (θ_1, θ_2) we can compute a set of e_t's from (5) which may then be used in (4) to calculate by simple least squares the conditional minimizing value $\hat{\varphi}|\theta_1, \theta_2$ and the associated conditional minimum sum of squares $S\{\hat{\varphi}|\theta_1, \theta_2\}$. Figure 4 shows the conditional sum of squares contours over the admissiable parameter space for the θ's. With the very wide availability of electronic computers, grid calculations and contour plots are not laborious and provide very valuable insight into the estimation situation.

Non-linear Estimation.

With appropriate caution the following iterative procedure is useful in determining maximum likelihood estimates and their approximate standard errors, particularly if the excellent starting values which can usually be obtained from sample autocorrelations are used.

Suppose we write the model

$$(6) \qquad a_t = \varphi(B)\theta^{-1}(B) w_t \quad \text{with} \quad w_t = \nabla^d z_t .$$

At the $(n-1)$th cycle of the iteration we have estimates $\varphi_1^{(n-1)}, \varphi_2^{(n-1)}, \ldots, \varphi_p^{(n-1)}; \theta_1^{(n-1)}, \theta_2^{(n-1)}, \ldots, \theta_q^{(n-1)}$ and denoting

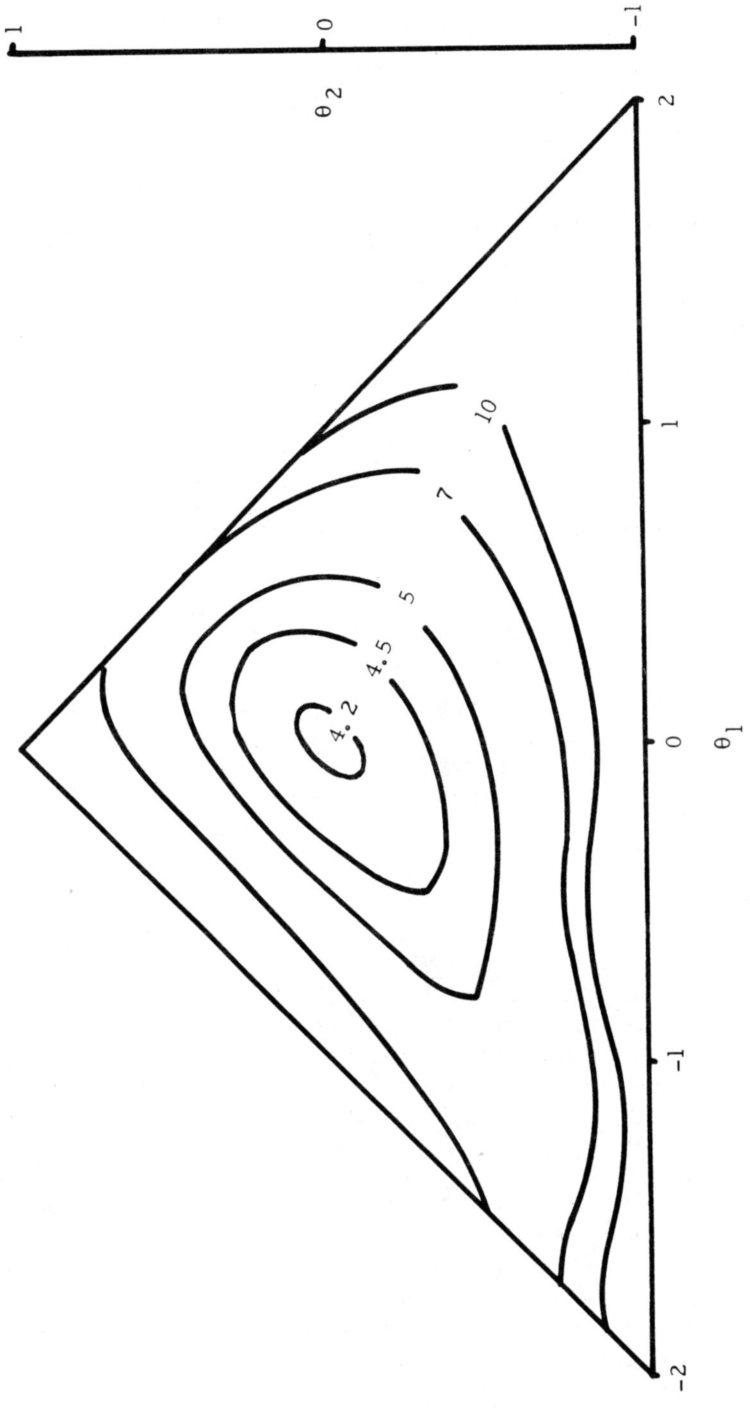

FIGURE 4. Contours of $S(\hat{\varphi}|\theta_1, \theta_2)$ plotted over the admissible parameter space for the θ's .

by $\varphi_{n-1}(B)$ and $\theta_{n-1}(B)$ the polynomials in B with these estimates substituted we have

$$a_t^{(n-1)} = \varphi_{n-1}(B)\theta_{n-1}^{-1}(B)w_t .$$

Now on linear expansion of (6) about the values $\varphi_1^{(n-1)}, \varphi_2^{(n-1)}, \ldots$ and after rearrangement we obtain approximately

$$a_t^{(n-1)} = \sum_{i=1}^{p} \{\varphi_i - \varphi_i^{(n-1)}\}\{\varphi_{n-1}^{-1}(B)a_{t-i}^{(n-1)}\} - \sum_{j=1}^{q} \{\theta_j - \theta_j^{(n-1)}\}\{\theta_{n-1}^{-1}(B)a_{t-j}^{(n-1)}\} + a_t$$

or

$$a_t^{(n-1)} = -\varphi(B)\{\varphi_{n-1}^{-1}(B)a_t^{(n-1)}\} + \theta(B)\{\theta_{n-1}^{-1}(B)a_t^{(n-1)}\} + a_t .$$

Fitting by linear least squares with $a_t^{(n-1)}$ as the "dependent" variable and $\varphi_{n-1}^{-1}(B)a_{t-1}^{(n-1)}, \varphi_{n-1}^{-1}(B)a_{t-2}^{(n-1)}, \ldots, \varphi_{n-1}^{-1}(B)a_{t-p}^{(n-1)}$; $\theta_{n-1}^{-1}(B)a_{t-1}^{(n-1)}, \ldots, \theta_{n-1}^{-1}(B)a_{t-q}^{(n-1)}$ as "independent" variables we obtain new estimates $\varphi_1^{(n)}, \varphi_2^{(n)}, \ldots, \varphi_p^{(n)}; \theta_1^{(n)}, \theta_2^{(n)}, \ldots, \theta_q^{(n)}$. These may now be used as the preliminary values in the nth cycle and so on. When reasonably good starting values are used, such as may be obtained from the sample autocorrelations, our experience has been that convergence to the least squares values is fast.

2.4. Diagnostic checks.

Two procedures we have used in checking the tentatively fitted model are (a) examination of residuals and (b) overfitting.

If the model is adequate and the number of fitted observations is not too small, then the values $(\hat{\varphi}, \hat{\theta})$ will be sufficiently close to the true values (φ, θ) so that the residuals $a_t(\hat{\varphi}, \hat{\theta})$ will be (very nearly) uncorrelated deviates. If they do not so behave, the nature of their dependence can provide clues as to how the model ought to be modified. It is a simple matter to calculate the theoretical autocorrelation function for the residuals which would occur with various types of model inadequacy. The sample autocorrelation function can thus be made to provide clues to necessary modifications. Particularly in the case of seasonal time series it is also valuable to inspect the cumulative periodogram of the residuals. We illustrate this check later in the analysis of a seasonal model. Checks of this kind on residuals have the advantage of generality. We do not need to have any very specific alternative model in mind. However, as always we pay for generality by lack of sensitivity. When there is a particular elaboration of the tentatively identified model which we fear, a much more sensitive check is provided by overfitting. That is by comparing the fits of the more elaborate and the less elaborate model. The process has already been illustrated in the case of series C in Figure 4. When the effect of non-linearities is not serious, as is usually the case if n is moderate or large, the usefulness of an additional set of parameters can be checked in a more formal manner by an approximate analysis of variance, where the reduction in the residual sum of squares accounted for by the additional parameters is compared with the residual sum of squares from the more elaborate model.

3. FORECASTING

3.1. General theory.

Suppose now that a satisfactory linear model of the form

$$\varphi(B)\nabla^d z_t = \theta(B) a_t \tag{7}$$

is exactly known. This may be written alternatively in the form

$$z_t = C(t) + \psi(B) a_t \tag{8}$$

where $C(t)$ is the complementary function of the difference equation (7). The first few coefficients $\psi_1, \psi_2, \ldots$ which are needed in what follows may be obtained by equating coefficients of $B, B^2, \ldots$ in

$$\varphi(B)\nabla^d \psi(B) = \theta(B) \,. \tag{9}$$

It is now immediate [7] that, given only the values of a series $z_t, z_{t-1}, z_{t-2}, \ldots$ for times up to an including t, the minimum mean square error estimate of some future value $z_{t+\ell}$ ($\ell \geq 1$) is given by $E_t(z_{t+\ell})$ where E_t is the conditional expectation at time t. That is

$$E_t(z_{t+\ell}) = \hat{z}_t(\ell) = C(t+\ell) + \psi_\ell a_t + \psi_{\ell+1} a_{t+1} + \ldots \,.$$

We refer to $\hat{z}_t(\ell)$ as the forecast for origin t and lead time ℓ. The corresponding forecast error for lead time ℓ is then

$$e_t(\ell) = z_{t+\ell} - \hat{z}_t(\ell) = a_{t+\ell} + \psi_1 a_{t+\ell-1} + \ldots + \psi_{\ell-1} a_{t+1}$$

and in particular

$$e_t(1) = a_{t+1} \,.$$

Hence the a_t's are in fact the one step ahead forecast errors and the variance of the forecast error for lead time ℓ is

$$V\{e(\ell)\} = (1 + \psi_1^2 + \psi_2^2 + \ldots + \psi_{\ell-1}^2)\sigma_a^2 . \tag{10}$$

3.2. Forecasting using the difference equation.

In practice the forecasts are most easily calculated directly from the stochastic difference equation. Writing for the overall autoregressive operator

$$\Phi_{p+d}(B) = \varphi_p(B)(1-B)^d = 1 - \Phi_1 B - \ldots - \Phi_{p+d} B^{p+d},$$

we can express the observation to be forecast in terms of previous z's and a's as follows

$$z_{t+\ell} = \Phi_1 z_{t+\ell-1} + \ldots + \Phi_{p+d} z_{t+\ell-p-d} + a_{t+\ell} - \theta_1 a_{t+\ell-1} \ldots - \theta_q a_{t+\ell-q} . \tag{11}$$

The forecast at origin t for lead time ℓ is now found by taking conditional expectations at time t on both sides of the equation using the relations

$$\left\{\begin{array}{ll} E_t z_{t-j} = z_{t-j}, \quad E_t a_{t-j} = a_{t-j} = z_{t-j} - \hat{z}_{t-j-1}(1), & (j = 0, 1, 2, \ldots) \\ E_t z_{t+j} = \hat{z}_t(j), \quad E_t a_{t+j} = 0, & (j = 1, 2, \ldots) . \end{array}\right. \tag{12}$$

Whence we obtain the forecast $\hat{z}_t(\ell)$ entirely in terms of previous forecasts and of known values of the series.

In practice we do not have the exact values of the parameters but only estimates which we substitute instead. Investigation shows that the forecast is rather robust to moderate changes in the parameters and that whenever the number of observations used to estimate the coefficients is reasonably large, the approximation is very good.

Example.

Series E (Figure 5) arising from a control study was closely represented by

$$\nabla^2 z_t = (1 - 0.9B + 0.5B^2) a_t$$

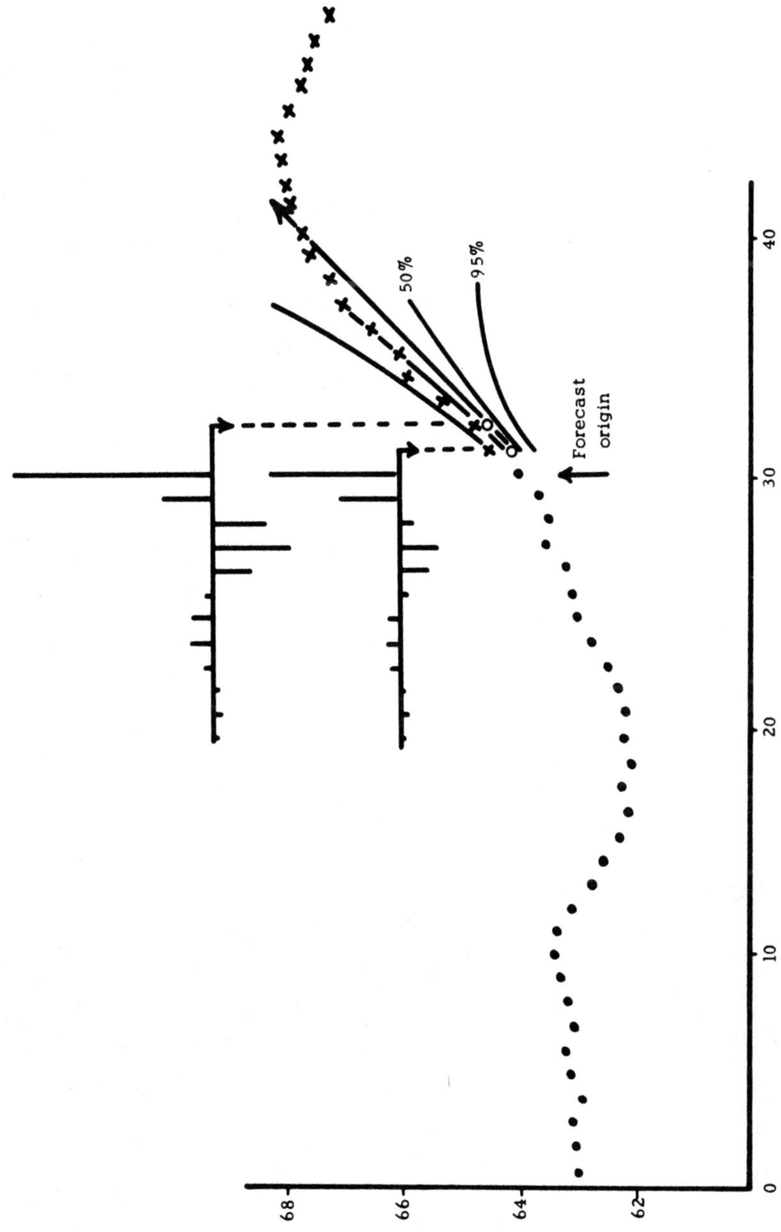

FIGURE 5. Part of a series arising in a control study (series E) with forecast function for origin t = 30 and weight function for forecasts $\hat{z}_t(1)$, $\hat{z}_t(2)$.

whence

$$z_{t+\ell} = 2z_{t+\ell-1} - z_{t+\ell-2} + a_{t+\ell} - 0.9\,a_{t+\ell-1} + 0.5\,a_{t+\ell-2}$$

and

$$\hat{z}_t(1) = 2z_t - z_{t-1} - 0.9\,a_t + 0.5\,a_{t-1}$$

$$\hat{z}_t(2) = 2\hat{z}_t(1) - z_t + 0.5\,a_t$$

$$\hat{z}_t(\ell) = 2\hat{z}_t(\ell-1) - \hat{z}_t(\ell-2) \qquad \ell > 2$$

with

$$a_t = z_t - \hat{z}_{t-1}(1) \ .$$

Thus the forecast function is a straight line determined by the first two forecasts $\hat{z}_t(1)$, $\hat{z}_t(2)$. Figure 5 shows a section of series E and forecast function at origin $t = 30$ together with 50% and 95% limits of error.

Updating the forecast.

From equation (8)

$$E_{t+1} z_{t+\ell+1} = E_t z_{t+\ell+1} + \psi_\ell a_{t+1}$$

so that

$$(13) \qquad \hat{z}_{t+1}(\ell) = \hat{z}_t(\ell+1) + \psi_\ell a_{t+1} \ .$$

For any model therefore of the type considered, the t-origin forecast of $z_{t+\ell+1}$ can be updated to become the t+1 origin forecast of the same observation by the addition of a constant multiple of the "one step ahead forecast error" $a_{t+1} = z_{t+1} - \hat{z}_t(1)$. This fact can still further simplify the practical calculation of forecasts.

3.3. The explicit forecast function.

Although for practical calculation of the forecasts it is simplest to use the difference equation directly in the manner already indicated, it is of interest to consider more explicitly the nature of the forecasts.

The rôle of the autoregressive operator.

By taking expectations on both sides of (11) with $\ell > q$ we obtain the difference equation

$$E_t\{\Phi_{p+d}(B)\, z_{t+\ell}\} = 0 \tag{14}$$

where

$$E_t z_{t+j} = \begin{cases} \hat{z}_t(j), & j = 1, 2, 3, \ldots \\ z_{t+j}, & j = 0, -1, -2, \ldots \end{cases}$$

We call the solution

$$E_t z_{t+j} = b_{1t} f_1(j) + b_{2t} f_2(j) + \ldots + b_{p+d,\,t} f_{p+d}(j), \qquad (j > q - p - d) \tag{15}$$

of this difference equation the "eventual" forecast function. The functions $f_1(j)$, $f_2(j)$, ..., $f_{p+d}(j)$ of the lead time j can include polynomials, exponentials, sines and cosines, and products of these functions.

For a given origin t the coefficients b_{it} are constants applicable for any lead time ℓ but change from one origin to the next and in fact adapt themselves appropriately to the particular part of the series being considered. In most practical cases $q \leq p + d$ so that the eventual forecast function applies to all the forecasts. If $q > p+d$, then it will apply for lead times $\ell > q - p - d$.

It will be seen that it is the autoregressive operator $\Phi_{p+d}(B)$ which completely determines the form of the eventual forecast function $\hat{z}_t(\ell)$.

Examples.

The model of order (1, 1, 0) fitted to series (B)

$$(1 - \varphi B)(1 - B)\, z_t = a_t$$

with $\varphi = 0.8$ yields the difference equation for $\ell > 0$

$$E_t\{z_{t+\ell} - (1+\varphi)\, z_{t+\ell-1} + \varphi z_{t+\ell-2}\} = 0 .$$

This has the solution

$$E_t(z_{t+j}) = \hat{z}_t(j) = b_{1t} - b_{2t}\varphi^{j+1}, \qquad j > -2$$

with

$$b_{1t} = z_t + \frac{\varphi}{1-\varphi}(z_t - z_{t-1}), \qquad b_{2t} = \frac{z_t - z_{t-1}}{1-\varphi} .$$

As the lead time is increased the exponential forecast approaches asymptotically the value b_{1t} with initial gradient $\hat{z}_t(1) - z_t = \varphi(z_t - z_{t-1})$.

Again the model of order $(0, 2, 2)$ fitted to series E

$$\nabla^2 z_t = (1 - 0.9B + 0.5B^2) a_t$$

yields the difference equation for $\ell > 2$

$$E_t\{\nabla^2 z_{t+\ell}\} = 0 ,$$

which has the solution

$$(16) \qquad E_t(z_{t+j}) = \hat{z}_t(j) = b_{1t} + b_{2t} j , \qquad j > 0$$

with

$$b_{1t} = 2\hat{z}_t(1) - \hat{z}_t(2) , \qquad b_{2t} = \hat{z}_t(2) - \hat{z}_t(1) .$$

For fixed origin t the forecasts for different lead times j lie along a straight line.

The rôle of the moving average operator.

While the autoregressive operator $\Phi_{p+d}(B)$ enitrely determines the form (15) of the eventual forecast function, the rôle of the moving average operator $\theta_q(B)$ is to decide how the coefficients b_{jt} are to be fitted and updated. The eventual forecast function is that unique curve of the form given by (15) which passes through the $p+d$ values $E_t z_{t+q}, E_t z_{t+q-1}, \ldots, E_t z_{t+q-p-d+1}$. In conjunction with $\Phi(B)$, $\theta(B)$ decides how the values (and hence the adaptive coefficients $b_{1t}, b_{2t}, \ldots, b_{p+d,t}$) depend on the available values $z_t, z_{t-1}, \ldots$ of the series.

To investigate this dependence, we write the model in the form

$$a_{t+1} = (1 - \pi_1 B - \pi_2 B^2 \ldots) z_{t+1} = \pi(B) z_{t+1}$$

where the π weights may be obtained explicitly by equating coefficients of $B, B^2, \ldots$ in

$$(17) \qquad \Phi(B) = \theta(B)\pi(B) .$$

Then taking conditional expectations at time t,

$$\hat{z}_t(1) = \sum_{j=1}^{\infty} \pi_j z_{t+1-j} \quad . \tag{18}$$

Now let us write

$$\hat{z}_t(\ell) = \sum_{j=1}^{\infty} \pi_j^{(\ell)} z_{t+1-j}$$

and call the $\pi_j^{(\ell)}$ the forecast weights at lead time ℓ. Then $\pi_j^{(1)} = \pi_j$ and the weights for longer lead times are readily obtained from the identity

$$\pi_j^{(\ell)} = \sum_{i=1}^{\ell} \psi_{\ell-i} \pi_{i+j-1}$$

which can be written as the convenient recurrence relation

$$\pi_j^{(\ell)} = \pi_{j+1}^{(\ell-1)} + \psi_{\ell-1}\pi_j \quad . \tag{19}$$

As an example Figure 5 shows part of series E which is fitted by $\nabla^2 z_t = (1 - 0.9B + 0.5B^2)a_t$. We have already seen that the fitted forecast function is a straight line which is updated as the origin is advanced. At origin t the line is determined by the forecasts $\hat{z}_t(1)$ and $\hat{z}_t(2)$. These values are in turn functions of the available values of the series $z_t, z_{t-1}, z_{t-2}, \ldots$ and have the weight functions indicated in the figure.

As we have emphasized, the actual process of forecasting and updating forecasts is best done directly in terms of the difference equation. However if desired the eventual forecast function could be used directly. The updating formulae for the coefficients b_{jt} can be found quite generally by noting that for $\ell > q - p - d$, equations (13) and (15) yield

$$\sum_{i=1}^{p+d} b_{it} f_i(\ell) = \sum_{i=1}^{p+d} b_{i,t-1} f_i(\ell+1) + \psi_\ell a_t \quad , \tag{20}$$

where $a_t = z_t - \hat{z}_{t-1}(1)$ is the lead-one forecast error. Solving $p + d$ such equations for different values of ℓ yields the constant d_{ih} in the updating formulae

$$b_{it} = \sum_{h=1}^{p+d} d_{ih} b_{i,t-1} + d_{i0} a_t \, .$$

For instance, for series E the straight line forecast (16)

$$\hat{z}_t(\ell) = b_{1t} + b_{2t}\ell$$

is updated in accordance with

$$b_{1t} = b_{1,t-1} + b_{2,t-1} + (1+\theta_2) a_t$$

$$b_{2t} = b_{2,t-1} + (1-\theta_1-\theta_2) a_t \, .$$

In matrix notation the p+d equations (20) may be written

$$\underset{\sim}{F}_\ell \underset{\sim}{b}_t = \underset{\sim}{F}_{\ell+1} b_{t-1} + \underset{\sim}{\psi} a_t$$

or

$$\underset{\sim}{b}_t = \underset{\sim}{L} \underset{\sim}{b}_{t-1} + \underset{\sim}{g} a_t$$

with

$$\underset{\sim}{L} = \underset{\sim}{F}_\ell^{-1} \underset{\sim}{F}_{\ell+1} \quad \text{and} \quad \underset{\sim}{g} = \underset{\sim}{F}_\ell^{-1} \underset{\sim}{\psi} \, .$$

These updating formulae are of the same form as those given by Brown and Meyer [8] and by Brown [9]. However these authors choose the functions $f_i(\ell)$ arbitrarily and fit them arbitrarily using "discounted least squares". This procedure will not in general lead to minimum mean square error properties nor any other optimal properties for the forecasts. In fact they can be very inefficient. For example in his book, Brown [9], illustrates his procedure by forecasting the I. B. M. stock price 3 days ahead using a quadratic forecast function. His forecast error variance is almost 3 times that of the optimal forecast. The latter is in turn almost equivalent to using todays price to forecast that 3 days ahead.

4. FORECASTING SEASONAL TIME SERIES

4.1. Multiplicative seasonal models.

We often need to analyze time series in which a recurrent pattern having known period s occurs. The pattern may change with time. For example the data of Table 3 quoted by Brown [9] are logarithms of totals of monthly international airline ticket sales. They show, as might be expected, a marked yearly seasonal pattern (s = 12). The general model (1) is entirely adequate to accomodate seasonal behavior. However with such behavior, there are known special features of the situation which can be exploited to give the model a parsimonious form.

Suppose for example that the observation z_t is made in the month of July. We expect that July sales of previous years will show simularities with this July. We can link the July figures for different years by a model of the form

(21) $$\Phi_P(B^s)\nabla_s^D z_t = \Theta_Q(B^s)e_t ,$$

where

$$\nabla_s = 1 - B^s .$$

In a similar way the sales in the month of June might be linked to those of previous June's by a model

(22) $$\Phi_P(B^s)\nabla_s^D z_{t-1} = \Theta_Q(B^s)e_{t-1} .$$

We should not however expect that e_t in (21) would be independent of e_{t-1} in (22) or of $e_{t-2}, e_{t-3} \ldots$ associated with May, April, etc. Instead it is reasonable to suppose that the e_t's might be linked month-wise by a second model of the form

(23) $$\varphi_p(B)\nabla^d e_t = \theta_p(B)a_t .$$

On multiplying (21) on both sides by the operator $\varphi_p(B)\nabla^d$ and using (23) we obtain the multiplicative model

(24) $$\Phi_P(B^s)\varphi_p(B)\nabla_s^D\nabla^d z_t = \Theta_Q(B^s)\varphi_q(B)a_t .$$

To represent some situations, for example data arising from weekly payments into banks, more than one seasonal component may be needed. The customary payment of bills at the end of each month would create a weekly seasonal pattern superimposed on the monthly pattern within years. For cases such as this a three-way model with multiplicative autoregressive, moving average, and difference components for periods s_2, s_1 and 1 could be used.

	Jan.	Feb.	Mar.	Apr.	May	June	July	Aug.	Sept.	Oct.	Nov.	Dec.
1949	4.718	4.771	4.883	4.860	4.796	4.905	4.997	4.997	4.913	4.779	4.644	4.771
1950	4.745	4.836	4.949	4.905	4.828	5.004	5.136	5.136	5.063	4.890	4.736	4.942
1951	4.977	5.011	5.182	5.094	5.147	5.182	5.293	5.293	5.215	5.088	4.984	5.112
1952	5.142	5.193	5.263	5.199	5.209	5.384	5.438	5.489	5.342	5.252	5.147	5.268
1953	5.278	5.278	5.464	5.460	5.434	5.493	5.576	5.606	5.468	5.352	5.193	5.303
1954	5.318	5.236	5.460	5.425	5.455	5.576	5.710	5.680	5.557	5.434	5.313	5.434
1955	5.489	5.451	5.587	5.595	5.598	5.753	5.897	5.849	5.743	5.613	5.468	5.628
1956	5.649	5.624	5.759	5.746	5.762	5.924	6.023	6.004	5.872	5.724	5.602	5.724
1957	5.753	5.707	5.875	5.852	5.872	6.045	6.142	6.146	6.001	5.849	5.720	5.817
1958	5.829	5.762	5.892	5.852	5.894	6.075	6.196	6.225	6.001	5.883	5.737	5.820
1959	5.886	5.835	6.006	5.981	6.040	6.157	6.306	6.326	6.138	6.009	5.892	6.004
1960	6.033	5.969	6.038	6.133	6.157	6.282	6.433	6.407	6.230	6.133	5.966	6.068

TABLE 3. Logarithms of totals of monthly international airline ticket sales.

4.2. Identification of seasonal models.

The autocorrelations for z, ∇z, $\nabla_{12} z$, and $\nabla\nabla_{12} z$ for the airline data are shown in Table 4. Whereas the autocorrelations for z, ∇z and $\nabla_{12} z$ had a complicated structure the double differences $\nabla\nabla_{12} z$ seem to possess a fairly simple correlation function with r_1 and r_{12} the largest values, suggesting a moving average formulation of the type

$$\nabla\nabla_{12} z_t = (1-\theta B)(1-\Theta B^{12}) a_t .$$

For this model it is easy to show that the only non-zero theoretical autocorrelations are

$$\rho_1 = -\theta/1+\theta^2$$

$$\rho_{11} = \rho_{13} = \theta\Theta/(1+\theta^2)(1+\Theta^2)$$

$$\rho_{12} = -\Theta/(1+\Theta^2) .$$

By equating the expressions for ρ_1 and ρ_{12} to their sample estimates we obtain rough values for the parameters of

$$\theta = 0.39 \qquad \Theta = 0.57 .$$

4.3. Fitting and checking seasonal models.

Fitting.

Using the above preliminary estimates for iterative least squares routine of equation (7) we obtain

Iteration	θ	Θ
Starting Values	.390	.570
1	.413	.569
2	.415	.567
3	.415	.568
4	.415	.568

	LAGS	AUTOCORRELATIONS											
z	1-12	.96	.91	.87	.83	.81	.79	.78	.78	.79	.81	.84	.85
	13-24	.81	.76	.72	.68	.66	.64	.63	.62	.64	.66	.69	.70
	25-36	.67	.62	.59	.55	.53	.52	.50	.50	.52	.55	.57	.59
	37-48	.56	.52	.48	.45	.42	.41	.39	.39	.41	.44	.47	.49
$\nabla_1 z$	1-12	.20	-.12	-.15	-.33	-.09	.03	-.12	-.36	-.12	-.12	.22	.92
	13-24	.24	-.15	-.13	-.31	-.06	.01	-.13	-.39	-.13	-.09	.24	.89
	25-36	.24	-.15	-.13	-.26	-.08	.02	-.15	-.37	-.17	-.05	.20	.88
	27-48	.46	-.18	-.08	-.23	-.08	.01	-.16	-.41	-.16	-.05	.19	.89
$\nabla_{12} z$	1-12	.72	.63	.49	.45	.40	.33	.25	.21	.16	-.01	-.13	-.27
	13-24	-.16	-.16	-.11	-.17	-.11	-.13	-.16	-.19	-.13	-.10	.00	-.06
	25-36	-.12	-.11	-.15	-.17	-.23	-.24	-.23	-.17	-.28	-.28	-.33	-.27
	37-48	-.21	-.19	-.16	-.10	-.02	.08	.11	.21	.29	.39	.51	.54
$\nabla_1 \nabla_{12} z$	1-12	-.34	.11	-.21	.02	.06	.03	-.06	-.00	.19	-.08	.07	-.43
	13-24	.17	-.06	.17	-.16	.08	.02	-.01	-.14	.05	-.11	.27	-.02
	25-36	-.12	.06	-.04	.06	-.02	-.07	-.07	.26	-.16	.11	-.21	-.01
	37-48	.07	.04	-.02	-.05	-.10	.14	-.13	.04	-.06	-.07	.17	-.08

TABLE 4. Sample autocorrelations for the airline data.

Figure 6 shows for this example a sum of squares grid and contours with the approximate 95% confidence region shaded. In particular it will be seen that there is very little dependence between θ and Θ.

Diagnostic checks.

The sample autocorrelations of the residuals from the least squares fit are shown in Table 5 with their approximate standard errors An approximate overall test that all $m = 48$ correlations are zero is supplied by referring

$$n \sum_{k=1}^{48} r_k^2 = 60.0$$

to the χ^2 table with $m = 48$ degrees of freedom. The value is very nearly significant at the 10% level thus casting some shadow of doubt on the model. In particular the occurrence of the correlation $r_{23}=0.26$ was somewhat disturbing.

Figure 7 shows the cumulative periodogram of the residuals. Here [10]

$$\sum_{i=1}^{j} I(i/n)/ns^2$$

is plotted against j with

$$I(i/n) = \frac{2}{n}[\{\sum_{t=1}^{n} \hat{a}_t \cos 2\pi \frac{it}{n}\}^2 + \{\sum_{t=1}^{n} \hat{a}_t \sin 2\pi \frac{it}{n}\}^2]$$

and

$$ns^2 = \sum_{i=1}^{q} I(i/n) \quad \text{and} \quad q = \begin{cases} \frac{1}{2}(n-2) & n \text{ even} \\ \frac{1}{2}\frac{(n-1)}{2} & n \text{ odd} \end{cases}.$$

The 95% and 75% Kolmogorov-Smirnov limits about the theoretical cumulative uniform distribution are also shown.

Our conclusion from this analysis was that while the proposed simple model described the series remarkably well it was possible that a more sensitive "overfitting" might show a means of further improvement.

Our model was therefore tentatively taken to be

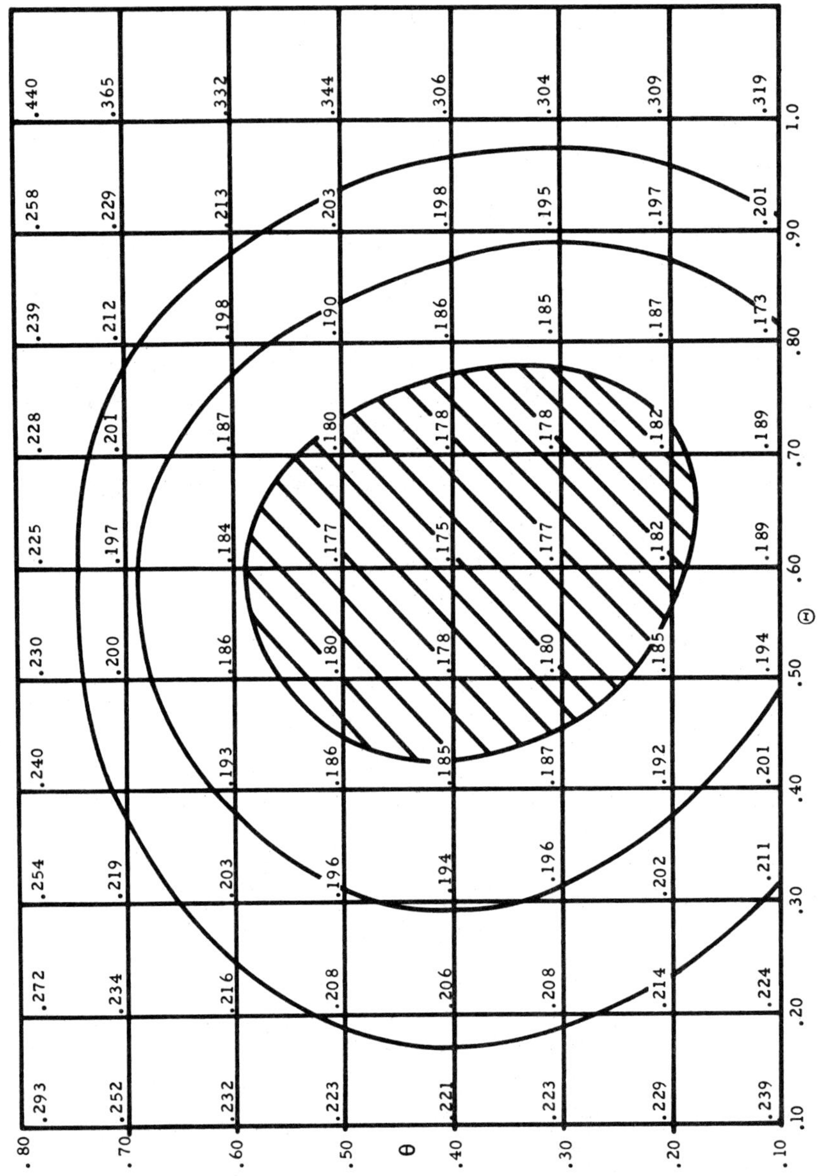

FIGURE 6. Sum of squares grid and contours of S(θ,Θ) for airline data.

Lag	Autocorrelations											
1-12	.02	.02	-.13	-.13	.05	.05	-.08	-.04	.11	-.06	.01	-.05
13-24	.00	.03	.06	-.19	.03	.00	-.13	-.12	-.06	-.06	.26	.02
25-36	.00	.08	-.03	-.08	-.10	-.11	-.06	.17	-.15	-.01	-.10	-.03
37-48	.16	.09	-.08	-.07	-.13	.01	-.04	.00	-.12	.03	.10	.11

Estimated standard error ± 0 0 8

TABLE 5. Autocorrelations of residuals $a_t(\hat{\theta},\hat{\Theta})$ for airline data.

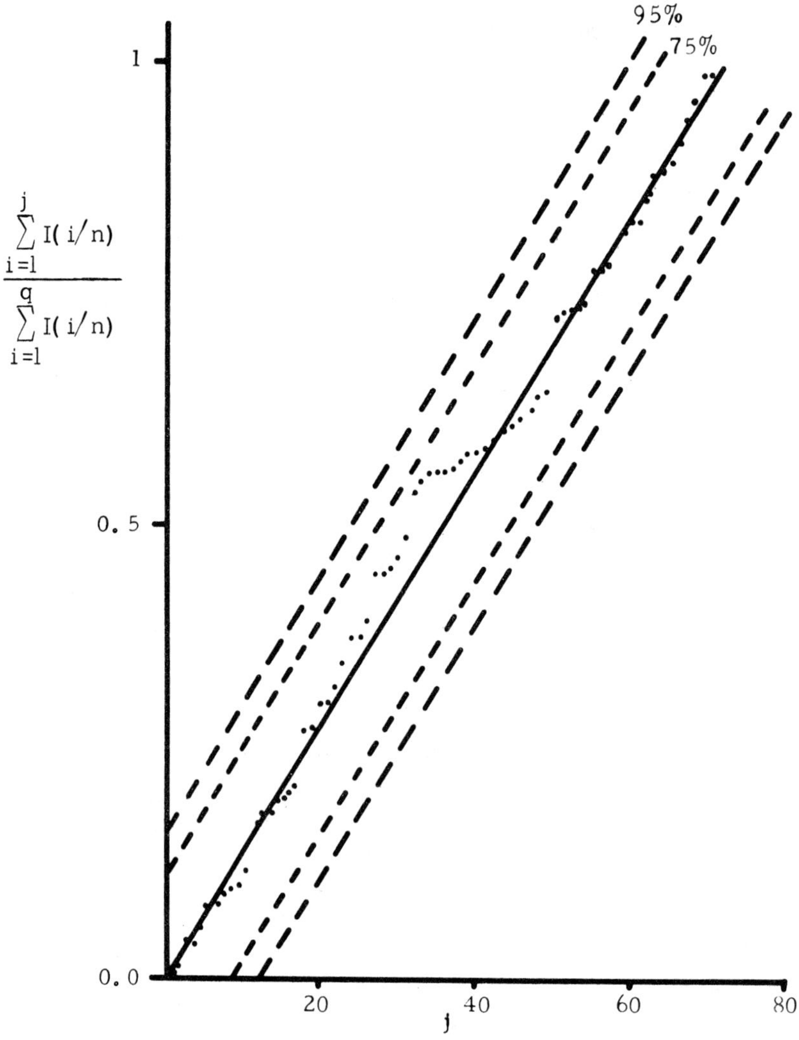

FIGURE 7. Cumulative periodogram for residuals from airline data.

$$(1-B)(1-B^{12})z_t = (1-0.4B)(1-0.6B^{12})a_t$$

with an estimate for σ_a^2 of $s_a^2 = 0.00124$.

4.4 Forecasts for the airline data.

Given the model, forecasts for any lead time can at once be written down. Thus, for example, the forecast for lead time 3 is obtained by taking expectations at time t on both sides of the equation

$$z_{t+3} = z_{t+2} + z_{t-9} - z_{t-10} + a_{t+3} - 0.4a_{t+2} - 0.6a_{t-9} + 0.24a_{t-10}$$

to give

$$\hat{z}_t(3) = \hat{z}_t(2) + z_{t-9} - z_{t-10} - 0.6a_{t-9} + 0.24a_{t-10} .$$

The airline series together with forecasts all made from origin July 1957 for lead times of 1, 2, . . . , 36 months are shown in Figure 8. For a lead time $\ell = 12r + m$ of r years and m months the forecast function takes the value

$$\hat{z}_t(12r+m) = A_t + rB_t + C_{m,t}$$

where A, B, and C are adaptive constants updated at each new origin t. A measures the initial level, B the yearly trend and C_m the monthly seasonal movement. Equating coefficients as in (9) and writing ψ_j as $\psi_{r,m}$, so that for example $\psi_{15} = \psi_{1,3}$, we find

$$\psi_{r,m} = \theta(1+r\Theta) + \delta\Theta$$

where

$$\delta = \begin{cases} 1 & \text{when } m = 12 \\ 0 & \text{when } m \neq 12 \end{cases} .$$

Setting $\theta = 0.4$, $\Theta = 0.6$, and $\hat{\sigma}_a^2 = 1.24 \times 10^{-3}$, the estimated standard deviations of the forecast errors (log data) for lead times 1 to 36 are shown in Table 6.

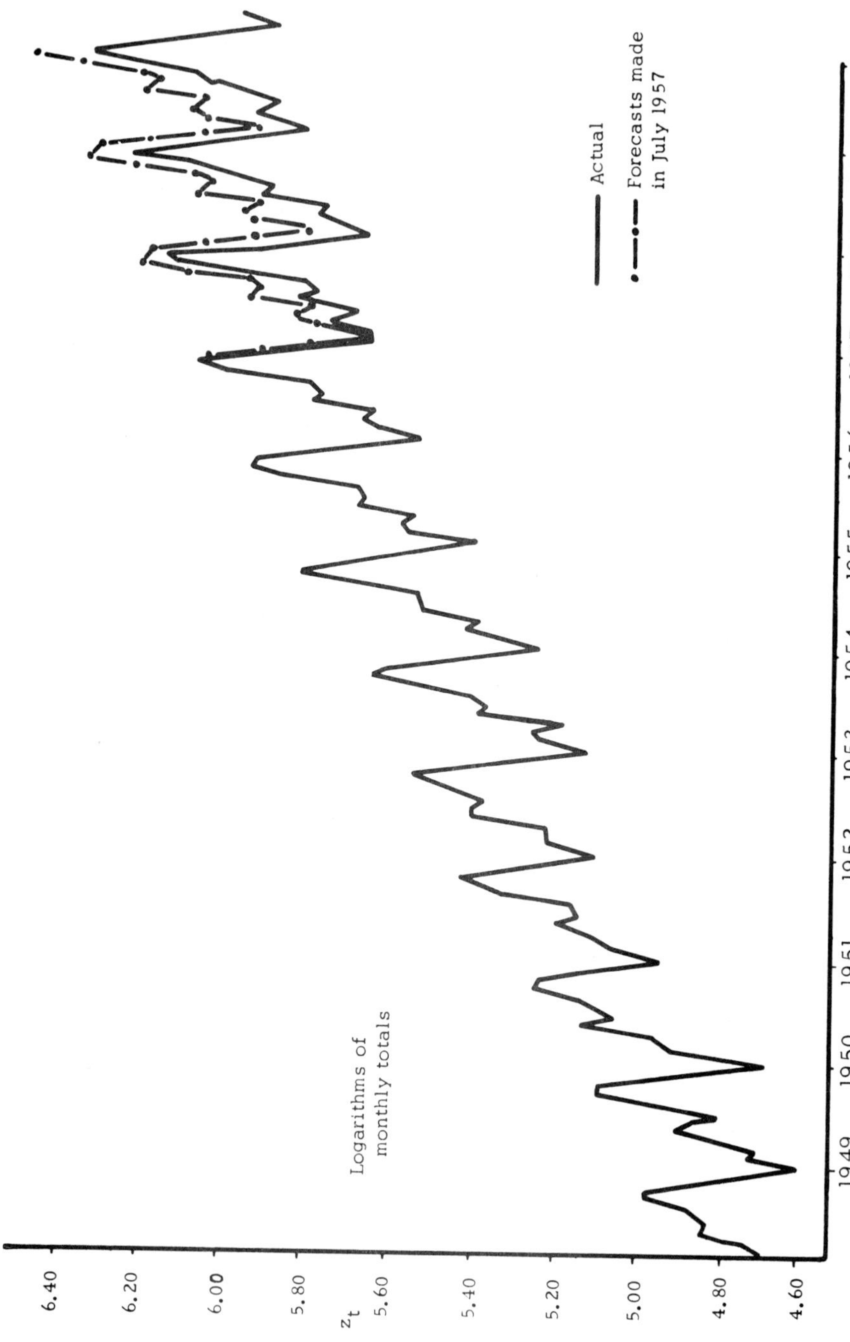

FIGURE 8. Logarithms of monthly totals of international airline passengers with forecasts made at origin July 1957 for 1, 2, 3, . . . , 36 months ahead.

Forecast lead times	1	2	3	4	5	6	7	8	9	10	11	12
S. E. $\times 10^{-2}$	3.5	4.1	4.6	5.1	5.5	5.9	6.1	6.6	6.9	7.3	7.6	7.8
Forecast lead times	13	14	15	16	17	18	19	20	21	22	23	24
S. E. $\times 10^{-2}$	8.6	9.1	9.6	10.0	10.4	10.9	11.2	11.6	12.0	12.4	12.7	13.0
Forecast lead times	25	26	27	28	29	30	31	32	33	34	35	36
S. E. $\times 10^{-2}$	13.8	14.3	14.8	15.3	15.7	16.2	16.6	17.0	17.5	17.9	18.3	18.7

TABLE 6. Estimated standard deviation of forecast errors for various lead times (logarithmic airline data).

Forecast weights.

Further insight into the nature of the forecasts is obtained by calculting the forecast weights (π) using (17) and (19). Those for the one step ahead forecast are shown in Figure 9. Suppose we denote exponentially weighted averages over months and over years by the symbols

$$\overset{\theta}{\bar{z}}_t = (1-\theta)\sum_{j=0}^{\infty} \theta^j z_{t-j}$$

$$\overset{\Theta}{\bar{z}}_t = (1-\Theta)\sum_{i=0}^{\infty} \Theta^i z_{t-12i} \, .$$

Then since $\hat{z}_t(1) = z_{t+1} - a_{t+1}$ and

$$a_{t+1} = \{1 - \frac{\theta B}{1-\theta B}\}\{1 - \frac{\Theta B^{12}}{1-\Theta B^{12}}\} z_{t+1}$$

it follows that

$$\hat{z}_t(1) = \overset{\theta}{\bar{z}}_t + \overset{\Theta}{\overline{z_{t-11} - \overset{\theta}{\bar{z}}_{t-12}}} \tag{25}$$

Thus the forecast is an exponentially weighted average (e. w. a.) taken over previous months modified by a second e. w. a. of discrepancies found between similar monthly e. w. a. 's and actual performance. For example, suppose we were attempting to predict December sales for a department store which would include a heavy component from Christmas buying. The first term on the right of (25) would be an e. w. a. taken over previous months the same year. However we know this will be an underestimate, so we correct it by taking a second e. w. a. over previous years of the discrepancies between actual December sales and the corresponding monthly e. w. a. taken over previous months.

As we have explained a more sensitive but less general diagnostic check is obtained by overfitting in the feared direction. In the course of such checking it was found for the present series that a slightly better representation resulting in an 8% reduction in the residual sum of squares was possible in terms of the model

$$\nabla\nabla_{12}^2 z_t = (1-\theta B)(1-\Theta_1 B \quad - \Theta_2 B^{24}) a_t \, .$$

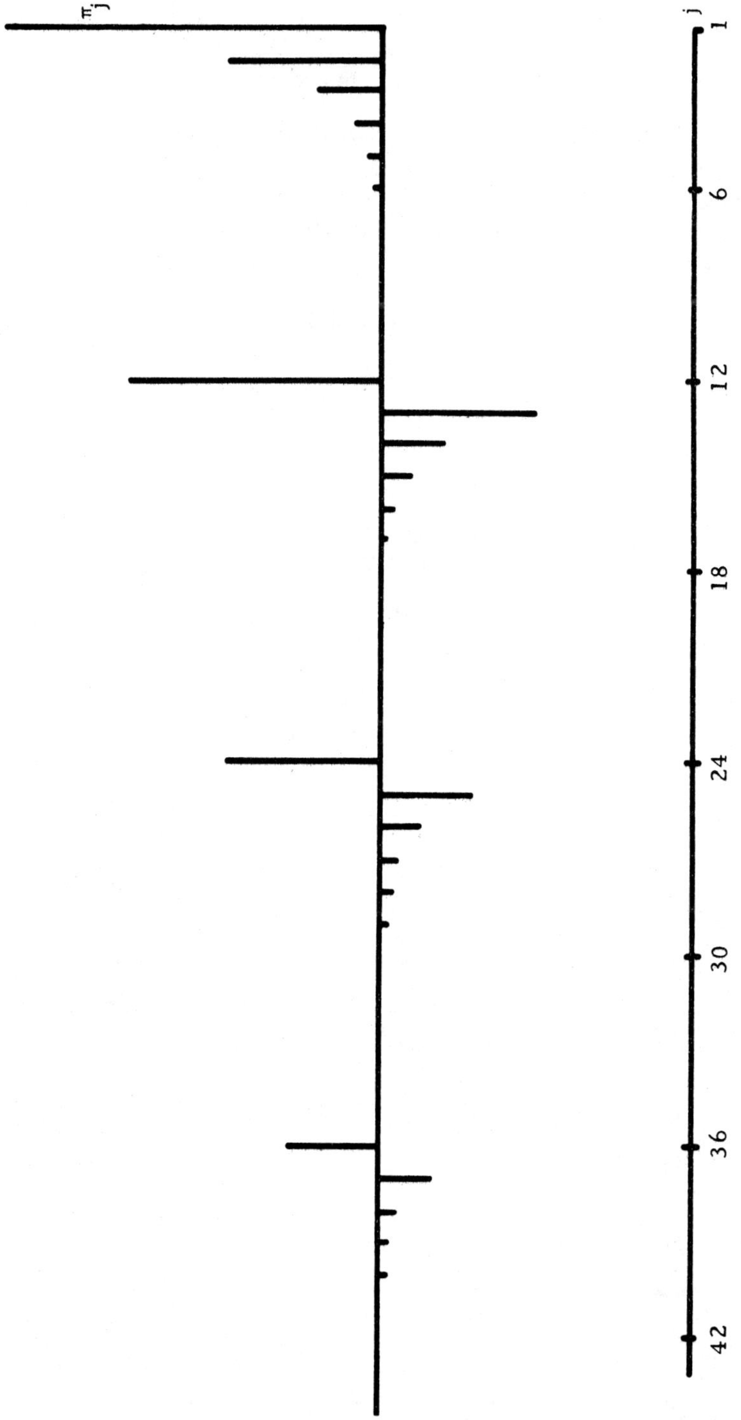

FIGURE 9. Lead-one forecast weights for airline model.

The forecasts provided by this model are very similar to those provided by the simpler model. The forecast function differs in containing an adaptive quadratic term in the yearly trend.

General procedure for identification of seasonal models.

Examination of the autocorrelation function of some suitable difference of the form $w_t = \nabla^d \nabla_s^D z_t$ of the series will usually indicate which model to fit in the first instance. This model can then be fitted, checked and modified if the fit is not adequate. This type of analysis has been used with success to study a variety of seasonal series [1]. In particular the simple models

$$w_t = (1 - \theta B)(1 - \Theta B^{12}) a_t$$

and

$$(1 - \Phi B^{12}) w_t = (1 - \theta B)(1 - \Theta B^{12}) a_t$$

with $w_t = \nabla \nabla_s z_t$ involving two and three parameters seem to provide adequate representations for many series.

The procedures for identifying, fitting, checking and forecasting these more elaborate models are similar to those set out above and are described in greater detail in [1].

REFERENCES

[1] Box, G. E. P. and Jenkins, G. M., Statistical Models for Forecasting and Control, Holden-Day, San Francisco (to appear).

[2] Box, G. E. P. and Cox, D. R., "An analysis of transformations", J. Roy. Stat. Soc., B, 26, 211 (1964).

[3] Barnard, G. A., Jenkins, G. M. and Winsten, C. B., "Likelihood, inference and time series", J. R. Stat. Soc., A, 125, 321 (1962).

[4] Box, G. E. P. and Jenkins, G. M., "Some statistical aspects of adaptive optimization and control", J. Roy. Stat. Soc., B, 24, 297 (1962).

[5] Box, G. E. P. and Jenkins, G. M., "Further contributions to adaptive quality control", Proc. of the 34th Session of I.S. I., Ottawa, Canada (1963).

[6] Box, G. E. P. and Jenkins, G. M., "Mathematical models for adaptive control and optimization", I. Chem. Eng. Symposium No. 4, 61 (1965).

[7] Whittle, P., Prediction and Regulation, Engligh U. P., London (1963).

[8] Brown, R. G. and Meyer, R. F., "The fundamental theorem of exponential smoothing", Operations Research, 9, 673 (1961).

[9] Brown, R. G., Smoothing, Forecasting and Prediction of Discrete Time Series, Prentice Hall, Inc., New Jersey (1962).

[10] Jenkins, G. M. and Watts, D. G., Applied Spectral Analysis, Holden Day, San Francisco (to appear).

3.4

Some Recent Advances in Forecasting and Control, Part I†

G. E. P. BOX
University of Wisconsin

G. M. JENKINS
University of Lancaster

1. INTRODUCTION

1.1. *Nature of Forecasting and Control Problems*

UNTIL fairly recently the word "control" has been principally associated in the statistician's mind with quality control, and especially with the quality control chart techniques developed originally by Shewhart in the United States and by Dudding and Jennett in Great Britain.

During the war the development of sequential inspection methods by Wald and Barnard gave new impetus to techniques in which sequential aspects were emphasised and led eventually to the introduction of cumulative sum charts by Page (1954, 1957) and by Barnard (1959).

The need for control implies the existence of an inherent disturbance in the process of one kind or another such as might be described by a time series. Thus, in recent years we find contributions to control problems from workers in stochastic processes such as Whittle (1963) and Bather (1963). Because one approach to control would be to forecast the deviation from target which would occur if no action were taken and then to act so as to cancel out that deviation, *forecasting* and *control* problems are closely linked together. However, we can forecast a time series in an optimal manner only if we have an adequate stochastic model for that series.

In the past a great deal of attention has been given to stationary time series models which have the property of remaining in equilibrium about a constant mean. However, forecasting has been of particular importance in business and economics where many series (for example, the monthly sales of an industrial product) are *non-stationary* and have no natural mean. It is not surprising, therefore, that the economic forecasting methods which have been proposed by such workers as Holt (1957, 1960), Winters (1960) and Brown (1962) and the control chart techniques proposed by Roberts (1959), all using the exponentially weighted moving averages, are appropriate for a particular type of *non-stationary* process. The fact that such methods have been successful supplies a clue to the *kind* of non-stationary model which might be useful in these problems.

† Originally presented at the European Meeting of Statisticians, held at Imperial College, London, in September 1966.

The Collected Works of George E. P. Box, 1984, Wadsworth, Inc., Belmont, CA 94002. Originally published in *Applied Statistics,* vol. 17, no. 2 (1968), pp. 91–109.

To a control engineer the word "control" has had a different connotation. He usually thinks in terms of feedback and feed forward control loops, the dynamics and stability of the system, and often of particular types of hardware to carry out the control action. In this paper we outline a statistical approach to optimal forecasting and to the optimal design of feedback and feedforward control schemes that we have developed in previous papers (Box and Jenkins, 1962, 1963, 1965; Box *et al.* 1967) and which will be described in a forthcoming book (Box and Jenkins, 1968).

The control techniques we discuss are different from those of standard quality control procedures, but this is because they have a different purpose. We certainly do not believe that the traditional quality control chart is unimportant or outmoded. Appropriate *display* of data on such a chart (rather than the burying of it in a process record book) ensures that changes that occur are regularly brought to the attention of those in charge of the process. They are thus nudged into seeking "assignable causes" for the changes and a continuous incentive for process improvement is achieved. This device is of enormous importance because it can stimulate new thinking about the process. However, in many situations a control scheme is required which adjusts some variable, whose precise effect on the quality characteristic is *known*, so as to minimize the variation of this quality characteristic about a target value. It is with such control problems that we are concerned here.

1.2. *An Outline of the Approach*

We suppose throughout that observations are available at *discrete* equispaced intervals of time. For example, in a sales forecasting problem, figures might be available every month and we might wish to forecast sales for 1, 2, 3, ..., 12 months ahead. Again, in a chemical process, observations and the opportunity to make control changes might occur every 5 minutes, every hour, or every shift depending on the rate at which the state of the system could change. In the case of a chemical process discrete observations might arise from a discrete or batch process, or a continuous record of the process characteristic might be "sampled" at equally spaced intervals. In practice, if the sampling interval is suitably chosen almost nothing will be lost by employing the discrete rather than the continuous record and there may be considerable gain in the simplicity of the analysis.

The optimal forecasts of future values of a time series are determined by the stochastic model that describes that series. Therefore the main object in statistical analysis directed to forecasting must be in obtaining a suitable stochastic model for the series in question. Therefore, we first develop a class of *stochastic models* which are capable of representing not only stationary behaviour but also non-stationary behaviour of the kind that we have encountered in practice. We show how models which satisfactorily describe a particular series may be derived and how they can be used to forecast seasonal as well as non-seasonal series. The same kind of stochastic model used in the forecasting problem may also be used to represent the disturbances which infect a system and which make control action necessary.

Now any control action which is taken will not be felt immediately but usually its effect will build up gradually because of the inertia of the system. Therefore we next describe *dynamic models* capable of representing the dynamic relationship between a controlling variable X and a controlled variable Y and we show how these dynamic models may be fitted to data obtained from the system.

An important principle in the choice of our models is that they should, whilst adequately representing the data, contain as few parameters as possible. Following Tukey we call this the *principle of parsimony.*

In Part II of this paper we shall describe how the stochastic and dynamic models may be brought together to design optimal feed-forward and feedback control schemes and also how the parameters in the stochastic and dynamic models may be simultaneously estimated from measurements made on the operating system.

2. Time Series Models

A criterion of great importance in discussing time series is *stationarity.* A series is strictly stationary if its properties are completely unaffected by a shift in the time origin. In particular, a stationary series varies about some *fixed* mean μ. It exhibits no change in mean and no drift.

2.1. *Autoregressive and Moving Average Models for Stationary Time Series*

Suppose we denote the values of a stationary series at equally spaced times $t, t-1, t-2, \ldots$ by $w_t, w_{t-1}, w_{t-2}, \ldots$ Let $a_t, a_{t-1}, a_{t-2}, \ldots$ be a "white noise" series consisting of uncorrelated random Normal deviates all having mean zero and variance σ_a^2. It is helpful to think of these a's as a series of random "shocks".

The time series model we employ, originally developed by Yule, is essentially a device for transforming the original series w_t, the observations of which are often highly correlated, into a series of uncorrelated component shocks a_t which can be thought of as generating the series. There are basically two different ways in which this is done.

The deviation $\dot{w}_t = w_t - \mu$ from the mean μ can be made linearly dependent on previous deviations $\dot{w}_{t-1} = w_{t-1} - \mu$, $\dot{w}_{t-2} = w_{t-2} - \mu$, etc., and on a_t. We then have what is called an *autoregressive* model. Thus

$$\dot{w}_t = \phi_1 \dot{w}_{t-1} + a_t, \tag{1}$$

$$\dot{w}_t = \phi_1 \dot{w}_{t-1} + \phi_2 \dot{w}_{t-2} + a_t, \tag{2}$$

are autoregressive models of orders 1 and 2, respectively.

Alternatively, we can make $\dot{w}_t$ linearly dependent on a_t and on one or more previous a's. We then have what is called a *finite moving average model.* Thus

$$\dot{w}_t = a_t - \theta_1 a_{t-1}, \tag{3}$$

$$\dot{w}_t = a_t - \theta_1 a_{t-1} - \theta_2 a_{t-2}, \tag{4}$$

are moving average models of orders 1 and 2, respectively. One might ask: can an autoregressive model be used to represent moving average behaviour? The answer is that this can be done but an infinite number of autoregressive terms are needed to represent a finite moving average model and vice versa.

To ensure parsimony we may need terms of both kinds and we are thus led to the *general mixed autoregressive-moving average model of order* (p, q), which may be written

$$\dot{w}_t - \phi_1 \dot{w}_{t-1} - \ldots - \phi_p \dot{w}_{t-p} = a_t - \theta_1 a_{t-1} - \ldots - \theta_q a_{t-q}, \tag{5}$$

where p and q would by 0, 1 or 2 in most applications. To manipulate models of this kind it is convenient to define a backward shift operator B such that

$$B w_t = w_{t-1}. \tag{6}$$

Using the operator B, (5) can be written

$$\phi_p(B)\dot{w}_t = \theta_q(B)a_t, \tag{7}$$

where

$$\phi_p(B) = 1-\phi_1 B-\phi_2 B^2 \ldots -\phi_p B^p,$$
$$\theta_q(B) = 1-\theta_1 B-\theta_2 B^2 \ldots -\theta_q B^q,$$

are polynomials in B of degree p and q respectively and $\phi_p(B)$ is called the *autoregressive* operator and $\theta_q(B)$ the *moving average* operator.

For example, the models of equations (1), (2), (3), and (4) could be written

A.R. 1: $(1-\phi_1 B)\dot{w}_t = a_t,$

A.R. 2: $(1-\phi_1 B-\phi_2 B^2)\dot{w}_t = a_t,$

M.A. 1: $\dot{w}_t = (1-\theta_1 B)a_t,$

M.A. 2: $\dot{w}_t = (1-\theta_1 B-\theta_2 B^2)a_t.$

Now consider the first order autoregressive model (1). The values of the series may be built up recursively as follows:

$$\dot{w}_1 = \phi_1 \dot{w}_0 + a_1,$$
$$\dot{w}_2 = \phi_1 \dot{w}_1 + a_2 = \phi_1^2 \dot{w}_0 + \phi_1 a_1 + a_2,$$
$$\dot{w}_3 = \phi_1 \dot{w}_2 + a_3 = \phi_1^3 \dot{w}_0 + \phi_1^2 a_1 + \phi_1 a_2 + a_3,$$
$$\dot{w}_t = \phi_1 \dot{w}_{t-1} + a_t = \phi_1^t \dot{w}_0 + \phi_1^{t-1} a_1 + \phi_1^{t-2} a_2 + \ldots + a_t. \tag{8}$$

We can ensure stationarity for this series by requiring that ϕ_1 lies between the values -1 and $+1$. If ϕ_1 lay outside these limits (if for example, ϕ_1 were equal to 2) then we can readily see from equation (8) that the deviation $\dot{w}_t$ would be dominated by remote events led by $\dot{w}_0$ and a_1 which would become more and more important as t became larger. On the other hand, if ϕ_1 lay between -1 and $+1$, as we require, the behaviour of $\dot{w}_t$ would be dominated by the most recent shock a_t, as is sensible.

A similar argument applied to the first order moving average model (3) leads to the conclusion that θ_1 must lie between -1 and $+1$ if a_t is not to be dominated by remote events. If this condition is satisfied the moving average model is said to be *invertible*.

Now one way of expressing the condition that ϕ_1 in the autoregressive operator $1-\phi_1 B$ lies between -1 and $+1$ is to say that the roots of the equation $1-\phi_1 B = 0$ (where B is regarded as a variable) lie *outside* the interval -1 to $+1$.

The corresponding condition for stationarity and invertibility of the general mixed autoregressive moving average model (5) is that the roots (which may be complex) of $\phi(B) = 0$ and $\theta(B) = 0$ must lie outside the unit circle and we shall suppose in all that follows that this condition is imposed.

With these conditions satisfied the model (5) turns out to be a valuable device for representing stationary time series. If the model is expressed in terms of the w_t's themselves, instead of deviations from the mean, the general form of the model may be written

$$\phi_p(B)w_t = \theta_0 + \theta_q(B)a_t, \tag{9}$$

where

$$\theta_0 = (1-\phi_1-\phi_2-\ldots-\phi_p)\mu.$$

2.2. *A General Model which can represent Stationary and Homogeneous Non-stationary Time Series*

Time series representing economic phenomena and disturbances in processes to be controlled are often best represented by non-stationary models. There is an unlimited number of ways in which a time series may be non-stationary. We now adapt our models to take account of the kinds of non-stationarity which we have frequently met in practice. Figure 1(a) shows one type of non-stationary series of common occurrence. This series is homogeneous except in its level. By this is meant that apart from a vertical translation, one part of the series looks much like another. A series z_t which is stationary in its *first difference*

$$\nabla z_t = z_t - z_{t-1} = (1-B)z_t$$

exhibits precisely this kind of behaviour. Again Figure 1(b) shows a second kind of non-stationarity which is frequently met. This series has neither a fixed level nor a fixed slope but is homogeneous if one allows for differences in these characteristics. We can reproduce such behaviour in a series z_t by a representation in which the *second difference*

$$\nabla^2 z_t = z_t - 2z_{t-1} + z_{t-2} = (1-B)^2 z_t$$

follows a stationary model.

Finally then, if z_t is the variable whose behaviour we wish to represent, it is assumed that its dth difference $\nabla^d z_t = w_t$ can be represented by the stationary and invertible model of equation (9). Since $\nabla^d = (1-B)^d$, the model for z_t becomes

$$\phi_p(B)(1-B)^d z_t = \theta_0 + \theta_q(B)a_t, \tag{10}$$

which will be non-stationary unless $d = 0$. The model is said to be of order (p,d,q) where p, d, and q are usually 0, 1, or 2.

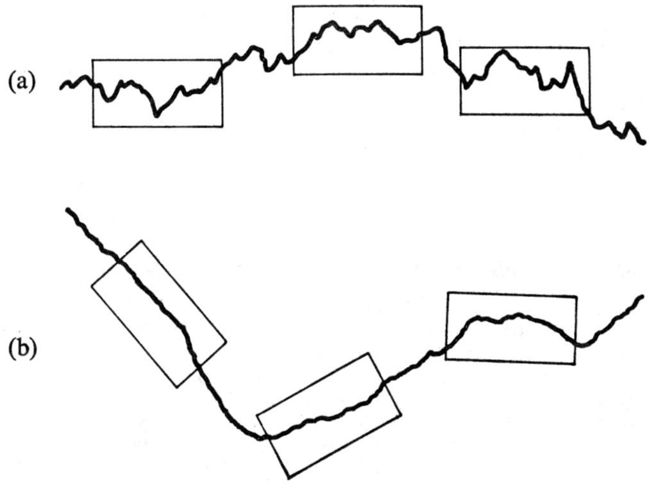

FIG. 1(a). A series showing non-stationarity in level such as can be represented by the model $\phi(B)\nabla z_t = \theta(B)a_t$.

FIG. 1(b). A series showing non-stationarity in level and in slope such as can be represented by the model $\phi(B)\nabla^2 z_t = \theta(B)a_t$.

The operator $\Phi_{p+d}(B) = \phi_p(B)(1-B)^d$ is called the *general* autoregressive operator. Since d of the roots of $\Phi_{p+d}(B) = 0$ are unity, this non-stationary operator will, of course, not satisfy the stationarity condition that all roots lie outside the unit circle. In many practical cases where differencing is needed to obtain stationarity (that is where $d \geqslant 1$), $\nabla^d z_t = w_t$ can be assumed to have a zero mean so that θ_0 in (10) can be set equal to zero.

Suppose we wish to determine a suitable model for a series for which observations $z_1, z_2, z_3, \ldots$ are available (where if possible there should be at least 50 and preferably more than 100 observations). In practice such model determination has to be done iteratively using a process of *identification, estimation, diagnostic checking, refitting and rechecking* until a satisfactory representation is found.

2.3. *Identification*

Equation (10) supplies too rich a class of models to permit immediate estimation. Therefore, using experience and the data we first identify a sub-class of models *worthy to be entertained.*

The primary data-analysis tool at this stage is the *sample autocorrelation function* of the original series and its differences. Suppose that n differences $w_1, w_2, \ldots, w_n$ are available. The sample autocorrelation coefficient at lag k for $w_t = \nabla^d z_t$ is

$$r_k(w) = c_k(w)/c_0(w),$$

where

$$c_k = \frac{1}{n}\sum_{t=1}^{n-k}(w_t - \bar{w})(w_{t+k} - \bar{w}) \quad \text{and} \quad \bar{w} = \frac{1}{n}\sum_{t=1}^{n} w_t.$$

We shall use $\rho_k(w)$ for the corresponding theoretical autocorrelation.

A suitable value for d may be inferred by finding the degree of differencing necessary to induce the sample autocorrelation function to damp out fairly quickly. For example, Table 1 shows the sample autocorrelation function of z, ∇z, and $\nabla^2 z$ for a series of IBM Common Stock Daily Closing Prices given by Brown (1962). While the sample autocorrelations for the original series are very slow to die out, indicating non-stationarity, its first and higher differences behave like those of a stationary series suggesting that we set $d = 1$.

Values to be entertained for p and q may usually be deduced by inspecting the sample autocorrelations using knowledge of the behaviour of the theoretical autocorrelation function ρ_k for various types of models. The characteristics of $\rho_k(w)$ for

TABLE 1

Sample autocorrelations for various differences of the IBM Common Stock Daily Closing Prices

Source: New York Stock Exchange, May 1961–November 1962 (369 observations)

		1	2	3	4	5	6	7	8	9	10
z	Lags 1–10	·99	·99	·98	·97	·96	·96	·95	·94	·93	·92
	11–20	·91	·91	·90	·89	·88	·87	·86	·85	·84	·83
∇z	Lags 1–10	·09	·00	−·05	−·04	−·02	·12	·07	·04	−·07	·02
	11–20	·08	·05	−·05	·07	−·07	·12	·12	·05	·05	·07
$\nabla^2 z$	Lags 1–10	−·45	−·02	−·04	·00	−·07	·11	−·01	·04	−·10	·02
	11–20	·04	·04	−·12	·13	−·17	·10	·05	−·04	−·01	·09

models of order $(1,d,0)$, $(2,d,0)$, $(0,d,1)$, $(0,d,2)$ and $(1,d,1)$ are shown in Table 2. The boundaries of the admissible parameter space are indicated by the inequalities. We see from Table 1 that the autocorrelations of ∇z are all small and appear consistent with a model of order $(0,1,0)$ or perhaps $(0,1,1)$.

Of considerable help in judging the reality of sample autocorrelations is the following approximate formula due to Bartlett for the standard error (S.E.) of r_k, namely

$$\text{S.E.}\,[r_k] \simeq \sqrt{\left\{\frac{1}{n}(1+2\rho_1^2+2\rho_2^2+\ldots)\right\}}. \tag{11}$$

Since we do not know the theoretical autocorrelations ρ_k, they have to be replaced by their sample estimates r_k.

Thus, under the assumption that the first difference of the IBM series is a moving average of order 1 (that is, the series is of order $(0,1,1)$)

$$\text{S.E.}\,[r_k] \simeq \sqrt{[\tfrac{1}{369}\{1+2(0{\cdot}09)^2\}]} = 0{\cdot}05.$$

Referring to Table 1, we see that only 3, that is 6 per cent of the sample autocorrelations of ∇z from the second onwards are greater than two standard deviations, confirming that a model of order $(0,1,1)$ is worthy to be entertained.

TABLE 2

Behaviour of theoretical autocorrelation function of dth difference of series for various simple (p,d,q) models

Order	$(1, d, 0)$	$(0, d, 1)$
Behaviour of ρ_k	$\rho_k = \phi^k$ decays exponentially	only ρ_1 non-zero
Preliminary estimates from	$\phi_1 = \rho_1$	$\rho_1 = \dfrac{-\theta_1}{1+\theta_1^2}$
Admissible region	$-1 < \phi_1 < 1$	$-1 < \theta_1 < 1$

Order	$(2, d, 0)$	$(0, d, 2)$
Behaviour of ρ_k	mixture of exponentials or damped sine wave	only ρ_1 and ρ_2 non-zero
Preliminary estimates from	$\phi_1 = \dfrac{\rho_1(1-\rho_2)}{1-\rho_1^2}$ $\quad \phi_2 = \dfrac{\rho_2-\rho_1^2}{1-\rho_1^2}$	$\rho_1 = \dfrac{-\theta_1(1-\theta_2)}{1+\theta_1^2+\theta_2^2}$ $\quad \rho_2 = \dfrac{-\theta_2}{1+\theta_1^2+\theta_2^2}$
Admissible region	$-1 < \phi_2 < 1$ $\phi_2+\phi_1 < 1$ $\phi_2-\phi_1 < 1$	$-1 < \theta_2 < 1$ $\theta_2+\theta_1 < 1$ $\theta_2-\theta_1 < 1$

Order	$(1, d, 1)$	
Behaviour of ρ_k	decays exponentially after first lag, $\rho_k = \phi\rho_{k-1}$ $(k \geqslant 2)$	
Preliminary estimates from	$\rho_1 = \dfrac{(1-\theta_1\phi_1)(\phi_1-\theta_1)}{1+\theta_1^2-2\phi_1\theta_1}$	$\rho_2 = \rho_1\phi_1$
Admissible region	$-1 < \phi_1 < 1$	$-1 < \theta_1 < 1$

By substituting sample estimates for ρ_k in Table 2, preliminary values for the model parameters (which, however, are in general not efficient estimates) may be obtained. For instance, in the case of the IBM Stock Price series suppose that we tentatively entertain the model $\nabla z_t = (1-\theta B)a_t$ of order (0, 1, 1). Then, because r_1 of ∇z_t is 0·09, a first guess for the parameter θ_1 is $-0{\cdot}09$ since this is the root of the equation $0{\cdot}09 = -\theta_1/1+\theta_1^2$ which lies within the admissible region $-1 < \theta_1 < 1$.

A complementary tool for identification called the *sample partial autocorrelation function* may also be used (see for example Box and Jenkins (1968)).

2.4. *Fitting*

Using efficient statistical methods we may now fit the tentatively identified model, or to be on the safe side, a slightly over-parameterized version of it.

On the assumption that the a's are Normally distributed, a close approximation to the maximum likelihood estimates of $\boldsymbol{\phi} = (\phi_1, \phi_2, \dots \phi_p)$ and $\boldsymbol{\theta} = (\theta_1, \theta_2, \dots \theta_q)$ will be obtained by minimizing the sum of squares

$$S(\boldsymbol{\phi}, \boldsymbol{\theta}) = \sum a_t^2(\boldsymbol{\phi}, \boldsymbol{\theta}).$$

The values a_t $(\boldsymbol{\phi}, \boldsymbol{\theta})$ for any $\boldsymbol{\phi}$ and $\boldsymbol{\theta}$ may readily be calculated recursively using

$$a_t = \theta_0 + \theta_1 a_{t-1} + \dots + \theta_q a_{t-q} + w_t - \phi_1 w_{t-1} - \dots - \phi_p w_{t-p}$$

with $w_t = \nabla^d z_t$ and when $d \neq 0$, θ_0 would often be set equal to zero. The process can be started off by commencing with a_{p+1} and setting $a_p, a_{p-1}, \dots a_{p-q+1}$ equal to their expected values of zero. This procedure is adequate for most purposes but a more exact calculation of the likelihood function will be described in Box and Jenkins (1968).

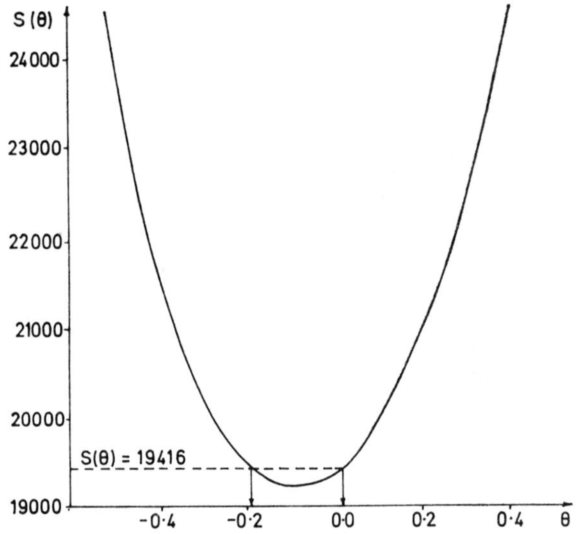

FIG. 2. Sum of squares function for I.B.M. data with approximate 95% confidence region for θ.

An approximate $1-\varepsilon$ confidence region for $\boldsymbol{\phi}$ and $\boldsymbol{\theta}$ is enclosed by the contour

$$S_{1-\varepsilon}(\boldsymbol{\phi},\boldsymbol{\theta}) = S(\hat{\boldsymbol{\phi}},\hat{\boldsymbol{\theta}})[1+\{\chi^2_{1-\varepsilon}(p+q)/\nu\}] \tag{12}$$

where $\chi^2_{1-\varepsilon}(m)$ is the upper $1-\varepsilon$ significance point of the chi-square distribution having m degrees of freedom and ν is equal to the number of a's in the sum $S(\boldsymbol{\phi},\boldsymbol{\theta})$ less the number of parameters fitted.

We illustrate again with the IBM data, using the tentatively entertained model $\nabla z_t = a_t - \theta a_{t-1}$. Figure 2 shows a plot of $S(\theta)$ against θ with a minimum at $\hat{\theta} = -0{\cdot}09$ and $S(\hat{\theta}) = 19{,}216$. The approximate 95 per cent confidence limits for θ of $-0{\cdot}19$ and $0{\cdot}03$ are those values for which

$$S(\theta) = 19{,}216\ (1+3{\cdot}84/367) = 19{,}417.$$

Complicating the model by adding an extra term on either side produced no appreciable reduction in the residual sum of squares. Hence the form of the model which was finally accepted is $\nabla z_t = a_t + 0{\cdot}1 a_{t-1}$. Least squares estimates and approximate confidence limits may be obtained without the use of graphical methods using iterative non-linear least squares procedures described in Box *et al.* (1967). However, in general graphs and contour plots of the sum of squares function $S(\boldsymbol{\phi},\boldsymbol{\theta})$, or of sections of it, are of great value in illuminating the estimation situation.

2.5. *Diagnostic Checks*

If the form of the model is correct and if $\hat{\boldsymbol{\phi}}$ and $\hat{\boldsymbol{\theta}}$ are close to their "true" values, then the estimated residuals $\hat{a}_t = a_t(\hat{\boldsymbol{\phi}},\hat{\boldsymbol{\theta}})$ will be (very nearly) uncorrelated random deviates. Inadequacies of the model may be shown up for example by examining the autocorrelation function of the residuals. A fuller discussion is given in Box and Jenkins (1968).

2.6. *Seasonal Models*

One often has to analyse time series in which recurrent patterns with *known period* s occur, for example, yearly patterns in monthly sales data ($s = 12$). Here parsimony can often be achieved using multiplicative models of the type

$$\phi_p(B)\Phi_P(B^s)(1-B)^d(1-B^s)^D z_t = \theta_q(B)\Theta_Q(B^s)a_t. \tag{13}$$

To see how this model is arrived at, suppose we are analysing a series of monthly sales data so that $s = 12$. Suppose we consider all the data at a fixed point in the period s. For example, suppose we consider the sequence of January sales figures. This series would be free of seasonality and might be described by a suitably chosen model of the general form given in equation (10). Bearing in mind that successive Januarys are $s = 12$ months apart and assuming that $\theta_0 = 0$, we would have

$$\Phi_P(B^s)\nabla_s{}^D z_t = \Theta_Q(B^s)e_t, \tag{14}$$

where

$$\nabla_s z_t = z_t - z_{t-s} \qquad \text{and} \qquad B^s z_t = z_{t-s}.$$

It could reasonably be assumed that February sales, March sales, etc. would follow precisely similar models *with the same parameters*. However, it could not be expected that the residuals e_{t+1} from February sales would be independent of the residuals e_t from January sales. To allow for this dependence a second model may be fitted to the "seasonal free" residuals e_t in the form

$$\phi_p(B)\nabla^d e_t = \theta_q(B)a_t. \tag{15}$$

On eliminating e_t between (14) and (15), we obtain (13). When $s = 12$ the model embodies parameters which describe month-to-month variation (little letters) and parameters which describe year-to-year variation (capital letters).

Procedures for identifying, fitting and checking such models closely follow those described above. For instance, it was shown in Box *et al.* (1967) that the airline passenger data of Fig. 4 was closely fitted by the model

$$(1-B)(1-B^{12})z_t = (1-0{\cdot}4B)(1-0{\cdot}6B^{12})a_t \tag{16}$$

corresponding to $p = 0$, $P = 0$, $s = 12$, $d = 1$, $D = 1$, $q = 1$, $Q = 1$, $\theta_1 = 0{\cdot}4$, and $\Theta_1 = 0{\cdot}6$. The sum of squares plot for this example is shown in Fig. 3.

2.7. *Forecasting*

Suppose now that we have determined an adequate model for a given series and we have new data $z_t, z_{t-1}, \ldots$ from the same series extending up to the present time t from which we wish to make a forecast l steps ahead. We call this an *origin t* forecast for *lead time l.*

It may be shown that the minimum mean square error forecast for any lead time is given by

$$\hat{z}_t(l) = \underset{t}{E}[z_{t+l}],$$

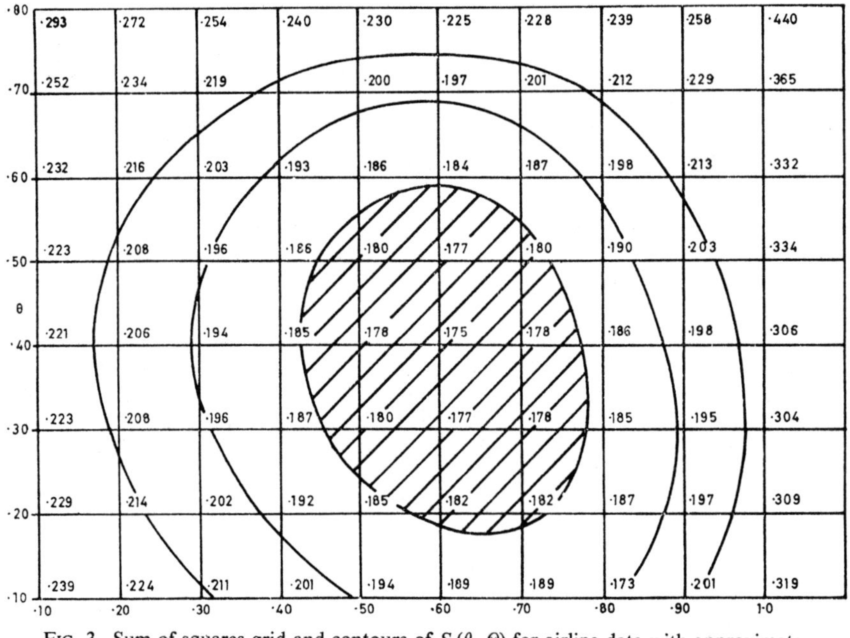

FIG. 3. Sum of squares grid and contours of $S(\theta, \Theta)$ for airline data with approximate 95% confidence region shaded

where $\underset{t}{E}$ is the conditional expectation given the z's up to time t. It follows, in particular, that

$$a_t = z_t - \hat{z}_{t-1}(1). \tag{17}$$

Thus the "shocks" a_t in the models (10) and (13) are in fact the forecast errors for unit lead time. That for an optimal forecast these "one step ahead" forecast errors ought to form an uncorrelated series is otherwise obvious. For suppose these forecast errors were autocorrelated; then it would be possible to forecast the next forecast error in which case the forecast could not be optimal.

The required expectations are easily found because

$$\underset{t}{E}[z_{t+j}] = \hat{z}_t(j), \qquad \underset{t}{E}[a_{t+j}] = 0, \qquad j = 1,2,3,\ldots$$

$$\underset{t}{E}[z_{t-j}] = z_{t-j}, \qquad \underset{t}{E}[a_{t-j}] = a_{t-j} = z_{t-j} - \hat{z}_{t-j-1}(1), \qquad j = 0,1,2,\ldots \quad (18)$$

For instance, to determine the 3-month ahead forecast for the airline series, we first use (16) to write down

$$z_{t+3} = z_{t+2} + z_{t-9} - z_{t-10} + a_{t+3} - 0{\cdot}4a_{t+2} - 0{\cdot}6a_{t-9} + 0{\cdot}24a_{t-10}.$$

Taking conditional expectations at time t,

$$\hat{z}_t(3) = \hat{z}_t(2) + z_{t-9} - z_{t-10} - 0{\cdot}6a_{t-9} + 0{\cdot}24a_{t-10},$$

and using (17),

$$\hat{z}_t(3) = \hat{z}_t(2) + z_{t-9} - z_{t-10} - 0{\cdot}6\{z_{t-9} - \hat{z}_{t-10}(1)\} + 0{\cdot}24\{z_{t-10} - \hat{z}_{t-11}(1)\},$$

that is

$$\hat{z}_t(3) = \hat{z}_t(2) + 0{\cdot}4z_{t-9} - 0{\cdot}76z_{t-10} + 0{\cdot}6\hat{z}_{t-10}(1) - 0{\cdot}24\hat{z}_{t-11}(1).$$

The forecast $\hat{z}_t(2)$ can be obtained in a similar way in terms of $\hat{z}_t(1)$ from $\underset{t}{E}[z_{t+2}]$.

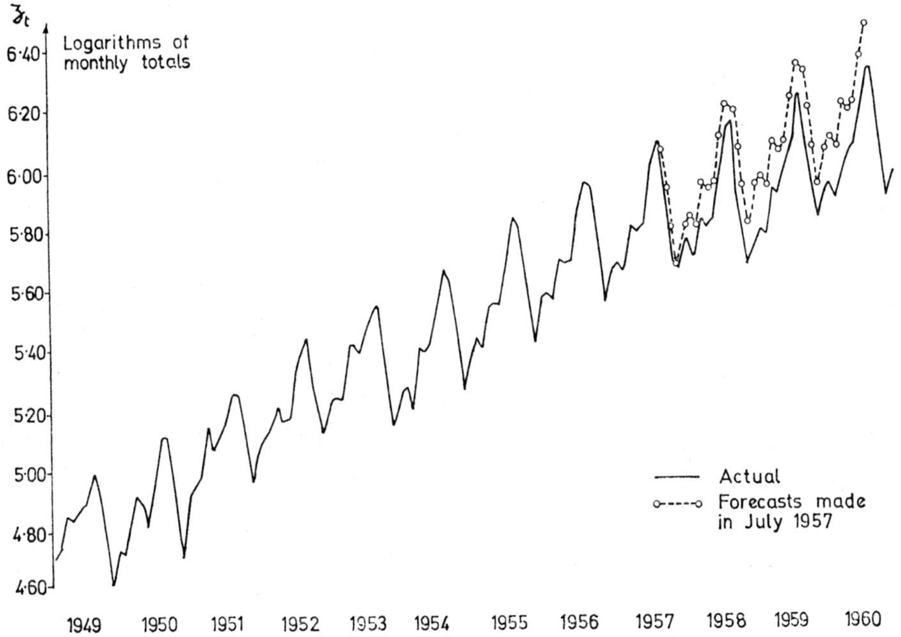

FIG. 4. Logarithms of monthly totals of international airline passengers with forecast made origin July 1957 for 1, 2, 3, . . . , 36 months ahead.

Similarly $\hat{z}_t(1)$ can be obtained from $\underset{t}{E}[z_{t+1}]$ which employs only values of the perviously occurring z's and a's. In practice then it is a very simple matter to compute the forecasts $\hat{z}_t(1)$, $\hat{z}_t(2)$, $\hat{z}_t(3)$, etc. recursively, using the forecast function

$$\underset{t}{E}[z_{t+l}] = \underset{t}{E}[z_{t+l-1}+z_{t+l-12}-z_{t+l-13}-0{\cdot}4a_{t+l-1}-0{\cdot}6a_{t+l-12}+0{\cdot}24a_{t+l-13}]$$

and (18). Note that this form of computation is ideally suited for use on an automatic computer. Using these methods, forecasts made at origin July 1957 for lead times 1, 2, 3, ..., 36 months ahead are shown in Fig. 4 where they may be compared with the values actually realised.

The procedure provides a very convenient and efficient method for industrial forecasting. In particular, it is ideally suited for forecasting sales or inventory on a large variety of products. Since only a very small amount of previous information need be stored for each product a computer with only modest storage capacity may be employed. In those cases where a past history of 50 or so observations is not available one can proceed by using experience and whatever past information *is* available to yield a preliminary model which may then be updated from time to time as more information becomes available.

3. Dynamic Models

In this section we consider the estimation of dynamic models which describe the relationship between a manipulated variable X and a controlled variable Y. Since the dynamic model describes how changes in X are transmitted into Y, it may be said to describe the *transfer function* between X and Y. Knowledge of the appropriate transfer function is essential for the design of control schemes. However, dynamic models of the type we now describe are also useful in forecasting a time series Y from past values of another time series X as well as from past values of Y.

3.1. *Linear Dynamic Models*

Suppose that in the study of the dynamic characteristics of some system, such as a chemical reactor, pairs of observations $(X_1, Y_1), (X_2, Y_2), \ldots$ are available of an input X, such as gas feed rate and an output Y, such as product viscosity. Suppose further that over the operating ranges of variation of Y and X there exists an approximately linear steady-state relationship

$$\dot{Y} = g\dot{X},$$

where $\dot{Y}, \dot{X}$ denote deviations from some average levels, and g is called the *steady state gain* of the system (or the linear regression coefficient between Y and X).

The *dynamic* characteristics of such systems can usually be represented parsimoniously by linear difference equations of the form

$$\xi(\nabla)\,\dot{Y}_{t+1} = g\eta(\nabla)\dot{X}_{t-b} \tag{19}$$

with

$$\xi(\nabla) = 1+\xi_1\nabla+\ldots+\xi_u\nabla^u,$$
$$\eta(\nabla) = 1+\eta_1\nabla+\ldots+\eta_v\nabla^v,$$

where b represents the number of whole intervals of pure dead time (delay) in the system. Most systems occurring in practice can be represented parsimoniously with u and v at most 2. For instance, the simple model

$$(1+\xi\nabla)\,\dot{Y}_{t+1} = g(1+\eta\nabla)\dot{X}_t \tag{20}$$

or

$$\dot{Y}_{t+1} = \frac{\xi}{1+\xi}\,\dot{Y}_t + g\frac{(1+\eta)}{1+\xi}\,\dot{X}_t - \frac{g\eta}{1+\xi}\,\dot{X}_{t-1} \tag{21}$$

can represent a system whose response to a step change of X_0 in the input is to produce an eventual change gX_0 in the output which is approached exponentially at a rate depending on ξ and delayed by an amount depending on η. Fig. 5 illustrates the model (20) with $\xi = 1$, $g = 4$ and $\eta = -0{\cdot}5$.

By solving the difference equation (19) the dynamic model can be written in the alternative form

$$\begin{aligned}\dot{Y}_{t+1} &= v_0\dot{X}_t + v_1\dot{X}_{t-1} + \ldots \\ &= V(B)\dot{X}_t, \end{aligned} \tag{22}$$

where the weights v_j applied to past inputs are called the *impulse response function* of the discrete system. The form (22) is not a parsimonious way of representing the dynamic model, but is useful in identifying the model (19) as will be shown in section 3.2.

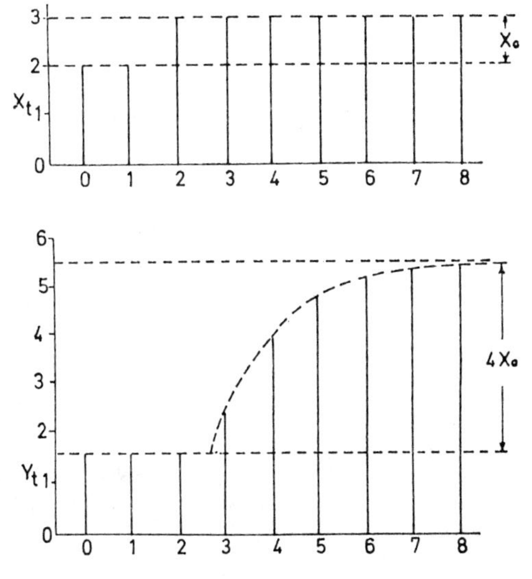

FIG. 5. Delayed exponential response to a step change produced by $(1+\nabla)(Y_{t+1}-1{\cdot}5) = 4(1-0{\cdot}5\nabla)(X_t-2)$ or $(1+\nabla)Y_{t+1} = -6{\cdot}5+4(1-0{\cdot}5\nabla)X_t$

Dynamic models with added noise

The relationship (19) between the input and output will usually be obscured by noise due to measurement error and variation in other variables not under one's control. In this case, we can write (19) as

$$\xi(\nabla)\,\dot{Y}_{t+1} = g\eta(\nabla)\dot{X}_{t-b}+E_{t+1} \tag{23}$$

and (22) as

$$\dot{Y}_{t+1} = V(B)\dot{X}_t+F_{t+1}, \tag{24}$$

where E_{t+1} and F_{t+1} are supposed not correlated with the input process $\dot{X}_t$.

3.2. *Identification of Dynamic Models*

In the same way that the sample autocorrelation function can be used to identify univariate time series models, the basic tool in the identification of dynamic models is the *sample cross correlation function*. To describe a pair of time series by their cross correlation function it is necessary to assume that both series are stationary. Hence it is first necessary to difference both input and output d times until the resulting input and output series are stationary.

If this differencing operation is applied to both sides of (23) and (24) the differenced dynamic models become

$$\xi(\nabla)y_{t+1} = g\eta(\nabla)x_{t-b}+e_t \tag{25}$$

and

$$y_{t+1} = V(B)x_t+f_t, \tag{26}$$

where $\quad y_t = \nabla^d \dot{Y}_t, \quad x_t = \nabla^d \dot{X}_t, \quad e_t=\nabla^d E_t, \quad f_t=\nabla^d F_t.$

Suppose that after differencing, n pairs of differences $(x_1,y_1), (x_2,y_2) \ldots (x_n,y_n)$ are available. Then the sample cross correlation function at lag $+k$ is defined by

$$r_{xy}(k) = \frac{c_{xy}(k)}{\sqrt{\{c_{xx}(0)\,c_{yy}(0)\}}}, \quad k = 0, +1, +2, \ldots \tag{27}$$

where

$$c_{xy}(k) = \frac{1}{n}\sum_{t=1}^{n-k}(x_t-\bar{x})(y_{t+k}-\bar{y}), \tag{28}$$

and at lag $-k$ by

$$c_{xy}(-k) = c_{yx}(k),$$

where $\bar{x}$, $\bar{y}$ are the means of the x and y series.

Prewhitening of the input series

Suppose that it is assumed that the input x_t is uncorrelated with the noise in (26). Then, on multiplying throughout in (26) by x_{t-k+1} and taking expectations,

$$\gamma_{xy}(k) = V(B)\gamma_{xx}(k-1), \tag{29}$$

where $\gamma_{xy}(k)$, $\gamma_{xx}(k)$ are the theoretical cross covariance function and input autocovariance function respectively, and B now operates on k.

Suppose now that we carry out the usual identification and estimation methods as described in section 2 to obtain a model

$$\phi(B)\theta^{-1}(B)x_t = x'_t \tag{30}$$

which transforms the correlated input series x_t to a white noise series x'_t. Suppose also that this transformation is now applied to both sides of (26), yielding,

$$y'_{t+1} = V(B)x'_t + f'_t, \tag{31}$$

where x'_t is white noise uncorrelated with f'_t. On multiplying throughout in (31) by x'_{t-k+1} and taking expectations, we obtain

$$\gamma_{x'y'}(k) = v_k \sigma^2_{x'}. \tag{32}$$

In terms of the cross correlation function, (32) may be rewritten

$$v_k = \rho_{x'y'}(k)\frac{\sigma_{y'}}{\sigma_{x'}}. \tag{33}$$

Hence after "prewhitening", the cross correlation function is directly proportional to the impulse function.

The presence of small initial values of v_k is indicative of pure delay or dead time. Thereafter the presence of values of v_k not following a pattern indicates that terms should be introduced on the *right hand side* of the model (19) and the presence of exponential decay or damped sine wave behaviour in v_k indicates that terms should be introduced on the *left hand side* of the model (19).

3.3. *An Example of identifying a Dynamic Model*

Fig. 6 shows continuous records of the input airfeed (X) and the output carbon dioxide concentration (Y) from a gas furnace. The input airfeed was deliberately varied so as to follow an autoregressive process and the input and output records read at 9-sec intervals resulting in 226 pairs of observations.

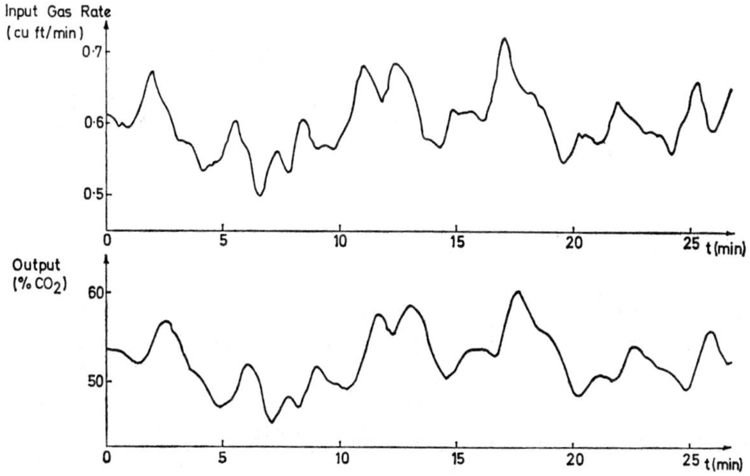

FIG. 6. Input and output records to a gas furnace.

The sample auto- and cross-correlation functions damped out fairly quickly indicating that no differencing was necessary. Hence $x_t = \dot{X}_t$, $y_t = \dot{Y}_t$. The usual identification and fitting procedure applied to the input indicates that it is a third order autoregressive process

$$(1-\phi_1 B-\phi_2 B^2-\phi_3 B^3)x_t = x_t'$$

with $\hat{\phi}_1 = 1{\cdot}97$, $\hat{\phi}_2 = -1{\cdot}37$, $\hat{\phi}_3 = 0{\cdot}34$ and $s_{x'}^2 = 0{\cdot}0353$.

Hence the transformations

$$x_t' = (1-1{\cdot}97B+1{\cdot}37B^2-0{\cdot}34B^3)x_t$$

$$y_t' = (1-1{\cdot}97B+1{\cdot}37B^2-0{\cdot}34B^3)y_t$$

were applied to the input and output series to yield the series x_t' and y_t' with $s_{x'} = 0{\cdot}188$, $s_{y'} = 0{\cdot}358$. The sample cross-correlation function between x' and y' is shown in Table 3 together with the estimate of the impulse response function obtained from (33), that is

$$v_k = \frac{0{\cdot}358}{0{\cdot}188} r_{x'y'}(k).$$

Fig. 7 shows the plot of v_k versus k and indicates that there are two whole periods of delay, then one or two preliminary values v_3 and v_4 which do not correspond to a pattern, followed by a decay pattern which could be first or second order.

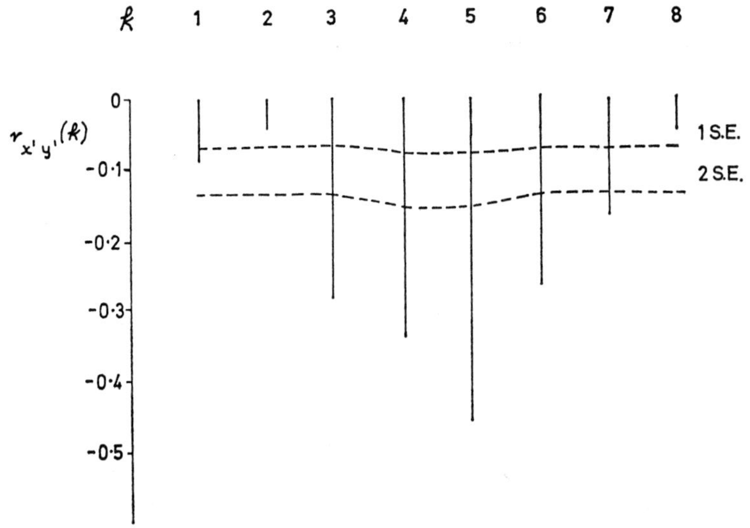

FIG. 7. Gas furnace data sample cross correlations after prewhitening.

TABLE 3

Cross correlation function and approximate impulse response function for gas furnace data

k	1	2	3	4	5	6	7	8
$r_{x'y'}(k)$	−0·05	−0·03	−0·28	−0·33	−0·46	−0·27	−0·17	−0·03
S.E. [r]	0·07	0·07	0·07	0·08	0·08	0·07	0·07	0·07
v_k	−0·09	−0·04	−0·53	−0·63	−0·88	−0·52	−0·32	−0·04

To help in the identification procedure, approximate standard errors for the sample cross correlations were computed using Bartlett's approximate formula

$$\text{cov}\,\{r_{x'y'}(k), r_{x'y'}(l)\} \approx \frac{1}{N}\sum_{j=-\infty}^{+\infty} \{\rho_{x'x'}(j)\,\rho_{y'y'}(j+l-k)+\rho_{x'y'}(j+l)\,\rho_{y'x'}(j-k)\} \tag{34}$$

for the covariance between two values of the sample cross correlation at different lags k and l. On writing $k = l$ in (34) and making use of the fact that the x' series is white noise, the variance of a single cross correlation coefficient is given approximately by

$$\text{var}\,\{r_{x'y'}(k)\} \approx \frac{1}{N}\left\{1+\sum_{j=-\infty}^{+\infty} \rho_{x'y'}(j+k)\,\rho_{y'x'}(-k)\right\}. \tag{35}$$

The standard errors given in Table 3 are based on the assumption that the cross correlations up to lag+2 and from lag+8 onwards are effectively zero. The one standard error and two standard error limits are plotted on Fig. 7 and confirm the identification of a dynamic model

$$(1+\xi_1\nabla+\xi_2\nabla^2)\,\dot{Y}_{t+1} = (1+\eta_1\nabla+\eta_2\nabla^2)\dot{X}_{t-b} \tag{36}$$

(probably with $b = 2$) or some simplification of it.

3.4. *Estimation of the Transfer Function*

In the first instance η_2 was set equal to zero in (36) and the model with added noise written as

$$\dot{Y}_{t+1} = \beta_1\,\dot{Y}_t+\beta_2\,\dot{Y}_{t-1}+\beta_3\dot{X}_{t-b}+\beta_4\dot{X}_{t-b-1}+E_{t+1}. \tag{37}$$

If the errors E_t were uncorrelated, then the parameters in (37) could be estimated by linear least squares. Under the added assumption that the E_t are Normal, these would also be maximum likelihood estimates. In practice the E_t's would rarely be uncorrelated, and hence it would be necessary to arrive at a model by iteration as is now illustrated for the gas furnace example.

Initially the model (37) was fitted by linear least squares for different values of the delay parameter b assuming that the errors E_t were uncorrelated. The minimum sum of squares was attained when $b = 1$, yielding the preliminary fitted model

$$\dot{Y}_{t+1} = 1{\cdot}39\,\dot{Y}_t-0{\cdot}55\,\dot{Y}_{t-1}-0{\cdot}14\dot{X}_{t-1}-0{\cdot}34\dot{X}_{t-2}+E_t,$$

where the dot notation is used to denote deviations of Y and X from their average values. The first ten autocorrelations of the residuals E_t from this model are given in Table 4.

TABLE 4

Autocorrelations of residuals from fitted dynamic model

k	1	2	3	4	5	6	7	8	9	10
r_k	·24	·18	·00	−·02	·01	·16	·08	·06	−·07	·00

These residuals might be explained by a first order autoregressive noise model

$$E_{t+1} = \phi E_t+a_{t+1}.$$

If so, then for an appropriate choice of ϕ we can rewrite the model (37) as

$$\tilde{Y}_{t+1} = \beta_1 \tilde{Y}_t + \beta_2 \tilde{Y}_{t-1} + \beta_3 \tilde{X}_{t-b} + \beta_4 \tilde{X}_{t-b-1} + a_{t+1} \tag{38}$$

where $\tilde{Y}_t = \dot{Y}_t - \phi \dot{Y}_{t-1}$, $\tilde{X}_t = \dot{X}_t - \phi \dot{X}_{t-1}$, and a_t is now white noise.

The model (38) was fitted to the transformed data $\tilde{Y}_{t+1} = \dot{Y}_{t+1} - \phi \dot{Y}_t$ and $\tilde{X}_{t+1} = \dot{X}_{t+1} - \phi \dot{X}_t$ for a grid of values of ϕ and b. The minimum sum of squares occurred at $b = 2$, $\phi = 0{\cdot}7$ yielding the model

$$\tilde{Y}_{t+1} = \underset{(\pm\cdot05)}{0{\cdot}90}\ \tilde{Y}_t - \underset{(\pm\cdot04)}{0{\cdot}19}\ \tilde{Y}_{t-1} - \underset{(\pm\cdot07)}{0{\cdot}48}\ \tilde{X}_{t-2} - \underset{(\pm\cdot10)}{0{\cdot}44}\ \tilde{X}_{t-3},$$

the figures in parentheses under the estimated parameters being their standard errors obtained from the usual least squares formula. The autocorrelations of the residuals a_{t+1} from this model were all small, confirming that the model is adequate.

Hence the final model is

$$(1-0{\cdot}7B)(1-0{\cdot}90B+0{\cdot}19B^2)\,\dot{Y}_{t+1} = -(1-0{\cdot}7B)(0{\cdot}48B^2+0{\cdot}44B^3)\dot{X}_t + a_{t+1}. \tag{39}$$

Rewriting (39) as

$$\dot{Y}_{t+1} = \frac{-(0{\cdot}48B^2+0{\cdot}44B^3)\dot{X}_t}{1-0{\cdot}90B+0{\cdot}19B^2} + \frac{a_{t+1}}{(1-0{\cdot}7B)(1-0{\cdot}90B+0{\cdot}19B^2)}, \tag{40}$$

we see that the fitted dynamic model is

$$(1-0{\cdot}90B+0{\cdot}19B^2)\,\dot{Y}_{t+1} = -(0{\cdot}48B^2+0{\cdot}44B^3)\dot{X}_t. \tag{41}$$

This model implies transfer function characteristics which agree very closely with those estimated in Jenkins and Watts (1968) using cross spectral analysis. In the control engineer's language it corresponds to a second order system with time constants $T_1 = 15{\cdot}8$ seconds and $T_2 = 8{\cdot}2$ seconds, and a pure delay or dead time of 22·8 seconds.

The model (40) also implies that the noise n_{t+1} at the output of the system is a third order autoregressive process

$$(1-0{\cdot}70B)(1-0{\cdot}90B+0{\cdot}19B^2)n_{t+1} = a_{t+1}. \tag{42}$$

A more direct fitting procedure which employs iterative non-linear least squares, and which is readily adapted to the analysis of multiple input data, is described in Box and Jenkins (1968).

References

Barnard, G. A. (1959). Control charts and stochastic processes. *J. R. Statist. Soc.* B, **21**, 239–271.

Bather, J. A. (1963). Control charts and the minimisation of costs. *J. R. Statist. Soc.* B, **25**, 49–80.

Box, G. E. P. and Jenkins, G. M. (1962). Some statistical aspects of adaptive optimisation and control. *J. R. Statist. Soc.* B, **24**, 297–343.

Box, G. E. P. and Jenkins, G. M. (1963). Further contributions to adaptive quality control: simultaneous estimation of dynamics: non-zero costs. *I.S.I. Bulletin*, 34th Session, Ottawa, Canada, 943–974.

Box, G. E. P. and Jenkins, G. M. (1965). Mathematical models for adaptive control and optimisation. *A.I.Ch.E.–I.Chem.E. Symposium* Series No. 4, 61–68.

Box, G. E. P., Jenkins, G. M., and Bacon, D. W. (1967). Models for forecasting seasonal and non-seasonal time series, pp. 271–311 of *Advanced Seminar on Spectral Analysis of Time Series.* Edited by B. Harris, New York: John Wiley.

BOX, G. E. P. and JENKINS, G. M. (1968). *Time Series Forecasting and Control*, San Francisco: Holden-Day. The original draft of this book has been issued in 1966 and 1967 as Technical Reports Nos. 72, 77, 79, 94, 95, 99, 103, 104, 116, 121 and 122 of the Department of Statistics, University of Wisconsin, Madison and simultaneously as Technical Reports Nos. 1, 2, 3, 4, 6, 7, 8, 9, 10, 11, 13 of the Department of Systems Engineering, University of Lancaster.

BROWN, R. G. (1962). *Smoothing, Forecasting and Prediction of Discrete Time Series.* New Jersey: Prentice-Hall.

HOLT, C. C. (1957). Forecasting trends and seasonals by exponentially weighted moving averages. O.N.R. Memorandum No. 52, Carnegie Institute of Technology.

HOLT, C. C., MODIGLIANI, F., MUTH, J. F., and SIMON, H. A. (1960). *Planning Production, Inventories and Work Force*, New Jersey: Prentice-Hall.

JENKINS, G. M. and WATTS, D. G. (1968). *Spectral Analysis and its Applications.* San Francisco: Holden-Day.

PAGE, E. S. (1954). Continuous inspection schemes. *Biometrika*, **41**, 100–114.

PAGE, E. S. (1957). On problems in which a change in a parameter occurs at an unknown point. *Biometrika*, **44**, 248–252.

ROBERTS, S. W. (1959). Control chart tests based on geometric moving averages. *Technometrics*, **1**, 239–250.

WHITTLE, P. (1963). *Prediction and Regulation*, London: English Universities' Press.

WINTERS, P. R. (1960). Forecasting sales by exponentially weighted moving averages. *Management Science*, **6**, 324–342.

3.5

Distribution of Residual Autocorrelations in Autoregressive-Integrated Moving Average Time Series Models

G. E. P. BOX and DAVID A. PIERCE*

Many statistical models, and in particular autoregressive—moving average time series models, can be regarded as means of transforming the data to white noise, that is, to an uncorrelated sequence of errors. If the parameters are known exactly, this random sequence can be computed directly from the observations; when this calculation is made with estimates substituted for the true parameter values, the resulting sequence is referred to as the "residuals," which can be regarded as estimates of the errors.

If the appropriate model has been chosen, there will be zero autocorrelation in the errors. In checking adequacy of fit it is therefore logical to study the sample autocorrelation function of the residuals. For large samples the residuals from a correctly fitted model resemble very closely the true errors of the process; however, care is needed in interpreting the serial correlations of the residuals. It is shown here that the residual autocorrelations are to a close approximation representable as a *singular* linear transformation of the autocorrelations of the errors so that they possess a singular normal distribution. Failing to allow for this results in a tendency to overlook evidence of lack of fit. Tests of fit and diagnostic checks are devised which take these facts into account.

1. INTRODUCTION

An approach to the modeling of stationary and non-stationary time series such as commonly occur in economic situations and control problems is discussed by Box and Jenkins [4, 5], building on the earlier work of several authors beginning with Yule [19] and Wold [17], and involves iterative use of the three-stage process of identification, estimation, and diagnostic checking. Given a discrete time series $z_t, z_{t-1}, z_{t-2}, \cdots$ and using B for the backward shift operator such that $Bz_t = z_{t-1}$, the general autoregressive—integrated moving average (ARIMA) model of order (p, d, q) discussed in [4, 5] may be written

$$\phi(B)\nabla^d z_t = \theta(B)a_t \tag{1.1}$$

where $\phi(B) = 1 - \phi_1 B - \cdots - \phi_p B^p$ and $\theta(B) = 1 - \theta_1 B - \cdots \theta_q B^q$, $\{a_t\}$ is a sequence of independent normal deviates with common variance σ_a^2, to be referred to as "white noise," and where the roots of $\phi(B)=0$ and $\theta(B)=0$ lie outside the unit circle. In other words, if $w_t = \nabla^d z_t = (1-B)^d z_t$ is the dth difference of the series z_t, then w_t is the stationary, invertible, mixed autoregressive (AR)—moving average (MA) process given by

$$w_t = \sum_{i=1}^{p} \phi_i w_{t-i} - \sum_{j=1}^{q} \theta_j a_{t-j} + a_t,$$

and permitting $d>0$ allows the original series to be (homogeneously) nonsta-

* G. E. P. Box is professor of statistics, University of Wisconsin. David A. Pierce is on leave from the Department of Statistics, University of Missouri, Columbia, as statistician, Research Department, Federal Reserve Bank of Cleveland. This work was supported jointly by the Air Force Office of Scientific Research under Grant AFOSR-69-1803 and by the U. S. Army Research Office under Grant DA-ARO-D-31-124-G917.

The Collected Works of George E. P. Box, 1984, Wadsworth, Inc., Belmont, CA 94002.
Originally published in *J. Amer. Stat. Assoc.*, vol. 65, no. 332 (1970), pp. 1509–1526.

tionary. In some instances the model (1.1) will be appropriate after a suitable transformation is made on z; in others z may represent the noise structure after allowing for some systematic model.

This general class of models is too rich to allow immediate fitting to a particular sample series $\{z_t\} = z_1, z_2, \cdots, z_n$, and the following strategy is therefore employed:

1. A process of identification is used to find a smaller subclass of models worth considering to represent the stochastic process.
2. A model in this subclass is fitted by efficient statistical methods.
3. An examination of the adequacy of the fit is made.

The object of the third or diagnostic checking stage is not merely to determine whether there is evidence of lack of fit but also to suggest ways in which the model may be modified when this is necessary. Two basic methods for doing this are suggested:

Overfitting. The model may be deliberately overparameterized in a way it is feared may be needed and in a manner such that the entertained model is obtained by setting certain parameters in the more general model at fixed values, usually zero. One can then check the adequacy of the original model by fitting the more general model and considering whether or not the additional parameters could reasonably take on the specified values appropriate to the simpler model.

Diagnostic checks applied to the residuals. The method of overfitting is most useful where the nature of the alternative feared model is known. Unfortunately, this information may not always be available, and less powerful but more general techniques are needed to indicate the way in which a particular model might be wrong. It is natural to consider the stochastic properties of the residuals $\hat{a} = (\hat{a}_1, \hat{a}_2, \cdots, \hat{a}_n)'$ calculated from the sample series using the model (1.1) with estimates $\hat{\phi}_1, \hat{\phi}_2, \cdots, \hat{\phi}_p; \hat{\theta}_1, \hat{\theta}_2, \cdots \hat{\theta}_q$ substituted for the parameters. In particular their autocorrelation function

$$\hat{r}_k = \sum \hat{a}_t\hat{a}_{t-k} / \sum \hat{a}_t^2 \tag{1.2}$$

may be studied.

Now if the model were appropriate and the a's for the particular sample series were calculated using the *true* parameter values, then these a's would be uncorrelated random deviates, and their first m sample autocorrelations $r = (r_1, r_2, \cdots, r_m)'$, where m is small relative to n and

$$r_k = \frac{\sum a_t a_{t-k}}{\sum a_t^2}, \tag{1.3}$$

would for moderate or large n possess a multivariate normal distribution [1]. Also it can readily be shown that the $\{r_k\}$ are uncorrelated with variances

$$V(r_k) = \frac{n-k}{n(n+2)} \approx 1/n, \tag{1.4}$$

from which it follows in particular that the statistic $n(n+2)\sum_{k=1}^{m}(n-k)^{-1}r_k^2$ would for large n be distributed as χ^2 with m degrees of freedom; or as a further approximation,

$$n \sum_{k=1}^{m} r_k^2 \sim \chi_m^2. \tag{1.5}$$

It is tempting to suppose that these same properties might to a sufficient approximation be enjoyed by the $\hat{r}$'s from the *fitted* model; and diagnostic checks based on this supposition were suggested by Box and Jenkins [4] and Box, Jenkins, and Bacon [6]. If this assumption were warranted, approximate standard errors of $1/\sqrt{n}$ [or more accurate standard errors of $\sqrt{n-k}/n(n+2)$] could be attached to the $\hat{r}$'s and a quality-control-chart type of approach used, with particular attention being paid to the $\hat{r}$'s of low order for the indication of possible model inadequacies. Also it might be supposed that Equation (1.5) with $\hat{r}$'s replacing r's would still be approximately valid, so that large values of this statistic would place the model under suspicion.

It was pointed out by Durbin [10], however, that this approximation is invalid when applied to the residual autocorrelations from a fitted autoregressive model. For example, he showed that $\hat{r}_1$ calculated from the residuals of a first order autoregressive process could have a much smaller variance than r_1 for white noise.

The present paper therefore considers in some detail the properties of the $\hat{r}$'s and in particular their covariance matrix, both for AR processes (Sections 2 and 3) and for MA and ARIMA processes (Section 5). This is done with the intention of obtaining a suitable modification to the above diagnostic checking procedures (Sections 4 and 5.3)

The problem of testing fit in time series models has been considered previously by several authors. Quenouille [14][1] developed a large-sample procedure for AR processes based on their sample partial autocorrelations, which possesses the same degree of accuracy as the present one.[2] Quenouille's test was subsequently extended [3, 15, 18] to cover MA and mixed models. Whittle [16] proposed tests based on the likelihood ratio and resembling the overfitting method above. The present procedure (a) is a unified method equally applicable to AR, MA, and general ARIMA models, (b) is motivated by the intuitive idea that the residuals from a correct fit should resemble the true errors of the process, and (c) can be used to suggest particular modifications in the model when lack of fit is found [5].

2. DISTRIBUTION OF RESIDUAL AUTOCORRELATIONS FOR THE AUTOREGRESSIVE PROCESS

In this section we obtain the joint large-sample distribution of the residual autocorrelations $\hat{\mathbf{r}} = (\hat{r}_1, \cdots, \hat{r}_m)'$ where $\hat{r}_k$ is given by (1.2), for an autoregressive process. This is done by first setting forth some general properties of AR processes, using these to obtain a set of linear constraints (2.9) satisfied by the $\{\hat{r}_k\}$, and then approximating $\hat{r}_k$ by a first order Taylor expansion (2.22) about the white noise autocorrelation r_k. Finally, these results are combined in matrix form to establish a linear relationship (2.27) between $\hat{\mathbf{r}}$ and $\mathbf{r}$ analogous to that between the residuals and true errors in a standard regression model, from which the distribution (2.29) of $\hat{\mathbf{r}}$ readily follows. Subsections 2.5–2.7 then discuss examples and applications of this distribution.

---

[1] See also [11].

[2] The authors are grateful to a referee for this observation.

2.1 The Autoregressive Process

The general AR process of order p,

$$\phi(B)y_t = a_t, \tag{2.1}$$

where B, $\phi(B)$, and $\{a_t\}$ are as in (1.1), can also be expressed as a moving average of infinite order by writing $\psi(B)=\phi^{-1}(B)=(1+\psi_1B+\psi_2B^2+\cdots)$ to obtain

$$y_t = \psi(B)a_t = \sum_{j=0}^{\infty} \psi_j a_{t-j}, \tag{2.2}$$

where $\psi_0=1$. By equating coefficients in the relation $\psi(B)\cdot\phi(B)=1$, it is seen that the ψ's and ϕ's satisfy the relation

$$\psi_\nu = \begin{cases} \phi_1\psi_{\nu-1} + \cdots + \phi_{\nu-1}\psi_1 + \phi_\nu, & \nu \le p \\ \phi_1\psi_{\nu-1} + \cdots + \phi_p\psi_{\nu-p}, & \nu \ge p. \end{cases} \tag{2.3}$$

Therefore by setting $\psi_\nu=0$ for $\nu<0$, we have

$$\psi_0 = 1; \quad \phi(B)\psi_\nu = 0, \qquad \nu \ne 0. \tag{2.4}$$

Suppose then we have a series $\{y_t\}$ generated by the model (2.1) or (2.2), where in general $y_t=\nabla^d z_t$ can be the dth difference ($d=0, 1, 2, \cdots$) of the actual observations. Then for given values $\dot{\boldsymbol{\phi}}=(\dot\phi_1, \cdots, \dot\phi_p)'$ of the parameters we can define

$$\dot a_t = a_t(\dot{\boldsymbol{\phi}}) = y_t - \dot\phi_1 y_{t-1} - \cdots - \dot\phi_p y_{t-p} = \dot\phi(B)y_t \tag{2.5}$$

and the corresponding autocorrelation

$$\dot r_k = r_k(\dot{\boldsymbol{\phi}}) = \frac{\sum \dot a_t \dot a_{t-k}}{\sum \dot a_t^2}. \tag{2.6}$$

Thus, in particular,

1. $a_t(\boldsymbol{\phi})=a_t$ as in (2.1), (2.2);
2. $a_t(\hat{\boldsymbol{\phi}})=\hat a_t$ are the residuals when (2.1) is fitted and least squares estimated $\hat{\boldsymbol{\phi}}$ obtained; and
3. $r_k(\hat{\boldsymbol{\phi}})$ and $r_k(\boldsymbol{\phi})$ are respectively the residual and white noise autocorrelations (1.2) and (1.3).

2.2 Linear Constraints on the $\hat r$'s

It is known that the residuals $\{\hat a_t\}$ above satisfy the orthogonality conditions

$$\sum_{t=p+1}^{n} \hat a_t y_{t-j} = 0, \qquad 1 \le j \le p. \tag{2.7}$$

Therefore if we let

$$\hat\psi(B) = \hat\phi^{-1}(B) = (1 - \hat\phi_1 B - \cdots - \hat\phi_p B^p)^{-1}, \tag{2.8}$$

then $y_t=\hat\psi(B)\hat a_t$, and from (2.7) we have

$$\begin{aligned} 0 &= \sum_t \sum_k \hat\psi_k \hat a_t \hat a_{t-k-j} \\ &= \sum_k \hat\psi_k \hat r_{k+j} \\ &= \sum \psi_k \hat r_{k+j} + O_p(1/n) \end{aligned} \tag{2.9}$$

where the symbol introduced in (2.9) denotes "order in probability" as defined in [13].

In leading up to (2.9) we have presumably summed an infinite number of autocorrelations from a finite series. However since $\{y_t\}$ is stationary we have $\psi_k \to 0$ as k becomes large; and unless ϕ is extremely close to the boundary of the stationarity region, this dying off of ψ_k is fast so that the summation can generally be stopped at a value of k much less than n. More precisely, we are assuming that n is larger than a fixed number N and for such n there exists a sequence of numbers m_n such that

(a) all ψ_j where $j \geq m_n - p$ are of order $1/\sqrt{n}$ or smaller, and
(b) the ratio m_n/n is itself of order $1/\sqrt{n}$.

Then in (2.9) and in all following discussion the error in stopping the summations at $k=m$ (we write m for m_n in the sequel) can to the present degree of approximation be ignored; and (b) also ensures that "end effects" (such as there being only $n-k$ terms summed in the numerator of $\hat{r}_k$ compared with n terms in the denominator) can also be neglected.

2.3 Linear Expansion of $\hat{r}_k$ about r_k

The root mean square error of $\hat{\phi}_j$, $1 \leq j \leq p$, defined by $\sqrt{E(\phi_j - \hat{\phi}_j)^2}$, is of order $1/\sqrt{n}$, and we can therefore approximate $\hat{r}_k$ by a first order Taylor expansion about $\hat{\boldsymbol{\phi}} = \boldsymbol{\phi}$ (evaluating the derivatives, however, at $\hat{\boldsymbol{\phi}}$ rather than $\boldsymbol{\phi}$ in order to obtain the simplification (2.12) below). Thus

$$\hat{r}_k = r_k + \sum_{j=1}^{p} (\phi_j - \hat{\phi}_j)\hat{\delta}_{jk} + O_p(1/n), \tag{2.10}$$

where

$$\hat{\delta}_{jk} = -\left.\frac{\partial \dot{r}_k}{\partial \dot{\phi}_j}\right|_{\dot{\boldsymbol{\phi}}=\hat{\boldsymbol{\phi}}}. \tag{2.11}$$

Now

$$\frac{\partial}{\partial \dot{\phi}_j}\left[\sum \dot{a}_t^2\right] = 0 \quad \text{at } \dot{\boldsymbol{\phi}} = \hat{\boldsymbol{\phi}}, \tag{2.12}$$

so that

$$\hat{\delta}_{jk} = -\left[\sum \hat{a}_t^2\right]^{-1} \left.\frac{\partial \dot{c}_k}{\partial \dot{\phi}_j}\right|_{\dot{\boldsymbol{\phi}}=\hat{\boldsymbol{\phi}}} \tag{2.13}$$

where

$$\begin{aligned} \dot{c}_k &= \sum \dot{a}_t \dot{a}_{t-k} = \sum [\dot{\phi}(B)y_t][\dot{\phi}(B)y_{t-k}] \\ &= \sum_t \sum_{i=0}^{p} \sum_{j=0}^{p} \dot{\phi}_i \dot{\phi}_j y_{t-i} y_{t-k-j}, \end{aligned} \tag{2.14}$$

where in (2.14) and below, $\phi_0 = \dot{\phi}_0 = -1$. From (2.13) and (2.14) it follows that

$$\begin{aligned} \hat{\delta}_{jk} &= -\frac{\sum y_t^2}{\sum \hat{a}_t^2} \sum_{i=0}^{p} \hat{\phi}_i [r^{(y)}{}_{k-i+j} + r^{(y)}{}_{k+i-j}] \\ &= -\frac{\sum_{i=0}^{p} \hat{\phi}_i [r^{(y)}{}_{k-i+j} + r^{(y)}{}_{k+i-j}]}{\sum_{i=0}^{p} \sum_{j=0}^{p} \hat{\phi}_i \hat{\phi}_j r^{(y)}{}_{i-j}}, \end{aligned} \tag{2.15}$$

where

$$r_\nu^{(y)} = \frac{\sum y_t y_{t-\nu}}{\sum y_t^2}\ .$$

Let us approximate $\hat{\delta}_{jk}$ by replacing $\hat{\phi}$'s and $r^{(y)}$'s in (2.15) by ϕ's and ρ's (the theoretical parameters and autocorrelations of the autoregressive process $\{y_t\}$) and denote the result by δ_{jk}. That is,

$$\delta_{jk} = \frac{\sum_{i=0}^{p} \phi_i [\rho_{k-i+j} + \rho_{k+i-j}]}{-\sum_{i=0}^{p} \sum_{j=0}^{p} \phi_i \phi_j \rho_{i-j}}\ . \tag{2.16}$$

Now from Bartlett's formula [2, Equation (7)] we have

$$r_k^{(y)} = \rho_k + O_p(1/\sqrt{n}), \tag{2.17}$$

and as in the discussion preceding (2.10), $\hat{\phi}_j = \phi_j + O_p(1/\sqrt{n})$;
thus

$$\hat{\delta}_{jk} = \delta_{jk} + O_p(1/\sqrt{n}), \tag{2.18}$$

so that equation (2.10) holds when $\hat{\delta}_{jk}$ is replaced by δ_{jk}.

By making use of the recursive relation which is satisfied by the autocorrelations of an autoregressive process, namely

$$\rho_\nu - \phi_1 \rho_{\nu-1} - \cdots - \phi_p \rho_{\nu-p} = \phi(B)\rho_\nu = 0, \qquad \nu \geq 1, \tag{2.19}$$

expression (2.16) can be simplified to yield

$$\delta_{jk} = \frac{\sum_{i=0}^{p} \phi_i \rho_{k-j+i}}{\sum_{i=0}^{p} \phi_i \rho_i}\ . \tag{2.20}$$

Thus δ_{jk} depends only on $(k-j)$, and we therefore write $\delta_{k-j} = \delta_{jk}$. Then it is straightforward to show that

(a) $\delta_0 = 1$

(b) $\delta_\nu = 0, \quad \nu < 0,$ and thus

(c) $\phi(B)\delta_\nu = \dfrac{\sum_{i=0}^{p} \phi_i [\phi(B)\rho_{\nu+i}]}{\sum_{i=0}^{p} \phi_i \rho_i} = 0, \qquad \nu \geq 1.$

Comparing (a), (b), and (c) with the corresponding results (2.4) for ψ_ν, we therefore have $\delta_\nu = \psi_\nu$, that is

$$\delta_{jk} = \psi_{k-j}, \tag{2.21}$$

whence, for $k = 1, 2, \cdots, m$,

$$\hat{r}_k = r_k + \sum_{j=1}^{p} (\phi_j - \hat{\phi}_j)\psi_{k-j} + O_p(1/n). \tag{2.22}$$

2.4 Representation of $\hat{r}$ as a Linear Transformation of r

We can now establish a relationship between the residual autocorrelations $\hat{\boldsymbol{r}}$ and the white noise autocorrelations $\boldsymbol{r}$. Let

$$X = \begin{bmatrix} 1 & 0 & \cdots & 0 \\ \psi_1 & 1 & \ddots & \vdots \\ \psi_2 & \psi_1 & \ddots & 0 \\ \vdots & \vdots & & 1 \\ \vdots & \vdots & & \vdots \\ \psi_{m-1} & \psi_{m-2} & \cdots & \psi_{m-p} \end{bmatrix} = [\mathbf{x}_1 \mid \mathbf{x}_2 \mid \cdots \mid \mathbf{x}_p]. \tag{2.23}$$

Then to $O_p(1/n)$ we can write (2.22) in matrix form as

$$\hat{\mathbf{r}} = \mathbf{r} + X(\boldsymbol{\phi} - \hat{\boldsymbol{\phi}}), \tag{2.24}$$

where from (2.9)

$$\hat{\mathbf{r}}'X = \mathbf{0}. \tag{2.25}$$

If we now multiply (2.24) on both sides by

$$Q = X(X'X)^{-1}X', \tag{2.26}$$

then using (2.25) we obtain

$$\hat{\mathbf{r}} = (I - Q)\mathbf{r}. \tag{2.27}$$

It is known [1] that $\mathbf{r}$ is very nearly normal for n moderately large. The vector of residual autocorrelations is thus approximately a linear transformation of a multi-normal variable and is therefore itself normally distributed. Specifically,

$$\mathbf{r} \sim N(\mathbf{0}, (1/n)I), \tag{2.28}$$

and hence

$$\hat{\mathbf{r}} \sim N(\mathbf{0}, (1/n)[I - Q]). \tag{2.29}$$

Note that the matrix $I-Q$ is idempotent of rank $m-p$, so that the distribution of $\hat{\mathbf{r}}$ has a p-dimensional singularity.

2.5 Further Consideration of the Covariance Structure of the $\hat{r}$'s

It is illuminating to examine in greater detail the covariance matrix of $\hat{\mathbf{r}}$, or equivalently the matrix Q. The latter matrix is idempotent of rank p, and its non-null latent vectors are the columns of X. Also,

$$X'X = \begin{bmatrix} \sum \psi_j^2 & \sum \psi_j\psi_{j-1} & \cdots & \sum \psi_j\psi_{j-p+1} \\ \sum \psi_j\psi_{j-1} & \sum \psi_j^2 & \cdots & \sum \psi_j\psi_{j-p+2} \\ \vdots & \vdots & & \vdots \\ \sum \psi_j\psi_{j-p+1} & \sum \psi_j\psi_{j-p+2} & \cdots & \sum \psi_j^2 \end{bmatrix} = \frac{\sigma_y^2}{\sigma_a^2} \begin{bmatrix} 1 & \rho_1 & \cdots & \rho_{p-1} \\ \rho_1 & 1 & \cdots & \rho_{p-2} \\ \vdots & \vdots & & \vdots \\ \rho_{p-1} & \rho_{p-2} & \cdots & 1 \end{bmatrix} \tag{2.30}$$

which when multiplied by σ_a^2 is the autocovariance matrix of the process itself. Let c^{ij} be the (ij)th element of $(X'X)^{-1}$ (given explicitly in [9]), and similarly q_{ij} for Q. If $\xi_j' = (\psi_{j-1}, \cdots, \psi_{j-p})$ denotes the jth row of X, then

$$\begin{aligned} q_{ij} &= \xi_i'(X'X)^{-1}\xi_j \\ &= \sum_{k=1}^{p} \sum_{\ell=1}^{p} \psi_{i-k} c^{k\ell} \psi_{j-\ell} \\ &= (-n)\operatorname{cov}[\hat{r}_i, \hat{r}_j] \quad \text{if } i \neq j. \end{aligned} \tag{2.31}$$

Since the elements of each column of X satisfy the recursive relation (2.4), we have $\phi(B)\xi_j = 0$, and hence

$$\phi(B) q_{ij} = 0, \tag{2.32}$$

where in (2.32) B can operate either on i or on j. This establishes an interesting recursive structure in the residual autocorrelation covariance matrix $(1/n) \cdot (I-Q)$ and provides an important clue as to how rapidly the covariances die out and the variances approach 1. Also, because of this property the entire covariance matrix is determined by specifying the elements

$$\begin{matrix} q_{11} & q_{12} & \cdots & q_{1p} \\ & q_{22} & \cdots & q_{2p} \\ & & \ddots & \vdots \\ & & & q_{pp} \end{matrix} \tag{2.33}$$

of Q, which are readily obtained by inverting the $X'X$ matrix (2.30).

2.6 Covariance Matrix of $\hat{r}$ for first and second order processes

Consider, for example, the first order autoregressive process $y_t = \phi y_{t-1} + a_t$, which in accordance with (2.2) we can write as

$$y_t = (1 - \phi B)^{-1} a_t = \sum_{j=0}^{\infty} \phi^j a_{t-j}. \tag{2.34}$$

For this process, $\psi_j = \phi^j$ and $(X'X)^{-1} = 1 - \phi^2$. From (2.31) the (ij)th element of Q is therefore $\phi^{i+j-2}(1-\phi^2)$, so that approximately the covariance matrix of the sample residual autocorrelations is

$$\Sigma_{\hat{r}} = (1/n)(I - Q) = 1/n \begin{bmatrix} \phi^2 & -\phi + \phi^3 & -\phi^2 + \phi^4 & \cdots \\ -\phi + \phi^3 & 1 - \phi^2 + \phi^4 & -\phi^3 + \phi^5 & \cdots \\ -\phi^2 + \phi^4 & -\phi^3 + \phi^5 & 1 - \phi^4 + \phi^6 & \cdots \\ \vdots & \vdots & \vdots & \end{bmatrix} \tag{2.35}$$

For the second order process

$$y_t = (1 - \phi_1 B - \phi_2 B^2)^{-1} a_t = \psi(B) a_t, \tag{2.36}$$

we have

$$X = \begin{bmatrix} 1 & 0 \\ \psi_1 & 1 \\ \psi_2 & \psi_1 \\ \vdots & \vdots \\ \cdot & \cdot \end{bmatrix}, \qquad X'X = \frac{\sigma_y^2}{\sigma_a^2} \begin{bmatrix} 1 & \rho_1 \\ \rho_1 & 1 \end{bmatrix},$$

$$(X'X)^{-1} = \frac{\sigma_a^2}{\sigma_y^2(1-\rho_1^2)}\begin{bmatrix} 1 & -\rho_1 \\ -\rho_1 & 1 \end{bmatrix}, \qquad \sigma_y^2 = \frac{(1-\phi_2)\sigma_a^2}{(1+\phi_2)[(1-\phi_2)^2-\phi_1^2]}.$$

Thus

$$q_{11} = 1-\phi_2^2, \qquad q_{12} = -\phi_1\phi_2(1+\phi_2), \qquad q_{22} = 1-\phi_2^2-\phi_1^2(1+\phi_2)^2,$$

from which Q and $\sum\hat{r} = 1/n(I-Q)$ may be determined using (2.32). In particular,

$$\left.\begin{aligned} V(\hat{r}_1) &= 1/n\cdot\phi_2^2, \\ V(\hat{r}_2) &= 1/n[\phi_2^2+\phi_1^2(1+\phi_2)^2], \quad \text{and} \\ V(\hat{r}_k) &= 1/n[1-\phi_1 q_{k,k-1}-\phi_2 q_{k,k-2}],\ k\geq 3. \end{aligned}\right\} \tag{2.37}$$

From these examples we can see a general pattern emerging. As in (2.33) the first p variances and corresponding covariances will be heavily dependent on the parameters $\phi_1, \cdots, \phi_p$ and in general can depart sharply from the corresponding values for white noise autocorrelations, whereas for $k \geq p+1$ a "1" is introduced into the expression for variances (as in (2.35) and (2.37)), and the recursion (2.32) ensures that as k increases the $\{\hat{r}_k\}$ behave increasingly like the corresponding $\{r_k\}$ with respect to both their variances and covariances.

2.7 The distribution of $n\sum_1^m \hat{r}_k^2$

We have remarked earlier that if the fitted model is appropriate and the parameters $\boldsymbol{\phi}$ are exactly known, then the calculated a_t's would be uncorrelated normal deviates, their serial correlations $\mathbf{r}$ would be approximately $N(\mathbf{0}, (1/n)I)$, and thus $n\sum_1^m r_k^2$ would possess a χ^2 distribution with m degrees of freedom. We now see that if m is taken sufficiently large so that the elements after the mth in the latent vectors of Q are essentially zero, then we should expect that to the order of approximation we are here employing, the statistic

$$n\sum_1^m \hat{r}_k^2, \tag{2.38}$$

obtained when estimates $\hat{\boldsymbol{\phi}}$ are substituted for the true parameters $\boldsymbol{\phi}$ in the model, will still be distributed as χ^2, only now with $m-p$ rather than m degrees of freedom. This result is of considerable practical interest because it suggests that an overall test of the type discussed in [4] can in fact be justified when suitable modifications coming from a more careful analysis are applied. Later we consider in more detail the use of this test, along with procedures on individual $\hat{r}$'s, in diagnostic checking.

3. MONTE CARLO EXPERIMENT

We have made certain approximations in deriving the distribution of the residual autocorrelations, and it is therefore of interest to investigate this distribution empirically through repeated sampling and to compare the results with (2.29). This was done for the first order AR process for $\phi=0, \pm.1, \pm.3, \pm.5, \pm.7, \pm.9$. For given ϕ, $s=50$ sets of $n=200$ random normal deviates were generated on the computer using a method described in [7], with separate aggregates of deviates obtained for each parameter value. For the jth set a series $\{y_t^{(j)}\}$ was generated using formula (2.34), $\hat{\phi}^{(j)}$ was estimated, $\{\hat{a}_t^{(j)}\}$ determined, and the quantities

$$\hat{r}_k^{(j)} = \frac{\sum \hat{a}_t^{(j)} \hat{a}_{t-k}^{(j)}}{\sum [\hat{a}_t^{(j)}]^2} \tag{3.1}$$

computed for $1 \leq k \leq m = 20$, $1 \leq j \leq s = 50$. This yielded sample variances and covariances

$$C_{k\ell} = \frac{1}{50} \sum_{j=1}^{50} \hat{r}_k^{(j)} \hat{r}_\ell^{(j)} \tag{3.2}$$

and sample correlations

$$R_{k\ell} = C_{k\ell} / \sqrt{C_{kk} C_{\ell\ell}} \,. \tag{3.3}$$

The results of this Monte Carlo sampling are set out in detail in [8] and in general confirm the adequacy of the approximations used. As an example of these calculations, Table 1 compares the empirical variances (3.2) of $\hat{r}_k$ and correlations (3.3) of $(\hat{r}_1, \hat{r}_k)$ with their theoretical counterparts obtained from (2.35). Allowing for the sampling error of the Monte Carlo estimates themselves, there is good agreement between the two sets of quantities, a phenomenon which occurred also for the other values of ϕ considered.

Since the large-sample variance ϕ^2/n of $\hat{r}_1$ departs the most from the common variance of $1/n$ for white noise autocorrelations, an examination of the empirical behavior of this quantity is of particular interest. Thus Figure 1 shows the sample variance of $\hat{r}_1$ for $\phi = 0, \pm.1, \pm.3, \pm.5, \pm.7, \pm.9$ in relation to the parabola $V(\hat{r}_1) = \phi^2/n$, with reasonable agreement between the two. (The coefficient of variation of the sample variance of $\hat{r}_k$ for $\phi \neq 0$ is approximately $\sqrt{2/s} = 1/5$, independent of k and n; at $\phi = 0$, $V(\hat{r}_1) = O(1/n^2)$.

Table 1. *THEORETICAL (AS IN (2.35)) AND EMPIRICAL (FROM MONTE-CARLO SAMPLING) VARIANCES AND CORRELATIONS OF SAMPLE RESIDUAL AUTOCORRELATIONS FROM FIRST-ORDER AR PROCESS WITH $\phi = .5$*

k	*Variance of $\hat{r}_k$ (multiplied by n)*		*Correlation between $\hat{r}_1$ and $\hat{r}_k$*	
	Theoretical	*Empirical*	*Theoretical*	*Empirical*
1	.250	.244	1.000	1.000
2	.813	.676	−.832	−.812
3	.953	.741	−.384	−.301
4	.988	.864	−.189	−.186
5	.997	1.240	−.094	−.366
6	.999	.967	−.047	−.221
7	1.000	.870	−.023	.083
8	1.000	1.203	−.012	−.148
9	1.000	.982	−.006	−.009
10	1.000	.881	−.003	−.080

Figure 1. THEORETICAL (LINE) AND EMPIRICAL (DOTS) VARIANCES OF $\hat{r}_1$

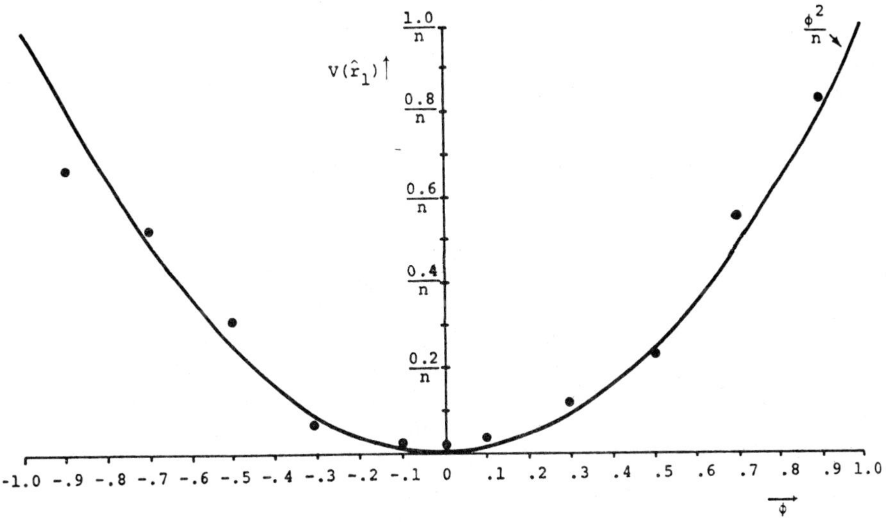

There are several additional comparisons which can be made based on certain functions of the $\hat{r}$'s. Thus we have seen that

$$\hat{l} = \sum \hat{\phi}^{k-1}\hat{r}_k = 0, \tag{3.4}$$

and in the course of our derivations we have had to make the approximation

$$l = \sum \phi^{k-1}\hat{r}_k = 0. \tag{3.5}$$

Some indication of the validity of this approximation is gained by examining the actual values of l from the sampling experiment, which were found to be distributed about zero with a variance of about one-hundredth that which would have been expected from the same linear form in white noise autocorrelations.

Of considerable importance because of its role in diagnostic checking is an examination of the quantity

$$n \sum_{k=1}^{m} \hat{r}_k^2 = 200 \sum_{k=1}^{20} \hat{r}_k^2, \tag{3.6}$$

which as in (2.38) should possess a χ^2—distribution with $\nu = m-1 = 19$ degrees of freedom. Such a distribution has a mean and variance of 19 and 38, respectively, with which the Monte Carlo values can be compared. When this was done, the overall or pooled empirical mean was found to be 18.1 and significantly different from 19. This difference is plausible, however, when it is realized that the statistic $n\sum_1^m \hat{r}_k^2$ possesses a χ^2_{m-p} distribution only insofar as the white noise autocorrelations $\mathbf{r} = (r_1, \cdots, r_m)'$ have a common variance of $1/n$; and from (1.4) it is seen that this approximation overestimates the true variance of a given r_k by a factor of $(n+2)/(n-k)$. In particular, for $n = 200$, $m = 20$, and a typical value of $k = 10$, the actual variance $V(r_k)$ is $190/202 \approx 94$ percent of the $1/n$ approximation. Since the residual autocorrelations $\hat{\mathbf{r}}$ are by (2.27) a linear transformation of $\mathbf{r}$, it is reasonable to expect that a comparable depression of the variances of $\{\hat{r}_k\}$ would occur, and this would account for the discrepancy between the theoretical and empirical means of the statistic $200 \sum_1^{20} \hat{r}_k^2$ encountered above. (This phenomenon would also explain the tendency for the empirical variances themselves, such as those in Table 1, to take on values averaging about 5 percent lower than those based on the matrix $(1/n)(I-Q)$ of (2.29).)

4. USE OF RESIDUAL AUTOCORRELATIONS IN DIAGNOSTIC CHECKING

We have obtained the large sample distribution of the residual autocorrelations $\hat{r}$ from fitting the correct model to a time series, and we have discussed the ways in which this distribution departs significantly from that of the white noise autocorrelations r. It is desirable now to consider the practical implications of these results in examining the adequacy of fit of a model.

First of all it appears that even though the $\hat{r}$'s have a variance/covariance matrix which can differ very considerably from that of the r's, the statistic $n\sum_{k=1}^{m} \hat{r}_k^2$ will (since the matrix $I-Q$ is idempotent) still possess a χ^2-distribution, only now with $m-p$ rather than m degrees of freedom. Thus the overall χ^2-test discussed in Section 1 may be justified to the same degree of approximation as before when the number of degrees of freedom is appropriately modified.

However, regarding the "quality-control-chart" procedure, that is the comparison of the $\{\hat{r}_k\}$ with their standard errors, some modification is clearly needed.

Figure 2 shows the straight-line standard error bands of width $1/\sqrt{n}$ associated with any set of white noise autocorrelations $\{r_k\}$. These stand in marked contrast to the corresponding bands for the residual autocorrelations $\{\hat{r}_k\}$, derived from their covariance matrix $(1/n)(I-Q)$ and shown in Figure 3 for selected first and second order AR processes. Since it is primarily the $\hat{r}$'s of small lags that are most useful in revealing model inadequacies, we see that the consequence of treating $\hat{r}$'s as r's in the diagnostic checking procedure can be a serious underestimation of significance, that is, a failure to detect lack of fit in the model when it exists. Of course, if the model would have been judged inadequate anyway, our conviction in this regard is now strengthened.

Suppose, for example, that we identify a series of length 200 as first order

Figure 2. STANDARD ERROR LIMITS FOR WHITE NOISE AUTOCORRELATIONS r_k

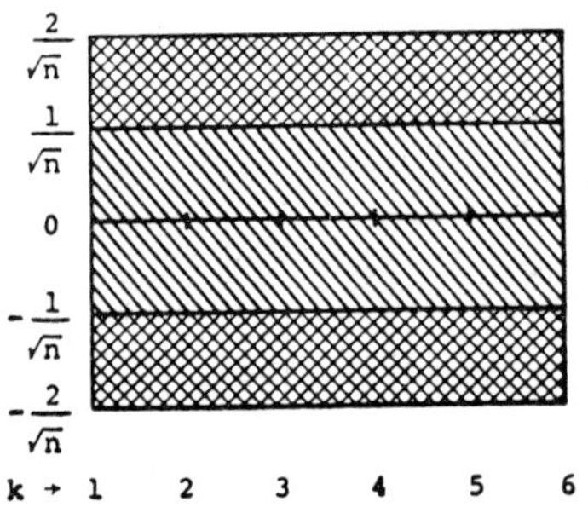

autoregressive and after fitting $\hat{\phi}=.5$. Suppose also that $\hat{r}_1=.10$. Now the standard error of r_1 for white noise is $1/\sqrt{n}=.07$, so that $\hat{r}_1$ is well within the limits in Figure 2. Therefore if we erroneously regarded these as limits on $\hat{r}_1$ we would probably not conclude that this model was inadequate. However, if the true process actually were first order autoregressive (say with $\phi=.5$), the standard error of $\hat{r}_1$ would be $|\phi|/\sqrt{n}=.035$; since the observed $\hat{r}_1=.10$ is almost three times this value, we should be very suspicious of the adequacy of this fit.

The situation is further complicated by the existence of rather high correlations between the $\hat{r}$'s, especially between those of small lags. For the first order process, the most serious correlation is

$$\rho[\hat{r}_1, \hat{r}_2] = -\frac{\phi}{|\phi|}\frac{1-\phi^2}{\sqrt{1-\phi^2+\phi^4}}$$

which, for example, approaches -1 as $\phi \to 0^+$ and is still as large as $-.6$ for $\phi=.7$. Correlation among the $\hat{r}$'s is even more prevalent in second and higher-order processes, where (as for variances) those involving lags up to $k=p$ can be particularly serious. From then on their magnitude is controlled by the recursive relationship (2.32); in particular, the closer ϕ is to the boundary of the stationarity region, the slower will be the dying out of $\text{cov}(\hat{r}_k, \hat{r}_l)$ or $\rho(\hat{r}_k, \hat{r}_l)$ although often in these situations the less serious will the initial correlations $\rho(\hat{r}_1, \hat{r}_2)$, $\rho(\hat{r}_2, \hat{r}_3)$, $\rho(\hat{r}_1, \hat{r}_3)$, etc., tend to be.

We have thus seen that the departure of the distribution of the residual autocorrelations $\hat{r}$ from that of white noise autocorrelations r is serious enough to

Figure 3. STANDARD ERROR LIMITS FOR RESIDUAL AUTOCORRELATIONS $\hat{r}_k$

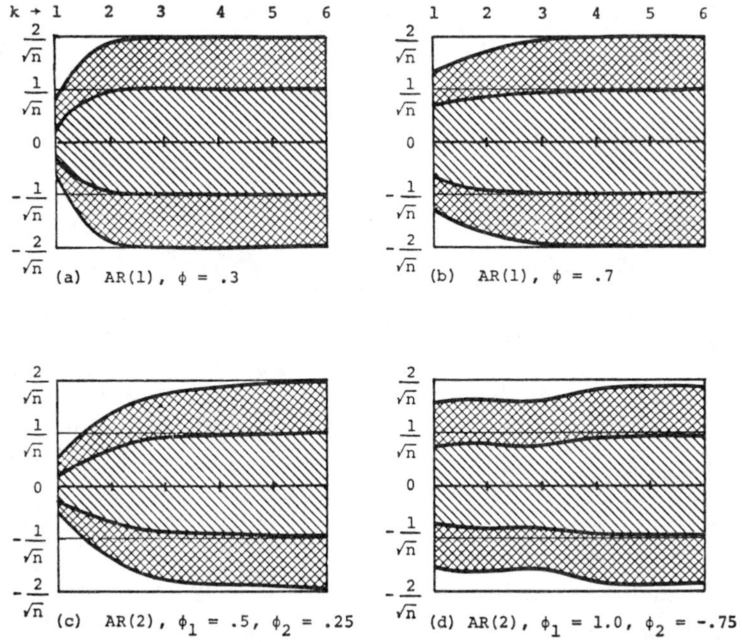

warrant some modifications in their use in diagnostic checking. The residual autocorrelation function, however, remains a powerful device for this purpose.

5. DISTRIBUTION OF RESIDUAL AUTOCORRELATIONS FOR THE MOVING AVERAGE AND GENERAL ARIMA PROCESSES

In obtaining the distribution of $\hat{r} = (\hat{r}_1, \cdots, \hat{r}_m)'$ for the pure autoregressive process in Section 2, considerable use was made of the recursive relation $\phi(B)\rho_k = 0$, which is not satisfied by moving average models $y_t = \theta(B)a_t$, or more generally by mixed models of the form (1.1) with $w_t = \nabla^d z_t$ denoting the stationary dth difference.

It is fortunate, therefore, that these models have in common with the pure AR models (2.1) an important property (derived in Section 5.1) because of which the distribution of their residual autocorrelations can be found as an immediate consequence of the autoregressive solution (2.29). This property is that if two time series, (a) the mixed autoregressive—moving average series (1.1), and (b) an autoregressive series

$$\pi(B)x_t = (1 - \pi_1 B - \cdots - \pi_{p+q}B^{p+q})x_t = a_t \tag{5.1}$$

are both generated from the *same set* of deviates $\{a_t\}$, and moreover if

$$\pi(B) = \phi(B)\theta(B), \tag{5.2}$$

then when these models are each fitted by least squares, their residuals, and hence also their residual autocorrelations, will be very nearly the same. Therefore if a mixed model of order (p, d, q) is correctly identified and fitted, its residual autocorrelations for n sufficiently large will be distributed as though the model had been of order $(p+q, d, 0)$ with the relations between the two sets of parameters given by (5.2). In particular the ψ's comprising the X-matrix (2.23) for the model (1.1) are the coefficients in $\psi(B) = [\phi(B)\theta(B)]^{-1}$.

5.1 Equality of Residuals in AR and ARIMA Models

Let w_t and x_t be as in (1.1) and (5.1); (5.2) then implies

$$w_t = \theta^2(B)x_t. \tag{5.3}$$

As in (2.5), define

$$\dot{a}_t^{\mathrm{AR}} = a_t^{\mathrm{AR}}(\dot{\pi}) = \dot{\pi}(B)x_t = -\sum_{j=0}^{p+q} \dot{\pi}_j x_{t-j} \tag{5.4}$$

where $\dot{\pi}_0 = -1$, and now also

$$\dot{a}_t^* = a_t^*(\dot{\phi}, \dot{\theta}) = \dot{\phi}(B)\dot{\theta}^{-1}(B)w_t = [\sum_{i=0}^{p} \dot{\phi}_i B^i][\sum_{j=0}^{q} \dot{\theta}_j B^j]^{-1} w_t, \tag{5.5}$$

where $\dot{\phi}_0 = \dot{\theta}_0 = -1$. We will expand these quantities about the true parameter values and go through a least squares estimation in each case which is analogous to writing the linear regression model $\mathbf{y} = X\boldsymbol{\beta} + \boldsymbol{\varepsilon}$ as

$$\dot{\mathbf{e}} = \mathbf{y} - \dot{\mathbf{y}} = X(\boldsymbol{\beta} - \dot{\boldsymbol{\beta}}) + \boldsymbol{\varepsilon} = X\boldsymbol{\delta} + \boldsymbol{\varepsilon}, \tag{5.6}$$

for fixed $\dot{\boldsymbol{\beta}}$, and then performing the regression directly on $\dot{\mathbf{e}}$ rather than on $\mathbf{y}$. The equality of the residuals in the two cases depends heavily on the fact that the derivatives in each expansion involve the same autoregressive variable x_t.

Thus

$$\frac{\partial \dot{a}_t^{\mathrm{AR}}}{\partial \dot{\pi}_j} = -x_{t-j}, \qquad 1 \le j \le p+q, \text{ irrespective of } \dot{\pi};$$

$$\frac{\partial \dot{a}_t^*}{\partial \dot{\phi}_j} = -\dot{\theta}^{-1}(B)w_{t-j}, \qquad 1 \le j \le p$$

$$= -\theta(B)x_{t-j} \qquad \text{at } (\dot{\boldsymbol{\phi}}, \dot{\boldsymbol{\theta}}) = (\boldsymbol{\phi}, \boldsymbol{\theta}); \quad \text{and}$$

$$\frac{\partial \dot{a}_t^*}{\partial \dot{\theta}_j} = \dot{\phi}(B)\dot{\theta}^{-2}(B)w_{t-j}, \qquad 1 \le j \le q$$

$$= \phi(B)x_{t-j} \qquad \text{at } (\dot{\boldsymbol{\phi}}, \dot{\boldsymbol{\theta}}) = (\boldsymbol{\phi}, \boldsymbol{\theta}).$$

Then

$$\dot{a}_t^{\mathrm{AR}} = a_t^{\mathrm{AR}} + \sum_{j=1}^{p+q} (\pi_j - \dot{\pi}_j)x_{t-j}, \tag{5.7}$$

and approximately

$$\dot{a}_t^* = a_t^* + \sum_{i=1}^{p} (\phi_i - \dot{\phi}_i)\theta(B)x_{t-i} - \sum_{j=1}^{q} (\theta_j - \dot{\theta}_j)\phi(B)x_{t-j} \tag{5.8}$$

$$\begin{aligned} &= a_t^* + \sum_{i=1}^{p} (\phi_i - \dot{\phi}_i)x_{t-i} - \sum_{j=1}^{q} (\theta_j - \dot{\theta}_j)x_{t-j} \\ &\quad + \sum_{i=1}^{p} \sum_{j=1}^{q} [\phi_i(\theta_j - \dot{\theta}_j) - \theta_j(\phi_i - \dot{\phi}_i)]x_{t-i-j} \\ &= a_t^* + \sum_{i=1}^{p} (\phi_i - \dot{\phi}_i)x_{t-i} - \sum_{j=1}^{q} (\theta_j - \dot{\theta}_j)x_{t-j} \\ &\quad + \sum_{i=1}^{p} \sum_{j=1}^{q} [\dot{\phi}_i(\theta_j - \dot{\theta}_j) - \dot{\theta}_j(\phi_i - \dot{\phi}_i)]x_{t-i-j} \end{aligned} \tag{5.9}$$

$$= a_t^* + \sum_{j=1}^{p+q} (\beta_j - \dot{\beta}_j)x_{t-j}.$$

Thus letting $\boldsymbol{\beta} = (\beta_1, \cdots, \beta_{p+q})'$ and $\boldsymbol{\lambda} = \begin{bmatrix}\boldsymbol{\phi}\\ \boldsymbol{\theta}\end{bmatrix}$, we see that

$$\dot{\boldsymbol{\beta}} = A\dot{\boldsymbol{\lambda}}, \tag{5.10}$$

where A is a $(p+q)$ − square matrix whose elements involve $\dot{\boldsymbol{\lambda}}$ but not the true parameter values $\boldsymbol{\lambda}$. For example, if $p = q = 1$, we would have

$$\begin{bmatrix}\dot{\beta}_1\\ \dot{\beta}_2\end{bmatrix} = \begin{bmatrix}1 & -1\\ -\dot{\theta} & \dot{\phi}\end{bmatrix}\begin{bmatrix}\dot{\phi}\\ \dot{\theta}\end{bmatrix} \tag{5.11}$$

Now equations (5.7) and (5.9) can be written as

$$\dot{\mathbf{a}}^{\mathrm{AR}} = \mathbf{a} + X(\boldsymbol{\pi} - \dot{\boldsymbol{\pi}}) \tag{5.12}$$

$$\dot{\mathbf{a}}^* = \mathbf{a} + X(\boldsymbol{\beta} - \dot{\boldsymbol{\beta}}) \tag{5.13}$$

where the error in (5.13) is $O(|\boldsymbol{\beta} - \dot{\boldsymbol{\beta}}|^2)$, and where we have made use of the fact that, at $\dot{\boldsymbol{\pi}} = \boldsymbol{\pi}$, $\dot{\boldsymbol{\theta}} = \boldsymbol{\theta}$, and $\dot{\boldsymbol{\phi}} = \boldsymbol{\phi}$,

$$a_t^{\mathrm{AR}} = a_t^* = a_t. \tag{5.14}$$

Thus in (5.12) the sum of squares

$$\mathbf{a}'\mathbf{a} = \sum a_t^2 = \sum [a_t^{\mathrm{AR}}(\boldsymbol{\pi})]^2$$

is minimized as a function of $\boldsymbol{\pi}$ when

$$\boldsymbol{\pi} - \dot{\boldsymbol{\pi}} = \hat{\boldsymbol{\pi}} - \dot{\boldsymbol{\pi}} = (X'X)^{-1}X'\dot{\mathbf{a}}^{\mathrm{AR}}, \tag{5.15}$$

while in (5.13) if we write

$$\mathbf{a}^* = \mathbf{a} + X[A(\boldsymbol{\lambda} - \dot{\boldsymbol{\lambda}})] = \mathbf{a} + Z(\boldsymbol{\lambda} - \dot{\boldsymbol{\lambda}}),$$

then the sum of squares

$$a'\mathbf{a} = \sum a_t^2 = \sum [a_t^*(\boldsymbol{\lambda})]^2$$

is minimized as a function of $\boldsymbol{\lambda}$ when

$$\boldsymbol{\lambda} - \dot{\boldsymbol{\lambda}} = \hat{\boldsymbol{\lambda}} - \dot{\boldsymbol{\lambda}} = (Z'Z)^{-1}Z'\dot{\mathbf{a}}^* = A^{-1}(\hat{\boldsymbol{\beta}} - \dot{\boldsymbol{\beta}});$$

that is,

$$\hat{\boldsymbol{\beta}} - \dot{\boldsymbol{\beta}} = (X'X)^{-1}X'\dot{\mathbf{a}}^*. \tag{5.16}$$

Then by setting $\dot{\mathbf{a}} = \mathbf{a}$ in (5.15) and (5.16), we have from (5.14) the important equality

$$\hat{\boldsymbol{\pi}} - \boldsymbol{\pi} = (X'X)^{-1}X'\mathbf{a} = \hat{\boldsymbol{\beta}} - \boldsymbol{\beta}; \tag{5.17}$$

and finally by setting "." = "^" in (5.12) and (5.13), it follows from (5.17) that to $O_p(1/n)$

$$\hat{\mathbf{a}}^{AR} = \mathbf{a} + X(\boldsymbol{\pi} - \hat{\boldsymbol{\pi}}) = \mathbf{a} + X(\boldsymbol{\beta} - \hat{\boldsymbol{\beta}}) = \hat{\mathbf{a}}^*, \tag{5.18}$$

and thus (to the same order) $\hat{r}^{AR} = \hat{r}^*$, as we set out to show.

5.2 Monte Carlo Experiment

The equality (5.18) between the residuals from the autoregressive and mixed models depends on the accuracy of the expansion (5.8), that is, on the extent of linearity in the moving average model, between the true and estimated values $\boldsymbol{\theta}$ and $\hat{\boldsymbol{\theta}}$. It is therefore worthwhile to confirm this model-duality by generating and fitting pairs of series of the form (1.1) and (5.1) and comparing their residuals, or more to our purpose, their residual autocorrelations. This was done for $p+q=1$ and $p+q=2$ for series of length 200. Some indication of the close-

Table 2. RESIDUAL CORRELATIONS FROM FIRST ORDER AR AND MA TIME SERIES GENERATED FROM SAME WHITE NOISE ($n=200$)

k	$\phi=\theta=.1$		$\phi=\theta=.5$		$\phi=\theta=.9$	
	$\hat{r}_k^{AR}$	$\hat{r}_k^{MA}$	$\hat{r}_k^{AR}$	$\hat{r}_k^{MA}$	$\hat{r}_k^{AR}$	$\hat{r}_k^{MA}$
1	−.029	−.010	.003	−.005	−.048	−.057
2	.164	.169	.044	.045	.157	.151
3	.096	.099	−.098	−.096	.008	.009
4	−.050	−.049	.014	.021	−.126	−.127
5	−.003	−.006	.057	.058	.034	.035
6	−.143	−.144	.010	.012	−.091	−.090
7	−.023	−.026	−.004	.001	−.001	−.000
8	−.040	−.041	−.054	−.046	−.038	−.035
9	.010	.009	.052	.052	−.004	.000
10	−.049	−.049	−.065	−.067	.113	.116
$\hat{\phi}$ or $\hat{\theta}\rightarrow$	.159	.057	.543	.451	.922	.870

ness of the agreement is obtained from the few results for first order AR and MA processes shown in Table 2, where it is seen that the residual autocorrelation $\hat{r}_k^{AR}$ and $\hat{r}_k^{MA}$ are equal or nearly equal to the second decimal place.

A sampling experiment of the type described in Section 3 was also performed for the first order MA process. The results were very similar, which is to be expected in view of (5.18).

5.3 Conclusions

We have shown above that to a close approximation the residuals from any moving average or mixed autoregressive-moving average process will be the same as those from a suitably chosen autoregressive process. We have further confirmed the adequacy of this approximation by empirical calculation. It follows from this that we need not consider separately these two classes of processes; more precisely,

1. We can immediately use the AR result to write down the variance/covariance matrix of $\hat{r}$ for any autoregressive-integrated moving average process (1.1) by considering the corresponding variance/covariance matrix of $\hat{r}$ from the pure AR process

$$\pi(B)x_t = \theta(B)\phi(B)x_t = a_t. \tag{5.19}$$

2. All considerations regarding the use of residual autocorrelations in tests of fit and diagnostic checking discussed in Section 4 for the autoregressive model therefore apply equally to moving average and mixed models.
3. In particular it follows from the above that a "portmanteau" test for the adequacy of any ARIMA process is obtained by referring $n\sum_{k=1}^{m} \hat{r}_k^2$ to a χ^2 distribution with ν degrees of freedom, where $\nu = m - p - q$.

REFERENCES

[1] Anderson, R. L., "Distribution of the Serial Correlation Coefficient," *The Annals of Mathematical Statistics*, 13 (March 1942), 1–13.

[2] Bartlett, M. S., "On the Theoretical Specification and Sampling Properties of Autocorrelated Time Series," *Journal of the Royal Statistical Society*, Series B, 8 (April 1946), 27–41.

[3] ——— and Diananda, P. H., "Extensions of Quenouille's Tests for Autoregressive Schemes," *Journal of the Royal Statistical Society*, Series B, 12 (April 1950), 108–15.

[4] Box, G. E. P. and Jenkins, G. M., *Statistical Models for Prediction and Control*, Technical Reports #72, 77, 79, 94, 95, 99, 103, 104, 116, 121, and 122, Department of Statistics, University of Wisconsin, Madison, Wisconsin, 1967.

[5] ———, *Time Series Analysis Forecasting and Control*, San Francisco: Holden-Day, Inc., 1970.

[6] ——— and Bacon, D. W., "Models for Forecasting Seasonal and Non-Seasonal Time Series," in B. Harris, ed., *Spectral Analysis of Time Series*, New York: John Wiley & Sons, Inc., 1967.

[7] Box, G. E. P. and Muller, M. E., "Note on the Generation of Random Normal Deviates," *The Annals of Mathematical Statistics*, 29 (June 1958), 610–11.

[8] Box, G. E. P. and Pierce, D. A., "Distribution of Residual Autocorrelations in Integrated Autoregressive-Moving Average Time Series Models," Technical Report #154, Department of Statistics, University of Wisconsin, Madison, April, 1968.

[9] Durbin, J., "Efficient Estimation of Parameters in Moving Average Models," *Biometrika*, 46 (December 1959), 306–16.

[10] ———, "Testing for Serial Correlation in Least-Squares Regression When Some of the Regressors are Lagged Dependent Variables," *Econometrica*, 38 (May 1970), 410–21.

[11] Grenander, U. and Rosenblatt, M., *Statistical Analysis of Stationary Time Series*, New York: John Wiley & Sons, Inc., 1957.

[12] Mann, H. B. and Wald, A., "On the Statistical Treatment of Linear Stochastic Difference Equations," *Econometrica*, 11 (July 1943), 173–220.

[13] ——— "On Stochastic Limit and Order Relationships," *The Annals of Mathematical Statistics*, 14 (September 1943), 217–26.

[14] Quenouille, M. H., "A Large-Sample Test for the Goodness of Fit of Autoregressive Schemes," *Journal of the Royal Statistical Society*, Series A, 110 (June 1947), 123–9.

[15] Walker, A. M., "Note on a Generalization of the Large-Sample Goodness of Fit Test for Linear Autoregressive Schemes," *Journal of the Royal Statistical Society*, Series B, 12 (April 1950), 102-7.

[16] Whittle, P., "Tests of Fit in Time Series," *Biometrika*, 39 (December 1952), 309–18.

[17] Wold, H., *A Study in the Analysis of Stationary Time Series*, Stockholm: Almquist and Wiksell, 1938.

[18] ———, "A Large-Sample Test for Moving Averages," *Journal of the Royal Statistical Society*, Series B, 11 (April 1949), 297–305.

[19] Yule, G. U., "On a Method of Investigating Periodicities in Disturbed Series, with Special Reference to Wolfer's Sunspot Numbers," *Philosophical Transactions*, (A) 226 (July 1927), 267–98.

3.6

Some Comments on a Paper of Coen, Gomme and Kendall

G. E. P. BOX[1] and PAUL NEWBOLD[2]
University of Wisconsin

[Received July 1970. Revised February 1971]

SUMMARY

The method of analysis used in a recent paper on economic forecasting is reviewed. Evidence is presented that what were believed to be highly significant relationships making possible the forecasting of the *Financial Times* share index arise because of the inflexibility of the assumed error structure.

1. INTRODUCTION

IN a recent publication, by Coen, Gomme and Kendall (1969) which for convenience we refer to as the C.G.K. paper, the forecasting of the *Financial Times* ordinary share index using various other lagged series is discussed. It is sufficient for illustration to consider relation (7) of the C.G.K. paper which we write as

$$Y_t = \alpha + \beta_1 X_{1,t-6} + \beta_2 X_{2,t-7} + n_t \tag{1}$$

which is projected to obtain forecasts. In this expression the "output" Y_t is the *Financial Times* ordinary share index. The two "inputs" are $X_{1,t}$, United Kingdom car production, and $X_{2,t}$, the *Financial Times* commodity index, and n_t is an error term. Quarterly data were employed yielding time series containing 51 successive observations. Of course this is only one of a number of such relationships which the authors postulate. However, it is their methods which we are doubtful about and our reservations would apply equally to their other analyses.

The authors built their model (1) by cross-correlating the detrended series which in this instance were the residuals remaining after fitting linear least-squares regressions on time to each series. For example, Fig. 1(a) shows a plot of the sample cross-correlation function† between the *Financial Times* share index (Y_t detrended) and United Kingdom car production ($X_{1,t}$ detrended). The authors display this cross-correlation for positive lags only (Y leading X) and note that a moderately large cross-correlation occurs near lag 5 or 6. The choice of lags for the independent variables in the linear regrassion was made by initially including among the regressors an independent variable at several different lags close to the value where the sample cross-correlation was a maximum in absolute value. A stepwise regression program was used to determine which lags should be included in the final equation. Values of Student's t-statistic of 14·1 and −9·9 were computed for the estimates of the parameters β_1 and β_2 in (1) and these appeared to be very highly significant. The authors were thus led to believe, for example, that car production six quarters previously and the *Financial Times* commodity index seven quarters previously could be used to forecast

[1] At University of Essex in 1971.
[2] Now at University of Nottingham.
† Our calculated cross-correlations differ somewhat from those given in the C.G.K. paper; however, the general pattern is similar.

The Collected Works of George E. P. Box, 1984, Wadsworth, Inc., Belmont, CA 94002.
Originally published in *J. Roy. Stat. Soc.*, Series A., vol. 134, no. 2 (1971), pp. 229–240.

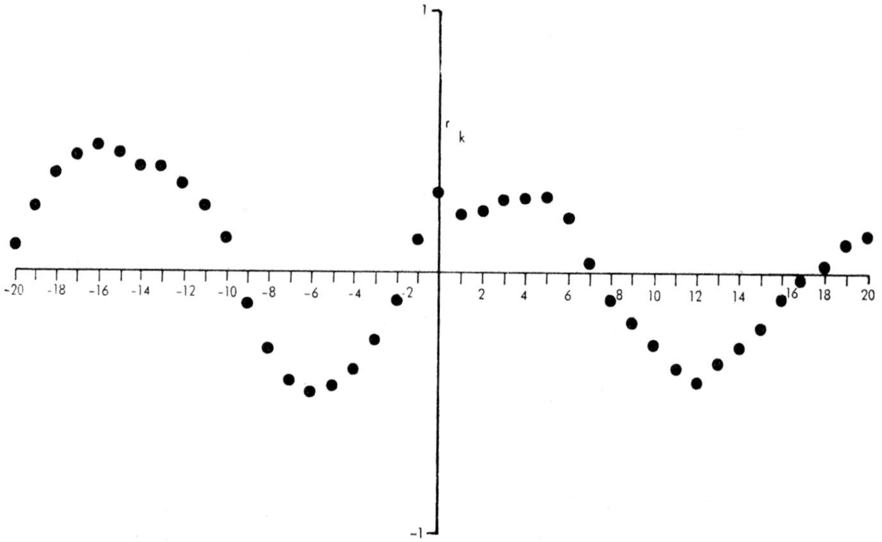

FIG. 1(a). Sample cross-correlations r_k between the *Financial Times* share index (detrended) and lagged values of United Kingdom car production (detrended). Fifty-one pairs of observations.

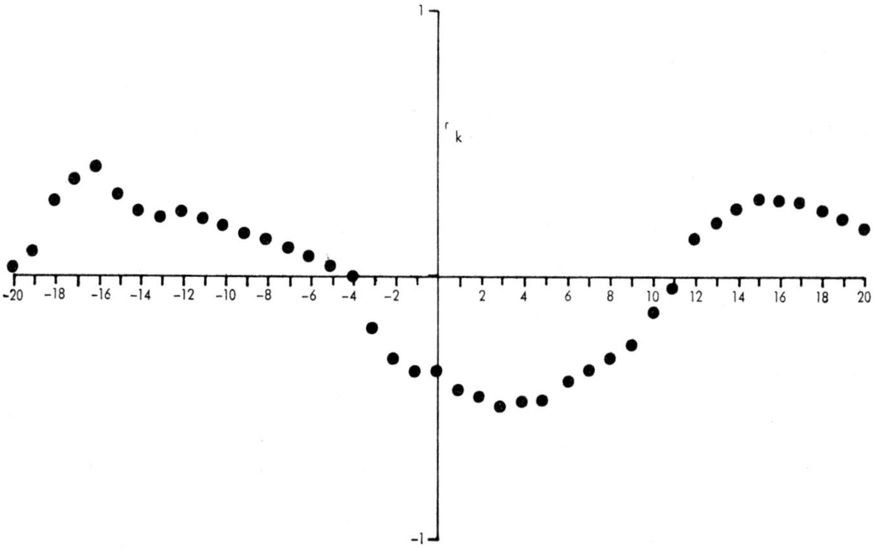

FIG. 1(b). Sample cross-correlations r_k between two unrelated detrended random walks generated from the first two columns of Wold's table of random normal deviates. Fifty pairs of observations.

the *Financial Times* ordinary share index. Because it was felt that the method of detrending itself might give rise to a cross-correlation effect the paper includes an appendix by E. M. L. Beale which shows that *if* individual series were of the form

$$X_t = \alpha + \beta t + a_t, \tag{2}$$

where a_t are assumed to be independent identically distributed random deviates, then the serial correlations of the residuals obtained when (2) is fitted by ordinary least squares would not differ much from those of a_t unless the sample size was small, thus indicating that sample cross-correlations such as those of Fig. 1 were not created by the detrending procedure. They, furthermore, conducted sampling experiments in which cross-correlations of residuals from detrended *random* series were plotted. They remark that the resulting diagrams are sufficiently unlike the *smooth* curves (see Fig. 1) which characterized the cross-correlations of the economic series as "to indicate that the latter are not artefacts created by the trend removal process, at least so far as concerns random residuals".

We believe that the authors were right to suspect that the apparent lagged relationships which they found might be produced by an artefact. The object of this report is to present evidence that this is so, and to show that this happens because of the inappropriateness of the error structure chosen.

2. Examination of the Proposed Model

In this paper we suppose throughout that $\ldots a_t, a_{t-1}, a_{t-2}, \ldots$ are a sequence of *independent* identically distributed random variables having means equal to zero. We shall subsequently call this a *white noise* process.

Let us consider equation (1). Allowing for detrending, the model may be written

$$Y_t = \alpha + \beta_0 t + \beta_1 X_{1,t-6} + \beta_2 X_{2,t-7} + n_t. \tag{3}$$

A critical assumption on which the C.G.K. analysis hangs is that the n_t's, representing error or "noise", are independent identically distributed random variables. Thus, by employing ordinary least squares, it is tacitly assumed that $n_t = a_t$.

However, it must surely be rare that the noise structure for economic models of this kind can be so represented. One might suspect instead that the n_t's were dependent and possibly best represented by some stable non-stationary noise model such as one in which the first difference of the noise was represented by a stationary process. A particular noise structure which has often proved useful in applications of this kind is the integrated moving average or "noisy random walk".

$$n_t = (1-\theta)\sum_{j=1}^{\infty} a_{t-j} + a_t, \quad -1 < \theta \leqslant 1, \tag{4}$$

where the a_t's are independent but for $\theta \neq 1$ the n_t's can be highly dependent.

Characteristics of the integrated moving average model of special interest are:

(i) It may alternatively be written as an infinite autogressive process in which the autogressive weights diminish exponentially

$$n_t = (1-\theta)(n_{t-1} + \theta n_{t-1} + \theta^2 n_{t-2} + \ldots) + a_t$$

with θ between zero and unity. The model thus exponentially discounts past information.

(ii) It follows, as was first shown in Muth (1960), that the model produces the widely used exponentially weighted average as an optimal forecast.
(iii) The model can be written in the convenient alternative form

$$\nabla n_t = a_t - \theta a_{t-1}$$

which implies that the first difference $\nabla n_t = n_t - n_{t-1}$ of the noise is a first-order moving average process and is therefore readily identifiable.
(iv) In the special case $\theta = 1$ we obtain the noise structure assumed in the C.G.K. paper.
(v) In the special case $\theta = 0$ we obtain a noise model which is a pure random walk.
(vi) For intermediate values we have the discounted disturbance structure of (i) which often provides a satisfactory representation of noise in economic series models.

We may now substitute the alternative noise structure (4) in (3) and regard θ as a parameter to be estimated along with the other parameters. If the estimate of θ is close to unity then the simpler error structure assumed in the C.G.K. paper will be vindicated.

After substituting the augmented error structure and differencing we obtain

$$y_t = \beta_0 + \beta_1 x_{1,t-6} + \beta_2 x_{2,t-7} + a_t - \theta a_{t-1},$$

where

$$y_t = Y_t - Y_{t-1}, \quad x_{1,t} = X_{1,t} - X_{1,t-1},$$

$$x_{2,t} = X_{2,t} - X_{2,t-1} \quad \text{and} \quad a_t - \theta a_{t-1} = n_t - n_{t-1}.$$

Now the model may be written in the form

$$a_t = \theta a_{t-1} + y_t - \beta_0 - \beta_1 x_{1,t-6} - \beta_2 x_{2,t-7},$$

whence we may recursively compute the quantities $a_t = (a_t | \theta, \beta_0, \beta_1, \beta_2)$ from $t = 8$ onwards for any given choice of parameters. It will make little practical difference if the starting value of a_7 in this recursion is set equal to its unconditional expected value of zero, or if it too is treated as a parameter to be estimated. In either case the least-squares estimates of the parameters obtained by minimizing $\sum (a_t | \theta, \beta_0, \beta_1, \beta_2)^2$, where summation extends over the whole sample, will closely approximate maximum-likelihood estimates (Barnard *et al.*, 1962).

The least-squares estimates with their approximate standard errors obtained from an iterative nonlinear squares fit are as follows:

$$\hat{\theta} = -0{\cdot}06 \pm 0{\cdot}15, \quad \hat{\beta}_0 = 1{\cdot}78 \pm 2{\cdot}71,$$

$$\hat{\beta}_1 = 0{\cdot}00016 \pm 0{\cdot}00009, \quad \hat{\beta}_2 = -1{\cdot}16 \pm 1{\cdot}18.$$

The analysis is remarkably revealing.

(1) The value of $\hat{\theta}$ is close to zero and not to unity implying that the noise structure is very different from that assumed in the C.G.K. paper and is in fact like a random walk.
(2) While one of the four estimates ($\hat{\beta}_1$) is 1·78 times its standard error, this can hardly be regarded as unusual. Thus with the less restrictive error structure there is no real evidence of any relation at all between the output on the one

hand and the two lagged inputs on the other hand. Thus among the class of models considered there is in fact little reason to question the unsophisticated model

$$\nabla Y_t = a_t,$$

which implies that Y_t is approximately a random walk and agrees with the frequently confirmed conclusion of Bachelier (1900) concerning the behaviour of stock prices.

(3) It would follow in particular if this model were appropriate that an observation at time $t+l$ could be expressed in terms of that at time t by the equation

$$Y_{t+l} = Y_t + \sum_{j=t+1}^{t+l} a_j.$$

Now (see, for example, Whittle, 1963) the minimum mean-square error (m.m.s.e.) forecast $\hat{Y}_t(l)$ of Y_{t+l} made at origin t for lead time l is given by taking expected values conditional on available knowledge at the time origin t. Hence the m.m.s.e. forecast for l steps ahead would be independent of the inputs $X_{1,t-6}$ and $X_{2,t-7}$ and would be simply the current value of the output

$$\hat{Y}_t(l) = Y_t. \tag{5}$$

The more general error structure proposed above might of course still be unduly restrictive. This could be checked in two ways:

(i) by considering other error structures;
(ii) by examining the behaviour of residuals from the fitted models.

A noise structure which presents a plausible alternative to the integrated moving average (4) is a low-order autoregressive process such as the second-order process

$$n_t = \phi_1 n_{t-1} + \phi_2 n_{t-2} + a_t. \tag{6}$$

It is to be noted that with $\phi_1 = 1$ and $\phi_2 = 0$ this coincides with the random walk model $\nabla n_t = a_t$. Thus the two classes of models represented by (4) and (6) intersect at the random walk.

Table 1 summarizes the results from fitting various models and it will be seen that the calculations confirm that the noise model is approximately a random walk. Furthermore, pronounced residual autocorrelations from the C.G.K. model are evident with a highly significant value for the Durbin and Watson (1951) statistic but, no such evidence of inadequacy is found for the other models.

3. CROSS-CORRELATION PATTERNS OF RANDOM WALKS

The apparent relationships in the C.G.K. paper between economic series have a number of puzzling aspects. For example, we see from our Fig. 1(a) that when cross-correlations with negative as well as positive lags are plotted one finds even larger cross-correlations existing at negative lags than those found in the C.G.K. paper at positive lags. This might suggest on the reasoning of that paper that the stock prices might be used to forecast car production instead of vice versa. And *a priori* this seems at least equally plausible.

The question thus arises of how cross-correlation patterns of this kind can have arisen. The answer involves the structure of the individual series which are not adequately represented by the classical regression model

$$X_t = \alpha + \beta t + a_t,$$

but rather require a model closer to the random walk form

$$\nabla X_t = \beta + a_t,$$

that is,

$$X_t = X_{t-1} + \beta + a_t$$

or equivalently

$$X_t = \alpha + \beta t + \sum_{j=0}^{\infty} a_{t-j}.$$

TABLE 1

Estimates and standard errors of coefficients in equation (3) *for various noise structures*

Assumed structure for noise in Y_t	*White noise (C.G.K.)*	*Integrated moving average*	*Second-order autoregressive*	*First-order autoregressive*	*Random walk*
	$n_t = a_t$	$n_t = a_t + (1-\theta)\sum_{j=1}^{\infty} a_{t-j}$	$n_t = \phi_1 n_{t-1} + \phi_2 n_{t-2} + a_t$	$n_t = \phi_1 n_{t-1} + a_t$	$n_t = \sum_{j=0}^{\infty} a_{t-j}$
α	653 ± 57		306 ± 108	318 ± 106	
β_0		1·78 ± 2·7	2·31 ± 1·0	2·04 ± 1·1	1·74 ± 2·6
β_1	0·00047 ± 0·00004	0·00016 ± 0·00009	0·00017 ± 0·00009	0·00018 ± 0·00009	0·00017 ± 0·00008
β_2	−6·13 ± 0·62	−1·16 ± 1·18	−1·76 ± 1·22	−1·87 ± 1·18	−1·27 ± 1·17
θ		−0·06 ± 0·15			
ϕ_1			0·93 ± 0·16	0·82 ± 0·10	
ϕ_2			−0·14 ± 0·16		
σ_a^2	497	321	299	298	315
Durbin–Watson statistic	Significant at 1 per cent	Not significant	Not significant	Not significant	Not significant

To gain preliminary insight into the behaviour of cross-correlations between series generated by models of this latter kind, a sampling experiment was performed as follows.

Each column of Wold's table of random normal deviates contains 50 entries. The first five columns on the first page of the table could therefore be used to generate five random walks of 50 observations by computing cumulative sums of column entries. The five independent random walk series so obtained were detrended as in the C.G.K. paper and the sample cross-correlations between each pair computed. Fig. 1(b) shows the cross-correlation function between the series generated from the first two columns of the random deviates. It has a pattern typical of those found for the other pairs.

Now persuasive features of the cross-correlation patterns in the C.G.K. paper, to which the authors have drawn attention are:

(i) their smoothness;
(ii) the large absolute magnitude of the biggest cross-correlation.

But it is exactly these features which are displayed by the cross-correlations of the independent random walks. In particular the largest in absolute value of the cross-correlations found and, in brackets, the lag at which this correlation appeared is shown in Table 2.

TABLE 2

Largest cross-correlations (with lag in brackets) found between independent random walks generated from the first five columns of Wold's table of random normal deviates

Column \ Column	2	3	4	5
1	−0·48(5)	−0·42(5)	0·50(−9)	−0·48(8)
2		0·51(1)	0·53(5)	0·54(4)
3			0·49(4)	−0·35(−16)
4				−0·79(18)
5				

Suppose we now treat one detrended random walk series as the detrended "input" X'_t and another as the detrended "output" Y'_t and following the C.G.K. paper fit by ordinary least squares the usual regression model

$$Y'_t = \alpha + \beta X'_{t-j} + a_t,$$

where j is chosen to give the maximum cross-correlation. Then applying the standard t-test it is readily confirmed that *every one* of the ten pairs of series yields a regression "significantly different" from zero *at least* at the 5 per cent point, even though the series are in fact independent.

In the C.G.K. paper the authors mention that much of the early work was done by graphing the series on transparencies to a roughly comparable scale, superposing them and sliding along the time axis to see whether there was any fairly obvious coincident variation. Their Fig. 1 is such a superposition which seemingly shows a remarkably close relationship. Where high cross-correlation is found at the particular lag we would expect to be able to visually demonstrate the relationship by this graphical technique and vice versa. It may be asked then whether a visual impression of a relationship is obtained when two unrelated random walks are treated in this way. Fig. 2 shows the plot for independent random walks 2 and 4 obtained from Wold's random deviates detrended, comparably scaled, and at lag 5 where maximum correlation is produced. The apparent relationship is partly due to the flexibility allowed in what is treated as similar—we can in effect adjust for location, spread, trend and lag before we need find similarity, partly due to the comparative smoothness of what is to be compared—to find a correlation only *a few* detrended rescaled and suitably lagged bumps have to roughly match, and partly due to selection process—among n series there are $\frac{1}{2}n(n-1)$ pairs of series that could show such an apparent relationship.

The reason for the smoothness and large absolute magnitudes of the cross correlations is readily explained from a theoretical viewpoint as follows.

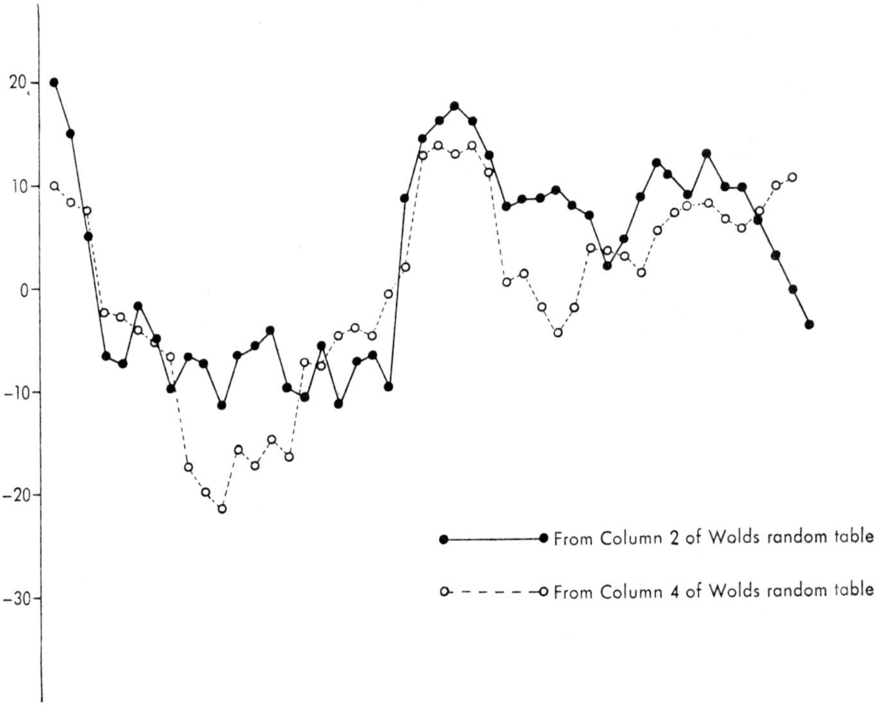

FIG. 2. A plot of comparably scaled and detrended independent random walks at lag 5.

Smoothness

Suppose we have a series of n values generated by the random walk process

$$(X_t - X_{t-1}) - \beta_1 = U_t$$

or

$$x_t - \beta_1 = U_t$$

and N values of Y generated similarly by

$$(Y_t - Y_{t-1}) - \beta_2 = V_t$$

or

$$y_t - \beta_2 = V_t$$

where U_t and V_t are independent white noise processes and let us define the cross-covariance between the differences x_t and y_t as

$$C_k^* = (N-1)^{-1} \sum_{t=2}^{N-k} (x_t - \bar{x})(y_{t+k} - \bar{y}).$$

Now, suppose we postulate trend relationships of the form

$$X_t = \alpha_1 + \beta_1 t + e_{1,t}$$

and

$$Y_t = \alpha_2 + \beta_2 t + e_{2,t},$$

and let $\hat{\alpha}_1$, $\hat{\alpha}_2$, $\hat{\beta}_1$ and $\hat{\beta}_2$ be *any* estimators of α_1, α_2, β_1 and β_2. The detrended series are then

$$X'_t = X_t - \hat{\alpha}_1 - \hat{\beta}_1 t,$$

$$Y'_t = Y_t - \hat{\alpha}_2 - \hat{\beta}_2 t,$$

and the sample cross-covariances for the detrended series are

$$C_k = N^{-1} \sum_{t=1}^{N-k} (X'_t - \bar{X}')(Y'_{t+k} - \bar{Y}'). \tag{7}$$

It is shown in the appendix that for moderate or large sample sizes to a close approximation

$$\nabla^2 C_{k+1} = -C_k^*. \tag{8}$$

Now (see, for example, Bartlett, 1955) the cross-covariances C_k^* between two independent white noise processes are independently distributed about zero with constant variance. Thus writing $e_k = -C_{k-1}^*$ the e_k's form a white noise process and the C_k's satisfy the difference equation

$$\nabla^2 C_k = e_k$$

the solution of which may be written

$$C_k = \sum_{i=0}^{\infty} \sum_{j=0}^{\infty} e_{k-i-j}.$$

Thus on the assumption made the cross-covariances C_k themselves follow a highly *non-stationary* stochastic process—the cumulative sum of a cumulative sum of random deviates. The appearance of any particular series of cross-covariances and hence of the corresponding cross-correlations is bound therefore to be smooth. Thus with the assumptions made, even though X and Y are generated by independent processes, their cross-covariances and hence their cross-correlations will wander about in a smooth pattern peculiar to each generating set of random numbers, in much the same way as was found for the economic series in the C.G.K. paper. This will be so irrespective of whether, or in which way, the series are detrended.

Size

The above can explain the smooth appearance of the cross-correlations, there remains the question of their large size. The variance of the sample cross-correlations r_k between two independent normal sequences X and Y is given (Bartlett, 1955) approximately by

$$\operatorname{var}(r_k) = (n-k)^{-1} \sum_{\nu=-\infty}^{+\infty} \rho_{xx}(\nu)\, \rho_{yy}(\nu),$$

where $\rho_{xx}(\nu)$ and $\rho_{yy}(\nu)$ are the theoretical autocorrelations. This variance is $(n-k)^{-1}$ for series which are not autocorrelated. However, it can be substantially inflated for correlated sequences. For example, suppose the series under study can be represented by unrelated first-order autoregressive processes each with parameter ϕ.

Then substitution yields

$$\text{var}(r_k) = (n-k)^{-1}(1+\phi^2)/(1-\phi^2).$$

The "inflation factor" $(1+\phi^2)/(1-\phi^2)$ becomes large as ϕ approaches unity and as the sequences approximate to the random walks we are considering.

Furthermore, as is noted, for example, by Hannan (1960), the variance of the regression coefficient between two such unrelated autoregressive sequences is inflated approximately by the same factor. The standard errors of the regression coefficients quoted in the C.G.K. paper could thus easily be underestimated by an order of magnitude. This possibly accounts for the high levels of significance obtained.

It is seen then that the observed cross-correlation phenomena are to be expected from unrelated but autocorrelated sequences.

Actual performance of the forecasts

To compare the C.G.K. forecasts with those obtained using equation (5) which totally ignores the inputs X_1 and X_2, forecasts were compared from one step, to six steps, ahead. The forecasting process was begun from the origin 1963/4—that is, the fourth quarter of 1963. All previous data were used to obtain forecasts

(i) from model (1) (C.G.K. forecast),
(ii) from model (5) (present price is forecast price)

The origin was then moved forward to 1964/1 and the whole process was repeated. The origin was moved forward one step at a time to 1967/3, thus producing 16 pairs of forecasts made one step ahead, 15 pairs two steps ahead, and so on. The averages for the squared errors of these forecasts are shown below

	Equation (1)	*Equation* (5)
One step ahead	969	386
Two steps ahead	1,164	894
Three steps ahead	1,264	1,301
Four steps ahead	1,279	1,270
Five steps ahead	1,274	739
Six steps ahead	1,500	375

Comparison of these results verifies that for these data equation (5) usually provides better forecasts.

CONCLUSIONS

Coen, Gomme and Kendall end their paper with the conclusion that their method deserves serious consideration for short-term economic forecasting. We have written this paper because on the contrary we believe this method should not be employed because of an innate and insidious capacity to mislead which we have discussed in some detail. The criticisms we have made are in the spirit of a recently published book (Box and Jenkins, 1970) which is, in turn, based on a number of previous reports and papers there referenced. These latter authors regard the process of model construction as involving first the consideration of an adequately flexible and theoretically sensible family of models followed by the iterative use of the sequence: model identification—model fitting—model diagnostic checking.

In that context we believe we have shown in this paper that:

(i) the *class of* C.G.K. *models* considered—linear multiple regression on lagged input variables with uncorrelated errors—is a demonstrably inadequate family. Adequacy would mean that transformations of the data of the kind

$$n_t = Y_t - \alpha - \beta_0 t - \beta_1 X_{1,t-s_1} - \beta_2 X_{2,t-s_2}$$

for suitable choice of α, β_0, β_1, s_1 and s_2 could produce uncorrelated noise n_t and this has not been found to be so;

(ii) the process of *identification* involving superposition of the highly autocorrelated time series backed by cross-correlation analysis invites the discovery of spurious relationships;

(iii) the process of *fitting* by ordinary least squares with implied uncorrelated errors is inappropriate and could lead to t values inflated by an order of magnitude;

(iv) no diagnostic *checking*, such as analysis of residuals, which would have pointed to these inadequacies, seems to have been attempted.

ACKNOWLEDGEMENT

We are grateful to Larry Haugh for carrying out additional calculations and for his help in the revision of this paper. This work was supported by the Air Force Office of Scientific Research and by the U.S. Army Research Office, Durham.

APPENDIX

CROSS-CORRELATION PROPERTIES OF DETRENDED RANDOM WALKS

We may establish the approximate relation between C_k and C_k^* of equation (8) as follows:

$$(N-1)\,C_k^* = \sum_{t=2}^{N-k}\left[\left\{X'_t-(N-1)^{-1}\sum_{t=2}^{N}X'_t\right\}-\left\{X'_{t-1}-(N-1)^{-1}\sum_{t=1}^{N-1}X'_t\right\}\right]$$
$$\times\left[\left\{Y'_{t+k}-(N-1)^{-1}\sum_{t=2}^{N}Y'_t\right\}-\left\{Y'_{t+k-1}-(N-1)^{-1}\sum_{t=1}^{N-1}Y'_t\right\}\right]$$

and approximately

$$(N-1)^{-1}\sum_{t=2}^{N}X'_t=(N-1)^{-1}\sum_{t=1}^{N-1}X'_t=\bar{X}';$$

$$(N-1)^{-1}\sum_{t=2}^{N}Y'_t=(N-1)^{-1}\sum_{t=1}^{N-1}Y'=\bar{Y}'.$$

Thus

$$(N-1)\,C_k^*\simeq\sum_{t=2}^{N-k}[(X'_t-\bar{X}')-(X'_{t-1}-\bar{X}')]\,[(Y'_{t+k}-\bar{Y}')-(Y'_{t+k-1}-\bar{Y}')],$$

therefore

$$(N-1)\,C_k^*\simeq\sum_{t=2}^{N-k}(X'_t-\bar{X}')(Y'_{t+k}-\bar{Y}')+\sum_{t=1}^{N-k-1}(X'_t-\bar{X}')(Y'_{t+k}-\bar{Y}')$$
$$-\sum_{t=2}^{N-k}(X'_t-\bar{X}')(Y'_{t+k-1}-\bar{Y}')-\sum_{t=1}^{N-k-1}(X'_t-\bar{X}')(Y'_{t+k+1}-\bar{Y}').$$

For moderate or large samples, on dividing by N the approximate relation (8) is now obtained.

REFERENCES

BACHELIER, L. (1900). Théorie de la spéculation. *Ann. Sci Éc. Norm. Sup., Paris*, Series 3, **17**, 21–86.

BARNARD, G. A., JENKINS, G. M. and WINSTEN, C. B. (1962). Likelihood inference and time series. *J.R. Statist. Soc.* A, **125**, 321–352.

BARTLETT, M. S. (1955). *Stochastic Processes*. Cambridge: Cambridge University Press.

BOX, G. E. P. and JENKINS, G. M. (1970). *Time Series Analysis Forecasting and Control.* San Francisco: Holden-Day.

COEN, P. G., GOMME, E. D. and KENDALL, M. G. (1969). Lagged relationships in economic forecasting. *J.R. Statist. Soc.* A, **132**, 133–152.

DURBIN, T. and WATSON, G. S. (1951). Testing for serial correlation in least squares regression. II. *Biometrika*, **38**, 159–178.

HANNAN, E. J. (1960). *Time Series Analysis*. London: Methuen.

MUTH, G. F. (1960). Optimal properties of exponentially weighted forecasts of time series with permanent and transitory components. *J. Amer. Statist. Ass.*, **55**, 299–306.

WHITTLE, P. (1963) *Prediction and Regulation by Linear Least-squares Methods*. London: English Universities Press.

3.7
The Analysis of Closed-Loop Dynamic-Stochastic Systems

G. E. P. BOX *University of Wisconsin*

JOHN F. MacGREGOR *McMaster University*

In the process industries data must often be obtained under conditions of closed-loop operation; that is, under conditions where feedback control is being applied. In the analysis of such data care is needed to properly take account of the manner of its generation. In particular, if standard open-looped procedures of model identification, estimation and diagnostic checking are applied to closed-loop data incorrect models may result and lack of fit not be detected.

This paper discusses the identification, estimation and diagnostic checking of closed-loop systems and illustrates the ideas on two real sets of data.

KEY WORDS

Identification
Estimation
Diagnostic Checking
Closed-Loop Data
Feedback Control
Cross-Correlation Function

1. INTRODUCTION

The building of a statistical model is characterized* by the iteration

IDENTIFICATION → FITTING → DIAGNOSTIC CHECKING (with feedback from DIAGNOSTIC CHECKING to IDENTIFICATION)

Usually the whole class of models which might be capable of representing the system under study is too extensive to be fitted directly to the data. Therefore

(i) *identification* (or specification) of a subclass of models worthy to be entertained is first achieved by a rough preliminary data analysis in the light of subject matter knowledge,

(ii) *fitting* of this subclass to the data is then carried out using efficient methods of parameter estimation conditioned on the truth of the tentative model.

(iii) *diagnostic checking*, by analyses of residuals of the tentatively fitted model, is directed not only to detection of lack of fit but also to recognition of its possible cause and remedy.

Indication of serious model inadequacy ordinarily leads to repetition of the iterative cycle until a suitable representation is obtained.

This research was supported by the Air Force Office of Scientific Research under Grant No. AFOSR 72-2363A.

* Terminology is important because engineers frequently refer to the whole model building process as "identification".

Received Dec. 1972; revised Nov. 1973

Considering systems in which data is available at discrete equispaced intervals of time, Box and Jenkins [1, 2, 3, 4] employed this iterative approach to build models characterizing dynamic and stochastic behaviour. Although for most of the systems they studied the data was collected under open-loop conditions, they also gave some attention to the problem of estimating parameters under closed-loop operation. Because when a full scale industrial process is under study operation under open-loop conditions could produce a dangerous situation or an unsatisfactory product, data for model building often must be obtained from closed-loop operation in which an input is varied by a feedback controller which attempts to eliminate at least major deviations of the output from target value. In this situation the input is no longer independent of the output and "open-loop" procedures may no longer be applicable. The situation at time t is shown diagrammatically in Figure 1. The system is described by the "open-loop" equation in which the error at the output at time t is given by

$$\epsilon_t = V(B)X_t + N_t$$

In this equation the backward shift operator B is such that $BX_t = X_{t-1}$, $V(B) = v_0 + v_1B + v_2B^2 + \cdots$ is the process impulse response function relating the output deviation from target ϵ_t to the deviation of the input variable X_t from its corresponding equilibrium value. The disturbance N_t, the existence of which is the reason that control is needed, represents the effect at the *output* of all sources of variation in the system in the absence of control action. It is represented by $N_t = \psi(B)a_t$, where $\psi(B) = 1 + \psi_1B + \psi_2B^2 + \cdots$ and a_t, a_{t-1}, a_{t-2}, $\cdots$ is a sequence of independent random shocks, roughly normally distributed with mean zero and variance σ_a^2. We shall refer to this sequence as "white noise". This open-loop equation can

The Collected Works of George E. P. Box, 1984, Wadsworth, Inc., Belmont, CA 94002.
Originally published in *Technometrics,* vol. 16, no. 3 (1974), pp. 391-398.

usually be characterized more parsimoniously in the form

$$\epsilon_t = \frac{\omega(B)B^{f+1}}{\delta(B)} X_t + \frac{\theta(B)}{\phi(B)\nabla^d} a_t \tag{1}$$

where $V(B) = \omega(B)B^{f+1}/\delta(B)$ is a transfer function representation of the dynamics, $\psi(B) = \theta(B)/\phi(B)\nabla^d$ is an autoregressive-integrated-moving-average (ARIMA) model for the noise, $\delta(B)$, $\omega(B)$, $\theta(B)$, and $\phi(B)$ are polynomials in B (which in practice are usually of low degree) having zeroes lying outside the unit circle, and $\nabla = 1 - B$ is the backward difference operator. The disturbance is frequently best represented by a non-stationary model in which $d > 0$.

In the situation depicted in Figure 1 some pilot feedback control scheme, typified by the transfer function $C(B) = c_0 + c_1B + c_2B^2 \cdots$, is in operation while the data is being collected. This pilot scheme might make no claim to optimality, but could at least prevent unacceptable digressions from the target output. The problem is to appropriately utilize data, obtained under such feedback control, to model $V(B)$ and $\psi(B)$ so that an appropriate near-optimal controller [1] may be designed.

As we see later to identify an unknown system, it is often desirable and sometimes necessary temporarily to collect data with an additional artificially generated noise source $\{D_t\}$ sometimes called a "dither" signal (see Figure 1) in the input which is uncorrelated with $\{\epsilon_t\}$ and with $\{a_t\}$. As in evolutionary operation [9] the sequence $\{D_t\}$ may be chosen so that consequent additional deviations at the output are of an acceptable magnitude. With the additional input noise source the feedback equation is

$$X_t = C(B)\epsilon_t \ (+D_t) \tag{2}$$

where here and in what follows brackets about the "dither" signal term indicates that it may or may not be included.

The object of this paper is to study problems of model building with closed-loop data and to consider whether, how and why, one can be misled if methods appropriate to open-loop data are employed. We shall discuss individually the procedures of identification, fitting and diagnostic checking as previously defined.

2. Identification of the Process Transfer Function

There are situations where the model *form* can be adequately guessed. In such cases we can proceed directly to the fitting stage. We shall, however, consider here some of the problems that arise when this is not possible and empirical identification of the model form must be carried out using operating data.

2.1. *Identification by cross-correlating the input and output sequences.*

When the disturbance N_t is non-stationary then the input series X_t and possibly* the output series ϵ_t will be non-stationary and their auto- and cross-correlations will fail to damp out quickly. In what follows we assume that it is possible by differencing d times to induce stationarity. Then adequate differencing is indicated when the estimated auto- and cross-correlations of $x_t = \nabla^d X_t$ and $e_t = \nabla^d \epsilon_t$ damp out quickly. In practice d is rarely greater than one. Equations (1) and (2) then become:

$$e_t = V(B)x_t + n_t \tag{3}$$

$$x_t = C(B)e_t(+d_t) \tag{4}$$

where $n_t = [\theta(B)/\phi(B)]a_t$ is stationary. If we suppose that x_t follows some stochastic process which can be "prewhitened" by fitting a stochastic model of the form

$$\phi_x(B)\theta_x^{-1}(B)x_t = x_t' \tag{5}$$

where $\{x_t'\}$ is a white noise sequence with mean zero and variance $\sigma_{x'}^2$, then, after applying this same transformation to e_t and n_t so that

$$e_t' = \phi_x(B)\theta_x^{-1}(B)e_t$$

$$n_t' = \phi_x(B)\theta_x^{-1}(B)n_t \ ,$$

the closed-loop equations (3) and (4) can be written

$$e_t' = V(B)x_t' + n_t' \tag{6}$$

$$x_t' = C(B)e_t'(+d_t'). \tag{7}$$

Multiplying (6) by x_{t-k}' and taking expectations we get after dividing by $\sigma_{x'}\sigma_{e'}$

$$\begin{aligned} \rho_{x'e'}(k) &= v_k \frac{\sigma_{x'}}{\sigma_{e'}} + \rho_{x'n'}(k) \frac{\sigma_{n'}}{\sigma_{e'}} \qquad k \geq 0 \\ &= \rho_{x'n'}(k) \frac{\sigma_{n'}}{\sigma_{e'}} \qquad k < 0 \end{aligned} \tag{8}$$

where $\rho_{x'e'}(k)$ and $\rho_{x'n'}(k)$ are cross-correlations at lag k. If the input x_t is uncorrelated with the disturbance n_t, then $\rho_{x'n'}(k) = 0$ and (8) reduces to

$$\rho_{x'e'}(k) = v_k \frac{\sigma_{x'}}{\sigma_{e'}} \qquad k \geq 0 \tag{9}$$

This is the relationship used by Box and Jenkins [1] for identifying the process transfer function from open-loop data. If one replaces $\sigma_{x'}$, $\sigma_{e'}$, and $\rho_{x'e'}(k)$, $k = 0, 1, 2, \cdots$ by their estimated values

* For any practical pilot control scheme ϵ_t would of course need to be stationary.

$s_{x'}$, $s_{e'}$, and $r_{x'e'}(k)$, $k = 0, 1, 2, \cdots$ respectively, equation (9) will yield rough estimates of the v_k's. From the pattern of these estimated v_k's one can often guess the number of periods of delay f, and the polynomial orders for $\omega(B)$ and $\delta(B)$ and so identify the transfer function model. However, the relationship in (9) will not hold under the feedback conditions of (3) and (4) since the input x_t is related to the disturbance n_t by

$$x_t = C(B)[1 - V(B)C(B)]^{-1}n_t \qquad (+ [1 - V(B)C(B)]^{-1}d_t) \qquad (10)$$

and therefore is highly dependent upon n_t (the extent of this dependency being determined by the relative magnitudes of the two terms in (10)). In such a situation trying to identify the process impulse response by using (9) will lead to an incorrect identification.

Under the ordinary conditions of feedback control where there is no added "dither" signal the feedback equation (7) becomes

$$x_t' = C(B)e_t'$$

which on inversion gives

$$e_t' = \frac{1}{C(B)} x_t' = (c_0' + c_1'B + c_2'B^2 + \cdots)x_t' \qquad (11)$$

Multiplying (11) by x_{t-k}', taking expectations, and dividing by $\sigma_{x'}\sigma_{e'}$, gives

$$\rho_{x'e'}(k) = c_k' \frac{\sigma_{x'}}{\sigma_{e'}} \quad k \geq 0; \qquad = 0 \quad k < 0 \qquad (12)$$

This shows that under normal feedback conditions the cross-correlation function $\rho_{x'e'}(k)$ can give no information whatsoever on the process transfer function. Rather it will reflect only the inverse of the controller transfer function as was shown by Akaike [5] using cross-spectral methods. Comparing this relationship for $\rho_{x'e'}(k)$ from the closed-loop with that given in (9) for the open-loop system it becomes apparent how important it is to determine whether one really has open- or closed-loop data. Since in most dynamic systems v_0 would be zero (no instantaneous transfer of the input to the output), it is clear from equation (12) that the presence of a substantial sample cross-correlation $r_{x'e'}(0)$ at lag zero would usually indicate the presence of a feedback loop.

2.2. *Identification techniques using an added "dither" signal D_t*

When during data gathering the process must be maintained under feedback control for safety or economic reasons, it is possible to deliberately add to the input a programmed noise sequence D_t which is uncorrelated with the disturbance N_t. The process transfer function can then be identified [5]. For, if we multiply equation (3) by D_{t-k} and take expectations, we get

$$\gamma_{De}(k) = v_0\gamma_{Dx}(k) + v_1\gamma_{Dx}(k-1) + \cdots \quad k \geq 0 \qquad (13)$$

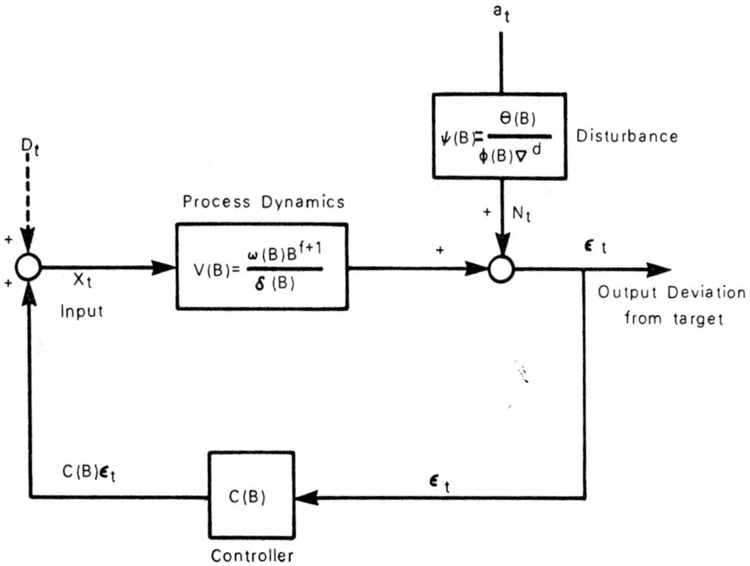

FIGURE 1—*Closed-loop System*

(Alternatively one could use any variate derived from D_t such as $d_t = \nabla^d D_t$). By assuming that the v_j weights are effectively zero beyond $k = K$, and substituting estimates for the two sets of cross-covariances $\gamma_{De}(k)$ and $\gamma_{Dx}(k)$ the first $K + 1$ equations could be solved for $\hat{v}_j$, $j = 0, 1 \cdots, K$. Note that equations (13) may also be written in the form

$$\gamma_{De}(B) = V(B)\gamma_{Dx}(B) \tag{14}$$

where $\gamma_{De}(B)$ and $\gamma_{Dx}(B)$ are cross-covariance generating functions. For example

$$\gamma_{De}(B) = \sum_{k=-\infty}^{\infty} \gamma_{De}(k)B^k.$$

Again by using estimates for the cross-covariances in (14) one may solve for $V(B)$ by equating coefficients of B or by long division.

This method would be available even if the controller $C(B)$ were unknown. Normally $C(B)$ will be known in which case $\gamma_{dx}(k)$ may be expressed as a function of the cross-covariances $\gamma_{de}(k)$ and the known autocovariances $\gamma_{dd}(k)$ of the programmed signal d_t. We can then identify $V(B)$ by using only estimates of the cross-correlations $\gamma_{de}(k)$.

To see this, if we multiply equation (4) by d_{t-k} and take expectations we obtain the set of equations defined by

$$\gamma_{dx}(B) = C(B)\gamma_{de}(B) + \gamma_{dd}(B) \tag{15}$$

Substitution of (15) in (14), where d_t has been used in place of D_t, yields

$$\gamma_{de}(B) = V(B)\{C(B)\gamma_{de}(B) + \gamma_{dd}(B)\} \tag{16}$$

whence by substituting estimates for $\gamma_{de}(k)$ and equating coefficients $V(B)$ may be identified.

Often the dither signal (D_t) would be chosen to be white noise in which case $\gamma_{dd}(B) = (1 - B)^d(1 - B^{-1})^d\sigma_D^2$. If the added signal is not white noise but is generated using the difference equation model

$$\phi_D(B)D_t = \theta_D(B)d_t^\circ$$

where $\{d_t^\circ\}$ is a white noise sequence, then we can substitute in (16)

$$\gamma_{dd}(B) = (1 - B)^d(1 - B^{-1})^d\theta_D(B) \cdot \theta_D(B^{-1})\phi_D^{-1}(B)\phi_D^{-1}(B^{-1})\sigma_{d^\circ}^2$$

Alternatively after applying the d-prewhitening transformation

$$\phi_D(B)e_t = (1 - B)^d\theta_D(B)e_t^\circ \text{ to the } e_t\text{'s we have}$$
$$\gamma_{d^\circ e^\circ}(B) = V(B)\{C(B)\gamma_{d^\circ e^\circ}(B) + \sigma_{d^\circ}^2\} \tag{17}$$

whence again $V(B)$ can be identified by substituting estimates of the cross-correlations $\gamma_{d^\circ e^\circ}(k)$ and equating coefficients for $k > 0$.

2.3 *Identification using the autocorrelations of the output errors.*

We have seen that, for closed-loop data and in the absence of an added noise signal D_t, identification cannot be accomplished by the cross-correlation of e_t and x_t. However, useful information could often be obtained in these circumstances by study of the autocorrelation of the output errors e_t

On substituting equation (4), with $d_t = 0$, into (3) we have

$$\{1 - V(B)C(B)\}e_t = \psi(B)a_t \tag{18}$$

where $n_t = \psi(B)a_t = \phi^{-1}(B)\theta(B)a_t$. In this equation $C(B)$ may be assumed to be known, for, even if it were not known a priori, it could be immediately deduced using equation (4) with the available data on e_t and x_t.

From equation (18) it is then immediately clear that if either $\psi(B)$ or $V(B)$ were already known then the sample autocorrelation function of the e_t's (or of any variate derived from them) could be used to identify the other. In practice there are examples in which $V(B)$ is known (for example from physical considerations) but $\psi(B)$ is not known, and other situations where $\psi(B)$ is known (for example from a previous open-loop study) but $V(B)$ is not.

If $V(B)$ is known then to identify $\psi(B) = \theta(B)/\phi(B)$ we can, for example, first compute the series $z_t = \{1 - V(B)C(B)\}\ e_t$ from the known e_t's. The stochastic model $\psi(B) = \theta(B)/\phi(B)$ is then identified by identifying the mixed autoregressive-moving-average process

$$\phi(B)z_t = \theta(B)a_t$$

by standard methods (see for example [1]).

If on the other hand $\psi(B) = \theta(B)/\phi(B)$ is known, then to identify $V(B)$ we may, for example, first compute the series $w_t = \theta^{-1}(B)\phi(B)e_t$. Then with $L(B) = 1 - V(B)C(B)$ equation (18) yields

$$L(B)w_t = a_t \tag{19}$$

Using (19) $L(B)$ may now be estimated from the data and $V(B)$ identified by equating coefficients in

$$V(B)C(B) = 1 - L(B)$$

If nothing whatever were known about $V(B)$ or about $\psi(B)$ then unambiguous identification would not in general be possible. For notice that any pair of dynamic and stochastic model forms $V_\circ(B)$ and $\psi_\circ(B)$, such that

$$\psi_\circ^{-1}(B)\{1 - V_\circ(B)C(B)\} = \psi^{-1}(B)\{1 - V(B)C(B)\}, \tag{20}$$

can exactly reproduce equation (18) and its white noise sequence $\{a_t\}$. Two examples are

$$\text{(i)}\ \psi_o(B) = 1 \text{ with } V_o(B) = \{V(B)C(B) + \psi(B) - 1\}/\psi(B)C(B) \quad 21(a)$$

$$\text{(ii)}\ \psi_o(B) = \psi(B)/\{1 - V(B)C(B)\} \text{ with } V_o(B) = 0. \quad 21(b)$$

This is not to say that the autocorrelation function of the e_t's is useless for the identification of closed-loop data. In practice one often has some notion as to the likely forms that $V(B)$ and $\psi(B)$ might take. Considered together with such practical knowledge, the autocorrelation function can be very helpful.

3. Estimation

As explained the purpose of identification is to select a parsimonious model *form* worthy to be entertained and fitted efficiently. Often the identification process also produces rough values for the parameters which may be used as starting values for the iterative fitting process.

We suppose now that an identification and analysis of the data, hopefully combined with some basic understanding of the system, has suggested appropriate degrees for the polynomials $\phi(B)$, $\theta(B)$, $\omega(B)$, and $\delta(B)$ in the general dynamic-stochastic model (1) and rough values for the parameters. Under appropriate conditions a close approximation to maximum likelihood estimates [1] may now be obtained by iteratively minimizing the sum of squares function $\sum_{t=1}^{n} a_t^2$ conditional upon selected starting values of the $\{e_t\}$ and $\{x_t\}$ sequences. Whether or not the system is under feedback control, in principle the required a_t's may be computed recursively for any given values of the parameters using equation (3), which may be written

$$a_t = \psi^{-1}(B)\{e_t - V(B)x_t\}. \quad (22)$$

For example, on page 453 of [1], Box and Jenkins discuss a control scheme for maintaining the error ϵ_t of polymer viscosity as small as possible by manipulating the gas rate X_t. Preliminary open-loop identification had suggested the model form

$$\epsilon_t = \frac{\omega_o B}{1 - \delta_o B} X_t + \frac{(1 - \theta_o B)}{1 - B} a_t \quad (23)$$

with preliminary values $\theta_o = 0$, $\omega_o = 0.10$ and $\delta_o = 0.50$. These values were used to design the pilot feedback control scheme

$$X_t = -\frac{(1 - \theta_o)(1 - \delta_o B)}{\omega_o(1 - B)} \epsilon_t = \frac{-10(1 - 0.5B)}{1 - B} \epsilon_t = C(B)\epsilon_t \quad (24)$$

which if the model were correct would give minimum mean square error control. Using the pilot feedback scheme hourly deviations from target ϵ_t and consequent adjustments $x_t = \nabla X_t$ were recorded for 13 days. These recorded data were then used to refit the model. Commencing the iteration with the preliminary pilot scheme values convergence to $\hat{\theta} = 0.18$, $\omega = 0.046$ and $\hat{\delta} = 0.78$ were obtained after 6 iterations of a nonlinear estimation routine.*

Preliminary identification should usually ensure that convergence to incorrect model forms such as (20) and (21) does not occur. However, those examples (equations (20) and (21)) make it clear that attempts at model fitting without preliminary identification by simply fitting a number of diverse model forms and comparing their residual sums of squares (as had been suggested [6] for open loop systems) may fail in closed-loop situations when d_t is absent.

However difficulties can arise associated with extreme correlation among parameters. In such cases the addition of a noise signal d_t can greatly improve the situation. For, if (4) is substituted in (3), we obtain

$$\{1 - V(B)C(B)\}e_t = V(B)d_t + \psi(B)a_t \quad (25)$$

From this equation it is clear that whereas if d_t is absent the nonlinear estimation routine is relying only on the autocorrelation properties of e_t to achieve the estimates, when d_t is present it tacitly uses also the cross-correlations between d_t and e_t. In doubtful cases a preliminary study of the system to be fitted written in the form of (25) may indicate what difficulties, if any, may arise with and without d_t.

4. Diagnostic Checking

Suppose an incorrect model using $V_o(B)$ and $\psi_o(B)$ has been employed. Then the a_{0t}'s which will be computed from the data will be such that

$$\{1 - V_o(B)C(B)\}e_t = V_o(B)d_t + \psi_o(B)a_{0t}$$

whence

$$a_{0t} = \frac{V(B) - V_o(B)}{\psi_o(B)\{1 - V(B)C(B)\}} d_t + \frac{\psi(B)}{\psi_o(B)} \frac{\{1 - V_o(B)C(B)\}}{\{1 - V(B)C(B)\}} a_t \quad (26)$$

If d_t is absent then autocorrelations in the a_{0t}'s could indicate inadequacy of $V_o(B)$, or $\psi_o(B)$, or both. On the other hand lack of significant autocorrelations need not of itself imply a correct model since, as is clear from equation (20) many incorrect models will make $a_{0t} = a_t$. Also since in these

* These estimates are based on a recalculation from the data printed at the end of [1] and differ slightly but not materially from those given earlier.

circumstances $x_t = C(B)\psi_0(B)\{1 - V_0(B)C(B)\}^{-1} a_{0t}$ is a known function of the a_{0t}'s no additional information is to be had by cross-correlating a_{0t} and x_t.

However, when d_t is present, significant cross-correlations between a_{0t} and d_t implies that $V_0(B) \neq V(B)$ and so yields a valuable check on the dynamic model.

5. Illustrative Examples

We now illustrate some of these points using two previously published examples.

5.1. *Polymer viscosity example*

Consider again the polymer viscosity-gas rate feedback scheme of [1]. As mentioned earlier preliminary identification suggested the model form (23) which in turn yielded the controller (24). The 312 observations collected under feedback conditions with this controller were correctly used in [1] to refit the previously identified model. However, this data may be used to illustrate some of the points made earlier in this paper.

First to illustrate the effect of cross-correlating the input and output data when a feedback scheme is in operation, Figure 2(a) shows the cross-correlations between x_t' and e_t' for the data from this example, while Figure 2(b) shows the theoretical cross-correlations induced by the controller and computed from (12) with c_k' the coefficient of B^k in $C^{-1}(B) = -(1 - B)/\{10(1 - 0.5B)\}$ and with $s_{x'} = 37.6$ and $s_{e'} = 5.0$ replacing $\sigma_{x'}$ and $\sigma_{e'}$ respectively. The agreement between Figures 2(a) and 2(b) is close and would be even closer if the control equation had been more exactly followed by the plant operators.

Next we illustrate the point that preliminary identification is an important first step in model building and we cannot rely on the mere fitting of a diversity of models and selection of that giving the smallest sum of squares. For example the following AR(3) model may be identified and fitted to the polymer viscosity data

$$(1 - 0.47B + 0.05B^2 + 0.17B^3)\epsilon_t = a_t$$

Although in the present context the model has little meaning it yields a residual sum of squares

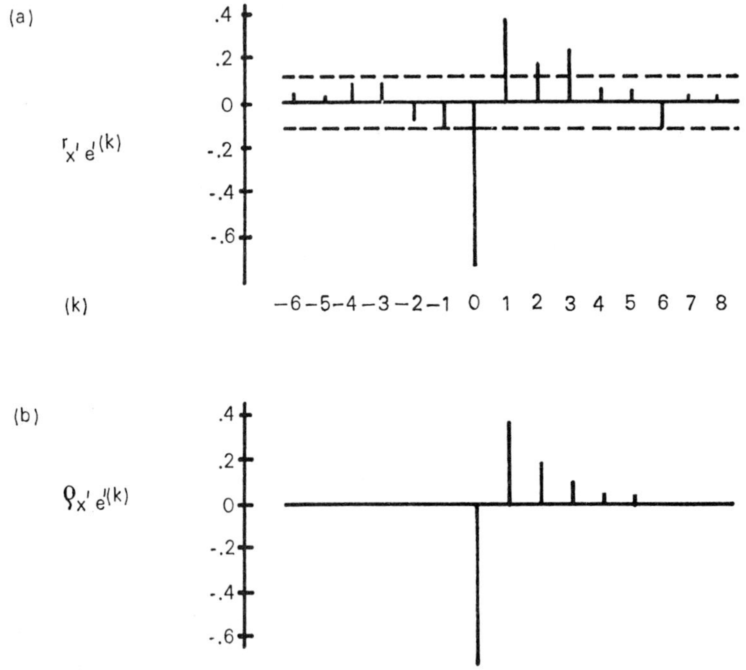

FIGURE 2—Polymer viscosity data.

(a) Estimated cross-correlations between the prewhitened input x_t' and the transformed output e_t'.

(b) Theoretical cross-correlations based on $\rho_{x'e'}(k) = c_k' s_{x'}/s_{e'}$ and the controller $C(B) = -10(1 - 0.5B)/(1 - B)$

which is actually smaller than that given by fitting the model (23) to the feedback data by computing the a_t's directly as in equation (22) from the x's and ϵ's to yield the revised estimates $\hat{\theta} = 0.18$, $\hat{\omega} = 0.046$, and $\hat{\delta} = 0.78$ referred to earlier. Furthermore, the autocorrelations of the residuals shown in Figure 3(a) and (b) from these two different models indicate no inadequacy in either model. (The cross-correlations between these residuals and the input sequence $\{x_t\}$ would therefore also fail to indicate inadequacy of either model as pointed out in section 4).

To examine the nature of this estimation procedure more closely equation (25) may be inspected for this example. In fact the control action (24) was not followed exactly, but if it had been, the substitution of (24) in (23) shows that the ϵ_t's would follow the process

$$[1 - \{1 + \delta - (1 - \theta_\circ)\omega/\omega_\circ\}B + \{\delta - \delta_\circ(1 - \theta_\circ) \cdot \omega/\omega_\circ\}B^2]\epsilon_t = (1 - \theta B)(1 - \delta B)a_t \quad (27)$$

which is an ARMA (2, 2) process whose parameters are functions of the parameters (θ, ω, δ) of the identified process (23). In this example substitution in (27) of ($\hat{\theta}$, $\hat{\omega}$, $\hat{\delta}$) for (θ, ω, δ) and of (0., 0.10, 0.50) for ($\theta_\circ$, $\omega_\circ$, $\delta_\circ$) yields

$$(1 - 1.32B + 0.55B^2)\epsilon_t = (1 - 0.96B + 0.14B^2)a_t$$

Division of the left hand polynomial by that on the right gives approximately

$$(1 - 0.36B + 0.06B^2 + 0.11B^3)\epsilon_t = a_t$$

which is close to the (meaningless) AR(3) model fitted directly.

5.2. *Paper machine example*

The second illustration is from an article [7] which considers the control of a papermaking process about which there is no prior information on the form of the dynamic model. In an attempt to identify the system 160 observations on ϵ_t and X_t were collected in a closed-loop situation in which an experienced operator regulated the stock gate opening, that is, changed X_t according to the observed deviation ϵ_t . The operator's feedback control was probably fairly efficient since he was experienced and the deviations from target which occurred as a result of his operation appear to be uncorrelated.

The authors of the article used* the open-loop

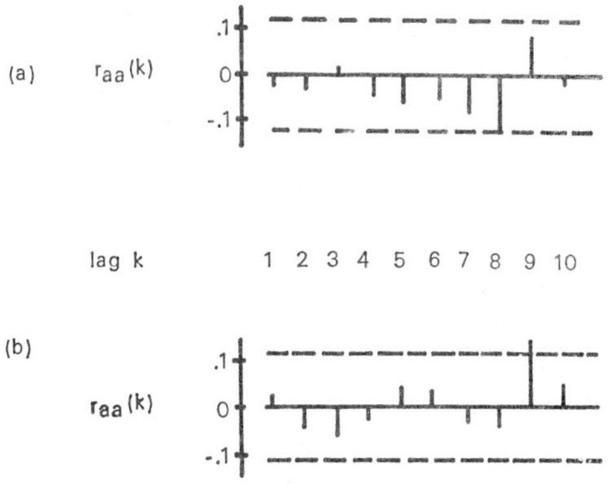

FIGURE 3—*Polymer viscosity data.*

Estimated autocorrelations of the residual a_t's. (a) from the model $(1 - 0.47B + 0.05B^2 + 0.17B^3)\ \epsilon_t = a_t$ (b) from the model $\epsilon_t = \{0.046B/(1 - 0.78B)\}X_t + \{(1 - 0.18B)/(1 - B)\}a_t$ (with $\pm 2/\sqrt{n}$ limits)

identification procedure in an attempt to identify the process transfer function $V(B)$. Prewhitening X_t with the AR(1) model $(1 - .90B)X_t = x_t'$ and applying this same transformation to ϵ_t the estimated cross-correlation function $r_{x'\epsilon'}(k)$ is as shown in Figure 4. The cross-correlations were calculated for positive lags only ($k \geq 1$) and since only one significant cross-correlation at lag $k = 1$ was found it was concluded that the process transfer function was of the form

$$V(B) = \omega_\circ B \quad (28)$$

The large negative cross-correlation at lag $k = 0$, which was ignored, is the clue to the presence of feedback in the system. The cross-correlations are

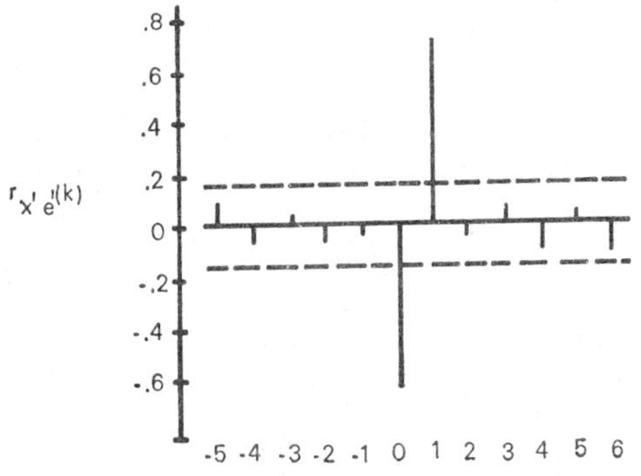

FIGURE 4—*Paper machine data.*

Estimated cross-correlations between the prewhitened input x_t' and the transformed output e_t'

* In recent letters to the editor the inappropriateness of this analysis has been pointed out [10] and acknowledged [11]. The latter reference sketches a proposal for a different analysis which, however, we believe, leads to questionable results.

of course principally reflecting the inverse of the controller transfer function and not as was supposed yielding information about $V(B)$. After fitting the model (28) for $V(B)$ together with a stationary AR(1) disturbance model for $\psi(B)$ it was concluded that this representation was adequate on the basis that, for positive lags, the residual cross-correlation with the prewhitened input and the residual auto-correlations were statistically insignificant. However, from [8] it appears that the first residual autocorrelation $r_{a_0a_0}(1)$ is significant indicating that either the transfer function model (28) or the AR(1) disturbance model or both are inadequate. Indeed the procedure outlined could not in general identify a correct model except by chance.

A further set of data presented in [7] to show the newly developed control scheme in actual operation tended to confirm its inadequacy for:

(1) the output deviations from target were not uncorrelated as those of the skilled operator were, and as those from an efficient scheme would be (for a process having less than one period of delay), but had a very significant autocorrelation at lag one (-0.31);
(2) the proposed scheme produced a variance about the target considerably larger than that obtained by the skilled operator.

6. Conclusions

When performing an iterative model building cycle of identification, estimation and diagnostic checking on a dynamic-stochastic system, it is important to determine whether the data is being collected under open-loop or closed-loop conditions. Sometimes this distinction is difficult to make because the feedback loop may be formed by an operator or by an unknown feedback element. If there is some doubt as to the type of data one has, a test for the presence of feedback can usually be obtained by testing for the presence of a significant cross-correlation $r_{x'e'}(k)$ (between the prewhitened input x_t' and the transformed output e_t') at lag $k = 0$.

The procedure for identifying the process transfer function in open-loop situations by cross-correlating the input X_t with the output deviation ϵ_t is inappropriate in closed-loop situations because of correlation between the disturbance N_t at the output and the input X_t. In cases of pure feedback control where the input is calculated completely from the output this cross-correlation technique identifies not the process transfer function but rather the inverse of the feedback controller transfer function. However, if one deliberately adds an independent programmed noise sequence D_t to the input, then the process transfer function can be identified by cross-correlating this independent noise sequence with the input and output or just with the output if the controller is known, as it usually would be. Useful information can also be obtained from the autocorrelation function of the output.

If there is negligible added noise, D_t, in the input, attempts at process identification merely by fitting competitive models of varying structure to closed-loop data will usually be unsuccessful because several different models can yield the same residual sequence. Furthermore, with such closed-loop data the standard open-loop diagnostic checks based on these residuals are not very useful for revealing model inadequacies because they cannot distinguish between these "equivalent" models. But, if an independent noise sequence D_t is present in the input, then the residual autocorrelation function and the cross-correlation function between this noise and the residuals will provide valuable diagnostic checks. This paper is illustrated with examples from the process industries. The conclusions apply of course with at least equal force to economic data. We are indebted to a referee for helpful comments.

References

[1] Box, G. E. P. and Jenkins, G. M. (1970), *Time Series Analysis, Forecasting and Control*, Holden Day, San Francisco.
[2] Box, G. E. P. and Jenkins, G. M., (1962). "Some Statistical Aspects of Adaptive Optimization and Control", *JRSS* (B), *24*, 297.
[3] Box, G. E. P. and Jenkins, G. M., (1963). "Further Contributions to Adaptive Quality Control: Simultaneous Estimation of Dynamics: Non-zero Costs", *Bull. Intl. Stat. Inst.*, 34th Session, 943, Ottawa, Canada.
[4] Box, G. E. P. and Jenkins, G. M., (1965). "Mathematical Models for Adaptive Control and Optimization", *AIChE-IChE Symp. Series*, *4*, 61.
[5] Akaike, H., (1967). "Some Problems in the Application of the Cross-Spectral Method", *Proceedings of an Advanced Seminar on Spectral Analysis of Time Series*, B. Harris editor, Wiley, New York.
[6] Astrom, K. J. and Bohlin, T., (1966). "Numerical Identification of Linear Dynamic Systems from Normal Operating Records", in *Theory of Self-Adaptive Control Systems*, P. H. Hammond (ed.), Plenum Press, New York.
[7] Tee, L. H. and Wu, S. M., (1972). "An Application of Stochastic and Dynamic Models for the Control of a Papermaking Process." *Technometrics*, *14*, 481.
[8] Tee, L. H., (1970). Ph.D Thesis, Department of Mechanical Engineering, University of Wisconsin.
[9] Box, G. E. P. and Draper, N. R., (1969). *Evolutionary Operation*, Wiley, New York.
[10] Box, G. E. P. and MacGregor, J. F. (1972), Letter to the Editor, *Technometrics*, *14*, No. 4, 985.
[11] Wu, S. M. (1973). Letter to the Editor, *Technometrics*, *15*, No. 1, 207.

3.8
Some Recent Advances in Forecasting and Control, Part II

G. E. P. BOX†
University of Wisconsin

J. F. MacGREGOR†
McMaster University

G. M. JENKINS
University of Lancaster

[Received January 1973. Revised May 1973]

SUMMARY

A brief discussion of Statistical Quality Control Charting procedures is first presented with special reference to the relevance of the objectives and assumptions. An approach to the design of discrete feedforward and feedback control schemes, which are of great importance for example, in the chemical industry, is then given. This approach to control employs discrete stochastic and dynamic models discussed in Part I of this paper (Box and Jenkins, 1968) and has a close link with the forecasting problems discussed there. The control algorithms obtained are ideally suited to discrete digital computer control. However, for common simple situations the algorithms may be represented by suitable charts or nomograms which may be employed to obtain improved manual control. The paper ends with a discussion of a problem typical of that arising in the parts manufacturing industry. Here, attention must be given to the cost of making an adjustment to the machine as well as to the cost of being off target and to the stochastic nature of the disturbance. An example is given where the appropriate form of action is like that required by Roberts's modification of a Shewhart chart. However, the justification required to make such action appropriate is very different from that previously given.

Keywords: FORECASTING; CONTROL CHARTS; FEEDFORWARD CONTROL; FEEDBACK CONTROL; COST OF ADJUSTMENT

1. INTRODUCTION

IN Part I of this paper (Box and Jenkins, 1968b) (we apologize for the delay in presenting this final part), we discussed a class of discrete time series and dynamic models together with the theory for identifying, fitting and checking them. The principal application there considered was *forecasting*. In this second part, which also relies on these models, we outline an approach to *control*. This approach is discussed in more generality and detail in the references (Box and Jenkins, 1962, 1963, 1965, 1968a, b) and in a book (1970). Opportunity is also taken here to correct a mistake which occurred in references (1968a, 1970) concerning optimal feedforward control (see Appendix II, Case 2).

In the past, the word "control" has usually meant to the statistician the quality control techniques developed originally by Shewhart (1931) in the United States, and

† Supported by the Air Force Office of Scientific Research (AFOSR–72–2363B).

The Collected Works of George E. P. Box, 1984, Wadsworth, Inc., Belmont, CA 94002. Originally published in *Applied Statistics,* vol. 23, no. 2 (1974), pp. 158–179.

by Dudding and Jennet (1942) in Great Britain. More recently, the sequential aspects of quality control have been emphasized, leading to the introduction of cumulative sum charts by Page (1957, 1961) and Barnard (1959) and the geometric moving average charts by Roberts (1959).

The word "control" has a different meaning to the control engineer. He thinks in terms of feedforward and feedback control, of the dynamics and stability of the system, and usually of particular types of hardware to carry out the control action. The control devices are automatic in the sense that information is fed to them automatically from instruments on the process and from them to adjust automatically the inputs to the process.

Mostly, the control techniques discussed here are, at least from the point of view of motivation, closer to those of the control engineer than are the quality control procedures most familiar to statisticians. This does not mean that we believe that the traditional quality control chart is unimportant, but rather that it usually performs a function different from that with which we are here concerned. However, as a convenient point of departure, we begin with a brief discussion of familiar quality control charts.

2. Quality Control Charts

Suppose that observations on an industrial process are being made at equispaced intervals of time to produce a time series $\{y_t\}$. Then, as Shewhart (1931) pointed out, it is very useful to plot the data as they come to hand on a chart which shows the target value T. Such a plotting procedure can first provide timely warning of a deviation from target and of possible need for *corrective action* and, second, it can provide clues as to possible *assignable causes* of variation which may subsequently be eliminated or compensated for. The first is a form of feedback control and the second a form of process improvement.

To assist judgment, Shewhart introduced limit lines on one or both sides of the target value. In computing these limits he assumed that $y_t = \mu + a_t$ where μ is a fixed mean and $\{a_t\}$ is what we shall call a "white noise" stochastic process, that is, a sequence of *independently* distributed errors having mean zero and variance σ_a^2. On this assumption, when the mean μ was in fact equal to the target value T, there was a small probability that a value would lie outside the limit lines. If, however, the mean changed from T to some other value, then the probability of a point lying outside the limit lines would be larger and, on the assumptions mentioned above, could be calculated.

If the referral of a point to such limit lines is thought of as a test of the hypothesis that $\mu = T$ then, given the above assumptions, such a test would not be very powerful. Indeed, since the introduction of sequential likelihood ratio tests by Wald and by Barnard during the Second World War, it has been known that on the stated assumptions a procedure of much greater power would employ, not the current deviation $y_t - T$, but the *cumulative sum* of the deviations $\sum_{j=1}^{t}(y_j - T)$. Following proposals by Page (1957, 1961) and by Barnard (1959), charts which used this cumulative sum have been introduced with considerable success into industry.

There are two important suppositions in the above discussion:

(1) that the observations are independently distributed about a fixed mean μ;
(2) that the object is to test the hypothesis that the mean $\mu = T$ against the alternative that μ has some other value.

Independent observations?

Industrial data occurring serially are very likely not to be independent but serially correlated. Suppose, for example, that the observations were generated by the first-order autoregressive process

$$\tilde{y}_t - \phi\tilde{y}_{t-1} = a_t, \tag{1}$$

where $\tilde{y}_t = y_t - \mu$, $-1 < \phi < 1$, μ is the mean and $\{a_t\}$ is a sequence of independent random Normal variables having mean zero and variance σ_a^2. If the problem is one of hypothesis testing, then the appropriate sequential likelihood ratio statistic is

$$\phi\{y_t - y_0\} + (1-\phi)\left\{\sum_{j=1}^{t}(y_j - T)\right\}.$$

If ϕ is not too close to one, and t is moderate in size, this expression is dominated by the cumulative sum in the second bracket. One might expect, therefore, that cumulative sum techniques would remain powerful in situations where autocorrelations of this kind occur and this has been demonstrated to be so by Goldsmith and Whitfield (1961). However, it is also true that if ϕ were very close to unity (that is, the disturbance was essentially a random walk), the first term would dominate, and assuming the initial value y_0 close to target, one would essentially be back to plotting

$$y_t - T,$$

the Shewhart statistic.

Another type of chart, suggested by Roberts (1959), is a modification of the Shewhart chart in which, instead of the current value y_t, the geometric moving average of the observations $\hat{y}_t$ is plotted, where

$$\begin{aligned}\hat{y}_t &= (1-\theta)(y_t + \theta y_{t-1} + \theta^2 y_{t-2} + \ldots) \\ &= \theta\hat{y}_{t-1} + (1-\theta)\,y_t.\end{aligned}$$

However, this intuitively sensible criterion is appropriate for testing for a change in level of a time series on the assumption that the stochastic process is defined by the *non-stationary* model

$$\nabla y_t = a_t - \theta a_{t-1}, \tag{2}$$

where

$$\nabla y_t = y_t - y_{t-1}.$$

It cannot be justified if one assumes, as does Roberts, that y_t is a *stationary* white noise process.

Hypothesis testing?

A second supposition underlying some uses of control charts, for example Truax (1961), has been that adjustment should only be made if a "significant" deviation is observed. Such a use implies important assumptions about the desired objectives. For optimal process control we should not need to be convinced of the "reality" of the difference $y_t - T$ to persuade us to take action. It is enough that a given control policy will lead to a desirable objective such as minimization of costs, or, in some instances, of the mean square deviation from target. To justify a "test like" procedure one normally needs the requirement that some cost is involved in making an adjustment, a point which is also made by Barnard (1959) and Bather (1963).

In the parts manufacturing industry, for which statistical control charts were originally devised, the making of an adjustment usually *does* have a specific cost. Later (in Section 4) we shall discuss briefly an example where a procedure, which closely resembles Shewhart's, results from taking appropriate account of costs and of the stochastic nature of the industrial time series. Before discussing that example we consider below a different and in some respects simpler situation typically met in the chemical and process industries which, in the past, has been more familiar to the control engineer than to the statistician. This is the design of feedback and feedforward control schemes when the extra cost of making an adjustment is zero.

3. Feedback and Feedforward Process Control

The process control schemes we discuss in this Section are appropriate for the periodic, optimal adjustment of a manipulated variable, whose effect on some quality characteristic is already known. They are designed to minimize the variation of that quality characteristic about some target value. We assume that data are available at discrete equispaced time intervals when opportunity can also be taken to make adjustments. It is assumed also that surveillance (by operator or computer) is needed in any case and so no appreciable cost is associated with corrective action (a situation commonly met, for example, in the chemical and process industries).

The reason control is necessary at all is that there are inherent disturbances or *noise* in the system. When we can measure these disturbances directly, the application of appropriate compensatory changes in some other variable to undo their effect is referred to as *feedforward* control. Alternatively, or in addition, use of the deviation from target or "error signal" of the output (quality) characteristic itself, to calculate appropriate compensatory changes, is referred to as *feedback* control. Feedback control can be employed even when the source of the disturbances is not accurately known or their magnitude measured. More generally, feedforward control can be used to compensate for those disturbances that can be measured and feedback control to compensate for the remainder.

The approach adopted here is to typify the disturbances by a suitable time series or *stochastic model* and the inertial (or dynamic) characteristics of the system by a suitable *transfer function model*. It is then possible to calculate a control equation which produces the smallest mean square error at the output, possibly subject to a constraint on the variance of the manipulated variable. Execution of the control action can then be accomplished at various levels of technical sophistication—by a digital computer linked directly to the process, by a pneumatic or electronic automatic controller or by manual manipulation by an operator using a suitable chart or nomogram.

It should perhaps be pointed out that, so far as the computer is concerned, *discrete* control is of great practical importance because of its economy. With the rather slow changes encountered, for example, in many chemical processes it would often be adequate to monitor and take whatever control action was necessary say, once every ten minutes, or for some processes, only once every hour. In these circumstances a single computer can monitor sequentially a large number of variables and analyse a number of different systems.

3.1. *Stochastic and Transfer Function Models*

For completeness, we now summarize the models discussed in Part I (1968b) which form the basic building blocks for the design of the feedforward and feedback control schemes. Just as linear differential equations represent continuous dynamic systems, so in the discrete case, linear difference equations may be employed.

(1) *The Transfer Function model*

It is supposed that the relationship between an output Y and an input X of a dynamic system may be represented by the difference equation

$$\tilde{Y}_t - \delta_1 \tilde{Y}_{t-1} - \ldots - \delta_r \tilde{Y}_{t-r} = \omega_0 \tilde{X}_{t-b} - \omega_1 \tilde{X}_{t-b-1} - \ldots - \omega_s \tilde{X}_{t-b-s}, \qquad (3)$$

where $\tilde{Y}_t$ is the output deviation from equilibrium of the system, $\tilde{X}_t$ is the corresponding input deviation, b determines the number of time intervals of pure delay between input and output and the δ's and ω's are constants. The equation is said to describe a linear filtering operation with X the input to the filter and Y the output. We shall assume that the ω's and δ's are chosen so that the filter is stable, which implies that a finite input change yields a finite change in the output. Provided that the ranges of variation of Y and X are not too great, this model is capable of representing many of the dynamic relationships met in practice.

(2) *The Univariate Stochastic model*

It is supposed that the disturbance N_t is described by a time series which is generated by passing white noise through a suitable linear filter. Specifically it is supposed that

$$w_t - \phi_1 w_{t-1} - \ldots - \phi_p w_{t-p} = a_t - \theta_1 a_{t-1} - \ldots - \theta_q a_{t-q}, \qquad (4)$$

where $w_t = \nabla^d N_t$, N_t is the observed disturbance, d is the degree of differencing required to induce stationarity and a_t is a sequence of random shocks each with zero mean and variance σ_a^2. The above autoregressive-integrated moving average (ARIMA) model of order (p, d, q) provides a representationally useful class of models for describing non-stationary time series. (Special forms useful for describing seasonal time series were described in Part I.) If $d = 0$ (i.e. no differencing is required to induce stationarity), then it is supposed that $w_t = \tilde{N}_t$, the deviation of N_t from its mean value. In many examples we have encountered, the disturbance could be equally well represented by a stationary or a non-stationary model. Thus, in Box and Jenkins (1970, p. 239) a disturbance sequence was equally well fitted by the stationary model

$$\tilde{N}_t - 0{\cdot}9\tilde{N}_{t-1} = a_t - 0{\cdot}6a_{t-1}$$

and by the closely related non-stationary model

$$\nabla N_t = a_t - 0{\cdot}7a_{t-1}.$$

We have usually preferred to employ the non-stationary disturbance model because it does not imply that the series has a fixed mean. Theoretically, the assumption of a non-stationary model implies that the adjustment could be infinite. In practice, such a model merely allows the control equation to compensate for large changes in level.

Difference and backward shift notation

For illustration, consider the transfer function model (3) with $r = 1$, $s = 0$, $b = 1$

$$\tilde{Y}_t - \delta \tilde{Y}_{t-1} = \omega \tilde{X}_{t-1}. \qquad (5)$$

With $-1<\delta<1$, equation (5) represents a discrete "first order" system which, if subjected, for example, to a step change of one unit in $\tilde{X}$, produces *eventually* a change of $g=\omega/(1-\delta)$ units in $\tilde{Y}$, the new equilibrium value being approached exponentially. If we write equation (5) in terms of the backward shift operator B, such that $BX_t = X_{t-1}$, it becomes

$$(1-\delta B)\tilde{Y}_t = \omega \tilde{X}_{t-1}$$

or

$$\tilde{Y}_t = \frac{\omega B}{1-\delta B}\tilde{X}_t. \tag{6}$$

Alternatively, since $\nabla X_t = (1-B)X_t = X_t - X_{t-1}$, the model can be written in terms of the difference operator ∇ as

$$(1+\xi\nabla)\tilde{Y}_t = g\tilde{X}_{t-1}$$

$$\tilde{Y}_t = \frac{g}{1+\xi\nabla}\tilde{X}_{t-1}, \tag{7}$$

where $\xi = \delta/(1-\delta)$.

The backward shift operator may also be used in the representation of stochastic models. Thus, the (1, 0, 0) model of equation (1) may be written

$$(1-\phi B)\tilde{y}_t = a_t \tag{8}$$

and the (0, 1, 1) model of equation (2) as

$$(1-B)y_t = (1-\theta B)a_t.$$

3.2. *Feedback Control*

Before describing a more general approach to feedback control, we consider two simple examples. Both of these occurred in the actual control of chemical processes.

A simple example (*integral control*)

In a scheme to control the viscosity Y of a polymer employed in the manufacture of a synthetic fibre, the controlled variable, viscosity, was checked every hour and adjusted by manipulating the catalyst formulation X. The desired target value for viscosity was 47 units. It was found that the dynamic characteristics relating Y and X were such that essentially all the effect of X on Y occurred within the one hour sampling interval. The transfer function model (3) was, therefore, of the form

$$\tilde{Y}_t = g\tilde{X}_{t-1}, \tag{10}$$

where $\tilde{Y}_t$ and $\tilde{X}_t$ are deviations from equilibrium values and g is called the *steady state gain.* The catalyst formulation changes were, by custom, scaled in terms of the effect they were expected to produce on viscosity. Thus one unit of formulation increase was such as would decrease viscosity by one unit. Hence the gain $g=-1{\cdot}0$.

To design a feedback controller, it is also necessary to take account of the nature of the disturbance which creates the need for control. We shall define N_t as the joint effect on the viscosity measurement of all unobserved disturbances occurring in the process. It is defined as the deviation from target viscosity that would occur at time t

if no control action were taken. In this example the disturbance was found to be adequately described by a stochastic model (4) of order (0, 1, 1)

$$\nabla N_t = (1-\theta B)\, a_t, \tag{11}$$

where $\theta = 0{\cdot}53$ and a_t is a sequence of random shocks with mean zero and variance σ_a^2. This stochastic model is non-stationary in its level (it has a tendency to wander with no fixed mean) and its minimum mean square error forecast l-steps ahead is the geometrically (exponentially) weighted moving-average of previous observations.

$$\hat{N}_t(l) = \theta \hat{N}_{t-1}(l) + (1-\theta)\, N_t = (1-\theta) \sum_{j=0}^{\infty} \theta^j N_{t-j}. \tag{12}$$

The aim of a feedback controller is to compensate for this disturbance in the output viscosity by making suitable changes in $\tilde{X}_t$. The total effect on the output viscosity at time $t+1$ of the disturbance is N_{t+1} and of any compensatory action at time t is $g\tilde{X}_t$. Thus the effect of the disturbance would be cancelled exactly if it were possible to set $\tilde{X}_t = -g^{-1} N_{t+1}$. Since N_{t+1} has not yet occurred, this is not possible, but we can obtain the minimum mean square control error by replacing N_{t+1} by its one step ahead forecast $\hat{N}_t(1)$, that is, by taking control action

$$\tilde{X}_t = -g^{-1} \hat{N}_t(1)$$

or, in terms of the adjustment to be made ($x_t = \nabla \tilde{X}_t = \tilde{X}_t - \tilde{X}_{t-1}$),

$$x_t = -g^{-1}\{\hat{N}_t(1) - \hat{N}_{t-1}(1)\}. \tag{13}$$

Using the forecasting theory in Part I (1968b) it can easily be shown that, for this disturbance model (11), the updating expression for the forecasts is given by

$$\hat{N}_t(1) - \hat{N}_{t-1}(1) = (1-\theta)\, a_t \tag{14}$$

and therefore the optimal control action (13) becomes

$$x_t = -\frac{(1-\theta)}{g} a_t.$$

With this adjustment the error in the output viscosity ε_t will simply be equal to the one step ahead forecast error a_t, and so we can write the optimal adjustment as

$$x_t = -\frac{(1-\theta)}{g} \varepsilon_t = -\frac{(1-0{\cdot}53)}{-1{\cdot}0} \varepsilon_t = 0{\cdot}47\, \varepsilon_t. \tag{15}$$

This controller is worthy of further discussion. As each new deviation ε_t becomes available, an adjustment is made to the input which corrects for our change in the forecast of the disturbance. Recalling that $\varepsilon_t = a_t$ is the one step ahead forecast error, we can see that the updating equation (15) implies that, because our previous forecast $\hat{N}_{t-1}(1)$ falls short of the realized value N_t by a_t, we should adjust it by an amount $(1-\theta)\, a_t = 0{\cdot}47 a_t$ which the model says from past experience is the amount of any shock which is permanently absorbed into the "level" of the process. A naive adjustment a_t to the forecast produced by control action $x_t = -(a_t/g)$ would result in over-correction, while no action would result in the viscosity drifting from target. The action $x_t = -(0{\cdot}47 a_t/g)$ produces an optimal compromise between these extremes, yielding minimum mean square error deviation from the target. The efficiency of control action of this kind is insensitive to moderate changes in parameter values and to a sufficient approximation we can take (15) to be

$$x_t = 0{\cdot}5\varepsilon_t.$$

A convenient chart for use when, as in this example, manual control action is employed, is shown in Fig. 1. On this chart the output (viscosity) scale and the action

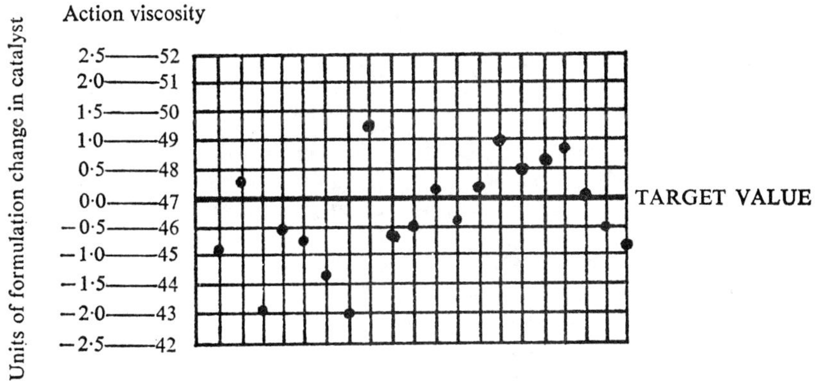

FIG. 1. A feedback chart for integral control action.

scale are arranged so that the output target is aligned with zero action, and so that one unit of output is matched by $-\{(1-\theta)/g\}$ units of action. To employ the chart, the plant operator simply plots the latest output (viscosity) value and reads off the appropriate adjustment on the action scale. For example, the fourth entry on Fig. 1 corresponds to 46 units of viscosity, so that the appropriate action, read off from the action scale, is to decrease the catalyst formulation by 0·5 units.

In terms of the actual catalyst level at the input, the control equation (15) becomes

$$\tilde{X}_t = 0{\cdot}47 \sum_{j=1}^{t} \varepsilon_j. \qquad (16)$$

This simple result is also worthy of further discussion. For many years automatic continuous time controllers, known as three-term controllers, have been used which base the control action empirically on a mixture of proportional, integral and derivative control. For instance, if $\varepsilon(t)$ were the continuous deviation of the output from target, the deviation of the input $\tilde{X}(t)$ from some equilibrium value would be calculated from

$$\tilde{X}(t) = k_D \frac{d\varepsilon(t)}{dt} + k_P\, \varepsilon(t) + k_I \int_0^t \varepsilon(u)\, du,$$

where k_D, k_P and k_I are constants. In some situations only one or two of these three modes of action need be used. The discrete analogue of this continuous control equation is

$$\tilde{X}_t = k'_D \nabla \varepsilon_t + k'_P\, \varepsilon_t + k'_I \sum_{j=1}^{t} \varepsilon_j.$$

It will be seen, therefore, that the above control action (16), appropriate to the control of viscosity, is simply the discrete analogue of integral control action. At first sight this appears to be similar to a cumulative sum procedure. However, it is important to notice that it is accumulating, not the deviations from target when no control action is being taken, but rather the deviations from target when action *is* being taken at every interval. It can be shown that this action is equivalent, not to taking a cumulative sum, but to taking an *exponentially weighted moving average* of past disturbances $N_t, N_{t-1}, N_{t-2}, \ldots$. However, the chart used here bears no resemblance to the exponentially weighted moving average control chart recommended by Roberts (1959).

Rounded control charts

Although the feedback control chart of Fig. 1 is extremely simple it was simplified further. For the particular process under study, control had previously been carried out using a chart based somewhat arbitrarily on a sequential significance testing scheme. It had turned out in this connection that it was convenient to add or subtract from the catalyst formulation in standard steps. Possible actions were: no action, $\pm$one step or $\pm$two steps of catalyst formulation and rules had been instituted on an *ad hoc* basis for taking each action.

Although difficult to justify, the previous scheme did have the advantages:

(1) that it had not been necessary to make changes every time; and
(2) when changes were called for, they were of one of five definite types, making the procedure easy to apply and supervise.

These features were readily included in the new control scheme, with very little increase in the error, by using a "rounded" action chart. A rounded chart is easily constructed from the original chart by dividing the action scale into bands. The adjustment made when an observation falls within the band is that appropriate to the middle point of the band on an ordinary chart. Fig. 2 shows the rounded chart used

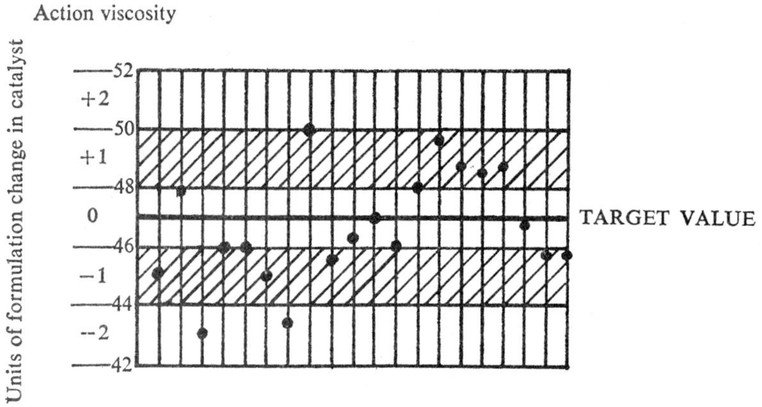

FIG. 2. A feedback chart for rounded integral control action.

for this application in which possible action was limited to -2, -1, 0, 1 or 2 catalyst formulation changes. Figs. 1 and 2 have been constructed by back calculating the values of a_t from a set of operating data and reconstructing the charts that would have resulted from using an unrounded and a rounded scheme. For manual control the increase in mean square error (less than 5 per cent for this example), which results from using the rounded scheme as compared to the unrounded scheme, is often outweighed by the convenience of working with a small number of standard adjustments.

A more elaborate example (proportional plus integral control)

To illustrate other possibilities, consider now the case where the dynamics of the system must be taken account of by a first-order transfer function model (7)

$$(1+\xi\nabla)\tilde{Y}_t = g\tilde{X}_{t-1}.$$

If we allow the disturbance model at the output to be represented again by the integrated-moving average (0, 1, 1) model

$$\nabla N_t = (1-\theta B)\,a_t$$

then we find (see Appendix I) that the optimal feedback controller (that which minimizes the mean square error at the output) is

$$\tilde{X}_t = -\frac{(1-\theta)}{g}\left\{\xi\varepsilon_t + \sum_{j=1}^{t}\varepsilon_j\right\} \tag{17}$$

which is the discrete analogue of proportional-integral control with the ratio of the amount of proportional to integral action being given by the dynamic parameter ξ. Again, it has been shown (Box and Jenkins, 1968a, 1970) that this form of control action can be represented by a simple control chart called a *projection chart*.

Similarly, with the same disturbance model as above, but a second-order transfer function model, we would find that the optimal controller is the discrete analogue of proportional-integral-derivative control. However, these actions are by no means the only kind of control to which this procedure can give rise. A general development is given in Appendix I.

Constrained control schemes

There are situations, particularly if dynamic changes are slow compared with the sampling rate, where minimum mean square error control gives undesirably large variations in the input X_t. It is then possible to introduce a constrained controller in which the mean square error of the output deviation from target ε_t is minimized subject to a constraint on the variance of the input changes. The remarkable feature of these constrained controllers is that by allowing only a very small increase in the mean square error of the output a very large reduction can usually be made in the variance of the input. An example of this behaviour, details of which have been given previously (Box and Jenkins, 1968a, 1970), is now given.

In a scheme to control the viscosity of the product of a chemical reaction by varying the input gas rate, the minimum mean square error control action was found to be given by

$$x_t = \nabla\tilde{X}_t = -10(\varepsilon_t - 0{\cdot}5\varepsilon_{t-1}). \tag{18}$$

If the fluctuations in x_t as a result of this scheme were unacceptably large, then a constrained scheme could be used. In particular, for a 10 per cent increase in the standard deviation of the output, the standard deviation of the input changes can be halved by using the control action

$$x_t = 0{\cdot}15x_{t-1} - 5{\cdot}5(\varepsilon_t - 0{\cdot}5\varepsilon_{t-1}). \tag{19}$$

Fig. 3 illustrates this point. A set of 24 successive observations showing the values of inputs (gas rate) and outputs (viscosity) are reproduced in the left-hand diagrams as they were actually recorded using the optimal unconstrained scheme (18). Also shown is the reconstructed disturbance. Supposing the scheme to be initially on

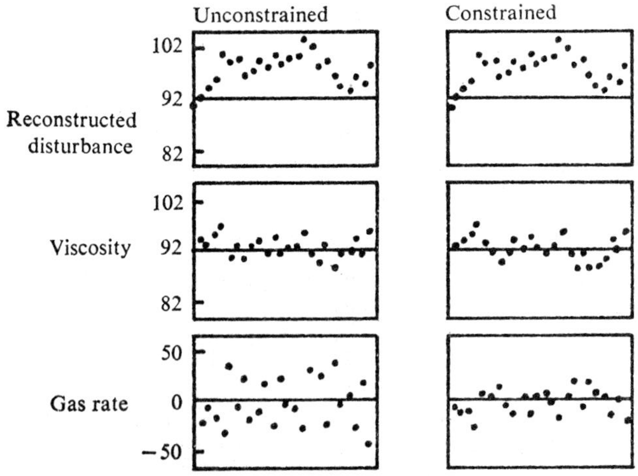

$$x_t = -10(\varepsilon_t - 0{\cdot}5\varepsilon_{t-1}), \quad x_t = 0{\cdot}15x_{t-1} - 5{\cdot}5(\varepsilon_t - 0{\cdot}5\varepsilon_{t-1})$$

FIG. 3. Behaviour of unconstrained and constrained control schemes for viscosity/gas rate example.

target, this reconstructed disturbance is the computed drift away from target that would have occurred if no control action had been taken. The right-hand diagrams show the calculated behaviour that would have occurred with the same disturbance if the constrained scheme of equation (19) had been used. Further information on the design of these optimal constrained schemes has been given by Box and Jenkins (1970), Wilson (1970) and MacGregor (1972).

3.3. *Feedforward Control*

Sometimes one or more major sources of disturbance can be located and measured. In feedforward control these measurements are used to calculate compensatory action which forestalls the effect of these disturbances on the output.

To illustrate the general principles of feedforward control, we consider a case based on an example different from the one given in Box and Jenkins (1970). The example is based on the manufacture of an intermediate product used for the production of a synthetic resin, where the specific gravity Y_t of the product had to be maintained as close as possible to the value 1·260. Specific gravity is known to vary in part due to fluctuations in the feed concentration z_t which can be observed, but not adjusted. The pressure $\tilde{X}_t$ of the reaction vessel is a control variable which is measured, can be manipulated and is potentially available to alter specific gravity by any desired amount. Hence it can compensate potential deviations from target caused by fluctuations in feed concentration z_t. Changes can be made in X at times $t, t-1, t-2, \ldots$ immediately after the observations $z_t, z_{t-1}, z_{t-2}, \ldots$ have come to hand, and then X held constant over the interval as in the feedback case.

Suppose that the transfer function model relating specific gravity and the observed disturbance (feed concentration) over the range of normal operation was estimated to be

$$(1-0{\cdot}2B)\,\tilde{Y}_t = 0{\cdot}0016\tilde{z}_t.$$

Suppose also that the transfer function model relating specific gravity and the compensating variable (pressure) was estimated to be

$$(1-0{\cdot}7B)\,\tilde{Y}_t = 0{\cdot}0024\tilde{X}_{t-1}.$$

Thus, at time t, the effect at the output of:

(1) the disturbance $\tilde{z}_t$ is

$$z'_t = \frac{0{\cdot}0016}{1-0{\cdot}2B}\tilde{z}_t;$$

(2) the compensation X_{t-1} is

$$\tilde{Y}_t = \frac{0{\cdot}0024}{1-0{\cdot}7B}\tilde{X}_{t-1}.$$

Thus, the effect of the observed disturbance will be exactly cancelled if we write

$$\tilde{Y}_t = -z'_t,$$

that is

$$\frac{0{\cdot}0024}{1-0{\cdot}7B}\tilde{X}_{t-1} = -z'_t$$

or

$$\tilde{X}_t = -\frac{(1-0{\cdot}7B)}{0{\cdot}0024}z'_{t+1}.$$

Since the effect z'_{t+1} of the disturbance at time $t+1$ is not known at time t, it is necessary to replace it by its forecast $\hat{z}'_t(1)$, and the minimum mean square error control action is then given by

$$\tilde{X}_t = -\frac{(1-0{\cdot}7B)}{0{\cdot}0024}\hat{z}'_t(1). \tag{20}$$

Study of the feed concentration showed that it could be represented by the linear stochastic model of order (0, 1, 1)

$$\nabla\tilde{z}_t = (1-0{\cdot}5B)\,a_t. \tag{21}$$

Therefore

$$z'_t = \frac{0{\cdot}0016}{1-0{\cdot}2B}\tilde{z}_t = \frac{0{\cdot}0016}{1-0{\cdot}2B}\frac{(1-0{\cdot}5B)}{(1-B)}a_t.$$

This can be arranged in the form

$$\begin{aligned} z'_{t+1} &= 0{\cdot}0016a_{t+1} + \hat{z}'_t(1) \\ &= 0{\cdot}0016a_{t+1} + \frac{0{\cdot}0016(0{\cdot}7-0{\cdot}2B)}{(1-0{\cdot}2B)(1-B)}a_t. \end{aligned}$$

Substituting for a_t in the above from (21) we obtain

$$\hat{z}_t'(1) = \frac{0{\cdot}0016(0{\cdot}7-0{\cdot}2B)}{(1-0{\cdot}2B)(1-B)}a_t = \frac{0{\cdot}0016(0{\cdot}7-0{\cdot}2B)}{(1-0{\cdot}2B)(1-0{\cdot}5B)}\tilde{z}_t.$$

Hence the optimal feedforward control equation (20) is given by

$$\tilde{X}_t = -\frac{(1-0{\cdot}7B)}{0{\cdot}0024}\frac{0{\cdot}0016(0{\cdot}7-0{\cdot}2B)}{(1-0{\cdot}2B)(1-0{\cdot}5B)}\tilde{z}_t$$

or

$$\tilde{X}_t = 0{\cdot}7\tilde{X}_{t-1} - 0{\cdot}1\tilde{X}_{t-2} - 0{\cdot}47\tilde{z}_t + 0{\cdot}46\tilde{z}_{t-1} - 0{\cdot}09\tilde{z}_{t-2}.$$

This control action requires the storage of the three most recent measurements of feed concentration (z) and the past two settings of the manipulated pressure level (X). It is easily handled by a mini-computer if direct digital control is being used or by the use of a nomogram or a small programmable desk calculator in the case of manual control. A general algorithm for feedforward control is given in Appendix II.

Further aspects of this approach to feedback and feedforward control are discussed elsewhere by Box and Jenkins (1970), Wilson (1970) and MacGregor (1972). Among the topics considered there are combined feedforward–feedback control, control schemes in which the variance of the manipulated variable is constrained, the choice of sampling interval and the modelling of transfer function-disturbance models from closed-loop operating data.

4. A Quality Control Problem Revisited

We now return to consider briefly a control problem typically met in the mass production industries where Statistical Quality Control was first employed. A machine mass produces components such as ball bearings. Some quality characteristic y such as weight, or diameter, is measured at discrete intervals of time. Because, for problems of this kind, the Shewhart chart has been used very successfully, one wonders whether such charts might not be justified on more plausible assumptions than those usually given.

Assumptions

Following an earlier paper (Box and Jenkins, 1963) a typical situation seems to be:

(a) the machine is liable to drift out of adjustment;
(b) the further the measurement is away from target the worse is the quality of the component;
(c) when deviations become sufficiently large it is necessary to stop the machine and reset it.

This situation might be approximately represented by the following set of assumptions:

(1) The integrated moving-average model of equation (11)

$$\nabla y_t = (1-\theta B)\, a_t$$

represents the behaviour of the quality characteristic y_t which is liable to drift away from the target value.

(2) The loss sustained through being δ units off target is proportional to the square of the deviation and is $k\delta^2$ pounds sterling (£$k\delta^2$).

(3) When the deviation is sufficiently serious, the mean level must be adjusted, but now in making each adjustment a fixed loss of £C is sustained.

(4) The dynamic characteristics associated with making the change are of no significance, that is, the adjustment is effective at once.

A control system in which the level is periodically adjusted and the controlled observations are considered in relation to a fixed target is equivalent to a system in which the uncontrolled observations y are considered in relation to a movable set point X to which the adjustments are applied with opposite sign (see Figs 4 and 5).

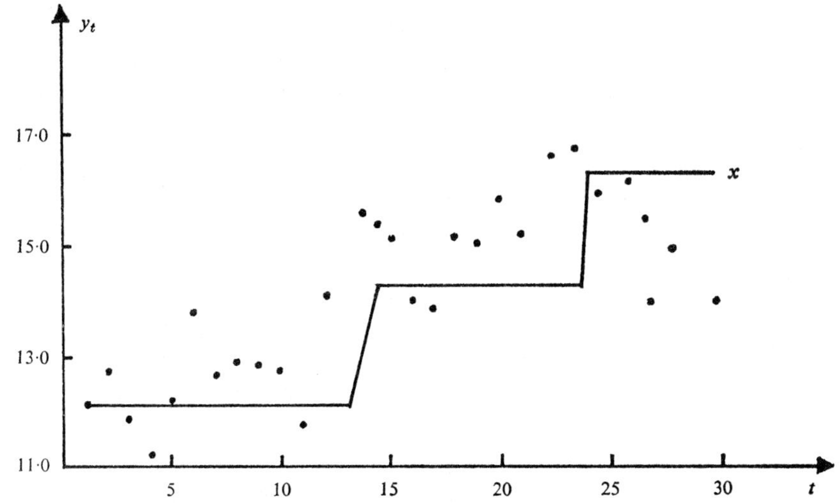

FIG. 4. Uncontrolled series with changes in set point X to achieve optimal control.

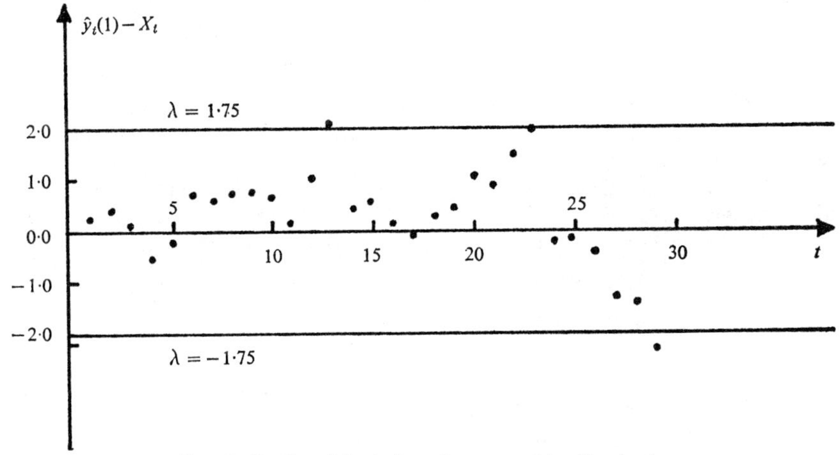

FIG. 5. Predicted deviations from target leading to the adjustments shown in Fig. 4.

Therefore, if X_{t+1} is the adjustable "set point" at time $t+1$, then the deviation from target at this time is $y_{t+1}-X_{t+1}$. If it costs nothing to make an adjustment ($C=0$), then we could minimize costs by making an adjustment $X_{t+1}=\hat{y}_t(1)$ at each stage. However, if we must pay £C to make an adjustment, the predicted deviation $\hat{y}_t(1)-X_t$ at time t must be such that $k(\hat{y}_t(1)-X_t)^2$ is sufficiently large to warrant paying £C for a change. Hence the set point will usually be kept at a constant level for considerable periods.

The control rule

Our problem is, knowing k, C, θ and σ_a^2, to choose an optimal policy that tells us (a) *when* to change, and (b) *by how much* to change so that the overall loss in running the control scheme is minimized. It is demonstrated in Box and Jenkins (1963) that the optimal control rule is to make an adjustment if $|\hat{y}_t(1)-X_t|\geqslant\lambda_{\text{opt}}$ and to remain at the current set point if $|\hat{y}_t(1)-X_t|<\lambda_{\text{opt}}$. Thus the optimal control rule leads to a chart which is very similar to Roberts's modification of a Shewhart chart. A quantity $\hat{y}_t(1)-X_t$, is plotted and referred to control lines $\pm\lambda_{\text{opt}}$ units above and below the target X_t. An adjustment $-\{\hat{y}_t(1)-X_t\}$ is then made if these control limits are exceeded.

A detailed analysis of this problem has been carried out previously (Box and Jenkins, 1963). The optimal value of the control limit λ_{opt} is summarized in Table 1

TABLE 1

$\lambda_{\text{opt}}/\{(1-\theta)\,\sigma_a\}$, *$E(n)$ and g_w* as a function of $c=(C/k)/\{(1-\theta)^2\,\sigma_a^2\}$

$c=\dfrac{C/k}{(1-\theta)^2\,\sigma_a^2}$	$\dfrac{\lambda_{\text{opt}}}{(1-\theta)\,\sigma_a}$	$E(n)$	g_w
20	2·6	10·7	1·5
50	3·5	17·5	2·5
100	4·3	24·5	3·8
200	5·3	34·2	5·6
500	6·8	55·4	8·9
1,000	8·2	78·2	12·7

in terms of the general parameters θ and σ_a^2 of the stochastic model and for a range of values of the ratio of the cost parameters C and k. In this Table we have also tabulated the expected run length $E(n)$ (the average number of observations taken before an adjustment is made) and a cost variable g_w which enables one to calculate the expected "within changes" loss per sampling interval $L_w=k\sigma_a^2+k(1-\theta)^2\,\sigma_a^2\,g_w$ due to being off target. The expected loss per sampling interval due to making a change is given by $L_c=C/E(n)$ and the expected overall loss per sampling interval L is the sum $L=L_w+L_c$.

An example

As an example, consider the case where $k=8$, $C=100$, $\theta=0{\cdot}5$, $\sigma_a=1{\cdot}0$ so that $c=(C/k)/(1-\theta)^2\,\sigma_a^2=50$. From Table 1 we find that $E(n)=17{\cdot}5$ observations and $\lambda_{\text{opt}}/(1-\theta)\,\sigma_a=3{\cdot}5$, so that the best rule is to change when

$$|\hat{y}_t(1)-X_t|>3{\cdot}5\times0{\cdot}5=1{\cdot}75.$$

The within-run loss is then $L_w = 8+2\times2\cdot5 = 13$, the loss due to changing $L_c = 100/17\cdot5 = 5\cdot8$ and hence the overall loss $L = 18\cdot8$. Fig. 4 shows a time series for which $\theta = 0\cdot5$ and $\sigma_a = 1\cdot0$ in its uncontrolled state and Fig. 5 shows the same series controlled in accordance with the above optimal rule. Plotted in Fig. 5 are the predicted deviations from target $\hat{y}_t(1) - X_t$ one step ahead. Appropriate adjustment is made when a point crosses the line. This in effect periodically refers the original series to the best current prediction at new origin as shown in Fig. 4. In Table 2 we give for this example the values of $E(n)$ and the losses for a range of values of λ.

TABLE 2

Expected losses as a function of λ when $k = 8$, $C = 100$, $\theta = 0\cdot5$, $\sigma_a = 1$

λ	$E(n)$	L_w	L_c	L	*Mean square error*
0·0	1·0	8·0	100·0	108·0	1·00
1·0	7·2	9·8	13·9	23·7	1·22
1·5	13·4	11·9	7·4	19·3	1·49
1·75	17·5	13·0	5·8	18·8	1·62
2·0	21·7	14·6	4·6	19·2	1·82
2·5	31·9	18·0	3·1	21·1	2·25
3·75	58·4	26·6	1·7	28·3	3·32
4·5	92·9	38·2	1·1	39·3	4·78

It may be seen from Table 2 that the overall loss L in the region of the minimum λ_{opt} is fairly flat. This suggests that the control scheme is remarkably robust to changes in the position of the "control lines".

In practice the ratio of costs C/k will not be precisely known. In this, as in many other instances, one can usually best proceed by presenting management with alternative possibilities. For instance, the expected mean square error L_w/k of the fluctuations about target can be tabulated as we have done in Table 2. This will increase as the spread of the control lines λ increases, but so will the average run length $E(n)$ between changes, and hence the expected loss due to making changes L_c will decrease. These figures could help in the balancing of costs subjectively. Conversely, if a particular value of λ is presently being used for the control lines, Table 1 would give the value of C/k to which this corresponds and enable one to examine whether this was sensible.

Relation to geometric moving average charts

The form of the control scheme arrived at in the above example is worth further consideration. As illustrated in Fig. 5, the one step ahead forecasts are plotted about a target X_t and referred to the control lines drawn at a distance $\pm\lambda_{opt}$ above and below the target. As long as the forecast falls within these control lines no change is made. This is similar to keeping a Shewhart chart on the *predicted* deviation from target one step ahead. Thus, a procedure very similar to one which has been found to be highly successful can be justified on assumptions which are probably more realistic than those originally adopted to justify it. By further noting that for the (0, 1, 1) process considered in this particular example the one step ahead forecast can be written in the form

$$\{\hat{y}_t(1)-X_t\} = \theta\{\hat{y}_{t-1}(1)-X_t\}+(1-\theta)(y_t-X_t), \tag{22}$$

we see that by plotting $\{\hat{y}_t(1)-X_t\}$ we have in fact the geometric moving average chart such as was suggested by Roberts (1959). However, the justification for it and the basis for choice of the "control limits" are quite different. We have not assumed that the series is a sequence of random independent deviates about a fixed mean but rather that it is highly dependent and has a tendency to drift. The "control limits" are seen to be related to the cost of making a change relative to that of being off target and to the parameters of the stochastic model rather than to any ideas of significance testing and probabilities of being outside control limits. In addition it should be noted that the "adjustable" parameter of the geometric moving average (22) rather than being selected in some arbitrary manner, is in our situation the θ parameter of the (0, 1, 1) process. As θ tends to zero, this stochastic process becomes a random walk and, as would be expected from the earlier discussion in Section 2 of this paper, the optimal control procedure then reduces to the standard Shewhart chart, but with different "control limits". It is worth noting also that, in this situation where we have introduced inertia into the system by associating a cost with making a change, we do not end up with a cumulative sum chart.

References

Barnard, G. A. (1959). Control charts and stochastic processes. *J. R. Statist. Soc.* B, **21**, 239–271.

Bather, J. A. (1963). Control charts and the minimization of costs. *J. R. Statist. Soc.* B, **25**, 49–80.

Box, G. E. P. and Jenkins, G. M. (1962). Some statistical aspects of adaptive optimization and control. *J. R. Statist. Soc.* B, **24**, 297–343.

—— (1963). Further contributions to adaptive quality control: simultaneous estimation of dynamics: non-zero costs. *Bull. Int. Statist. Inst.*, 34th Session, 943–974.

—— (1965). Mathematical models for adaptive control and optimization. *A.I.Ch.E.–J.Ch.E.Symp.*, Series **4**, 61–68.

—— (1968a). Discrete models for feedback and feedforward control. In *The Future of Statistics*, 201–240. (D. G. Watts, ed.) New York: Academic Press.

—— (1968b). Some recent advances in forecasting and control. I. *Appl. Statist.*, **17**, 91–109.

—— (1970). *Time Series Analysis, Forecasting and Control.* San Francisco: Holden Day.

Dudding, B. P. and Jennet, W. J. (1942). Quality control charts. British Standard 600R. London: British Standards Institution.

Goldsmith, P. L. and Whitfield, H. (1961). Average run lengths in cumulative sum quality control schemes. *Technometrics*, **3**, 11–20.

MacGregor, J. F. (1972). Topics in the control of linear processes with stochastic disturbances. Ph.D. Thesis, University of Wisconsin (also Department of Statistics Technical Reports).

Page, E. S. (1957). On problems in which a change in a parameter occurs at an unknown point. *Biometrika*, **44**, 249–252.

—— (1961). Cumulative sum charts. *Technometrics*, **3**, 1–10.

Roberts, S. W. (1959). Control chart tests based on geometric moving averages. *Technometrics*, **1**, 239–250.

Shewhart, W. A. (1931). *The Economic Control of the Quality of Manufactured Product.* New York: Macmillan.

Truax, H. M. (1961). Cumulative sum charts and their application to the chemical industry. *Ind. Qual. Contr.*, **17**, 18.

Wilson, G. T. (1970). Modelling linear systems for multivariate control. Ph.D. Thesis, Department of System Engineering, University of Lancaster, England.

APPENDIX I

A General Algorithm for Feedback Control

To derive this algorithm, we assume that the transfer function model relating the output deviation from equilibrium $\tilde{Y}_t$ to the manipulated variable deviation $\tilde{X}_t$ is

$$\tilde{Y}_t = L_1^{-1}(B)\, L_2(B)\, B^{f+1}\, \tilde{X}_t, \tag{A1}$$

where $L_1(B)$ is a polynomial in B of degree r, $L_2(B)$ is a polynomial in B of degree s and where f is the number of complete intervals of pure delay before an adjustment to the input X_t begins to affect the output Y_t. Suppose also that the disturbance may be represented by the ARIMA (p, d, q) model

$$\phi_p(B)\, \nabla^d N_t = \theta_q(B)\, a_t,$$

where $\phi_p(B)$ is a polynomial in B of degree p and $\theta_q(B)$ is a polynomial in B of degree q, or equivalently by its expanded form

$$N_t = a_t + \psi_1 a_{t-1} + \psi_2 a_{t-2} + \ldots,$$

where a_t is a random series or white noise. Then, referring to Fig. A1, we have at the point P for time t:

$$\text{total effect of disturbance} = N_t,$$

$$\text{total effect of adjustment} = L_1^{-1}(B)\, L_2(B)\, \tilde{X}_{t-f-1}.$$

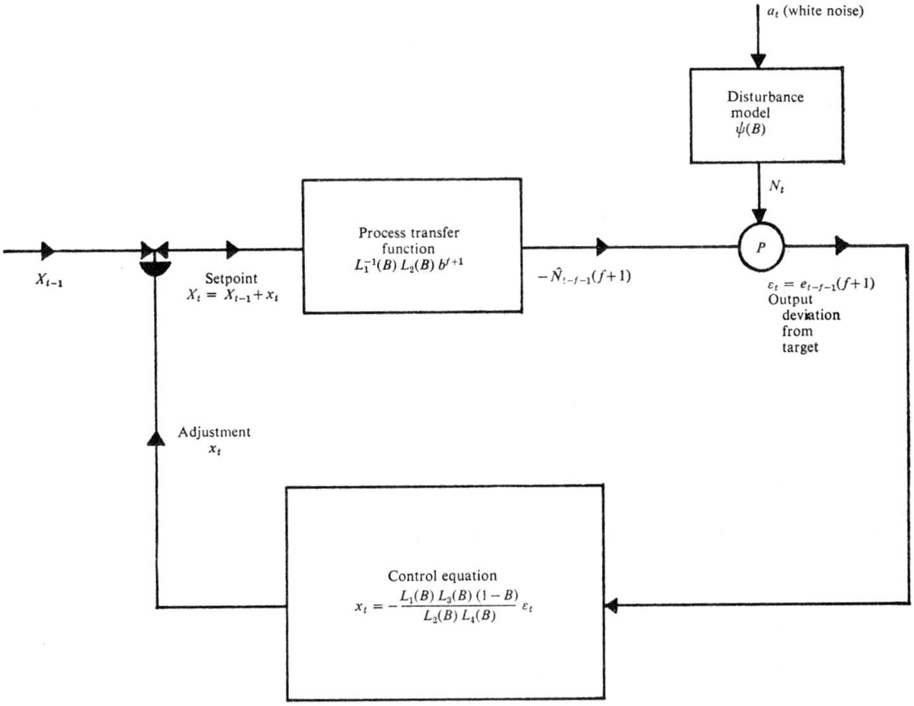

FIG. A1. Feedback control scheme at time t.

Hence the effect of the disturbance would be exactly cancelled if it were possible to set

$$\tilde{X}_t = -L_1(B)\,L_2^{-1}(B)\,N_{t+f+1}.$$

Since $f+1$ is positive, this control action is not realizable. However, we can minimize the mean square error of the deviation of the output from its target value by replacing N_{t+f+1} by its forecast $\hat{N}_t(f+1)$ made at time t, that is

$$\tilde{X}_t = -L_1(B)\,L_2^{-1}(B)\,\hat{N}_t(f+1). \tag{A2}$$

Hence the error, or deviation from target, at time $(t+f+1)$ is the forecast error $e_t(f+1) = N_{t+f+1} - \hat{N}_t(f+1)$ made $f+1$ steps ahead at time t.

To write the control equation in terms of the error $\varepsilon_t = e_{t-f-1}(f+1)$ observed at time t, we note that

$$\hat{N}_t(f+1) = (\psi_{f+1}\,a_t + \psi_{f+2}\,a_{t-1} + \ldots) = L_3(B)\,a_t$$

and

$$\varepsilon_t = e_{t-f-1}(f+1) = a_t + \psi_1\,a_{t-1} + \ldots + \psi_f\,a_{t-f} = L_4(B)\,a_t.$$

Hence

$$\hat{N}_t(f+1) = \frac{L_3(B)}{L_4(B)}\,\varepsilon_t \tag{A3}$$

and on substituting (A3) in (A2) the control equation becomes

$$\tilde{X}_t = -\frac{L_1(B)\,L_3(B)}{L_2(B)\,L_4(B)}\,\varepsilon_t \tag{A4}$$

or in terms of the adjustment to the manipulated variable,

$$x_t = \tilde{X}_t - \tilde{X}_{t-1} = -\frac{L_1(B)\,L_3(B)}{L_2(B)\,L_4(B)}(1-B)\,\varepsilon_t. \tag{A5}$$

An example

Returning to the example of proportional plus integral control given in Section 3.2 we have for the transfer function model

$$(1+\xi\nabla)\,\tilde{Y}_t = g\tilde{X}_{t-1},$$

$L_1(B) = 1+\xi(1-B)$, $L_2(B) = g$, $f = 0$. Similarly, for the disturbance model

$$\nabla N_t = (1-\theta B)\,a_t,$$

$$\hat{N}_t(1) = \frac{1-\theta}{1-B}a_t = \frac{1-\theta}{1-B}\varepsilon_t$$

so that $L_4(B) = 1$, $L_3(B) = (1-\theta)/(1-B)$. Hence the control equation (A4) becomes

$$\tilde{X}_t = -\frac{(1-\theta)}{g}\left[\frac{1}{1-B} + \xi\right]\varepsilon_t$$

and

$$\tilde{X}_t - \tilde{X}_0 = -\frac{(1-\theta)}{g}\left\{\sum_{j=1}^{t}\varepsilon_j + \xi\varepsilon_t\right\}$$

which is the same as the control equation (17) given in the main text if we take X_0 to be the origin.

Appendix II

A General Algorithm for Feedforward Control

Suppose, as in Fig. A2, that the transfer function model which relates the observed disturbance z_t and the output Y_t is

$$\tilde{Y}_t = \delta^{-1}(B)\,\omega(B)\,B^b\,\tilde{z}_t$$

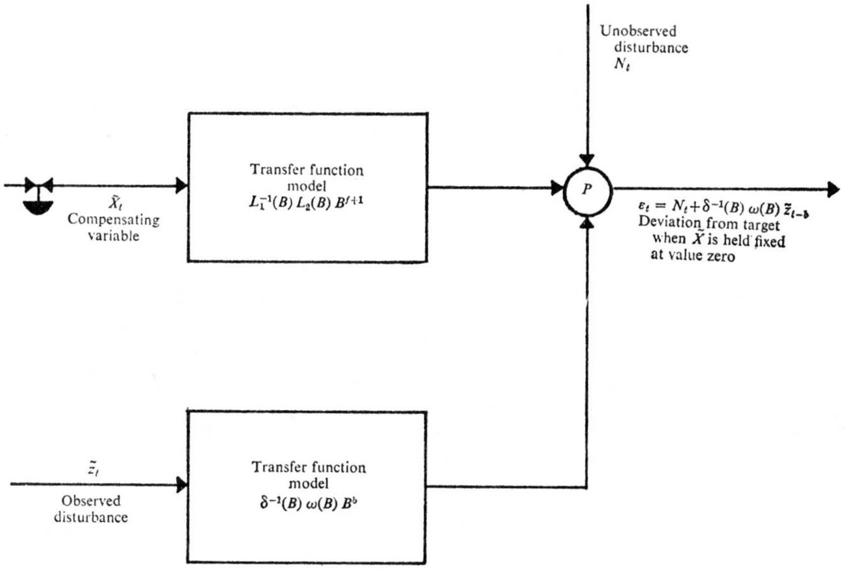

Fig. A2. A system at time t subject to an observed disturbance z_t and unobserved disturbance N_t, with potential compensating variable $\tilde{X}_t$ held fixed at $\tilde{X}_t = 0$.

and that the transfer function model relating the compensating variable X_t and the output is

$$\tilde{Y}_t = L_1^{-1}(B)\,L_2(B)\,B^{f+1}\,\tilde{X}_t,$$

where f is the number of whole periods of pure delay. Suppose also that the total effect in the output of all other unmeasured sources of disturbance at time t is denoted by N_t. Then if no control is exerted ($\tilde{X}_t$ is held fixed at $\tilde{X}_t = 0$), the total error in the output at point P in Fig. A2 will be

$$\varepsilon_t = N_t + \delta^{-1}(B)\,\omega(B)\,\tilde{z}_{t-b}. \qquad \text{(A6)}$$

Clearly, it ought to be possible to compensate the effect of the measured parts of the overall disturbance by manipulating X_t. Now, at time t and at point P in Fig. A2:

(1) the total effect of the disturbance (z) is $\delta^{-1}(B)\,\omega(B)\,\tilde{z}_{t-b}$;
(2) the total effect of the compensation (X) is $L_1^{-1}(B)\,L_2(B)\,\tilde{X}_{t-f-1}$.

Thus the effect of the observed disturbance will be cancelled if we set

$$L_1^{-1}(B)\,L_2(B)\,\tilde{X}_{t-f-1} = -\,\delta^{-1}(B)\,\omega(B)\,\tilde{z}_{t-b}.$$

Thus the control action at time t should be such that

$$L_1^{-1}(B)\,L_2(B)\,\tilde{X}_t = -\,\delta^{-1}(B)\,\omega(B)\,\tilde{z}_{t-(b-f-1)}. \quad \text{(A7)}$$

Case 1: $b \geqslant f+1$.

At time t, the values $z_{t+1}, z_{t+2}, \ldots$ are unknown. The control action (A7) is directly realizable then only if $(b-f-1) \geqslant 0$ in which case the desired control action at time t is to set the manipulated variable $\tilde{X}_t$ to the level

$$\tilde{X}_t = -\frac{L_1(B)\,\omega(B)}{L_2(B)\,\delta(B)}\,\tilde{z}_{t-(b-f-1)}.$$

With this control action the component at the output (point P in Fig. A2) of the deviation from target due to z_t is (theoretically at least) exactly eliminated at the observation times, and only the component N_t due to unobserved disturbances remains.

Case 2: $b < f+1$†

Suppose now that $f+1 > b$. This means that an observed disturbance reaches the output before it is possible for compensating action to become effective. In this case the action of equation (A7) is not realizable because at time t, when action is to be taken, the relevant value $z_{t-(b-f-1)}$ of the disturbance is not yet available. One would usually avoid this situation if one could (if, for example, some quicker acting compensating variable could be used), but sometimes such an alternative is not available.

If the disturbance z_t can be represented by the linear stochastic model

$$\varphi(B)\,\tilde{z}_t = \theta(B)\,a_t \quad \text{(A8)}$$

then we can express $z_t' = \delta^{-1}(B)\,\omega(B)\,\tilde{z}_t$ as the stochastic model

$$\varphi'(B)\,z_t' = \theta'(B)\,a_t,$$

where $\varphi'(B) = \varphi(B)\,\delta(B)$ and $\theta'(B) = \theta(B)\,\omega(B)$ and $\{a_t\}$ is the same white noise sequence as in (A8). This can be equivalently expressed in the form

$$z_t' = \left\{1 + \sum_{i=1}^{\infty} \psi_i\, B^i\right\} a_t.$$

Then

$$z'_{t+f+1-b} = \hat{z}_t'(f+1-b) + e_t'(f+1-b),$$

where in this expression

$$e_t'(f+1-b) = a_{t+f+1-b} + \psi_1\, a_{t+f-b} + \ldots + \psi_{f-b}\, a_{t+1}$$

is the $(f+1-b)$-step ahead forecast error at origin t and $\hat{z}_t'(f+1-b)$ is the forecast. Then we can write equation (A7) in the form

$$L_1^{-1}(B)\,L_2(B)\,\tilde{X}_t = -\hat{z}_t'(f+1-b) - e_t'(f+1-b).$$

† The solution given for this case was incorrect in the first three printings of Box and Jenkins's book *Time Series Analysis, Forecasting and Control* (1970).

Now $e'_t(f+1-b)$ is a function of the uncorrelated random deviates $a_{t+h}(h \geqslant 1)$ which have not yet occurred at time t (and are therefore unforecastable). It follows that the optimal action is achieved by setting

$$\tilde{X}_t = -\frac{L_1(B)}{L_2(B)}\hat{z}'_t(f+1-b) \tag{A9}$$

or by making a change in the compensating variable at time t equal to

$$x_t = -\frac{L_1(B)}{L_2(B)}\{\hat{z}'_t(f+1-b) - \hat{z}'_{t-1}(f+1-b)\}.$$

The needed forecast $\hat{z}'_t(f+1-b)$, obtained as in Part I of this paper (Box and Jenkins, 1968b), can then be written conveniently in terms of the previous z_t's (viscosity measurements) and a_t's. (Recall that these a_t's are common to both the z'_t and z_t series.)

This control scheme results in an additional component in the deviation ε_t from the target, which now becomes

$$\varepsilon_t = N_t + e'_{t-f-1}(f+1-b).$$

Note that in both the above cases of feedforward control the output deviation ε_t from target still includes the disturbance N_t representing the effect in the output at time t of all other disturbances in the system. These disturbances can often be substantial and in particular can result in uncontrolled drift. Hence feedback control will often have to be applied to the output simultaneously with feedforward control to the input.

3.9

Intervention Analysis with Applications to Economic and Environmental Problems

G. E. P. BOX and G. C. TIAO*

This article discusses the effect of interventions on a given response variable in the presence of dependent noise structure. Difference equation models are employed to represent the possible dynamic characteristics of both the interventions and the noise. Some properties of the maximum likelihood estimators of parameters measuring level changes are discussed. Two applications, one dealing with the photochemical smog data in Los Angeles and the other with changes in the consumer price index, are presented.

1. INTRODUCTION

Data of potential value in the formulation of public and private policy frequently occur in the form of time series. Questions of the following kind often arise: "Given a known intervention,[1] is there evidence that change in the series of the kind expected actually occurred, and, if so, what can be said of the nature and magnitude of the change?"

For example, in early 1960 two events occurred, here referred to jointly as the intervention, which might have been expected to reduce the oxidant (denoted by O_3) pollution level in downtown Los Angeles. These events were the diversion of traffic by the opening of the Golden State Freeway and the coming into effect of a new law (Rule 63) which reduced the allowable proportion of reactive hydrocarbons in the gasoline sold locally. The expected effect of this intervention would be to produce a more or less immediate reduction (i.e., a step change) in the oxidant level in early 1960. Figure A shows the monthly averages of oxidant concentration level from 1955–72 in downtown Los Angeles [6]. Using this highly variable and seasonal time series, is there evidence for a change in level and, if so, what is its magnitude?

Many other problems of this kind have come to our attention in recent years. These have included the possible effect of the opening of a nuclear power station on measurements made on river samples, the possible effect of the Nixon Administration's Phases I and II on an economic indicator, and the possible effect of promotions, advertising campaigns and price changes on the sale of a product.

Available procedures such as Student's t test for estimating and testing for a change in mean have played an important role in statistics for a very long time.

A. Monthly Average of Hourly Readings of O_3 (pphm) in Downtown Los Angeles (1955–1972)[a]

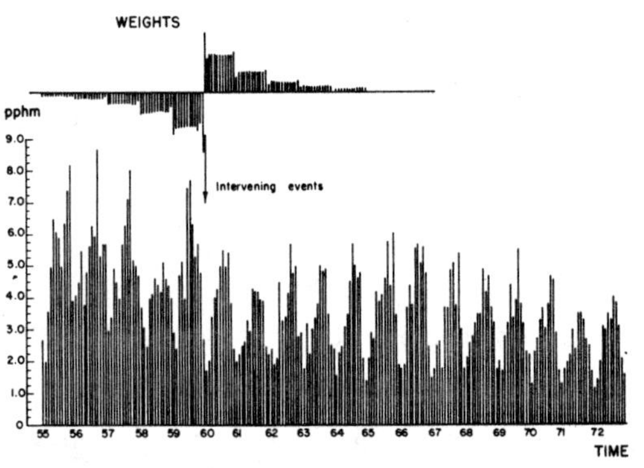

[a] With the weight function for estimating the effect of intervening events in 1960.

However, the ordinary t test would be valid only if the observations before and after the event of interest varied about means μ_1 and μ_2, not only normally and with constant variance but *independently*. In the examples quoted, however, the data are in the form of time series in which successive observations are usually serially dependent and often nonstationary, and there may be strong seasonal effects. Thus the ordinary parametric or nonparametric statistical procedures which rely on independence or special symmetry in the distribution function are not available nor are the blessings endowed by randomization.

An approach we initiated earlier [2] was to build a stochastic model which included the possibility of change of the form expected. Such model building is necessarily iterative and, as discussed, e.g., in [3], involves inferences from a tentatively entertained model alternating with criticism of the appropriate tentative analysis. The process proceeds [1] by successive use of Identification (tentative specification of the model form), Fitting, and Diagnostic Checking. Using these ideas in the present context, we come to the following general strategy:

1. Frame a model for change which describes what is expected to occur given knowledge of the known intervention;

* G.E.P. Box is R. A. Fisher professor and G. C. Tiao is professor and chairman, Department of Statistics, University of Wisconsin, Madison, Wis. 53706. This research was supported in part by a grant from the American Petroleum Institute and in part by U.S. Army Research Office under Grant DA-ARO-D-31-124-72-G162.

[1] A term introduced in [5], based on our earlier work [2].

The Collected Works of George E. P. Box, 1984, Wadsworth, Inc., Belmont, CA 94002.
Originally published in *J. Amer. Stat. Assoc.*, vol. 70, no. 349 (1975), pp. 70–79.

2. Work out the appropriate data analysis based on that model;
3. If diagnostic checks show no inadequacy in the model, make appropriate inferences; if serious deficiencies are uncovered, make appropriate model modification, repeat the analysis, etc.

Suppose the data . . . Y_{t-1}, Y_t, Y_{t+1}, . . . are available as a series obtained at equal time intervals. Following, e.g., [1], we will employ models of the general form

$$y_t = f(\kappa, \xi, t) + N_t \tag{1.1}$$

where:

$y_t = F(Y_t)$ is some appropriate transformation of Y_t, say $\log Y_t$, $(Y_t)^{\frac{1}{2}}$ or Y_t itself;
$f(\kappa, \xi, t)$ can allow for deterministic effects of time, t, the effects of exogenous variables, ξ, and in particular, interventions;
N_t represents stochastic background variation or noise;
κ is a set of unknown parameters.

In Section 2 we discuss a general integrated mixed autoregressive moving average model for representing the noise N_t. A class of general dynamic models capable of representing the effect of interventions is given in Section 3. The associated parameter estimation procedures are given in Section 4. In Section 5 two illustrative examples of intervention analysis are presented. The first concerns the Los Angeles oxidant data, and the second considers possible effects on the consumer price index of recent government actions. Finally, in Section 6, the nature of the maximum likelihood estimators for some specific level-change parameters is discussed in some detail.

2. A STOCHASTIC MODEL FOR THE NOISE

We suppose that the noise $N_t = y_t - f(\mathbf{k}, \xi, t)$ may be modeled by a mixed autoregressive moving average process

$$\varphi(B)N_t = \theta(B)a_t \tag{2.1}$$

where:

1. B is the backshift operator such that $By_t = y_{t-1}$;
2. $\ldots a_{t-1}, a_t, a_{t+1}, \ldots$ is a sequence of independently distributed normal variables having mean zero and variance $(\sigma_a)^2$ which for brevity we refer to as "white" noise;
3. $\theta(B) = 1 - \theta_1 B - \theta_2 B^2 \cdots - \theta_q B^q$, $\varphi(B) = 1 - \varphi_1 B - \varphi_2 B^2 \cdots - \varphi_p B^p$ are "moving average" and "autoregressive" polynomials in B of degrees q and p, respectively;
4. the roots of $\theta(B)$ lie outside, and those of $\varphi(B)$ lie on or outside the unit circle.

For the representation of certain kinds of homogeneous nonstationary series, the operator $\varphi(B)$ is factored so that

$$\varphi(B) = (1 - B)^d \phi(B) \tag{2.2}$$

where the roots of $\phi(B)$ all lie outside the unit circle. This corresponds to the use of a stationary model in the dth difference. Also, for seasonal data with period s (e.g., monthly data with $s = 12$), it is often helpful to write $\varphi(B) = \varphi_1(B)\varphi_2(B^s)$ and $\theta(B) = \theta_1(B)\theta_2(B^s)$ with $\varphi_2(B^s) = (1 - B^s)^D \phi_2(B^s)$ to allow for seasonal nonstationarity.

Finally, we entertain a class of noise model of the form

$$\phi_1(B)\phi_2(B^s)(1 - B)^d(1 - B^s)^D N_t = \theta_1(B)\theta_2(B^s)a_t \tag{2.3}$$

where the polynomials $\phi_1(B)$, $\phi_2(B^s)$, $\theta_1(B)$, $\theta_2(B^s)$ are of degrees p_1, p_2, q_1, q_2, respectively.

3. A DYNAMIC MODEL FOR INTERVENTION

Frequently the effects of exogenous variables ξ can be represented by a dynamic model of the form

$$f(\delta, \omega, \xi, t) = \sum_{j=1}^{k} \mathcal{Y}_{tj} = \sum_{j=1}^{k} \{\omega_j(B)/\delta_j(B)\}\xi_{tj} \tag{3.1}$$

where:

1. The $\mathcal{Y}_{tj}$ represent the dynamic transfer from ξ_{tj};
2. The parameters κ previously lumped together are now denoted by δ and ω;
3. The polynomials in B

$$\delta_j(B) = 1 - \delta_{1j}B - \cdots - \delta_{r_j j}B^{r_j} \quad \text{and}$$
$$\omega_j(B) = \omega_{0j} - \omega_{1j}B - \cdots - \omega_{s_j j}B^{s_j}$$

are of degrees r_j and s_j, respectively;
4. We shall normally assume that $\omega_j(B)$ has roots outside, and $\delta_j(B)$, outside or on, the unit circle.

In general, the individual ξ_{tj} could be exogenous time series whose influence needs to be taken into account. For the present purpose, however, some or all of them will be indicator variables taking the values 0 and 1 to denote the nonoccurrence and occurrence of intervention.

For illustration, suppose for a single exogenous variable ($k = 1$) the model is

$$y_t = \mathcal{Y}_t + N_t = (\omega(B)/\delta(B))\xi_t + (\theta(B)/\varphi(B))a_t \;; \tag{3.2}$$

then the transfer $\mathcal{Y}_t$ to the output from ξ_t is generated by the linear difference equation

$$\delta(B)\mathcal{Y}_t = \omega(B)\xi_t \; .$$

Figures B(a), B(b) and B(c) show the response $\mathcal{Y}_t$ transmitted to the output for various simple dynamic systems by an indicator variable representing a step. We can denote such an indicator by $\xi_t = S_t^{(T)}$ where

$$S_t^{(T)} = \begin{cases} 0, & t < T \\ 1, & t \geq T \end{cases} . \tag{3.3}$$

Similarly, we use $P_t^{(T)}$ for a pulse indicator where

$$P_t^{(T)} = \begin{cases} 0, & t \neq T \\ 1, & t = T \end{cases} . \tag{3.4}$$

Referring to the figure for the case we have discussed for the Los Angeles 1960 intervention, we would expect that the change could be modelled as in Figure B(a), so that immediately following the known step change in the input, an output step change of unknown magnitude would be produced according to

$$\mathcal{Y}_t = \omega B S_t^{(T)} \; .$$

Sometimes a step change would not be expected to produce an immediate response but rather a "first order" dynamic response like that in Figure B(b). The appropriate transfer function model is then

$$\mathcal{Y}_t = \{\omega B/(1 - \delta B)\}S_t^{(T)} \; ,$$

($\delta < 1$). It is readily shown that the time constant of this system is estimated by $\{-\log_e \delta\}^{-1}$ and the steady state gain is $\omega/(1-\delta)$. When δ approaches the value unity, we have the transfer function model

$$\mathcal{Y}_t = \{\omega B/(1-B)\} S_t^{(T)}$$

in which a step change in the input produces a "ramp" response in the output (Figure B(c)).

Note that since

$$(1-B)S_t^{(T)} = P_t^{(T)} , \qquad (3.5)$$

any of these transfer functions could equally well be discussed in terms of the unit pulse $P_t^{(T)}$, and sometimes matters are best thought of directly in terms of $P_t^{(T)}$. Thus, suppose we have monthly sales data and wish to represent the effect of a promotion or advertising campaign lasting less than a month. The simple first order model

$$\mathcal{Y}_t = \{\omega_1 B/(1-\delta B)\} P_t^{(T)}$$

might do this (Figure B(d)) with ω_1 indicating the initial increase in sales immediately following the intervention and δ representing the rate of decay of this increase.

This particular model implies that no lasting effect will occur as a result of the intervention. When this might not be so, the model $B(e)$

$$\mathcal{Y}_t = \{(\omega_1 B/(1-\delta B)) + (\omega_2 B/(1-B))\} P_t^{(T)}$$

could be used in which the possibility is entertained that a residual gain (or loss) in sales ω_2 persists.

B. Responses to a Step and a Pulse Input[a]

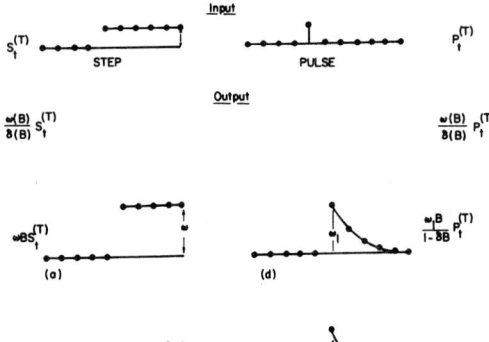

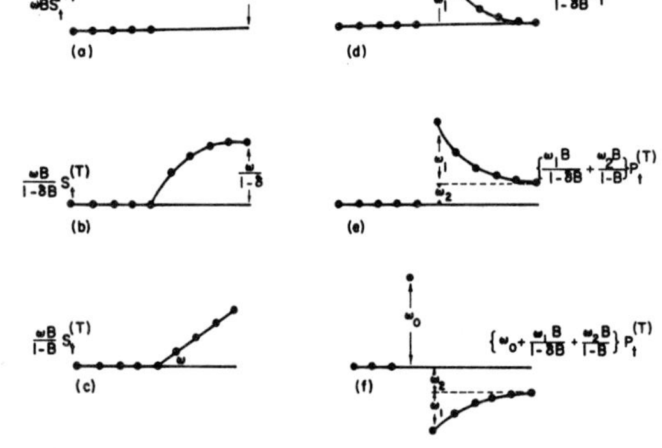

[a] (a), (b), (c) show the response to a step input for various simple transfer function models; (d), (e), (f) show the response to a pulse for some models of interest.

If it were believed that the full impact of intervention might not be felt until the second month, after which there would be a decay and possibly a residual effect as in the previous case, the model

$$\mathcal{Y}_t = \{\omega_0 B + (\omega_1 B^2/(1-\delta B)) + (\omega_2 B^2/(1-B))\} P_t^{(T)}$$

might be appropriate. This would insert a preliminary value ω_0 into the output (which in the preceding context would usually be less than ω_1). The same form of model, shifted forward and with some sign changes in the parameters, could be useful to represent the effect of price changes. In the application shown in Figure B(f), ω_0 would be positive and would represent an immediate rush of buying when a prospective price change was announced. The reduction in buying immediately after the change occurred would be represented by $\omega_1 + \omega_2$ and the final effect of the change would be represented by ω_2 which is shown as negative but, of course, could have a zero or positive value.

Obviously, these difference equation models may be readily extended to represent many situations of potential interest.

The following points are worthy of note:

(i) The function $\mathcal{Y}_t$ represents the *additional* effect of the intervention over the noise. In particular, when N_t is nonstationary, large changes could occur in the output even with no intervention. Fitting the model can make it possible to distinguish between what can and what cannot be explained by the noise.

(ii) Intervention extending over several time intervals can be represented by a series of pulses. A three month advertising campaign might be represented, for example, by three pulses whose magnitude might represent expenditure in the three months.

4. CALCULATIONS BASED ON THE LIKELIHOOD

Suppose we entertain a model of the form

$$y_t = \sum_{j=1}^{k} \mathcal{Y}_{tj} + N_t \qquad (4.1)$$

where $\sum_{j=1}^{k} \mathcal{Y}_{tj}$ is the transfer function given in (3.1) associated with known interventions, N_t assumes the form in (2.3), and a time series is available of length $n + d + sD$. Then the likelihood may be obtained in terms of an n dimensional vector $\mathbf{w}$ whose tth element is $w_t = (1-B)^d(1-B^s)^D(y_t - \sum_{j=1}^{k} \mathcal{Y}_{tj})$. The corresponding model for w_t,

$$w_t = \{\theta_1(B)\theta_2(B^s)/\phi_1(B)\phi_2(B^s)\} a_t , \qquad (4.2)$$

is stationary. Thus, following the argument given, e.g., in [1, p. 273], and with the vector $\boldsymbol{\beta}$ having for its g elements the stochastic and dynamic parameters in the model, the likelihood function may be written

$$L(\boldsymbol{\beta}, (\sigma_a)^2 | \mathbf{y}) = (2\pi(\sigma_a)^2)^{-(n/2)} |\mathbf{M}|^{\frac{1}{2}} \cdot \exp\{-S(\boldsymbol{\beta})/2(\sigma_a)^2\} \qquad (4.3)$$

where $\mathbf{M}^{-1}(\sigma_a)^2$ is the covariance matrix of the vector w and

$$S(\boldsymbol{\beta}) = \mathbf{w}'\mathbf{M}\mathbf{w} = \sum_{t=-\infty}^{n} [a_t | \mathbf{y}, \boldsymbol{\beta}]^2 \qquad (4.4)$$

with $[a_t | \mathbf{y}, \boldsymbol{\beta}]$ as the expected value of a_t conditional on $\boldsymbol{\beta}$ and $\mathbf{y}$.

If none of the roots in (4.2) is close to the unit circle, then for moderate and large n, the likelihood is dominated by the exponent. The values of the elements of $\boldsymbol{\beta}$ minimizing (4.4), which we shall call the *least squares* values, are to a close approximation also the maximum likelihood values. Alternatively, if we introduce a prior distribution such that in the neighborhood where the likelihood is nonnegligible $p(\boldsymbol{\beta}, \sigma_a) \propto p(\boldsymbol{\beta})(\sigma_a)^{-1}$, we obtain the posterior distribution

$$p(\boldsymbol{\beta} | \mathbf{y}) \propto p(\boldsymbol{\beta}) |\mathbf{M}|^{\frac{1}{2}} \{S(\boldsymbol{\beta})\}^{-(n/2)} . \qquad (4.5)$$

Again for moderate or large samples and for a non-informative distribution $p(\boldsymbol{\beta})$, the term involving $S(\boldsymbol{\beta})$ dominates and approximately

$$p(\boldsymbol{\beta} | \mathbf{y}) \dot{\propto} \{S(\boldsymbol{\beta})\}^{-(n/2)} \qquad (4.6)$$

so that the least square estimates correspond with the point of maximum posterior density.

Now if, over the region where the density is appreciable, $S(\boldsymbol{\beta})$ is approximately quadratic (and in any given case it is easy to check this numerically), then the posterior distribution is approximately a multivariate t. Then,

$$p(\boldsymbol{\beta} | \mathbf{y}) \dot{\propto} \{1 + (\sum_{ij} S_{ij}(\beta_i - \hat{\beta}_i)(\beta_j - \hat{\beta}_j)/ (n - g)(s_a)^2)\}^{-(n/2)} \qquad (4.7)$$

where

$$S_{ij} = \tfrac{1}{2}\partial^2\{S(\boldsymbol{\beta})\}/\partial\beta_i\partial\beta_j|_{\beta=\hat{\beta}}$$

and $(s_a)^2 = S(\hat{\boldsymbol{\beta}})/(n - g)$. Thus, for moderate or large n, $\boldsymbol{\beta}$ is approximately distributed as multivariate normal with mean $\hat{\boldsymbol{\beta}}$ and covariance matrix

$$V(\boldsymbol{\beta}) = (s_a)^2\{S_{ij}\}^{-1} .$$

The square roots of the diagonal elements of $V(\boldsymbol{\beta})$ will be referred to as standard errors (S.E.).

In practice we may obtain $\hat{\boldsymbol{\beta}}$, $V(\boldsymbol{\beta})$ and $(s_a)^2$ using a standard nonlinear least squares computer program for the numerical minimization of $S(\boldsymbol{\beta})$. To do this we need only to be able to compute the quantities $[a_t | \mathbf{y}, \boldsymbol{\beta}]$ for any $\boldsymbol{\beta}$ and we may proceed as follows. Since the model for w_t is stationary, $[a_t | \mathbf{y}, \boldsymbol{\beta}]$ will be negligible for values $t \leq -Q$ where Q is some suitably chosen positive number. We, therefore, replace $S(\boldsymbol{\beta})$ by the finite sum $\sum_{t=-Q}^{n} [a_t | \mathbf{y}, \boldsymbol{\beta}]^2$. It is shown in [1] that the initial values $[a_0]$, $[a_{-1}]$, $\cdots$, $[a_{-Q}]$ may often be obtained conveniently by a process of "back forecasting" which also indicates an appropriate value for Q.

5. TWO ILLUSTRATIVE EXAMPLES

The theory developed here is illustrated in this section by two examples, one employing the Los Angeles oxidant data and the other, the rate of change in the United States consumer price index, to determine the effect of known interventions.

5.1 Example 1: The Los Angeles Oxidant Data

Monthly averages of the oxidant (O_3) level in Downtown Los Angeles from January 1955 to December 1972 are shown in Figure A.

Identification (Specification) of the Model. The periods 1955–60 and 1960–65 were regarded as containing no major intervention which would affect the O_3 level. The series themselves and the sample autocorrelation functions within these periods suggest nonstationary and highly seasonal behavior. The autocorrelation functions of such differences $(1 - B^{12})y_t$ taken twelve months apart show significant correlations only at lags 1 and 12. This suggests the following model for the noise N_t:

$$(1 - B^{12})N_t = (1 - \theta_1 B)(1 - \theta_2 B^{12})a_t . \qquad (5.1)$$

Interventions I_1 and I_2 of potential major importance are:

I_1: In 1960 the opening of the Golden State Freeway and the coming into effect of a new law (Rule 63) reducing the allowable proportion of reactive hydrocarbons in locally sold gasoline.

I_2: From 1966 onwards regulations required engine design changes in new cars which would be expected to reduce the production of O_3.

As already argued, I_1 might be expected to produce a step change in the O_3 level at the beginning of 1960. The effect of I_2 might be most accurately represented if we knew, for example, the proportion of new cars having specified engine changes which were in the pool of all cars driven at any point in time. Unfortunately, such data are not available to us presently. We have, therefore, represented the possible effect of intervention as a constant intervention change from year to year reflecting the increased proportion of "new design vehicles" in the car population. As explained more fully in [6], the engine changes would be expected to slow down the photochemical reactions which produce O_3 and, because of the summer-winter atmospheric temperature inversion differential and the difference in the intensity of sunlight, the net effect would be different in winter when oxidant pollution is low from that in summer when it is high.

A model form was, therefore, tentatively entertained for all the available monthly O_3 data from January 1955 to December 1972, which may be conveniently written as:

$$y_t = \omega_{01}\xi_{t1} + \omega_{02}\frac{\xi_{t2}}{1 - B^{12}} + \omega_{03}\frac{\xi_{t3}}{1 - B^{12}} + \frac{(1 - \theta_1 B)(1 - \theta_2 B^{12})}{(1 - B^{12})} a_t \qquad (5.2)$$

where

$$\xi_{t1} = \begin{cases} 0, & t < \text{January, 1960} \\ 1, & t \geq \text{January, 1960} \end{cases}$$

$$\xi_{t2} = \begin{cases} 1, & \text{"summer" months June–October beginning 1966} \\ 0, & \text{otherwise} \end{cases}$$

$$\xi_{t3} = \begin{cases} 1, & \text{"winter" months November–May beginning 1966} \\ 0, & \text{otherwise.} \end{cases}$$

This allows for a step change in the level of O_3 beginning in 1960 of size ω_{01} associated with I_1 and for progressive yearly increments in the O_3 level beginning 1966 of ω_{02} and ω_{03} units, respectively, for the summer and the winter months. This representation is admittedly somewhat crude, and we hope to improve on it as more data become available.

Estimation Results. The maximum likelihood estimates and the associated standard errors are as follows:

Parameter	*MLE*	*S.E.*
ω_{01}	−1.09	.13
ω_{02}	−0.25	.07
ω_{03}	−0.07	.06
θ_1	−0.24	.03
θ_2	0.55	.04

Since examination of residuals $\hat{a}_t$ fails to show any obvious inadequacies in the model, we interpret the results as follows. The marginal distributions *a posteriori* of ω_{01}, ω_{02} and ω_{03} are very nearly normal and centered at the maximum likelihood estimate values with the approximate standard deviations shown.

Thus, there is evidence that

(i) associated with I_1 is a step change of approximately $\hat{\omega}_0 = -1.09$ units in the level of O_3;

(ii) associated with I_2 there is a progressive reduction in O_3. Over the period studied, there is a yearly increment of approximately $\hat{\omega}_{02} = -.25$ in the summer months, but the increment (if any) in the winter is slight.

5.2 Example 2: The Rate of Change in the U.S. Consumer Price Index

A second example supplies further intuitive appreciation for the kind of calculations being performed.

Figure C shows the latter part of a record of the monthly rate of change in the consumer price index (CPI) given more completely in [4]. The complete (July 1953 to December 1972) data include 234 successive values, 218 of which occurred prior to the institution of controls in August 1971. As indicated in the figure, in the three months beginning September 1971, Phase I control was applied; and after that to the end of the recorded period, Phase II was in effect.

C. Monthly Rate of Inflation of the U.S. Consumer Price Index: January 1964–December 1972

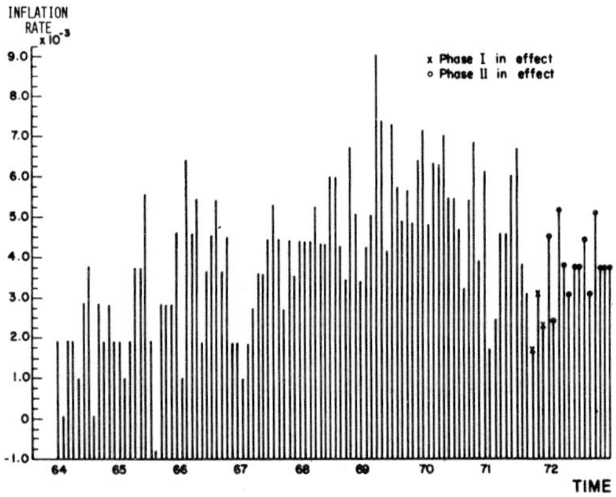

Inspection of the autocorrelation functions of the first 218 observations and their differences prior to Phase I suggests a noise model of the form

$$(1 - B)N_t = (1 - \theta B)a_t \ . \quad (5.3)$$

The maximum likelihood values for the parameters are:

Parameter	*MLE*	*S.E.*
θ	0.84	.04
σ_a	0.0019	

Inspection of the residuals and their autocorrelations reveals no obvious inadequacies of this model, so we adopt it.

We now ask the question, "What are the possible effects of Phases I and II?" To answer, we suppose:

(i) that Phases I and II can be expected to produce changes in level of the rate of change of the CPI,

(ii) that the form of the noise model remains essentially the same.

On these assumptions, the approximate model (ignoring estimation errors in the noise structure) is

$$y_t = \omega_{01}\xi_{t1} + \omega_{02}\xi_{t2} + \{(1 - .84B)/(1 - B)\}a_t \quad (5.4)$$

where

$$\xi_{t1} = \begin{cases} 1, & t = \text{September, October and November 1971} \\ 0, & \text{otherwise} \end{cases}$$

$$\xi_{t2} = \begin{cases} 1, & t \geq \text{December 1971} \\ 0, & \text{otherwise} \end{cases}$$

which may be written

$$z_t = \omega_{01}x_{t1} + \omega_{02}x_{t2} + a_t \ . \quad (5.5)$$

The sequences $\{z_t\}$, $\{x_{t1}\}$, $\{x_{t2}\}$ may be readily calculated from the equations

$$(1 - .84B)z_t = (1 - B)y_t$$
$$(1 - .84B)x_{t1} = (1 - B)\xi_{t1}$$
$$(1 - .84B)x_{t2} = (1 - B)\xi_{t2}$$

using, e.g., the initial approximation $z_1 = x_{11} = x_{12} = 0$.

Also, since

$$(1 - B)/(1 - \theta B) = 1 - B(1 - \theta)(1 + \theta B + \theta^2 B^2 + \cdots) \ ,$$

we have

$$z_t = y_t - \bar{y}_{t-1} \ , \quad x_{t1} = \xi_{t1} - \bar{\xi}_{t-1,1} \ , \quad x_{t2} = \xi_{t2} - \bar{\xi}_{t-1,2}$$

where $\bar{y}_{t-1}$, $\bar{\xi}_{t-1,1}$ and $\bar{\xi}_{t-1,2}$ are exponentially weighted moving averages of values prior to time t, e.g.,

$$\bar{y}_{t-1} = (1 - \theta)(y_{t-1} + \theta y_{t-2} + \theta^2 y_{t-3} + \cdots) \ .$$

We see that (5.5) is very much like the regression equations we are all familiar with in which the deviation of y_t from its average is related to the deviations of ξ_{t1} and ξ_{t2} from their averages. Notice, however, that the model copes with nonstationarity by using not the usual arithmetic averages, but local exponentially weighted averages which change as the series progresses.

Using (5.5), the constants ω_{01} and ω_{02} may now be estimated by ordinary linear least squares as

Parameter	*MLE*	*SE*
ω_{01}	−0.0022	0.0010
ω_{02}	−0.0007	0.0009

Alternatively, a nonlinear least squares program may be employed to estimate ω_{01}, ω_{02} and θ simultaneously from the complete set of 234 data values yielding the estimates (essentially as before):

Parameter	*MLE*	*SE*
θ	0.85	.05
ω_{01}	−0.0022	0.0010
ω_{02}	−0.0008	0.0009

The analysis suggests that a real drop in the rate of increase of the CPI is associated with Phase I, but the effect of Phase II is less certain.

6. NATURE OF THE MAXIMUM LIKELIHOOD ESTIMATORS FOR SOME LEVEL CHANGE PARAMETERS

The maximum likelihood estimators of parameters such as ω_{01}, ω_{02} and ω_{03} in (5.2) and (5.4) which measure level changes are functions of the data. It is instructive to consider the nature of these functions. Several results in the summation of series useful in the following discussion are given in the appendix.

6.1 One Parameter "Linear" Dynamic Model

Consider first the dynamic model in (3.2). Formally, it can be written

$$Q(B)y_t = (\varphi(B)/\theta(B))(\omega(B)/\delta(B))\xi_t + a_t \quad (6.1)$$

where $Q(B) = \varphi(B)/\theta(B)$, even though in practice the y_t are only available for $t = 1, \cdots, n$. Since the roots of $\theta(B)$ all lie outside the unit circle, $Q(B)$ can be expressed as a power series in B which converges for $|B| = 1$.

Here we discuss the situation where

$$(\varphi(B)/\theta(B))(\omega(B)/\delta(B)) = \beta R(B) \quad (6.2)$$

and investigate the nature of the maximum likelihood estimator of β, assuming that (i) the coefficients in $Q(B)$ and $R(B)$ are known and (ii) the power series $R(B)$ converges for $|B| = 1$.

Letting

$$z_t = Q(B)y_t \quad \text{and} \quad x_t = R(B)\xi_t \ ,$$

we can write (6.1) in the form of the usual linear model

$$z_t = \beta x_t + a_t \quad (6.3)$$

so that the maximum likelihood estimator of β is

$$\hat{\beta} = \sum_{t=1}^{n} z_t x_t / \sum_{t=1}^{n} (x_t)^2$$

with (6.4)

$$\mathrm{Var}\,(\hat{\beta}) = (\sigma_a)^2 (\sum_{t=1}^{n} (x_t)^2)^{-1}.$$

For large n, we apply the results (A.6) and (A.7) in the appendix to obtain

$$\sum_{t=1}^{\infty} z_t x_t = \sum_{t=1}^{\infty} Q(B)y_t R(B)\xi_t = \sum_{t=1}^{\infty} \xi_t R(F)Q(B)y_t = R(F)Q(B)C_{\xi y}(0)$$

where $F = B^{-1}$ and

$$\sum_{t=1}^{n} (x_t)^2 = \sum_{t=1}^{\infty} R(B)\xi_t R(B)\xi_t = R(F)R(B)C_{\xi\xi}(0) \ ,$$

where

$$C_{\alpha\beta}(k) = \sum_{t=1}^{\infty} \beta_t \alpha_{t-k} \ , \quad k = 0, \pm 1, \pm 2, \cdots ,$$

and for a given k

$$B^l C_{\alpha\beta}(k) = C_{\alpha\beta}(k - l) \ , \quad l = 0, \pm 1, \pm 2, \cdots .$$

Thus,

$$\hat{\beta} = R(F)Q(B)C_{\xi y}(0)/R(F)R(B)C_{\xi\xi}(0) \quad (6.5)$$

and

$$\mathrm{Var}\,(\hat{\beta}) = (\sigma_a)^2/R(F)R(B)C_{\xi\xi}(0) \ .$$

Making use of (A.10) in the appendix, we can write $R(B)R(F)$ as

$$R(B)R(F) = r_0 + \sum_{l=1}^{\infty} r_l(B^l + F^l) \ . \quad (6.6)$$

Suppose that $\xi_t = P_t^{(T)}$ is a pulse at time T, and a large number of observations are available before and after T. In this case

$$C_{\xi\xi}(k) = \begin{cases} 1, & k = 0 \\ 0, & k \neq 0 \end{cases} \quad \text{and} \quad C_{\xi y}(k) = y_{T-k} \ , \quad (6.7)$$

so that

$$\hat{\beta} = (r_0)^{-1}R(F)Q(B)y_T \quad \text{and} \quad \mathrm{Var}\,(\hat{\beta}) = (\sigma_a)^2(r_0)^{-1} \quad (6.8)$$

where it is understood that B is operating on T.

Now, nonstationarity in time series data can often be removed by differencing. In what follows we suppose that the polynomial $\varphi(B)$ in (6.1) is divisible by $(1 - B)$. We consider two special cases of interest.

Case (i). $\quad \omega(B)/\delta(B) = \beta B \ , \quad (6.9)$

that is, the pulse input $P_t^{(T)}$ gives rise to a response at time $(T + 1)$ measured by β which dissipates completely after the $(T + 1)$th period. It should be noted that with any number of periods of pure delay, the response will follow the same pattern but be appropriately shifted. In this case, $Q(B) = R(B)F$ so that, from (6.6) and (6.8),

$$\hat{\beta} = y_{T+1} - \tfrac{1}{2} \sum_{l=1}^{\infty} \lambda_l(y_{T+1+l} + y_{T+1-l}) \ , \quad (6.10)$$

where $\lambda_l = -2r_l/r_0$. Also, since $\varphi(B)$ is assumed divisible by $(1-B)$, $r_0 + 2\sum_{l=1}^{\infty} r_l = 0$, and hence $\sum_{l=1}^{\infty} \lambda_l = 1$.

As an example, consider the integrated moving average model of order one for the noise term N_t for which

$$\varphi(B) = 1 - B \quad \text{and} \quad \theta(B) = 1 - \theta B \ . \tag{6.11}$$

Since

$$R(B)R(F) = \frac{(1-B)(1-F)}{(1-\theta B)(1-\theta F)}$$

$$= (1+\theta)^{-1}\cdot[2 - (1-\theta)\sum_{l=1}^{\infty} \theta^{l-1}(B^l + F^l)] \ ,$$

we find that

$$\lambda_l = (1-\theta)\theta^{l-1} \ . \tag{6.12}$$

Thus, $\hat{\beta}$ represents a comparison between y_{T+1} and the mean of two exponentially weighted averages, one of the observations before time $(T+1)$ and the other after, with the magnitude of the weights $(1-\theta)\theta^{l-1}$ monotonically decreasing as l increases.

This formulation is applicable to situations where the response to the pulse input is expected to be short-lived, e.g., the effect on the demand for electricity during a sudden heat wave in the summer or the sale of beer in Wisconsin should the Packers win the Super Bowl. Essentially, we are comparing the observation y_{T+1} with the neighboring ones to determine if y_{T+1} is an "aberrant" or "outlying" observation. The results in (6.10) and (6.12) are appealing since, in forming the comparison, more weight is given to observations close to the intervening event and less and less weight to observations remote from the time of the event.

Case (ii). $\omega(B)/\delta(B) = \beta B/(1-B)$. (6.13)

Here, the response to the pulse $P_t^{(T)}$ is a step change in the level of the observations measured by β. Thus

$$Q(B) = (1-B)R(B)F \tag{6.14}$$

and, from (6.6), (6.8) and (A.11), we have that

$$\hat{\beta} = (r_0)^{-1}R(B)R(F)(1-B)y_{T+1}$$

$$= \sum_{l=0}^{\infty} \alpha_l y_{T+1+l} - \sum_{l=0}^{\infty} \alpha_l y_{T-l} \tag{6.15}$$

where $\alpha_l = (r_0)^{-1}(r_l - r_{l+1})$ so that $\sum_{l=0}^{\infty} \alpha_l = 1$.

The quantity $\hat{\beta}$ is, therefore, a contrast between two weighted averages, one of observations before the intervening pulse $P_t^{(T)}$ and the other afterward, where the weights are symmetrical.

As a first example, consider again the integrated moving average model in (6.11). We find

$$\hat{\beta} = (1-\theta)\sum_{l=0}^{\infty} \theta^l y_{T+1+l} - (1-\theta)\sum_{l=0}^{\infty} \theta^l y_{T-l} \tag{6.16}$$

as obtained in [2].

As a second example, we return to the model in (5.2) for the monthly averages of ozone in downtown Los Angeles. For illustration, we shall ignore the effect of interventions after 1966 and discuss the step change

$$(\beta B/(1-B))P_t^{(T)} = \omega_{01}\xi_{t1} \ , \quad T = \text{December 1959}$$

in the level of the series due to the intervening events around that time. In this case, the noise model is such that

$$\varphi(B) = (1 - B^{12})$$

and

$$\theta(B) = (1-\theta_1 B)(1-\theta_2 B^{12}) \ .$$

Thus,

$$R(B)R(F) = \frac{(\sum_{j=0}^{11} B^j)(\sum_{j=0}^{11} F^j)}{(1-\theta_1 B)(1-\theta_2 B^{12})(1-\theta_1 F)(1-\theta_2 F^{12})}$$

$$= (\sum_{j=0}^{\infty} \pi_j B^j)(\sum_{j=0}^{\infty} \pi_j F^j) \tag{6.17}$$

so that from (A.10),

$$r_l = \sum_{j=0}^{\infty} \pi_j \pi_{j+l} \ .$$

The π_j can be obtained from the relationship

$$(1-\theta_1 B)(1-\theta_2 B^{12})\sum_{j=0}^{\infty} \pi_j B^j = \sum_{j=0}^{11} B^j \ .$$

By writing $\pi_j = 12n + m$, we find

$$\pi_{12n+m} = (1-\theta_1)^{-1}(\phi-\theta_2)^{-1}[(\theta_1)^{m+1}\{(1-\phi)\phi^n - (1-\theta_2)(\theta_2)^n\} + (\phi-\theta_2)(\theta_2)^n] \ ,$$
$$m = 0, \cdots, 11; n = 0, \cdots, \infty \tag{6.18}$$

where $\phi = (\theta_1)^{12}$.

From (6.18) and after some algebraic reduction, we obtain, on setting $l = 12k + s$,

$$r_{12k+s} = (1-\theta_1)^{-2}(1-(\theta_2)^2)^{-1} \cdot\left[12 - s(1-\theta_2) + \frac{\theta_1(1-\theta_2)^2}{1-(\theta_1)^2} \cdot\left(\frac{\phi(\theta_1)^{-s}}{1-\phi\theta_2} - \frac{(\theta_1)^s}{\phi-\theta_2}\right)\right](\theta_2)^k + (1-\theta_1)^{-2} \cdot(\phi-\theta_2)^{-1}(1-\phi\theta_2)^{-1}(1-(\theta_1)^2)^{-1} \cdot(1-\phi)^2(\theta_1)^{s+1}\phi^k \ , \tag{6.19}$$

$$s = 0, \cdots, 11; \quad k = 0, \cdots, \infty \ .$$

The resulting weight function for the Los Angeles data is shown in Figure A above the observations.

6.2 The General "Linear" Dynamic Model

The result in (6.5) can be readily extended to the case of more than one parameter. In the general dynamic model with k inputs in (4.1), letting

$$(\varphi(B)/\theta(B))(\omega_j(B)/\delta_j(B)) = \beta_j R_j(B) \tag{6.20}$$

we can write

$$Q(B)y_t = \sum_{j=1}^{k} \beta_j R_j(B)\xi_{tj} + a_t, \quad t = 1, \cdots, n \tag{6.21}$$

where, as before in (6.1), $Q(B) = \varphi(B)/\theta(B)$. Assuming that all the coefficients in $Q(B)$ and $R_j(B)$ are known and these $k+1$ power series converge for $|B| = 1$, the model is then linear in the k parameters $\boldsymbol{\beta} = (\beta_1, \cdots, \beta_k)'$. It readily follows that, for large n, the maximum likelihood estimator $\hat{\boldsymbol{\beta}}$ satisfies the normal equations

$$\mathbf{A}\hat{\boldsymbol{\beta}} = \mathbf{b} \qquad (6.22)$$

where $\mathbf{A}$ is a $k \times k$ matrix and $\mathbf{b}$ a $k \times 1$ vector such that

$$\mathbf{A} = [a_{ij}] \ , \quad a_{ij} = R_i(F)R_j(B)C_{\xi_i\xi_j}(0)$$
$$\mathbf{b} = (b_1, \cdots, b_k)'$$

with

$$b_j = R_j(F)Q(B)C_{\xi_j y}(0) \ ; \quad i, j = 1, \cdots, k \ .$$

In what follows, we investigate the special case having two parameters,

$$y_t = \{\beta_1\eta(B)B + \beta_2(1-B)^{-1}B\}P_t^{(T)} + (\theta(B)/\varphi(B))a_t \ . \qquad (6.23)$$

In this model, $\beta_1\eta(B)BP_t^{(T)}$, where $\eta(B)$ is assumed to converge for $|B| = 1$, measures the transient effect, and β_2 represents the eventual change in the level of the observations induced by the pulse input $P_t^{(T)}$ (see Figure B(e) for the special case $\eta(B) = (1-\delta B)^{-1}$). When $\beta_1 = 0$, the model reduces to that considered in (6.13). It is, therefore, of particular interest to know to what extent the nature and precision of the estimator of β_2 is affected by the presence of β_1. We again suppose that the noise term is nonstationary so that $\varphi(B)$ is divisible by $(1-B)$.

To facilitate comparison with the model (6.13) we again define a quantity $R(B)$ such that

$$Q(B) = (1-B)R(B)F \ ,$$

so that in (6.22)

$$R_1(B) = Q(B)\eta(B)B = R(B)\eta(B)(1-B)$$

and

$$R_2(B) = R(B) \ .$$

It follows that, provided $|\mathbf{A}| \neq 0$,

$$\hat{\beta}_1 = |\mathbf{A}|^{-1}\{a_{22}b_1 - a_{12}b_2\} \ , \qquad \hat{\beta}_2 = |\mathbf{A}|^{-1}\{a_{11}b_2 - a_{12}b_1\} \qquad (6.24)$$

where

$$|\mathbf{A}| = a_{11}a_{22} - (a_{12})^2 \ ,$$
$$b_1 = R(B)R(F)(1-F)\eta(F)(1-B)y_{T+1} \ ,$$
$$b_2 = R(B)R(F)(1-B)y_{T+1} \ ,$$

and a_{11}, a_{12} and a_{22} are, respectively, the coefficients of B^0 in the power series

$$R(B)R(F)\eta(B)\eta(F)(1-B)(1-F) \ ,$$
$$R(B)R(F)\eta(B)(1-B) \ ,$$
$$R(B)R(F) \ .$$

Some Properties of $\hat{\beta}_1$ and $\hat{\beta}_2$.

(i) Both b_1 and b_2 are linear functions of the observations y_t. By setting $B = F = 1$, the sum of the coefficients associated with y_t is zero for both of these functions. Thus, $\hat{\beta}_1$ and $\hat{\beta}_2$ are linear contrasts in y_t.

(ii) The estimator $\hat{\beta}_2$ can be expressed in the form

$$\hat{\beta}_2 = \sum_{l=0}^{\infty} \alpha_{1l}y_{T+1+l} - \sum_{l=0}^{\infty} \alpha_{2l}y_{T-l}$$

where

$$\sum_{l=0}^{\infty} \alpha_{1l} = \sum_{l=0}^{\infty} \alpha_{2l} = 1 \ , \qquad (6.25)$$

i.e., a contrast between two weighted averages, one of observations on or before the pulse input and the other afterward. To see this, since $\hat{\beta}_2$ is a linear contrast, it suffices to show that $\sum_{l=0}^{\infty} \alpha_{1l} = 1$.

From the expression for b_2 in (6.24), letting

$$G(B) = R(B)R(F) \ , \quad H(B) = 1 - B$$

and

$$b_2 = \sum_{l=-\infty}^{\infty} d_l y_{T+1-l}$$

it follows from (A.11) that $\sum_{l=-\infty}^{0} d_l = a_{22}$.

Further, making use of (A.12) and (A.13), we see that a_{12} in (6.24) is also the coefficient of B^0 in $R(B)R(F) \cdot (1-F)\eta(F)$. If we now set

$$G_1(B) = R(B)R(F)(1-F)\eta(F) \ , \quad H_1(B) = 1 - B$$

and

$$b_1 = \sum_{l=-\infty}^{\infty} d_l^* y_{T+1-l} \ ,$$

we then have $\sum_{l=-\infty}^{0} d_l^* = a_{12}$. The desired result follows since

$$\sum_{l=0}^{\infty} \alpha_{1l} = |\mathbf{A}|^{-1}\{a_{11}\sum_{l=-\infty}^{0} d_l - a_{12}\sum_{l=-\infty}^{0} d_l^*\} = 1 \ .$$

This property is similar to that of $\hat{\beta}$ in (6.15) for the model (6.13), except that the weight functions are no longer symmetrical. From least squares theory, we have

$$\hat{\beta}_2 = \hat{\beta} - (a_{12}/|\mathbf{A}|)(b_1 - a_{12}\hat{\beta}) \ , \qquad (6.26)$$

and the second term on the right side measures the effect of the presence of the term $\beta_1\eta(B)BP_t^{(T)}$ in the model.

(iii) One would expect that addition of the parameter β_1 to the model would reduce the precision with which β_2 could be estimated. A useful measure of the loss of information is the variance ratio $\mathrm{Var}(\hat{\beta}_2)/\mathrm{Var}(\hat{\beta})$ where it is understood that the denominator corresponds to the model in (6.13). Now

$$\mathrm{Var}(\hat{\beta}_2)/\mathrm{Var}(\hat{\beta}) = (1-\rho^2)^{-1}$$

where

$$\rho = a_{12}/(a_{11}a_{22})^{\frac{1}{2}} \ . \qquad (6.27)$$

We illustrate these results in terms of a specific example. Consider the case of (6.23) in which

$$\eta(B) = (1-\delta B)^{-1}, \ \varphi(B) = 1 - B \text{ and } \theta(B) = 1 - \theta B.$$

We find

$$\hat{\beta}_2 = \hat{\beta} - \frac{(1-\theta)(1+\delta)}{(\theta-\delta)} \sum_{l=0}^{\infty} [(1-\delta)\delta^l - (1-\theta)\theta^l] y_{T+1+l} \ , \quad (6.28)$$

where $\hat{\beta}$ is given in (6.16). In this case only the weights associated with the observations after the intervening pulse $P_t^{(T)}$ are affected by the presence of $\beta_1(1-\delta B)^{-1}BP_t^{(T)}$ in the model. The weight function is shown in Figure D for $\theta = .5$ and $\delta = .25$.

D. Comparison of Weights Associated with y_{T+1+l} $\hat{\beta}_2$ and for $\hat{\beta}$ ($\theta = .05$, $\delta = .25$, $l = 0, 1, 2, \ldots$)

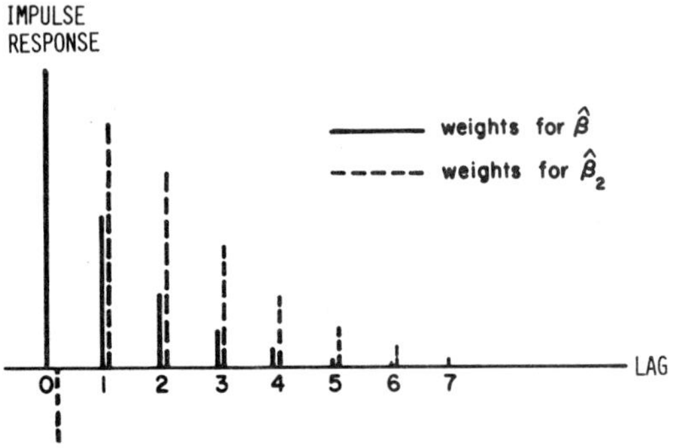

Also, for this model the variance ratio is

$$V = \text{Var}\,(\hat{\beta}_2)/\text{Var}\,(\hat{\beta})$$
$$= 1 + ((1-\theta)(1+\delta)/(1+\theta)(1-\delta)) \ . \quad (6.29)$$

The value of this ratio for various values of θ and δ is shown in the following tabulation:

	δ				
θ	$-.5$	$-.25$	0	$.25$	$.5$
$-.5$	2.00	2.80	4.00	6.00	10.00
$-.25$	1.56	2.00	2.67	3.78	6.00
0	1.33	1.60	2.00	2.67	4.00
.25	1.20	1.36	1.60	2.00	2.80
.5	1.11	1.20	1.33	1.56	2.00

Thus, the presence of β_1 in the model can cause large increases in the variance of $\hat{\beta}_2$, compared with $\hat{\beta}$, when θ is negative and δ is positive.

7. CONCLUDING REMARKS

In the past, much attention has been given to statistical analysis linking phenomena which are coincidental in time. In practice, it is perhaps more often the case that a response at a given point of time depends on events, both known and unknown, which have occurred not necessarily coincidentally but over the recent past. Statistical methods have, in a word, "lacked memory." The dynamic characteristics of both the transfer function and the noise parts of the model have tended to be ignored. The application of time series methods can amend this situation. This is illustrated in this article in the particular case where the object is to study the possible effect of interventions in the presence of dependent noise structure.

APPENDIX

We here state some useful results in the summation of series.

Lemma 1: Let $\{v_k\}_0^\infty$ be a sequence of numbers and let $\{x_t\}_{-\infty}^\infty$ and $\{y_t\}_{-\infty}^\infty$ be two sequences of numbers such that $x_t \equiv y_t \equiv 0$ for $t \leq 0$. If one of the following three double sums is absolutely convergent,

$$S_1 = \sum_{t=1}^{\infty} \sum_{k=0}^{\infty} x_t v_k y_{t-k} \ , \quad S_2 = \sum_{u=1}^{\infty} \sum_{k=0}^{\infty} y_u v_k x_{u+k} \ , \quad (A.1)$$

$$S_3 = \sum_{k=0}^{\infty} \sum_{u=1}^{\infty} v_k y_u x_{u+k},$$

the other two are absolutely convergent and

$$S_1 = S_2 = S_3 \ .$$

Proof of the lemma can be found in any standard text on infinite series.

It is convenient to express S_1, S_2 and S_3 in terms of the backshift operator B and its reciprocal, the forward shift operator $F = B^{-1}$. Letting

$$V(B) = \sum_{k=0}^{\infty} v_k B^k \quad \text{and} \quad V(F) = \sum_{k=0}^{\infty} v_k F^k \quad (A.2)$$

we can then write

$$S_1 = \sum_{t=1}^{\infty} x_t V(B) y_t \quad \text{and} \quad S_2 = \sum_{t=1}^{\infty} y_t V(F) x_t \ . \quad (A.3)$$

Further, suppose we define

$$C_{xy}(k) = \sum_{t=1}^{\infty} y_t x_{t-k} \ , \quad C_{yx}(k) = \sum_{t=1}^{\infty} x_t y_{t-k} \ , \quad k = 0, \pm 1, \pm 2, \cdots$$

so that

$$C_{xy}(k) = C_{yx}(-k) \ . \quad (A.4)$$

The quantity S_3 in (6.1) can be expressed as

$$S_3 = \sum_{k=0}^{\infty} v_k C_{xy}(-k) \ ,$$

and, by letting $C_{xy}(-k) = B^k C_{xy}(0)$, we have

$$S_3 = V(B) C_{xy}(0) \quad (A.5)$$

It follows that when the conditions of Lemma 1 hold,

$$\sum_{t=1}^{\infty} x_t V(B) y_t = \sum_{t=1}^{\infty} y_t V(F) x_t = V(B) C_{xy}(0) \ . \quad (A.6)$$

This result can be readily extended to the following:

Lemma 2: Suppose $W(B) = V_1(B) + V_2(F)$ where $V_1(B)$ and $V_2(F)$ are two power series in B and F, respectively, such that the sum $\sum_{t=1}^{\infty} x_t W(B) y_t$ is absolutely convergent. Then

$$\sum_{t=1}^{\infty} x_t W(B) y_t = W(B) C_{xy}(0) \ . \quad (A.7)$$

Lemma 3: Let $G(B) = \sum_{j=-\infty}^{\infty} g_j B^j$ and $H(B) = \sum_{k=-\infty}^{\infty} h_k B^k$ be two power series in B and converge for $|B| = 1$, and let $D(B)$

$= G(B)H(B)$. Then

$$D(B) = \sum_{l=-\infty}^{\infty} d_l B^l \tag{A.8}$$

where $$d_l = \sum_{j=-\infty}^{\infty} g_j h_{l-j} \ .$$

In particular

(i) if $g_j = g_{-j}$ and $h_k = h_{-k}$, then

$$d_l = d_{-l} = \sum_{u=0}^{\infty} h_u g_{u+l} + \sum_{u=1-l}^{\infty} g_u h_{u+l} \ , l = 0, \cdots, \infty \ ; \tag{A.9}$$

(ii) if $g_j = 0, j \leq -1$ and $H(B) = G(F)$, then

$$d_l = d_{-l} = \sum_{j=0}^{\infty} g_j g_{j+l} \ , \quad l = 0, \cdots, \infty \ ; \tag{A.10}$$

(iii) if $H(B) = 1 - B$, then

$$d_l = g_l - g_{l-1} \qquad l = 0, \pm 1, \cdots, \pm\infty \ , \tag{A.11}$$

so that $\sum_{l=1}^{\infty} d_l = -g_0$ and $\sum_{l=-\infty}^{0} d_l = g_0$;

(iv) if $g_j = g_{-j}$ and $h_j = 0 \quad j \leq -1$, then

$$d_0 = \sum_{j=0}^{\infty} h_j g_j \ ; \tag{A.12}$$

(v) if $g_j = g_{-j}$ and $h_j = 0 \quad j \geq 1$, then

$$d_0 = \sum_{j=-\infty}^{0} h_j g_j \ . \tag{A.13}$$

[Received October 1973. Revised August 1974.]

REFERENCES

[1] Box, G.E.P. and Jenkins, G.M., *Time Series Analysis, Forecasting and Control*, San Francisco: Holden-Day, Inc., 1970.

[2] ——— and Tiao, G.C., "A Change in Level of a Non-stationary Time Series," *Biometrika*, 52 (June 1965), 181–92.

[3] ——— and Tiao, G.C., *Bayesian Inference in Statistical Analysis*, Reading, Mass.: Addison-Wesley Publishing Co., 1973.

[4] Feige, E. and Pearce, D.K., "Inflation and Income Policy: An Application of Time Series Models," Report #7318, Social Systems Research Institute, University of Wisconsin, Madison, 1973.

[5] Glass, G.V., "Estimating the Effects of Intervention into a Nonstationary Time Series," *American Educational Research Journal*, 9, No. 3 (1972), 463–77.

[6] Tiao, G.C., Box, G.E.P. and Hamming, W.J., "Analysis of Los Angeles Photochemical Smog Data: A Statistical Overview," *Journal of Air Pollution Control Association*, 25 (March 1975).

3.10

Identification of Dynamic Regression (Distributed Lag) Models Connecting Two Time Series

LARRY D. HAUGH and G. E. P. BOX*

A methodology is introduced for identifying dynamic regression or distributed lag models relating two time series. First, specification of a bivariate time-series model is discussed, and its relationship to the usual dynamic regression model is indicated. Then, a two-stage identification procedure is presented which involves fitting univariate time-series models to each series, and identifying a dynamic shock model relating the two univariate model innovation series. The models obtained at these two stages are combined to identify a dynamic regression model, which may then be fitted in the usual ways. Two systems of economic time series illustrate the methodology.

KEY WORDS: Rational distributed lag; Identification of distributed lag model; Bivariate time-series model; Autoregressive moving average model; Cross-correlation function; Univariate innovations.

1. INTRODUCTION

In time-series forecasting work there arises the question as to whether it would be useful to consider a related series x when forecasting a series y (e.g., is x a leading indicator of y?). More generally the question arises as to how two time series are related to each other. If the two series can be transformed (e.g., by differencing) in such a way that they are jointly covariance stationary, then their interrelationships can most often be usefully described by either their cross-correlation function or cross spectrum. It will be assumed in this discussion that both of the series x and y are observed at regular sampling intervals and can be adequately described by mixed autoregressive integrated moving average models of the sort described, for example, in [3].

If it is felt that x leads y, then one may attempt to build a dynamic regression model relating the two series with y as the dependent variable, so that the fitted model could be used to forecast y. By a dynamic regression model, or distributed lag model, we mean a regression of y_t, y at time t, on the present and past values of x, x_s, for $s \leq t$. The model form of primary interest is that of a rational distributed lag function of y on x plus an independent noise term of the general mixed autoregressive moving average type [3]:

$$\begin{aligned} y_t &= \omega(B)\delta(B)^{-1}x_t + \theta(B)\phi(B)^{-1}a_t \\ &= (\omega_0 - \omega_1 B - \ldots - \omega_s B^s)(1 - \delta_1 B - \ldots - \delta_r B^r)^{-1}x_t \\ &\quad + (1 - \theta_1 B - \ldots - \theta_q B^q) \\ &\quad \cdot (1 - \phi_1 B - \ldots - \phi_p B^p)^{-1}a_t \qquad (1.1) \end{aligned}$$

where B is the lag operator $Bx_t = x_{t-1}$, and the roots of the polynomials $\delta(B)$, $\theta(B)$, and $\phi(B)$ lie outside the unit circle. The mean values μ_x and μ_y, of x and y, will often be ignored for convenience in this paper, but in practice the y and x values in (1.1) may be deviations from their mean values.

1.1 The Usual Cross-Correlation Estimator

Given that the two series are jointly, and hence individually, covariance stationary there still exists a wide range of possible relationships between the series. Traditionally the most common method of describing the interrelationship has been through the lagged cross-correlation function.

The cross-correlation function $\rho_{xy}(\cdot)$ at lag k is defined as

$$\rho_{xy}(k) = E\{(x_t - \mu_x)(y_{t+k} - \mu_y)\}/\sigma_x\sigma_y \qquad (1.2)$$

where μ and σ denote mean and standard deviation. In practice $\rho_{xy}(k)$ can be estimated by

$$r_{xy}(k) = c_{xy}(k)c_x(0)^{-\frac{1}{2}}c_y(0)^{-\frac{1}{2}} \qquad (1.3)$$

where

$$\begin{aligned} c_{xy}(k) &= N^{-1} \sum_{t=1}^{N-k} (x_t - \bar{x})(y_{t+k} - \bar{y}) , \quad k \geq 0 \\ &= N^{-1} \sum_{t=1-k}^{N} (x_t - \bar{x})(y_{t+k} - \bar{y}) , \quad k < 0 , \end{aligned}$$

$$c_x(0) = N^{-1} \sum_{t=1}^{N} (x_t - \bar{x})^2 ,$$

$$c_y(0) = N^{-1} \sum_{t=1}^{N} (y_t - \bar{y})^2 .$$

* Larry D. Haugh is Assistant Professor, Statistics Program, University of Vermont, Burlington, VT 05401. G.E.P. Box is R.A. Fisher Professor, Statistics Department, University of Wisconsin, Madison, WI 53706. Much of the research by the first author was done at the Universities of Wisconsin and Essex. The authors wish to acknowledge financial support from a Fulbright-Hays Grant, the General Electric Foundation, and the Air Force Office of Scientific Research (AF-AFOSR-69-1803). The detailed referee reports were helpful in revision and are also gratefully acknowledged.

The Collected Works of George E. P. Box, 1984; Wadsworth, Inc., Belmont, CA 94002.
Originally published in *J. Amer. Stat. Assoc.,* vol. 72, no. 357 (1977), pp. 121–130.

However, in practice it can be difficult to interpret such an estimate of the cross-correlation function. For example, when each of the series x and y are themselves autocorrelated, then the lagged cross-correlation estimates can have high variance and the estimates at different lags can be highly correlated with one another [1]. One may be misled in such situations by attributing some significance to apparent patterns in the cross-correlation function, which in fact are a result of the sampling properties of the estimates used (e.g., [4]). This can happen even if the two series are in fact independent, as will be shortly discussed. The approach developed here will help to alleviate these interpretive problems of cross-correlation analysis.

1.2 Motivation for a Univariate Innovations Approach

For motivation consider the problem of forecasting y_{N+k} with $k > 0$ given $\{y_N, y_{N-1}, \ldots\}$. Assume that we may construct a univariate forecasting model for y of the form

$$\phi_y(B)(1 - B)^{d_2}y_t = \theta_y(B)u_{yt} \tag{1.4}$$

where d_2 is the order of differencing required for stationarity, u_{yt} is white noise,

$$\begin{aligned}\phi_y(B) &= 1 - \phi_{y1}B - \ldots - \phi_{yp}B^p , \\ \theta_y(B) &= 1 - \theta_{y1}B - \ldots - \theta_{yq}B^q .\end{aligned} \tag{1.5}$$

In this model the univariate innovations (or shocks) u_{yt} are also the theoretical one-step-ahead prediction errors made when forecasting y_t from its own past $\{y_{t-1}, y_{t-2}, \ldots\}$.

To decide when the x series as well as the past history of y will be useful in forecasting y_{N+k}, it is natural to consider whether or not x and u_y are related. If x_t and $u_{y(t+k)}$ are correlated for some positive k, then the forecast of y_{N+k} should be improved by taking account of this relationship—if the forecast error can itself be forecast, then prediction may be improved.

Just as it was useful above to think in terms of the white-noise or driving series u_y instead of the series y itself, we may also consider the white-noise series u_x driving x. The word "driving" is now being used in the sense that x can be described as y was in (1.4):

$$\phi_x(B)(1 - B)^{d_1}x_t = \theta_x(B)u_{xt} \tag{1.6}$$

where the parameters involved are defined as in (1.4). Equivalently we may write

$$x_t = V_x(B)u_{xt} \tag{1.7}$$

where

$$V_x(B) = \theta_x(B)/(1 - B)^{d_1}\phi_x(B) . \tag{1.8}$$

In this sense x is the output of a linear system (described by $V_x(B)$) having input or driving force u_x. For example, if the u_x series were "shut off," or equivalently if there were no forecast errors to be made in predicting x_{N+k} from its own past, then x would always stay at its mean value. In summary we have the two univariate time-series models

$$u_{xt} \rightarrow V_x(B) = \theta_x(B)/(1 - B)^{d_1}\phi_x(B) \rightarrow x_t ,$$

and

$$u_{yt} \rightarrow V_y(B) = \theta_y(B)/(1 - B)^{d_2}\phi_y(B) \rightarrow y_t ,$$

where reference to the series means is suppressed.

Although we desire to relate x with u_y to determine the value of using x in predicting y, it would be more convenient in fact to relate u_x to u_y. This is because the distribution of cross-correlation estimator $r_{u_xu_y}(\cdot)$ is simpler than that of $r_{xu_y}(\cdot)$ when x and y are independent and hence the interpretation of such estimates will be simpler.

1.3 Cross Correlation of the Univariate Innovation Series

To describe the relationship existing between x and y we will consider the cross-correlation function, $\rho_{u_xu_y}(\cdot)$, of u_x and u_y instead of considering $\rho_{xy}(\cdot)$. In practice the series u_x and u_y are unobservable and we only have available the two residual series $\hat{u}_x$ and $\hat{u}_y$, which can be obtained by efficiently fitting the two univariate models (1.4) and (1.6) to the data available. Then $\rho_{u_xu_y}(k)$ is estimated by $r_{\hat{u}_x\hat{u}_y}(k)$.

This idea of considering the associated univariate white-noise series in identifying time-series relationships is not entirely new (see [3, 7, 12, 14, 18, 19]) although the following method (more completely described in [9]) of identifying a model relating x to y is different than those considered before.

1.4 Asymptotic Distribution of Cross-Correlation Estimates $r_{xy}(\cdot)$

It is known that a set of cross-correlation estimates $\{r_{xy}(k)\}$ at a fixed number of chosen lags k are consistent estimators for $\{\rho_{xy}(k)\}$ and are asymptotically normally distributed under reasonable assumptions on x and y (e.g., [8]).

Assuming that x and y are jointly covariance stationary and normally distributed, Bartlett [2] presents the asympotic covariance between two cross-correlation estimates:

$$\begin{aligned}&\operatorname{cov}\{r_{xy}(k), r_{xy}(k + \ell)\} \\ &\cong (N - k)^{-1} \sum_{\nu=-\infty}^{\infty} \{\rho_x(\nu)\rho_y(\nu + \ell) \\ &\quad + \rho_{xy}(-\nu)\rho_{xy}(\nu + 2k + \ell) + \rho_{xy}(k)\rho_{xy}(k + \ell) \\ &\quad \cdot[\rho_{xy}(\nu)^2 + \tfrac{1}{2}\rho_x(\nu)^2 + \tfrac{1}{2}\rho_y(\nu)^2] \\ &\quad - \rho_{xy}(k)[\rho_x(\nu)\rho_{xy}(\nu + k + \ell) \\ &\quad + \rho_{xy}(-\nu)\rho_y(\nu + k + \ell)] - \rho_{xy}(k + \ell) \\ &\quad \cdot[\rho_x(\nu)\rho_{xy}(\nu + k) + \rho_{xy}(-\nu)\rho_y(\nu + k)]\} .\end{aligned} \tag{1.9}$$

This relation indicates that the covariance pattern of the cross correlations can be quite complicated, depending both on $\rho_x(\cdot)$ and $\rho_y(\cdot)$. In particular if one were to judge the size of individual estimates by comparison to a standard error of $N^{-\frac{1}{2}}$ (or $(N-k)^{-\frac{1}{2}}$), one could be greatly misled.

When the series are in fact independent, the normality assumption is no longer required, and (1.9) simplifies somewhat, but it will still depend on $\rho_x(\cdot)$ and $\rho_y(\cdot)$:

$$\operatorname{cov}\{r_{xy}(k), r_{xy}(k+\ell)\} \cong (N-k)^{-1} \sum_{\nu=-\infty}^{\infty} \rho_x(\nu)\rho_y(\nu+\ell) \quad . \tag{1.10}$$

Alternatively, if the series x and y are each white noise, the formula again simplifies but will still involve $\rho_{xy}(\cdot)$.

Therefore it will be quite easy to visually interpret the cross-correlation estimates only in the situation of two independent white-noise series, in which case we have

$$\begin{aligned} \operatorname{cov}\{r_{xy}(k), r_{xy}(k+l)\} &\cong o(N^{-1}) \ , \\ \operatorname{var}\{r_{xy}(k)\} &\cong (N-k)^{-1} \ . \end{aligned} \tag{1.11}$$

The estimates at various lags in this case may be judged individually against an approximate standard error of $N^{-\frac{1}{2}}$ (or $(N-k)^{-\frac{1}{2}}$).

Fortunately, this latter situation is of practical importance because at the identification stage of the model building process one is often interested in comparing these cross-correlation estimates with benchmarks appropriate to the null hypothesis of series independence. For example, when identifying a univariate time-series model by its sample autocorrelation function one initially judges the size of these estimates against an approximate (asymptotic) standard deviation of $N^{-\frac{1}{2}}$, which is appropriate for a white-noise series. (Of course one would also be interested in any overall pattern that may emerge from the plot of these estimates.)

1.5 Asymptotic Distribution of Cross-Correlation Estimates $r_{\hat{u}_x\hat{u}_y}(\cdot)$

Using (1.11) it is known that $\{r_{u_xu_y}(k)\}$ are asymptotically uncorrelated and normally distributed with variance N^{-1}, but what of $\{r_{\hat{u}_x\hat{u}_y}(k)\}$ which are the estimates that must be used in practice? It is established in [9, 11] that if x and y are independent, $\{r_{\hat{u}_x\hat{u}_y}(k)\}$ has the same asymptotic distribution as $\{r_{\hat{u}_x\hat{u}_y}(k)\}$. Further, in these references some Monte Carlo work indicates the practical utility of the asymptotic distributional results for various first-order autoregressive and moving average models (for x and y) and for sample sizes of $N = 50{,}100{,}200$.

1.6 A Two-Stage Identification Procedure

It is proposed to identify the relationship between two series, x and y, by characterizing separately each of their univariate models, and the relationship between the two univariate residual series driving each time series. Thus at the first stage, appropriate autoregressive integrated moving average processes are fitted to each of x and y. The residual series $\hat{u}_x$ and $\hat{u}_y$ from these fits are then cross correlated in the usual way. Based on the appearance of this cross-correlation function, $r_{\hat{u}_x\hat{u}_y}(\cdot)$, a tentative dynamic shock model is identified which relates u_x to u_y. By recombining the two univariate models for x and y with the identified model connecting u_x and u_y, a distributed lag model relating x to y may be identified. Once a dynamic regression model is identified, it may be fitted and checked using the methodology, e.g., of [3].

2. SPECIFICATION OF BIVARIATE MODELS RELATING X AND Y

Since x and y are assumed to be jointly covariance stationary, we may consider their combined covariance generating function $\mathbf{\Gamma}(B)$ where

$$\begin{aligned} \mathbf{\Gamma}(B) &= \sum_{k=-\infty}^{\infty} \mathbf{\Gamma}_k B^k \\ &= \Big(\sum_{k=-\infty}^{\infty} \gamma_{\ell m}(k) B^k\Big)_{2\times 2} \\ \mathbf{\Gamma}_k &= (\gamma_{\ell m}(k)) \ , \quad \ell, m = 1, 2 \ , \\ \gamma_{11}(k) &= E(x_t - \mu_x)(x_{t+k} - \mu_x) \ , \\ \gamma_{12}(k) &= \gamma_{21}(-k) = E(x_t - \mu_x)(y_{t+k} - \mu_y) \ , \\ \text{and} \quad \gamma_{22}(k) &= E(y_t - \mu_y)(y_{t+k} - \mu_y) \ . \end{aligned} \tag{2.1}$$

We now assume that $\mathbf{\Gamma}$ will be rational; i.e., we suppose that each element of the 2×2 matrix $\mathbf{\Gamma}$ is a rational function of B. In the univariate case this is equivalent to assuming that the autocorrelation function of a time series can be modeled exactly by an appropriate mixed autoregressive moving average process. We also assume that det $(\mathbf{\Gamma}(B))$ has zeroes lying on or outside the unit circle; which assures that the joint (x, y) process is of full rank. In practice this is a very reasonable restriction, eliminating from consideration, for example, processes in which both series are transformations of the same white-noise process.

The key result allowing unique specification of a model relating x to y in terms of the two series' joint covariance function is the factorization of $\mathbf{\Gamma}$ [20]. For example, we may factorize $\mathbf{\Gamma}$ as

$$\mathbf{\Gamma} = \mathbf{\Phi}(F)\mathbf{\Sigma}_\epsilon \mathbf{\Phi}(B)^T \tag{2.2}$$

where $F = B^{-1}$, $\mathbf{\Phi}(0) = \mathbf{I}_{2\times 2}$, and $\mathbf{\Sigma}_\epsilon$ is the covariance matrix of one-step-ahead forecast errors obtained with optimal mean square prediction. The factorization is unique if we suppose that the denominator polynomials (in B) of each element of $\mathbf{\Phi}$ have zeroes lying outside the unit circle, if the processes are jointly of full rank, and if det $(\mathbf{\Phi}(B))$ has zeroes lying on or outside the unit circle.

In the univariate case this means that we may uniquely factorize the covariance generating function of a rational process as

$$\Gamma(B) = V(F)V(B)\sigma^2 \tag{2.3}$$

where $V(B) = \theta(B)\phi(B)^{-1}$, $\theta(B) = 1 - \theta_1 B - \ldots - \theta_q B^q$, and $\phi(B) = 1 - \phi_1 B - \ldots - \phi_p B^p$. The uniqueness conditions mentioned imply here that $\phi(B)$ should have zeroes lying outside the unit circle, and that the zeroes of $\theta(B)$ should lie outside or on the unit circle.

We still further restrict ourselves to invertible processes, in which the zeroes of $\theta(B)$, or of det $(\mathbf{\Phi}(B))$, in general, may not lie on the unit circle. Thus we have exactly the conditions of invertibility and stationarity discussed more extensively in [3, Ch. 3] for univariate series.

The factorization of $\mathbf{\Gamma}$ produces a unique $\mathbf{\Phi}$ and hence a unique process model generating the observed covariance structure

$$\begin{pmatrix} x_t \\ y_t \end{pmatrix} = \mathbf{\Phi}(B) \begin{pmatrix} \epsilon_{1t} \\ \epsilon_{2t} \end{pmatrix} \tag{2.4}$$

with cov $(\boldsymbol{\varepsilon}_t) = \mathbf{\Sigma}_\epsilon$.

Although this form of specification is probably the most commonly used one in multiple time-series analysis, another form is most convenient for specifying the usual dynamic regression model. We would like the covariance matrix of $\boldsymbol{\varepsilon}_t$ to be diagonal.

$$\text{cov}\,(\boldsymbol{\varepsilon}_t) = \begin{pmatrix} \sigma_1^2 & 0 \\ 0 & \sigma_2^2 \end{pmatrix}. \tag{2.5}$$

With this error specification we no longer will have a joint transfer function $\mathbf{\Phi}(B)$ with $\mathbf{\Phi}(0) = \mathbf{I}_2$. Instead we may choose to make a nonsingular transformation specified by $\mathbf{C}$ such that

$$\begin{pmatrix} x_t \\ y_t \end{pmatrix} = \mathbf{\Phi}(B)\boldsymbol{\varepsilon}_t = \mathbf{\Phi}(B)\mathbf{C}\mathbf{C}^{-1}\boldsymbol{\varepsilon}_t \tag{2.6}$$

and $\mathbf{C}^{-1}\boldsymbol{\varepsilon}_t$ has a diagonal covariance matrix.

Lemma 1: Given the positive definite matrix $\mathbf{\Sigma}_\epsilon$ there exists a unique nonsingular matrix $\mathbf{C}$, which is lower triangular with one's on the diagonal, such that $\mathbf{C}\mathbf{\Sigma}_\epsilon\mathbf{C} = \mathbf{\Sigma}$, where $\mathbf{\Sigma}$ is diagonal. (The proof is straightforward, e.g., [9].)

Therefore we may obtain a unique representation for x and y as

$$\begin{pmatrix} x_t \\ y_t \end{pmatrix} = \mathbf{V}(B)\mathbf{a}_t \tag{2.7}$$

where $\mathbf{V}(0) = \mathbf{V}_0$ is lower triangular with one's on the diagonal and $\mathbf{C}^{-1}\boldsymbol{\varepsilon}_t = \mathbf{a}_t$ so that cov $(\mathbf{a}) = \mathbf{\Sigma}_a$ is diagonal. In turn, $\mathbf{V}(B)$ may be described uniquely as

$$\begin{aligned} \mathbf{V}(B) &= \begin{pmatrix} V_{11}(B) & V_{12}(B) \\ V_{21}(B) & V_{22}(B) \end{pmatrix} \mathbf{a}_t \\ &= \begin{pmatrix} \theta_{11}(B)/\phi_{11}(B) & C_{12}\theta_{12}(B)/\phi_{12}(B) \\ C_{21}\theta_{21}(B)/\phi_{21}(B) & \theta_{22}(B)/\phi_{22}(B) \end{pmatrix} \mathbf{a}_t \end{aligned} \tag{2.8}$$

where $\{\phi_{ij}(B)\}$ are polynomials in B with leading term 1 and roots outside the unit circle, $\theta_{11}(B)$ and $\theta_{22}(B)$ have leading terms 1 while $\theta_{12}(0) = 0$, and the roots of det $(\mathbf{V}(B))$ lie outside the unit circle. This parameterization has the advantage that all changes in variable scales will be absorbed by the $\{C_{ij}, \sigma_{a_i}^2\}$, leaving the structural parameters $\{\phi_{ij}(B), \theta_{ij}(B)\}$ unchanged.

2.1 Covariance Structure of the Bivariate Model

The covariance structure of (2.7) will be seen to take an appealing form if use is made of the following lemmas, whose proofs are straightforward [9].

Lemma 2: Consider x and y as jointly covariance stationary time series, where $x_t = V(B)a_t$, $y_t = W(B)\epsilon_t$, and both a and ϵ are white-noise processes. Then $\gamma_{xy}(k) = V(F)W(B)\gamma_{a\epsilon}(k)$, where B and F now operate on the lag k.[1]

Lemma 3: With x, y as in Lemma 2, $V(B)\gamma_{xy}(k) = V(F)\gamma_{yx}(-k)$.

Using Lemma 2 and the fact that the innovations of the specification (2.7) are independent, the covariance structure of x and y may be conveniently written

$$\begin{aligned} \gamma_x(k) &= V_{11}(F)V_{11}(B)\gamma_{a_1}(k) + V_{12}(F)V_{12}(B)\gamma_{a_2}(k)\,, \\ \gamma_y(k) &= V_{21}(F)V_{21}(B)\gamma_{a_1}(k) + V_{22}(F)V_{22}(B)\gamma_{a_2}(k)\,, \\ \gamma_{xy}(k) &= V_{11}(F)V_{21}(B)\gamma_{a_1}(k) + V_{12}(F)V_{22}(B)\gamma_{a_2}(k)\,, \end{aligned} \tag{2.9}$$

where $\gamma_{a_i}(k) = 0$ for $k \neq 0$, and $\gamma_{a_i}(0) = \sigma_{a_i}^2$.

2.2 Univariate Model Stage of Identification

With the assumptions made concerning x and y it is possible to model each series individually as

$$\begin{aligned} x_t &= V_x(B)u_{xt} = \theta_x(B)\phi_x(B)^{-1}u_{xt} \\ y_t &= V_y(B)u_{yt} = \theta_y(B)\phi_y(B)^{-1}u_{yt} \end{aligned}$$

which may be written as

$$\begin{pmatrix} x_t \\ y_t \end{pmatrix} = \begin{pmatrix} V_x(B) & 0 \\ 0 & V_y(B) \end{pmatrix} \begin{pmatrix} u_{xt} \\ u_{yt} \end{pmatrix}. \tag{2.10}$$

However, the joint process (u_x, u_y) is not bivariate white noise, since u_x and u_y may be cross correlated at nonzero lags.

2.3 Bivariate Model for u_x and u_y

The general bivariate model for u_x and u_y is just a special case of the general model (2.7, 2.8). Of course it is a very interesting case in that the autocorrelation functions of the white-noise series u_x and u_y are particularly simple. We may write

$$\mathbf{u}_t = \begin{pmatrix} u_{xt} \\ u_{yt} \end{pmatrix} = \mathbf{W}(B)\mathbf{a}_t = \begin{pmatrix} W_{11}(B) & W_{12}(B) \\ W_{21}(B) & W_{22}(B) \end{pmatrix} \begin{pmatrix} a_{1t} \\ a_{2t} \end{pmatrix} \tag{2.11}$$

where as for (2.8) we may take, for i, $j = 1$, 2 and $\theta_{12}(0) = 0$,

$$W_{ii}(B) = \theta_{ii}(B)/\phi_{ii}(B)\,,$$

and

$$W_{ij}(B) = C_{ij}\theta_{ij}(B)/\phi_{ij}(B)\,. \tag{2.12}$$

The cross-covariance structure is like that of (2.9),

$$\begin{aligned} \gamma_{u_x u_y}(k) = W_{11}(F)W_{21}(B)\gamma_{a_1}(k) \\ + W_{12}(F)W_{22}(B)\gamma_{a_2}(k)\,, \end{aligned} \tag{2.13}$$

[1] An application of this lemma has appeared in [19], there providing a justification for the method of covariance contraction.

but the autocovariance structure is simply $\gamma_{u_x}(k) = \gamma_{u_y}(k) = 0$ for $k \neq 0$.

2.4 Combination of the Univariate and Joint Univariate Shock Models

A complete model for x and y is formed by combining (2.10) and (2.11).

$$\begin{pmatrix} x_t \\ y_t \end{pmatrix} = \begin{pmatrix} V_x(B) & 0 \\ 0 & V_y(B) \end{pmatrix} \begin{pmatrix} u_{xt} \\ u_{yt} \end{pmatrix}$$
$$= \begin{pmatrix} V_x(B) & 0 \\ 0 & V_y(B) \end{pmatrix} \begin{pmatrix} W_{11}(B) & W_{12}(B) \\ W_{21}(B) & W_{22}(B) \end{pmatrix} \begin{pmatrix} a_{1t} \\ a_{2t} \end{pmatrix}$$
$$= \begin{pmatrix} V_x(B)W_{11}(B) & V_x(B)W_{12}(B) \\ V_y(B)W_{21}(B) & V_y(B)W_{22}(B) \end{pmatrix} \begin{pmatrix} a_{1t} \\ a_{2t} \end{pmatrix}.$$

No claim is made that the factorization of $\mathbf{V}(B)$ as $\begin{pmatrix} V_x(B) & 0 \\ 0 & V_y(B) \end{pmatrix} \mathbf{W}(B)$ will always be the most parsimonious for identification or structural description, but it is believed that this two-stage analysis will lead to a more easily interpretable cross-correlation analysis in many situations. It has been our experience in fact, as with others [17], that after we cross correlate different macroeconomic series with varying univariate model complexity, the cross-correlation function of the univariate residual series typically reveals a quite simple structure for $\mathbf{W}(B)$.

3. IDENTIFICATION OF THE JOINT UNIVARIATE SHOCK MODEL

To check how u_x and u_y may be cross correlated, one could compute $r_{\hat{u}_x\hat{u}_y}(\cdot)$. Hopefully any pattern appearing in $r_{\hat{u}_x\hat{u}_y}(\cdot)$ will lead one to identify a bivariate model of the form (2.11). The identification process in practice is very similar to that found to be useful in [3]. By knowing the cross-correlation patterns appropriate to various bivariate models of interest one may match $r_{\hat{u}_x\hat{u}_y}(\cdot)$ to the true cross-correlation patterns $\rho_{u_xu_y}(\cdot)$ for such models.

Our chief interest in this discussion is that of dynamic regression (x leads to y), although some discussion is given in [9] for more general bivariate feedback models in which x leads y and also y leads x. A result of great aid in identifying models in the dynamic regression case is the fact that y may be expressed as a distributed lag function of present and past values of x when $V_{12}(B) = 0$. In terms of $\rho_{u_xu_y}(\cdot)$ this is equivalent to having $\rho_{u_xu_y}(k) = 0$ for negative lags k.

Thus when $r_{\hat{u}_x\hat{u}_y}(\cdot)$ seems to indicate no feedback effect, in that no significant cross correlation occurs at negative lags, a dynamic regression model may be identified as

$$\begin{pmatrix} x_t \\ y_t \end{pmatrix} = \begin{pmatrix} V_x(B) & 0 \\ 0 & V_y(B) \end{pmatrix} \begin{pmatrix} W_{11}(B) & 0 \\ W_{21}(B) & W_{22}(B) \end{pmatrix} \mathbf{a}_t$$
$$= \begin{pmatrix} V_x(B)W_{11}(B) & 0 \\ V_y(B)W_{21}(B) & V_y(B)W_{22}(B) \end{pmatrix} \mathbf{a}_t \, . \quad (3.1)$$

Hence in this case $W_{11}(B) = 1$, $a_{1t} = u_{xt}$, and $a_{2t} = a_t$, so that

$$\begin{pmatrix} x_t \\ y_t \end{pmatrix} = \begin{pmatrix} V_x(B) & 0 \\ V_y(B)W_{21}(B) & V_y(B)W_{22}(B) \end{pmatrix} \begin{pmatrix} u_{xt} \\ a_t \end{pmatrix}$$

and

$$y_t = V_y(B)W_{21}(B)V_x(B)^{-1}x_t + V_y(B)W_{22}(B)a_t \, . \quad (3.2)$$

The dynamic regression model (3.2) is of course the usual rational distributed lag model with mixed autoregressive moving average noise (1.1),

$$y_t = \omega(B)\delta(B)^{-1}x_t + \theta(B)\phi(B)^{-1}a_t \quad (3.3)$$
$$= V(B)x_t + \psi(B)a_t \, .$$

Therefore the distributed lag transfer function, $\omega(B)/\delta(B)$, for such a model form can be identified as $V_y(B)W_{21}(B)V_x(B)^{-1}$ and the noise model, $\theta(B)\phi(B)^{-1}$, can be identified as $V_y(B)W_{22}(B)$.

The fact that when there is no feedback effect from y to x, $\rho_{u_xu_y}(k) = 0$ for negative lags k, points out another advantage that the proposed two-stage identification approach has. Often it is the case that *a priori* one is not able to exclude the possibility of feedback effects from y to x. If one looked at $r_{xy}(\cdot)$ in an attempt to check feedback, one could easily be misled, since $\rho_{xy}(\cdot)$ can have nonzero values at negative lags even when there is no feedback present. This can be seen from the cross-covariance function $\gamma_{xy}(\cdot)$, now ignoring mean levels,

$$\gamma_{xy}(k) = Ex_t(\omega(B)\delta(B)^{-1}x_{t+k} + \theta(B)\phi(B)^{-1}a_{t+k})$$
$$= \omega(B)\delta(B)^{-1}\gamma_x(k)$$
$$= \theta_x(F)\phi_x(F)^{-1}\omega(B)\delta(B)^{-1}\theta_x(B)\phi_x(B)^{-1}\gamma_{u_x}(k) \quad (3.4)$$

since

$$\gamma_x(k) = \theta_x(F)\phi_x(F)^{-1}\theta_x(B)\phi_x(B)^{-1}\gamma_{u_x}(k) \, . \quad (3.5)$$

So unless $\theta_x(B) = \phi_x(B) = 1$, $\rho_{xy}(\cdot)$ can have nonzero values for $k < 0$, as is indicated by the factor $\theta_x(F)\phi_x(F)^{-1}$ in (3.4).

An alternative approach to remedying the defect in $\rho_{xy}(\cdot)$ just mentioned concerning feedback detection, is that of just prewhitening the x series. One could then consider $\rho_{u_xy}(\cdot)$ to check feedback effects, although there is some advantage in cross correlating u_x with a transformed version of y instead, $\phi_x(B)\theta_x(B)^{-1}y_t$, so as to more clearly identify $\omega(B)/\delta(B)$ [3, Ch. 11].

3.1 Four Cases of Primary Interest

Depending upon the type of pattern in $\rho_{u_xu_y}(\cdot)$ four cases of primary interest in model identification can arise [10, 18].

a. The two series are uncorrelated at all lags k (when the two series are normally distributed, they will be independent in this case), and the x series will not aid in the forecasting of y.
b. $\rho_{u_xu_y}(0) \neq 0$, but the two series are uncorrelated at all other nonzero lags k. This is a peculiar case in that one could build a dynamic regression model for y on x as well as a dynamic regression model of x on y. However, there will be no improvement in our ability to forecast y using the past of x beyond that already achieved by using the past of y.

c. There is at least one nonzero cross correlation $\rho_{u_x u_y}(k)$ for $k > 0$, and there is no cross correlation at negative lags k. One may build a dynamic regression model for y on x and thereby improve the forecastability of y.
d. There exist nonzero cross-correlation coefficients $\rho_{u_x \rho_{u_y}}(k)$ for some positive k as well as for some negative k. This would indicate feedback from y to x, and it would not be possible to build a dynamic regression model of y on the present and past of x. Of course more general bivariate time-series models may be fitted, if desired, but identification of such models will not be considered here.

3.2 Identification of a Dynamic Shock Model

In either (b) or (c) we may attempt to identify a model connecting u_x and u_y which will lead to a dynamic regression model (1.1) connecting x and y. In this case $\rho_{u_x u_y}(\cdot)$ will have the following structure:

$$\begin{aligned}\gamma_{u_x u_y}(k) &= Eu_{xt}(\phi_y(B)\theta_y(B)^{-1}y_{t+k}) \\ &= Eu_{xt}(\phi_y(B)\theta_y(B)^{-1}\omega(B)\delta(B)^{-1}x_{t+k} \\ &\qquad + \phi_y(B)\theta_y(B)^{-1}\theta(B)\phi(B)^{-1}a_t) \\ &= \phi_y(B)\theta_y(B)^{-1}\omega(B)\delta(B)^{-1}\phi_x(B)^{-1}\theta_x(B)\gamma_{u_x}(k) \qquad (3.6)\end{aligned}$$

and

$$\rho_{u_x u_y}(k) = \gamma_{u_x u_y}(k)\sigma_{u_x}^{-1}\sigma_{u_y}^{-1} .$$

Hence the values $\rho_{u_x u_y}(k)$ are not proportional to the transfer function (or impulse response) weights V_k, where

$$V(B) = \omega(B)/\delta(B) = \sum_{k=0}^{\infty} V_k B^k .$$

We may consider instead, however, the dynamic shock model

$$\begin{aligned}u_{yt} &= \omega'(B)\delta'(B)^{-1}u_{xt} + \theta'(B)\phi'(B)^{-1}a_t \\ &= V'(B)u_{xt} + \psi'(B)a_t\end{aligned} \qquad (3.7)$$

which is analagous to the usual dynamic regression model (3.3). Now $\rho_{u_x u_y}(\cdot)$ has the simpler form

$$\begin{aligned}\rho_{u_x u_y}(k) &= (\sigma_{u_x}\sigma_{u_y})^{-1}V'(B)\gamma_{u_x}(k) \\ &= (\sigma_{u_x}\sigma_{u_y})^{-1}V_k'\end{aligned} \qquad (3.8)$$

so that $\rho_{u_x u_y}(\cdot)$, as identified by $r_{\hat{u}_x \hat{u}_y}(\cdot)$, is directly indicative of the transfer function $V'(B)$.

Note that in either (b) or (c) one would be led to such a dynamic shock form (3.7) anyway, because when $\gamma_{u_x u_y}(k) = 0$ for $k < 0$ we have

$$\begin{pmatrix}u_{xt}\\u_{yt}\end{pmatrix} = \begin{pmatrix}W_{11}(B) & 0\\ W_{21}(B) & W_{22}(B)\end{pmatrix}\mathbf{a}_t .$$

Since u_{xt} is white noise it must also be that $W_{11}(B) = 1$, and hence $u_{xt} = a_{1t}$. Therefore we have

$$u_{yt} = W_{21}(B)u_{xt} + W_{22}(B)a_{2t}$$

which leads to (3.7).

3.3 Identification of $V'(B)$

Various patterns of transfer function weights V_k corresponding to the forms of $V(B) = \omega(B)/\delta(B)$ that often occur in practice are discussed in [3, Ch. 10]. Fortunately, the analogy existing between (3.7) and (3.3) permits us to carry over this discussion of $V(B)$ to $V'(B)$. For example, if it appeared that $\rho_{u_x u_y}(\cdot)$ had nonzero values at lags $k = 0$ and 1 only, $V'(B) = \omega'(B)/\delta'(B)$ could be identified as $V'(B) = \omega_0' - \omega_1'B$. Similarly a cyclic pattern in $\rho_{u_x u_y}(\cdot)$ may indicate that $\delta'(B)$ is at least of second order in B.

Initial parameter estimates for the parameters appearing in the identified forms of $\delta'(B)$ and $\omega'(B)$ may be obtained, for example, as those of $\delta(B)$ and $\omega(B)$ are obtained in [3, Ch. 11].

3.4 Identification of $\psi'(B)$

The cross-correlation function $r_{\hat{u}_x \hat{u}_y}(\cdot)$ gives an indication of $\rho_{u_x u_y}(\cdot)$, and hence $V'(B)$; but also we may easily identify the form of $\psi'(B) = \theta'(B)/\phi'(B)$ by making use of the fact that u_x is white noise. It will be indicated next that $\phi'(B) = \delta'(B)$ and that $\theta'(B)$ is at most of order r' or s', where r' and s' are the orders of $\delta'(B)$ and $\omega'(B)$, respectively.

To establish the form of $\psi'(B)$, consider the covariance structure of $z_t = \psi'(B)a_t$:

$$\begin{aligned}\gamma_z(k) &= E(u_{yt} - V'(B)u_{xt})(u_{yt+k} - V'(B)u_{xt+k}) \\ &= \gamma_{u_y}(k) - V'(F)V'(B)\gamma_{u_x}(k) .\end{aligned} \qquad (3.9)$$

To show that $\phi'(B) = \delta'(B)$, consider $\delta'(B)z_t = z_t'$ say,

$$\begin{aligned}\gamma_{z'}(k) &= \delta'(F)\delta'(B)\gamma_z(k) \\ &= \delta'(F)\delta'(B)\gamma_{u_y}(k) - \omega'(F)\omega'(B)\gamma_{u_x}(k) .\end{aligned} \qquad (3.10)$$

Hence $z_t' = \delta'(B)z_t$ has the covariance structure associated with a moving average process of order q', MA (q'), where $q' \leq \max(r', s')$. Since

$$z_t' = \delta'(B)\theta'(B)\phi'(B)^{-1}a_t$$

is MA (q'), then $\phi'(B)$ must be a factor of $\delta'(B)$. Similarly one can show that $\delta'(B)$ must be a factor of $\phi'(B)$, so that $\delta'(B) = \phi'(B)$.

Since $\delta'(B) = \phi'(B)$, $z_t' = \theta'(B)a_t$ and the coefficients of $\theta'(B)$ may be determined as the parameters having covariance structure (3.10). The determination of initial parameter estimates for $\theta'(B)$ will be illustrated in the following discussions of two examples.

4. AN EXAMPLE OF DYNAMIC REGRESSION MODEL IDENTIFICATION

To illustrate the identification methodology just mentioned consider two macroeconomic indicators recently discussed by Bray [5]. They are the first differences of U.K. Gross Domestic Product (GDP) and U.K. Unemployment. The original series are given quarterly for 1955I-1968IV ($N = 56$). Let $X =$ GDP and $Y = \log_{10}$ (unemployment/1000), while $y_t = Y_t - Y_{t-1}$ and $x_t = X_t - X_{t-1}$. Note that the data have already been seasonally corrected and otherwise adjusted before appearing in [5].

Although there appears to be some cycling in y, the following model gives a reasonable univariate fit to the y series.[2]

$$(1 - .63B)y_t = \hat{u}_{yt} \qquad (4.1)$$

with a standard error of .11 attached to .63 and $\hat{\sigma}_{u_y}{}^2 = .80\ E - 3$.

An adequate model for x was similarly obtained as

$$x_t = .66 + \hat{u}_{xt} \qquad (4.2)$$

with a standard error of .12 attached to .66 and $\hat{\sigma}_{u_x}{}^2 = .80$.

4.1 Dynamic Shock Model Identification

At this point the residual series obtained above ($\hat{u}_y$ and $\hat{u}_x$) are cross correlated. Figure A illustrates $r_{\hat{u}_x\hat{u}_y}(\cdot)$. It is concluded that any feedback effect is of secondary importance, as evidenced by the small cross correlations at negative lags. The correlations at lags 1 and 2 are also small, with the strongest link being at $k = 0$; but to better illustrate the distributed lag identification procedure we will also tentatively consider $\rho_{u_xu_y}(1)$ and $\rho_{u_xu_y}(2)$ to be nonzero in identifying a model connecting u_x to u_y. This direction of causation from x to y agrees with that considered by Bray [5].

A. Estimated Univariate Residual Cross-Correlation Function $r_{\hat{u}_x\hat{u}_y}(\cdot)$ for Bray Data ($1/\sqrt{N} = .13$)

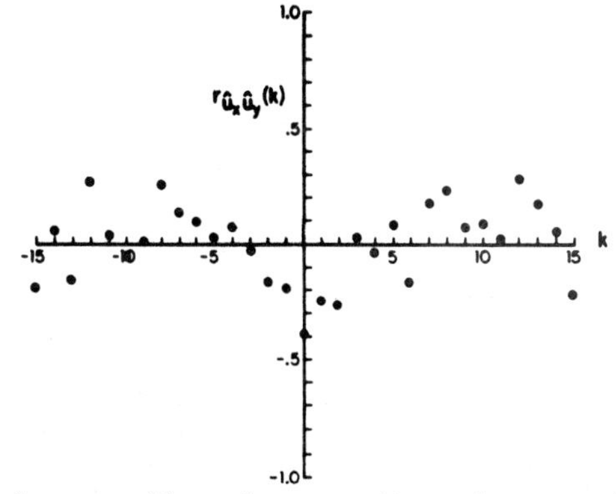

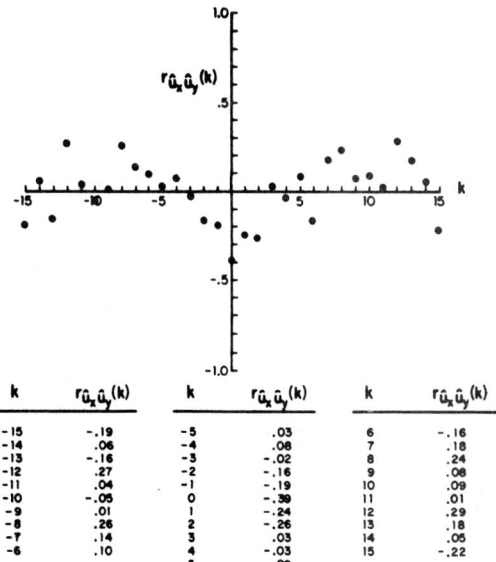

k	$r_{\hat{u}_x\hat{u}_y}(k)$	k	$r_{\hat{u}_x\hat{u}_y}(k)$	k	$r_{\hat{u}_x\hat{u}_y}(k)$
-15	-.19	-5	.03	6	-.16
-14	.06	-4	.08	7	.18
-13	-.16	-3	-.02	8	.24
-12	.27	-2	-.16	9	.08
-11	.04	-1	-.19	10	.09
-10	-.05	0	-.39	11	.01
-9	.01	1	-.24	12	.29
-8	.26	2	-.26	13	.18
-7	.14	3	.03	14	.05
-6	.10	4	-.03	15	-.22
		5	.09		

A pattern in which $\rho_{u_xu_y}(\cdot)$ is nonzero for $k = 0$, 1, and 2 leads one to the dynamic shock model

$$u_{yt} = (\omega_0' - \omega_1'B - \omega_2'B^2)u_{xt} + (1 - \theta_1'B - \theta_2'B^2)a_t\ . \qquad (4.3)$$

[2] A general nonlinear least-squares program was used to obtain the parameter estimates reported here and elsewhere. A standard error is attached to each parameter estimate, presuming model linearity in the region of parameter convergence as necessary.

Here, $\theta'(B)$ is chosen to be of the same order as $\omega'(B)$, to insure that u_y can be specified as white noise (see Section 3).

Initial parameter values for $\omega'(B)$, $\theta'(B)$, and $\sigma_a{}^2$ may be obtained, in general, by solving a system of moment equations determined by $\gamma_{u_y}(\cdot)$ and $\gamma_{u_xu_y}(\cdot)$. We have from (4.3),

$$\begin{aligned}
\sigma_{u_y}{}^2 &= (\omega_0'^2 + \omega_1'^2 + \omega_2'^2)\sigma_{u_x}{}^2 + (1 + \theta_1'^2 + \theta_2'^2)\sigma_a{}^2 \\
0 &= -\omega_1'(\omega_0' - \omega_2')\sigma_{u_x}{}^2 - \theta_1'(1 - \theta_2')\sigma_a{}^2 \qquad (4.4) \\
0 &= -\omega_0'\omega_2'\sigma_{u_x}{}^2 - \theta_2'\sigma_a{}^2
\end{aligned}$$

and from the form of $\gamma_{u_xu_y}(\cdot)$ we have,

$$\begin{aligned}
\gamma_{u_xu_y}(0) &= \omega_0'\sigma_{u_x}{}^2\ , & \rho_{u_xu_y}(0) &= \omega_0'\sigma_{u_x}/\sigma_{u_y}\ , \\
\gamma_{u_xu_y}(1) &= -\omega_1'\sigma_{u_x}{}^2\ , & \rho_{u_xu_y}(1) &= -\omega_1'\sigma_{u_x}/\sigma_{u_y}\ , \qquad (4.5) \\
\gamma_{u_xu_y}(2) &= -\omega_2'\sigma_{u_x}{}^2\ , & \rho_{u_xu_y}(2) &= -\omega_2'\sigma_{u_x}/\sigma_{u_y}\ .
\end{aligned}$$

Hence the equations (4.5) to be solved for ω_0', ω_1', and ω_2' are in fact linear. Substituting estimates of σ_{u_y}, σ_{u_x}, and $\rho_{u_xu_y}(0)$, $\rho_{u_xu_y}(1)$, and $\rho_{u_xu_y}(2)$ in (4.5) yields the estimates $\tilde{\omega}_0' = -.012$, $\tilde{\omega}_1' = .0074$, and $\tilde{\omega}_2' = .0080$.

The nonlinear equations (4.4) to be solved for θ_1', θ_2', and $\sigma_a{}^2$ can be viewed as the moment equations usually solved in obtaining moment estimates for parameters of an MA (2) process [3, Ch. 6]:

$$\begin{aligned}
(1 + \theta_1'^2 + \theta_2'^2)\sigma_a{}^2 &= \sigma_{u_y}{}^2 - (\omega_0'^2 + \omega_1'^2 + \omega_2'^2)\sigma_{u_x}{}^2 \\
&= \sigma_{u_y}{}^2(1 - \rho_0{}^2 - \rho_1{}^2 - \rho_2{}^2)\ , \\
-\theta_1'(1 - \theta_2')\sigma_a{}^2 &= \omega_1'(\omega_0' - \omega_2')\sigma_{u_x}{}^2 \qquad (4.6) \\
&= -\sigma_{u_y}{}^2\rho_1(\rho_0 + \rho_2)\ , \\
-\theta_2'\sigma_a{}^2 &= \omega_0'\omega_2'\sigma_{u_x}{}^2 \\
&= -\sigma_{u_y}{}^2\rho_0\rho_2\ ,
\end{aligned}$$

where, for convenience, we let $\rho_k = \rho_{u_xu_y}(k)$.

Such equations may be solved iteratively, in general, or one may graphically obtain estimates in the MA (2) case [3, p. 519, Chart 6]. To use the graphical procedure one would calculate from (4.6) the corresponding first- and second-order autocorrelations $\rho_{z'}(1)$ and $\rho_{z'}(2)$, where $z_t' = \theta'(B)a_t$. For example, $\rho_{z'}(1)$ would just be the quotient of the right sides of the first two equations in (4.6). For these data we find $\tilde{\rho}_{z'}(1) = -.22$ and $\tilde{\rho}_{z'}(2) = -.14$. Using the chart mentioned, one finds $\tilde{\theta}_1' = .28$ and $\tilde{\theta}_2' = .15$. Using the third equation, for example, one can also estimate the noise variance as $\tilde{\sigma}_a{}^2 = .54E - 3$.

The identified dynamic shock model is hence

$$u_{y_t} = -.012(1 + .62B + .67B^2)u_{x_t} + (1 - .28B - .15B^2)a_t \qquad (4.7)$$

with $\sigma_a{}^2$ about $.54E - 3$.

4.2 Completing Identification

Substitution of the identified univariate models into the identified dynamic shock model gives the dynamic regression form,

$$y_t = .049 - .012(1 + .62B + .67B^2)(1 - .63B)^{-1}x_t + (1 - .28B - .15B^2)(1 - .63B)^{-1}a_t\ . \qquad (4.8)$$

However, since

$$(1 - .28B - .15B^2)/(1 - .63B) \doteq (1 - .55B)(1 + .27B)/(1 - .63B) ,$$

it will be useful at this stage to eliminate the nearly common factors $(1 - .55B)$ and $(1 - .63B)$ leaving[3]

$$y_t = .049 - .012(1 + .62B + .67B^2)(1 - .63B)^{-1}x_t + (1 + .27B)a_t . \quad (4.9)$$

4.3 Estimation

Nonlinear estimation of the parameters in a model of the form (4.9), using a program of the Marquardt type [13], yielded the following fitted model.

$$y_t = .047 - .013(1 + .79B + .81B^2)(1 - .43B)^{-1}x_t + (1 + .27B)a_t \quad (4.10)$$

with the standard errors attached to .047, − .79, − .81, .43, − .27 being, respectively, .008, .44, .47, .16, .14 and $\hat{\sigma}_a^2 = .54E - 3$.

The usual residual auto- and cross-correlation checks [3, Ch. 11] indicate a good data fit. As can be seen from (4.10) the identified values provided good initial estimates for the iterative estimation program. It may be possible to further simplify the model form slightly by considering whether $\omega(B)$ and $\theta(B)$ may not be altered, but such possibilities will not be developed here.

Recall that the univariate model (4.1), for y, had a white-noise variance of $\hat{\sigma}_{u_y}^2 = .80E - 3$, so that the dynamic model (4.10) has reduced the unexplained noise variance by about 30 percent. Thus the differenced (log) GDP series seems to be of some value in explaining the stochastic structure of the differenced unemployment series. The proposed identification methodology was seen to be successful in arriving at a dynamic regression model relating the two time series.

5. ANOTHER EXAMPLE OF DYNAMIC REGRESSION IDENTIFICATION

As another example consider the possible relationship between the interest rate on AAA corporate bonds and the interest rate on commercial paper. Pierce [15], for example, discusses these two series as part of a study involving two additional series as well. Both series are given quarterly from 1953I to 1970IV ($N = 72$), and a listing of the data was kindly provided to the authors by Pierce. Let $y_t = Y_t - Y_{t-1}$ and $x_t = X_t - X_{t-1}$ where Y is the commercial paper interest rate and X is the AAA corporate bond interest rate.

The y series (with $\hat{\sigma}_y^2 = .20$) appears to be fitted well by either an AR (2) or ARMA (1, 1) univariate time series model. For example, the AR (2) model is

$$(1 - .76B + .39B^2)y_t = \hat{a}_{yt} \quad (5.1)$$

[3] The proposed simplification would normally be detected in a more straightforward manner by checking the division of $(1 - .63B)$ into $(1 - .28B - .15B^2)$ yielding $1 + .35B + .07B^2$. The small second-order term is a candidate for deletion.

with the standard errors .13 and .13 attached to the estimates .76 and −.39 and $\hat{\sigma}_{u_y}^2 = .14$.

The x series (with $\hat{\sigma}_x^2 = .028$) appears to be adequately fitted by an MA (1) model with nonzero mean.

$$x_t = .069 + (1 + .55B)\hat{a}_{xt} \quad (5.2)$$

with the standard errors .027 and .10 attached to the estimates .069 and −.55 and $\hat{\sigma}_{u_x}^2 = .021$.

5.1 Dynamic Shock Model Identification

The cross-correlation function $r_{\hat{a}_x\hat{a}_y}(\cdot)$ between the two univariate residual series above is given in Figure B. It

B. Estimated Univariate Residual Cross-Correlation Function $r_{\hat{a}_x\hat{a}_y}(\cdot)$ for Interest Rate Data ($1/\sqrt{N} = .12$)

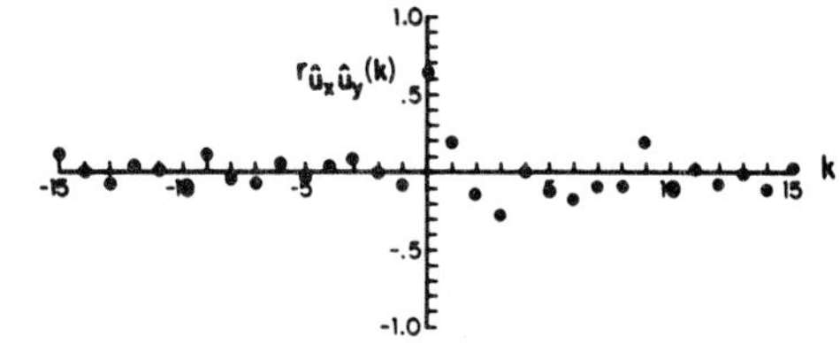

k	$r_{\hat{u}_x\hat{u}_y}(k)$	k	$r_{\hat{u}_x\hat{u}_y}(k)$	k	$r_{\hat{u}_x\hat{u}_y}(k)$
-15	.12	-5	-.03	6	-.15
-14	.01	-4	.04	7	-.08
-13	-.08	-3	.10	8	-.08
-12	.04	-2	.00	9	.20
-11	.02	-1	-.08	10	-.08
-10	-.08	0	.64	11	.02
-9	.12	1	.20	12	-.06
-8	-.04	2	-.13	13	-.01
-7	-.06	3	-.26	14	-.10
-6	.05	4	.01	15	.03
		5	-.10		

does not appear that any feedback effect exists, judging from the insignificant cross correlations at negative lags. As in our first example (Bray data) the strongest link appears at lag 0, but there appears to be an interesting second order pattern in the cross correlations at small positive lags. A dynamic shock model relating u_x and u_y which may explain the second order decay pattern in $\rho_{u_xu_y}(\cdot)$ is

$$u_{yt} = \omega_0'(1 - \delta_1'B - \delta_2'B^2)^{-1}u_{xt} + (1 - \theta_1'B - \theta_2'B^2)(1 - \delta_1'B - \delta_2'B^2)^{-1}a_t . \quad (5.3)$$

Moment estimates of ω_0', δ_1', and δ_2' may be obtained from $r_{\hat{a}_x\hat{a}_y}(0)$, $r_{\hat{a}_x\hat{a}_y}(1)$, and $r_{\hat{a}_x\hat{a}_y}(2)$ since

$$\gamma_{u_xu_y}(0) = \omega_0'\sigma_{u_x}^2 , \qquad \rho_0 = \omega_0'\sigma_{u_x}/\sigma_{u_y} ,$$
$$\gamma_{u_xu_y}(1) = \delta_1'\gamma_{u_xu_y}(0) , \qquad \rho_1 = \delta_1'\rho_0 , \quad (5.4)$$
$$\gamma_{u_xu_y}(2) = \delta_1'\gamma_{u_xu_y}(1) + \delta_2'\gamma_{u_xu_y}(0) ,$$
$$\rho_2 = \delta_1'\rho_1 + \delta_2\rho_0 ,$$

where again $\rho_k = \rho_{u_xu_y}(k)$.

Hence as in the previous example, the equations (5.4) to be solved are linear. Using estimates of σ_{u_y} and σ_{u_x}, in addition to $r_{\hat{a}_x\hat{a}_y}(0)$, $r_{\hat{a}_x\hat{a}_y}(1)$, and $r_{\hat{a}_x\hat{a}_y}(2)$, we find $\tilde{\omega}_0' = 1.63$, $\tilde{\delta}_1' = .32$, and $\tilde{\delta}_2' = -.30$.

Again the equations to be solved for θ_1', θ_2', and σ_a^2 are nonlinear.

$$\begin{aligned}
&(1 + \theta_1'^2 + \theta_2'^2)\sigma_a^2 \\
&\quad = (1 + \delta_1'^2 + \delta_2'^2)\sigma_{u_y}^2 - \omega_0'^2\sigma_{u_x}^2 \\
&\quad = \sigma_{u_y}^2\rho_0^{-2}(\rho_0^2 + \rho_1^2 + (\rho_2 - \rho_1^2/\rho_0)^2) - \sigma_{u_y}^2\rho_0^2 , \\
&-\theta_1'(1 - \theta_2')\sigma_a^2 \\
&\quad = -\delta_1'(1 - \delta_2')\sigma_{u_y}^2 \qquad (5.5)\\
&\quad = -\sigma_{u_y}^2\rho_1\rho_0^{-2}(\rho_0 - \rho_2 + \rho_1^2/\rho_0) , \\
&-\theta_2'\sigma_a^2 \\
&\quad = -\delta_2'\sigma_{u_y}^2 \\
&\quad = -\sigma_{u_y}^2\rho_0^{-2}(\rho_2 - \rho_1^2/\rho_0) .
\end{aligned}$$

We may estimate θ_1' and θ_2' graphically as before, using the relevant autocorrelations of an MA (2) process, $\tilde{\rho}_{z'}(1) = -.53$, and $\tilde{\rho}_{z'}(2) = .39$. We find $\tilde{\theta}_1 = .58$, $\tilde{\theta}_2 = -.73$, and $\tilde{\sigma}_a^2 = .057$. Hence the identified model is

$$\begin{aligned}
u_{y_t} &= 1.63(1 - .32B + .30B)^{-1}u_{x_t} \\
&\quad + (1 - .58B + .73B^2)(1 - .31B + .30B^2)^{-1}a_t \qquad (5.6)
\end{aligned}$$

with σ_a^2 about .057.

5.2 Completing Identification

A substitution of the identified univariate models for x and y into the preceding relationship leads to the dynamic regression model

$$\begin{aligned}
y_t = 1.63&(1 - .76B + .39B^2)^{-1} \\
&\cdot(1 - .32B + .30B^2)^{-1}(1 + .55B)^{-1}(x_t - .069) \\
&+ (1 - .58B + .73B^2)(1 - .76B + .39B^2)^{-1} \\
&\qquad \cdot(1 - .32B + .30B^2)^{-1}a_t . \qquad (5.7)
\end{aligned}$$

Expansion of the polynomials in B represented by each of the denominators in the two preceding transfer functions, and omission of the very small fourth and fifth order terms, leaves us

$$\begin{aligned}
y_t &= 1.63(1 - .53B + .34B^2 + .16B^3)^{-1}(x_t - .069) \\
&+ (1 - .58B + .73B^2)(1 - 1.08B + .93B^2 - .35B^3)^{-1}a_t .
\end{aligned}$$

One may simplify the identified noise model by noting that $(1 - 1.08B + .93B^2 - .35B^3)(1 - .58B + .93B^2)^{-1} \doteq 1 - .50B - .09B^2$, so that we may consider instead

$$\begin{aligned}
y_t &= 1.63(1 - .53B + .34B^2 + .16B^3)^{-1}(x_t - .069) \\
&\qquad + (1 - .50B - .09B^2)^{-1}a_t \\
&= -.071 + 1.63(1 - .53B + .34B^2 + .16B^3)^{-1}x_t \\
&\qquad + (1 - .50B - .09B^2)^{-1}a_t . \qquad (5.8)
\end{aligned}$$

5.3 Estimation

Estimation of the parameters in (5.8) reveals an adequate model fit (with $\hat{\sigma}_a^2 = .073$), but the third-order term (of $\delta(B)$) in the transfer function relating x to y is not significantly different from zero. Hence the following fitted model will be considered instead.

$$\begin{aligned}
y_t &= -.10 + 1.51(1 - .65B + .39B^2)^{-1}x_t \\
&\qquad + (1 - .71B + .19B^2)^{-1}a_t \qquad (5.9)
\end{aligned}$$

with standard errors of .07, .21, .11, .11, .13, .13 attached to the estimates −.10, 1.51, .65, −.39, .71, −.19, respectively, and $\hat{\sigma}_a^2 = .072$.

Further simplification to a first-order autoregressive noise model was contraindicated by a significance test of the sum of squared residual autocorrelations from the fitted model [16].

In this example the unexplained variance remaining in the univariate model fit ($\hat{\sigma}_{u_y}^2 = .137$) has been reduced to an unexplained variance of $\hat{\sigma}_a^2 = .072$, a reduction of 47 percent.

6. SUMMARY REMARKS

Concerning the two economic systems briefly discussed, our analysis clearly depends on the assumption of joint covariance stationarity, which some econometricians may not feel is appropriate for such series. Our analyses should be viewed as an attempt at improving the current identification methodology, which has often been used with the same assumptions.

Once u_x and u_y have been obtained at the first stage of identification, the bivariate dynamic shock identification which follows depends in no way on the univariate models employed. Hence seasonal univariate models of various types may usefully be employed as well. Of course the seasonality may appear in the cross-correlation function $\rho_{u_x u_y}(\cdot)$, but more experience needs to be gained as to how real data behave in this regard.

The specification arguments of Section 2 are useful in that an alternative to the more usual multiple time-series specification, in which the noise is correlated at zero lag, has been found to be helpful in identifying models. The specification provides a generalization of the usual dynamic regression model, and gives a clearer indication of the nature of feedback, which occurs when the bivariate transfer function is not lower triangular in form. Also the independent shock specification was found to be helpful in calculating the forms of cross correlation that may occur in various cases (e.g., note the additive form of (2.9)).

The identification methodology will hopefully provide a useful alternative to the usual cross-correlation techniques because

a. the distribution of cross-correlation estimates $r_{\hat{u}_x\hat{u}_y}(\cdot)$ is simpler in mathematical form than that of $r_{xy}(\cdot)$, particularly when x and y are independent;
b. the identification of feedback between two series should be easier;
c. the analysis in two stages may provide a useful visualization of the bivariate process;
d. the identification of dynamic regression models at the second stage of univariate residual cross correlation follows the already developed techniques of [3].

It is probably useful to briefly compare this two-stage identification method with the related technique discussed in [3, Ch. 11]. The latter procedure involves at the first stage a univariate modelling of x alone, $x_t - \mu_x = V_x(B)u_{xt}$, and y is then transformed using the same filter to obtain

$$\beta_t = V_x(B)^{-1}y_t \; . \tag{6.1}$$

Therefore, the lagged cross-correlation function between u_{xt} and β_t is [3]

$$\rho_{u_x\beta}(k) = V_k\sigma_\alpha/\sigma_\beta \tag{6.2}$$

where $V(B) = \sum_{k\geq 0} V_kB^k$. This procedure has the advantage that the pattern in the transfer function weights V_k is seen directly in the cross-correlation function $\rho_{u_x\beta}(\cdot)$, estimated by $r_{\hat{a}_x\beta}(\cdot)$. Recall that what is seen in $\rho_{u_xu_y}(\cdot)$ is the transfer function $V'(B)$ relating u_x to u_y.

It should be noted though that, in general, β_t will not be white noise, so that these cross-correlation estimates are themselves correlated with one another (Section 1). In particular then, these estimates do not provide as convenient a check for independence of the two series. It can be speculated as well that for many real series the double prewhitening would reduce the deviance of the estimates' covariance matrix, from that appropriate when correlating two independent white-noise series, beyond that afforded by a prewhitening of x alone and even further beyond that if no prewhitening at all is done. Recall too that an identification of the noise model is provided directly upon identification of $V'(B)$, since u_y is white noise.

A problem that requires further investigation is the distribution of $r_{\hat{u}_x\hat{u}_y}(\cdot)$ when the series x and y are not independent. It is assumed, for example, in the methodology employed that the pattern in $r_{\hat{u}_x\hat{u}_y}(\cdot)$ can be used to identify the true form of $\rho_{u_xu_y}$ when $\rho_{u_xu_y} \not\equiv 0$. In most examples considered to date, the magnitude of the cross correlations involved is small, and thus hopefully any required adjustments arising from a knowledge of the distribution just mentioned will not significantly deter a proper identification of $\rho_{u_xu_y}$.

Another question deserving of future research is the effect on the univariate residual cross-correlation function of possible misspecification of the univariate models in the first stage of identification.

[*Received April 1974. Revised July 1976.*]

REFERENCES

[1] Bartlett, M.S., "Some Aspects of the Time-Correlation Problem in Regard to Tests of Significance," *Journal of the Royal Statistical Society*, 98 (1935), 536–43.

[2] ———, *An Introduction to Stochastic Processes*, 2nd ed., Cambridge: Cambridge University Press, 1966.

[3] Box, G.E.P., and Jenkins, G.M., *Time Series Analysis, Forecasting and Control*, rev. ed., San Francisco: Holden-Day, Inc., 1976.

[4] ———, and Newbold, P., "Some Comments on a Paper of Coen, Gomme, and Kendall," *Journal of the Royal Statistical Society, Ser. A*, 134, Part 2 (1971), 229–40.

[5] Bray, J., "Dynamic Equations for Economic Forecasting with the G.D.P.—Unemployment Relation and the Growth of G.D.P. in the U.K. as an Example," *Journal of the Royal Statistical Society, Ser. A*, 134 (1971), 167–227.

[6] Granger, C.W.J., "Investigating Causal Relations by Econometric Models and Cross-Spectral Methods," *Econometrica*, 37 (1969), 424–38.

[7] Granger, C.W.J., and Newbold, P., "Identification of Two-Way Causal Systems," in M.D. Intriligator, ed., *Frontiers of Quantitative Economics, Vol. 3*, New York: American Elsevier Publishing Co., 1976.

[8] Hannan, E.J., *Multiple Time Series*, New York: John Wiley & Sons, Inc., 1970.

[9] Haugh, L.D. "The Identification of Time Series Interrelationships with Special Reference to Dynamic Regression Models," unpublished Ph.d. dissertation, Statistics Department, University of Wisconsin-Madison, 1972.

[10] ———, "Checking Time Series Interrelationships and Identifying Dynamic Regression Models for Short-Term Forecasting," *1972 Proceedings of the Business and Economic Statistics Section-American Statistical Association*, 325–30.

[11] ———, "Checking the Independence of Two Covariance-Stationary Time Series: A Univariate Residual Cross Correlation Approach," *Journal of the American Statistical Association*, 71 (June 1976), 378–85.

[12] Jenkins, G.M., and Watts, D.G., *Spectral Analysis and Its Applications*, San Francisco: Holden-Day, Inc., 1968.

[13] Marquardt, D.W., "An Algorithm for Least Squares Estimation of Non-Linear Parameters," *Journal of the Society of Industrial and Applied Mathematics*, 11, No. 2 (June 1963), 431–41.

[14] Parzen, E., "Multiple Time Series Modelling," in P.R. Krishnaiah, ed., *Multivariate Analysis-II*, New York: Academic Press, Inc., 1969, 389–409.

[15] Pierce, D.A., "Fitting Dynamic Time Series Models: Some Considerations and Examples," in *Proceedings of the Fifth Annual Symposium on the Interface of Computer Science and Statistics*, 1971, 45–53.

[16] ———, "Residual Correlations and Diagnostic Checking in Dynamic-Disturbance Time Series Models," *Journal of the American Statistical Association*, 67 (September 1972), 636–40.

[17] ———, "Relationships—and the Lack Thereof—Between Economic Time Series, with Special Reference to Money, Reserves, and Interest Rates," *Journal of the American Statistical Association*, 72 (1977), 11–22.

[18] ———, and Haugh, L.D., "Causality in Temporal Systems: Characterizations and a Survey," *Journal of Econometrics*, to appear.

[19] Priestley, M.B., "Fitting Relationships Between Time Series," *Bulletin of the International Statistical Institute*, 34 (1971), 1–27.

[20] Rozanov, Y.A., *Stationary Random Processes*, San Francisco: Holden-Day, Inc., 1967.

3.11
Analysis and Modeling of Seasonal Time Series

G. E. P. BOX, STEVEN C. HILLMER, and GEORGE C. TIAO
University of Wisconsin

INTRODUCTION

It is frequently desired to obtain certain derived quantities from economic time series. These include smoothed values, deseasonalized values, forecasts, trend estimates, and measurements of intervention effects.

The form that these derived quantities should take is clearly a statistical question. As with other statistical questions, attempts to find answers follow two main routes: An empirical approach and a model-based approach.

The first appeals directly to practical considerations, the latter to theory. We shall argue that best results are obtained by an iteration between the two and further show how this process has led to a useful class of time series models that may be used to study the questions mentioned above.

A SIMPLE EXAMPLE

To illustrate with an elementary example, consider the classical problem of selecting a measure of location M for a set of repeated measurements $Z_1, Z_2, \ldots, Z_n$. Arguing empirically: (1) If it were postulated that a change in any observation Z_j to $Z_j+\delta$ should have a linear effect on the location measure changing M to $M+c_j\delta$, this would imply that M was of the form

$$M=\sum_j w_j Z_j \qquad (1)$$

(2) Suppose it was further postulated that if the set of observations all happened to be equal to some value Z, then M should also be equal to Z. This would imply that

$$\sum_j w_j=1 \qquad (2)$$

(3) Finally, if there were no reason to believe that any single observation supplied more information about M than any other, then the w_j's might be taken equal so that

$$w_j=n^{-1},\ (j=1, 2, \ldots, n) \qquad (3)$$

Thus, from this empirical argument, M would be the arithmetic average, $\bar{Z}$, of the data.

From a theoretical or model-based viewpoint, the same quantity $\bar{Z}$ might be put forward if it were believed that the generation of the measurements was realistically simulated by random sampling from a Normal population. From this premise, well-known mathematical argument would lead to $\bar{Z}$.

Iteration Between Theory and Practice

The empirical method and the model-based method of attack are each employed by knowledgeable statisticians and are sometimes thought of as rivals. But they are, we believe, only rivals in the same sense that the two sexes are rivals. In both cases, isolation is necessarily sterile, while interaction can be fruitful.

The model-based approach works only if we can postulate a realistic class of models. But, from where are such models to come? One important source is from empirical procedures that have been found to behave satisfactorily in practice. Suppose, for a particular type of data, practical experience shows that the arithmetic average measures location well. Then, we can be led to the standard Normal model by asking what assumptions would make $\bar{Z}$ a good measure of location.

But, if $\bar{Z}$ can be arrived at empirically, where is the need for a model? The answer is, of course, that the existence of a model acts as a catalyst to further development. In particular, a model allows—

1. Constructive criticism of the original empirical idea. For instance, model-based arguments that recommend $\bar{Z}$ on Normal theory assumptions can also warn of consequences, not easily foreseen from an empirical standpoint, if the population is Cauchy-like or contains an outlier. It can also suggest better functions of the observations in these latter circumstances.
2. Generalization of the idea. For instance, in cases where the Normal theory assumptions are sensible for estimating a mean, they are equally sensible for more general models, leading in particular to the method of least squares estimation.

This research was supported, in part, by a grant from the Wisconsin Alumni Research Foundation and, in part, by the U.S. Army Research Office under grant No. DAAG29-76-G-0304.

The Collected Works of George E. P. Box, 1984, Wadsworth, Inc., Belmont, CA 94002.
Originally published in *Proceedings of Conference on Seasonal Analysis of Economic Time Series* (1976), pp. 309–344.

History seems to show that most rapid progress occurs when theory and practice are allowed to confront, criticize, and stimulate each other. A brief sketch of some historical developments in time series analysis illustrates this point.

ITERATIVE DEVELOPMENT OF SOME IMPORTANT IDEAS IN TIME SERIES ANALYSIS

Visual inspection of economic time series, such as annual wheat prices, suggests the existence of cycles. When Beveridge [2] fitted empirical models containing linear aggregates of cosine waves plus Normally distributed independent errors, statistically significant cycles, indeed, appeared. However, these cycles had exotic and mysterious periods for which no direct cause could be found. This led Yule [35] to propose a revolutionary kind of model in which it was supposed that time series could be represented as the output from a dynamic system excited by random shocks. The dynamic system was represented by a difference equation, and the random shocks were represented by independent drawings from a Normal distribution.

For illustration, suppose the sequence $\{Z_t\}$ is a time series with mean μ and $\{a_t\}$ is a sequence of independent drawings from the distribution $N(O, \sigma^2)$, which we shall call white noise. Then, an important model proposed by Yule was the second-order autoregressive process

$$z_t=\phi_1 z_{t-1}+\phi_2 z_{t-2}+a_t, \text{ where } z_t=Z_t-\mu$$

With B the back-shift operator, such that $Bz_t=z_{t-1}$, this model may be written

$$(1-\phi_1 B-\phi_2 B^2)z_t=a_t \quad (4)$$

If in (4) the second-degree polynomial (with B treated temporarily as a variable) has complex zeros, then the impulse response of (4) can be oscillatory, and the generated series can exhibit pseudocyclic behavior, like that of the economic data.

More general dynamic characteristics may be obtained with a model of the form

$$\varphi(B)z_t=\theta(B)a_t \quad (5)$$

where the polynomial $\varphi(B)=1-\varphi_1 B-\varphi_2 B^2-\ldots-\varphi_P B^P$ is called an autoregressive operator of order P and $\theta(B)=1-\theta_1 B-\theta_2 B^2-\ldots-\theta_q B^q$ is called a moving average operator of order q. The resulting model (5) is called an autoregressive moving average process of order (P, q), or simply an ARMA (P, q) process.

In the decades that followed Yule's proposal, evidence was obtained for the practical usefulness of models of the form of equation (5), and much study was given to them.

Authors, notably Bartlett, Durbin, Hannan, Jenkins, Kendall, Quenouille and Wald, studied their properties including their autocorrelation functions, methods for their fitting, and tests for their adequacy.

Heavy emphasis had, up to this time, been placed on the stationarity of time series, and it was known that, for a stationary process, the zeros of $\varphi(B)$ in (5) had to lie outside the unit circle. However, in the 1950's, operations research workers, such as Holt [14; 23] and Winters [33], required methods for smoothing and forecasting nonstationary economic and business series. Their need sparked a return to empiricism, resulting in the development and generalization of exponential weighting for smoothing and forecasting.

Suppose it is desired to measure the location at current time t of a nonstationary economic time series $\{Z_t\}$. For this purpose, the first two postulates advanced before, resulting in equations (1) and (2), would seem appropriate,[1] so that

$$M_t=\sum_{j=0}^{\infty} w_j Z_{t-j}, \text{ with } \sum_{j=0}^{\infty} w_j=1$$

However, it is sensible to require that the current value Z_t should be given more weight than the penultimate value Z_{t-1}, which, in turn, should have more weight than Z_{t-2}, etc. If the weights are made to fall off exponentially, so that $w_{j+1}=\theta w_j$, where $0<\theta<1$ is a smoothing constant, then, since $\sum w_j=1$, it follows that $w_j=(1-\theta)\theta^j$. The measure M_t was called an exponentially weighted average and has the very convenient property that it can be updated by the formula $M_{t+1}=(1-\theta)Z_{t+1}+\theta M_t$.

The practical usefulness of this measure soon became apparent, and it was developed and generalized by Brown [11; 12] and by Brown and Meyer [13]. One important application was to employ M_t as the one step ahead (lead one) forecast $\hat{Z}_t(1)$ of Z_{t+1}, where the notation $\hat{Z}_t(l)$ refers to a forecast made l steps ahead from time origin t. Thus, in this application

$$\hat{Z}_t(1)=(1-\theta)\sum_{j=0}^{\infty} \theta^j Z_{t-j} \quad (6)$$

The practice-theory iteration proceeded through one more important step when Muth [27] asked what stochastic process would be such that (6) provided a forecast having minimal mean square error. He showed that the required stochastic process was of the form

$$(1-B)Z_t=(1-\theta B)a_t, \ -1<\theta<1 \quad (7)$$

This is a nonstationary process of the form of (5) with $P=1$, $q=1$, and $\varphi_1=1$. Thus, a root of $\varphi(B)$ is on the unit circle.

The account of further developments follows the approach adopted by Box and Jenkins [3; 4; 5; 6; 7; 8; 9].

Empirical evidence from control engineering also pointed to the importance of nonstationary stochastic models of the form of (5), having roots on the unit circle.

---

[1] The expressions that follow imply an infinite series of past data. However, since in practice the weights quickly decay towards zero, M_t can be calculated to required accuracy from a fairly short series.

Long before the introduction of James Watt's governor, empirical methods of feedback control were being developed; Mayr [26]. The earliest forms used control in which the adjustment x_t at time t was made proportional to the deviation e_t from target T of some objective function $y_t=T+e_t$. The adjustment function, or controller, was, thus, of the form $x_t=-ke_t$. Adjustments with such controllers lagged behind when trends occurred in the disturbance, and it was soon realized that control could often be greatly improved by adding an integral term. Equivalently, for discrete control with observations and adjustments made at equally spaced times, a summation terms Se_t such that $Se_t=\sum_{i=0}^{\infty} e_{t-j}$ was added yielding a controller

$$x_t=-(k_1e_t+k_2Se_t) \tag{8}$$

Since such proportional plus integral controllers have been eminently successful and continue to be widely used throughout the process industries, it is natural to ask "What theoretical setup would make such a controller optimal?" Supposing, as would often be the case, that the dynamic relation between x_t and y_t could be roughly approximated by a first-order system, modeled by the difference equation $y_t=\delta y_{t-1}+\omega x_{t-1}$. It is easily shown that the control equation (8) would be optimal for a disturbance modeled by the stochastic process (7).

Thus, successful empiricism in two widely different fields point to the usefulness of models in which a first difference $\nabla Z_t=(1-B)Z_t$, or possibly a higher difference $\nabla^d Z_t=(1-B)^d Z_t$, could be represented by a stationary model. It, subsequently, turned out that such a class of models had, in fact, been proposed by Yaglom [34]. Such models are, thus, of the form

$$\phi(B)\nabla^d z_t=\theta(B)a_t \tag{9}$$

where $\varphi(B)=\phi(B)(1-B)^d$ and with $P=p+d$. In this model, $\phi(B)$ is a stationary autoregressive operator of degree p, having zeros outside the unit circle, $\nabla^d z_t$ is the dth difference of the series, and $\theta(B)$ is of degree q and has zeros outside the unit circle. Such a model is called an autoregressive integrated moving average process or ARIMA of order (p, d, q). More generally, these developments point to the possible usefulness of models in which one or more of the zeros of $\varphi(B)$ in (5), although not necessarily unity, lie on the unit circle. In what follows, $z_t=Z_t-\mu$, where μ is the mean of the series if stationary and, otherwise, is any convenient origin.

Properties of the ARIMA Models

It now becomes appropriate to test whether these generalizations of models arising from successful empiricism are practical.

Two relevant questions are—

1. What kinds of dependence of a current value z_t on past history can be represented by the model?
2. What kinds of projection or forecast of a time series are possible with the model?

Dependence of z_t on Past Values

Any model of the form of (5) can be thought of as an autoregressive model of possibly infinite order. Thus,

$$z_{t+1}=\sum_{j=0}^{\infty} \pi_j z_{t-j}+a_{t+1} \tag{10}$$

In this model, the one-step-ahead forecast $\hat{z}_t(1)=\sum_{j=0}^{\infty} \pi_j z_{t-j}$ is a linear aggregate of previous observations. Using parlance popular in econometrics, the forecast is obtained by applying a distributed lag weight function[2] to previous observations. Alternatively, the weights π_j can be thought of as defining the memory of the past, contained in the current value z_t.

Now, many forms have been proposed by econometricians for distributed lag-weight functions. How general are those implied by (5)?

This equation may be written

$$\varphi(B)\theta^{-1}(B)z_t=a_t \tag{11}$$

which may be compared with (10), written as

$$\pi(B)z_t=a_t \tag{12}$$

where

$$\pi(B)=1-\pi_1B-\pi_2B^2-\dots$$

By equating coefficients in the identity $\varphi(B)=\theta(B)\pi(B)$, the π weights may be calculated correspondingly for any choice of the polynomials $\varphi(B)$ and $\theta(B)$. Also, the nature of these weights may be deduced, and it can be shown that using this form we can represent—

1. A finite set of π weights that are functionally unrelated when we have a pure autoregressive model in which $\theta(B)=1$.
2. Of more interest, a convergent series of π weights that, after any desired number of initial, unrelated values, follows a function which can be any mixture of damped exponentials, sine and cosine waves satisfying the difference equation $\theta(B)\pi_j=0$. Convergence is assured by the requirement that the zeros of $\theta(B)$ lie outside the unit circle.

Convergence of the weights seems essential for any sensible memory function. The class of functions included is sufficiently rich to be capable of representing a very wide variety of practical situations.

[2] Distributed lag models are usually credited to Irving Fisher [18] but were used earlier by R. A. Fisher [19].

Forms of Projection Implied by the Model

The model in the form (10) for z_{t+l} is

$$z_{t+l}=\sum_{j=0}^{\infty}\pi_j z_{t+l-j}+a_{t+l} \tag{13}$$

Multiplying both sides by $(1+\psi_1 B+\psi_2 B^2+\ldots+\psi_{l-1}B^{l-1})$ and choosing the l-1 coefficients $\psi_1, \psi_2, \ldots, \psi_{l-1}$ so that the coefficients of $z_{t+l-1}, z_{t+l-2}, \ldots, z_{t+1}$ are all zero, we can write (13) as

$$z_{t+l}=\sum_{j=0}^{\infty}\pi_j^{(l)}z_{t-j}+\sum_{i=0}^{l-1}\psi_i a_{t+l-i} \tag{14}$$

where $\psi_0=1$. In fact, the ψ's are obtained by equating coefficients in $\pi(B)\psi(B)=1$, where $\psi(B)$ is the infinite series $1+\psi_1 B+\psi_2 B^2+\ldots$.

Now, suppose that (14) is written in the form

$$z_{t+l}=\hat{z}_t(l)+e_t(l) \tag{15}$$

Then $\hat{z}_t(l)$ is the conditional expectation of z_{t+l} given all past history up to time t which we will also write as $[z_{t+l}]_t$. It may be shown (see, e.g., [32]) that this conditional expectation is the minimum mean square error forecast of z_{t+l}, made at origin t.

Consider the forecasts $\hat{z}_t(1), \hat{z}_t(2), \hat{z}_t(3), \ldots, \hat{z}_t(l), \ldots$, all made at origin t. The values together yield a forecast function $f(l, \underline{b}^{(t)})$ that constitutes the appropriate projection of the series.

Specifically, (15) may be written

$$z_{t+l}=f(l,\underline{b}^{(t)})+e_t(l) \tag{16}$$

where $f(l, \underline{b}^{(t)})=\hat{z}_t(l)=[z_{t+l}]_t$ is some function of l, with $\underline{b}^{(t)}$ a vector of coefficients that are fixed for any origin t but are updated when the origin is advanced.

For example, if the model (5) were of the form

$$(1-B)^2 z_t=(1-\theta_1 B-\theta_2 B^2)a_t \tag{17}$$

it is readily shown that the forecast function at origin t would be a straight line

$$\hat{z}_t(l)=f(l, \underline{b}^{(t)})=b_0^{(t)}+b_1^{(t)}l \tag{18}$$

Any series that could be regarded as an outcome of the model (17) should, thus, be projected or forecasted along a straight line.

Writing the model for z_{t+l} in the form of (5) and taking conditional expectations

$$\varphi(B)[z_{t+l}]_t=\theta(B)[a_{t+l}]_t \tag{19}$$

where

$$[z_{t+j}]_t=\begin{cases}\hat{z}_t(j) & \text{if } j>0\\ z_{t+j} & \text{if } j\le 0\end{cases} \quad\text{and}\quad [a_{t+j}]_t=\begin{cases}0 & \text{if } j>0\\ z_{t+j}-\hat{z}_{t+j-1}(1) & \text{if } j\le 0\end{cases}$$

so that for $l>q$,

$$\varphi(B)[z_{t+l}]_t=0 \tag{20}$$

It is, thus, easily seen that possible forecast functions are—

1. A set of unrelated values, followed by a fixed value equal to the mean when we have a pure moving average model with $\varphi(B)=1$.
2. Of more interest, any mixture of polynomials, damped exponentials and damped sine and cosine waves, possibly preceded by one or more unrelated values.

We see, therefore, that the forecast functions implied by the model are also sensible and of sufficient variety to satisfy many practical needs.

The relationship (16) has a form that is of interest quite apart from forecasting. The function $f(l, \underline{b}^{(t)})$ supplies all the information about z_{t+l} that is available up to time t, while $e_t(l)$ represents information that enters the system at a later time. If, in (5), $P\le q$, then $e_t(l)$ is a particular integral, and $f(l, \underline{b}^{(t)})$ is the complementary function of (5), i.e., it is the solution of

$$\varphi(B)f(l, \underline{b}^{(t)})=0 \tag{21}$$

Adaptive Updating of Forecast Function

It may be shown, by taking conditional expectations, that the coefficients $\underline{b}^{(t)}$ in the forecast function are automatically updated as the origin for the forecast is advanced. Thus, for the model $(1-B)^2 z_t = (1-\theta_1 B-\theta_2 B^2)a_t$, for which the forecast function is the straight line $\hat{z}_t(l)=b_0^{(t)}+b_1^{(t)}l$, the updating formulas[3] are

$$b_0^{(t+1)}-(b_0^{(t)}+b_1^{(t)})=\lambda_0 a_{t+1} \tag{22}$$

and

$$b_1^{(t+1)}-b_1^{(t)}=\lambda_1 a_{t+1} \tag{23}$$

where

$$\lambda_0=1+\theta_2,\ \lambda_1=1-\theta_1-\theta_2 \text{ and } a_{t+1}=z_{t+1}-\hat{z}_{t}(1) \tag{24}$$

Such formulas can be obtained for any stochastic model of the form of (5) and its associated forecast function.

From this, we reach the following conclusions: Time series models of the form of (5)—

1. Yield a sensible and rich class of memory functions, relating the dependence of present on the past.
2. Yield a sensible and rich class of forecast (complementary) functions for describing the future behavior of the series, which depends on the past.

[3] Obviously, $b_0^{(t)}+b_1^{(t)}$ would be the updated intercept if no adjustments in the coefficients were made. The constants λ_0 and λ_1, which are functions of the model memory parameters, determine the changes necessary in the coefficients in the light of the discrepancy $a_{t+1}=z_{t+1}-\hat{z}_t(1)$ between prediction and actuality.

3. Yield a sensible procedure for updating forecast functions.

This serves to show that the proposed models are not arbitrary. Indeed, if the desirable characteristics listed were set out as requirements for a model, it can be shown that we would inevitably be led to the form (5) for the generating stochastic process. There is, thus, a strong prima facie case in favor of this form of model.

Further properties, however, are needed for success, since—

1. It should be possible to build a model from available data. When an actual time series is under study, the appropriate form of the model within the general class and appropriate values for the memory parameters $\underset{\sim}{\theta}=(\theta_1, \ldots, \theta_q)'$ and $\underset{\sim}{\varphi}=(\varphi_1, \ldots, \varphi_p)'$ should be deducible from the time series itself. The deduced stochastic model will then automatically determine the form of the forecast and memory functions.
2. Evidence from many applications should show that the models do adequately represent real series, and application to problems, such as forecasting and intervention analysis, Box and Tiao [10], ought to have yielded satisfactory results with reasonable consistency.

Model Building

Models have been built using the following three-stage iterative procedure (identification-fitting-diagnostic checking):

1. Identification seeks a tentative model form, worthy to be entertained. In particular, it should suggest appropriate orders for the polynomials $\phi(B)$ and $\theta(B)$, indicate needed transformation of the series, and indicate appropriate orders of differencing. It is usually accomplished chiefly by graphical inspection of the time series and of computed auxiliary sample functions, such as the autocorrelation function, partial autocorrelation function, and, in some cases, the spectrum.
2. Fitting involves the estimation, by maximum likelihood, of the memory parameters $\underset{\sim}{\phi}$ and $\underset{\sim}{\theta}$ and, where necessary, the parameters of the transformation.
3. Diagnostic checking is intended to show up inadequacies in the model and to suggest remedies. It is usually accomplished by inspection of residuals and of their computed auxiliary functions. When inadequacies are found, a further iterative cycle is initiated.

Evidence From Applications

In recent years, applications of these methods have become increasingly common not only in economics and business but in widely diverse areas, such as environmental studies and educational psychology. Indeed, it is nearly impossible to keep up with the literature. Up to now, these applications have dealt with forecasting and with intervention analysis. This literature seems to show that models of this general class have usually worked successfully over a very wide field of subject matter.

Seasonality

Seasonality is a phenomenon commonly found in economic, environmental, and other time series. Models of the form (5) are, in principle, sufficiently general to represent such series, but, to allow representation in a most parsimonious fashion, special forms of (5) have been worked out and have proved useful.

Seasonal series are such that similarities occur at equivalent parts of a cycle. As an example, consider monthly data for department store sales that might have a seasonal pattern because of Christmas, Easter, holidays, summer vacations, etc. Now, sales for a particular month, e.g., December, might be related from year to year by a model like (2.2) in which B is replaced by B^{12}. Thus,

$$\Phi(B^{12})z_t=\Theta(B^{12})u_t \tag{25}$$

We may suppose that a similar model applies to the other months. However, the residual u_t's from such a model would be expected to be dependent from month to month. If they can be related by a model

$$\varphi(B)u_t=\theta(B)a_t \tag{26}$$

then, on substituting (26) in (25), we obtain the seasonal model

$$\Phi(B^{12})\varphi(B)z_t=\Theta(B^{12})\Theta(B)a_t \tag{27}$$

For illustration, a typical form of a seasonal nonstationary model that has been identified and used successfully in many economic series arises when the functional form (7) represents both monthly and yearly components. It is, thus,

$$(1-B)(1-B^{12})z_t=(1-\theta_1B)(1-\theta_2B^{12})a_t \tag{28}$$

Figure 1a shows the π weights for this model when it is written as an infinite autoregressive process of the form of (10)

$$z_{t+1}=\sum_{j=0}^{\infty}\pi_jz_{t-j}+a_{t+1} \tag{29}$$

To see the implications, note that if we use $z_t^{(\theta)}$ to mean an exponential average of the form of (6), then the model may also be written

$$z_{t+1}=z_t^{(\theta_1)}+\{z_{t-11}-z_{t-12}^{(\theta_1)}\}^{(\theta_2)}+a_{t+1} \tag{30}$$

Thus, the contribution to z_{t+1} of all previous data values

Figure 1a. π WEIGHTS FOR MODEL (2.25)

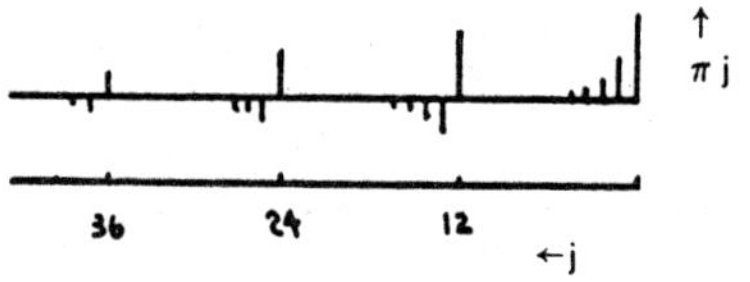

Figure 1b. FORECAST FUNCTION FOR MODEL (2.25)

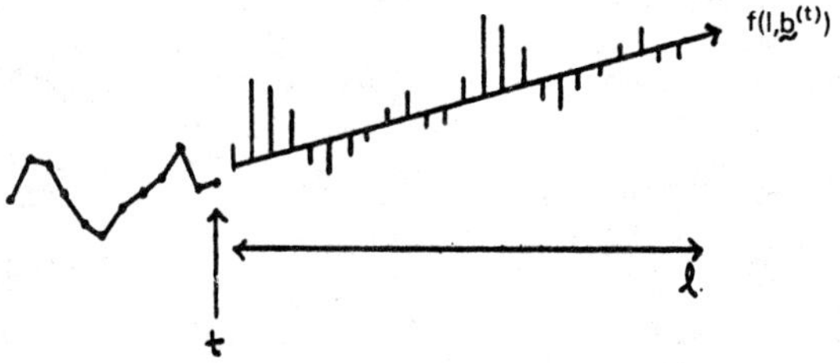

Figure 1c. SEASONAL COMPONENTS OF $(1-B^{12})$ ASSOCIATED WITH ROOTS ON THE UNIT CIRCLE

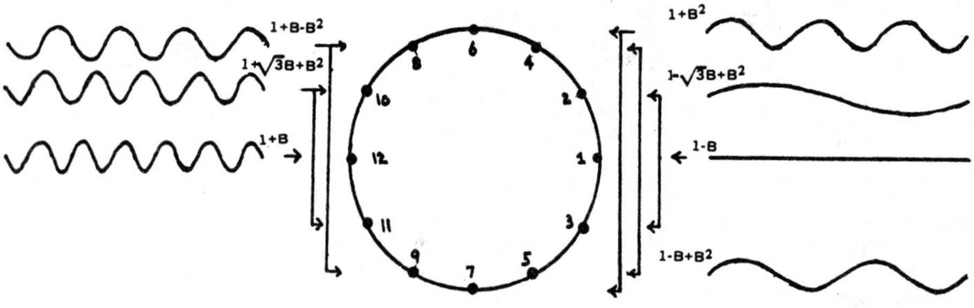

takes the appealingly sensible form of a crude forecast $z_t^{(\theta_1)}$, which is an exponential average taken over previous months, corrected by an exponential average of discrepancies between similar crude forecast and actuality for the same month, discounted over previous years.

More generally, any model of the form of (27) may be written

$$\frac{\varphi(B)}{\theta(B)}\frac{\Phi(B^{12})}{\Theta(B^{12})}z_t=a_t \tag{31}$$

or

$$R(B)Q(B^{12})z_t=a_t$$

where

$R(B)=1-R_1B-R_2B^2-\ldots$ and $Q(B^{12})=1-Q_1B^{12}-Q_2B^{24}-\ldots$

Now write

$z_t^{(R)}$ for the weighted sum $R_1z_t+R_2z_{t-1}+\ldots$

and

$z_t^{(Q)}$ for the weighted sum $Q_1z_t+Q_2z_{t-12}+\ldots$

Then

$$z_{t+1}=z_t^{(R)}+\{z_{t-11}-z_{t-12}^{(R)}\}^{(Q)}+a_{t+1} \tag{32}$$

which may be interpreted as before but with monthly and seasonal weights following a more general, and not necessarily, exponential pattern.

The particular model (28) written in the form of (16) becomes

$$z_{t+l}=b_0^{(t)}+b_1^{(t)}l+b_{0,m}^{(t)}+e_t(l) \tag{33}$$

The typical appearance of the forecast function is sketched in figure 1b. In this expression, the forecast function consists of an updated straight line plus seasonal adjustment factors $b_{0,m}^{(t)}$, such that $\sum_{m=1}^{12} b_{0,m}^{(t)}\neq 0$. These factors are automatically adjusted as each new piece of data comes to hand and are weighted averages of past data.

Alternatively, (33) may be written

$$z_{t+l}=b^{(t)}+b_1^{(t)}l+\sum_{j=1}^{5}\left\{b_{1j}^{(t)}\cos\left(\frac{2j\pi}{12}\right)+b_{2j}^{(t)}\sin\left(\frac{2j\pi}{12}\right)\right\}+e_{t+l} \tag{34}$$

In this form, the seasonal component contains a complete set of undamped sinusoids, adaptive in amplitude and phase with frequencies 0, 1, 2, ..., 6 cycles per year. These components are associated with the 12 roots of unity on the unit circle produced by the operator $(1-B^{12})$ and indicated in figure 1c. Thus, the complementary function is a solution of

$$(1-B)(1-B^{12})f(l,\ b^{(t)})=0$$

or equivalently of

$$(1-B^2)(1+B+B^2+\ldots+B^{11})f(l,\ b^{(t)})=0$$

or equivalently of

$$(1-B)^2(1-\sqrt{3B}+B^2)(1-B+B^2)(1+B^2)(1+B+B^2)(1+\sqrt{3B}+B^2)(1+B)f(l,\ b^{(t)})=0,\ l>13 \tag{35}$$

More elaborate models produce appropriately more elaborate weight functions and forecast functions.

Summary

We have attempted to show that, as a result of the practice-theory iteration extended over many decades and carried on by many different investigators, a class of stochastic models capable of representing nonstationary and seasonal time series has evolved. When these models have been employed for forecasting and intervention analysis, they have worked well. There is no reason to believe that they would be any less useful for a model-based attack on problems, such as smoothing and seasonal adjustment. These problems are now considered.

SMOOTHING AND SEASONAL ADJUSTMENT

Like other problems, smoothing and seasonal adjustment can be tackled from either an empirical or a model-based point of view. Also, like other problems, it is fairly certain that an iteration between the two approaches in which each stimulates the other is likely to be most fruitful. Inspired empiricism, as we have seen, first produced exponential smoothing and its generalizations. It has also produced valuable methods for seasonal adjustment, exemplified, in particular, by the census X-11 procedure, Shiskin et al [29].

The additive version of the census procedure assumes that an observed time series $\{z_t\}$ can be written

$$z_t=S_t+p_t+e_t \tag{36}$$

where S_t is the seasonal component, p_t is the trend component, and e_t is the noise component. Specific symmetric filters of the form

$$M(\delta,k)=\sum_{j=-k}^{k}\delta_j z_{t-j} \tag{37}$$

with $\delta_j=\delta_{-j}$, are employed to produce estimates $\hat{S}_t$, $\hat{p}_t$ and $\hat{e}_t$ of these unobserved components. For the majority of economic data met in practice, the weights δ_j's used for computing $\hat{S}_t$ and $\hat{p}_t$ are shown in figure 3, given later in this section. The procedure has been widely used in Government and industry and found to produce sensible results.

The empirical success of this procedure motivated Cleveland and Tiao [16] to ask the question: Are there stochastic models for S_t, p_t and e_t for which the census procedure would be optimal? In general, if each of these components follows a model of the form (5), then, the minimum mean square error estimates of p_t and S_t are, respectively, the conditional expectations $E(p_t|z)$ and $E(S_t|z)$. They showed that if the components follow the models

$$(1-B)^2p_t=(1+0.49B-0.49B^2)c_t$$

$$(1-B^{12})S_t=(1+0.64B^{12}+0.83B^{24})b_t \tag{38}$$

and $\{c_t\}$, $\{b_t\}$, and $\{e_t\}$ are three independent Gaussian white-noise processes, such that $\sigma_b^2/\sigma_c^2=1.3$ and $\sigma_e^2/\sigma_b^2=14.4$, then the conditional expectations $E(p_t|z)$ and $E(S_t|z)$ will be of the symmetric form (37), with weights almost identical to the corresponding weights of $\hat{p}_t$ and $\hat{S}_t$ for the census procedure. This finding makes it possible to assess, at least partially, the appropriateness of the census procedure in a given instance. Specifically, expression (38) implies that the overall model for z_t is

$$(1-B)(1-B^{12})z_t=\theta(B)a_t \tag{39}$$

where $\theta(B)$ is strictly a polynomial in B of degree 25, but the two largest coefficients are $\theta_1=0.34$ and $\theta_{12}=0.42$. This model is broadly similar to the one in (28), although the moving average structure is different. The implication is that the use of the census procedure for seasonal adjustment would be justified in situations where the series can be adequately represented by the model (39). On the other hand, if the model for a series $\{z_t\}$ were found to be vastly different from (39), then the appropriateness of the census decomposition would be in doubt.

The significance of this consideration is that it links the empirically successful census decomposition procedure to a model (39), for which this appropriateness can be verified for any given set of data. It should be borne in mind that, since only the series $\{z_t\}$ is available, any smoothing or seasonal adjustment procedure in the framework of (36) is necessarily arbitrary. On the other hand, whatever procedure one uses must at least be consistent with a model of z_t which can be built from the data. A procedure satisfying this minimum requirement will be called a model-based decomposition procedure.

We now consider what procedures would be produced using the stochastic model in (5). We will suppose that, by using past values of the series, a model has been carefully built in exactly the same way as for any other time series application, i.e., by an iterative sequence involving identification of a model worthy to be entertained, efficient fitting of the tentative model, and diagnostic checking of the fitted tentative model. Based on such a model for the observed series $\{z_t\}$, we shall then derive smoothing and seasonal adjustment procedures, illustrate how these derived procedures behave with actual data, and compare the results with other methods.

It is helpful for the development of ideas to first consider the simpler smoothing problem when there are no seasonal components and then to build onto this to develop seasonal adjustment methods.

Trend Plus Noise

We make the following assumptions, referred to as assumption I:

1. $\{T_t\}$ is an observed time series following the known model $\varphi(B)T_t=\eta(B)d_t$, with d_t being independent drawings from $N(O, \sigma_d^2)$.
2. $\varphi(B)$ is a polynomial in B of degree p with zeros on the outside of the unit circle, $\eta(B)$ is a polynomial of degree u with zeros outside the unit circle, and $\varphi(B)$ and $\eta(B)$ has no common zeros.
3. $T_t=p_t+e_t$, with $\{p_t\}$ and $\{e_t\}$ being independent of each other.
4. $\{p_t\}$ is a stochastic process following some ARIMA model.
5. $\{e_t\}$ is a Gaussian white-noise process, with mean O and variance σ_e^2.

An estimate of p_t is required from the time series $\{T_t\}$. When the models for p_t and e_t are given, the solution to this problem has been derived by Weiner [30], Kolmogorov [24], and Cleveland and Tiao [16]. In the analysis of economic time series, however, it is reasonable to assume that only the model for T_t is known. It is important, therefore, to consider how the models for p_t and e_t are restricted by this knowledge. On assumption I, the following results are readily proved (see, e.g., Cleveland [15]):

1. The autoregressive part of the model for p_t is the polynomial $\varphi(B)$.
2. The model for p_t is $\varphi(B)p_t=\alpha(B)c_t$, where $\alpha(B)$ is a polynomial in B of degree less than or equal to $\max(p, u)$, and c_t are independently distributed as $N(O, \sigma_c^2)$.
3. $$\sigma_d^2\eta(B)\eta(F)=\sigma_c^2\alpha(B)\alpha(F)+\sigma_e^2\varphi(B)\varphi(F), \tag{40}$$

   where $F=B^{-1}$. Obviously, numerous combinations of σ_c^2, σ_e^2 and $\alpha(B)$ will satisfy equation (40).

In the remainder of this section, we sketch the results set out in more detail in Hillmer [22]. A model for p_t is called an acceptable model if, given the model for T_t—

1. $\alpha(B)$ satisfies equation (40) for some $\sigma_c^2\geq 0$ and $\sigma_e^2\geq 0$.
2. The zeros of $\alpha(B)$ lie on or outside the unit circle.

It is easy to obtain the following results:

1. Every given model for T_t has at least one acceptable model for p_t.
2. Given the model for T_t the possible values of σ_e^2 are bounded. We call this bound K^*.

3. Given the model for T_t, then every σ_e^2 in the range $0 \le \sigma_e^2 \le K^*$ determines a unique acceptable model for p_t.

Result (1) follows from letting $\sigma_e^2=0$ in which case $p_t=T_t$ with probability one. Result (2) follows from the fact that, for a model to be acceptable, we require $\sigma_c^2\alpha(B)\alpha(F)\ge 0$ for all $|B|=1$. Then, from equation (40), $\sigma_e^2 \le \dfrac{\sigma_a^2\eta(B)\eta(F)}{\varphi(B)\varphi(F)}$ for all $|B|=1$. For result (3), if $0\le \sigma_e^2 \le K^*$, then $\sigma_a^2\eta(B)\eta(F)-\sigma_e^2(B)\,(F)=g(B)$ is nonnegative for B on the unit circle. Therefore, $g(B)$ determines a unique moving average polynomial $\alpha(B)$.

When $\sigma_e^2=0$, $T_t=p_t$, and there is no smoothing. When $\sigma_e^2=K^*$, on the other hand, the variance of the added white noise is maximized.

An illustrative example—For illustration, consider the rate of change in the consumer price index. Box and Tiao [10] have studied this time series and found that the model

$$(1-B)T_t=(1-0.84B)d_t,\ \sigma_a^2=0.0019$$

adequately describes its behavior from 1953 to 1971. If we assume that this time series follows assumption I, then the model for p_t must be of the form

$$(1-B)p_t=(1-\alpha B)c_t$$

Figure 2 shows the original and smoothed processes for various values of σ_e^2 and α. Also shown are the weight functions, ω_j, implicitly employed when the smoothed value is written in the form

$$\hat{p}_t=\sum_{-\infty}^{\infty}\omega_j T_{t-j}$$

The functions illustrate the case in which the time series available is $T_1, T_2, \ldots T_n$, and t is not close to an end value. The details of how the smoothing was carried out for this example are given later in this section.

While it seems natural in the practical circumstance when σ_e^2 is unknown to choose the variance of the added noise as large as possible, it is also to be noted, as in this example, that a wide range of models for p_t correspond to approximately the same σ_e^2. Furthermore, for this example, models for p_t with α in the range $-1\le\alpha\le 0$ will imply almost identical smoothed estimates.

Smoothing with maximum σ_e^2—We proceed then on the basis that the smoothing to be used should maximize σ_e^2. It is readily shown that—

1. The bound for σ_e^2, K^*, is attainable.
2. The bound K^* occurs when the moving average polynomial of the model for p_t has a zero on the unit circle.

To see (1),

$$K^*=\min_{|B|=1}\left\{\frac{\sigma_a^2\eta(B)\eta(F)}{\varphi(B)\varphi(F)}\right\} \qquad (41)$$

which can be calculated given the model for T_t. Now for (2), if we let $g(B)=\sigma_a^2\eta(B)\eta(F)-K^*\varphi(B)\varphi(F)$, it is then easily seen that $g(B)$ is a covariance generating function that uniquely determines a $\sigma_{c_*}^2$ and $\alpha_*(B)$ such that

$$\sigma_a^2\eta(B)\eta(F)=\sigma_{c_*}^2\alpha_*(B)\alpha_*(F)+K^*\varphi(B)\varphi(F)$$

Furthermore, $\alpha_*(B)$ is uniquely determined such that it has at least one zero on the unit circle and the remainder of its zeros on or outside the unit circle.

Deviation of the smoothing weights and calculation of end values—Therefore, given any model for T_t, we can calculate the maximum σ_e^2 that is consistent with the model for T_t. We shall see that knowing σ_e^2 is sufficient to be able to carry out the smoothing. Cleveland and Tiao [10] have established the following results:

1. When t is not close to the beginning or end of a time series, the smoothed estimate, $\hat{p}_t$, is a symmetric moving average of T_t where the weights, ω_j, are given by the coefficients of B in the generating function
$$\omega(B)=\frac{\sigma_c^2}{\sigma_a^2}\,\frac{\alpha(B)\alpha(F)}{\eta(B)\eta(F)}$$
However, from equation (40), it follows that
$$\omega(B)=1-\frac{\sigma_e^2}{\sigma_a^2}\,\frac{\varphi(B)\varphi(F)}{\eta(B)\eta(F)} \qquad (42)$$
so that knowledge of the model for T_t together with σ_e^2 will enable us to perform the smoothing. A method for computing the weights ω_j from (42) is described in the appendix.
2. For values of $\hat{p}_t$ near the end of the observed time series the following modification is appropriate. Conditional expectations (minimum mean square error forecasts) given the observed series, of a sufficient number of future values are first obtained. The method to obtain the forecasts is given in Box and Jenkins [3]. Then, the forecasted values are used as observations in the formula appropriate for the middle of the series. Consequently, we only need one set of weights. Precisely similar procedures are used for t near the beginning of the series.

Examples—The next two examples will help clarify the ideas that have been presented.

First, suppose T_t follows the model

$$(1-B)T_t=(1-\eta B)d_t$$

Under assumption I, the model for p_t must be of the form

$$(1-B)p_t=(1-\alpha B)c_t$$

for some α, and

$$\sigma_a^2(1-\eta B)(1-\eta F)=\sigma_c^2(1-\alpha B)(1-\alpha F)+\sigma_e^2(1-B)(1-F) \qquad (43)$$

Figure 2. SOME POSSIBLE DECOMPOSITIONS INTO TRENDS AND ERROR, CONSUMER PRICE INDEX EXAMPLE

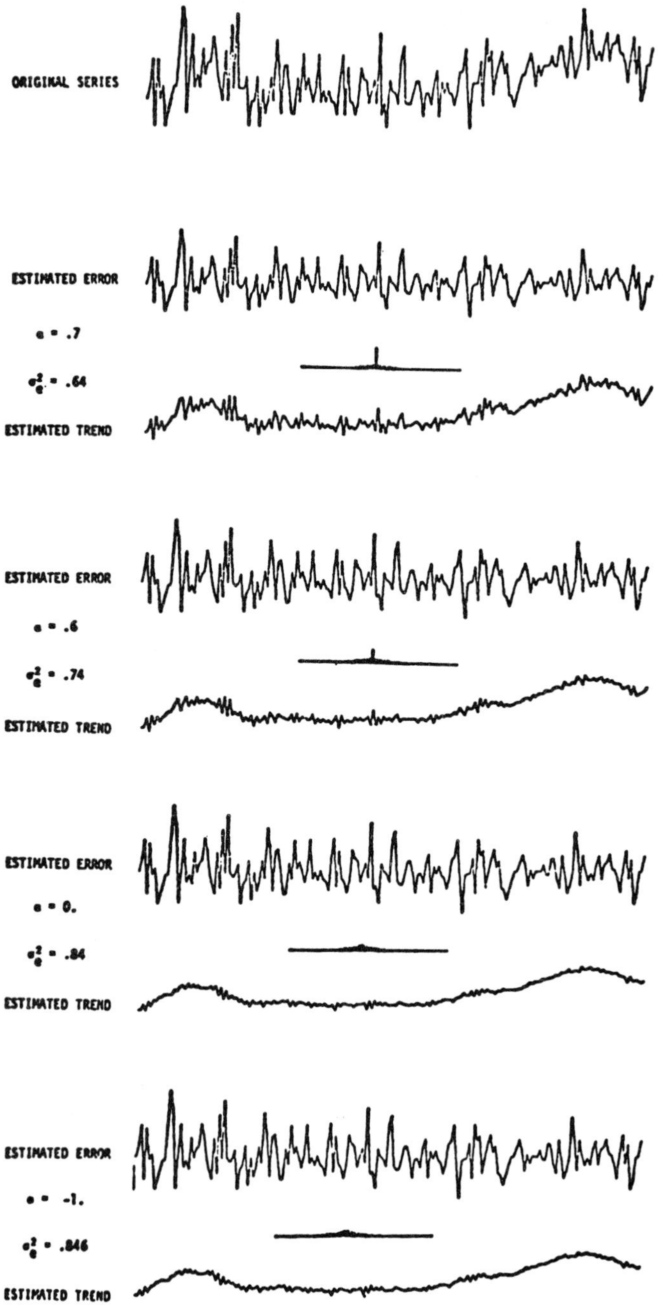

By setting $B=1/\alpha$ in equation (43) and solving for σ_e^2, we obtain

$$\sigma_e^2=\frac{\sigma_d^2(\alpha-\eta)(1-\alpha\eta)}{(1-\alpha)^2}$$

It is easy to see that the maximum possible σ_e^2 occurs when $\alpha=-1$. This agrees with the result that the maximum σ_e^2 is attained when the zero of $(1-\alpha B)$ is on the unit circle. Consequently, the model for p_t, corresponding to the largest possible σ_e^2, is

$$(1-B)p_t=(1+B)c_t$$

and

$$\sigma_e^2=\frac{(1+\eta)^2}{4}\sigma_d^2$$

From (42), the appropriate smoothing weights are then given by the generating function

$$\omega(B)=1-\frac{(1+\eta)^2}{4}\frac{(1-B)(1-F)}{(1-\eta B)(1-\eta F)} \quad (44)$$

Using the method given in the appendix, we find $\omega_0=\frac{1-\eta}{2}$, $\omega_1=\omega_{-1}=\frac{1-\eta^2}{4}$, and $\omega_j=\omega_{-j}=\eta\omega_{j-1}$ for $j=2, 3, \ldots$. Consequently, for estimating p_t in the middle of the observed T_t series, the smoothed estimate is

$$\hat{p}_t=\frac{1-\eta}{2}T_t+\frac{1-\eta^2}{4}\{(T_{t+1}+T_{t-1})+\eta(T_{t+2}+T_{t-2})+\ldots\} \quad (45)$$

Now, to smooth the observed series at the end, one can proceed to first calculate the conditional expectations of $T_{n+1}, T_{n+2}, \ldots$, given $T_n, T_{n-1}, \ldots$. These forecasts are: $\hat{T}_{n+1}=(1-\eta)T_n+\eta(1-\eta)T_{n-1}+\eta^2(1-\eta)T_{n-2}+\ \ldots$ and $\hat{T}_{n+j}=\hat{T}_{n+1}$ for $j=2, 3, \ldots$. Then,

$$\hat{p}_n=\frac{1-\eta}{2}T_n+\frac{1-\eta^2}{4}\{(\hat{T}_{n+1}+T_{n-1})+\eta(\hat{T}_{n+2}+T_{n-2})+\ldots\}$$

or

$$\hat{p}_n=\frac{(3-2\eta-\eta^2)}{4}T_n+\frac{1-\eta^2}{4}\{(1-\eta)T_{n-1}+\eta(1-\eta)T_{n-2}+\ldots\} \quad (46)$$

Smoothed values for other observations near the end of the observed series can be obtained in a similar manner.

Second, suppose the model for T_t is

$$(1-\phi_1B-\phi_2B^2)T_t=(1-\eta_1B-\eta_2B^2)d_t$$

Then, the model for p_t must be of the form

$$(1-\phi_1B-\phi_2B^2)p_t=(1-\alpha_1B-\alpha_2B^2)c_t$$

for some unknown $\alpha_1, \alpha_2, \sigma_c^2$ and σ_e^2. From (41), the maximum possible σ_e^2 consistent with the given model for T_t is

$$K^*=\sigma_d^2\min_{|B|=1}\frac{(1-\eta_1B-\eta_2B^2)(1-\eta_1F-\eta_2F^2)}{(1-\phi_1B-\phi_2B^2)(1-\phi_1F-\phi_2F^2)}$$

or, by letting $B=e^{i\omega}$ for $-0\le\omega\le\pi$, we have that $K^*=\sigma_d^2\min_{0\le\omega\le\pi}\{f(\omega)\}$, where

$$f(\omega)=\frac{(1-\eta_1e^{i\omega}-\eta_2e^{2i\omega})(1-\eta_1e^{-i\omega}-\eta_2e^{-2i\omega})}{(1-\phi_1e^{i\omega}-\phi_2e^{2i\omega})(1-\phi_1e^{-i\omega}-\phi_2e^{-2i\omega})}$$

$$=\frac{1+\eta_1^2+\eta_2^2-(\eta_1-\eta_1\eta_2)2\cos(\omega)-2\eta_2\cos(2\omega)}{1+\phi_1^2+\phi_2^2-(\phi_1-\phi_1\phi_2)2\cos(\omega)-2\phi_2\cos(2\omega)}$$

It is straightforward to numerically minimize $f(\omega)$, given specific values of η_1, η_2, ϕ_1 and ϕ_2. Once this is done, K^* can be calculated, and the appropriate smoothing weights can be calculated from the equation

$$\omega(B)=1-\frac{K^*}{\sigma_d^2}\frac{(1-\phi_1B-\phi_2B^2)(1-\phi_1F-\phi_2F^2)}{(1-\eta_1B-\eta_2B^2)(1-\eta_1F-\eta_2F^2)} \quad (47)$$

Seasonal Adjustment

For a nonseasonal T_t, the smoothing method described may be used directly to estimate the trend component p_t in $T_t=p_t+e_t$. Basically, the idea is to choose $\hat{p}_t$ to extract as much white noise from the observed time series as possible.

It is now supposed that the model contains a seasonal component S_t so that, as in (36),

$$z_t=S_t+T_t=S_t+p_t+e_t \quad (48)$$

The problem is to estimate p_t in the presence of the seasonal component S_t and the noise component e_t. To do this, we must ask what the properties of the seasonal component should be. The concept of a seasonal component and, hence, its definition are, to some extent, arbitrary, but it seems reasonable that it satisfies the following conditions:

1. It should be capable of evolving over time.
2. It should be such that for monthly data the sum of twelve consecutive components varies about zero with minimum variance. The minimum variance requirement in (2) arises because variation greater than minimal variation in the twelve monthly sums should properly be reflected in the trend component p_t or the noise component e_t.

To illustrate how the requirements for a seasonal component can be incorporated into a model-based decomposition procedure, we make a preliminary study in this section by considering the particular stochastic model (28)

$$(1-B)(1-B^{12})z_t=(1-\theta_1 B)(1-\theta_2 B^{12})a_t$$

which has been found to provide an adequate representation of many seasonal time series.

We proceed by making the following assumptions:

1. An observed time series is well represented by (28), with θ_1, θ_2 and σ_a^2 known.
2. S_t and T_t in (48) are independent and follow models of the ARIMA class. Then, necessarily (Cleveland [15]), the product of their autoregressive operators is $(1-B)(1-B^{12})$.
3. $T_t=p_t+e_t$ follows assumption I.

As we have seen, the complementary function for this model satisfies

$$(1-B)(1-B^{12})\hat{z}_t(l)=(1-B)^2(1+B+\ldots+B^{11})\hat{z}_t(l)=0$$

or

$$\hat{z}_t(l)=b_0^{(t)}+b_1^{(t)}l+b_{0,m}^{(t)} \text{ with } \sum_{m=1}^{12} b_{0,m}^{(t)}=0 \tag{49}$$

The adaptive trend term, $b_0^{(t)}+b_1^{(t)}l$, satisfies

$$(1-B)^2(b_0^{(t)}+b_1^{(t)}l)=0$$

and the seasonal component satisfies

$$(1+B+\ldots+B^{11})b_{0,m}^{(t)}=0$$

Thus, an appropriate model for the seasonal component is

$$(1+B+\ldots+B^{11})S_t=(1-\psi_1 B-\ldots-\psi_{11}B^{11})b_t \tag{50}$$

and the corresponding model for the trend component is

$$(1-B)^2T_t=(1-\eta_1 B-\eta_2 B^2)d_t$$

where $\{b_t\}$ and $\{d_t\}$ are two independent Gaussian white-noise processes with zero means and variances σ_b^2 and σ_d^2, respectively.

Letting $\psi(B)=(1-\psi_1 B-\ldots-\psi_{11}B^{11})$, $\eta(B)=(1-\eta_1 B-\eta_2 B^2)$, $\theta(B)=(1-\theta_1 B)(1-\theta_2 B^{12})$, and $U(B)=(1+B+\ldots+B^{11})$, we now observe that

$$(1-B)(1-B^{12})z_t=(1-B)(1-B^{12})S_t+(1-B)(1-B^{12})T_t$$

It follows that

$$\theta(B)a_t=(1-B)^2\psi(B)b_t+U(B)\eta(B)d_t$$

and

$$\sigma_a^2\theta(B)\theta(F)=\sigma_b^2(1-B)^2\psi(B)(1-F)^2\psi(F) + \sigma_d^2U(B)\eta(B)U(F)\eta(F) \tag{51}$$

Minimum variance solution—We require a solution such that $\hat{S}_t=U(B)S_t$ has the smallest variance. Let $\sigma_b^2\geq 0$, $\sigma_d^2\geq 0$, $\eta(B)$ and $\psi(B)$ be an acceptable solution, i.e. one which satisfies equation (3.16). Consider the polynomial

$$g(B)=\sigma_b^2\psi(B)\psi(F)-\epsilon^*U(B)U(F)$$

where

$$\epsilon^*=\min_{|B|=1}\left|\frac{\sigma_b^2\psi(B)\psi(F)}{U(B)U(F)}\right| \tag{52}$$

It follows that $g(B)$ is nonnegative for $|B|=1$ and hence there exists a unique $\sigma_{b_*}^2>0$ and $\psi^*(B)$ such that $\sigma_{b_*}^2\psi^*(B)\psi^*(F)=g(B)$. Now, (51) can be written as

$$\sigma_a^2\theta(B)\theta(F)=\sigma_{b_*}^2(1-B)^2(1-F)^2\psi^*(B)\psi^*(F)+U(B)U(F)H(B)$$

where

$$H(B)=\sigma_d^2\eta(B)\eta(F)+\epsilon^*(1-B)^2(1-F)^2$$

Clearly, $H(B)>0$ for $|B|=1$ so that we can determine a unique $\sigma_{d_*}^2>0$ and $\eta_*(B)$ such that

$$\sigma_{d_*}^2\eta_*(B)\eta_*(F)=H(B) \tag{53}$$

Thus,

$$\sigma_a^2\theta(B)\theta(F)=\sigma_{b_*}^2(1-B)^2(1-F)^2\psi_*(B)\psi_*(F) +\sigma_{d_*}^2 U(B]U(F)\eta_*(B)\eta_*(F) \tag{54}$$

It is shown in Hillmer [22] that the seasonal and trend component model, corresponding to (54), has the desired property that the sum $\hat{S}_t$ has the smallest variance and, further, that the solution is unique.

Thus, to obtain the minimum variance solution, we must first find an acceptable model. As we shall see, for estimating the seasonal component S_t and the trend T_t, it is only necessary to know $\eta_*(B)$. From (51), we see that an acceptable solution is one for which $\sigma_d^2>0$ and

$$\sigma_a^2\theta(B)\theta(F)-\sigma_d^2U(B)U(F)\eta(B)\eta(F) \tag{55}$$

is nonnegative for $|B|=1$ and having four zeros at $B=1$. For the given $\sigma_a^2\theta(B)\theta(F)$, we can then employ numerical methods to obtain an σ_d^2 and an $\eta(B)=1-\eta_1 B-\eta_2 B^2$ having this property. To find the desired $\eta_*(B)$, note that (52) can be alternatively written as

$$\epsilon_*=\min_{|B|=1}\left\{\frac{\sigma_a^2\theta(B)\theta(F)U^{-1}(B)U^{-1}(F)-\sigma_d^2\eta(B)\eta(F)}{(1-B)^2(1-F)^2}\right\} \tag{56}$$

which can calculated using numerical methods. Once ϵ_* is obtained, the required $\eta_*(B)$ can be determined from (53).

Determination of the weight functions for the seasonal and trend components—It remains to obtain the smoothing weights when the models for S_t and T_t are determined. Let the estimated seasonal component be $\hat{S}_t$ and the estimated trend plus noise be $\hat{T}_t$. To estimate a component in the middle of the time series, we have, from Cleveland and Tiao [10], that

$$\hat{S}_t=\sum_{j=-\infty}^{\infty} w_j z_{t-j}=w(B)z_t$$

and

$$\hat{T}_t=\sum_{j=-\infty}^{\infty} h_j z_{t-j}=h(B)z_t \tag{57}$$

The generating function for the seasonal component is

$$w(B)=\frac{\sigma_b^2}{\sigma_a^2}\,\frac{(1-B)^2\psi(B)(1-F)^2\psi(F)}{\theta(B)\theta(F)} \tag{58}$$

However, since $z_t=\hat{S}_t+\hat{T}_t$, we have that $1=w(B)+h(B)$, so that

$$h(B)=\frac{\sigma_d^2}{\sigma_a^2}\,\frac{U(B)\eta(B)U(F)\eta(F)}{\theta(B)\theta(F)} \tag{59}$$

We do not need to know $\frac{\sigma_d^2}{\sigma_a^2}$, since this ratio can be obtained from the fact that $h(1)=1$. We suggest calculating $h(B)$, which can then be used to calculate $\hat{T}_t$, then $\hat{S}_t=z_t-\hat{T}_t$.

The smoothed values near the ends of the series are obtained as before, by forecasting the required unobserved values of z_t and using these forecasts in the formula for the center of the series.

The estimated trend plus noise component T_t can now be used to compute the estimate of the trend p_t. Noting that $T_t=p_t+e_t$ and the model for T_t is $(1-B)\,T_t=\eta(B)d_t$, it is readily verified from the results in the subsection on trend plus noise that

$$\hat{p}_t=\left[1-\frac{\sigma_e^2(1-B)^2(1-F)^2}{\sigma_d^2\eta(B)\eta(F)}\right]\hat{T}_t \tag{60}$$

The smoothest estimate of p_t is obtained when σ_e^2 is maximized, i.e., when

$$\sigma_e^2=\min_{|B|=1}\left|\frac{\sigma_d^2\eta(B)\eta(F)}{(1-B)^2(1-F)^2}\right| \tag{61}$$

Application to the Times Series of Monthly U.S. Unemployed Males, 20 Years Old and Over

These ideas will be illustrated by applying them to the time series of monthly U.S. unemployed males, 20 years old and over. This series is the largest component of the total unemployed series. The political and economic impact of this series has been especially important recently, and we consider it an important series in which to try out our methods.

By following the model building procedure, sketched in the subsection on model building, we find that the model

$$(1-B)(1-B^{12})z_t=(1-0.75B^{12})a_t,\ \sigma_a^2=0.0037 \tag{62}$$

adequately describes the behavior of the log of the observed series. Therefore, we can apply the seasonal adjustment and smoothing procedures previously outlined to this data. We find—

1. An acceptable solution results when $\eta_1=0.9251$, $\eta_2=0.05$ and $\sigma_d^2=0.700\sigma_a^2$.

2. $\min_{|B|=1}\left\{\frac{\sigma_a^2\theta(B)\theta(F)U^{-1}(B)U^{-1}(F)-\sigma_d^2\eta(B)\eta(F)}{(1-B)^2(1-F)^2}\right\}$ $=0.0376\sigma_a^2$.

3. $\eta^*(B)=(1-0.9798B+0.0034B^2)$, $\sigma_{a_*}^2=0.7780\sigma_a^2$.

4. $\min_{|B|=1}\left\{\frac{\sigma_{a_*}^2\eta_*(B)\eta_*(F)}{(1-B)^2(1-F)^2}\right\}=0.1915\sigma_a^2$

The seasonal and trend weights obtained by our procedure and by the census procedure are plotted in figure 3. It is interesting that the seasonal weights of both procedures, although arrived at quite differently, behave in a similar manner. Notice, however, that the trend weights for the two procedures are different. It should be remembered that our weight functions will adjust appropriately depending on the parameters of the particular series. The census weight function, however, is fixed for all series. The observed time series and the three estimated components derived from our procedure are plotted in figure 4a; the analogous series for the census procedure are plotted in figure 4b. Observe that the estimated seasonal components appear similar to each other. The estimated trend component from the census procedure is smoother than the trend component from our procedure, but this may reflect oversmoothing.

Generalization to other models—We have considered seasonal adjustment when the observed time series follows the model (28). Examination of the complementary function (49) led us to models for the seasonal and trend components that seemed sensible. Because the class of ARIMA models is more general than the model (28), it is doubtful whether all seasonal models in the ARIMA class could be treated in a precisely similar manner. If this is so, it will be because of the too limited nature of the seasonal adjustment concept; this concept may need to be widened. Every model in the ARIMA class will have a corresponding complementary function. Appropriate adjustment, we feel, may turn on suitable factorization of this function. The problem is an important one for future consideration.

Figure 3. TREND AND SEASONAL WEIGHTS FOR THE BUREAU OF THE CENSUS AND MODEL-BASED PROCEDURES

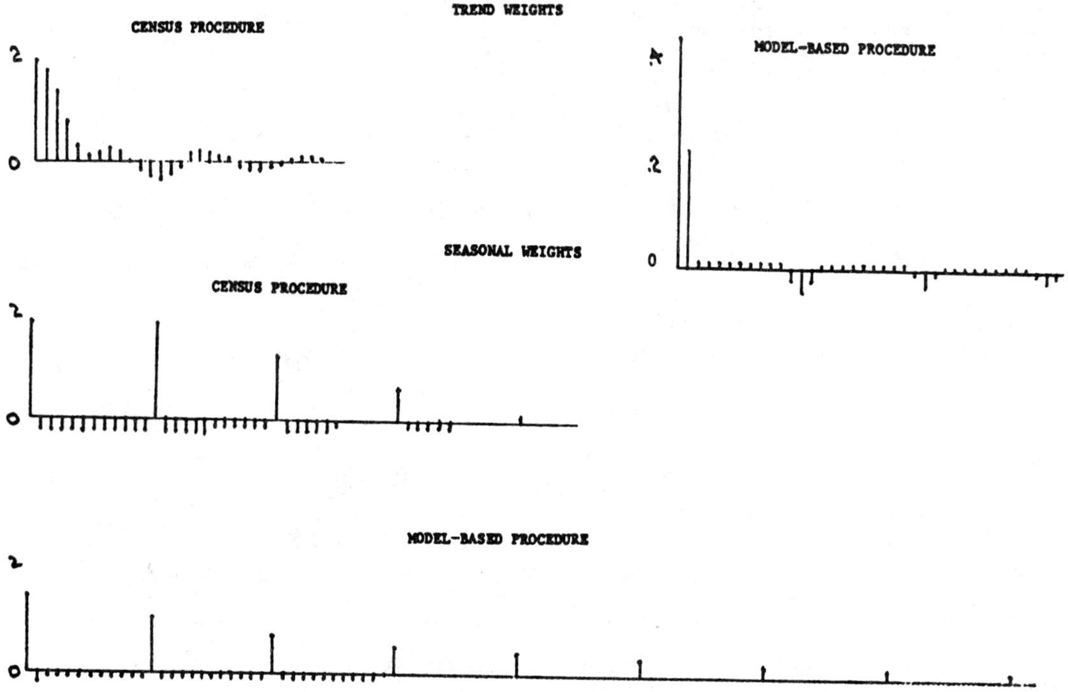

Figure 4a. THE ESTIMATED COMPONENTS FROM THE MODEL-BASED PROCEDURES FOR THE TIME SERIES OF UNEMPLOYED MALES, 20 YEARS OLD AND OVER

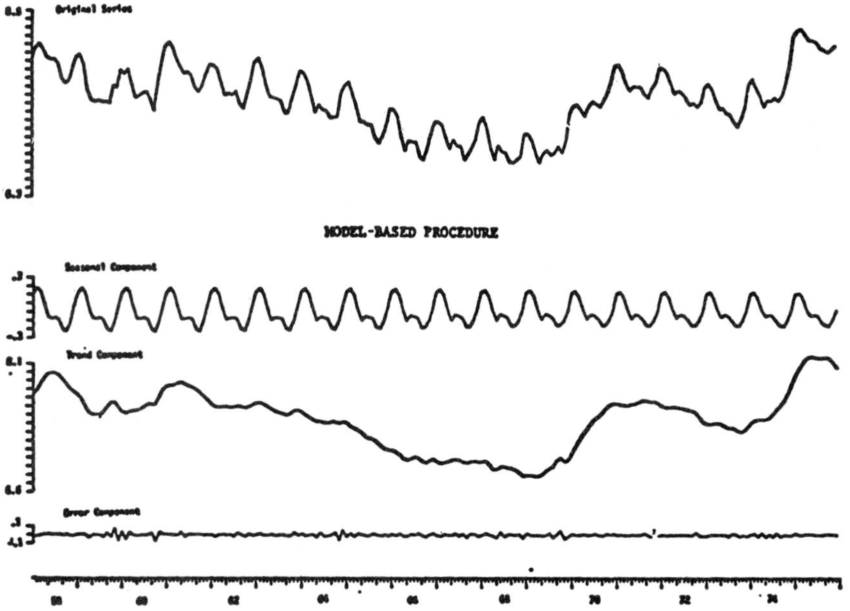

(a)

Figure 4b. THE ESTIMATED COMPONENTS FROM THE CENSUS PROCEDURE FOR THE TIME SERIES OF UNEMPLOYED MALES, 20 YEARS OLD AND OVER

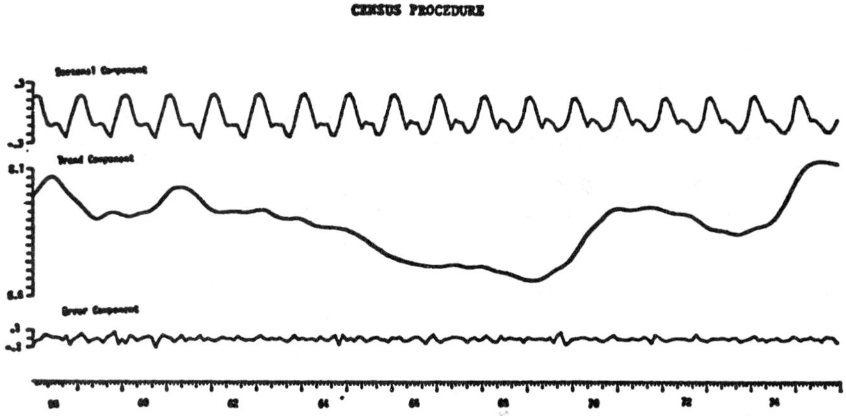

(b)

MULTIVARIATE GENERALIZATION OF THE MODELS

When k series $\{z_{1t}\}$, $\{z_{2t}\}$, ..., $\{z_{kt}\}$ are considered simultaneously, it is necessary to allow for dynamic relationships that may exist between the series, for possibilities of feedback and for correlations between the shocks affecting the series. A useful class of models is obtained by direct generalization of (5) to

$$\Phi(B)\underset{\sim}{z}_t=\Theta(B)\underset{\sim}{a}_t \qquad (63)$$

In this model,

$$\underset{\sim}{z}'_t=(z_{1t}, \ldots, z_{kt}),\ \underset{\sim}{a}'_t=(a_{1t}, \ldots, a_{kt})$$

$$\Phi(B)=I-\Phi_1B-\Phi_2B^2-\ldots-\Phi_pB^p$$

$$\Theta(B)=I-\Theta_1B-\Theta_2B^2-\ldots-\Theta_qB^q$$

and the Φ_i's and Θ_j's are $k \times k$ matrices of autoregressive and moving average parameters. $\{\underset{\sim}{a}_t\}$ is a sequence of vector-valued independent random shocks, distributed as multivariate normal $N(O, \Sigma)$ that allows for contemporaneous correlation between the elements $a_{1t}, \ldots, a_{kt}$.

While multiple time series models are naturally more complicated to handle and experience in their use is more limited, they provide a potential means of improving on results from univariate models. For example, information about z_{1t}, in addition to that contained in its own past, may be available from other related series $\{z_{2t}\}$, $\{z_{3t}\}$, etc. When this is so, improved forecasts, smoothed values, seasonal adjustments, etc., should be possible.

One difficulty that has previously impeded progress has been the estimation of parameters contained in the models. Initially, therefore, our attack has been directed at this problem, and, in the next section, we describe a practical means of computing exact maximum likelihood estimates.

Estimation

Whittle [31] and Hannan and Dunsmuir [21] have shown, for multivariate ARIMA models with normally distributed errors, that maximum likelihood estimates have desirable asymptotic properties. In particular, the estimates are asymptotically consistent and efficient. In addition, a number of authors have preformed simulations which indicate that maximum likelihood estimates have desirable small sample properties. Several examples are reported in Hillmer [22]. We will proceed to find maximum likelihood estimates of the parameters, assuming that the $\underset{\sim}{a}_t$'s are normally distributed.

Most time series estimation procedures that have been proposed to date are motivated by first considering the likelihood function and, then, by making some simplifying approximations to this function. (See, e.g., Anderson [1], Box and Jenkins [9], and Hannan [20].) While most of the simplifying approximations to the likelihood function have no effect upon the asymptotic estimates, these approximations can have an effect upon the estimates in small or even moderately large samples. In particular, the approximations can have a significant effect when estimating seasonal multiplicative ARIMA models. The likelihood function is usually simplified by ignoring the effect of the ends of the time series and by ignoring the changes that occur in the normalizing determinant. For univariate time series, Box and Jenkins [9] overcame the first problem by proposing an exact method for computing the exponent in the likelihood.

Recent papers, have indicated that further worthwhile improvement can be made in the estimation procedure if the terms in the determinant are not ignored. In the case of univariate time series several procedures have been recently proposed to obtain exact maximum likelihood estimates, see in particular, Ljung [25] and Dent [17].

The problem of estimating the parameters in a multivariate time series model is more difficult than the univariate problem. However, Hillmer [22] has developed procedures that give maximum likelihood estimates for multivariate ARIMA models. In order to illustrate those ideas, we give the details for a first-order moving average model.

First Order Moving Average Model

The multivariate MA(1) model is

$$\underset{\sim}{z}_t=\underset{\sim}{a}_t-\theta\underset{\sim}{a}_{t-1},\ t=1, \ldots n \qquad (64)$$

where $\underset{\sim}{z}_t$ is a vector of k-observed time series, $\underset{\sim}{a}_t$ is a vector of unobserved errors following a multivariate normal distribution with mean vector $\underset{\sim}{0}$ and unknown covariance matrix Σ, and θ is a $k \times k$ matrix of parameters. Our objective will be to estimate θ and Σ.

First we derive the likelihood function of θ and Σ given the observations $\underset{\sim}{z}_1, \ldots, \underset{\sim}{z}_n$. Consider the transformation

$$\begin{aligned}\underset{\sim}{a}_0&=\underset{\sim}{a}_0\\ \underset{\sim}{z}_1&=\underset{\sim}{a}_1-\theta\underset{\sim}{a}_0\\ &\ \vdots\\ \underset{\sim}{z}_n&=\underset{\sim}{a}_n-\theta\underset{\sim}{a}_{n-1}\end{aligned} \quad \text{or} \quad \begin{bmatrix}\underset{\sim}{a}_0\\ \underset{\sim}{z}\end{bmatrix}=\begin{bmatrix}I_k & 0\\ D & C\end{bmatrix}\begin{bmatrix}\underset{\sim}{a}_0\\ \underset{\sim}{a}\end{bmatrix} \qquad (65)$$

where $\underset{\sim}{z}=(\underset{\sim}{z}'_1, \ldots, \underset{\sim}{z}'_n)'$ $\underset{\sim}{a}=(\underset{\sim}{a}'_1, \ldots, \underset{\sim}{a}'_n)'$,

$$D=\begin{bmatrix}-\theta\\ 0\\ \cdot\\ \cdot\\ \cdot\\ 0\end{bmatrix} \quad \text{and } C=\begin{bmatrix}I_k & 0 & \cdots & & 0\\ -\theta & I_k & & & \cdot\\ 0 & -\theta & I_k & & \cdot\\ \cdot & & & \ddots & \cdot\\ \cdot & & & & 0\\ 0 & \cdots & 0 & -\theta & I_k\end{bmatrix}$$

Equation (65) implies that

$$\begin{bmatrix}\underset{\sim}{a}_0\\ \underset{\sim}{a}\end{bmatrix}=\begin{bmatrix}I_k & 0\\ -C^{-1}D & C^{-1}\end{bmatrix}\begin{bmatrix}\underset{\sim}{a}_0\\ \underset{\sim}{z}\end{bmatrix}$$

$$=\begin{bmatrix}I_k\\ -C^{-1}D\end{bmatrix}\underset{\sim}{a}_0+\begin{bmatrix}0\\ C^{-1}\end{bmatrix}\underset{\sim}{z} \qquad (66)$$

Because the a_t are independently distributed as $N_k(0, \Sigma)$, the joint probability density function of a_0 and a is

$$p(a_0, a)=\left(\frac{1}{2\pi}\right)^{\frac{k(n+1)}{2}}|\Sigma|^{-\frac{n+1}{2}}$$

$$\exp\left\{-\frac{1}{2}[a_0', a'][I_{n+1}\otimes\Sigma^{-1}]\begin{bmatrix}a_0\\ a\end{bmatrix}\right\} \quad (67)$$

where $\otimes$ denotes the Kronecker product of two matrices. Noting that the Jacobian of the transformation (66) is unity, expression (67) implies that the joint probability density function of a_0 and z is

$$p(a_0, z)=\left(\frac{1}{2\pi}\right)^{\frac{k(n+1)}{2}}|\Sigma|^{-\frac{n+1}{2}}\exp\left\{-\frac{1}{2}S(a_0, z|\theta, \Sigma)\right\} \quad (68)$$

where

$$S(a_0, z|\theta, \Sigma)=\left\{\begin{bmatrix}I\\ -C^{-1}D\end{bmatrix}a_0+\begin{bmatrix}0\\ C^{-1}\end{bmatrix}z\right\}$$

$$\{I_{n+1}\otimes\Sigma^{-1}\}\left\{\begin{bmatrix}I\\ -C^{-1}D\end{bmatrix}a_0+\begin{bmatrix}0\\ C^{-1}\end{bmatrix}z\right\} \quad (69)$$

By completing the square for a_0 in the quadratic form $S(a_0, z|\theta, \Sigma)$, we obtain

$$S(a_0, z|\theta, \Sigma)=S(\hat{a}_0, z|\theta, \Sigma)+(a_0-\hat{a}_0)'A_n(\hat{a}_0-a_0) \quad (70)$$

where

$$A_n=[I_k, -D'C'^{-1}][I_{n+1}\otimes\Sigma^{-1}]\begin{bmatrix}I_k\\ -C^{-1}D\end{bmatrix} \quad (71)$$

and

$$a_0=-A_n^{-1}[I_k, -D'C'^{-1}][I_{n+1}\otimes\Sigma^{-1}]\begin{bmatrix}0\\ C^{-1}\end{bmatrix}z \quad (72)$$

Consequently, we have that

$$p(a_0, z)=\left(\frac{1}{2\pi}\right)^{\frac{k(n+1)}{2}}|\Sigma|^{-\frac{n+1}{2}}$$

$$\exp\left\{-\frac{1}{2}S(\hat{a}_0, z|\theta, \Sigma)\right\}\exp\left\{-\frac{1}{2}(a_0-\hat{a}_0)'A_n(a_0-\hat{a}_0)\right\}$$

By integrating out a_0 in $p(a_0, z)$, we obtain the probability density function of z. By treating this probability density function as a function of θ and Σ, we obtain the likelihood function of θ and Σ given z as

$$L(\theta, \Sigma|z)=|\Sigma|^{-\frac{n+1}{2}}\cdot|A^n|^{-\frac{1}{2}}\exp\left\{-\frac{1}{2}S(\hat{a}_0, z|\theta, \Sigma)\right\} \quad (73)$$

Observe, from equation (69), that

$$S(a_0, z|\theta, \Sigma)=\sum_{t=0}^{n}a_t'\Sigma^{-1}a_t$$

Furthermore, given θ, a_0, z_t for $t=1, \ldots n$, we can calculate the a_t's from the recursive relationship

$$a_t=z_t+\theta a_{t-1}$$

which is defined by the transformation in equation (65). Similarly, we can calculate $S(a_0, z|\theta, \Sigma)$, given θ, z and a_0, by calculating

$$\hat{a}_t=z_t+\theta\hat{a}_{t-1} \text{ for } t=1, \ldots n$$

and then

$$S(\hat{a}_0 z|\theta, \Sigma)=\sum_{t=0}^{n}\hat{a}_t'\Sigma^{-1}\hat{a}_t \quad (74)$$

The important point to observe from this discussion is that the likelihood function can be evaluated for any given θ and Σ very quickly once A_n and $\hat{a}_0$ have been calculated. Consequently, we focus our attention upon the calculation of A_n and a_0. Note that

$$C^{-1}=\begin{bmatrix}I_k & & & & 0\\ \theta & I_k & & & \\ . & \theta & . & & \\ . & & . & . & \\ . & & & . & . \\ \theta^{n-1} & \theta^{n-2} & \ldots & \theta & I_k\end{bmatrix}$$

then, from equation (71), we obtain

$$A_n=\Sigma^{-1}+\theta'\Sigma^{-1}\theta+\ldots+\theta'^n\Sigma^{-1}\theta^n \quad (75)$$

One easy way to perform the calculation of A_n is by the recursive calculation

$$A_0=\Sigma^{-1} \text{ and } A_j=\Sigma^{-1}+\theta'A_{j-1}\theta \text{ for } j=1, \ldots n \quad (76)$$

Calculation of A_n can be quickly performed on a computer with a minimum of storage. Next, consider the calculation of $\hat{a}_0$. From (72),

$$\hat{a}_0=A_n^{-1}D'C'^{-1}\{I_n\otimes\Sigma^{-1}\}C^{-1}z$$

By multiplying out the matrices, we have that

$$D'C'^{-1}\{I_n\otimes\Sigma^{-1}\}C^{-1}z=\theta'A_{n-1}z_1+\theta'^2A_{n-2}z_2+\ldots+\theta'^nA_0z_n$$

where the A_j matrices are defined by equation (76). The vector $D'C'^{-1}\{I_n\otimes\Sigma^{-1}\}C^{-1}z$ can be calculated recursively as follows:

let $g_0=0$ and $A_0=\Sigma^{-1}$; let $g_j=\theta'\{g_{j-1}+A_{j-1}z_{n-j+1}\}$ for $j=1, \ldots n$

then

$$g_n = \theta' A_{n-1} z_1 + \theta'^2 A_{n-2} z_2 + \ldots + \theta'^{n-1} A_0 z_n$$

To summarize, we have the following algorithm for the calculation of A_n and a_0:

1. Let $A_0 = \Sigma^{-1}$ and $g_0 = 0$.
2. For $j = 1, \ldots, n$, let $g_j = \theta'\{g_{j-1} + A_{j-1} z_{n-j+1}\}$ and let $A_j = \Sigma^{-1} + \theta' A_{j-1}\theta$.
3. $\hat{a}_0 = -A_n^{-1} g_n$.

For any given θ and Σ, we now have a way to calculate A_n and $\hat{a}_0$. Therefore, we can use these values to calculate the exact likelihood function.

We desire to maximize $L(\theta, \Sigma | z)$ with respect to the parameter matrices θ and Σ. We have illustrated a way to evaluate $L(\theta, \Sigma | z)$ for any particular values of θ and Σ. What is needed is some maximization algorithm that will systematically search over the parameter values and find the θ and Σ that will maximize $L(\theta, \Sigma | z)$. There are numerous optimization algorithms that can be used. We have used a nonlinear regression algorithm to estimate the parameters in the multivariate ARIMA examples given in the next section.

Application: The Durable Shipments, Durable New Orders, and Durable Inventories Series

We shall use the three series—

1. Shipments of durable goods
2. New orders of durable goods
3. Inventories of durable goods

to illustrate an analysis of seasonal multivariate models. A useful identification step is to build univariate models for the three series; they are:

1. $(1-B)(1-B^{12})z_{1t} = (1-B)(1-0.75B^{12})a_{1t}$

   $\sigma^2_{a_1} = 0.00093$ with z_{1t} = log-durable shipments.

2. $(1-B)(1-B^{12})z_{2t} = (1-0.26B)(1-0.80B^{12})a_{2t}$ (77)

   $\sigma^2_{a_2} = 0.00179$ with z_{2t} = log-durable new orders.

3. $(1-0.85B)(1-B^{12})z_{3t} = (1-0.36B)(1-0.73B^{12})a_{3t}$

   $\sigma^2_{a_3} = 0.0000268$ with $z_{3t} = (1-B)$ log-durable inventories.

From consideration of the univariate models we tentatively entertained the multivariate seasonal model

$$(I-\Phi_1)(I-\Phi_{12}B^{12})z_t = (I-\theta_1 B)(I-\theta_{12}B^{12})a_t \quad (78)$$

where

$z_t' = (z_{1t}, z_{2t}, z_{3t})$, $a_t' = (a_{1t}, a_{2t}, a_{3t})$ and Φ_1, Φ_{12}, θ_1 and θ_{12} are 3 x 3 matrices. The parameters in this model are estimated as follows:

$$\Phi_1 = \begin{bmatrix} 0.36 & 0.53 & 0.75 \\ (0.12) & (0.09) & (0.79) \\ -0.29 & 1.30 & -2.00 \\ (0.15) & (0.11) & (0.83) \\ -0.05 & 0.04 & 0.79 \\ (0.01) & (0.01) & (0.07) \end{bmatrix}$$

$$\Phi_{12} = \begin{bmatrix} 0.93 & 0.03 & 0.03 \\ (0.06) & (0.06) & (0.21) \\ 0.20 & 0.75 & 0.20 \\ (0.07) & (0.08) & (0.27) \\ 0.01 & -0.01 & 0.93 \\ (0.01) & (0.01) & (0.03) \end{bmatrix}$$

$$\hat{\theta}_1 = \begin{bmatrix} -0.22 & 0.31 & -3.22 \\ (0.16) & (0.11) & (0.90) \\ -0.25 & 0.53 & -3.62 \\ (0.24) & (0.16) & (1.07) \\ -0.05 & 0.02 & 0.50 \\ (0.02) & (0.03) & (0.10) \end{bmatrix}$$

$$\hat{\theta}_{12} = \begin{bmatrix} 0.54 & 0.08 & 0.24 \\ (0.11) & (0.08) & (0.43) \\ -0.02 & 0.57 & 0.58 \\ (0.11) & (0.12) & (0.65) \\ 0.04 & -0.01 & 0.76 \\ (0.02) & (0.01) & (0.06) \end{bmatrix}$$

$$\Sigma = \begin{bmatrix} 0.000548 & 0.000596 & -0.000010 \\ & 0.001464 & -0.000014 \\ & & 0.000021 \end{bmatrix} \quad (79)$$

The numbers in the parentheses are the estimated standard errors of the parameter estimates. The residuals from this model are plotted in figure 5, and the autocorrelation and cross-correlation functions of the residuals are plotted in figure 6. Examination of these plots does not suggest any inadequacies in the model.

Inspection of the fitted model reveals a structure that seems to be readily capable of interpretation. This is made clearer if we simplify the model by setting to zero coefficients that are small, compared with their standard errors, and make other minor adjustments.

A valuable feature of the multivariate maximum likelihood program is that it permits any chosen coefficient to

Figure 5. **RESIDUALS FROM THE MULTIVARIATE FIT IN THE DURABLES EXAMPLE**

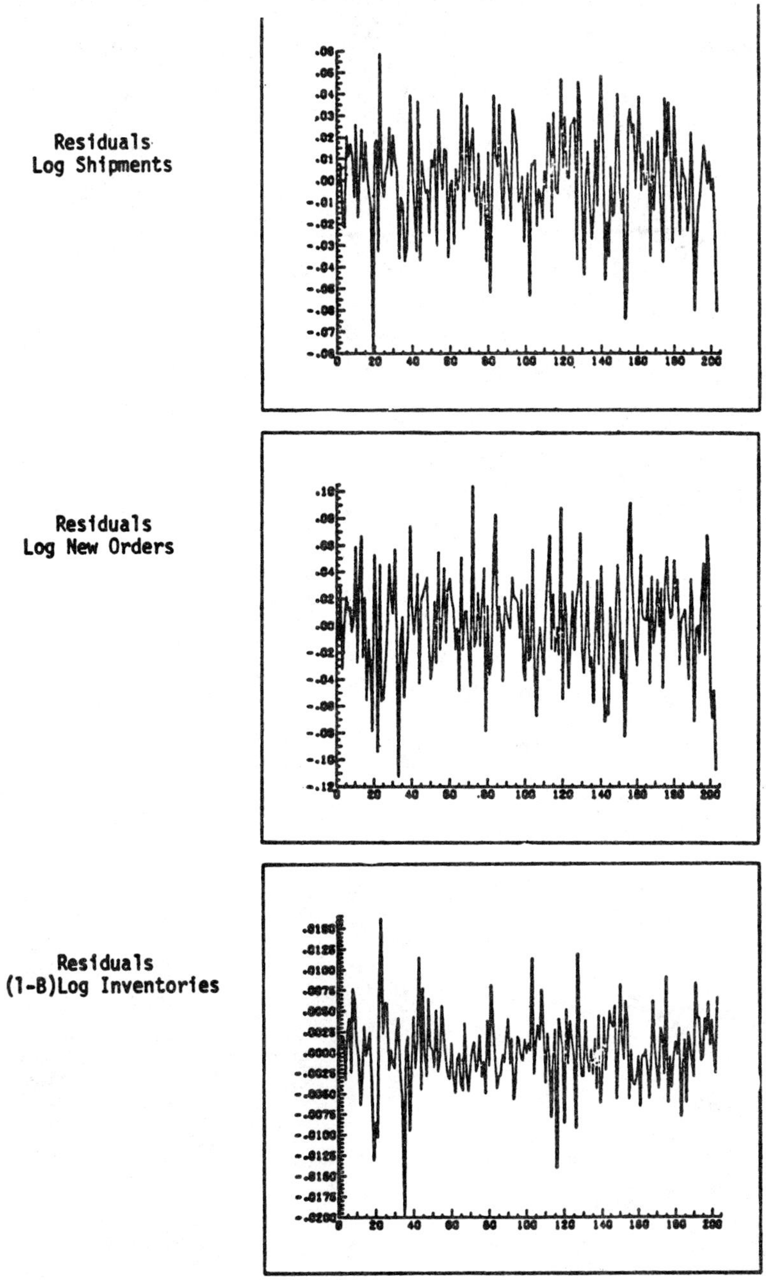

Figure 6. AUTOCORRELATIONS AND CROSSCORRELATIONS OF THE RESIDUALS FROM THE ESTIMATED MULTIVARIATE MODEL FOR THE DURABLES SERIES

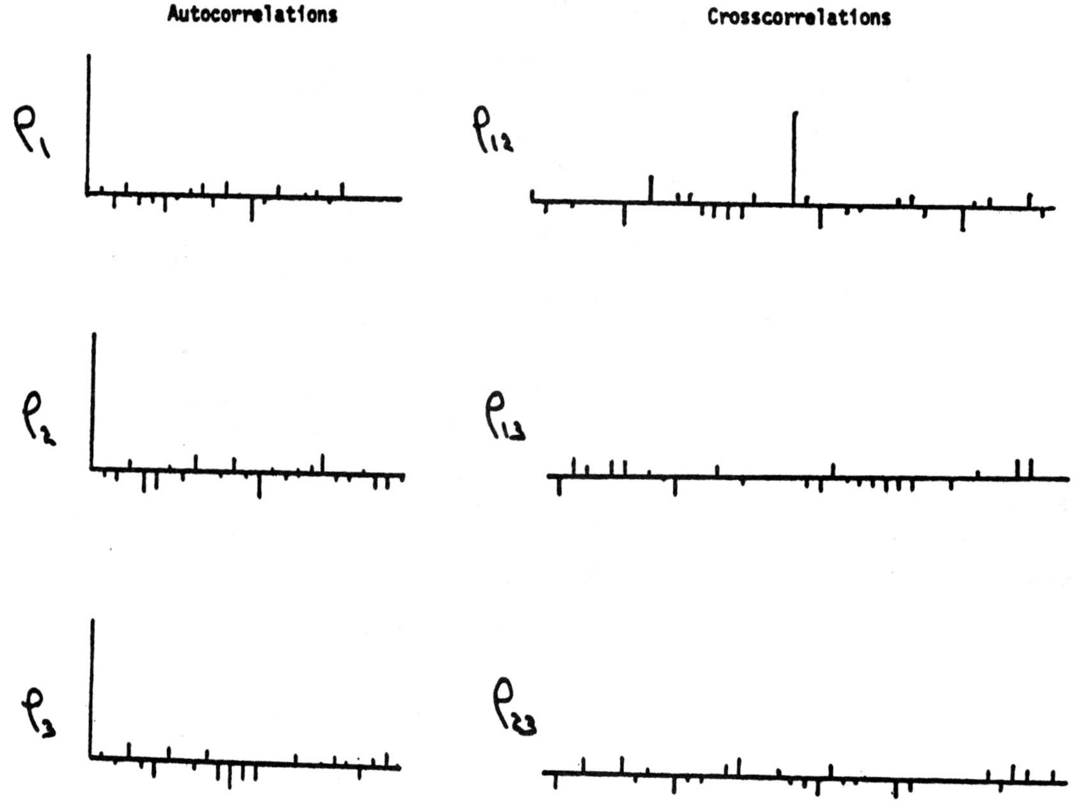

be tentatively fixed at any desired value and, in particular, to be set equal to zero. It also allows coefficients to be estimated under a given constraint. For example, two coefficients may be estimated subject to the constraint that they be equal. This permits any desired simplification to be explored and makes it easier to see the implications of a given model. In the present example, the matrices $\hat{\Phi}_{12}$ and $\hat{\theta}_{12}$ are, apart from elements that are not large compared with their standard errors, very nearly diagonal. Also, the diagonal elements in $\hat{\Phi}_{12}$ are large, two of them being close to unity. Now, moderate changes made in $\hat{\Phi}_{12}$ can be very nearly compensated by appropriate changes in $\hat{\theta}_{12}$. In refitting, therefore, Φ_{12} was simplified to be an identity matrix, and θ_{12} was made diagonal. Furthermore, corresponding elements in (1) the first row and second column and (2) the second row and third column of $\hat{\Phi}_1$ and $\hat{\theta}_1$ are not significantly different. This has a particular interpretation that forecasted values may be used directly in the equations. Thus, in refitting, each of these two pairs of elements were tentatively set equal. The results are given in (80). It will be noted that the diagonal elements of $\hat{\Sigma}$, associated with the one step ahead forecast errors of the refitted model, are somewhat larger than those in (79) but still much smaller than the corresponding variances in (77) obtained for the univariate models. No strong evidence of lack of fit was shown in the residual analysis.

$$\hat{\Phi}_1=\begin{bmatrix} \underset{(0.05)}{0.79} & \underset{(0.05)}{0.20} & \cdot \\ \cdot & \underset{(0.06)}{0.97} & \underset{(0.36)}{-0.69} \\ \cdot & \cdot & \underset{(0.04)}{0.90} \end{bmatrix}, \quad \hat{\Phi}_{12}=\begin{bmatrix} 1. & \cdot & \cdot \\ \cdot & 1. & \cdot \\ \cdot & \cdot & 1. \end{bmatrix}$$

$$\hat{\theta}_1=\begin{bmatrix} \cdot & \underset{(0.05)}{0.20} & \underset{(0.25)}{-1.72} \\ \cdot & \underset{(0.80)}{0.36} & \underset{(0.36)}{-0.69} \\ \cdot & \cdot & \underset{(0.08)}{0.40} \end{bmatrix}, \quad \hat{\theta}_{12}=\begin{bmatrix} \underset{(0.04)}{0.77} & \cdot & \cdot \\ \cdot & \underset{(0.05)}{0.74} & \cdot \\ \cdot & \cdot & \underset{(0.05)}{0.75} \end{bmatrix}$$

$$\hat{\Sigma}=\begin{bmatrix} 0.000644 & 0.000711 & -0.000007 \\ \cdot & 0.001655 & -0.000002 \\ \cdot & \cdot & 0.000027 \end{bmatrix} \qquad (80)$$

Implications of the model—If we write

$$w_t=(I-\theta_{12}B^{12})^{-1}(I-\Phi_{12}B^{12})z_t \qquad (81)$$

then

$$w_{1t}=z_{1t}-\tilde{z}_{1,\,t-12},\ w_{2,\,t}=z_{2t}-\tilde{z}_{2,\,t-12},\ w_{3t}=z_{3t}-\tilde{z}_{3,\,t-12}$$

where, for example,

$$\tilde{z}_{1,\,t-12}=0.23(z_{1,\,t-12}+0.77z_{1,\,t-24}+0.77^2z_{1,\,t-36}+\ldots)$$

Thus, the w's are deviations from seasonal exponentially discounted averages and have the effect of removing the seasonal component. The relationships between these deviations are now approximately as follows:

$$w_{1t}-0.8w_{1,\,t-1}=a_{1t}+0.2(w_{2,\,t-1}-a_{2,\,t-1})+1.7a_{3,\,t-1} \qquad (82)$$

$$w_{2t}-w_{2,\,t-1}=a_{2t}-0.4a_{2,\,t-1}-0.7(w_{3,\,t-1}-a_{3,\,t-1}) \qquad (83)$$

$$w_{3t}-0.9w_{3,\,t-1}=a_{3t}-0.4a_{3,\,t-1} \qquad (84)$$

We will now consider the implications of these equations.

Inventories—Inspection of $\hat{\Sigma}$ in (80) shows that there is no evidence that a_{3t} is correlated with a_{1t} and a_{2t}; also, (84) does not contain w_{1t} or w_{2t}. Thus, rather remarkably, the inventory series behaves independently of the other two series, and its complementary (forecast) function is very nearly such that

$$(1-B)^3(1+B+\cdots+B^{11})\hat{x}_t(l)=0$$

where $\hat{x}_t(l)$ is the log of the (undifferenced) inventories. The solution, thus, involves a quadratic trend component plus a seasonal component. This holds out the possibility that upturns and downturns in inventories might be forecast.

New orders—Again, using a notation adopted earlier, we write $x_t^{(\theta)}$ to mean an exponentially smoothed value with smoothing coefficient θ, $x_t^{(\theta)}=(1-\theta)\sum_{j=0}^{\infty}\theta^j x_{t-j}$.

Thus, equation (83) may be written aproximately as

$$w_{2t}=w_{2,\,t-1}^{(.4)}-\hat{w}_{3,\,t-2}^{(.4)}(1)+a_{2t},\ \sigma_{a_2}^2=0.00166$$

The implication is that the change in new orders from last month's (smoothed) value is just such as will balance the (smoothed) value forecast for last month's change in inventories. Since we are dealing with logged data, it should be remembered that change refers to percentage change.

Shipments—Equation (82) may be written

$$w_{1t}=w_{1,\,t-1}+0.2\{\hat{w}_{2,\,t-2}(1)-w_{1,\,t-1}\}$$
$$+1.7\{w_{3,\,t-1}-\hat{w}_{3,\,t-2}(1)\}+a_{1t}$$

Now, whereas a_{3t} appears to be independent of a_{1t} and a_{2t}, a_{1t} and a_{2t} have a correlation of 0.69. Correspondingly, if we write

$$a_{1t}=0.43a_{2t}+a_{1t}',\ \text{with } \sigma_{a_1'}^2=0.000339$$

Figure 7. FORECASTS FROM THE UNIVARIATE AND MULTIVARIATE MODELS FOR THE DURABLES SERIES

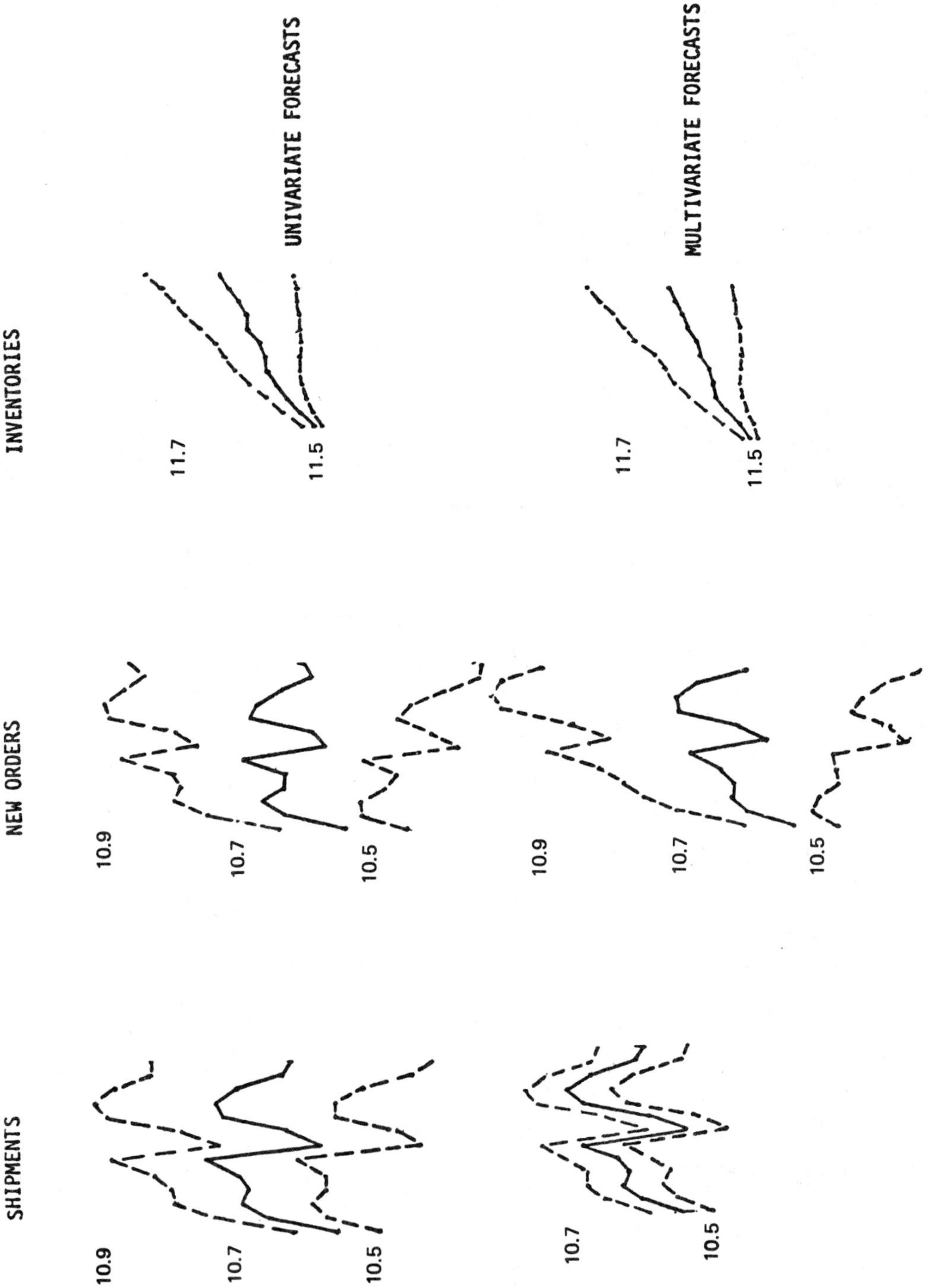

then a'_{1t} is uncorrelated with a_{2t}. Thus, we can obtain three equations that have independent errors by taking the third to be

$$w_{1t}=w_{1,\,t-1}+0.2\{\hat{w}_{2,\,t-2}(1)-w_{1,\,t-1}\}+1.7\{w_{3t-1}-\hat{w}_{3,\,t-2}(1)\}$$
$$+0.4\{w_{2t}-\hat{w}_{2,\,t-1}(1)\}+a'_{1t} \quad \sigma^2_{a'_1}=0.00034$$

Thus, this month's shipments are increased over last month's when—

1. Forecast orders for last month exceeded shipments.
2. Increases in inventories for last month exceeded that forecasted.
3. Current new orders exceed last month's forecast.

It is gratifying to see that this model makes economic sense.

Forecasting—As already noted, there is a substantial reduction in the variance of the shipments and new order series when the multivariate model is used. A corresponding increase is found in the accuracy of the forecasts, as shown in figure 7. Therefore, by multivariate extensions of the arguments used in this paper, it is to be expected that even more precise methods for smoothing and seasonal adjustment should be derivable for sets of related series.

SUMMARY

It is argued that most rapid progress in statistical methods occurs when the empirical approach and the model-based approach iteratively interact. Such progress has led to the useful and rich class of ARIMA models that may be fitted to a wide variety of time series. We illustrate, in this paper, how a fitted model can then determine appropriate techniques for smoothing and seasonal adjustment of a particular series.

The development of methods for convenient computation of exact maximum likelihood estimates for multivariate extensions of these models makes the multivariate ARIMA models more accessible. Using the shipments-order-inventories series, we have illustrated how such multivariate models permit the analysis of complex relationships, allow more accurate forecasts, and have the potential for improving smoothing and seasonal adjustment methods still further.

REFERENCES

1. Anderson, T. W. "Maximum Likelihood Estimation of Parameters of Autoregressive Processes with Moving Average Residuals and Other Covariance Matrices with Linear Structure." *Annals of Statistics*. 3 (1975): 1283–1304.

2. Beveridge, W. H. "Wheat Prices and Rainfall in Western Europe." *Journal of the Royal Statistical Society* 85 (1922): 412–459.

3. Box, G. E. P., and Jenkins, G. M. "Some Statistical Aspects of Adaptive Optimization and Control." *Journal of the Royal Statistical Society*, ser. B, 24 (1962): 297–331.

4. ———. "Further Contributions to Adaptive Control; Simultaneous Estimation of Dynamics: Non-Zero Costs," *Bulletin of the International Statistical Institute 34th Session* (1963): 943–974.

5. ———. "Mathematical Models for Adaptive Control and Optimization." *American Institute of Chemical Engineering Institute of Chemical Engineering Symposium Series* 4 (1965): 61–68.

6. ——— and Bacon, D. W. "Models for Forecasting Seasonal and Non-Seasonal Time Series." *Advanced Seminar on Spectral Analysis of Time Series*. Edited by B. Harris. New York: John Wiley and Sons, Inc., 1967.

7. ———. "Some Recent Advances in Forecasting and Control I." *Applied Statistics* 17 (1968): 91–109.

8. ———. "Discrete Models for Forecasting and Control," *Encyclopedia of Linguistics, Information and Control*. Calcutta: Pergamon Press, 1969, pp. 162–167.

9. ———. *Time Series Analysis: Forecasting and Control*. San Francisco: Holden Day, 1970.

10. Box, G. E. P., and Tiao, G. C. "Intervention Analysis with Applications to Economic and Environmental Problems." *Journal of the American Statistical Association* 70 (1975): 70–79.

11. Brown, R. G. *Statistical Forecasting for Inventory Control*. New York: McGraw-Hill, 1959.

12. ———. *Smoothing, Forecasting and Prediction of Discrete Time Series*. Englewood Cliffs, N.J.: Prentice-Hall, 1963.

13. ——— and Meyer, R. F. "The Fundamental Theory of Exponential Smoothing." *Operations Research* 9 (1961): 673–685.

14. Carnegie Institute of Technology. "Forecasting Trends and Seasonals by Exponentially Weighted Moving Averages." Office of Naval Research Memorandum No. 52, by C. C. Holt, 1957.

15. Cleveland, W. P. "Analysis and Forecasting of Seasonal Time Series" Ph.D. dissertation, University of Wisconsin, Statistics Department, 1972.

16. ——— and Tiao, G. C. "Decomposition of Seasonal Time Series: A Model for the Census X–11 Program." *Journal of the American Statistical Association* 71 (1976): 581–587.

17. Dent, W. "Computation of the Exact Likelihood Function of an ARIMA Process." Technical Report No. 45, University of Iowa, Statistics Department, 1975.

18. Fisher, I. "On Unstable Dollar and the So-Called Business Cycle." *Journal of the American Statistical Association*, 20 (1925): 179.

19. Fisher, R. A. "The Influence of Rainfall on the Yield of Wheat at Rothamsted." *Philosophical Transactions of the Royal Society of London*, B, 213 (1924): 89–142.

20. Hannan, E. J. *Multiple Time Series*. New York: John Wiley, 1970.

21. ——— and Dunsmuir, W. "Vector Linear Time Series Models," 1975. (Unpublished manuscript.)

22. Hillmer, S. C. "Time Series: Estimation, Smoothing and Seasonal Adjusting," Ph.D. dissertation, University of Wisconsin, Statistics Department, 1976.

23. Holt, C. C., Modigliani, F., Muth, J. F., and Simon, H. A. *Planning Production, Inventories and Work Force*. Englewood Cliffs, N.J.: Prentice-Hall, 1960.

24. Kolmogorov, A. "Interpolation and Extrapolation von Stationären Zufälligen Folgen." *Bulletin Academy of Science, U.S.S.R. Series Math* 5 (1941): 3–14.

25. Ljung, G. "Studies in the Modeling of Discrete Time Series." Ph.D. dissertation, University of Wisconsin, Statistics Department, 1976.

26. Mayr, O. *The Origins of Feedback Control*. Cambridge, Mass.: M.I.T. Press, 1970.

27. Muth, J. F. "Optimal Properties of Exponentially Weighted Forecasts of Time Series with Permanent and Transitory Components." *Journal of the American Statistical Association* 55 (1960): 299–306.

28. Nelder, J. A. and Mead, R. "A Simplex Method for Function Minimization." *The Computer Journal* 7 (1965): 308–313.

29. U.S. Department of Commerce, Bureau of the Census. *The X–11 Variant of Census Method II Seasonal Adjustment Program,* by Julius Shiskin, Allan H. Young, and J. C. Musgrave. Technical Paper 15. Washington, D.C.: Government Printing Office, 1967.

30. Weiner, N. *Extrapolation, Interpolation and Smoothing of Stationary Time Series.* New York: John Wiley, 1949.

31. Whittle, P. "Gaussian Estimation in Stationary Time Series" *Bulletin Institute International Statistics* 33 (1961): 1–26.

32. ———. *Prediction and Regulation.* New York: D. Van Nostrand, 1963.

33. Winters, P. R. "Forecasting Sales by Exponentially Weighted Moving Averages." *Management Science* 6 (1960): 324–342.

34. Yaglom, A. M. "The Correlation Theory of Processes Whose nth Difference Constitute a Stationary Process." *American Mathematical Society Translations,* ser. 2 and 8 Providence, R.I.: American Mathematical Society, 1955.

35. Yule, G. U. "On the Method of Investigating Periodicities in Disturbed Series, with Special Reference to Wolfer's Sunspot Numbers." *Philosophical Transactions,* A, 226 (1927): 267–298.

APPENDIX

EXPANSION OF THE GENERATING FUNCTION $\phi(B)\phi(F/\eta(B)\eta(F)$

For the convenience of the reader, we will sketch a method by which the weight functions ω_j's in (42) may be determined. Note that $\phi(B)$ and $\eta(B)$ are polynomials in degree p and u, respectively. With no loss in generality we may write

$$\phi(B)=1-\phi_1 B-\ldots-\phi_r B^r,\ \eta(B)=1-\eta_1 B-\ldots-\eta_r B^r$$

where $r=\max(p, u)$. First set

$$\frac{\phi(B)}{\eta(B)}=C(B), \text{ where } C(B)=1+C_1B+C_2B^2+\ldots$$

and solve for the C_j's by matching coefficients of B^j in

$$\phi(B)=\eta(B)C(B) \qquad (A.1)$$

Specifically,

$$C_1=\eta_1-\phi_1,\ C_j=\eta_j-\phi_j+\sum_{i=1}^{j-1} C_{j-i}\eta_i,\ j=2,\ \ldots,\ r \qquad (A.2)$$

and for $j>r$, the C_j's can be recursively computed from the relation

$$C_j=\sum_{i=1}^{r} C_{j-i}\eta_i \qquad (A.3)$$

Next, set

$$\frac{\phi(F)}{\eta(F)}C(B)=X(B, F)$$

where

$$X(B, F)=X_0+\sum_{j=0}^{\infty} X_j(B^j+F^j)$$

Equivalently,

$$\phi(F)C(B)=\eta(F)X(B, F) \qquad (A.4)$$

Note that the largest degree of F on the left hand side of (A.4) is r. By matching coefficients of F^j in (A.4), we find that

1. For $j=0, 1, \ldots, r$

$$\underset{\sim}{\eta}\underset{\sim}{X}=\underset{\sim}{C}\underset{\sim}{\phi} \qquad (A.5)$$

where

$$\underset{\sim}{X}'=(X_0, \ldots, X_r),\ \phi'=(1,-\phi_1, \ldots, -\phi_r)$$

$$\underset{\sim}{C}=\begin{bmatrix} 1 & C_1 & \cdot & \cdot & C_r \\ & \cdot & \cdot & & \cdot \\ & & \cdot & \cdot & \\ & & & & \cdot \\ & & & \cdot & C_1 \\ \underset{\sim}{0} & & & & 1 \end{bmatrix} \quad \underset{\sim}{\eta}=\begin{bmatrix} 1 & -\eta_1 \ldots & -\eta_r \\ -\eta_1 & -\eta_r & \\ \cdot & \cdot & \\ \cdot & \cdot & \\ \cdot\ \cdot & & \underset{\sim}{0} \\ -\eta_r & & \end{bmatrix}$$

$$+\begin{bmatrix} 0 & & & & \underset{\sim}{0} & \\ \cdot & 1 & & & & \\ \cdot & -\eta_1 & \cdot & & & \\ \cdot & \cdot & \cdot & \cdot & & \\ \cdot & \cdot & & \cdot & \cdot & \\ 0 & -\eta_{r-1} & \cdot & \cdot & -\eta_1 & 1 \end{bmatrix}$$

which define a set of r+1 linear equations in $r+1$ unknowns $X_0, \ldots, X_r$, so that

$$\underset{\sim}{X}=\underset{\sim}{\eta}^{-1}\underset{\sim}{C}\underset{\sim}{\phi} \qquad (A.6)$$

2. For $j>r$, the X_j's can be computed recursively from the relation

$$X_j=\sum_{i=1}^{r} X_{j-i}\eta_i \qquad (A.7)$$

Finally, the weights ω_j's in (42) are

$$\omega_0=1-\frac{\sigma_e^2}{\sigma_a^2}X_0 \text{ and } \omega_j=-\frac{\sigma_e^2}{\sigma_a^2}X_j,\ j=1, \ldots \qquad (A.8)$$

COMMENTS ON "ANALYSIS AND MODELING OF SEASONAL TIME SERIES" BY GEORGE E. P. BOX, STEVEN C. HILLMER, AND GEORGE C. TIAO

George A. Barnard
University of Essex

I wish to make clear, at the outset, that I have no claim to expertise in the matter of time series analysis. On the rare occasions when I have referred to stochastic processes in published papers, I have taken care to base my remarks on models to which none of the existing theory applies. As the domain of existing theory extends, it becomes a little more difficult to do this—but only a little more difficult. It is, perhaps, worthwhile to point out that, since the universe, in its evolution, can be regarded as the exemplification of a stochastic process, a full treatment of the theory of such processes still has a very long way to go before it can be said to be complete. Such a reflection is not merely philosophical. It is relevent to the issues we are here to discuss in that attempts to claim universality for any one method of analysis are bound to fail. The iterative process of empirical analysis, interacting with model building so well emphasized by the authors of this paper, should converge when related to a particular case, on the basis of given data; but, in relation to the set of all possible cases, it will clearly have an open-ended character.

Such knowledge as I have gained over the years in connection with time series has been almost wholly derived from conversations with George Box and with Gwilym Jenkins. These began in the mid-fifties, when I had the privilege of hearing Box's first thoughts on the subject of nonstationary series. Perhaps he found me a sympathetic listener, because one of the difficulties I had had with the subject, up to that time, was with the near universality of the assumption of stationarity, which conflicted fundamentally with my own evolutionary view of nature and society. I record this here, because it is not clear, from the sketch of history given in the paper, that although it has subsequently turned out that their work was in some respects anticipated by others, such as Yaglom, Box and Jenkins developed their theoretical and practical approach almost entirely independently. I believe that the coherence of the various aspects and methods, which is such a strong feature of their work and that of their school, is, to a large extent, due to this.

Up to now, the lessons in time series analysis that I have had from Box and Jenkins have come in doses sufficiently small and well expounded that I have been able to digest them without undue effort. Perhaps, because the present instalment represents the work of several of Box's coworkers at Madison, it comes here as an advance in the theory of such magnitude that its implications will take me, at least, a considerable time to assimilate. The traditional approach to seasonality and smoothing has been to process the series in question through a set of filters to remove, as far as possible, the high frequency noise, and the excess amplitude of frequencies associated with seasonality and, then, to set up a model for the residual trend. What the authors propose here is to turn this process upside down—to model the process as it stands and, then, express the model as the sum of the traditional three components. That something of this kind needs to be done, at least in some cases, especially in economic contexts, is indicated by the experience of most workers in the field, that even the highly sophisticated seasonality adjustments, such as the census X–II program or the mixed-multiplicative and additive-adaptive methods studied on the other side of the Atlantic, always seem to leave some element of seasonal pattern in each series to which they are applied. And, it is common sense to think that the way in which each series reacts to and remembers seasonal factors is bound to have some degree of individuality about it, so that methods which, in principle, assume the same type of reaction to seasonal influences are bound to be limited in application.

In so far as our current ideas and practice with seasonal adjustments are based on dating, we are, in effect, assuming that the factors that are influencing the series we are concerned with and whose effect we wish to separate out, are directly and simply related to the position of the earth in its orbit round the sun. This will be true of series in which the date, by itself, has major significance—one thinks primarily here of the influence of, e.g., Christmas on sales figures. But, already with the date of Easter, we have to consider not only the position of the earth relative to the sun, but also its position in relation to the Paschal moon, so that series (such as the Irish consumption of Guinness Stout) for which Easter is an important festival will exhibit peaks that sometimes occur in March and sometimes in April. And, in the many series for which the major influence of this kind is the weather, the patterns will still be more complicated. Most series will be influenced by a mixture of such factors, in proportions peculiar to the particular series, and, perhaps more important, the degree to which these influences are remembered by the system in question will vary, so that an approach, such as that taken in this paper, would certainly have advantages.

Whether the advantages would, in all cases, be worth the extra effort, involved in dealing with each series individually, is a point worth discussion.

My own understanding of the history of the exponentially weighted moving average is that the idea occurred to my friend Arthur A. Brown, working for Arthur D. Little, who was working on stock control problems. It occurred to him, waiting for a plane at O'Hare International Airport, that the EWMA would require, in a computerised setup, only two storage locations per series to be forecast. I say this, partly by way of comment on the historical section of the author's paper, but more especially because the subsequent history of the associated model would suggest that an optimistic attitude to the simplicity of the real world (something that one is trained as a mathematician to despise—mostly to one's detriment) has a lot to be said in its favour. Correspondingly, while, as I have said, I have little doubt that the approach to seasonality, proposed by the authors, will have, in principle, many advantages, the differences between the final results of the census treatment of the unemployed males series and that of the authors' is perhaps arguably small enough to be ignored for most "practical" purposes.

I put "practical" in quotation marks, because I wish it to be understood in the sense in such contexts where there is usually an element of the short run involved in the "practical." In the long run, the difference of principle involved in the difference of approach can have enormous practical consequences. Because, as the authors note, it raises the whole question of why we attempt to deseasonalise our series, and what we understand ourselves to be doing when we so treat them.

One approach to this question would suggest that we can draw a distinction between factors, such as the motion of the earth around the sun or variations in the weather, that are exogenous to the social system and that are altogether beyond our control, and, on the other hand, factors, such as tax rates, bank rates, and perhaps social attitudes, that also are exogenous to most aspects of the social system but which are, to some extent, within our control. Usually, it is assumed that the factors beyond our control will, in some sense, average themselves out in the future, and what planners and politicians need to concern themselves with are the controllable factors. The recent drought in Western Europe is exceptional in that it has brought into prominence, for practical politicians, the fact that most of their policies are based on an extrapolation of present weather conditions that may, in fact, be unjustified. To help with this, they need to have a model of the times series they are concerned with that has the effects of the uncontrollable factors removed or averaged out. If we understand the purpose of deseasonalisation, in this sense, we shall want to further develop the ideas expressed in this paper, especially in relation to the sketch of the approach to multiple time series. We shall also want to develop the ideas of intervention analysis that Box and Tiao have put forward but which they have not dealt with in the present contribution.

But, perhaps, a simpler notion will serve. It is that those who use deseasonalised and smoothed series regard such series as representing the average, relatively long-term behaviour of the series in question. Smoothing gets rid of the month-to-month random fluctuations, while deseasonalising enables one to get an idea of the movement of the yearly average. If this point of view is adopted, then instead of focussing attention, as we do now, on the periodic motion of the earth, we may begin to pay attention to what perhaps in some countries, is becoming a more important length of time—the 4-, 5-, or 6-year term between elections. Would it be too much to hope that one day statisticians will be able to discount the effects of the loosening of credit and other inflationary measures that are becoming all too common, even in advanced countries in preelection periods, so that politicians will be judged more on the long-term effects of their policies? The common man has much more sense than the credit he is usually given. If he is given the information that he needs to judge the long-term effects of political and economic policies, he can be relied upon to look to the reasonable future, instead of the immediate present.

What I am suggesting is that instead of producing, along with the raw data, the smoothed and deseasonalised values, we should think in terms of producing (for instance) a forecast of the discounted future values of the series that we are concerned with in the notation of their paper, e.g.,

$$\hat{\hat{Z}} = \sum_{l=1}^{\infty} \hat{z}_{t+l}(1-\delta)^{l-1}\,\delta$$

where δ is a discounting factor that might be around 0.008 for a monthly series. Such forecasts, updated from month to month, would surely be a better guide to policy and its effects than the values we now use. It is a feature of the methods proposed by the present authors that they provide a reasonably objective procedure for doing this. The necessity of having such an objective procedure is, perhaps, illustrated by some of the U.K. Government's recent *White Papers,* where assumptions about, e.g., growth rates are embodied without proper stress being laid upon the possibility that such assumptions could prove wrong.

To return to the paper and its details, it is perhaps worth emphasising that one aspect of the model-building process that can make it an essential adjunct to the empirical approach is its capacity to bring to bear information different, in kind, from the basic numerical data. As an example, it could be that someone who knew the processes used in stores in adjusting their prices could bring evidence that the extreme value $\alpha = -1$ for the coefficient of B in the forcing function for the smoothed model of the consumer price index was too large in absolute value, and a value near to $-1/2$ would better accord with practical experience. Examining the curves, given for $\alpha = 0$ and for $\alpha = -1$, shows that the data, by themselves, throw very little light on the values of α within this range.

The authors' somewhat apologetic tone in referring to their use of the Nelder-Mead procedure in solving the equations for the coefficients of the seasonal component of the male unemployment series is surely not appropriate. Methods of function minimisation, developed by Powell and others in recent years, have become so powerful that even for the solution of large sets of linear equations it may be preferable to minimise directly the sum of a residuals rather than use the (unstable) algorithms of pivotal condensation, etc. If I may suggest it, perhaps the authors should draw their problems to the attention of a numerical analyst working in this area; one gets the impression that theory has run ahead of practical problems.

Finally, in reference to the remarks concerning the calculation of the likelihood function and the virtues of maximum likelihood estimates, perhaps, the name of Ian McLeod should be added to that of Kang as having provided a practicable procedure for obtaining more exact maximums and showing, by simulation, that the improvement is worthwhile. And, it should, perhaps, be mentioned that Barnard and Winsten showed, using the data cited by Whittle in the paper referred to, that examination of the whole likelihood function threw a much clearer light on the estimate and its error than would come from calculations based on asymptotic results. The problem of exhibiting, in a usable form, the likelihood function for many parameters is not yet solved; but, good progress is being made. There really is no substitute for this, in the case of most economic time series, since the effective lengths of series are typically much too small for asymptotic results to be of relevance.

Since the form and content of my remarks is typical to an opening of discussion on my home round at the Royal Statistical Society, may I follow the custom there and conclude by moving, with much pleasure, a hearty vote of thanks to our three authors for a most rich and stimulating paper.

COMMENTS ON "ANALYSIS AND MODELING OF SEASONAL TIME SERIES" BY GEORGE E. P. BOX, STEVEN C. HILLMER, AND GEORGE C. TIAO

Emanuel Parzen
State University of New York at Buffalo

I would like to express my thanks to the authors for their important and seminal paper that will contribute greatly towards clarifying what is meant by seasonality in time series and developing useful methods for its analysis and modeling.

Given observed time series (or, in general, statistical data) that we seek to model, we are confronted with the basic questions: What do we mean by a model? and where do models come from? One can distinguish between two types of models that can be given names as follows:

Type I: Fundamental or structural
Type II: Technical or synthetic

I will use the adjectives "structural" and "synthetic."

A structural model comes from subject-matter theoretical considerations, and the statistician's role is usually to estimate its parameters or test its fit to data.

A synthetic model comes from statistical and probabilistic considerations, either from empirical data analysis procedures that have been found satisfactory, in practice, or from representation theory of random variables.

In time series analysis, both statistical and probabilistic considerations yield a simple general definition of what we mean by a model: It is a transformation (often a linear filter) F on the data $\{z_t\}$ that whitens the time series $\{z_t\}$. In symbols, let $\{a_t\}$ be a white-noise series such that

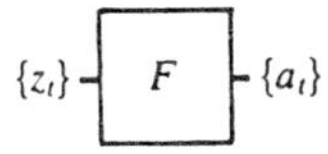

then the transformation F is the model.

The essence of the successful Box-Jenkins approach to time series analysis seems to be to develop empirical procedures for discovering a filter F (in the univariate case of the form $\theta^{-1}(B)\varphi(B)\nabla^d$) for which one can write conclusions similar to that which the authors write following their model in equation (79): "The residuals from this model are plotted in figure 5 and the autocorrelation and cross-correlation functions of the residuals are plotted in figure 6. Examination of these plots does not suggest any inadequacies in the model."

The models (means) to be used in time series analysis should depend on the intended ends or applications (thus, the eternal conundrum: Can one find means robust against ends?). I believe one may distinguish six basic types of applications of time series analysis:

1. Forecasting (or extrapolation)
2. Spectral analysis (or interpolation, by harmonics)
3. Parametrization (or data compression)
4. Intervention analysis (significant changes in forecasts or parameters)
5. Signal plus noise decomposition
6. Control

There is no doubt that ARIMA models are important and can provide means for all these ends. However, other ways of formulating equivalent models should not be discarded from (or fail to be incorporated into) the time series analyst's bag of tools. In particular, spectral and state space representations are often indispensable means.

I believe there can be no disagreement with the conclusion stated at the end of the section on the iterative development of some important ideas in time series analysis that ARIMA models are "a class of stochastic models capable of representing nonstationary and seasonal time series" and with the assertion that a successful seasonal nonstationary model for economic time series with both monthly and yearly components is given by their equation (28), which we shall call model I:

$$(1-B)(1-B^{12})\,z_t=(1-\theta_1 B)(1-\theta_2 B^{12})\,a_t$$

A question that I believe should be investigated is whether model I, by itself, can yield seasonal adjustment procedures, or is it necessary to pass (algorithmically and conceptually) from model I to the traditional representation of a time series z_t as having a seasonal component S_t and trend-noise component T_t, which we call model II:

$$z_t=S_t+T_t=S_t+p_t+e_t$$

where $T_t=p_t+e_t$ is the sum of a trend p_t and white noise e_t. The authors' main aim in the section on smoothing and seasonal adjustment is to outline approaches to pass from model I to model II. Whether this trip is necessary will be discussed.

Some technical comments on the section on smoothing and seasonal adjustment are the following: In estimating p_t from T_t, the authors consider only two-sided filters that use both past and future values of T_t; would it not be more appropriate to use one-sided filters that use only past values? Then, to estimate p_t from T_t, one would use Kalman filtering techniques. In regard to estimating models for p_t and e_t, an important reference seems, to me, to be Pagano [1].

The aim of the section on multivariate generalization of the models is to develop multivariate seasonal models such as that given by equation (75), which we call model III:

$$(I-\Phi_1 B)(I-\Phi_{12} B^{12}) z_t = (I-\theta_1 B)(I-\theta_{12} B^{12}) a_t$$

where z_t is a k vector and Φ_i and θ_j are $k \times k$ matrices. The time available to me for discussion prevents my commenting further on the section concerning multivariate generalization of the models other than to note my view that its results are pioneering and impressive and should stimulate much further research.

I believe this paper is important and seminal, because the problems considered by Box, Hillmer, and Tiao in the third and fourth sections are the problems at the heart of the problem of seasonal adjustment. In the remainder of my discussion, I would like to outline, from the point of view of my own approach to empirical time series analysis, why and how it might suffice to obtain seasonal adjustment procedures directly from a suitable reinterpretation of models, such as I (and, more generally, III).

A fundamental decomposition of a time series $\{z_t\}$ can be given in terms of its one-step-ahead infinite-memory predictors

$$z_t^{\mu} = E[z_t | z_{t-1}, z_{t-2}, \ldots]$$

and its one-step-ahead infinite-memory prediction errors or innovations

$$z_t^{\nu} = z_t - z_t^{\mu}$$

Then

$$z_t = z_t^{\mu} + z_t^{\nu}$$

The innovation time series $\{z_t^{\nu}\}$ is white noise and, indeed, $a_t = z_t^{\nu}$

For a nonstationary time series, the modeling problem is first to find the whitening filter (that transforms $\{z_t\}$ to $\{z_t^{\nu}\}$) and, second, to interpret it as several filters in tandem:

D_0: A detrending filter that, in the spectral domain, eliminates the low-frequency components corresponding to trend.

D_λ: A deseasonal filter that, in the spectral domain, eliminates the components corresponding to a periodic component with period λ or to the harmonics with frequencies that are multiples of $\frac{2\pi}{\lambda}$.

g_∞ or Π: An innovations filter that transforms the series $z_t^{(\mathrm{stat})} = D_0 D_\lambda z_t$, representing a transformation of z_t to a stationary series, to white noise.

The time series modeling problem is, thus, to find the filter representation

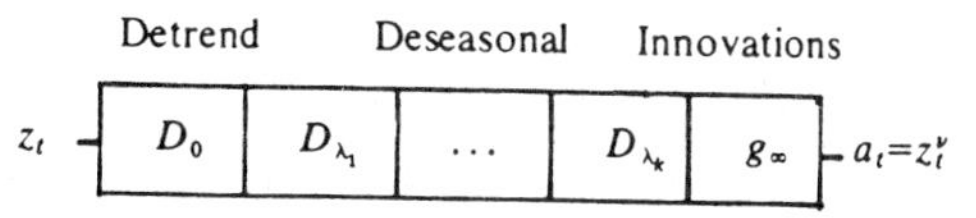

where we admit the possibility of several different periods $\lambda_1, \ldots, \lambda_k$ (e.g., in monthly data λ values are often 12 and 3, in daily data λ values are often 7 and 365, and in hourly data λ values are often 24 and 168).

Given this decomposition, one can form various derived series:

$z_t^{(0)} = D_0 z_t$, the detrended series.
$z_t^{(\lambda)} = D_\lambda z_t$, the seasonally adjusted series.
$z_t^{(0,\lambda)} = z_t^{(\mathrm{stat})} = D_0 D_\lambda z_t = D_\lambda D_0 z_t$, the detrended seasonally adjusted series.
$z_t^{\nu} = z_t^{(\mathrm{white})} = g_\infty D_0 D_\lambda z_t$, the innovations series.

Instead of universal detrending and seasonal adjustment procedures, what is being suggested are filters accomplishing the same ends that are custom tailored for each series.

Such decompositions seem to be crucial to the study of the relations between time series $Y_1(\cdot)$ and $Y_2(\cdot)$. To study their relations, it seems clear that if one relates $Y_1(\cdot)$ and $Y_2(\cdot)$ without filtering, one will often find spurious relationships. It has been suggested, therefore, that one attempt to relate $Y_1^{\nu}(\cdot)$ and $Y_2^{\nu}(\cdot)$, the individual innovations of each series. What remains to be examined is the insight to be derived from relating $Y_1^{(\lambda)}(t)$ and $Y_2^{(\lambda)}(t)$, the seasonally adjusted series or $Y_1^{(\mathrm{stat})}(t)$ and $Y_2^{(\mathrm{stat})}(t)$, the detrended and seasonally adjusted series.

The question remains of how to find, in practice, the detrending and deseasonal filters. To seasonally adjust for a period λ in data, several possibilities are available that may be interpreted as seasonal adjustment filters.

A filter with the same zeroes in the frequency domain as some usual procedures, which is recursive (acts only on past values) and yields a variety of filter shapes (in the frequency domain) between a square wave and a sinusoid, is the one-parameter family of filters

$$D_\lambda(\theta) = \frac{I - B^\lambda}{I - \theta B^\lambda}$$

where the parameter θ is chosen (usually by an estimation procedure) between 0 and 1. When $\theta = 0$, the filter is denoted ∇_λ and called λ-th difference.

To understand the role of the filter $D_\lambda(\theta)$, denote it for brevity by D and rewrite it, writing $I-B^{\lambda}=I-\theta B^{\lambda}-(1-\theta)B^{\lambda}$, we obtain

$$D=I-\frac{(1-\theta)B^{\lambda}}{I-\theta B^{\lambda}}=I-\{(1-\theta)(B^{\lambda}+\theta B^{2\lambda}+\theta^2 B^{3\lambda}+\dots)\}$$

Then, the output $z_t^{(I)}=Dz_t$ of a filter D with input z_t can be written

$$z_t^{(I)}=z_t-(1-\theta)\{z_{t-\lambda}+\theta z_{t-2\lambda}+\dots\}$$

In words, $z_t^{(I)}$ is the result of subtracting, from z_t, the exponentially weighted average of $z_{t-\lambda}$, $z_{t-2\lambda}$, ...

It seems to me open to investigation whether the filter D (of mixed autoregressive moving average type) is superior to the approximately equivalent autoregressive filter

$$D'=(I-B^{\lambda})(I+\theta B^{\lambda})=I-(1-\theta)B^{\lambda}-\theta B^{2\lambda}$$

whose output

$$z_t^{(I')}=D'z_t$$

can be written

$$z_t^{(I')}=z_t-(1-\theta)\,z_{t-\lambda}-\theta z_{t-2\lambda}$$

It appears to me that the role of moving averages in Box-Jenkins ARIMA models is to build filters of the type $D_{\lambda}(\theta)$. Thus, the ARIMA model

$$(I-B)(I-B^{12})z_t=(I-\theta_1 B)(I-\theta_{12}B^{12})a_t$$

should be viewed as the whitening filter

$$D_1(\theta_1)\,D_{12}(\theta_{12})\,z_t=\left(\frac{I-B}{I-\theta_1 B}\right)\left(\frac{I-B^{12}}{I-\theta_{12}B^{12}}\right)z_t=a_t$$

I should like to emphasize that the output of the filter $D_1(\theta_1)\;D_{12}(\theta_{12})$ is often not white noise but is only a stationary time series. For purposes of one-step-ahead prediction, it is often not important to differentiate between the case that $D_{12}D_1z_t$ is white noise or not, since most of the predictability is obtained by finding a suitable transformation to stationarity of the form D_1D_{12}. (Next, we will discuss naive prediction as a transformation to stationarity.)

A moral can be drawn from the foregoing considerations. To find a transformation of a nonstationary time series to stationarity, it may suffice to apply pure differencing operators, such as $I-B$ and $I-B^{12}$. However, the transformation of the residuals to the innovations series should be expressed, if possible, in terms of factors corresponding to the filters $I-\theta_1 B$ and $I-\theta_{12}B^{12}$, since such factors enable us to interpret the overall whitening filter

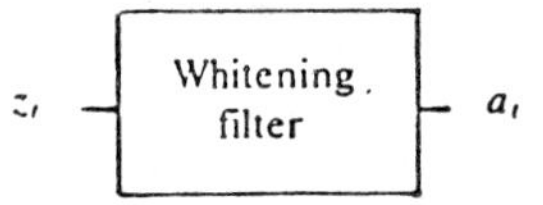

as a series of filters in tandem

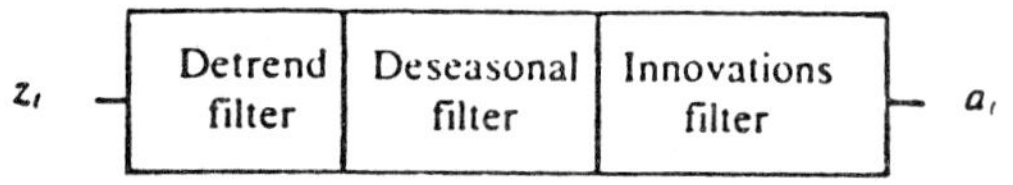

which can be interpreted as helping to provide solutions to the seasonal adjustment problem.

NAIVE PREDICTION AND TRANSFORMATIONS TO STATIONARITY

To predict a time series z_t, one can often suggest a naive predictor of the form

$$z_t^{\text{naive}}=z_{t-\lambda_1}+z_{t-\lambda_2}-z_{t-\lambda_1-\lambda_2}$$

The prediction error of this predictor is given by

$$\tilde{z}_t=z_t-z_t^{\text{naive}}=z_t-z_{t-\lambda_1}-z_{t-\lambda_2}+z_{t-\lambda_1-\lambda_2}=(I-B^{\lambda_1})(I-B^{\lambda_2})z_t$$

In words, taking λ_1-th and λ_2-th differences is equivalent to forming the naive prediction errors.

A criterion that z_t be nonstationary is that it be predictable (in the sense that the ratio of the average square of $\tilde{z}_t$ to the average square of z_t is of the order of $1/T$). When $\tilde{z}_t$ is stationary (nonpredictable), one models it (see app.) by an approximate autoregressive scheme

$$\hat{g}_{\hat{m}}(B)\tilde{z}_t=a_t$$

which can be used to form $\tilde{z}_t^{\mu}$, the best one-step-ahead predictor of $\tilde{z}_t$.

The best one-step-ahead predictor of z_t is given by

$$z_t^{\mu}=z_t^{\text{naive}}+\tilde{z}_t^{\mu}$$

to prove this, note the identity

$$z_t=z_t^{\text{naive}}+\tilde{z}_t$$

and form the conditional expectation of both sides of this identity with respect to z_{t-1}, z_{t-2}, ...

A remarkable fact is the equality of the prediction errors of z_t and $\tilde{z}_t$

$$z_t^{\nu}=z_t-z_t^{\mu}=\tilde{z}_t-\tilde{z}_t^{\mu}=\tilde{z}_t^{\nu}$$

It follows that, to find the whitening filter

for a nonstationary time series z_t (which includes almost all time series with seasonal components), it suffices to apply any one-sided filter (which, in practice, would be suggested by an ad hoc deseasonalizing procedure) whose output $\tilde{z}_t$ is stationary. The tandem filter

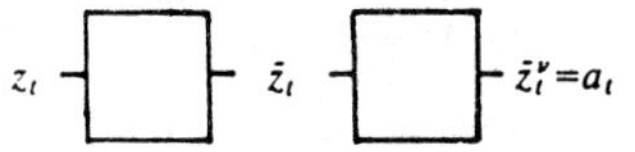

then yields the whitening filter. While the filter leading to $\tilde{z}_t$ is not unique, the overall filter leading to a_t is unique.

The final seasonal adjustment procedure is a filter D_λ that comes from interpreting the overall whitening filter as a series of filters in tandem, which can be interpreted as detrending and deseasonalizing filters. An illustration of the application of this approach to real data is contained in [2].

REFERENCES

1. Pagano, M. "Estimation of Autoregressive Signals Plus White Noise." *Annals of Statistics* 2 (1974): 99–108.

2. Parzen, Emmanuel. "An Approach to Time Series Modeling and Forecasting Illustrated by Hourly Electricity Demands," *Proceedings on Forecasting Methodology for Time-of-Day and Seasonal Electric Utility Loads.* Special Report 31. Washington, D.C.: Electric Power Research Institute, 1976.

3. ———. "Multiple Time Series: Determining the Order of Approximating Autoregressive Schemes." In *Multivariate Analysis IV*. Edited by P. Krishnaiah. New York: Academic Press, 1976.

APPENDIX

ESTIMATION OF THE WHITENING FILTER OF A STATIONARY TIME SERIES

A rigorous definition of the whitening filter can be given when the time series $\{z_t\}$ is stationary zero mean and has a continuous spectral density function $f(\omega)$, where, for mathematical convenience, we use the definition

$$f(\omega)=\frac{1}{2\pi}\sum_{\nu=-\infty}^{\infty} e^{-i\nu\omega}R(\nu),\ -\pi\leq\omega\leq\pi$$

$$R(\nu)=E[z_t z_{t+\nu}]$$

Assume that $\log f(\omega)$ and $f^{-1}(\omega)$ are integrable; then there is a frequency transfer function

$$g_\infty(z)=1+\alpha_1 z+\alpha_2 z^2+\ldots+\alpha_m z^m+\ldots$$

such that

$$f(\omega)=\frac{1}{2\pi}\sigma_\infty^2|g_\infty(e^{i\omega})|^{-2}$$

Further, g_∞ is a whitening filter

$$g_\infty(B)\,z_t=a_t$$

where $\{a_t\}$ is the innovation series (white noise) with variance

$$\sigma_\infty^2=E[|a_t|^2]$$

We call g_∞ the ARTF (autoregressive transfer function). It is the same as the transfer function $\Pi(B)$, defined in the authors' equation (12).

To every finite memory $\underline{m}$, one can define finite memory one-step-ahead prediction errors

$$z_t^{\nu,m}=g_m(B)z_t$$

where

$$g_m(z)=1+\alpha_{1,m}z+\ldots+\alpha_{m,m}z^m$$

is the polynomial of degree $\underline{m}$ minimizing

$$\int_{-\pi}^{\pi}|g_m(e^{i\omega})|^2 f(\omega)\,d\omega$$

among all polynomials of degree $\underline{m}$ with constant coefficient equal to 1. The memory $\underline{m}$ mean square prediction error is denoted by

$$\sigma_m^2=E[|z_t^{\nu,m}|^2]$$

The autoregressive spectral approximator

$$f_m(\omega)=\frac{1}{2\pi}\sigma_m^2|g_m(e^{i\omega})|^{-2}$$

may be shown to converge to $f(\omega)$ as $\underline{m}$ tends to ∞, as does σ_m^2 to σ_∞^2 and $g_m(z)$ to $g_\infty(z)$.

Estimators of these quantities from a finite sample $\{z_t, t=1,\ldots,T\}$ can be constructed as follows: Define

1. The sample spectral density

$$f_T(\omega)=\frac{1}{2\pi T}\left|\sum_{t=1}^{T} z_t e^{it\omega}\right|^2$$

2. Sample covariances

$$R_T(\nu)=\int_{-\pi}^{\pi} e^{i\nu\omega}f_T(\omega)\,d\omega$$

3. Sample order m autoregressive coefficients $\hat{\alpha}_{j,m}$, $j=1,\ldots,m$, as the solution of the normal (or Yule-Walker) equations

$$\sum_{j=1}^{m}\hat{\alpha}_{j,m}R_T(k-j)=-R_T(k),\ k=1,\ldots,m$$

4. Sample memory m mean square prediction error

$$\hat{\sigma}_m^2=R_T(0)+\sum_{j=1}^{m}\hat{\alpha}_{j,m}R_T(j)$$

5. Sample order m autoregressive transfer function

$$\hat{g}_m(z)=1+\hat{\alpha}_{1,m}z+\ldots+\hat{\alpha}_{m,m}z^m$$

6. Sample order m autoregressive spectral estimator

$$\hat{f}_m(\omega)=\frac{1}{2\pi}\hat{\sigma}_m^2|\hat{g}_m(e^{i\omega})|^{-2}$$

Finally, we estimate $\sigma_\infty^2, g_\infty$, and $f(\omega)$ by $\hat{\sigma}_{\hat{m}}^2$, $\hat{g}_{\hat{m}}$ and $\hat{f}_{\hat{m}}(\omega)$, respectively, where $\hat{m}$ is chosen by an order-determination criterion. In [3], I have proposed choosing $\hat{m}$ as the value of m, minimizing the criterion function CAT (criterion autoregressive transfer function)

$$\mathrm{CAT}(m)=\frac{1}{T}\sum_{j=1}^{m}\bar{\sigma}_j^{-2}-\bar{\sigma}_m^{-2}$$

where

$$\hat{\hat{\sigma}}^2_m = \left(1 - \frac{m}{T}\right)^{-1} \hat{\sigma}^2_m$$

and

$$\text{CAT}(0) = -1 - (1/T)$$

When $\hat{m}=0$, we say that the time series is white noise.

Having determine the maximum order $\hat{m}$, to help interpret $\hat{g}_{\hat{m}}(z)$, it is useful to use stepwise regression techniques to determine the significantly nonzero autoregressive coefficients. As an example on monthly data, if one had determined that $\hat{m}=13$, it would be of interest to determine whether $\hat{g}_{13}(z)$ were approximately of the form

$$\hat{g}_{13}(z) = (1-\theta_1 z)(1-\theta_{12} z^{12})$$

In my approach to empirical time series analysis, the identification stage is not accomplished chiefly by graphical inspection of the time series and of computed auxiliary sample functions, such as the autocorrelation function, partial autocorrelation function, and spectrum. Rather, the infinite parametric ARTF g_∞ is directly estimated and parsimoniously parametrized.

3.12
A Canonical Analysis of Multiple Time Series

G. E. P. BOX and G. C. TIAO
University of Wisconsin

SUMMARY

This paper proposes a canonical transformation of a k-dimensional stationary autoregressive process. The components of the transformed process are ordered from least to most predictable. The least predictable components are often nearly white noise which can reflect stable contemporaneous relationships among the original variables. The most predictable can be nearly nonstationary representing the dynamic growth characteristic of the series. The method is illustrated with a series with five variables.

Some key words: Autoregressive process; Canonical variable; Eigenvalue; Eigenvector; Multiple time series; Variance component.

1. INTRODUCTION

Data frequently occur in the form of k related time series simultaneously observed at some constant interval. In particular, economic, industrial and ecological data are often of this kind. Much work has been done on the problem of detecting, estimating and describing relationships of various kinds among such series; see, for example, Quenouille (1957), Hannan (1970), Box & Jenkins (1970), and Brillinger (1975). In this paper we shall consider a particular method for characterizing structure.

Consider a $k \times 1$ vector process $\{\mathscr{Z}_t\}$ and let $z_t = \mathscr{Z}_t - \mu$, where μ is a convenient $k \times 1$ vector of origin which is the mean if the process is stationary. Suppose z_t follows the pth order multiple autoregressive model

$$z_t = \hat{z}_{t-1}(1) + a_t, \tag{1·1}$$

where

$$\hat{z}_{t-1}(1) = E(z_t | z_{t-1}, z_{t-2}, \ldots) = \sum_{l=1}^{p} \pi_l z_{t-l}$$

is the expectation of z_t conditional on past history up to time $t-1$, the π_l are $k \times k$ matrices, $\{a_t\}$ is a sequence of independently and normally distributed $k \times 1$ vector random shocks with mean zero and covariance matrix Σ, and a_t is independent of $\hat{z}_{t-1}(1)$. The model (1·1) can be written

$$\left(I - \sum_{l=1}^{p} \pi_l B^l\right) z_t = a_t, \tag{1·2}$$

where I is the identity matrix and B is the backshift operator such that $Bz_t = z_{t-1}$. The process $\{z_t\}$ is stationary if the determinantal polynomial in B, det $(I - \Sigma \pi_l B^l)$, has all its zeros lying outside the unit circle, and otherwise the process will be called nonstationary.

Now suppose $k = 1$. Then, if the process is stationary,

$$E(z_t^2) = E\{\hat{z}_{t-1}(1)\}^2 + E(a_t^2),$$

that is

$$\sigma_z^2 = \sigma_{\hat{z}}^2 + \sigma_a^2.$$

We can define a quantity λ measuring the predictability of a stationary series from its past as $\lambda = \sigma_{\hat{z}}^2 \sigma_z^{-2} = 1 - \sigma_a^2 \sigma_z^{-2}$.

The Collected Works of George E. P. Box, 1984, Wadsworth, Inc., Belmont, CA 94002.
Originally published in *Biometrika,* vol. 64, no. 2 (1977), pp. 355–365.

Suppose that we are considering k different stock market indicators such as the Dow Jones Average, the Standard and Poors index, etc., all of which exhibit dynamic growth. It is natural to conjecture that each might be represented as some aggregate of one or more common inputs which may be nearly nonstationary, together with other stationary or white noise components. This leads one to contemplate linear aggregates of the form $u_t = m'z_t$, where m is a $k \times 1$ vector. The aggregates which depend most heavily on the past, namely having large λ, may serve as useful composite indicators of the overall growth of the stock market. By contrast, the aggregates with λ nearly zero may reflect stable contemporaneous relationships among the original indicators. The analysis given in this paper yields k 'canonical' components from least to most predictable. The most predictable components will often approach nonstationarity and the least predictable will be stationary or independent. Thus we may usefully decompose the k-dimensional space of the observation z_t into independent, stationary and nonstationary subspaces. Variables in the nonstationary space represent dynamic growth while those in the stationary and independent spaces can reflect relationships which remain stable over time.

2. Choice of the canonical variables

2·1. *General considerations*

Suppose that z_t is stationary. Let $\Gamma_j(z) = E(z_{t-j} z_t')$ be the lag j autocovariance matrix of z_t. It follows from (1·1) that

$$\Gamma_0(z) = \sum_{l=1}^{p} \pi_l \Gamma_l(z) + \Sigma = \Gamma_0(\hat{z}) + \Sigma, \tag{2·1}$$

say, where $\Gamma_0(\hat{z})$ is the covariance matrix of $\hat{z}_{t-1}(1)$. Until further notice, we shall assume that Σ, and therefore $\Gamma_0(z)$, are positive-definite.

Now, consider the linear combination $u_t = m'z_t$. We have that $u_t = \hat{u}_{t-1}(1) + v_t$, where $\hat{u}_{t-1}(1) = m'\hat{z}_{t-1}(1)$ and $v_t = m'a_t$. The predictability of u_t from its past is therefore measured by

$$\lambda = \sigma_{\hat{u}}^2 \sigma_u^{-2} = \{m'\Gamma_0(\hat{z})\, m\}/\{m'\Gamma_0(z)\, m\}. \tag{2·2}$$

It follows that for maximum predictability, λ must be the largest eigenvalue of $\Gamma_0^{-1}(z)\,\Gamma_0(\hat{z})$ and m the corresponding eigenvector. Similarly, the eigenvector corresponds to the smallest eigenvalue will yield the least predictable combination of z_t.

2·2. *The canonical transformation*

Let $\lambda_1, \ldots, \lambda_k$ be the k real eigenvalues of the matrix $\Gamma_0^{-1}(z)\,\Gamma_0(\hat{z})$. Suppose that the λ_j are ordered with λ_1 the smallest, and that the k corresponding linearly independent eigenvectors, $m_1', \ldots, m_k'$, form the k rows of a matrix M. Then, we can construct a transformed process $\{y_t\}$, where

$$y_t = \hat{y}_{t-1}(1) + b_t, \tag{2·3}$$

with

$$y_t = Mz_t, \quad b_t = Ma_t, \quad \hat{y}_{t-1}(1) = \sum_{l=1}^{p} \tilde{\pi}_l y_{t-l}, \quad \tilde{\pi}_l = M\pi_l M^{-1}.$$

Corresponding to (2·1), we now have

$$\Gamma_0(y) = \Gamma_0(\hat{y}) + \tilde{\Sigma}, \tag{2·4}$$

where $\Gamma_0(y) = M\Gamma_0(z)\,M'$, $\Gamma_0(\hat{y}) = M\Gamma_0(\hat{z})\,M'$ and $\tilde{\Sigma} = M\Sigma M'$.

Note that: (i)

$$M'^{-1}\Gamma_0^{-1}(z)\,\Gamma_0(\hat{z})\,M' = \Lambda, \quad M'^{-1}\Gamma_0^{-1}(z)\,\Sigma M' = I - \Lambda, \tag{2·5}$$

where Λ is the $k \times k$ diagonal matrix with elements $(\lambda_1, \ldots, \lambda_k)$; (ii) $0 \leqslant \lambda_j < 1$ $(j = 1, \ldots, k)$; and (iii) for $i \neq j$, $m_i' \Sigma m_j = m_i' \Gamma_0(\hat{z}) m_j = 0$. In other words, $M\Sigma M'$, $M\Gamma_0(\hat{z})M'$ and, therefore, $M\Gamma_0(z)M'$ are all diagonal. Thus, the transformation (2·3) produces k components series $\{y_{1t}, \ldots, y_{kt}\}$ which (i) are ordered from least predictable to most predictable, (ii) are contemporaneously independent, (iii) have predictable components $\{\hat{y}_{1(t-1)}(1), \ldots, \hat{y}_{k(t-1)}(1)\}$ which are contemporaneously independent, and (iv) have unpredictable components $\{b_{1t}, \ldots, b_{kt}\}$ which are contemporaneously and temporally independent.

2·3. *Zero roots*

Special interest may attach to situations where certain of the λ_j approach zero. When the k_1 roots, $\lambda_1, \ldots, \lambda_{k_1}$, are zero, the matrix $\Gamma_0(\hat{y})$ in (2·4) then can be written

$$\Gamma_0(\hat{y}) = \begin{bmatrix} 0 & 0 \\ 0 & D \end{bmatrix}, \tag{2·6}$$

where D is an $k_2 \times k_2$ diagonal matrix. With

$$y_t' = [y_{1t}' \vdots y_{2t}'], \quad b_t' = [b_{1t}' \vdots b_{2t}'], \tag{2·7}$$

where y_{1t} and b_{1t} are $k_1 \times 1$ vectors, it readily follows that, with probability one, $y_{1t} = b_{1t}$. For $l = 1, \ldots, p$, partitioning the $k \times k$ matrix $\tilde{\pi}_l$ into

$$\tilde{\pi}_l = \begin{bmatrix} \tilde{\pi}_{11}^{(l)} & \tilde{\pi}_{12}^{(l)} \\ \tilde{\pi}_{21}^{(l)} & \tilde{\pi}_{22}^{(l)} \end{bmatrix}, \tag{2·8}$$

where $\tilde{\pi}_{11}^{(l)}$ is a $k_1 \times k_1$ matrix, we have the transformed series $\{y_t\}$ expressed as

$$\begin{bmatrix} y_{1t} \\ y_{2t} \end{bmatrix} = \sum_{l=1}^{p} \begin{bmatrix} 0 & 0 \\ \tilde{\pi}_{21}^{(l)} & \tilde{\pi}_{22}^{(l)} \end{bmatrix} B^l \begin{bmatrix} y_{1t} \\ y_{2t} \end{bmatrix} + \begin{bmatrix} b_{1t} \\ b_{2t} \end{bmatrix}. \tag{2·9}$$

Thus, the canonical transformation decomposes the original $k \times 1$ vector process $\{z_t\}$ into two parts: (i) a part $\{y_{1t}\}$ which follows a k_1-dimensional white noise process, and (ii) a part $\{y_{2t}\}$ which is stationary but whose predictable part depends on both $y_{1(t-l)}$ and $y_{2(t-l)}$ for $l = 1, \ldots, p$.

The practical importance of (2·9) is that it implies that there are k_1 relationships between the original variables of the 'static' form

$$m_{j1}\mathscr{Z}_{1t} + \ldots + m_{jk}\mathscr{Z}_{kt} = \eta_j + b_{jt} \quad (j = 1, \ldots, k_1),$$

where the b_{jt} are contemporaneously and temporally independent. We shall later illustrate this situation with an example.

3. The first-order autoregressive process

3·1. *The canonical model*

In this section we discuss some properties of the canonical transformation when $\{z_t\}$ follows an k-dimensional autoregressive process of order one. Thus with $p = 1$ and $\pi_1 = \phi$, (1·1) yields

$$z_t = \hat{z}_{t-1}(1) + a_t = \phi z_{t-1} + a_t. \tag{3·1}$$

Since $\Gamma_1'(z) = \phi\Gamma_0(z)$ it follows from (2·1) that $\Gamma_0(z) = \phi\Gamma_0(z)\phi' + \Sigma$ and the required roots λ_j and vectors m_j are the k eigenvalues and eigenvectors of the matrix

$$Q = \Gamma_0^{-1}(z)\phi\Gamma_0(z)\phi'. \tag{3·2}$$

If $\tilde{\phi} = M\phi M^{-1}$ the transformed process can now be written

$$y_t = \tilde{\phi} y_{t-1} + b_t. \tag{3·3}$$

3·2. *Nonstationary series and unit roots*

In the above we have assumed that z_t is stationary. In practice, many time series exhibit nonstationary behaviour. A useful class of models to represent nonstationary series may be obtained by allowing the zeros of the $\det(I-\Sigma \pi_l B^l)$ of (1·2) to lie on the unit circle. For the model (3·1), let $\alpha_1, \ldots, \alpha_k$ be the eigenvalues of the matrix ϕ. Then

$$\det(I-\phi B) = \prod_{j=1}^{k}(1-\alpha_j B),$$

so that the zeros of $\det(I-\phi B)$ are simply $\alpha_1^{-1}, \ldots, \alpha_k^{-1}$. If one of more of the α_j are on the unit circle, then $\Gamma_0(z)$ does not exist and the canonical transformation method will break down. However, it is of interest to study the limiting situation when k_2 of the α_j approach values on the unit circle. Letting

$$y_t' = [y_{1t}' \vdots y_{2t}'], \quad b_t' = [b_{1t}' \vdots b_{2t}'], \quad \tilde{\phi} = \begin{bmatrix} \tilde{\phi}_{11} & \tilde{\phi}_{12} \\ \tilde{\phi}_{21} & \tilde{\phi}_{22} \end{bmatrix},$$

where y_1 and b_{1t} are $k_1 \times 1$ vectors and $\tilde{\phi}_{11}$ is a $k_1 \times k_1$ matrix with $k_1 = k - k_2$, we show in the Appendix that:

(i) if, and only if, k_2 of the eigenvalues α_j of ϕ approach values on the unit circle, then k_2 of the eigenvalues λ_j of Q in (3·2) approach unity;

(ii) the transformed model for y_t is, in the limit,

$$\begin{bmatrix} y_{1t} \\ y_{2t} \end{bmatrix} = \begin{bmatrix} \tilde{\phi}_{11} & 0 \\ \tilde{\phi}_{21} & \tilde{\phi}_{22} \end{bmatrix} B \begin{bmatrix} y_{1t} \\ y_{2t} \end{bmatrix} + \begin{bmatrix} b_{1t} \\ b_{2t} \end{bmatrix}. \tag{3·4}$$

The canonical transformation therefore decomposes z_t into two parts:

(i) a part y_{1t} which follows a stationary first-order autoregressive process, and

(ii) a part y_{2t} which is approaching nonstationarity and also depends on $y_{1(t-1)}$.

The practical significance of this result is that the components y_{2t} can serve as useful composite indicators of the overall dynamic growth of the original series.

3·3. *Zero and unit roots*

For the model (3·1), suppose that k_1 of the λ_j are zero, k_3 of them approach unity and the remaining $k_2 = k - k_1 - k_3$ are intermediate in size. Then, from the results in (2·9) and (3·4), and upon partitioning y_t, b_t and $\tilde{\phi}$ into

$$y_t' = [y_{1t}' \vdots y_{2t}' \vdots y_{3t}'], \quad b_t' = [b_{1t}' \vdots b_{2t}' \vdots b_{3t}'],$$

$$\tilde{\phi} = \begin{bmatrix} \tilde{\phi}_{11} & \tilde{\phi}_{12} & \tilde{\phi}_{13} \\ \tilde{\phi}_{21} & \tilde{\phi}_{22} & \tilde{\phi}_{23} \\ \tilde{\phi}_{31} & \tilde{\phi}_{32} & \tilde{\phi}_{33} \end{bmatrix}, \tag{3·5}$$

where y_{1t} and b_{1t} are $k_1 \times 1$ vectors, y_{2t} and b_{2t} are $k_2 \times 1$ vectors, and $\tilde{\phi}_{11}$ and $\tilde{\phi}_{22}$ are, respectively, $k_1 \times k_1$ and $k_2 \times k_2$ matrices, the transformed process $\{y_t\}$ takes the form

$$\begin{bmatrix} y_{1t} \\ y_{2t} \\ y_{3t} \end{bmatrix} = \begin{bmatrix} 0 & 0 & 0 \\ \tilde{\phi}_{21} & \tilde{\phi}_{22} & 0 \\ \tilde{\phi}_{31} & \tilde{\phi}_{32} & \tilde{\phi}_{33} \end{bmatrix} B \begin{bmatrix} y_{1t} \\ y_{2t} \\ y_{3t} \end{bmatrix} + \begin{bmatrix} b_{1t} \\ b_{2t} \\ b_{3t} \end{bmatrix}. \tag{3·6}$$

Thus there are: (i) a k_1-dimensional white noise process $\{y_{1t}\}$, (ii) a k_2-dimensional stationary process $\{y_{2t}\}$ such that the predictable part of y_{2t} depend only on $y_{1(t-1)}$ and $y_{2(t-1)}$, and (iii) a k_3-dimensional near nonstationary process $\{y_{3t}\}$ such that the predictable part of y_{3t} depends on $y_{1(t-1)}$, $y_{2(t-1)}$ and $y_{3(t-1)}$.

3·4. *Variance components for the first-order process*

Whatever the scaling of the transformed process $\{y_t\}$ in (3·3), since the jth element y_{jt} is $y_{jt} = \Sigma_i \tilde{\phi}_{ji} y_{i(t-1)} + b_{jt}$, where $(\tilde{\phi}_{j1}, \ldots, \tilde{\phi}_{jk})$ is the jth row of $\tilde{\phi}$, and since $y_{1(t-1)}, \ldots, y_{k(t-1)}$ and b_{jt} are independent, it follows that $\sigma^2_{y_j} = \Sigma_i \tilde{\phi}^2_{ji} \sigma^2_{y_i} + \sigma^2_{b_j}$. The contributions of $y_{1(t-1)}, \ldots, y_{k(t-1)}$ and b_{jt} to the variance of y_{jt} are, therefore, $\tilde{\phi}^2_{j1} \sigma^2_{y_1}, \ldots, \tilde{\phi}^2_{jk} \sigma^2_{y_k}$ and $\sigma^2_{b_j}$, respectively. It is convenient to consider these variance components in terms of their proportional contribution to $\sigma^2_{y_j}$, that is to consider $(\phi_{ji}\sigma^2_{y_i})/\sigma^2_{y_j}$ and $\sigma^2_{b_j}/\sigma^2_{y_j} = 1 - \lambda_j$. This can be done conveniently by arranging the canonical variables with scaling such that the variances of y_{jt} are all unity.

For the general process (1·1), to arrange for this scaling the matrix M must be chosen such that $M\Gamma_0(z) M' = I$. Corresponding to (2·3), let the transformed series in this scaling be written as

$$x_t = \hat{x}_{t-1}(1) + d_t. \tag{3·7}$$

Then, $\Gamma_0(x) = \Gamma_0(\hat{x}) + \bar{\Sigma}$, where $\Gamma_0(x) = I$, $\Gamma_0(\hat{x}) = \Lambda$, $\bar{\Sigma} = I - \Lambda$ and Λ is the diagonal matrix in (2·5). For the process (3·1), $\hat{x}_{t-1}(1) = \bar{\phi} x_{t-1}$, and hence

$$\bar{\phi}^2_{ji} = (\tilde{\phi}^2_{ji} \sigma^2_{y_i})/\sigma^2_{y_j}, \quad \bar{\phi}\bar{\phi}' = \Lambda. \tag{3·8}$$

In this scaling, then, the rows of $\bar{\phi}$ are orthogonal and the sum of squares of the jth row is λ_j.

The preceding canonical analysis will now be illustrated by an example.

4. An example

4·1. *U.S. hog, corn and wage series*

Quenouille (1957, pp. 88–101) studied a time series with 5 variates and containing 82 yearly observations from 1867–1948. He made adjustments where necessary, logarithmically transformed each variate and then linearly coded the logs, so as to produce numbers of comparable magnitude in the different series. His resulting five series denoted by $\mathscr{H}_{1t}, \ldots, \mathscr{H}_{5t}$ are plotted in Fig. 1 (*a*) and are identified in Table 4·1.

4·2. *The first-order autoregressive model*

Quenouille fitted the data to a first-order autoregressive process but was doubtful as to the adequacy of the model. We found, however, that the fit can be improved by appropriately shifting series 2 and 5 backward by one period as indicated in Table 4·1.

With the model $z_t = \phi z_{t-1} + a_t$ in (3·1), where $z_t = \mathscr{Z}_t - \mu$, the sample means $\hat{\mu}$ of $\mathscr{Z}_t$ and sample cross-covariance matrices C_j needed in our analysis are as follows:

$$10^{-3}\hat{\mu}' = (0·6989,\ 0·8949,\ 0·7714,\ 1·3281,\ 0·9956),$$

$$10^{-4}C_0 = \begin{bmatrix} 0·6831 & 1·2523 & 0·6535 & 0·9533 & 1·5224 \\ & 6·1939 & 3·7845 & 2·0209 & 5·5708 \\ & & 3·6877 & 0·2633 & 3·4746 \\ & & & 2·1407 & 2·1925 \\ & & & & 5·7206 \end{bmatrix},$$

$$10^{-4}C_1 = \begin{bmatrix} 0·5864 & 1·3670 & 0·7513 & 0·8632 & 1·5151 \\ 1·2038 & 5·2334 & 3·1639 & 1·8849 & 5·0392 \\ 0·4616 & 3·5820 & 2·7173 & 0·5605 & 3·0633 \\ 1·0108 & 1·8972 & 0·8338 & 1·6260 & 2·2508 \\ 1·3993 & 5·1586 & 3·2153 & 1·9817 & 5·3246 \end{bmatrix}.$$

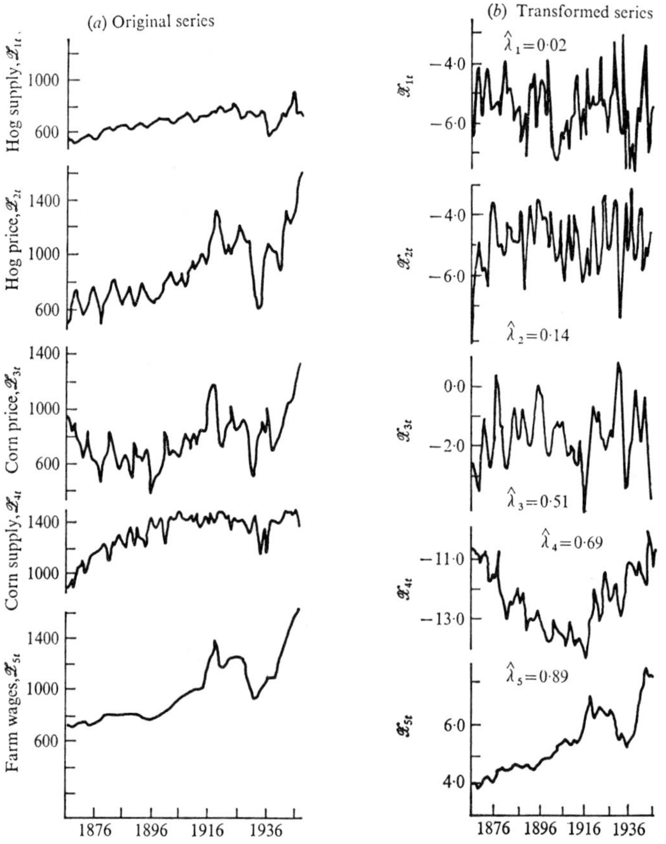

Fig. 1. U.S. hog data. (*a*) Original series, (*b*) transformed series.

Table 4·1. *Notation for Quenouille's U.S. hog series*

Variate	Symbol	As logged and linearly coded by Quenouille	Used in our analysis
Hog supply	H_s	$\mathscr{H}_{1t}$	$\mathscr{X}_{1t} = \mathscr{H}_{1t}$
Hog price	H_p	$\mathscr{H}_{2t}$	$\mathscr{X}_{2t} = \mathscr{H}_{2(t+1)}$
Corn price	R_p	$\mathscr{H}_{3t}$	$\mathscr{X}_{3t} = \mathscr{H}_{3t}$
Corn supply	R_s	$\mathscr{H}_{4t}$	$\mathscr{X}_{4t} = \mathscr{H}_{4t}$
Farm wages	W	$\mathscr{H}_{5t}$	$\mathscr{X}_{5t} = \mathscr{H}_{5(t+1)}$

Table 4·2. *Estimated eigenvalues and eigenvectors for the hog data*

j	λ_j	H_s	H_p	R_p	R_s	W		
1	0·0232	(1·0000	0·3876	−0·2524	−0·5896	−0·2665)	×	0·0284
2	0·1421	(0·2080	1·0000	−0·8614	−0·3382	−0·3655)	×	0·0111
3	0·5061	(0·8925	−0·6433	−0·8277	−0·4784	1·0000)	×	0·0074
4	0·6901	(−0·9358	−0·2410	−0·4391	−0·5614	1·0000)	×	0·0129
5	0·8868	(0·6687	−0·1206	−0·0134	0·0396	1·0000)	×	0·0039

Table 4·2 gives the estimated eigenvalues and eigenvectors of Q in (3·2). The latter are scaled according to (3·7) so that all the components of the transformed process $\{x_t\}$ have unit estimated variances. The transformed process is $x_t = \bar{\phi} x_{t-1} + d_t$ with the estimated $\bar{\phi}$ given by

$$\begin{bmatrix} 0\cdot1213 & -0\cdot0778 & 0\cdot0465 & -0\cdot0110 & 0\cdot0113 \\ 0\cdot2215 & 0\cdot2766 & -0\cdot1241 & -0\cdot0309 & 0\cdot0119 \\ -0\cdot0321 & 0\cdot3167 & 0\cdot6334 & 0\cdot0444 & -0\cdot0404 \\ 0\cdot0885 & -0\cdot0025 & -0\cdot0492 & 0\cdot8235 & 0\cdot0416 \\ -0\cdot0801 & 0\cdot0378 & 0\cdot0396 & -0\cdot0363 & 0\cdot9360 \end{bmatrix},$$

and the resulting series $\mathscr{X}_t = \hat{M}\mathscr{Z}_t$ are shown in Fig. 1 (*b*).

The estimated proportional contributions of $x_{1(t-1)}, \ldots, x_{5(t-1)}$ and d_{jt} to the variance of x_{jt} are given in Table 4·3.

Table 4·3. *Component variances of the transformed series*

	$x_{1(t-1)}$	$x_{2(t-1)}$	$x_{3(t-1)}$	$x_{4(t-1)}$	$x_{5(t-1)}$	d_{jt}
x_{1t}	0·015	0·006	0·002	0·000	0·000	0·977
x_{2t}	0·049	0·077	0·015	0·001	0·000	0·858
x_{3t}	0·001	0·100	0·401	0·002	0·002	0·494
x_{4t}	0·008	0·000	0·002	0·678	0·002	0·310
x_{5t}	0·006	0·001	0·002	0·001	0·876	0·113

We see from the above calculations that there is very little contribution to x_{1t} and x_{2t} from history. These two transformed series are essentially white noise. The remarkable feature of x_{3t}, x_{4t} and x_{5t} is their heavy dependence on their own past, and this is especially so for the latter two components. It is almost true that x_{4t} and x_{5t} can be expressed as two independent univariate first-order autoregressive processes

$$x_{4t} = 0\cdot82 x_{4(t-1)} + d_{4t}, \quad x_{5t} = 0\cdot94 x_{5(t-1)} + d_{5t}. \tag{4·1}$$

4·3. *Interpretation*

In terms of the original observations $\mathscr{Z}_t$, the model for the most predictable component $\mathscr{X}_{5t}$ is approximately,

$$\mathscr{X}_{5t} - 0\cdot94\mathscr{X}_{5t-1} = 0\cdot35 + d_{5t}. \tag{4·2}$$

The autoregressive parameter is close to unity, indicating that the series is nearly nonstationary Also, it is readily seen that the estimated standard deviation of the mean of $\mathscr{X}_{5t} - 0\cdot94\mathscr{X}_{5t-1}$ is 0·04 so that the term 0·35 on the right-hand side of (4·2) is real. Thus, what we have is nearly a random walk with a fixed increment of 0·35 per year. Now $x_{5t} = m_5' z_t$ and from the estimated eigenvector $\hat{m}_5$, x_{5t} is essentially a linear combination of the farm wages, W, and hog supply, H_s,

$$x_{5t} \simeq 0\cdot0039\,(z_{5t} + 0\cdot67 z_{1t}). \tag{4·3}$$

This is then the linear combination which serves as an indicator of the overall dynamic growth pattern in the original series.

The nearly random components x_1 and x_2, omitting the subscript t, associated with small values of λ are also of considerable interest. Their existence implies that any linear combination of the component series in the hyperplane

$$Z = \alpha x_1 + \beta x_2 = c_1 z_1 + c_2 z_2 + c_3 z_3 + c_4 z_4 + c_5 z_5 \tag{4·4}$$

varies nearly independently about fixed means. In choosing the component it is natural to seek combinations which are scientifically meaningful.

Now the dollar value of the hogs sold is proportional to $H_p H_s$ and the dollar value of the corn needed to feed them is $R_p R_s$. If then a Z exists involving these dollar values it will be such that approximately $c_1 = c_2$ and $c_3 = c_4$. By least squares or otherwise it is easy to find the linear combination for which this is nearly true. Specifically, by setting $\alpha = 30{\cdot}01$ and $\beta = 59{\cdot}51$ we obtain the relationship

$$Z = z_1 + z_2 - 0{\cdot}78z_3 - 0{\cdot}73z_4 - 0{\cdot}48z_5. \tag{4·5}$$

That is to say Z in (4·5) is approximately independently distributed about a fixed mean.

Now the average estimated variance of $(z_{1t}, \ldots, z_{5t})$ is $3{\cdot}69 \times 10^4$. For comparison, we normalize the linear combination (4·5) by letting $u = l'z$, where $l = (1{\cdot}84)^{-1}[1, 1, -0{\cdot}78, -0{\cdot}73, -0{\cdot}48]$, so that $l'l = 1$. Since x_1 and x_2 have unit variance and are independent, the variance of u is $0{\cdot}1326 \times 10^4$. Compared with the average variability of the original series, we have thus obtained a remarkably stable contemporaneous relationship among the 5 original variables.

Taking antilogs of (4·5) this implies that

$$I_1 = \frac{H_p H_s}{(R_p R_s)^{0{\cdot}75} W^{0{\cdot}50}} \tag{4·6}$$

is approximately constant. The numerator is obviously a measure of return to the farmer and the denominator a measure of his expenditure. The analysis points to the near constancy of this relation reminding us of the 'economic law' that a viable business must operate so as to balance expenditure and income.

Suppose we choose $\alpha = 46{\cdot}51$ and $\beta = -137{\cdot}80$, we then obtain

$$Z = 1{\cdot}00z_1 - 1{\cdot}02z_2 + 0{\cdot}98z_3 - 0{\cdot}26z_4 + 0{\cdot}21z_5. \tag{4·7}$$

Again, if we normalized the combination by expressing $u = l'z$ with $u = (1{\cdot}76)^{-1}Z$ such that $l'l = 1$, the variance of u would be $0{\cdot}68 \times 10^4$.

Upon taking antilogs, this implies that, very approximately,

$$I_2 = H_s R_p / H_p \tag{4·8}$$

is constant, indicating that a stable relationship existed between hog supply and the price ratio (Wallace & Bressman, 1937, p. 342–50).

In addition, we note that the percentage coefficients of variation of I_1 and I_2, given approximately by $100 \log\{10\sigma(Z)\}$ are 16 % and 18 %, respectively. Thus, both indices are remarkably stable when it is remembered that over the time period studied, the individual elements in the indices underwent massive changes. For example, hog prices increased tenfold.

4·4. *Differencing of the data*

For the hog data, since each of the original series exhibit a growth pattern, questions might be raised as to whether one should difference the data first and then perform a canonical analysis of the differenced series. Indeed, if one were to analyze these series individually, one would be led to consider differencing z_1, z_2, z_4 and z_5. However, in analyzing multiple time series of this kind, it is useful to entertain the possibility that the dynamic pattern in the data may be due to a small subset of nearly nonstationary components and that there may exist stable contemporaneous linear relationships among the variables. If this is so, then differencing all the original series could lead to complications in the analysis. To illustrate, suppose we have the bivariate model

$$z_{1t} = z_{1(t-1)} + a_{1t}, \quad z_{2t} = \beta z_{1t} + a_{2t}, \tag{4·9}$$

so that each series individually will be nonstationary. If we considered the two differenced series, $w_{1t} = (1-B)z_{1t}$ and $w_{2t} = (1-B)z_{2t}$, we would have

$$w_{1t} = a_{1t}, \quad w_{2t} = \beta a_{1t} + a_{2t} - a_{2(t-1)}. \tag{4·10}$$

It is readily shown that while (4·9) can be expressed in the form of a bivariate first-order autoregressive process, the differenced series cannot be put into the autoregressive form (1·1) making the analysis more complicated.

5. Further considerations

5·1. *Singularity of the matrix* Σ

So far it has been assumed that the covariance matrix of a_t, Σ, in (2·1) is positive-definite. Situations occur when $\Gamma_0(z)$ is positive-definite but Σ is singular. Specifically, suppose that the rank of Σ is $k_1 < k$. This means that the $k \times 1$ vector process $\{z_t\}$ is in fact driven by a k_1-dimensional nonsingular random shock process. Then, it is readily seen that the k_2 roots $\lambda_{k_1+1}, \ldots, \lambda_k$ of $\Gamma_0^{-1}(z)\,\Gamma_0(\hat{z})$ are exactly equal to one, and the transformed covariance $\tilde{\Sigma}$ matrix of b_t in (2·3) takes the form

$$\tilde{\Sigma} = \begin{bmatrix} D_1 & 0 \\ 0 & 0 \end{bmatrix}, \tag{5·1}$$

where D_1 is an $k_1 \times k_1$ diagonal matrix with positive elements. Partitioning y_t, b_t and $\tilde{\pi}_l$ as given in (2·7) and (2·8), we see that $b_{2t} = 0$ with probability one. Thus the transformed model is

$$\begin{bmatrix} y_{1t} \\ y_{2t} \end{bmatrix} = \sum_{l=1}^{p} \begin{bmatrix} \tilde{\pi}_{11}^{(l)} & \tilde{\pi}_{12}^{(l)} \\ \tilde{\pi}_{21}^{(l)} & \tilde{\pi}_{22}^{(l)} \end{bmatrix} B^l \begin{bmatrix} y_{1t} \\ y_{2t} \end{bmatrix} + \begin{bmatrix} b_{1t} \\ 0 \end{bmatrix}. \tag{5·2}$$

In other words, the k_2-dimensional vector y_{2t} is completely predictable from the past values $y_{1(t-l)}$ and $y_{2(t-l)}$ $(l = 1, \ldots, p)$.

In practice, situations may occur where Σ is nearly singular. From the results here and those discussed earlier in § 3·2, we see that for the first-order autoregressive process, certain of the roots λ_j will be nearly equal to one either when some of the eigenvalues of ϕ approach values on the unit circle or when Σ is nearly singular. The problem of how to distinguish between these two cases is currently being investigated.

5·2. *Singularity of the matrix* $\Gamma_0(z)$

Examples can also occur when $\Gamma_0(z)$ is singular. It is not unusual to find exact and quite complex linear relationships imposed by the method in which the data is put together so that $\Gamma_0(z)$ will necessarily be singular (Box *et al.*, 1973). Two situations can occur depending on whether or not the nature of any exact linear relationships existing in the data is already known. If known, then the problem may be avoided by eliminating, in advance of the analysis, any dependencies and applying the analysis to a linearly independent subset of r of the k series. When the nature of exact relationships in the data which might exist are not known, a principle component analysis of the estimate $C_0(z)$ of $\Gamma_0(z)$ should be first conducted. The existence of $k-r$ roots which are nearly zero indicates the existence of $k-r$ linearly independent exact relationships which define a hyperplane in the k space given by the $k-r$ corresponding eigenvectors. The canonical analysis of this paper may now be carried out on any subset of r linearly independent series which lie in the nonsingular space.

The work was supported by U.S. Army Research Office and the Wisconsin Alumni Research Foundation.

Appendix

Eigenvalues of first-order autoregressive process

We here sketch the proof of the results given in § 3·1 concerning the situation where k_2 of the eigenvalues of ϕ approach values on the unit circle.

Theorem. *Suppose that z_t follows the stationary model in* (3·1), *where the covariance matrix Σ of a_t is positive-definite. A sufficient and necessary condition for k_2 of the eigenvalues of*

$$\Gamma_0(z)^{-1}\phi\Gamma_0(z)\phi'$$

to tend to unity is that k_2 of the eigenvalues of ϕ approach values on the unit circle.

Proof. Let $k = k_1+k_2$ and the eigenvalues of ϕ be divided into two sets $\alpha_1^* = (\alpha_1, \ldots, \alpha_{k_1})$ and $\alpha_2^* = (\alpha_{k_1+1}, \ldots, \alpha_k)$ with no common element and such that α_j and its complex conjugate belong to the same set. The characteristic polynomial of ϕ can be written as the product

$$f(\alpha) = f_{k_1}(\alpha)\, f_{k_2}(\alpha), \tag{A 1}$$

where

$$f_{k_1}(\alpha) = \alpha^{k_1} - \gamma_1 \alpha^{k_1-1} - \ldots - \gamma_{k_1}, \quad f_{k_2}(\alpha) = \alpha^{k_2} - s_1 \alpha^{k_2-1} - \ldots - s_{k_2}$$

are real polynomials of degrees k_1 and k_2 with roots α_1^* and α_2^*, respectively. Now there exists a $k \times k$ real nonsingular matrix C such that $C\phi C^{-1}$ is of the block diagonal form

$$C\phi C^{-1} = \begin{bmatrix} R & 0 \\ 0 & S \end{bmatrix}, \tag{A 2}$$

where R and S are, respectively, $k_1 \times k_1$ and $k_2 \times k_2$ matrices such that

$$R = \left[\begin{array}{c|c} 0 & I \\ \hline \gamma_{k_1} & \gamma_{k_1-1} \cdots \gamma_1 \end{array}\right], \quad S = \left[\begin{array}{c|c} 0 & I \\ \hline s_{k_2} & s_{k_2-1} \cdots s_1 \end{array}\right].$$

Letting $V = C\Gamma_0(z)C'$ and $W = C\Sigma C'$ and partitioning V and W correspondingly, so that, for example, V_{11} and W_{11} are $k_1 \times k_1$ matrices, we obtain

$$V_{11} = RV_{11}R' + W_{11}, \quad V_{12} = RV_{12}S' + W_{12}, \quad V_{22} = SV_{22}S' + W_{22}, \tag{A 3}$$

where we use the relation $\Gamma_0(z) = \phi\Gamma_0(z)\phi' + \Sigma$. By writing $V_{22} = AA'$, it is readily seen from (3·2) that the λ_j are the roots of the determinantal polynomial

$$\det\left\{(1-\lambda)\begin{bmatrix} V_{11} & V_{12}A'^{-1} \\ A^{-1}V'_{12} & I \end{bmatrix} - \begin{bmatrix} W_{11} & W_{12}A'^{-1} \\ A^{-1}W'_{12} & A^{-1}W_{22}A'^{-1} \end{bmatrix}\right\} = 0. \tag{A 4}$$

To prove sufficiency, we need to show that if the k_2 eigenvalues

$$\alpha_j \to e^{i\omega_j} \quad (j = k_1+1, \ldots, k), \tag{A 5}$$

then (A 4) will tend to

$$(1-\lambda)^{k_2} \det\{(1-\lambda)V_{11} - W_{11}\} = 0. \tag{A 6}$$

It suffices to prove that (A 5) implies that

$$A^{-1} \to 0, \quad A^{-1}V'_{12} \to 0. \tag{A 7}$$

From (A 3)

$$V_{22}^{-1}SV_{22}S' = \{I + S'^{-1}V_{22}^{-1}S^{-1}W_{22}\}^{-1} \tag{A 8}$$

so that $\det S^2 = \det(I + S'^{-1}V_{22}^{-1}S^{-1}W_{22})^{-1}$. When (A 5) holds, $\det(S^2) = s_{k_2}^2 \to 1$. Since S is nonsingular and W_{22} is positive-definite, it follows that $V_{22}^{-1} \to 0$ and hence $A^{-1} \to 0$.

To show $A^{-1}V'_{12} \to 0$, we have, from (A 3),

$$I = PP' + A^{-1}W_{22}A'^{-1}, \tag{A 9}$$

$$A^{-1}V'_{12} = PA^{-1}V'_{12}R' + A^{-1}W'_{12}, \tag{A 10}$$

where $P = A^{-1}SA$. Thus, when (A 5) holds, in (A 9), $P \to P_0$ where P_0 is an orthogonal matrix, and hence (A 10) becomes

$$P_0' A^{-1} V_{12}' = A^{-1} V_{12}' R. \qquad \text{(A 11)}$$

Since by supposition, S and R have no common eigenvalues, it follows (Gantmacher, 1959, p. 220) that $A^{-1}V_{12}' \to 0$. This completes the proof of the sufficiency.

Next to show necessity, recall that $|\alpha_j| < 1$ and $0 \leqslant \lambda_j < 1$, for $j = 1, \ldots, k$. Thus, if k_2 of the λ_j tends to one, then exactly k_2 of the α_j must approach values on the unit circle. For, if otherwise, and suppose $k' \neq k_2$ of the α_j approach values on the unit circle, then from the sufficiency part of the theorem which we have just proved, k' of the λ_j must approach one, which contradicts the supposition. The theorem thus follows.

To study the eigenvectors and the transformed matrix $\tilde{\phi}$, it is easy to see that the systems of equations $(Q - \lambda I)\, m = 0$ is equivalent to

$$\{(1-\lambda) I - V^{-1} W\} h = 0, \qquad \text{(A 12)}$$

where $C'h = m$. When (A 5) holds, by using (A 7) it is straightforward to verify that the matrix of eigenvectors M' must be of the form

$$C'^{-1} M' = \begin{bmatrix} H_{11}' & -W_{11}^{-1} W_{12} \\ 0 & I \end{bmatrix}, \qquad \text{(A 13)}$$

where the columns of H_{11}' are the eigenvectors of $V_{11}^{-1} W_{11}$.

It follows from (A 13) that the transformed matrix $\tilde{\phi}$ takes the form

$$\tilde{\phi} = M \phi M^{-1} = \begin{bmatrix} \tilde{\phi}_{11} & 0 \\ \tilde{\phi}_{21} & \tilde{\phi}_{22} \end{bmatrix}, \qquad \text{(A 14)}$$

where $\tilde{\phi}_{11} = H_{11} R H_{11}^{-1}$, $\tilde{\phi}_{22} = S_0$, $\tilde{\phi}_{21} = (S_0 W_{12}' W_{11}^{-1} - W_{11}^{-1} R) H_{11}^{-1}$ and S_0 is the limiting matrix of S when all its roots approach values on the unit circle.

References

Box, G. E. P., Erjavec, J., Hunter, W. G. & MacGregor, J. F. (1973). Some problems associated with the analysis of multiresponse data. *Technometrics* **15**, 33–51.

Box, G. E. P. & Jenkins, G. M. (1970). *Time Series Analysis Forecasting and Control*. San Francisco: Holden-Day.

Brillinger, D. R. (1975). *Time Series Data Analysis and Theory*. New York: Holt, Rinehart and Winston.

Gantmacher, F. R. (1959). *The Theory of Matrices*, Vol. I. New York: Chelsea.

Hannan, E. J. (1970). *Multiple Time Series*. New York: Wiley.

Quenouille, M. H. (1957). *The Analysis of Multiple Time Series*. London: Griffin.

Wallace, H. A. & Bressman, E. N. (1937). *Corn and Corn Growing*, 4th edition. New York: Wiley.

[*Received November* 1975. *Revised November* 1976]

3.13
On a Measure of Lack of Fit in Time Series Models

G. M. LJUNG
University of Denver

G. E. P. BOX
University of Wisconsin

SUMMARY

The overall test for lack of fit in autoregressive-moving average models proposed by Box & Pierce (1970) is considered. It is shown that a substantially improved approximation results from a simple modification of this test. Some consideration is given to the power of such tests and their robustness when the innovations are nonnormal. Similar modifications in the overall tests used for transfer function-noise models are proposed.

Some key words: Autoregressive-moving average model; Residual autocorrelation; Test for lack of fit; Transfer function-noise model.

1. Introduction

Consider a discrete time series $\{w_t\}$ generated by a stationary autoregressive-moving average model

$$\phi(B)\, w_t = \theta(B)\, a_t,$$

where $\phi(B) = 1-\phi_1 B - \ldots - \phi_p B^p$, $\theta(B) = 1-\theta_1 B - \ldots - \theta_q B^q$, $B^k w_t = w_{t-k}$, and $\{a_t\}$ is a sequence of independent and identically distributed $N(0,\sigma^2)$ random deviates. The w_t's can in general represent the d-th difference or some other suitable transformation of a non-stationary series $\{z_t\}$.

After a model of this form has been fitted to a series $w_1, \ldots, w_n$, it is useful to study the adequacy of the fit by examining the residuals $\hat{a}_1, \ldots, \hat{a}_n$ and, in particular, their autocorrelations

$$\hat{r}_k = \sum_{t=k+1}^{n} \hat{a}_t \hat{a}_{t-k} \Big/ \sum_{t=1}^{n} \hat{a}_t^2 \quad (k = 1, 2, \ldots).$$

An informal graphical analysis of these quantities combined with overfitting (Box & Jenkins, 1970, § 8.1) usually proves most effective in detecting possible deficiencies in the model. In addition, however, it is often worthwhile to look at an overall criterion of adequacy of fit. Box & Pierce (1970) noted that if the model were appropriate and the parameters were known, the quantity

$$\tilde{Q}(r) = n(n+2) \sum_{k=1}^{m} (n-k)^{-1} r_k^2, \tag{1·1}$$

where

$$r_k = \sum_{t=k+1}^{n} a_t a_{t-k} \Big/ \sum_{t=1}^{n} a_t^2,$$

would for large n be distributed as χ^2_m since the limiting distribution of $r = (r_1, \ldots, r_m)'$ is multivariate normal with mean vector zero (Anderson, 1942; Anderson & Walker, 1964),

The Collected Works of George E. P. Box, 1984, Wadsworth, Inc., Belmont, CA 94002.
Originally published in *Biometrika,* vol. 65, no. 2 (1978), pp. 297–303.

$\text{var}(r_k) = (n-k)/\{n(n+2)\}$ and $\text{cov}(r_k, r_l) = 0\ (k \neq l)$. Using the further approximation $\text{var}(r_k) = 1/n$, Box & Pierce (1970) suggested that the distribution of

$$Q(r) = n \sum_{k=1}^{m} r_k^2 \tag{1·2}$$

could be approximated by that of χ_m^2. Furthermore, they showed that when the $p+q$ parameters of an appropriate model are estimated and the $\hat{r}_k$'s replace the r_k's, then

$$Q(\hat{r}) = n \sum_{k=1}^{m} \hat{r}_k^2$$

would for large n be distributed as χ_{m-p-q}^2 yielding an approximate test for lack of fit.

In applications of this test, suspiciously low values of $Q(\hat{r})$ have sometimes been observed, and studies by the present authors, reported in a University of Wisconsin technical report, and by Davies, Triggs & Newbold (1977) have verified that the distribution of $Q(\hat{r})$ can deviate from χ_{m-p-q}^2. This observation was also made by Prothero & Wallis (1976) in the discussion of their paper. The observed discrepancies could be accounted for by several factors, for instance departures from normality of the autocorrelations. It appears, however, that the main difficulty is caused by the approximation of (1·1) by (1·2). A modified test based on the criterion

$$\tilde{Q}(\hat{r}) = n(n+2) \sum_{k=1}^{m} (n-k)^{-1} \hat{r}_k^2$$

was recommended by the present authors but its usefulness was questioned by Davies *et al.* (1977) on the ground that the variance of $\tilde{Q}(\hat{r})$ exceeds that of the χ_{m-p-q}^2 distribution. Our studies show however that the modified test provides a substantially improved approximation that should be adequate for most practical purposes.

2. MEANS AND VARIANCES OF $Q(r)$ AND $\tilde{Q}(r)$

To examine the overall test, it is useful to consider initially the quantities $Q(r)$ and $\tilde{Q}(r)$ which involve the white noise autocorrelations r. Since the limiting distribution of r is $N(0, n^{-1} I_m)$, $Q(r)$ and $\tilde{Q}(r)$ are asymptotically distributed as χ_m^2 and have expectation m and variance $2m$. For finite values of n, $\tilde{Q}(r)$ has expectation m, whereas

$$E\{Q(r)\} = n \sum_{k=1}^{m} E(r_k^2) = \frac{mn}{n+2}\left(1 - \frac{m+1}{2n}\right). \tag{2·1}$$

Clearly, unless n is large relative to m, $E\{Q(r)\}$ can be much smaller than m.

The variances are

$$\text{var}\{Q(r)\} = n^2 \sum_{k=1}^{m} \text{var}(r_k^2) + 2n^2 \sum_{k=1}^{m-1} \sum_{l=k+1}^{m} \text{cov}(r_k^2, r_l^2),$$

$$\text{var}\{\tilde{Q}(r)\} = n^2(n+2)^2 \sum_{k=1}^{m} (n-k)^{-2} \text{var}(r_k^2) + 2n^2(n+2)^2 \sum_{k=1}^{m-1} \sum_{l=k+1}^{m} (n-k)^{-1}(n-l)^{-1} \text{cov}(r_k^2, r_l^2), \tag{2·2}$$

where, for fixed n, $\text{cov}(r_k^2, r_l^2)$ is nonzero. The univariate and bivariate moments of the r_k's needed to evaluate (2·2) can be obtained using the identity

$$E(r_k^i r_l^j) = \frac{E\{(\Sigma a_t a_{t-k})^i (\Sigma a_t a_{t-l})^j\}}{E\{(\Sigma a_t^2)^{i+j}\}}, \tag{2·3}$$

which follows from independence of the r_k's and Σa_t^2 (Anderson, 1971, p. 304). Taking

$\text{var}(a_t) = 1$ without loss of generality, we have that Σa_t^2 is distributed as χ_n^2 and $E(\Sigma a_t^2)^{i+j} = n(n+2)\ldots(n+2i+2j-2)$. The term in the numerator of (2·3) can be evaluated by multiplying term by term and taking the expected value. It can thus be verified that for $k < \frac{1}{2}n$

$$\text{var}(r_k^2) = \frac{6(3n-5k)+3(n-k)^2}{n(n+2)(n+4)(n+6)} - \frac{(n-k)^2}{n^2(n+2)^2},$$

$$\text{cov}(r_k^2, r_l^2) = \frac{(n-k)(n-l)+4(n-l)+8(n-k-l)}{n(n+2)(n+4)(n+6)} - \frac{(n-k)(n-l)}{n^2(n+2)^2}. \qquad (2\cdot4)$$

The exact variances of $Q(r)$ and $\tilde{Q}(r)$ are readily evaluated using (2·2) and (2·4). By ignoring terms of order higher than $1/n$ it may be shown that approximately, for n large relative to m,

$$\text{var}\{Q(r)\} = 2m\left(1+\frac{m-10}{n}\right), \quad \text{var}\{\tilde{Q}(r)\} = 2m\left(1+\frac{2m-5}{n}\right).$$

The variance of $\tilde{Q}(r)$ exceeds $2m$ but the absence of a location bias makes its distribution much closer to χ_m^2 than that of $Q(r)$. This is illustrated in Fig. 1 which compares Monte Carlo distributions of $Q(r)$ and $\tilde{Q}(r)$ based on 1000 replications to the χ_m^2 distribution for $m = 30$ and $n = 100$. The observed distribution of $Q(r)$ has mean 24·97 and variance 60·47; $\tilde{Q}(r)$ has mean 30·17 and variance 88·25. These values agree quite closely with the theoretical values 24·85, 63·15, 30·00 and 91·48, respectively. Also shown by dashed lines in Fig. 1 is a distribution of the form $a\chi_b^2$ for which both the mean and variance are adjusted to correspond with those of $\tilde{Q}(r)$. There is perhaps somewhat better agreement in the upper tail but the main improvement results from adjusting the mean.

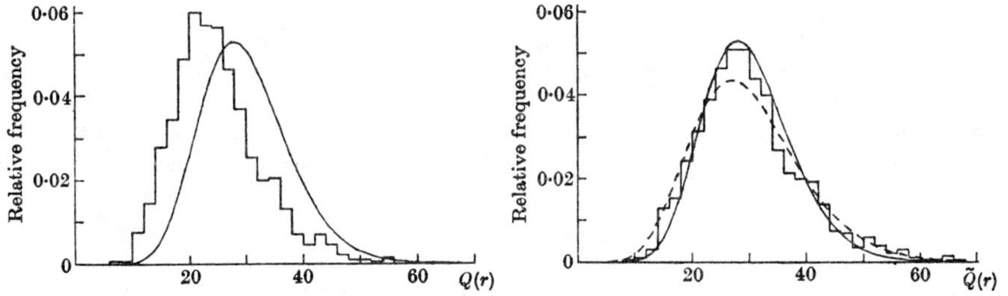

Fig. 1. Monte Carlo distributions of $Q(r)$ and $\tilde{Q}(r)$ and approximations; 1000 replications, $n = 100$ and $m = 30$; solid line, χ_{30}^2; dashed line, $a\chi_b^2$ ($a = 1{\cdot}52$, $b = 19{\cdot}68$).

3. The test statistics $Q(\hat{r})$ and $\tilde{Q}(\hat{r})$

Box & Pierce (1970) showed that the residual autocorrelations $\hat{r} = (\hat{r}_1, \ldots, \hat{r}_m)'$ from a correctly identified and fitted model can to a close approximation be represented as

$$\hat{r} \simeq (I-D)r,$$

where $I-D$ is an idempotent matrix of rank $m-p-q$. With this relationship the expectation of $Q(\hat{r})$ is

$$E\{Q(\hat{r})\} \simeq E\{nr'(I-D)r\} = \text{tr}\{n(I-D)C\},$$

where C is the exact covariance matrix of r. The matrix D has its largest elements in the upper left corner with the remaining elements d_{ij} decreasing to zero as i and/or j increases. The matrix DC is therefore nearly equal to $n^{-1}D$. Using this approximation and noting that

$E\{Q(r)\} = \text{tr}\,(nC)$, we have

$$E\{Q(\hat{r})\} \simeq E\{Q(r)\} - p - q. \tag{3·1}$$

On combining (2·1) and (3·1), the expected value of $Q(\hat{r})$ is approximately

$$E\{Q(\hat{r})\} \simeq \frac{mn}{n+2}\left(1 - \frac{m+1}{2n}\right) - p - q, \tag{3·2}$$

which indicates that the distribution of $Q(\hat{r})$ can deviate markedly from χ^2_{m-p-q} unless n is large relative to m. However, using the same approximations it can be shown that

$$E\{\tilde{Q}(\hat{r})\} \simeq E\{\tilde{Q}(r)\} - p - q = m - p - q.$$

It may be expected therefore that the distribution of $\tilde{Q}(\hat{r})$ might be approximated by the χ^2_{m-p-q} distribution.

The adequacy of this approximation was questioned by Davies *et al.* (1977) on the ground that the variance of $\tilde{Q}(\hat{r})$ exceeds $2(m-p-q)$. However, results from a simulation study reported in the next section suggest that the reduction in the location bias results as before in a markedly improved approximation that should be adequate for most practical purposes It also appears that the expression for the variance given by Davies *et al.*, which is not exact, overestimates the variance of $\tilde{Q}(\hat{r})$. For example, for fitting a first-order autoregressive model to white noise, Davies *et al.* obtain for $m = 20$ and $n = 50$, 100 and 200, $\text{var}\{\tilde{Q}(\hat{r})\} = 58·80$, 50·08 and 44·20, respectively, while our study gives $\text{var}\{\tilde{Q}(\hat{r})\} = 46·84$, 43·20 and 41·97, respectively.

4. Some numerical results

4·1. *Comparison of the overall tests*

A Monte Carlo study was conducted by generating 4000 sets of observations $\{w_1, \ldots, w_n\}$ from the first-order autoregressive model $w_t - \phi w_{t-1} = a_t$, estimating ϕ by the approximate maximum likelihood estimator

$$(n-2)\,(n-1)^{-1} \sum_{t=2}^{n} w_t w_{t-1} \Big/ \sum_{t=2}^{n-1} w_t^2$$

(Box & Jenkins, 1970, p. 279), and calculating autocorrelations of the residuals $\hat{a}_1 = (1-\hat{\phi}^2)\, w_1$, $\hat{a}_t = w_t - \hat{\phi} w_{t-1}$ $(t = 2, \ldots, n)$. The statistics $Q(\hat{r})$ and $\tilde{Q}(\hat{r})$ were then calculated.

Table 1 shows the proportion of $Q(\hat{r})$ and $\tilde{Q}(\hat{r})$ values exceeding the upper 5, 10 and 25 percentage points of the χ^2_{m-1} distribution for a few combinations of n and m and for $\phi = 0·5$. The table also gives the means and variances of the observed distributions. It seems clear that although the variance of $\tilde{Q}(\hat{r})$ exceeds $2(m-1)$ a test based on this statistic would for smaller sample sizes provide a considerable improvement over the previously used $Q(\hat{r})$ test.

Table 1. *Empirical means, variances and significance levels of the statistics $Q(\hat{r})$ and $\tilde{Q}(\hat{r})$; data generated from the model $w_t - \frac{1}{2}w_{t-1} = a_t$*

		$Q(\hat{r})$					$\tilde{Q}(\hat{r})$				
				% level					% level		
n	m	Mean	Var.	5	10	25	Mean	Var.	5	10	25
50	10	7·48	13·79	2·3	4·7	13·4	8·82	19·11	5·3	9·5	23·0
	20	13·96	27·50	1·3	2·3	6·4	18·58	47·76	6·1	10·4	23·2
100	10	8·14	16·04	3·4	7·0	18·2	8·83	18·88	5·0	9·9	23·1
	20	16·26	35·45	2·5	5·0	13·1	18·63	46·46	5·8	10·2	22·8
	30	23·53	55·74	1·7	3·6	9·1	28·58	81·71	7·2	11·6	23·4
200	10	8·57	16·76	4·2	8·3	21·5	8·92	18·16	5·0	9·8	23·9
	20	17·46	36·36	3·5	6·9	17·6	18·66	41·51	5·4	10·0	22·7
	30	26·11	56·01	2·9	5·6	14·2	28·66	67·37	5·9	10·5	23·8

4·2. *An alternative test based on* $Q(\hat{r})$

The above results suggest that a closer approximation to the distribution of $Q(\hat{r})$ should be obtainable by appropriate adjustment of the mean of the approximating distribution. Furthermore, Table 1 shows values of var $\{Q(\hat{r})\}$ which are nearly twice the mean, suggesting the approximation $Q(\hat{r}) \sim \chi^2_{E\{Q(\hat{r})\}}$ with $E\{Q(\hat{r})\}$ given by (3·2). Empirical significance levels obtained using this approximation and the criterion $\tilde{Q}(\hat{r})$ are compared in Table 2. The agreement is quite close. It may however be more convenient generally to use $\tilde{Q}(\hat{r})$, since the test based on $Q(\hat{r})$ will have noninteger degrees of freedom.

Table 2. *Empirical significance levels based on the approximations* $Q(\hat{r}) \sim \chi^2_{E\{Q(\hat{r})\}}$ *and* $\tilde{Q}(\hat{r}) \sim \chi^2_{m-1}$; *data generated from the model* $w_t - \phi w_{t-1} = a_t$

		$Q(\hat{r}) \sim \chi^2_{E\{Q(\hat{r})\}}$						$\tilde{Q}(\hat{r}) \sim \chi^2_{m-1}$					
		$m = 10$			$m = 20$			$m = 10$			$m = 20$		
		% level						% level					
n	ϕ	5	10	25	5	10	25	5	10	25	5	10	25
50	0·1	4·1	8·3	21·2	4·6	8·1	20·9	4·7	9·3	21·4	5·9	10·1	22·5
	0·4	4·3	8·5	22·1	4·6	8·6	21·6	5·1	9·3	22·8	6·0	10·3	23·0
	0·7	4·7	9·5	23·3	5·1	9·6	22·6	5·4	10·1	23·6	6·7	11·3	24·0
100	0·1	4·3	8·8	23·4	5·1	9·3	22·2	4·7	9·3	23·5	5·9	10·0	22·7
	0·4	4·4	8·5	23·5	5·3	9·1	22·7	4·8	9·3	23·5	6·0	10·0	23·0
	0·7	4·7	9·0	24·1	5·6	9·6	22·7	4·9	9·4	24·0	6·2	10·3	23·2
200	0·1	5·0	9·6	24·1	5·2	9·8	22·7	5·2	9·9	24·2	5·5	10·2	23·2
	0·4	4·8	9·5	23·8	5·1	9·6	22·5	5·1	9·8	24·0	5·4	10·1	22·8
	0·7	4·8	9·9	24·1	4·9	10·0	22·5	5·0	10·1	24·2	5·3	10·5	22·8

4·3. *A power calculation*

The two criteria $Q(\hat{r})$ and $\tilde{Q}(\hat{r})$ differ in the weighting which is applied to the autocorrelations $\hat{r}_k$ with $\tilde{Q}(\hat{r})$ giving more emphasis to later autocorrelations than $Q(\hat{r})$. This would perhaps be an advantage if serial correlation occurs at high lags k. However, for large n this difference should be rather small. If the type of discrepancies to be expected is known, tests specifically aimed at detecting these discrepancies should be used. Such specific tests will of course be much more powerful. This point is illustrated in Table 3 which empirically compares the power of the overall tests and the method of "overfitting" (Box & Jenkins, 1970). The results are based on data generated from a second-order autoregressive model, with a first-order model being fitted to obtain $Q(\hat{r})$ and $\tilde{Q}(\hat{r})$. As might be expected, the overall tests

Table 3. *Empirical power of the overall tests and the method of overfitting for* $n = 100$. *Assumed model*: $w_t - \phi w_{t-1} = a_t$; *true model*: $(1-0{\cdot}7B)(1-G_2 B)w_t = a_t$. *Nominal significance level:* 5%

Test	m	$G_2 = 0$	$G_2 = 0{\cdot}1$	$G_2 = 0{\cdot}3$	$G_2 = 0{\cdot}5$	$G_2 = 0{\cdot}7$	$G_2 = 0{\cdot}9$
Overfitting		5·3	12·0	59·7	93·8	99·7	99·1
$Q(\hat{r}) \sim \chi^2_{E\{Q(\hat{r})\}}$	10	4·7	6·7	28·6	72·0	96·6	99·9
	20	5·6	7·3	24·4	62·8	93·7	99·7
	30	6·0	7·7	22·9	58·1	91·7	99·5
$\tilde{Q}(\hat{r}) \sim \chi^2_{m-1}$	10	4·9	7·0	28·9	71·6	96·2	99·9
	20	6·2	8·0	24·7	61·7	93·2	99·6
	30	7·0	9·0	23·7	57·0	90·5	99·3

are much less powerful than overfitting which tests the hypothesis that the second-order autoregressive coefficient is zero. A smaller value of m improves the power of the overall tests for this particular alternative.

4·4. *Effect of nonnormality of the a_t's*

In developing the overall test, it is assumed that the innovations a_t in the model are normally distributed. Circumstances occur where this assumption is not true. For example, it is known that stock price innovations often have highly leptokurtic distributions. Results by Anderson & Walker (1964) show that the asymptotic normality of the r_k's does not require normality of the a_t's, only that var (a_t) is finite. The overall test might therefore be expected to be insensitive to departures from normality of the a_t's. This is supported by Table 4, which shows the behaviour of $\tilde{Q}(\hat{r})$ when the a_t's have a double exponential and a uniform distribution. The results agree closely with those obtained under the normality assumption in Table 1.

Table 4. *Empirical means, variances and significance levels of $\tilde{Q}(\hat{r})$ when the innovations a_t have* (i) *a double exponential and* (ii) *a uniform distribution; data generated from the model*

$$w_t - \tfrac{1}{2}w_{t-1} = a_t$$

		(i) $a_t \sim$ double exponential					(ii) $a_t \sim$ uniform				
					% level					% level	
n	m	Mean	Var.	5	10	25	Mean	Var.	5	10	25
50	10	8·50	13·59	4·7	8·6	20·7	9·01	19·35	5·6	10·0	24·4
	20	17·77	47·00	5·4	8·8	19·6	18·95	52·39	7·3	12·1	24·3
100	10	8·80	18·70	5·0	9·1	22·4	9·11	19·41	5·5	10·8	25·3
	20	18·37	43·62	4·8	9·2	22·0	19·00	47·52	6·4	11·5	25·7
	30	27·94	76·60	6·3	10·1	21·9	28·98	81·72	7·5	12·4	25·3

5. EXTENSION TO TRANSFER FUNCTION NOISE MODELS

To check the adequacy of the transfer function in the model

$$w_t = \frac{\omega(B)}{\delta(B)}\alpha_t + \frac{\theta(B)}{\phi(B)}a_t,$$

where

$$\omega(B)/\delta(B) = (\omega_0 - \omega_1 B - \ldots - \omega_u B^u)/(1-\delta_1 B - \ldots - \delta_v B^v)$$

and where the input series $\{\alpha_t\}$ is assumed to be white noise and independent of $\{a_t\}$, it is useful to examine the cross-correlations between $\{\alpha_t\}$ and the residuals $\{\hat{a}_t\}$

$$\hat{r}_k^* = \sum_{t=k+1}^{n} \alpha_{t-k}\hat{a}_t \bigg/ \left(\sum_{t=1}^{n}\alpha_t^2 \sum_{t=1}^{n}\hat{a}_t^2\right)^{1/2} \quad (k = 0, 1, \ldots).$$

D. A. Pierce in a University of Wisconsin technical report, Box & Jenkins (1970, § 11.3) and Pierce (1972) propose an overall test for lack of fit based on approximating the distribution of

$$S(\hat{r}^*) = n \sum_{k=0}^{m} (\hat{r}_k^*)^2$$

by the χ^2_{m-v-u} distribution. However, on arguing as above, it appears that a criterion of the form

$$\tilde{S}(\hat{r}^*) = n^2 \sum_{k=0}^{m} (n-k)^{-1}(\hat{r}_k^*)^2$$

would be more appropriate. The criterion $S(\hat{r}_k^*)$ is obtained by approximating the variance of the k-th sample cross-correlation between $\{\alpha_t\}$ and $\{a_t\}$ by $1/n$, while the actual variance is $(n-k)/n^2$.

The modification considered in the previous sections applies to the overall test for lack of fit in the noise model $\theta(B)/\phi(B)$ discussed by Box & Jenkins (1970, § 11.3).

This work was sponsored by the United States Army Research Office and the Air Force Office of Scientific Research.

References

Anderson, R. L. (1942). Distribution of the serial correlation coefficients. *Ann. Math. Statist.* **13**, 1–13.

Anderson, T. W. (1971). *The Statistical Analysis of Time Series*. New York: Wiley.

Anderson, T. W. & Walker, A. M. (1964). On the asymptotic distribution of the autocorrelations of a sample from a linear stochastic process. *Ann. Math. Statist.* **35**, 1296–303.

Box, G. E. P. & Jenkins, G. M. (1970). *Time Series Analysis Forecasting and Control*. San Francisco: Holden-Day.

Box, G. E. P. & Pierce, D. A. (1970). Distribution of residual autocorrelations in autoregressive-integrated moving average time series models. *J. Am. Statist. Assoc.* **65**, 1509–26.

Davies, N., Triggs, C. M. & Newbold, P. (1977). Significance levels of the Box–Pierce portmanteau statistic in finite samples. *Biometrika* **64**, 517–22.

Pierce, D. A. (1972). Residual correlations and diagnostic checking in dynamic-disturbance time series models. *J. Am. Statist. Assoc.* **67**, 636–40.

Prothero, D. L. & Wallis, K. F. (1976). Modelling macroeconomic time series (with discussion). *J. R. Statist. Soc.* A **139**, 468–500.

[*Received September* 1977. *Revised January* 1978]

3.14

The Likelihood Function of Stationary Autoregressive-Moving Average Models

GRETA M. LJUNG
University of Denver

G. E. P. BOX
University of Wisconsin

SUMMARY

This paper examines the likelihood function of a stationary autoregressive-moving average model of order (p,q) and presents a method for its evaluation. Numerical results illustrating the computational efficiency of the method are given.

Some key words: Autoregressive-moving average model; Likelihood function; Time series estimation.

1. INTRODUCTION

An autoregressive-moving average process of order (p,q) is defined by

$$w_t - \phi_1 w_{t-1} - \ldots - \phi_p w_{t-p} = a_t - \theta_1 a_{t-1} - \ldots - \theta_q a_{t-q}, \tag{1·1}$$

where the a_t's are independent random variables with mean zero and variance σ^2. The process may be written more compactly as $\phi(B)w_t = \theta(B)a_t$, where $\phi(B) = 1 - \phi_1 B - \ldots - \phi_p B^p$, $\theta(B) = 1 - \theta_1 B - \ldots - \theta_q B^q$ and B is the backshift operator $B^k w_t = w_{t-k}$. The process $\{w_t\}$ is stationary if $\phi(B) = 0$ has all roots outside the unit circle. A similar condition on the moving average operator $\theta(B)$ ensures that the model is invertible.

Assuming that the a_t's and hence the w_t's have a normal distribution, the likelihood function for the parameters $\phi = (\phi_1, \ldots, \phi_p)'$, $\theta = (\theta_1, \ldots, \theta_q)'$ and σ is

$$L(\phi, \theta, \sigma \mid w) = (2\pi\sigma^2)^{-\frac{1}{2}n} |\Sigma|^{-\frac{1}{2}} \exp\{-(2\sigma^2)^{-1} w' \Sigma^{-1} w\}, \tag{1·2}$$

where $w = (w_1, \ldots, w_n)'$ and $\Sigma = \sigma^{-2} \operatorname{cov}(w)$. The mean of the w_t's is assumed to be zero for convenience. The likelihood may be maximized with respect to ϕ and θ by minimizing the function

$$L_0(\phi, \theta \mid w) = (w' \Sigma^{-1} w) |\Sigma|^{1/n}.$$

The exact form of (1·2) is complicated and many of the estimation methods proposed in the literature are based on approximating or modifying the likelihood; for references see Shaman (1975) and Anderson (1975, 1977). A commonly used modification is to treat the initial values $w_{1-p}, \ldots, w_0$ and $a_{1-q}, \ldots, a_0$ in the model as fixed constants. Maximum likelihood is then equivalent to conditional least squares. Box & Jenkins (1970) obtain approximate maximum likelihood estimates by minimizing $w' \Sigma^{-1} w$. This quadratic form is written as a sum of squares and is minimized using a nonlinear least squares procedure.

These estimates usually closely approximate the exact maximum likelihood estimates. However, appreciable differences can occur if the sample is small or the moving average parameters are near or on the boundary of the invertibility region. Results reported in an unpublished report of the Australian Bureau of Census and Statistics by K. M. Kang and by

The Collected Works of George E. P. Box, 1984, Wadsworth, Inc., Belmont, CA 94002. Originally published in *Biometrika,* vol. 66, no. 2 (1979), pp. 265–270.

others suggest that the exact maximum likelihood method in these cases gives superior estimates.

The form of the exact likelihood function has been studied by several authors. Siddiqui (1958) gave expressions for $w'\Sigma^{-1}w$ and $|\Sigma|$ for a pure autoregressive model. Box & Jenkins (1970) derived the likelihood function for the moving average model. This derivation was extended to the general case by Newbold (1974) and Galbraith & Galbraith (1974). The likelihood function for the general model has been studied more recently by Dent (1977), Ali (1977) and Ansley (1979). Ansley presents a method for evaluating the likelihood which is shown to be more efficient than other methods described in the literature.

The present paper examines further the likelihood function for the general model (1·1). Various closed form expressions for $w'\Sigma^{-1}w$ are given and it is illustrated how this quadratic form and the determinant $|\Sigma|$ can be evaluated. The method for evaluating $w'\Sigma^{-1}w$ is based on estimating the $p+q$ initial values in (1·1) or $m = \max(p,q)$ linear combinations of these initial values. Some numerical results indicate that the efficiency of this method is similar to the efficiency of the method proposed by Ansley.

2. THE LIKELIHOOD FUNCTION FOR THE GENERAL MODEL

It is convenient to write the model for the $n\times 1$ observation vector w in matrix form as

$$L_1 w = L_2 a + V u_*, \tag{2·1}$$

where $a = (a_1, \ldots, a_n)'$ is the $n \times 1$ vector of random normal deviates,

$$u_* = (w_{1-p}, \ldots, w_0, a_{1-q}, \ldots, a_0)'$$

is the $(p+q)\times 1$ vector of initial values, L_1 and L_2 are simple $n\times n$ matrices with 1's down the leading diagonal, $-\phi_1$ and $-\theta_1$ down the subdiagonal, and so on. Further V is the $n\times(p+q)$ matrix

$$V = \begin{bmatrix} V_1 \\ \hdashline 0 \end{bmatrix},$$

with

$$V_1 = \begin{bmatrix} \phi_p & \phi_{p-1} & \cdots & \phi_1 & -\theta_q & -\theta_{q-1} & \cdots & -\theta_1 \\ & \phi_p & \cdots & \phi_2 & & -\theta_q & \cdots & -\theta_2 \\ & & & \vdots & & & & \vdots \\ & & & \phi_p & & & & -\theta_q \end{bmatrix}$$

The order of the matrix V_1 is $m\times(p+q)$. Clearly, the two sections of this matrix have different dimensions when $p \neq q$.

The joint density function of w and u_* can be written as

$$p(w, u_* \mid \phi, \theta, \sigma) = p(w \mid u_*, \phi, \theta, \sigma)\, p(u_* \mid \phi, \theta, \sigma),$$

where $p(u_* \mid \phi, \theta, \sigma)$ is a normal density with mean vector 0 and covariance matrix $\sigma^2 \Sigma_*$. The conditional density $p(w \mid u_*, \phi, \theta, \sigma)$ is readily obtained from the density function of a by the transformation

$$a = Mw - Xu_*, \tag{2·2}$$

where, from (2·1), $M = L_2^{-1} L_1$ and $X = L_2^{-1} V$.

The joint density of w and u_* may be factorized as

$$p(w, u_* \mid \phi, \theta, \sigma) = p(w \mid \phi, \theta, \sigma)\, p(u_* \mid w, \phi, \theta, \sigma)$$

to give expressions for $w'\Sigma^{-1}w$ and $|\Sigma|$ in (1·2). By applying generalized least squares theory we can show that

$$w'\Sigma^{-1}w = \hat{a}'\hat{a} + \hat{u}_*'\Sigma_*^{-1}\hat{u}_*, \tag{2·3}$$

where

$$\hat{u}_* = E(u_*\,|\,w) = (\Sigma_*^{-1}+X'X)^{-1}X'Mw, \quad \hat{a} = E(a\,|\,w) = Mw - X\hat{u}_*.$$

Furthermore,

$$|\Sigma| = |\Sigma_*|\,|\Sigma_*^{-1}+X'X|. \tag{2·4}$$

If the expressions for $\hat{u}_*$ and $\hat{a}$ are substituted in (2·3), we obtain

$$w'\Sigma^{-1}w = w'M'\{I_n - X(\Sigma_*^{-1}+X'X)^{-1}X'\}Mw, \tag{2·5}$$

which shows that

$$\Sigma^{-1} = M'\{I_n - X(\Sigma_*^{-1}+X'X)^{-1}X'\}M.$$

This expression for the inverse agrees with those obtained by Galbraith & Galbraith (1974) and Dent (1977).

From (2·2), Mw is the vector a corresponding to the initial values $u_* = 0$. Defining $\hat{a}_0 = Mw$, we can write (2·5) as

$$w'\Sigma^{-1}w = \hat{a}_0'\hat{a}_0 - \hat{a}_0'X(\Sigma_*^{-1}+X'X)^{-1}X'\hat{a}_0 = \hat{a}_0'\hat{a}_0 - \hat{u}_*'(\Sigma_*^{-1}+X'X)\hat{u}_*. \tag{2·6}$$

We observe that the first term on the right-hand side of (2·6) is the conditional sum of squares, given $u_* = 0$. Minimization of this sum of squares maximizes the conditional likelihood $L(\phi, \theta, \sigma\,|\,w, u_* = 0)$; see, for example, Box & Jenkins (1970, Chapter 7).

The expression (2·5) can be written in the equivalent form

$$w'\Sigma^{-1}w = \hat{a}_0'\hat{a}, \tag{2·7}$$

since $\hat{a} = Mw - X\hat{u}_* = \{I_n - X(\Sigma_*^{-1}+X'X)^{-1}X'\}Mw$. The elements of the vector $\hat{a}_0$ can be calculated recursively from the general model (1·1) setting the initial values equal to zero. The elements of $\hat{a}$ can be obtained in a similar manner using the estimates $\hat{u}_*$ as initial values. Thus, by (2·7), the problem of evaluating $w'\Sigma^{-1}w$ is reduced to that of calculating $\hat{u}_*$.

For the general autoregressive-moving average model, Box & Jenkins (1970) show that

$$w'\Sigma^{-1}w = \sum_{t=-\infty}^{n} \hat{a}_t^2, \tag{2·8}$$

where $\hat{a}_t = E(a_t\,|\,w)$. Comparing with (2·3), we see that

$$\hat{u}_*'\Sigma_*^{-1}\hat{u}_* = \sum_{t=-\infty}^{0} \hat{a}_t^2.$$

For $t \leqslant -q$, the $\hat{a}_t$'s satisfy the relation $\hat{a}_t = \phi_1\hat{a}_{t+1} + \ldots + \phi_p\hat{a}_{t+p}$. This may be verified by multiplying the model $a_t = \theta^{-1}(B)\phi(B)w_t$ by $\phi(F) = 1 - \phi_1F - \ldots - \phi_pF^p$, where $F^k w_t = w_{t+k}$, and taking the conditional expectation on both sides. Thus, once $\hat{u}_*$ and $\hat{a}$ are given, the remaining terms in (2·8) may be obtained by a simple recursive calculation.

3. Estimation of the initial values

Defining $h = X'Mw = X'\hat{a}_0$, we have from (2·3) that $\hat{u}_* = (\Sigma_*^{-1}+X'X)^{-1}h$. To evaluate $\hat{u}_*$ we thus need to compute Σ_*, $X'X$ and h.

Since $u_* = (w_*', a_*')'$, where $w_*' = (w_{1-p}, \ldots, w_0)$ and $a_*' = (a_{1-q}, \ldots, a_0)$, the matrix Σ_* has the form

$$\Sigma_* = \sigma^{-2}\operatorname{cov}(u_*) = \begin{bmatrix} \Gamma & \Psi \\ \Psi' & I_q \end{bmatrix},$$

where $\Gamma = \sigma^{-2} E(w_* w_*')$, $\Psi = \sigma^{-2} E(w_* a_*')$ and $I_q = \sigma^{-2} E(a_* a_*')$. The elements of Ψ are defined by

$$\sigma^{-2} E(w_t a_k) = \begin{cases} \psi_{t-k} & (k < t), \\ 1 & (k = t), \\ 0 & (k > t), \end{cases}$$

where the ψ's satisfy the difference equation $\phi(B)\psi_i = -\theta_i$, with $\psi_0 = 0$, $\psi_i = 0$, for $i < 0$, and $\theta_i = 0$, for $i > q$. The elements $\gamma_k = \sigma^{-2} E(w_t w_{t-k})$ of the matrix Γ satisfy the equation

$$\gamma_k - \phi_1 \gamma_{k-1} - \ldots - \phi_p \gamma_{k-p} = E(a_t w_{t-k}) - \theta_1 E(a_{t-1} w_{t-k}) - \ldots - \theta_q E(a_{t-q} w_{t-k}).$$

Substitution of $k = 0, 1, \ldots, p$ in this expression gives the equations $A\gamma = v$, where

$$A = \begin{bmatrix} 1 & 0 & \ldots & 0 & 0 \\ -\phi_1 & 1 & \ldots & 0 & 0 \\ \vdots & \vdots & & \vdots & \vdots \\ -\phi_{p-1} & -\phi_{p-2} & \ldots & 1 & 0 \\ -\phi_p & -\phi_{p-1} & \ldots & -\phi_1 & 1 \end{bmatrix} + \begin{bmatrix} 0 & -\phi_1 & -\phi_2 & \ldots & -\phi_p \\ 0 & -\phi_2 & -\phi_3 & \ldots & 0 \\ \vdots & \vdots & \vdots & & \vdots \\ 0 & -\phi_p & 0 & \ldots & 0 \\ 0 & 0 & 0 & \ldots & 0 \end{bmatrix},$$

$\gamma = (\gamma_0, \ldots, \gamma_p)'$ and v is a $(p+1) \times 1$ vector whose elements satisfy the relation

$$v_j = \psi_{1-j} - \sum_{i=1}^{q} \theta_i \psi_{i+1-j} \quad (j = 1, \ldots, p+1),$$

with $\psi_0 = 1$ and $\psi_l = 0$, for $l < 0$. The elements of Γ are given by the solution $\gamma = A^{-1} v$.

Next, we need to compute the matrix product $X'X$. Since $X = L_2^{-1} V$, the elements of X are

$$x_{ij} = \begin{cases} \theta_1 x_{i-1,j} + \ldots + \theta_q x_{i-q,j} + \phi_{p+i-j} & (j = 1, \ldots, p), \\ \theta_1 x_{i-1,j} + \ldots + \theta_q x_{i-q,j} - \theta_{q+i-j+p} & (j = p+1, \ldots, p+q). \end{cases}$$

for $i = 1, \ldots, n$, where $x_{ij} = 0$, for $i < 1$, $\phi_k = 0$, for $k > p$, and $\theta_l = 0$, for $l > q$. However, for high order models, the computation of $X'X$ becomes more efficient if we write $X'X = V' L_2'^{-1} L_2^{-1} V = V_1' \Pi' \Pi V_1$ and first evaluate $\Pi' \Pi$. The matrix Π consists of the first $m = \max(p, q)$ rows of L_2^{-1} and has the form $\Pi = (\pi_1, \ldots, \pi_m)$, where $\pi_1 = (1, \pi_1, \ldots, \pi_{n-1})'$, $\pi_2 = (0, 1, \ldots, \pi_{n-2})'$, $\ldots$, $\pi_m = (0, 0, \ldots, \pi_{n-m})'$. The elements π_i satisfy the difference equation $\theta(B)\pi_i = 0$, with initial values $\pi_0 = 1$ and $\pi_i = 0$, for $i < 0$. Having computed the vector π_1, we can obtain the elements $\{p_{i1}\}$ of the first column of $\Pi' \Pi$ using the equation $\theta(F) p_{i1} = \pi_{i-1}$ and setting $p_{i1} = 0$, for $i > n$. The remaining diagonal and subdiagonal elements of $\Pi' \Pi$ are given by $p_{ij} = p_{i-1,j-1} - \pi_{n-i+1} \pi_{n-j+1}$.

Finally, to evaluate the elements $\{h_j\}$ of the $(p+q) \times 1$ vector $h = X' \hat{a}_0$ it is convenient to write $h = V' g$, where $g = L_2'^{-1} \hat{a}_0$. Then

$$h_j = \begin{cases} \sum_{i=1}^{j} \phi_{p-j+i} g_i & (j = 1, \ldots, p), \\ \sum_{i=1}^{j-p} (-\theta_{q-j+p+i}) g_i & (j = p+1, \ldots, p+q), \end{cases}$$

where the g's satisfy the relation $\theta(F) g_i = \hat{a}_{0i}$, with $g_i = 0$, for $i > n$.

We can now calculate the estimates $\hat{u}_*$ in (2·3) and evaluate $w' \Sigma^{-1} w$ using (2·7). The determinant in the likelihood function is also readily calculated using (2·4).

Estimation of u_* requires the inversion of the $(p+q)\times(p+q)$ matrix Σ_* and solution of the $p+q$ equations $(\Sigma_*^{-1}+X'X)\hat{u}_* = h$. If p and q are large, the efficiency of the computations may be increased by directly evaluating the m linear combinations $\hat{d} = V_1\hat{u}_*$ appearing in the expression for the vector $\hat{a}$. Using well-known matrix results, it may be verified that

$$\hat{d} = V_1\hat{u}_* = (\Sigma_d^{-1}+\Pi'\Pi)^{-1}g_m,$$

where $\Sigma_d = V_1\Sigma_* V_1' = \sigma^{-2}\operatorname{cov}(V_1 u_*)$ and $g_m = (g_1,\ldots,g_m)'$. The calculation of $\hat{d}$ gives as a byproduct the components of the determinant (2·4), which can be written in the form $|\Sigma| = |\Sigma_d|\,|\Sigma_d^{-1}+\Pi'\Pi|$.

4. Simplifications for the autoregressive model

If the time series is generated by an autoregressive model of order p, then $u_* = w_*$, and $L_2 = I_n$, so that $M = L_1$ and $X = V$. Hence, from (2·3)

$$\hat{w}_* = (\Sigma_*^{-1}+V'V)^{-1}V'L_1 w = (\Sigma_*^{-1}+V_1'V_1)^{-1}V_1'L_{11}w_p^\dagger, \qquad (4\cdot1)$$

where $w_p^\dagger = (w_1,\ldots,w_p)'$ and L_{11} is the $p\times p$ submatrix in the upper left corner of L_1.

It may be shown that for this model

$$\Sigma_*^{-1} = L_{11}'L_{11}-V_1V_1' = L_{11}L_{11}'-V_1'V_1. \qquad (4\cdot2)$$

By substituting this expression in (4·1) and observing that V_1' and L_{11} commute we obtain $\hat{w}_* = L_{11}'^{-1}V_1'w_p^\dagger$. Equivalently, $L_{11}'\hat{w}_* = V_1'w_p^\dagger$, which shows that the estimates satisfy the relation

$$\hat{w}_t = \phi_1\hat{w}_{t+1}+\ldots+\phi_p\hat{w}_{t+p} \quad (t = 0,-1,\ldots,1-p), \qquad (4\cdot3)$$

with $\hat{w}_t = w_t$, for $t\geqslant 1$. The estimates $\hat{w}_*$ are thus obtained by a backward recursive calculation. We note that since, from (2·3), $\hat{w}_* = E(w_*\,|\,w)$, the result (4·3) could be obtained directly by taking the conditional expectation on both sides of the backward form of the autoregressive model $w_t = \phi_1 w_{t+1}+\ldots+\phi_p w_{t+p}+e_t$, where $E(e_t\,|\,w) = 0$, for $t\leqslant 0$.

Equations (4·3) and (2·7) define a simple procedure for evaluating $w'\Sigma^{-1}w$. It is also easy to evaluate $w'\Sigma^{-1}w$ using (2·6). For the autoregressive model, this expression simplifies to

$$w'\Sigma^{-1}w = \hat{a}_0'\hat{a}_0 - w_p^{\dagger\prime}V_1V_1'w_p^\dagger, \qquad (4\cdot4)$$

where $w_p^{\dagger\prime}V_1V_1'w_p^\dagger = \phi_p^2w_1^2+(\phi_{p-1}w_1+\phi_pw_2)^2+\ldots+(\phi_1w_1+\ldots+\phi_pw_p)^2$.

Since, in (4·4), $\hat{a}_0 = L_1 w$ and $V_1'w_p^\dagger = V'w$, we see that $\Sigma^{-1} = L_1'L_1 - VV'$. For the special case when $n = p$, this expression agrees with (4·2).

The determinant in the likelihood function of the autoregressive model is

$$|\Sigma| = |\Sigma_*| = |L_{11}'L_{11}-V_1V_1'|^{-1},$$

which follows from combining (2·4) and (4·2).

5. Some numerical results

Table 1 shows approximate CPU times required to evaluate the conditional likelihood $L(\phi,\theta,\sigma\,|\,w,u_* = 0)$, or equivalently the sum of squares $\hat{a}_0'\hat{a}_0$, and the exact likelihood (1·2) on a Burroughs 6700 computer. For the moving average and mixed models, the quadratic form $w'\Sigma^{-1}w$ was evaluated by directly computing the linear combinations $\hat{d} = V_1\hat{u}_*$. The m equations $(\Sigma_d^{-1}+\Pi'\Pi)\hat{d} = g_m$ were solved by the IMSL subroutine LEQT1P, which uses a Cholesky decomposition of the matrix $(\Sigma_d^{-1}+\Pi'\Pi)$. For the autoregressive models, $w'\Sigma^{-1}w$ was evaluated using the result (4·3). Table 1 compares the ratio of the CPU times for the exact and conditional likelihood calculations with those reported by Ansley (1979). The agreement is generally close, suggesting similarity in the efficiency of the two methods.

Table 1. *Comparison of approximate* CPU *times in milliseconds for computation of conditional and exact likelihoods for various models; series length* $n = 100$

Model	Conditional	Exact	Exact/conditional Present algorithm	Exact/conditional Ansley's algorithm
ARMA(1, 0)	16	18	1·1	1·1
ARMA(0, 1)	19	63	3·3	3·1
ARMA(2, 0)	24	29	1·2	1·2
ARMA(0, 2)	24	115	4·8	4·8
ARMA(1, 1)	25	102	4·1	3·8
ARMA$(1, 0).(1, 0)_{12}$	31	102	3·3	7·9
ARMA$(0, 1).(1, 0)_{12}$	26	410	15·8	15·2
ARMA$(1, 0).(0, 1)_{12}$	28	400	14·3	14·5
ARMA$(0, 1).(0, 1)_{12}$	30	225	7·5	8·7

Times for Fortran programs on the Burroughs 6700 computer at the University of Denver.

The present method assumes unlike Ansley's algorithm that the model (1·1) is invertible. If any of the zeros of $\theta(B)$ lie inside the unit circle, computer overflow problems may occur in the calculation of g_m and $\Pi'\Pi$. This difficulty may, however, be overcome by simply inverting any zero inside the unit circle prior to the computation of the likelihood. In the one parameter case this involves replacing θ by $1/\theta$, if $|\theta| > 1$. The present method has an advantage over Ansley's method in that it directly gives the residual vector $\hat{a}$, which is needed for diagnostic checking of a fitted model and for computation of minimum mean square error forecasts of future observations.

This work was supported by the U.S. Army Research Office and by the Air Force Office of Scientific Research.

REFERENCES

ALI, M. M. (1977). Analysis of autoregressive-moving average models: Estimation and prediction. *Biometrika* **64**, 535–45.

ANDERSON, T. W. (1975). Maximum likelihood estimation of parameters of autoregressive processes with moving average residuals and other covariance matrices with linear structure. *Ann. Statist.* **3**, 1283–1304.

ANDERSON, T. W. (1977). Estimation for autoregressive moving average models in the time and frequency domains. *Ann. Statist.* **5**, 842–65.

ANSLEY, C. F. (1979). An algorithm for the exact likelihood of a mixed autoregressive moving average process. *Biometrika* **66**, 59–65.

BOX, G. E. P. & JENKINS, G. M. (1970). *Time Series Analysis: Forecasting and Control.* San Francisco: Holden-Day.

DENT, W. T. (1977). Computation of the exact likelihood function for an ARIMA process. *J. Statist. Comp. & Simul.* **5**, 193–206.

GALBRAITH, R. F. & GALBRAITH, J. I. (1974). On the inverses of some patterned matrices arising in the theory of stationary time series. *J. Appl. Prob.* **11**, 63–71.

NEWBOLD, P. (1974). The exact likelihood function for a mixed autoregressive-moving average process. *Biometrika* **61**, 423–6.

SHAMAN, P. (1975). An approximate inverse for the covariance matrix of moving average and autoregressive processes. *Ann. Statist.* **3**, 532–8.

SIDDIQUI, M. M. (1958). On the inversion of the sample covariance matrix in a stationary autoregressive process. *Ann. Math. Statist.* **29**, 585–8.

[*Received July* 1978. *Revised November* 1978]

3.15

Modeling Multiple Time Series with Applications

G. C. TIAO and G. E. P. BOX*

An approach to the modeling and analysis of multiple time series is proposed. Properties of a class of vector autoregressive moving average models are discussed. Modeling procedures consisting of tentative specification, estimation, and diagnostic checking are outlined and illustrated by three real examples.

KEY WORDS: Multiple time series; Vector autoregressive moving average models; Cross-correlations; Partial autoregression; Intervention analysis; Transfer function.

1. INTRODUCTION

Business, economic, engineering and environmental data are often collected in roughly equally spaced time intervals, for example, hour, week, month, or quarter. In many problems, such time series data may be available on several related variables of interest. Two of the reasons for analyzing and modeling such series jointly are

1. To understand the dynamic relationships among them. They may be contemporaneously related, one series may lead the others or there may be feedback relationships.
2. To improve accuracy of forecasts. When there is information on one series contained in the historical data of another, better forecasts can result when the series are modeled jointly.

Let

$$\{Z_{1t}\}, \ldots, \{Z_{kt}\}, \quad t = 0, \pm 1, \pm 2, \ldots \tag{1.1}$$

be k series taken in equally spaced time intervals. Writing

$$\mathbf{Z}_t = (Z_{1t}, \ldots, Z_{kt})', \tag{1.2}$$

we shall refer to the k series as a k-dimensional vector of multiple time series. Models that are of possible use in representing such multiple time series, considerations of their properties, and methods for relating them to actual data have been extensively discussed in the literature. See in particular Quenouille (1957), Whittle (1963), Hannan (1970), Zellner and Palm (1974), Brillinger (1975), Dunsmuir and Hannan (1976), Box and Haugh (1977), Granger and Newbold (1977), Parzen (1977), Wallis (1977), Chan and Wallis (1978), Deistler, Dunsmuir, and Hannan (1978), Hallin (1978), Jenkins (1979), Hsiao (1979), Akaike (1980), Hannan (1980), Hannan, Dunsmuir, and Deistler (1980), and Quinn (1980). There are, however, considerable divergences of view. The object of this article is to describe an approach to the modeling and analysis that we have developed over a considerable period of time and that we are finding effective. Our main emphasis will be on motivating, describing, and illustrating the various methods used in an iterative model building process. Much, if not all, of the underlying theory can be found in the references given and, therefore, will not be repeated. Section 2 presents a short review of the widely used univariate ($k = 1$) time series and transfer function models as developed in Box and Jenkins (1970). Section 3 discusses a class of vector autoregressive moving average models. Model building procedures are discussed in Section 4 and applied to two actual examples in Section 5. A comparison with some alternative approaches and some concluding remarks pertaining to the analysis of fitting results are given in Section 6.

* G.C. Tiao is Professor of Statistics and Business, and G.E.P. Box is Vilas Research Professor, Department of Statistics, University of Wisconsin, Madison, WI 53706. The authors are grateful to W.R. Bell, I. Chang, M.R. Grupe, G.B. Hudak, and R.S. Tsay for computing assistance. This research was partially supported by the U.S. Bureau of the Census under JSA 80-10, the Army Research Office, Durham, NC under Grant No. DAAG29-78-G00166, and the Alcoa Foundation.

2. UNIVARIATE TIME SERIES AND TRANSFER FUNCTION MODELS

When $k = 1$ we shall write $\mathbf{Z}_t = Z_t$ in (1.2). An important class of models for discrete univariate series originally proposed by Yule (1927) and Slutsky (1937) and developed by such authors as Bartlett, Kendall, Walker, Wold, and Yaglom are stochastic difference equations of the form

$$\phi_p(B) z_t = \theta_q(B) a_t, \tag{2.1}$$

where $\phi_p(B) = 1 - \phi_1 B - \cdots - \phi_p B^p$ and $\theta_q(B) = 1 - \theta_1 B - \cdots - \theta_q B^q$. In (2.1) the a_t's are independently identically and normally distributed random shocks (or white noise) with zero mean and variance σ^2; B is the back-shift operator such that $BZ_t = Z_{t-1}$; and $z_t = Z_t - \eta$ is the deviation of the observation Z_t from some convenient location η.

Relationships between k series $\{z_{1t}\}, \ldots, \{z_{kt}\}$ can sometimes be represented by linear *transfer function* models of the form

$$\begin{aligned} z_{ht} = & \sum_{i \in k(h)} [\omega_{s_{hi}}(B) B^{b_{hi}} / \delta_{r_{hi}}(B)]\, z_{it} \\ & + [\theta_{q_h}(B)/\varphi_{p_h}(B)]\, a_{ht}, \quad (h = 1, 2, \ldots, k) \end{aligned} \tag{2.2}$$

where $z_{0t} \equiv 0$, $k(h)$ is the set $(1, \ldots, h - 1)$; $\omega_{s_{hi}}(B)$, $\delta_{r_{hi}}(B)$, $\varphi_{p_h}(B)$, and $\theta_{q_h}(B)$ are polynomials in B;

The Collected Works of George E. P. Box, 1984, Wadsworth, Inc., Belmont, CA 94002.
Originally published in *J. Amer. Stat. Assoc.,* vol. 76, no. 376 (1981), pp. 802–816.

the b_{hi}'s are nonnegative integers; and $\{a_{1t}\}, \ldots, \{a_{kt}\}$ are k independent Gaussian white-noise processes with zero means and variances $\sigma_1^2, \ldots, \sigma_k^2$. In particular, intervention models of this form with one or more of the z_h's indicator variables have proved useful (Box and Tiao 1975; Abraham 1980).

Transfer function models of the form (2.2), however, assume that the series, when suitably arranged, possess a triangular relationship, implying for example that z_1 depends only on its own past; z_2 depends on its own past and on the present and past of z_1; z_3 on its own past and on the present and past of z_2 and z_1; and so on. On the other hand, if z_1 depends on the past of z_2, and also z_2 depends on the past of z_1, then we must have a model that allows for this *feedback*.

3. MULTIPLE STOCHASTIC DIFFERENCE EQUATION MODELS

3.1 The Vector ARMA Model

A useful class of models obtained by direct generalization of the Yule-Slutsky ARMA models that allow for feedback relationships among the k series is obtained from (2.2) by letting $k(h)$ be the set $(1, \ldots, k)$ excluding h. These models can be alternatively expressed as the vector autoregressive moving average ARMA models (Quenouille 1957),

$$\boldsymbol{\varphi}_p(B)\mathbf{z}_t = \boldsymbol{\theta}_q(B)\mathbf{a}_t, \tag{3.1}$$

where

$$\boldsymbol{\varphi}_p(B) = \mathbf{I} - \boldsymbol{\varphi}_1 B - \ldots - \boldsymbol{\varphi}_p B^p,$$

$$\boldsymbol{\theta}_q(B) = \mathbf{I} - \boldsymbol{\theta}_1 B - \ldots - \boldsymbol{\theta}_q B^q$$

are matrix polynomials in B, the $\boldsymbol{\varphi}$'s and $\boldsymbol{\theta}$'s are $k \times k$ matrices, $\mathbf{z}_t = \mathbf{Z}_t - \boldsymbol{\eta}$ is the vector of deviations from some origin $\boldsymbol{\eta}$ that is the mean if the series is stationary, and $\{\mathbf{a}_t\}$ with $\mathbf{a}_t = (a_{1t}, \ldots, a_{kt})'$ is a sequence of random shock vectors identically independently and normally distributed with zero mean and covariance matrix $\boldsymbol{\Sigma}$. We shall suppose that the zeros of the determinantal polynomials $|\boldsymbol{\varphi}_p(B)|$ and $|\boldsymbol{\theta}_q(B)|$ are on or outside the unit circle. The series $\mathbf{z}_t$ will be stationary when the zeros of $|\boldsymbol{\varphi}_p(B)|$ are all outside the unit circle, and will be invertible when those of $|\boldsymbol{\theta}_q(B)|$ are all outside the unit circle. Properties of such models have been discussed by, for example, Hannan (1970), Anderson (1971), and Granger and Newbold (1977).

Some Simple Examples. To illustrate the behavior of observations from these models, Figure 1 shows two series with 250 observations generated from the bivariate ($k = 2$) first order moving average [MA(1)] model, $\mathbf{z}_t = (\mathbf{I} - \boldsymbol{\theta}B)\mathbf{a}_t$, with

$$\boldsymbol{\theta} = \begin{bmatrix} .2 & .3 \\ -.6 & 1.1 \end{bmatrix} \quad \text{and} \quad \boldsymbol{\Sigma} = \begin{bmatrix} 4 & 1 \\ 1 & 1 \end{bmatrix}. \tag{3.2}$$

Figure 2 shows two series with 150 observations generated from the bivariate first order autoregressive [AR(1)] model, $(\mathbf{I} - \boldsymbol{\varphi}B)\mathbf{z}_t = \mathbf{a}_t$, with

Figure 1. Data Generated From a Bivariate MA(1) Model With Parameter Values in (3.2)

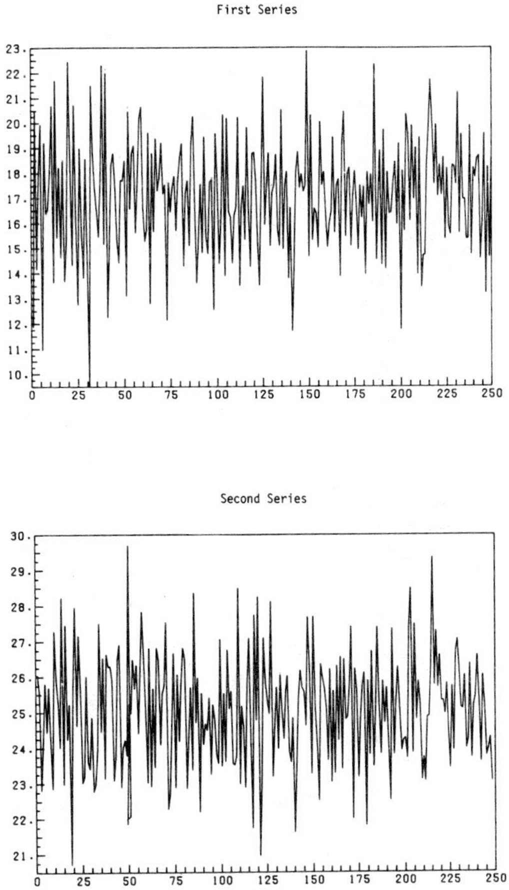

$$\boldsymbol{\varphi} = \begin{bmatrix} .2 & .3 \\ -.6 & 1.1 \end{bmatrix} \quad \text{and} \quad \boldsymbol{\Sigma} = \begin{bmatrix} 4 & 1 \\ 1 & 1 \end{bmatrix}. \tag{3.3}$$

While in both cases the series are seen to be stationary, observations from the autoregressive model are seen to have more "momentum" than those from the moving average model.

In practice, time series often exhibit nonstationary behavior. When several such series are considered jointly, nonstationarity may be modeled by allowing the zeros of $|\boldsymbol{\varphi}(B)|$ in (3.1) to lie on the unit circle. A particular example is the model $(1 - B)\mathbf{z}_t = (\mathbf{I} - \boldsymbol{\theta}B)\mathbf{a}_t$, that is, after differencing each series we obtain a vector MA(1) model. This is a vector analog of the commonly used univariate nonstationary model $(1 - B)z_t = (1 - \theta B)a_t$. However,

Figure 2. Data Generated From a Bivariate AR(1) Model With Parameter Values in (3.3)

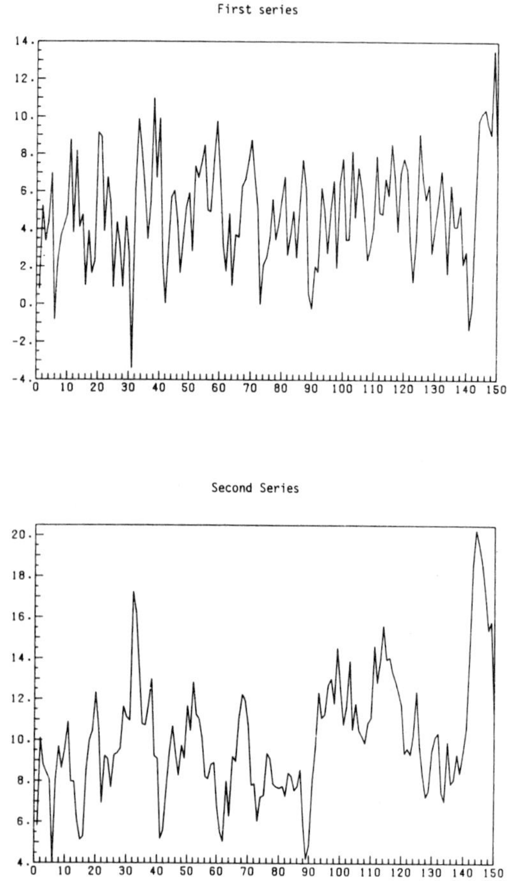

it should be noted here that for vector time series, linear combinations of the elements of $\mathbf{z}_t$ may often be stationary, and simultaneous differencing of all series can lead to unnecessary complications in model fitting. See, for example, the discussion in Box and Tiao (1977) and Hillmer and Tiao (1979).

Tranfer Function Model. For the vector model in (3.1), in general, all elements of $\mathbf{z}_t$ are related to all elements of $\mathbf{z}_{t-j}$ ($j = 1, 2, \ldots$) and there can be feedback relationships between all the series. However, if the $\mathbf{z}_t$'s can be arranged so that the coefficient matrices $\boldsymbol{\varphi}$'s and $\boldsymbol{\theta}$'s are all lower triangular, then (3.1) can be written as a transfer function model of the form (2.2). More generally, if the $\boldsymbol{\varphi}$'s and $\boldsymbol{\theta}$'s are all lower block triangular, then we obtain a generalization of the transfer function form of (2.2) in which both the input vector series and the output vector series are allowed to have feedback relationships. Furthermore, relationships between the vector transfer function model and the econometric linear simultaneous equation model have been discussed in Zellner and Palm (1974) and Wallis (1977).

3.2 Cross-Covariance and Cross-Correlation Matrices

For a stationary vector time series $\{\mathbf{Z}_t\}$ with mean vector $\boldsymbol{\eta}$, let $\boldsymbol{\Gamma}(l)$ be the lag l cross-covariance matrix

$$\begin{aligned}\boldsymbol{\Gamma}(l) &= E(\mathbf{z}_{t-l}\mathbf{z}'_t)\\ &= \{\gamma_{ij}(l)\}, \quad l = 0, \pm 1, \pm 2, \ldots \qquad (3.4)\\ &\qquad i, j = 1, \ldots, k\end{aligned}$$

and let $\boldsymbol{\rho}(l) = \{\rho_{ij}(l)\}$ be the corresponding cross-correlation matrix.

When the vector ARMA model in (3.1) is stationary, it is well known that

$$\boldsymbol{\Gamma}(l) = \begin{cases} \sum_{j=l-r}^{l-1} \boldsymbol{\Gamma}(j)\boldsymbol{\varphi}'_{l-j} - \sum_{j=0}^{r-l} \boldsymbol{\psi}_j \boldsymbol{\Sigma} \boldsymbol{\theta}'_{j+l}, & l = 0, \ldots, r \\ \sum_{j=1}^{r} \boldsymbol{\Gamma}(l-j)\boldsymbol{\varphi}_j', & l > r, \end{cases} \qquad (3.5)$$

where the $\boldsymbol{\psi}_j$'s are obtained from the relationship

$$\boldsymbol{\psi}(B) = \boldsymbol{\varphi}^{-1}(B)\boldsymbol{\theta}(B) = (\mathbf{I} + \boldsymbol{\psi}_1 B + \cdots),$$

$\boldsymbol{\theta}_0 = -\mathbf{I}$, $r = \max(p, q)$, and it is understood that (*a*) if $p < q$, $\boldsymbol{\varphi}_{p+1} = \cdots = \boldsymbol{\varphi}_r = 0$, and (b) if $q < p$, $\boldsymbol{\theta}_{q+1} = \cdots = \boldsymbol{\theta}_r = \mathbf{0}$.

In particular, when $p = 0$, that is, we have a vector MA(q) model, then

$$\boldsymbol{\Gamma}(l) = \begin{cases} \sum_{j=0}^{q-l} \boldsymbol{\theta}_j \boldsymbol{\Sigma} \boldsymbol{\theta}'_{j+l}, & l = 0, \ldots, q \\ \mathbf{0} & l > q. \end{cases} \qquad (3.6)$$

Thus, all auto- and cross-correlations are zero when $l > q$. On the other hand, for a vector autoregressive model the auto- and cross-correlations in general will decay gradually to zero as $|l|$ increases.

3.3 A Determinantal Criterion for ARMA Models and the Partial Autoregression Matrices

From the moment equations in (3.5) for a stationary ARMA (p, q) model, we see that the autocovariance matrices $\boldsymbol{\Gamma}(l)$'s and the autoregressive coefficient matrices $\boldsymbol{\varphi}_1, \ldots, \boldsymbol{\varphi}_p$ are related as follows:

$$\begin{bmatrix} \mathbf{A}(p, m) & \mathbf{b}(p, m) \\ \mathbf{g}'(p, m) & \boldsymbol{\Gamma}(m) \end{bmatrix} \begin{bmatrix} \boldsymbol{\Phi}_{p-1} \\ \boldsymbol{\varphi}'_p \end{bmatrix} = \begin{bmatrix} \mathbf{c}(p, m) \\ \boldsymbol{\Gamma}(p + m) \end{bmatrix}, \quad m = q, q + 1, \ldots, \qquad (3.7)$$

where

$$\mathbf{A}(p, m) = \begin{bmatrix} \Gamma(m)\ \Gamma(m-1) & \cdots & \Gamma(m-p+2) \\ \Gamma(m+1) & \ddots & \vdots \\ \vdots & \ddots & \Gamma(m-1) \\ \Gamma(m+p-2) \ldots \Gamma(m+1) & & \Gamma(m) \end{bmatrix},$$

$$\mathbf{b}(p, m) = \begin{bmatrix} \Gamma(m-p+1) \\ \vdots \\ \Gamma(m-1) \end{bmatrix},$$

$$\mathbf{c}(p, m) = \begin{bmatrix} \Gamma(m+1) \\ \vdots \\ \Gamma(m+p-1) \end{bmatrix},$$

$\mathbf{g}'(p, m) = [\Gamma(m + p - 1), \ldots, \Gamma(m + 1)]$, and $\mathbf{\Phi}'_{p-1} = [\varphi_1, \ldots, \varphi_{p-1}]$. Consider now the $k \times k$ matrix

$$\mathbf{D}(l, m) = [d_{ij}(l, m)] \quad \begin{array}{l} l = 1, 2, \ldots \\ m = 0, 1, \ldots, \end{array} \tag{3.8}$$

where $d_{ij}(l, m)$ is the determinant

$$d_{ij}(l, m) = \det \begin{bmatrix} \mathbf{A}(l, m) & \mathbf{c}_j(l, m) \\ \mathbf{g}'_i(l, m) & \gamma_{ij}(l + m) \end{bmatrix},$$

$$i, j = 1, \ldots, k$$

$\mathbf{c}_j(l, m)$ is the jth column of $\mathbf{c}(l, m)$, $\mathbf{g}'_i(l, m)$ is the ith row of $\mathbf{g}'(l, m)$, and $\gamma_{ij}(l + m)$ is the (i, j)th element of $\Gamma(l + m)$. It follows from (3.7) that for an ARMA (p, q) model

$$\mathbf{D}(l, m) = \mathbf{0} \quad \text{for} \quad l > p \quad \text{and} \quad m \geq q. \tag{3.9}$$

This provides a multivariate generalization of the results in Gray, Kelley, and McIntire (1978) for univariate ARMA models.

In the special case $m = q = 0$, (3.7) is a multivariate generalization of the Yule-Walker equations for autoregressive models in univariate time series. Analogous to the partial autocorrelation function for the univariate case, we may define a *partial autoregression matrix function* $\underset{\sim}{\mathcal{P}}(l)$ having the property that if the model is AR(p), then

$$\underset{\sim}{\mathcal{P}}(l) = \begin{cases} \varphi_l, & l = p \\ \mathbf{0}, & l > p \end{cases}. \tag{3.10}$$

From (3.7), we define $\underset{\sim}{\mathcal{P}}(l)$ as

$$\underset{\sim}{\mathcal{P}}'(l) = \begin{cases} \Gamma^{-1}(0)\Gamma(1), \; l = 1 \\ [\Gamma(0) - \mathbf{b}'(l, 0)\mathbf{A}^{-1}(l, 0)\mathbf{b}(l, 0)]^{-1} \\ [\Gamma(l) - \mathbf{b}'(l, 0)\mathbf{A}^{-1}(l, 0)\mathbf{c}(l, 0)], \; l > 1. \end{cases} \tag{3.11}$$

4. MODEL BUILDING STRATEGY FOR MULTIPLE TIME SERIES

The models in (3.1) contain a dauntingly large number $\{k^2(p + q) + \frac{1}{2}k(k + 1)\}$ of parameters, complicating methods for model building. It is natural that attempts have been made to simplify the general form in the model building process, for example by Granger and Newbold (1977) and Wallis (1977). While we sympathize with this aspiration, we feel that so far at least these attempts have not been successful. In some comparisons made later in Section 6, we argue that they do not result in genuine simplification, nor do they provide feasible methods when k is greater than 2 or 3. We see no alternative but to provide for direct initial fitting of models of the form (3.1). It must, however, be added

1. that often models of rather low order (p and q small) provide adequate approximation,
2. that occasionally knowledge of the system might allow simplification a priori, although even here prudent checking of the adequacy of the simplifcation would be necessary (see Zellner and Palm 1974),
3. that considerable simplification is almost invariably possible after an initial model has been fitted,
4. that 2 and 3 imply that provision should be made to allow models to be fitted in which certain parameters are fixed or constrained in some other way,
5. that other methods of seeking simplifications, for example principal component analysis or canonical analysis (see Box and Tiao 1977), will often prove effective.

In brief, we feel that although the full form (3.1) needs to be fitted initially, subsequent iterations will usually lead to simplification.

In what follows we sketch an iterative approach consisting of (a) tentative specification (identification), (b) estimation, and (c) diagnostic checking for the vector ARMA models in (3.1). A computer package to carry out this analysis has been completed (Tiao et al. 1979) consisting of three main programs: (a) Preliminary Analysis, (b) Stepwise Autoregression, and (c) Estimation and Forecasting.

4.1 Tentative Specification

The aim here is to employ statistics (a) that can be readily calculated from the data and (b) that facilitate the choice of subclass of models worthy of further examination.

Sample Cross-Correlations. The sample cross-correlations $\hat{\rho}_{ij}(l)$,

$$\hat{\rho}_{ij}(l) = \sum(Z_{it} - \bar{Z}_i)(Z_{j(t+l)} - \bar{Z}_j)/ \{\sum(Z_{it} - \bar{Z}_i)^2 \sum(Z_{jt} - \bar{Z}_j)^2\}^{1/2}$$

where $\bar{Z}_i$ is the sample mean of the ith component series of $\mathbf{Z}_t$, are particularly useful in spotting low order vector moving average models, since from (3.6) $\rho_{ij}(l) = 0$ for $l > q$.

For the data shown in Figure 1, which were generated from a bivariate MA(1) model, Figures 3(a)–(c) show, respectively, the sample autocorrelations $\hat{\rho}_{11}(l)$ and $\hat{\rho}_{22}(l)$, and the sample cross-correlations $\hat{\rho}_{12}(l)$. The large values occurring at $|l| = 1$ would lead to tentative specification of the model as an MA(1). However, graphs of this kind become increasingly cumbersome as the number of series is increased. Furthermore, identification is not easy from a listing of sample cross-correlation matrices $\hat{\boldsymbol{\rho}}(l)$ like that in Table 1(a), particularly when k is greater than 4 or 5.

In this circumstance, we have found the following simple device of great practical value. Instead of the numerical values, a plus sign is used to indicate a value greater than $2n^{-1/2}$, a minus sign a value less than $-2n^{-1/2}$, and a dot to indicate a value inbetween $-2n^{-1/2}$ and $2n^{-1/2}$. The motivation is that if the series were white noise, for large n the $\hat{\rho}_{ij}(l)$'s would be normally distributed with mean 0 and variance n^{-1}. The symbols can be arranged either as in Table 1(b) or as in Table 1(c). We realize that the variances of the $\hat{\rho}_{ij}(l)$'s can be considerably greater than $n^{-1/2}$ when the series are highly autocorrelated, so that these indicator symbols, if taken literally, can lead to overparameterization. However, we do not interpret these indicator symbols in the sense of a formal significance test, but as a rather crude "signal-to-noise ratio" guide. Taken together they can give useful and assimilable indicators of the general correlation pattern.

Table 2 shows sample cross-correlation matrices in terms of these indicator symbols for the series in Figure 2 generated from an AR(1) model. The persistence of large correlations suggests the possibility of autoregressive behavior. In general, the pattern of indicator symbols for the cross-correlation matrices makes it very easy to identify a low order moving average model.

Figure 3. Sample Auto- and Cross-Correlations for the Data in Figure 1

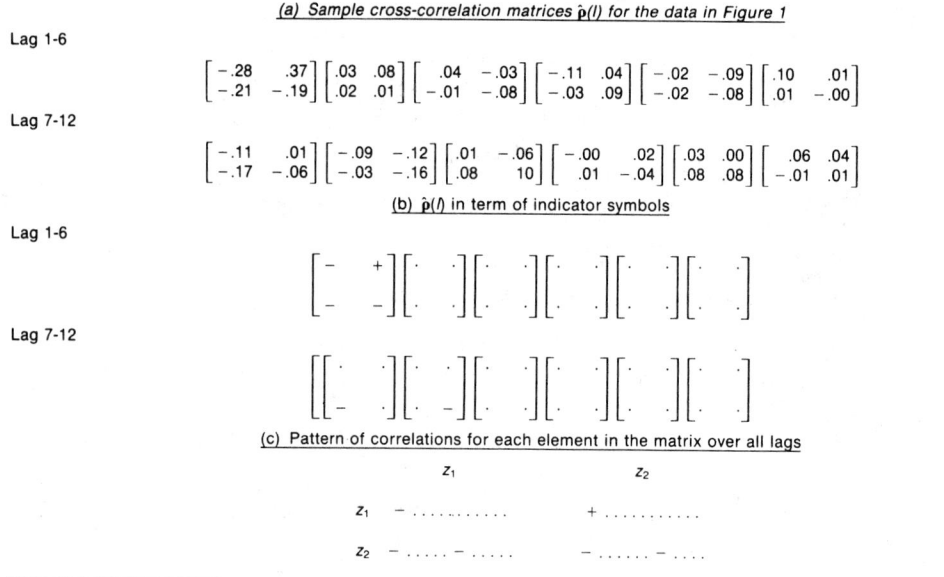

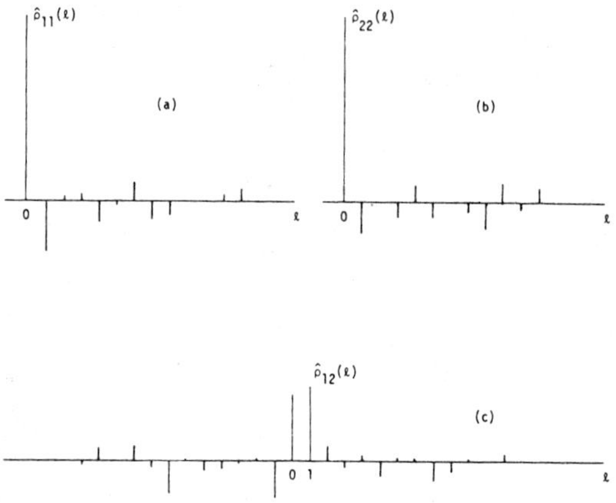

Sample Partial Autoregression and Related Summary Statistics. For an AR(p) process, the partial autoregression matrices $\mathcal{P}(l)$ in (3.11) are zero for $l > p$. They are therefore particularly useful for identifying an autoregressive model. Estimates of $\mathcal{P}(l)$ and their standard er-

Table 1. Cross-Correlations Matrices $\hat{\boldsymbol{\rho}}$ (l) for the Data in Figure 1

(a) Sample cross-correlation matrices $\hat{\boldsymbol{\rho}}(l)$ for the data in Figure 1

Lag 1-6

$$\begin{bmatrix} -.28 & .37 \\ -.21 & -.19 \end{bmatrix} \begin{bmatrix} .03 & .08 \\ .02 & .01 \end{bmatrix} \begin{bmatrix} .04 & -.03 \\ -.01 & -.08 \end{bmatrix} \begin{bmatrix} -.11 & .04 \\ -.03 & .09 \end{bmatrix} \begin{bmatrix} -.02 & -.09 \\ -.02 & -.08 \end{bmatrix} \begin{bmatrix} .10 & .01 \\ .01 & -.00 \end{bmatrix}$$

Lag 7-12

$$\begin{bmatrix} -.11 & .01 \\ -.17 & -.06 \end{bmatrix} \begin{bmatrix} -.09 & -.12 \\ -.03 & -.16 \end{bmatrix} \begin{bmatrix} .01 & -.06 \\ .08 & 10 \end{bmatrix} \begin{bmatrix} -.00 & .02 \\ .01 & -.04 \end{bmatrix} \begin{bmatrix} .03 & .00 \\ .08 & .08 \end{bmatrix} \begin{bmatrix} .06 & .04 \\ -.01 & .01 \end{bmatrix}$$

(b) $\hat{\boldsymbol{\rho}}(l)$ in term of indicator symbols

Lag 1-6

$$\begin{bmatrix} - & + \\ - & - \end{bmatrix} \begin{bmatrix} \cdot & \cdot \\ \cdot & \cdot \end{bmatrix} \begin{bmatrix} \cdot & \cdot \\ \cdot & \cdot \end{bmatrix} \begin{bmatrix} \cdot & \cdot \\ \cdot & \cdot \end{bmatrix} \begin{bmatrix} \cdot & \cdot \\ \cdot & \cdot \end{bmatrix} \begin{bmatrix} \cdot & \cdot \\ \cdot & \cdot \end{bmatrix}$$

Lag 7-12

$$\begin{bmatrix} \cdot & \cdot \\ - & \cdot \end{bmatrix} \begin{bmatrix} \cdot & \cdot \\ \cdot & - \end{bmatrix} \begin{bmatrix} \cdot & \cdot \\ \cdot & \cdot \end{bmatrix} \begin{bmatrix} \cdot & \cdot \\ \cdot & \cdot \end{bmatrix} \begin{bmatrix} \cdot & \cdot \\ \cdot & \cdot \end{bmatrix} \begin{bmatrix} \cdot & \cdot \\ \cdot & \cdot \end{bmatrix}$$

(c) Pattern of correlations for each element in the matrix over all lags

	z_1	z_2
z_1	−	+ . . . .
z_2	− . . . . . −	− . . . . . − . . . .

Table 2. Sample Cross-Correlation Matrices $\hat{\boldsymbol{\rho}}(l)$ for the Data in Figure 2 in Terms of Indicator Symbols

Lag 1-6

$$\begin{bmatrix} + & \\ + & + \end{bmatrix} \begin{bmatrix} \cdot & \cdot \\ + & + \end{bmatrix} \begin{bmatrix} \cdot & \cdot \\ + & + \end{bmatrix} \begin{bmatrix} \cdot & - \\ + & + \end{bmatrix} \begin{bmatrix} \cdot & \cdot \\ + & + \end{bmatrix} \begin{bmatrix} \cdot & \cdot \\ + & \cdot \end{bmatrix}$$

Lag 7-12

$$\begin{bmatrix} - & - \\ \cdot & \cdot \end{bmatrix} \begin{bmatrix} - & \cdot \\ \cdot & \cdot \end{bmatrix} \begin{bmatrix} \cdot & \cdot \\ \cdot & \cdot \end{bmatrix} \begin{bmatrix} \cdot & \cdot \\ \cdot & \cdot \end{bmatrix} \begin{bmatrix} \cdot & \cdot \\ \cdot & \cdot \end{bmatrix} \begin{bmatrix} \cdot & \cdot \\ \cdot & \cdot \end{bmatrix}$$

rors can be obtained by fitting autoregressive models of successively high order $l = 1, 2, \ldots$ by standard multivariate least squares.

It is well known (see, e.g., Anderson 1971) that for a stationary AR(p) model asymptotically the estimates $\varphi'_1, \ldots, \varphi'_p$ are jointly normally distributed. A useful summary of the pattern of the partials is obtained by listing indicator symbols, assigning a plus (minus) sign when a coefficient in $\hat{\mathcal{P}}(l)$ is greater (less) than 2 (-2) times its estimated standard errors, and a dot for values in between.

To help tentatively determine the order of an autoregressive model, we may also employ the likelihood ratio statistics corresponding to testing the null hypotheses $\boldsymbol{\varphi}_l = \mathbf{0}$ against the alternative $\boldsymbol{\varphi}_l \neq \mathbf{0}$ when an AR(l) model is fitted. Let

$$\mathbf{S}(l) = \sum_{t=l+1}^{n} (\mathbf{z}_t - \hat{\boldsymbol{\varphi}}_1 \mathbf{z}_{t-1} - \ldots - \hat{\boldsymbol{\varphi}}_l \mathbf{z}_{t-l}) \times (\mathbf{z}_t - \hat{\boldsymbol{\varphi}}_1 \mathbf{z}_{t-1} - \ldots - \hat{\boldsymbol{\varphi}}_l \mathbf{z}_{t-l})' \qquad (4.1)$$

Table 3. Indicator Symbols for Partial Autoregression and Related Statistics for Data in Figure 2

Lag l	*Indicator symbols*	*M (l)*[a] $\stackrel{a}{\sim} \chi_4^2$	*Diagonal elements of* $\hat{\boldsymbol{\Sigma}}$
1	$\cdot$ + − +	356.96	5.30 1.08
2	$\cdot$ $\cdot$ $\cdot$ +	7.04	5.16 1.03
3	$\cdot$ $\cdot$ $\cdot$ $\cdot$	2.63	5.07 1.03
4	$\cdot$ $\cdot$ $\cdot$ $\cdot$	4.38	5.01 1.02
5	$\cdot$ $\cdot$ $\cdot$ $\cdot$	2.42	4.95 1.01

[a] $\stackrel{a}{\sim}$ means approximately distributed as.

be the matrix of residual sum of squares and cross products after fitting an AR(l). The likelihood ratio statistic is the ratio of the determinants

$$U = |\mathbf{S}(l)| / |\mathbf{S}(l-1)|. \qquad (4.2)$$

Using Bartlett's (1938) approximation, the statistic

$$M(l) = -(N - \tfrac{1}{2} - l \cdot k)\log_e U \qquad (4.3)$$

is, on the null hypothesis, asymptotically distributed as χ^2 with k^2 degrees of freedom, where $N = n - p - 1$ is the effective number of observations, assuming that a constant term is included in the model.

Finally, a measure of the extent to which the fit is improved as the order is increased is provided by the diagonal elements of the residual covariance matrices $\hat{\boldsymbol{\Sigma}}$ corresponding to the successive AR models.

For illustration, the matrices of summary symbols, the $M(l)$ statistics, and the diagonal elements of the residual covariance matrices for the series in Figure 2 are shown in Table 3 for $l = 1, \ldots, 5$. They indicate that an AR(1) or at most an AR(2) would be adequate for the data.

For the series shown in Figure 1, the pattern of the partials and related statistics are given in Table 4. Notice here that if we had confined attention to autoregressive models as is advocated in Parzen (1977), we would have needed p to be as high as 7. This is not surprising since with the MA(1) model of (3.2) written in the autoregressive form $\mathbf{z}_t = \boldsymbol{\pi}_1 \mathbf{z}_{t-1} + \boldsymbol{\pi}_2 \mathbf{z}_{t-2} + \ldots + \mathbf{a}_t$, we find

$$\boldsymbol{\pi}_1 = \begin{bmatrix} -.2 & -.3 \\ .6 & -1.1 \end{bmatrix}, \quad \boldsymbol{\pi}_2 = \begin{bmatrix} .14 & -.39 \\ .78 & -1.03 \end{bmatrix}, \ldots,$$

$$\boldsymbol{\pi}_6 = \begin{bmatrix} .23 & -.25 \\ .49 & -.51 \end{bmatrix}, \qquad (4.4)$$

$$|\boldsymbol{\pi}_1| = .4, \quad |\boldsymbol{\pi}_2| = .16, \quad \ldots, \quad |\boldsymbol{\pi}_6| = .0041.$$

Thus, although the determinants $|\boldsymbol{\pi}_j|$ decrease rapidly towards zero as j increases, the elements of $\boldsymbol{\pi}_j$ converge to zero very slowly so that many autoregressive terms would be needed to provide an adequate approximation.

In general, the pattern of the partial autoregression matrices, the $M(l)$ statistic, and the diagonal elements of the residual covariance matrix are useful to distinguish between moving average and low order autoregressive models and to select tentatively the appropriate order for the latter.

Table 4. Pattern of Partial Autoregression and Related Statistics for Data in Figure 1

Lag	Pattern of $\hat{\mathcal{P}}(l)$		$M(l) \,\dot{\sim}\, \chi_4^2$	‡
1	−	−	123.2	4.78
	+	−		1.88
2	·	·	75.9	4.75
	+	−		1.43
3	+	·	35.2	4.63
	+	−		1.23
4	·	·	27.5	4.63
	+	·		1.08
5	·	·	16.6	4.61
	+	·		1.04
6	·	·	13.5	4.53
	+	·		.98
7	·	−	16.5	4.38
	+	·		.94
8	·	·	8.1	4.31
	·	−		.91

Sample Residual Cross-Correlation Matrices After AR Fit. After each AR(l) fit, $l = 1, \ldots, p$, cross-correlation matrices of the residuals $\hat{\mathbf{a}}_t$'s may be readily obtained. Table 5 shows indicator symbols for residual correlations after fitting AR(1) and AR(2) to the AR data plotted in Figure 2. Again a plus sign is used to indicate values greater than $2n^{-1/2}$, a minus sign for values less than $-2n^{-1/2}$, and a dot for in-between values. They verify that there is no need to go beyond an AR(2) model.

It is perhaps worth emphasizing here again that these indicator symbols are proposed as a rough preliminary device to help arrive at an initial model. They should not be treated as "exact significance testing." In a recent paper by Li and McLeod (1980), expressions have been obtained for the asymptotic distributions of the residual autocorrelations. As in the univariate case, the low order autocorrelations have variance considerably less than $n^{-1/2}$.

For mixed vector autoregressive moving average models in general, however, both the population cross-correlation matrices $\boldsymbol{\rho}(l)$ and the partial autoregression matrices $\mathcal{P}(l)$ decay only gradually toward $\mathbf{0}$. In some situations, the order of mixed models may be tentatively identified by inspection of patterns in residual cross-correlations after the AR fit, but in others study of residual correlations could be misleading. For illustration, consider the case of a stationary ARMA(1, 1) model

$$(\mathbf{I} - \boldsymbol{\varphi} B)\mathbf{z}_t = (\mathbf{I} - \boldsymbol{\theta} B)\mathbf{a}_t. \tag{4.5}$$

If an AR(1) model is fitted to $\{\mathbf{z}_t\}$, then the estimate $\hat{\boldsymbol{\varphi}}$ will be biased. In fact, asymptotically $\hat{\boldsymbol{\varphi}}$ converges in probability to

$$\hat{\boldsymbol{\varphi}} \rightarrow \boldsymbol{\varphi}_0 = \boldsymbol{\Gamma}'(1)\boldsymbol{\Gamma}(0)^{-1}. \tag{4.6}$$

Thus the residuals $\hat{\mathbf{a}}_t = \mathbf{z}_t - \boldsymbol{\varphi}_0 \mathbf{z}_{t-1}$ approximately follow the model

$$\hat{\mathbf{a}}_t = (\mathbf{i} - \boldsymbol{\varphi}_0 B)(\mathbf{I} - \boldsymbol{\varphi} B)^{-1}(\mathbf{I} - \boldsymbol{\theta} B)\mathbf{a}_t. \tag{4.7}$$

Table 5. Indicator Symbols for Residual Cross Correlations for the AR (1) Data of Figure 2

AR (1) Lag 1-6

$$\begin{bmatrix} \cdot & \cdot \\ \cdot & - \end{bmatrix}\begin{bmatrix} \cdot & \cdot \\ \cdot & \cdot \end{bmatrix}\begin{bmatrix} \cdot & \cdot \\ \cdot & \cdot \end{bmatrix}\begin{bmatrix} \cdot & \cdot \\ \cdot & \cdot \end{bmatrix}\begin{bmatrix} \cdot & \cdot \\ \cdot & \cdot \end{bmatrix}\begin{bmatrix} + & \cdot \\ \cdot & \cdot \end{bmatrix}$$

Lag 7-12

$$\begin{bmatrix} - & \cdot \\ - & \cdot \end{bmatrix}\begin{bmatrix} \cdot & \cdot \\ \cdot & \cdot \end{bmatrix}\begin{bmatrix} \cdot & \cdot \\ \cdot & \cdot \end{bmatrix}\begin{bmatrix} \cdot & \cdot \\ \cdot & \cdot \end{bmatrix}\begin{bmatrix} \cdot & \cdot \\ \cdot & \cdot \end{bmatrix}\begin{bmatrix} \cdot & \cdot \\ \cdot & \cdot \end{bmatrix}$$

AR (2) Lag 1-6

$$\begin{bmatrix} \cdot & \cdot \\ \cdot & \cdot \end{bmatrix}\begin{bmatrix} \cdot & \cdot \\ \cdot & \cdot \end{bmatrix}\begin{bmatrix} \cdot & \cdot \\ \cdot & \cdot \end{bmatrix}\begin{bmatrix} \cdot & \cdot \\ \cdot & \cdot \end{bmatrix}\begin{bmatrix} \cdot & \cdot \\ \cdot & \cdot \end{bmatrix}\begin{bmatrix} + & \cdot \\ \cdot & \cdot \end{bmatrix}$$

Lag 7-12

$$\begin{bmatrix} \cdot & \cdot \\ \cdot & \cdot \end{bmatrix}\begin{bmatrix} \cdot & \cdot \\ \cdot & \cdot \end{bmatrix}\begin{bmatrix} \cdot & \cdot \\ \cdot & \cdot \end{bmatrix}\begin{bmatrix} \cdot & \cdot \\ \cdot & \cdot \end{bmatrix}\begin{bmatrix} \cdot & \cdot \\ \cdot & \cdot \end{bmatrix}\begin{bmatrix} \cdot & \cdot \\ \cdot & \cdot \end{bmatrix}$$

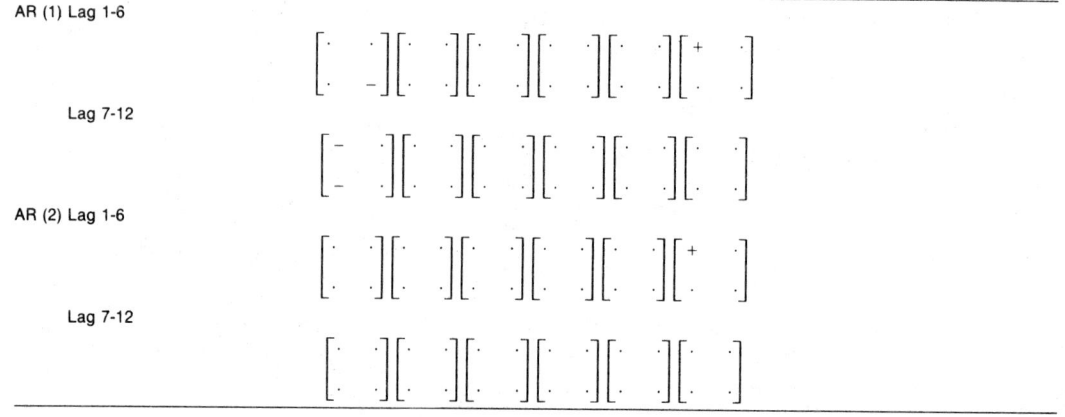

For $k = 1$, $\{\hat{a}_t\}$ follows an ARMA(1, 2) model so that the autocorrelations of $\hat{a}_t$ are

$$\rho_{\hat{a}}(j) = \varphi\rho_{\hat{a}}(j - 1), j > 2 \quad (4.8)$$

and $\rho_{\hat{a}}(1)$ and $\rho_{\hat{a}}(2)$ are functions of φ and θ. Table 6 gives values of $\rho_{\hat{a}}(1)$ and $\rho_{\hat{a}}(2)$ for various combinations of values of φ and θ. For each combination, the first value is $\rho_{\hat{a}}(1)$ and the second $\rho_{\hat{a}}(2)$.

We see that if the true value of φ is large in magnitude, residual autocorrelations would lead to the choice of an MA(1) model for $\hat{a}_t$ and therefore the correct identification. For intermediate values of φ, a moving average of order 2 or higher might be selected, resulting in overparametrization.

In Gray, Kelley, and McIntire (1978) and Beguin, Gourieroux, and Monfort (1980), methods have been proposed to determine the order of univariate ARMA model. These methods are essentially equivalent to estimating, for $k = 1$, the determinant $D(l, m)$ in (3.8) using sample estimates of the autocovariances and selecting the orders of autoregressive and moving average polynomials on the basis of the property in (3.9). We are currently studying sampling properties of estimates of appropriate functions of $\mathbf{D}(l, m)$ in the vector case.

4.2 Estimation

Once the order of the model in (3.1) has been tentatively selected, efficient estimates of the associated parameter matrices $\boldsymbol{\varphi} = (\boldsymbol{\varphi}_1, \ldots, \boldsymbol{\varphi}_p)$, $\boldsymbol{\theta} = (\boldsymbol{\theta}_1, \ldots, \boldsymbol{\theta}_q)$, and $\boldsymbol{\Sigma}$ are determined by maximizing the likelihood function. Approximate standard errors and correlation matrix of the estimates of elements of the $\boldsymbol{\varphi}_j$'s and $\boldsymbol{\theta}_j$'s can also be obtained.

Conditional Likelihood. For the ARMA (p, q) model, we can write

$$\mathbf{a}_t = \mathbf{z}_t - \boldsymbol{\varphi}_1\mathbf{z}_{t-1} - \cdots - \boldsymbol{\varphi}_p\mathbf{z}_{t-p} + \boldsymbol{\theta}_1\mathbf{a}_{t-1} + \cdots + \boldsymbol{\theta}_q\mathbf{a}_{t-q}. \quad (4.9)$$

As in the univariate case discussed in Box and Jenkins (1970), the likelihood function can be approximated by a "conditional" likelihood function as follows. The series is regarded as consisting of the $n - p$ vector observations $\mathbf{z}_{p+1}, \ldots, \mathbf{z}_n$. The likelihood function is then determined from $\mathbf{a}_{p+1}, \ldots, \mathbf{a}_n$, using the preliminary values $\mathbf{z}_1, \ldots, \mathbf{z}_p$ and conditional on zero values for $\mathbf{a}_p, \ldots, \mathbf{a}_{p-q-1}$. Thus, as shown in Wilson (1973),

$$l_c(\boldsymbol{\varphi}, \boldsymbol{\theta}, \boldsymbol{\Sigma} \mid \mathbf{z}) \propto |\boldsymbol{\Sigma}|^{-(n-p)/2}\exp\{-\tfrac{1}{2}\operatorname{tr}\boldsymbol{\Sigma}^{-1}\mathbf{S}(\boldsymbol{\varphi}, \boldsymbol{\theta})\}, \quad (4.10)$$

where $\mathbf{S}(\boldsymbol{\varphi}, \boldsymbol{\theta}) = \sum_{t=p+1}^{n} \mathbf{a}_t\mathbf{a}'_t$. Properties of the maximum likelihood estimates obtained from (4.10) have been discussed in Nicholls (1976, 1977) and Anderson (1980).

It has been shown in Hillmer and Tiao (1979) that this approximation can be seriously inadequate if n is not sufficiently large and one or more zeros of $|\boldsymbol{\theta}_q(B)|$ lie on or close to the unit circle. Specifically, this would lead to estimates of the moving average parameters with large bias.

Exact Likelihood Function. For univariate ARMA models, the exact likelihood function has been considered by Tiao and Ali (1971), Newbold (1974), Dent (1977), Ansley (1979), and others. For vector models, this function has been studied by Osborn (1977) for the pure moving average case and by Phadke and Kedem (1978), Nicholls and Hall (1979), and Hillmer and Tiao (1979). It takes the form

$$l(\boldsymbol{\varphi}, \boldsymbol{\theta}, \boldsymbol{\Sigma} \mid \mathbf{z}) \propto l_c(\boldsymbol{\varphi}, \boldsymbol{\theta}, \boldsymbol{\Sigma} \mid \mathbf{z})l_1(\boldsymbol{\varphi}, \boldsymbol{\theta}, \boldsymbol{\Sigma} \mid \mathbf{z}), \quad (4.11)$$

where l_1 depends (a) only on $\mathbf{z}_1, \ldots, \mathbf{z}_p$ if $q = 0$ and (b) on all the data vectors $\mathbf{z}_1, \ldots, \mathbf{z}_n$ if $q \neq 0$. Estimation algorithms have been developed and incorporated in our computer package for the vector MA(q) model. For the general ARMA(p, q) model, it has been shown that a close approximation to the exact likelihood can be obtained by considering the transformation

$$\mathbf{w}_t = (\mathbf{I} - \boldsymbol{\varphi}_1 B - \cdots - \boldsymbol{\varphi}_p B^p)\mathbf{z}_t$$

so that (4.12)

$$\mathbf{w}_t = \boldsymbol{\theta}_q(B)\mathbf{a}_t$$

and then applying the results for MA(q) to $\mathbf{w}_t$, $t = p + 1, \ldots, n$.

Because estimation of moving average parameters using the exact likelihood is rather slow, we presently employ the conditional method in the preliminary stages of iterative model building and switch to the exact method towards the end.

4.3 Diagnostic Checking

To guard against model misspecification and to search for directions of improvement, a detailed diagnostic analysis of the residual series $\{\hat{\mathbf{a}}_t\}$, where

$$\hat{\mathbf{a}}_t = \mathbf{z}_t - \hat{\boldsymbol{\varphi}}_1\mathbf{z}_{t-1} - \cdots - \hat{\boldsymbol{\varphi}}_p\mathbf{z}_{t-p} + \hat{\boldsymbol{\theta}}_1\hat{\mathbf{a}}_{t-1} + \cdots + \hat{\boldsymbol{\theta}}_q\hat{\mathbf{a}}_{t-q}, \quad (4.13)$$

is performed. Useful diagnostic checks include (a) plots of standardized residual series against time and/or other variables and (b) cross-correlation matrices of the resid-

Table 6. Asymptotic Values of $\rho_{\hat{a}}$ (1) and $\rho_{\hat{a}}$ (2)

φ \ θ	−.95	−.50	.50	.95
−.95	—	.265 .085	−.381 −.03	−.481 −.036
−.50	.049 −.222	—	−.223 −.201	−.321 −.267
.50	.321 −.267	.223 −.201	—	−.049 −.222
.95	.481 −.036	.381 −.03	−.265 .085	—

Table 7. Pattern of Sample Cross-Correlations for the SCC Data

	Z_1 Stocks	Z_2 Cars	Z_3 Commodities
Z_1 Stocks	+ + + + + + + + ...	− − − − − − − − − − − − − − − − − − − −	− − − − − − − − − − − − − − − − − − − −
Z_2 Cars	. . . . + + +	+ + + + + + + + + + + + + + + ...	+ + + + + + + + + + + + + + ...
Z_3 Commodities	− − − − − − − . . . . + + + + +	+ + + + + + + + + + + + + + + + + + . .	+ + + + + + + + + + + + + + + + + . . .

uals $\hat{\mathbf{a}}_t$. As before, the structures of the correlations are summarized by indicator symbols. Overall χ^2 tests based on the sample cross correlations of the residuals have been proposed in recent papers by Hosking (1980) and Li and McLeod (1980). However, as is noted in Box and Jenkins (1970), such overall tests are not substitutes for more detailed study of the correlation structure.

5. ANALYSES OF TWO EXAMPLES

We now apply the model building approach introduced in the preceding section to the following sets of data:

1. The Financial Time Ordinary Share Index, U.K. Car Production and the Financial Time Commodity Price Index: Quarterly Data 3/1952–4/1967, obtained from Coen, Gomme, and Kendall (1969). This will be referred to as the SCC data.
2. The Gas Furnace Data given in Box and Jenkins (1970).

5.1 The SCC Data

The three series are

Z_{1t}: Financial Time Ordinary Share Index
Z_{2t}: U.K. Car Production
Z_{3t}: Financial Time Commodity Price Index

The authors of the original study were interested in the possibility of predicting Z_{1t} from lagged values of Z_{2t} and Z_{3t} using a standard regression analysis in which Z_{1t} was treated as a dependent variable and $Z_{2(t-6)}$ and $Z_{3(t-7)}$ as regressors or independent variables. For a critical evaluation of this approach, see Box and Newbold (1970). Here we consider what structure is revealed by the present multiple time series analysis, in which the three series are jointly modeled.

Tentative Specification. We see in Table 7 that the original series show high and persistent auto- and cross-correlations. Examination of the partials and related statistics in Table 8 shows that for $l > 2$ most of the elements of $\mathcal{P}(l)$ are small compared with their estimated standard errors and the $M(l)$ statistic fails to show significant improvement. Table 9 shows that the pattern of the cross-correlations of the residuals after AR(2) is consonant with estimated white noise. However, note that there is one large residual correlation at lag 1 after the AR(1) fit, suggesting also the possibility of an ARMA(1, 1) model.

Estimation. Both an AR(2) and an ARMA(1, 1) model were fitted using the exact likelihood method* but results are given only for the ARMA(1, 1) model, which produced a marginally better representation. For this model,

$$(\mathbf{I} - \boldsymbol{\varphi} B)\mathbf{Z}_t = \boldsymbol{\theta}_0 + (\mathbf{I} - \boldsymbol{\theta} B)\mathbf{a}_t, \qquad (5.1)$$

where $\boldsymbol{\theta}_0$ is a vector of constants, Table 10 shows the initial unrestricted fit and also the fits for two simpler models obtained by setting to zero those coefficients whose estimates were small compared to their standard errors.

Diagnostic Checking. Table 11 suggests that the restricted ARMA(1, 1) model provides an adequate representation of the data.

Implication of the Model. The final model implies that the system is approximated by

$$(1 - .98B)Z_{1t} = a_{1t} \qquad (5.2a)$$

$$(1 - .93B)Z_{2t} = .2 + a_{2t} \qquad (5.2b)$$

$$(1 - .83B)Z_{3t} = 2.8 + .40a_{1(t-1)} + (1 + .41B)a_{3t}. \qquad (5.2c)$$

Upon substituting (5.2a) into (5.2c), we get

$$(1 - .83B)Z_{3t} = 2.8 + .40(1 - .98B)Z_{1(t-1)} + (1 + .41B)a_{3t}. \qquad (5.2d)$$

Thus all three series behave approximately as random walks with slightly correlated innovations. From the point of view of forecasting, (5.2d) is of some interest since it implies that ordinary share $Z_{1(t-1)}$ is a *leading indicator* at lag 1 for the commodity index Z_{3t}. Its effect is small, however, as can be seen for example by the improvement achieved over the corresponding best fitting univariate model, which was

$$(1 - .78B)Z_{3t} = 3.63 + (1 + .53B)a_t, \sigma^2 = .151 \qquad (5.3)$$

The residual variance of .151 from the univariate model is not much larger than the value .134 for a_{3t} obtained

* For this example, estimates from the conditional likelihood for the ARMA(1, 1) case are very close to the exact results.

Table 8. Partial Autoregression and Related Statistics: SCC Data

Lag	Indicator Symbols for Partials	$M(l)$ Statistic $\dot\sim \chi_9^2$	Diagonal Elements of $\hat{\Sigma} \times 10$
	+ · ·		.44
1	· + ·	301.3	.89
	· · +		1.62
	– · ·		.40
2	· · ·	18.6	.84
	– + –		1.23
	· · ·		.37
3	· · ·	9.6	.81
	· · ·		1.21
	· · ·		.36
4	· · ·	3.6	.79
	· · ·		1.19
	· + ·		.32
5	· + ·	11.9	.70
	· · ·		1.11

from the final vector model. Although the multiple time series analysis fails to reveal anything very surprising for this example, it shows what is there and does not mislead.

5.2 The Gas Furnace Data

The two series consist of (a) input gas rate and (b) output as CO_2 concentration at 9-second intervals from a gas furnace. We shall let Z_{1t} = gas rate + .057 and $Z_{2t} = CO_2 - 5.35$. This set of data was employed in Box and Jenkins (1970) to illustrate a procedure of identification, fitting, and checking of a transfer function model of the form (2.3) for $k = 2$ relating two time series one of which is *known* to be input for the other. Using this approach, the following models were found for the input Z_{1t} and the output Z_{2t};

$$(1 - 1.97B + 1.37B^2 - .34B^3)Z_{1t} = a_{1t},$$
$$\hat\sigma^2_{a_1} = .0353 \quad (5.4a)$$

$$Z_{2t} = \frac{\omega(B)}{\delta(B)} B^b Z_{1t} + \varphi(B)^{-1} a_{2t},\ \hat\sigma^2_{a_2} = .0561, \quad (5.4b)$$

Table 9. Pattern of Cross-Correlation Matrices of Residuals: SCC Data

	Lag 1	2	3	4	5	6	7	8
(a) AR(1) model								
	· · +	· · ·	· · ·	· · ·	· · ·	· · ·	· · ·	· · ·
	· · ·	· · ·	· · ·	· · ·	· · ·	· · ·	· · ·	· · ·
	· · ·	· · ·	· · ·	· · ·	· · ·	· · ·	· · +	· · ·
(b) AR(2) model								
	· · ·	· · ·	· · ·	· · ·	· · ·	· · ·	· · ·	· · ·
	· · ·	· · ·	· · ·	– · ·	· · ·	· · ·	· · ·	· · ·
	· · ·	· · ·	· · ·	· · ·	· · ·	· · ·	· · +	· · ·

Table 10. Estimation Results for the Model (5.1): SCC Data (exact likelihood)

	$\hat{\theta}_0$	$\hat{\varphi}$			$\hat{\theta}$			$\hat{\Sigma}$		
(1) Full Model	1.11	.81	.15	−.06	−.29	.23	.06	.037		
	(.64)	(.08)	(.07)	(.04)	(.15)	(.11)	(.07)			
	1.74	−.07	.98	−.09	−.45	.20	−.15	.022	.078	
	(.82)	(.10)	(.10)	(.05)	(.22)	(.17)	(.11)			
	4.08	−.32	.30	.76	−.79	.57	−.44	.013	.022	.129
	(1.47)	(.18)	(.17)	(.08)	(.28)	(.21)	(.13)			
(2) Restricted Model (intermediate)	.13	.90	.08					.042		
	(.09)	(.06)	(.06)	·	·	·	·			
	.59		.92	−.02				.022	.079	
	(.05)	·	(.04)	(.04)	·	·	·			
	2.48			.85	−.40		−.41	.017	.021	.131
	(1.10)	·	·	(.07)	(.23)	·	(.12)			
(3) Restricted Model (final)	.12	.98						.045		
	(.08)	(.03)		·	·	·	·			
	.24		.93					.024	.085	
	(.10)	·	(.04)	·	·	·	·			
	2.76			.83	−.40		−.41	.019	.023	.134
	(1.07)	·	·	(.06)	(.23)	·	(.12)			

Table 11. Pattern of Residual Cross-Correlations After Final Restricted ARMA(1,1) Model Fit: SCC Data

	$\hat{a}_1$	$\hat{a}_2$	$\hat{a}_3$
$\hat{a}_1$	........–.	....	....
$\hat{a}_2$	....	....	....
$\hat{a}_3$	....	........–.	....

where $\omega(B) = -(.53 + .37B + .51B^2)$, $\delta(B) = 1 - .57B$, $\varphi(B) = 1 - 1.53B + .63B^2$, and the $\{a_{1t}\}$ and $\{a_{2t}\}$ series are assumed independent.

Particularly when we are dealing with econometric rather than engineering models, feedback relationships may not be known a priori; it is of interest, therefore, to analyze the data using the present approach where no distinction is made between an input and output variable and the fact that no feedback could occur in the system is not used.

Tentative Specification. In Table 12, we see that the auto- and cross-correlations of the original data in part (a) are persistently large in magnitude, ruling out low order moving average models; the $M(l)$ statistic (χ_4^2) in part (b) suggests that an AR(6) model might be appropriate; and the residual cross correlation pattern after an AR(6) fit in part (c) seems to verify the appropriateness of this model.

Estimation Results. Estimation results corresponding to an unrestricted AR(6) model

$$(\mathbf{I} - \varphi_1 B - \cdots - \varphi_6 B^6)\mathbf{Z}_t = \mathbf{a}_t \qquad (5.5)$$

are as follows:

$$\hat{\varphi}_1 = \begin{bmatrix} 1.93 & -.05 \\ (.06) & (.05) \\ .06 & 1.55 \\ (.08) & (.06) \end{bmatrix} \quad \hat{\varphi}_2 = \begin{bmatrix} -1.20 & .10 \\ (.13) & (.08) \\ -.14 & -.59 \\ (.16) & (.11) \end{bmatrix}$$

$$\hat{\varphi}_3 = \begin{bmatrix} .17 & -.08 \\ (.15) & (.09) \\ -.44 & -.17 \\ (.19) & (.11) \end{bmatrix} \quad \hat{\varphi}_4 = \begin{bmatrix} -.16 & .03 \\ (.15) & (.09) \\ .15 & .13 \\ (.19) & (.11) \end{bmatrix}$$

$$\hat{\varphi}_5 = \begin{bmatrix} .38 & -.04 \\ (.14) & (.08) \\ -.12 & .06 \\ (.18) & (.10) \end{bmatrix} \quad \hat{\varphi}_6 = \begin{bmatrix} -.22 & .03 \\ (.08) & (.03) \\ .25 & -.04 \\ (.11) & (.04) \end{bmatrix}$$

$$\hat{\mathbf{\Sigma}} = \begin{bmatrix} .0345 & \\ -.0023 & .0566 \end{bmatrix}, \quad \hat{\rho}(a_1, a_2) = .045 \qquad (5.6)$$

If we let

$$\hat{\varphi}_l = \{\hat{\varphi}_{ij}^{(l)}\},$$

Table 12. Tentative Identification for the Gas Furnace Data

(a) Pattern of cross-correlations of the original data

	Z_{1t}	Z_{2t}
Z_{1t}	+ + + + + + + + + + + +	– – – – – – – – – – – –
Z_{2t}	– – – – – – – – – – . .	+ + + + + + + + + + + +

(b) M statistic for partial autoregression

Lag l	1	2	3	4	5	6	7	8	9	10	11
$M(l)$	1650	665	31.7	22.5	5.6	12.9	1.8	8.0	3.5	0	2.0

(c) Pattern of cross-correlations of the residuals after AR(6) fit

	$\hat{a}_{1t}$	$\hat{a}_{2t}$
$\hat{a}_{1t}$	...........–	....
$\hat{a}_{2t}$	....	....

we see that $\hat{\varphi}_{12}^{(l)}$ are small compared with their standard errors over all lags, confirming (as in this case is known from the physical nature of the apparatus generating the data) that there is a unidirectional relationship between Z_{1t} and Z_{2t} involving no feedback. Also, $\hat{\varphi}_{21}^{(l)}$ is small for $l = 1, 2$, and the residuals $\hat{a}_{1t}$ and $\hat{a}_{2t}$ are essentially uncorrelated, implying a delay of 3 periods. It should be noted also that the variances for a_{1t} and a_{2t} are very close to those for a_{1t} and a_{2t} in (5.4), and their correlation is negligible.

To facilitate comparison with (5.4), we set $\varphi_{11}^{(l)} = 0$ for $l > 3$, $\varphi_{12}^{(l)} = 0$ for all l, $\varphi_{21}^{(l)} = 0$ for $l = 1, 2$, and $\varphi_{22}^{(l)} = 0$ for $l = 5, 6$. Estimation results for this restricted AR(6) model are then

$$\hat{\varphi}_1 = \begin{bmatrix} 1.98 & \cdot \\ (.06) & \\ \cdot & 1.53 \\ & (.06) \end{bmatrix} \quad \hat{\varphi}_2 = \begin{bmatrix} -1.38 & \cdot \\ (.10) & \\ \cdot & -.58 \\ & (.11) \end{bmatrix} \quad \hat{\varphi}_3 = \begin{bmatrix} .35 & \cdot \\ (.06) & \\ -.53 & -.14 \\ (.07) & (.10) \end{bmatrix}$$

$$\hat{\varphi}_4 = \begin{bmatrix} \cdot & \cdot \\ .11 & .12 \\ (.16) & (.04) \end{bmatrix} \quad \hat{\varphi}_5 = \begin{bmatrix} \cdot & \cdot \\ -.04 & \cdot \\ (.17) & \end{bmatrix} \quad \hat{\varphi}_6 = \begin{bmatrix} \cdot & \cdot \\ .21 & \cdot \\ (.11) & \end{bmatrix}$$

$$\hat{\mathbf{\Sigma}} = \begin{bmatrix} .0359 & -.0029 \\ & .0561 \end{bmatrix}, \quad \hat{\rho}(a_1, a_2) \doteq 0 \qquad (5.7)$$

Examination of the pattern of the cross-correlations of the residuals suggests that the model is adequate.

Implication of the Bivariate Model. The final AR(6) model (5.7) can be written

$$\begin{bmatrix} \varphi_{11}(B) & \\ \varphi_{21}(B) & \varphi_{22}(B) \end{bmatrix} \begin{bmatrix} Z_{1t} \\ Z_{2t} \end{bmatrix} = \begin{bmatrix} a_{1t} \\ a_{2t} \end{bmatrix}, \qquad (5.8)$$

where $\varphi_{11}(B) = 1 - 1.98B + 1.38B^2 - .35B^3$, $\varphi_{21}(B) = (.53 - .11B - .21B)B^3$, and $\varphi_{22}(B) = (1 - 1.53B + .58B^2 + .14B^3 - .12B^4)$. Assuming a_{1t} and a_{2t} are uncorrelated, the input model $\varphi_{11}(B)Z_{1t} = a_{1t}$ with $\text{Var}(a_{1t}) = .0359$ is essentially the same as (5.4a). Now the model relating the output Z_{2t} to the input Z_{1t} is

$$Z_{2t} = -\frac{\varphi_{21}(B)}{\varphi_{22}(B)}Z_{1t} + \frac{1}{\varphi_{22}(B)}a_{2t} \qquad (5.9)$$

with $\text{Var}(a_{2t}) = .0561$. The noise model $\varphi_{22}^{-1}(B)a_{2t}$ is not very different from the corresponding one $\varphi^{-1}(B)a_{2t}$ in (5.4b), but the dynamic model $-\varphi_{21}(B)\varphi_{22}^{-1}(B)Z_{1t}$ at first sight appears markedly different from the first term on the right side of (5.4b). The reason is that in the form (5.9) the denominators of the dynamic model and of the noise model are constrained to be identical. This restriction is not present in the transfer function model (5.4b). The less restrictive form can however be written in the form of (5.9) if we set $\varphi_{22}(B) = \varphi(B)$ and $-\varphi_{21}(B) = \omega(B)B^b\{\varphi(B)\delta^{-1}(B)\}$. For this example, the factor $\varphi(B)\delta^{-1}(B) \doteq 1 - .96B$, and it is then seen that the models are in fact very similar. This may be confirmed by comparing the impulse response weights in Table 13, where $\omega(B)B^b\delta^{-1}(B) = \sum_{j=0}^{\infty}v_jB^j$ and $-\varphi_{21}(B)\varphi_{22}^{-1}(B) = \sum_{j=0}^{\infty}v^*_jB^j$.

Further Analysis of Stepwise AR Results. It is instructive to examine for this data the changes in the fitted autoregressive models as the order is increased. Using indicator symbols (and omitting the dots) Table 14 shows the situation for $p = 1, \ldots, 6$. The residual covariance matrix for each order is also given. The following observations may be made.

1. If only AR(1) or AR(2) were considered, one might be led to believe mistakenly that there was a feedback relationship between these two series.
2. The unidirectional dynamic relationship becomes clear when the order of the model, p, is increased to three. Since the input series Z_{1t} essentially follows a univariate AR(3) model, this suggests that the present procedure will correctly identify the one-sided causal dynamic relationship once the input model is appropriately selected.
3. The delay $b = 3$ emerges when the order p is increased to 4. Since only very marginal improvement in the fit occurs for $p > 4$, this is saying that the delay is correctly identified only when the model is specified essentially correctly.

Implications on General Time Series Model Building. The relative merit of the present procedure and more direct modeling of the system will depend on how much is known or how much we are prepared to assume. In some applications, particularly in engineering and most examples of intervention analysis, an adequate initial specification may be possible from knowledge of the nature of the problem. This may allow a flow diagram showing the feedback structure to be drawn and likely orders to be guessed for the various dynamic components. The resulting models can then be directly *fitted* in the manner described and illustrated in Box and MacGregor (1974, 1976) and Box and Tiao (1975). For a single input with feedback known to be absent, a prewhitening method is given in Box and Jenkins (1970) for *identifying* an unknown dynamic system, but extension of this identification method to multiple inputs is rather complex.

Particularly for economic and business examples, however, the feedback structure and orders of the multiple system are often unknown. The present multiple time series procedure has the great advantage that it allows *identification* of the feedback and dynamic structure. Furthermore,

1. A one-sided causal relationship, if it exists, will emerge in the identification process, and the stochastic structures of the input as well as the transfer function relationship between input and output will be modeled simultaneously.
2. Stochastic multiple input and multiple output situations are readily handled.
3. A useful method is provided for seeking leading indicators in economic and business applications. In this context it should be noted that a unidirectional dynamic relationship may not exist between two time series even when one variable is known to be the input for the other. One reason for this phenomenon is the effect of temporal aggregation. As shown in Tiao and Wei (1976), pseudo-feedback relationships could occur because of this temporal aggregation effect, and it would be a mistake to impose a transfer function model in such a situation.
4. However, when a simple transfer function structure of the form (2.2) *is* appropriate, the present multiple time series approach could rarely reproduce it directly—see, for example, (5.4b) and (5.9)—and some analysis of the fitted form might be necessary to reveal a more parsimonious and more easily understood structure.

6. COMPARISON WITH SOME OTHER APPROACHES AND CONCLUDING REMARKS

We have discussed various tools used in an iterative approach to modeling multiple time series and illustrated

Table 13. Impulse Response Weights for the Gas Furnace Data

	j												
	0	1	2	3	4	5	6	7	8	9	10	11	12
v_j	·	·	·	−.53	−.67	−.89	−.51	−.29	−.17	−.09	−.05	−.03	−.02
v^*_j	·	·	·	−.53	−.70	−.77	−.48	−.26	−.09	−.01	.01	.00	−.01

Table 14. Successive AR Fitting Results for the Gas Furnace Data

Order of AR	φ_1	φ_2	φ_3	φ_4	φ_5	φ_6	$\boldsymbol{\Sigma}$
1	$\begin{bmatrix} + & + \\ - & + \end{bmatrix}$						$\begin{bmatrix} .102 & .090 \\ & .346 \end{bmatrix}$
2	$\begin{bmatrix} + & - \\ + & + \end{bmatrix}$	$\begin{bmatrix} - & + \\ - & - \end{bmatrix}$					$\begin{bmatrix} .037 & -.004 \\ & .069 \end{bmatrix}$
3	$\begin{bmatrix} + & \\ & + \end{bmatrix}$	$\begin{bmatrix} - & \\ - & - \end{bmatrix}$	$\begin{bmatrix} + & \\ & + \end{bmatrix}$				$\begin{bmatrix} .036 & -.002 \\ & .063 \end{bmatrix}$
4	$\begin{bmatrix} + & \\ & + \end{bmatrix}$	$\begin{bmatrix} - & \\ & - \end{bmatrix}$	$\begin{bmatrix} & \\ - & \end{bmatrix}$	$\begin{bmatrix} & \\ + & + \end{bmatrix}$			$\begin{bmatrix} .036 & -.003 \\ & .059 \end{bmatrix}$
5	$\begin{bmatrix} + & \\ & + \end{bmatrix}$	$\begin{bmatrix} - & \\ & - \end{bmatrix}$	$\begin{bmatrix} & \\ - & \end{bmatrix}$	$\begin{bmatrix} & \\ & \end{bmatrix}$	$\begin{bmatrix} & \\ & \end{bmatrix}$		$\begin{bmatrix} .035 & -.003 \\ & .058 \end{bmatrix}$
6	$\begin{bmatrix} + & \\ & + \end{bmatrix}$	$\begin{bmatrix} - & \\ & - \end{bmatrix}$	$\begin{bmatrix} & \\ - & \end{bmatrix}$	$\begin{bmatrix} & \\ & \end{bmatrix}$	$\begin{bmatrix} + & \\ & \end{bmatrix}$	$\begin{bmatrix} - & \\ + & \end{bmatrix}$	$\begin{bmatrix} .035 & -.002 \\ & .057 \end{bmatrix}$

how they work in practice. Much further work is needed, especially in the identification of mixed autoregressive moving average models and in developing faster estimation algorithms and better tools for diagnostic checking. In spite of the imperfections of the present tools and the preliminary nature of the approach, we have felt it appropriate to present them here in order to (a) illustrate the potential usefulness of vector autoregressive moving average models in characterizing dynamic structures in the data and (b) stimulate further development of modeling procedures. Several alternative approaches to modeling multiple time series have been proposed in the literature. It may be of interest to discuss briefly those proposed by Granger and Newbold (1977), Wallis (1977), and Chan and Wallis (1978).

In the Granger and Newbold approach, one begins by fitting univariate ARMA models to each series,

$$\varphi_{p_j}(B)Z_{jt} = \theta_{q_j}(B)C_{jt}, \quad j = 1, \ldots, k \tag{6.1}$$

and then attempts to identify the dynamic structure of the k white noise residual series $\{C_{jt}\}$ by examination of their cross-correlations. A model of the form (2.2) with $k(h)$ being the set $(1, \ldots, k)$ excluding h is then fitted to the k residual series. This model and the prewhitening transformations (6.1) then determine the model for the original vector series. As the authors themselves pointed out, the procedure is complex and difficult to apply for $k > 2$. One major difficulty arises from the fact that the parameters in the model for the residuals are subject to various complicated nonlinear constraints. Also, it can be readily shown that even if the vector series $\{\mathbf{Z}_t\}$ follows a low order ARMA model (3.1), the corresponding model for the residual vector $\{\mathbf{C}_t\}$ where $\mathbf{C}'_t = (C_{1t}, \ldots, C_{kt})$ can be complex and difficult to identify in practice.

The Wallis and Chan approach uses the form (6.1) for each individual series and the fact that the model (3.1) can be written as

$$|\boldsymbol{\varphi}(B)|\,\mathbf{Z}_t = \mathbf{H}(B)\mathbf{a}_t, \tag{6.2}$$

where $\mathbf{H}(B) = \mathbf{A}(B)\boldsymbol{\theta}_q(B)$, $\mathbf{A}(B)$ is the adjoint matrix and $|\boldsymbol{\varphi}_p(B)|$ the determinant of $\boldsymbol{\varphi}_p(B)$. As in the G and N approach, an individual model is first constructed for each series. From the degrees of the moving average polynomials $\theta_{q_j}(B)$ of these individual models, the degree of $\mathbf{H}(B)$ is determined. Next, models of the form

$$\mathbf{D}_l(B)\mathbf{Z}_t = \mathbf{H}(B)\mathbf{a}_t, \tag{6.3}$$

where $\mathbf{D}_l(B)$ is a diagonal matrix polynomial in B of degree l, are fitted successively for $l = r, r-1, \ldots$, where r is some specified maximum order, to determine an appropriate value for l. A likelihood ratio test is then performed to check whether the diagonal elements of $\mathbf{D}_l(B)$ are identical, that is, of the form (6.2). Finally, from the fitted $\mathbf{H}(B)$ and $|\varphi(B)|$ or $\mathbf{D}_l(B)$, one guesses at the values of p and q in (3.1) and then proceeds to estimate the parameters in $\boldsymbol{\varphi}_p(B)$ and $\boldsymbol{\theta}_q(B)$. The efficacy of this approach is open to question on several grounds.

1. The degree of the polynomial $\mathbf{H}(B)$ in (6.2) can be higher than the maximum degree of $\theta_{q_j}(B)$ for the individual series. For example, suppose $k = 2$,

$$\mathbf{H}(B) = \begin{pmatrix} 1 & -h_1B \\ -h_2B & 1 \end{pmatrix}$$

and the two elements of $\mathbf{a}_t$ are independent. Then $q_1 = q_2 = 0$, but it would be a mistake to infer that $\mathbf{H}(B)$ is of degree zero.

2. For vector AR or ARMA models, the representation (6.2) is certainly nonparsimonious. Apart from the covariance matrix $\boldsymbol{\Sigma}$, for k series the maximum number of parameters in the original form (3.1) is $k^2(p + q)$, while the maximum number of parameters in the form (6.2) is $kp + [(k-1)p + q]k^2$, representing an increase

of $pk(k-1)^2$ parameters. The increase could be even greater if the diagonal form (6.3) is employed. Thus, assuming the degree of $\mathbf{H}(B)$ is correctly specified, even for k as low as 3 or 4, a very large number of additional parameters will have to be estimated merely to identify correctly a low order vector AR model, say $p = 1$ or 2.

3. Since the correspondence between the degrees of the determinantal polynomial $|\varphi(B)|$ and $\mathbf{H}(B)$ and the values of (p, q) is not necessarily one to one, it is not clear how one determines p and q in (3.1) from the form (6.2).

4. The approach is made even more computationally burdensome because the authors propose to employ the exact likelihood method for moving average parameters throughout the processes of model building. Our experience, however, suggests that because this method converges relatively slowly it is better to use it only in the final stage of the estimation process.

The chief distinction between our approach and the two alternatives just discussed is that we believe it better to tackle the dynamic relationships of the k series in their entirety, employing tools such as the estimates of cross-correlation matrices and partial autoregression matrices to shed light directly on the structure. Simplifications of one kind or another will then often follow. At least for the tentative specification of the vector autoregressive or the vector moving average model, our procedures seem far simpler to use in practice and do not require the multitude of steps these alternative approaches need to arrive at even a simple model.

To illustrate these points, we briefly consider the mink-muskrat example which Chan and Wallis used to illustrate their methods. They treat two series $Y_{1t}{}^*$ and $Y_{2t}{}^*$ obtained after "detrending" the muskrat and mink series by first and second degree polynomials respectively. Proceeding through the various steps outlined above, they eventually arrive at an AR(1) model. However, it will be seen that this same model is suggested immediately by the simple procedures we propose. Table 15(a) shows the partial autoregression results for $l = 1, 2$ and Table 15(b) the residual cross correlations after the AR(1) fit. A very similar analysis of this set of data is given in Ansley and Newbold (1979). For various reasons, we do not wish to sanctify this AR(1) model. These include the question of whether any linear structural model is adequate for these series (see Tong and Lim 1980). Also, the validity of the detrending procedures and the suspicious behavior of a high autocorrelation at lag 10 occurring in the residuals seem suspect. Our only point is to show that the circuitous route adopted by Chan and Wallis to arrive at this model is unnecessary.

Before concluding this paper, it is worth noting that in modeling as well as analysis of vector time series one often finds it useful to perform various eigenvalue and eigenvector analyses. Specifically, writing (3.1) in the form

$$\mathbf{z}_t = \hat{\mathbf{z}}_{t-1}(1) + \mathbf{a}_t, \qquad (6.4)$$

Table 15. Identification of Muskrat-Mink Data

(a) Partial Autoregression and Related Statistics

Lag	*Partials*		$M(l) \stackrel{.}{\rightarrow} \chi_4^2$	*Diagonal elements of* $\hat{\Sigma}$
1	+	−	111.7	.062
	+	+		.059
2	−	·	4.8	.0571
	·	·		.0572

(b) Cross-correlations of Residuals After AR(1) Fit

	a_1	a_2
a_1	. . . .	. − . . . .
a_2	. . . .	. . . . + . .

where $\hat{\mathbf{z}}_{t-1}(1)$ is the one step ahead forecast of $\mathbf{z}_t$ made at time $t-1$, and denoting, for stationary series,

$$\boldsymbol{\Gamma}_{\mathbf{z}}(0) = E(\mathbf{z}_t\mathbf{z}'_t)$$

and

$$\boldsymbol{\Gamma}_{\hat{\mathbf{z}}}(0) = E(\hat{\mathbf{z}}_{t-1}(1)\hat{\mathbf{z}}_{t-1}(1)'),$$

it will often be informative to compute eigenvalues and eigenvectors of estimates of the following matrices:

$$(a)\ \boldsymbol{\Gamma}_{\mathbf{z}}(0), \quad (b)\ \boldsymbol{\Sigma},$$

$$(c)\ \boldsymbol{\Gamma}_{\mathbf{z}}(0)^{-1}\boldsymbol{\Gamma}_{\hat{\mathbf{z}}}(0), \quad (d)\ \varphi_l, \text{ and } \boldsymbol{\theta}_l.$$

Such analyses are described in Quenouille (1957), Box and Tiao (1977), and Tiao et al. (1979). Also, the eigenvalues and eigenvectors of the spectral density matrix of the model should also be considered (see Brillinger 1975). These techniques are useful in (a) detecting exact concurrent or lagged linear relations between series, and (b) facilitating understanding and interpretation of the fitted model. In our opinion, this is one of the most important and challenging topics for further research.

[*Received January 1981. Revised June 1981.*]

REFERENCES

ABRAHAM, B. (1980), "Intervention Analysis and Multiple Time Series," *Biometrika*, 67, 73–78.

ABRAHAM, B., and BOX, G.E.P. (1978), "Deterministic and Forecast-Adaptive Time Dependent Models," *Applied Statistics*, 27, 120–130.

AKAIKE, H. (1980), "On the Identification of State Space Models and Their Use in Control," in *Directions in Time Series*, eds. D.R. Brillinger and G.C. Tiao, Institute of Mathematical Statistics, 175–187.

ANDERSON, T.W. (1971), *The Statistical Analysis of Time Series*, New York: John Wiley.

——— (1980), "Maximum Likelihood Estimation for Vector Autoregressive Moving Average Models," in *Directions in Time Series*, eds. D.R. Brillinger and G.C. Tiao, Institute of Mathematical Statistics, 49–59.

ANSLEY, C. (1979), "An Algorithm for the Exact Likelihood of a Mixed Autoregressive Moving Average Process," *Biometrika*, 66, 59–65.

ANSLEY, C.F., and NEWBOLD, P. (1979), "Multivariate Partial Autocorrelations," *Proceedings of Business and Economic Statistics Section*, American Statistical Association, 349–353.

BARTLETT, M.S. (1938), "Further Aspects of the Theory of Multiple Regression," *Proceedings of the Cambridge Philosophical Society*, 34, 33–40.

BEGUIN, J.M., GOURIEROUX, C., and MONFORT, A. (1980), "Identification of a Mixed Autoregressive-Moving Average Process: The Corner Method," in *Time Series*, ed. O.D. Anderson, Amsterdam: North-Holland, 423–436.

BOX, G.E.P, and HAUGH, L. (1977), "Identification of Dynamic Regression Models Connecting Two Time Series," *Journal of the American Statistical Association*, 72, 121–130.

BOX, G.E.P., and JENKINS, G.M. (1970), *Time Series Analysis—Forecasting and Control*, San Francisco: Holden-Day.

BOX, G.E.P., and MACGREGOR, J.F. (1974), "The Analysis of Closed Loop Dynamic Stochastic Systems," *Technometrics*, 16, 391–398.

——— (1976), "Parameter Estimation with Closed-Loop Operating data," *Technometrics*, 18, 371–380.

BOX, G.E.P., and NEWBOLD, P. (1970), "Some Comments on a Paper by Coen, Gomme and Kendall," *Journal of the Royal Statistical Society*, Ser. A, 134, 229–240.

BOX, G.E.P., and TIAO, G.C. (1975), "Intervention Analysis with Applications to Environmental and Economic Problems," *Journal of the American Statistical Association*, 70, 70–79.

——— (1977), "A Canonical Analysis of Multiple Time Series," *Biometrika*, 64, 355–365.

BRILLINGER, D.R. (1975), *Time Series Data Analysis and Theory*, New York: Holt, Rinehart, and Winston.

CHAN, W.Y.T., and WALLIS, K.F. (1978), "Multiple Time Series Modelling: Another Look at the Mink-Muskrat Interaction," *Applied Statistics*, 27, 168–175.

COEN, P.G., GOMME, E.D., and KENDALL, M.G. (1969), "Lagged Relationships in Economic Forecasting," *Journal of the Royal Statistical Society*, Ser. A, 132, 133–163.

DEISTLER, M., DUNSMUIR, W., and HANNAN, E.J. (1978), "Vector Linear Time Series Models: Corrections and Extensions," *Advances in Applied Probability*, 10, 360–372.

DENT, W. (1977), "Computation of the Exact Likelihood Function of an ARIMA Process," *Journal of Statistical Computation and Simulation*, 5, 193–206.

DUNSMUIR, W., and HANNAN, E.J. (1976), "Vector Linear Time Series Models," *Advances in Applied Probability*, 8, 339–364.

GRANGER, C.W.J., and NEWBOLD, P. (1977), *Forecasting Economic Time Series*, New York: Academic Press.

GRAY, H.L., KELLEY, G.D., and McINTIRE, D.D. (1978), "A New Approach to ARMA Modelling," *Communications in Statistics*, B7, 1–77.

HALLIN, M. (1978), "Mixed Autoregressive-Moving Average Multivariate Processes with Time-Dependent Coefficients," *Journal of Multivariate Analysis*, 8, 567–572.

HANNAN, E.J. (1970), *Multiple Time Series*, New York: John Wiley.

——— (1980), "The Estimation of the Order of an ARMA Process," *Annals of Statistics*, 8, 1071–1081.

HANNAN, E.J., DUNSMUIR, W.T.M., and DEISTLER, M. (1980), "Estimation of Vector ARMAX Models," *Journal of Multivariate Analysis*, 10, 275–295.

HILLMER, S.C., and TIAO, G.C. (1979), "Likelihood Function of Stationary Multiple Autoregressive Moving Average Models," *Journal of the American Statistical Association*, 74, 652–660.

HOSKING, J.R.M. (1980), "The Multivariate Portmanteau Statistic," *Journal of the American Statistical Association*, 75, 602–607.

HSIAO, C. (1979), "Autoregressive Modeling of Canadian Money and Income Data," *Journal of the American Statistical Association*, 74, 553–560.

JENKINS, G.J. (1979), *Practical Experiences with Modelling and Forecasting Time Series*, Channel Islands: GJP Ltd.

LI, W.K., and McLEOD, A.I. (1980), "Distribution of the Residual Autocorrelations in Multivariate ARMA Time Series Models," TR-80-03, University of Western Ontario.

NEWBOLD, P. (1974), "The Exact Likelihood Function for a Mixed Autoregressive Moving Average Process," *Biometrika*, 61, 423–427.

NICHOLLS, D.F. (1976), "The Efficient Estimation of Vector Linear Time Series Models," *Biometrika*, 63, 381–390.

——— (1977), "A Comparison of Estimation Methods for Vector Linear Time Series Models," *Biometrika*, 64, 85–90.

NICHOLLS, D.F., and HALL, A.D. (1979), "The Exact Likelihood of Multivariate Autoregressive-Moving Average Models," *Biometrika*, 66, 259–264.

OSBORN, D.R. (1977), "Exact and Approximate Maximum Likelihood Estimators for Vector Moving Average Processes," *Journal of the Royal Statistical Society*, Ser. B, 39, 114–118.

PARZEN, E. (1977), "Multiple Time Series: Determining the Order of Approximating Autoregressive Schemes," in *Multivariate Analysis-IV*, ed. P. Krishnaiah, Amsterdam: North-Holland, 283–295.

PHADKE, M.S., and KEDEM, G. (1978), "Computation of the Exact Likelihood Function of Multivariate Moving Average Models," *Biometrika*, 65, 511–519.

QUENOUILLE, M.H. (1957), *The Analysis of Multiple Time Series*, London: Griffin.

QUINN, B.G. (1980), "Order Determination for a Multivariate Autoregression," *Journal of the Royal Statistical Society*, Ser. B, 42, 182–185.

SLUTSKY, E. (1937), "The Summation of Random Causes as the Source of Cyclic Processes," *Econometrika*, 5, 105–146.

TIAO, G.C., and ALI, M.M. (1971), "Analysis of Correlated Random Effects: Linear Model with Two Random Components," *Biometrika*, 58, 37–51.

TIAO, G.C., BOX, G.E.P., GRUPE, M.R., HUDAK, G.B., BELL, W.R., and CHANG, I. (1979), "The Wisconsin Multiple Time Series (WMTS-1) Program: A Preliminary Guide," Department of Statistics, University of Wisconsin, Madison.

TIAO, G.C., and WEI, W.S. (1976), "Effect of Temporal Aggregation on the Dynamic Relationship of Two Time Series Variables," *Biometrika*, 63, 513–523.

TONG, H., and LIM, K.S. (1980), "Threshold Autoregression, Limit Cycles and Cyclical Data," *Journal of the Royal Statistical Society*, Ser. B, 42, 245–292.

WALLIS, K.F. (1977), "Multiple Time Series Analysis and the Final Form of Econometric Models," *Econometrika*, 45, 1481–1497.

WHITTLE, P. (1963), "On the Fitting of Multivariate Autoregressions, and the Approximate Canonical Factorization of a Spectral Density Matrix," *Biometrika*, 50, 129–134.

WILSON, G.T. (1973), "The Estimation of Parameters in Multivariate Time Series Models," *Journal of the Royal Statistical Society*, Ser. B, 35, 76–85.

YULE, G.U. (1927), "On a Method of Investigating Periodicities in Disturbed Series, With Special Reference to Wolfer's Sunspot Numbers," *Philosophical Transactions of the Royal Society of London*, Ser. A, 226, 267–298.

ZELLNER, A., and PALM, F. (1974), "Time Series Analysis and Simultaneous Equation Econometric Models," *Journal of Econometrics*, 2, 17–54.

3.16

Gwilym Jenkins, Experimental Design and the Time Series

G. E. P. BOX
University of Wisconsin

We are here to honor this remarkable man: Gwilym Meirion Jenkins. He was my friend, and the times when we worked together were some of the happiest and most exciting in my life.

Gwilym was Welsh. Wales, you remember, is that bit of Britain on the left looking north that sticks out into the Irish Sea. People sometimes think about Britain in much the same way as they think about Spain, and I am ashamed to confess that many people do not know that Catalonia is a distinct entity in itself with its own customs, its own proud history, and its own beautiful language that is very much alive. The same is true, of, course for Wales. In fact, Gwilym Jenkins spoke only Welsh up until the time he was seven years old. His grandmother, who lived to be 106, did not learn English until she was in her sixties because it was not until then that English was much spoken in her village. Gwilym told me that he sometimes thought in Welsh. Once, for example, when he lived in London, he went on an errand for his wife, Meg, to buy a colander but returned empty handed because he could only remember the Welsh but not the English word for "colander."

The Welsh are famous for their poetry and their song. It is said that a Welsh congregration sings automatically not only in tune but also in four-part harmony. Certainly you will find throughout British history that all minstrels that are mentioned -- in Shakespeare and elsewhere -- are all Welsh. I take that to mean, in modern terms, that they are a race with a very active right brain. Certainly Gwilym possessed unusual inventiveness and intuition as well as remarkable analytical power.

He took a bachelor's degree in mathematics in 1953 at University College of London with first class honors, and he completed his Ph.D. there in 1956. His disser-

The Collected Works of George E. P. Box, 1984, Wadsworth, Inc., Belmont, CA 94002.
To be published in *Qüestiió.*

tation was on time series analysis.

After this, he served for two years as a junior fellow at the Royal Aircraft Establishment at Farnborough, where he helped to design aircraft. One of the problems he told me about that illustrates the kind of work he did there concerned the design of an aircraft undercarriage. He explained how by running a little wheel along the runway, its ups and downs could be recorded as a time series and the spectrum estimated. Since it was also possible to compute transfer functions for any design of undercarriage, a design could be selected for which low transmission occurred at frequencies where the spectral power was high -- so ensuring that the plane did not shake to pieces on takeoff. It was this kind of useful application of theory that interested Gwilym. He was equally at home with time domain as with frequency domain analysis, and he was sometimes asked when one or the other was appropriate.

One reply he gave was simply that when, as in the aircraft undercarriage problem, you are directly interested in frequencies, it is natural to work with frequencies, but when you are concerned with matters such as forecasts occurring in real time, then you would usually want to work in real time. There would be cases, of course, where switching from one to the other would greatly assist mathematical reasoning.

It seems that strong forces are always at work which try to divide the world of statistics into two non-overlapping territories -- theory and application. Gwilym fought constantly against those forces. He believed that not much was to be expected from theory without practice, or from practice without theory, and that the two were inextricably wedded, with most useful theory arising from sensible practice and most useful practice based on sound theory in a never-ending iteration. The results of this mode of thinking come through strongly, for example, in his book with Don Watts on spectral analysis (1), which draws on practical experience and is genuinely concerned with the problem of how you do spectral analysis.

It was at the Royal Aircraft Establishment that he met his wife, Margaret Bellingham. From that day onwards Meg's loving enterprise and devotion were to be his unfailing support. George Barnard quickly recognized Gwilym's ability, and in 1957 he was appointed lecturer at Imperial College. His publications on time series also brought rapid recognition on the other side of the Atlantic, and in 1959 he came as a visitor to Stanford University and later to Princeton.

I remember very well his visit to the Statistical Techniques Research Group at Princeton. My friend George Barnard had written me a letter by way of introduction to Gwilym, in which he said something to the effect that "on matters concerning time series he would value Gwilym's judgment before anyone, even before John Tukey." I was almost as intrigued by this letter as was John Tukey, and we quickly invited him to come.

His arrival in Princeton marked the beginning of a long and happy collaboration between us, which later resulted in much visiting between England and the United States. It was during one such visit to Madison in 1964-65 that the seriousness of his medical condition was first realized. From now on he would fight a slowly losing battle against Hodgkin's disease.

For the next seventeen years, his condition fluctuated unpredictably and disappointingly between moderately well and desperately ill. In circumstances that would have undermined the courage of a hero, I have seen him continue to work at a pace which many a healthy person would have found impossible, and to somehow still maintain his buoyant optimism and sense of humor.

I want now to say something about the work we did together. Before I met Gwilym, I regarded time series as a very boring subject. I'm sure this was because I had never really used it. Indeed, I think that Gwilym was the first person I'd ever met who talked coherently about time series in terms of actually doing something with it.

But in the beginning our discussions weren't about time series at all, but about a problem in experimental design.

It can be interesting to consider by what strange routes real investigations sometimes proceed: starting off where you would not expect them to, making detours which later turn out to be profitless, and so forth. Certainly what is finally published seldom provides much idea of the rather haphazard and messy route which has been taken. So I thought it might be of interest to try to reconstruct how, so far as I can remember, some of our work progressed.

Let me first explain about the experimental design problem that we began with. It concerned an idea for making an industrial process track a moving optimum. For example, suppose for some chemical catalytic process, yield y was a function of temperature x. Then as the catalyst decayed with time, the process curve representing the relation between y and x might drift, as indicated in Figure 1, in a manner not knowable in advance. One way to make the process continuously adapt to such a moving optimum was to perturb temperature (2,3,4,5) sinusoidally and to use the observed variation in response to apply a correction to the mean level of temperature, as in Figure 2.

Gwilym agreed that the thing to do was to try to get such an apparatus actually working where we could see it, run it, and get data from it. We didn't have any luck with the chemical engineers at Princeton. However, a few months later I moved to Wisconsin and the chemical engineers there, especially Olaf Hougen, greeted the project with enthusiasm, and an apparatus of this kind was indeed eventually built (6). (In particular, the "gas-furnace data" in our book (7) came from that investigation.)

The analogy with standard experimental design and analysis and with the philosophy of evolutionary operation (8) was obvious (Figure 3). But Gwilym emphasized that if the continuous scheme of Figure 2 was going to work properly, then we should need to pay careful attention to the dynamics of the system that was being perturbed and to the correlation structure of the noise we encountered. So we began to work on this. Also, he found considerable entertainment (9) working out appropriate frequencies for sine wave designs when several factors were to be perturbed and optimized. In particular, for second-order designs you had to watch for confounding of frequencies arising from harmonics produced by second- and possible higher-order effects. In addition, it was necessary to worry about these frequencies in relation to the power spectrum of the noise.

The work which eventually resulted in our book took place roughly between 1960 and 1970. It evolved in this way: Having started off thinking about automatic optimization and the importance of dynamics, we realized that what we were really involved with was a

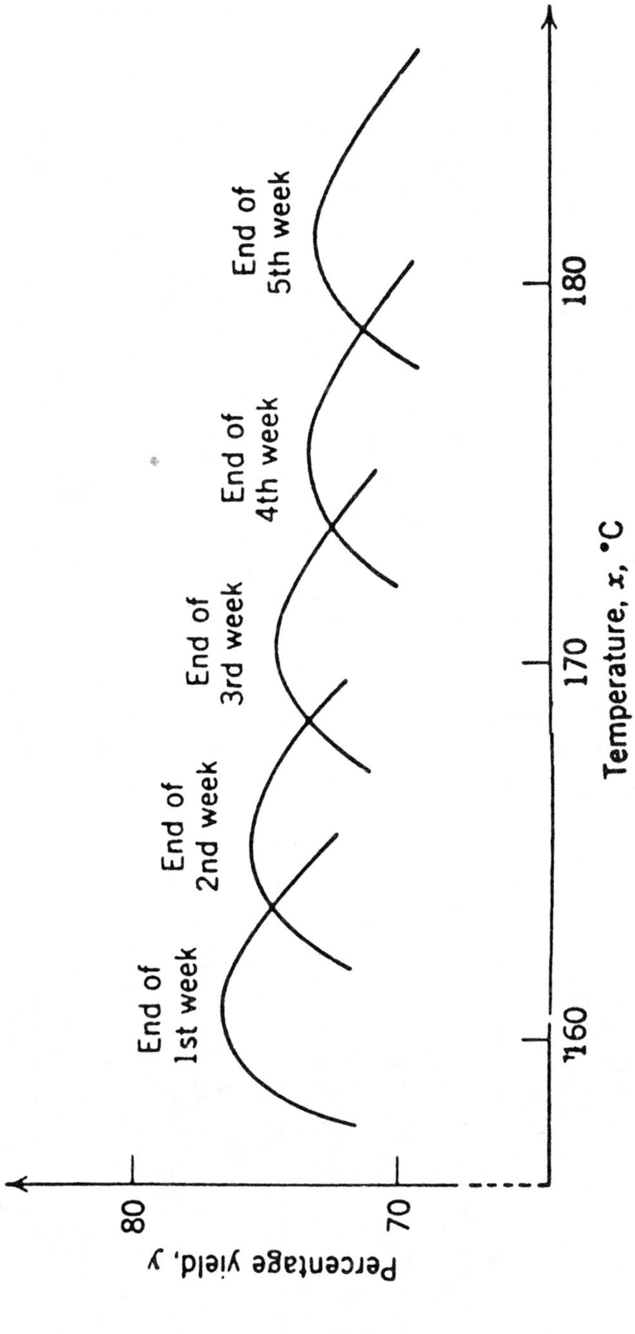

Figure 1. Typical drift in process curve caused by decaying catalyst.

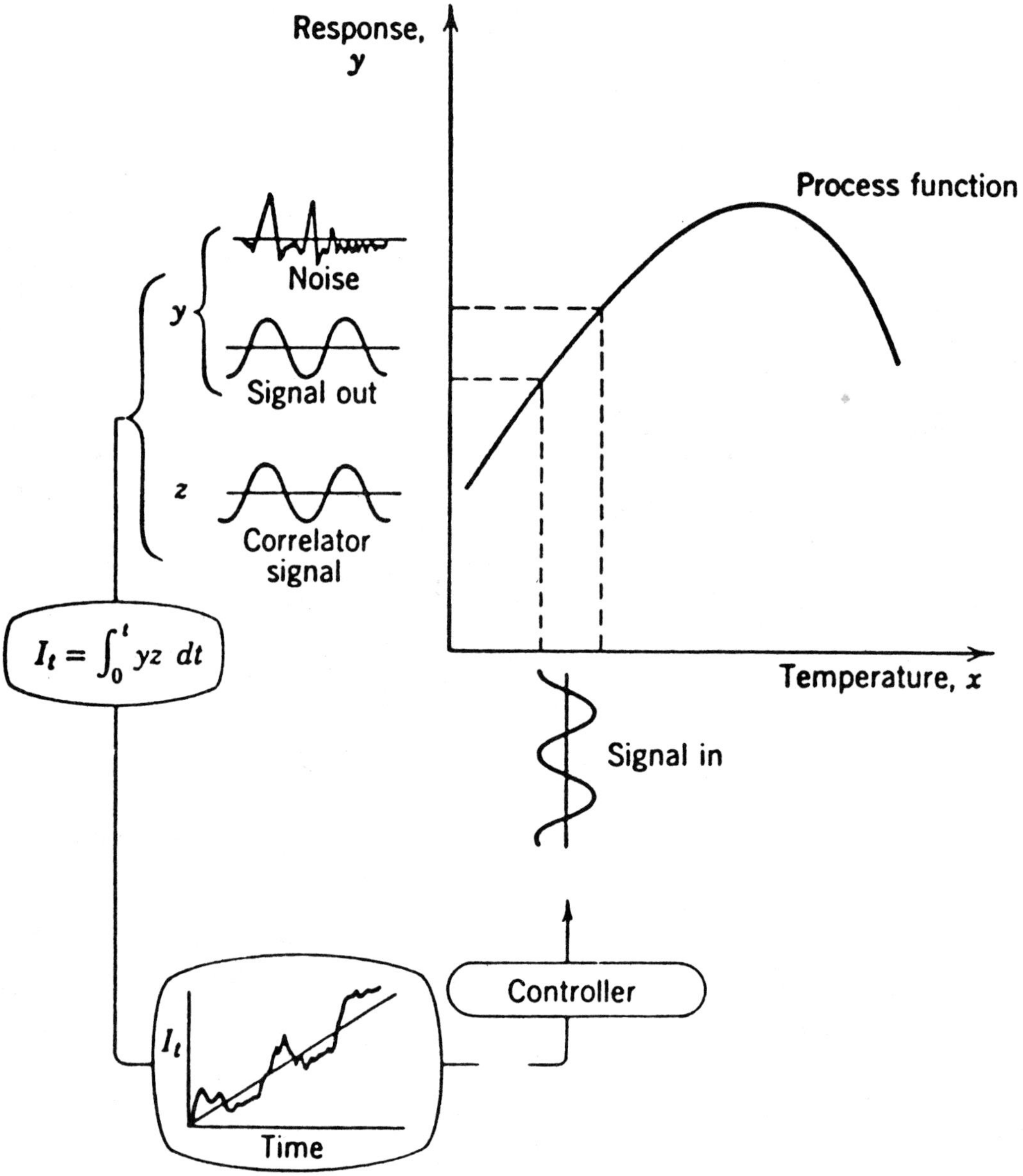

Figure 2. Automatic optimisation for a continuous chemical process.

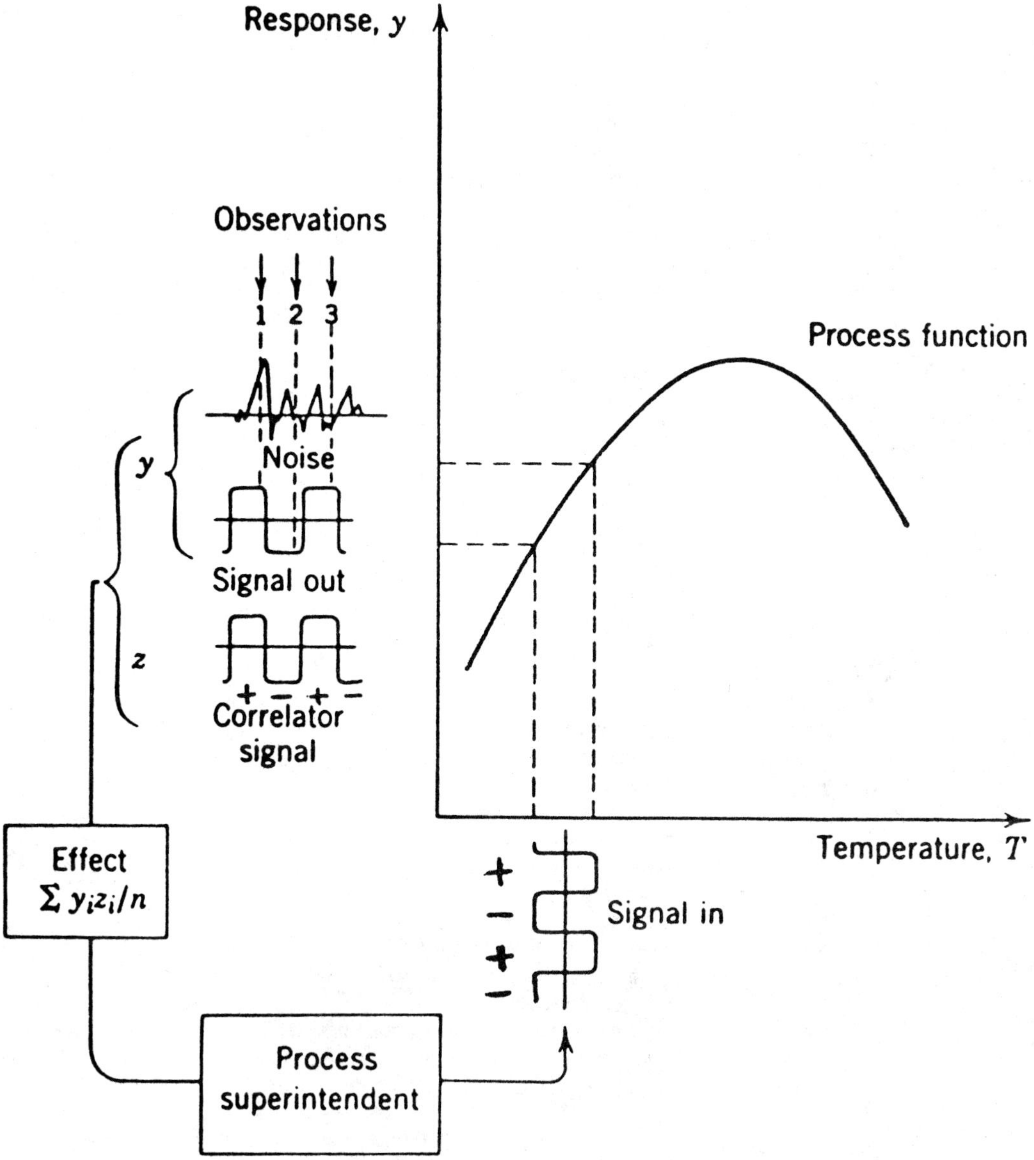

Figure 3. Evolutionary operation.

control problem of a very special kind (10). This meant that we started to work on more general problems of discrete control. In particular, we tried to better understand the relation of what we were doing to other kinds of statistical control.

A study that didn't appear in our book (11) but kept us intrigued for some time was nicknamed the "machine tool" problem. The idea was that in, say, a machining operation the natural tendency was for the measured characteristic not to vary about a fixed mean but to drift away from target in a manner which might be represented by a nonstationary process such as a (0,1,1) ARIMA process. We further supposed that the loss incurred by being a distance δ off target was $(a\delta^2)$ dollars. When this deviation became sufficiently serious, the machine could be stopped and the mean level readjusted, but this would cost C dollars. The problem was to design a strategy which minimized overall loss. We were a little surprised when the answer was equivalent to plotting deviations from target on a chart with two parallel action lines. This was like a Shewhart chart -- but with a totally different justification and a totally different basis for setting the "action lines." Another difference, but often in practice not an important one, was that it was a one-step-ahead forecast (an exponential average of past data) that was actually plotted. Such a chart using exponential averages had earlier been recommended on empirical grounds by Roberts (12). An example is shown in Figure 4.

Our early papers together emphasized the iterative and adaptive nature of experimentation itself and pointed parallels to evolutionary operation, adaptive optimization, and feedback control. But certain problems of feedback control can be thought of in terms of forecasting -- one can control by acting in such a manner that a forecasted discrepancy is canceled. So we became interested in forecasting also. Interwoven with all of this was the question of the type of models that were appropriate.

A general principle that Gwilym found very appealing was that if something worked and had withstood the test of time, then there must be a good reason for it. One entity that qualified in the area of control was the proportional plus integral controller, which in one form or another had been used successfully in industry for over a hundred years. For discrete data, such a device requires that compensatory action X_t at

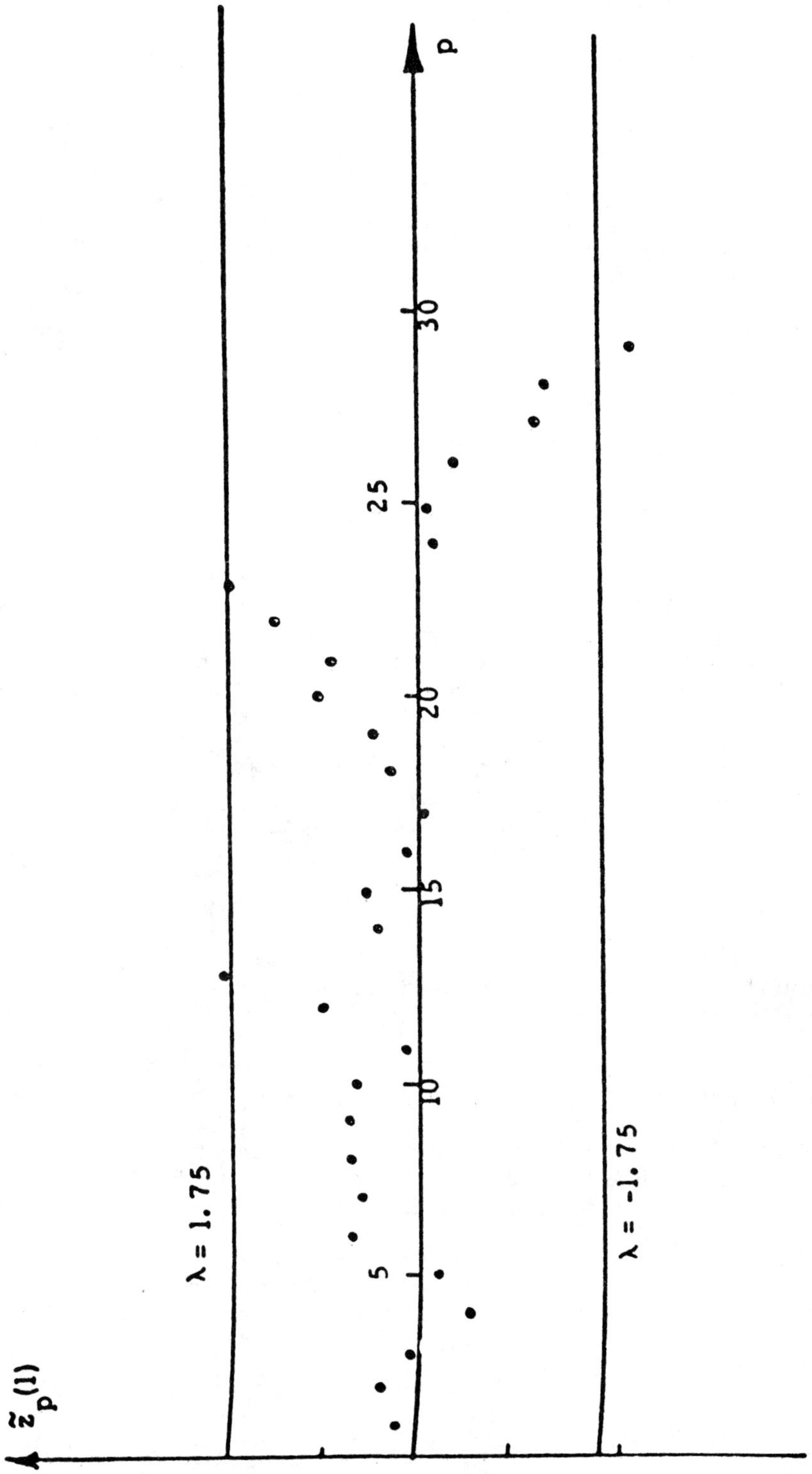

Figure 4. Predictions $\tilde{z}_p(1)$ of process deviations with action lines.

time t is of the form

$$(a) \quad X_t = k_1 e_t + k_2 \sum^{t} e_i$$

where e_i is the deviation from target of the controlled variable at time i, and k_1 and k_2 are constants.

Another such device that qualified in the area of forecasting a discrete series $\{z_t\}$ was the exponentially weighted moving average of past data

$$(b) \quad Z_{t+1} = (1 - \theta) \sum_{j=0}^{\infty} \theta^j z_{t-j}$$

which had been introduced more or less on empirical grounds by Holt (13), Winters (14), and others. For many business and economic series, this kind of forecast worked surprisingly well. At least it was capable of tracking nonstationary phenomena.

It can be shown that both equations (a) and (b) are optimal if the disturbance to be controlled in the first case and the series to be forecast in the second (15) are both members of nonstationary autoregressive-moving average (ARMA) processes whose autoregressive polynomial has one or more zeros on the unit circle. Some time later, it became clear that seasonal processes with period s could often be parsimoniously accommodated by employing models which included a backshift operator B^S which related items s intervals apart (16).

When models of this kind were appropriate, simple linear operations such as differencing could produce stationary autoregressive-moving average processes the properties of which had been extensively studied (17, 18, 19, 20, 21, 22, 23, 24, 25).

We soon became adherents of the view that whatever you wanted to do with a time series, whether it was to control it, forecast it, or (later) seasonally adjust it, you first needed to build a model for it using available data and some common sense. Once a satisfactory model had been built, then the right way of doing whatever you wanted to do would be made manifest. The model-building process was thus of central importance. While potential ingredients for model building were available, such as spectral analysis, autocorrelation analysis, various procedures for estimating parameters, and various tests of fit, we were not sure how, or if at all, these fitted together into a coherent system.

Gwilym's inclination was to try things and see if they worked, letting success or failure indicate in which direction we should go. I liked this way of working, too.

Modern time series analysis is possible because of the electronic computer. Gwilym had already realized this before the 1960s and worked very hard to have programs written that would enable us to experiment easily. Sam Weller said something to the effect that "nothing clarifies the thoughts of a man so much as the knowledge that he is going to be hung tomorrow morning." With this in mind, we set about attempting to model as many real time series as possible. By trial and error, this led to a three-stage system of model building that seemed to work and which we eventually adopted. This was of the form

$$\left\{\begin{matrix}\text{Identification}\\ \text{Specification}\end{matrix}\right\} \rightarrow \left\{\begin{matrix}\text{Estimation}\\ \text{Fitting}\end{matrix}\right\} \rightarrow \left\{\begin{matrix}\text{Diagnostic}\\ \text{Checking}\end{matrix}\right\}$$

The three ingredients were: <u>identification</u> -- getting an idea of what the general form of the model might be, using visual displays of the data and of minimally parametric identifiers, such as the autocorrelation and partial autocorrelation functions; <u>estimation</u> -- having got an idea of what kind of model might be worth trying, to act temporarily as if we believed it and to estimate its parameters using likelihood; <u>diagnostic checks</u> -- criticizing the fitted model by visual checks on residuals, their autocorrelations, etc., and by more formal checks of fit, leading in some cases to modification (reidentification) of the model.

We thought that it was particularly important <u>not</u> to try to make the model-building process automatic and entirely controlled by the computer, but to ensure that the human brain intervened and controlled, particularly at the identification and the diagnostic checking/model modification stages. Subsequent experience has, I believe, demonstrated the rightness of this idea.

We found the problem of model building philosophically puzzling because on the one hand you have a model which is very specific, and on the other you know that it must be false, since models are, at best, approximations. When, for example, you write down the likelihood, the model must necessarily be treated as true -- not nearly true, but exactly true. When you are checking fit, however, you are obviously acting as if you no longer believed in the model's necessary truth. To be a good model builder, it seemed essential to be a bit schizophrenic. One must be prepared to be a wholehearted sponsor for the model on one leg of the iteration and its wholehearted critic on the other. One of the ideas that later came, at least partially, out of this was the implied need for two different kinds of inference (34).

We came to think: (a) of the model-building process as the iterative building of a filter which transformed data to white noise which appeared to be independent of any known input; (b) that this could be accomplished by the mind and the computer appropriately combining their talents in an iteration involving identification, estimation, and diagnostic checking; (c) that probably this was the kind of procedure by which all statistical models ought to be built, whether time series or not.

Concerning estimation: At the time when we were writing our book, Gwilym favored the likelihood approach (26), although he was happy to indulge my wish to introduce a little Bayes as well. In fact, for samples of the size we usually encountered, the two approaches usually gave results that were essentially identical.

When you argued with Gwilym, you argued with someone who was constructive, friendly, sympathetic but firm, and I found our discussions highly educational. They were our chief way of "sorting things out," as Gwilym used to say. When we were together and Gwilym was well enough, we used to go on walks where we discussed things. I remember, for example, a problem we called the "golf course" problem because it was while walking on the golf course that we first thought of it. Another problem involved the "jam jar" model because we had an analog about a jam jar filling up with water. Discussion of a problem could be taken up at any time, and we would see if, by kicking it around between us a bit, some further progress could be made. During the long periods when we weren't together, we would send tapes to each other, usually wrapped in a piece of paper with the equations written on it. I saved some of the tapes, and one devastating thing I discovered was how idiotic I sounded as I listened to an old tape of myself discussing twelve months previously a problem which we now knew the answer to.

It was Gwilym's idea to write the book. I remember his saying to me something like: "Look, George, we can go on writing papers about this and that aspect, but we seem to have got ourselves involved in trying to sort out a philosophy of how to go about building and actually using time series models. We've got a lot of explaining to do -- so let's write a book."

Gwilym's work with me was, of course, but a small part of his contribution to statistics and still less of his overall scientific contribution. By 1964 he had been promoted to reader in statistics at Imperial College, but he was beginning to see that statistics, even when properly applied, could not of itself solve the problems he wanted to tackle. His consulting work concerned systems which, whether they were hospitals, government departments, or chemical plants, contained many interdependent subsystems, each of which had to be viewed in relation to the other. A neat statistical solution of one particular aspect of a problem might result in no benefit overall.

When he became professor of systems engineering at Lancaster in 1965, he was anxious to study further how statistical methods fitted into a wider system and also to train his students so that they should be able to go out into the world and do useful things. The master of science degree that he devised required not only knowledge of systems and of statistics but also the successful undertaking, under faculty supervision, of a major project in industry, government, or some other suitable field. This plan guaranteed close contact between the university and outside areas of application, wedding theory to practice, and tended to ensure the relevance and originality of the department's research.

I wish that more schemes of this sort could be instituted for training statisticians. This would certainly improve the students' knowledge of statistics. But it might have other desirable consequences. It might, indirectly and in time, improve curriculum and the teaching and competence of the professors. Perhaps paramount, in a few rare but important instances, it would inevitably lead to genuinely new and exciting research in statistics -- experience certainly shows that fresh theoretical innovation frequently originates from thoughtful practice.

Many of the cooperative projects at Lancaster were highly successful, and industrial and government clients became anxious to support further joint endeavor. It was to meet this need that in 1974 Gwilym Jenkins founded and became managing director of ISCOL, a consulting enterprise wholly owned by the University of Lancaster.

In these days, when the mutual benefit which can flow from interdependence between university and industry is better accepted, such an initiative might be welcomed unreservedly. In 1974, however, it generated political problems which finally proved unmanageable, and Jenkins left the university to form his own company and at the same time to take up a visiting professorial appointment at the London Business School.

The satisfaction of the absorbing work in which he was soon engaged and the international success of his company might have been enough for most men. But he saw his projects, whether they concerned coping with pollution of the Rhine, planning electricity generation, or forecasting employment for the European Economic Community, as important not only in themselves but also as case studies from which others could learn.

We are told that students and teachers of medicine in the Middle Ages did no dissections and had little practical knowledge of the functioning of the human body, but instead formulated their art entirely in terms of theory. Statistics is still sometimes taught in a similar way. Perhaps because of this, although hundreds of thousands of words are written each year in books and journals devoted to statistics, few of these concern the course of actual investigations in which statistics has been employed. For this reason, the two volumes of case studies which he gave us, Practical Experiences in Modelling and Forecasting Time Series (27) and Case Studies in Time Series Analysis (28), are gems of especial value for researcher, student, and teacher alike.

In 1969 Jenkins had founded, and was coordinating editor for, the Journal of Systems Engineering. Furthermore, in 1971, with his friend and longtime collaborator Philip Youle, he wrote an excellent book for the layman on systems engineering (29). These authors believed that systems methods were important not just to the chemical industry, the hospital, and the local bus company but, for example, to secretaries, housewives, and clergymen. The uninstructed might confuse systems analysis with regimentation. They showed how the ideas of systems analysis, far from restricting his options, could liberate the worker from the common frustrations and discouragements which arose from lack of a rational approach.

In recent years, the effectiveness of foreign industrial competition has all too clearly demonstrated the superiority of this kind of thinking over traditional European and American management practice, which seems to have adopted accounting rather than scientific method as its inspiration. In particular, the United States and the countries of Western Europe have found it increasingly difficult to match the quality and price of products from Japan. It is well known that an important reason for Japanese superiority is their application of statistical methods to the management of production.

The methods employed are not new and, indeed, were mostly invented in the West. What they have done is essentially to use the systems approach so that these methods are appropriately integrated and made to suffuse the whole structure of management. Experimental design, particularly using orthogonal arrays (30, 31, 32), is a statistical technique that is used extensively both in raising the mean level of quality and in reducing its variation.

The real problems of experimental design in quality improvement cannot be appreciated if we adopt a model that assumes that errors are random drawings from some distribution having fixed mean and variance. For if statistical design is to be used to move an unsatisfactory process to new operating conditions where it will be in a state of control, we should not assume that it is there already.

Fortunately, the basis of statistical design was laid by a very practical man -- R. A. Fisher (33), who worked in an industry (agriculture) where his experimental material was never in a state of control. His ideas using replication, randomization, and confounding to produce small blocks were fashioned for systems with disturbances that could be autocorrelated and even nonstationary. This is illustrated in Figure 5, which shows a nonrandom disturbance such as might be encountered in relation to a 2^3 design randomized in small blocks of four.

With this we come full circle to the link between time series models and experimental design which Gwilym and I found so intriguing.

Gwilym's life was an inspiring one. His boundless optimism and cheerfulness made possible his great achievements even in the face of pain and recurring disappointment. I count myself especially fortunate to have been his friend.

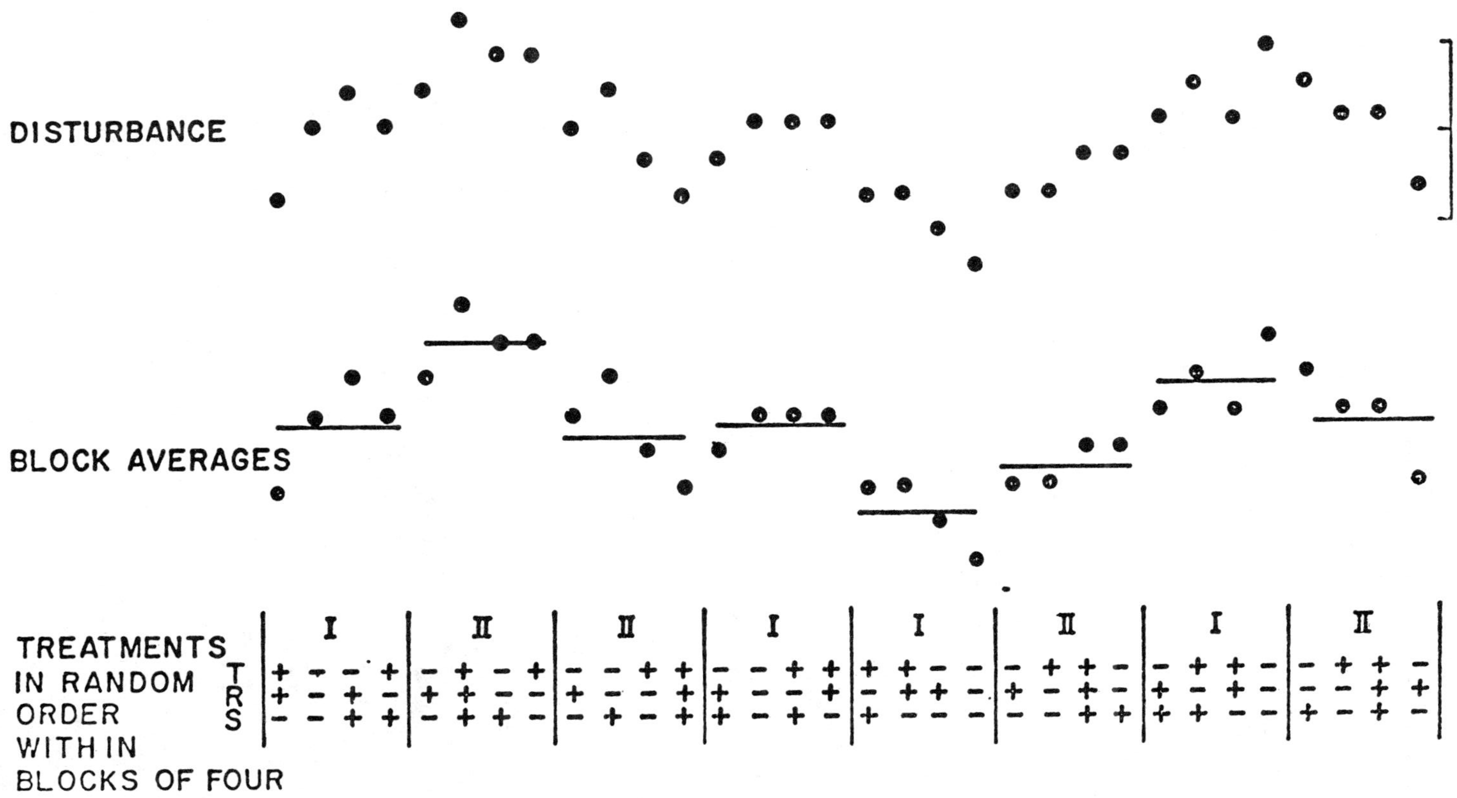

Figure 5. Autocorrelated disturbance. Averages for blocks of four. Replicated 2^3 design, randomised within blocks with TRS interaction confounded.

REFERENCES

(1) Jenkins, G. M., and D. G. Watts (1968). Spectral Analysis and Its Applications. Holden-Day, San Francisco.

(2) Draper, C. S., and Y. T. Li (1951). Principles of Optimizing Controls and an Application to the Internal Combustion Engine. ASME, New York.

(3) Tsien, H. S. (1954). Engineering Cybernetics, pp. 214-230. McGraw-Hill Book Company, Inc., New York.

(4) Box. G. E. P., and G. M. Jenkins (1962). Some statistical aspects of adaptive optimization and control. J. Roy. Stat. Soc., Ser. B 24: 297-343.

(5) Eveleigh, Virgil W. (1963). Adaptive control systems. In Electro-Technology, Science and Engineering Series, April 1963, pp. 79-98.

(6) Kotnour, K. D., G. E. P. Box, and R. J. Altpeter (1966). A discrete predictor-controller applied to sinusoidal perturbation adaptive optimization. Instr. Soc. Amer. Trans. 5: 225.

(7) Box, G. E. P., and G. M. Jenkins (1976). Time Series Analysis: Forecasting and Control. Holden-Day, San Francisco.

(8) Box, G. E. P., and N. R. Draper (1969). Evolutionary Operation. John Wiley & Sons, New York.

(9) Chanmugam, J., and G. M. Jenkins (1963). Optimum experimentation in the process industries. Proceedings Amer. Inst. Chem. Eng. Symposium ("Statistics and Numerical Methods in Chemical Engineering," Symposium series 42, volume 59, pp. 108-117).

(10) Box, G. E. P., and G. M. Jenkins (1962). Some statistical aspects of adaptive optimization and control. J. Roy. Stat. Soc., Ser. B 24: 297.

(11) Box, G. E. P., and G. M. Jenkins (1963). Further contributions to adaptive quality control: Simultaneous estimation of dynamics: Nonzero costs. Bull. Intl. Stat. Inst., 34th

Session, 943, Ottawa, Canada.

(12) Roberts, S. W. (1959). Control chart tests based on geometric moving averages. *Technometrics* 1: 239-250.

(13) Holt, C. C. (1957). Forecasting trends and seasonals by exponentially weighted moving averages. *O.N.R.Memorandum*, No. 52, Carnegie Institute of Technology.

(14) Winters, P. R. (1960). Forecasting sales by exponentially weighted moving averages. *Management Sci.* 6: 324.

(15) Muth, J. F. (1960). Optimal properties of exponentially weighted forecasts of time series with permanent and transitory components. *J. Amer. Stat. Assoc*. 55: 299.

(16) Box, G. E. P., G. M. Jenkins, and D. W. Bacon (1967). Models for forecasting seasonal and non-seasonal time series. In *Advanced Seminar on Spectral Analysis of Time Series*, ed. B. Harris, p. 271. John Wiley & Sons, New York.

(17) Yule, G. U. (1927). On a method of investigating periodicities in disturbed series, with special reference to Wolfer's sunspot numbers. *Phil. Trans*. A226, 267.

(18) Slutsky, E. (1937). The summation of random causes as the source of cyclic processes (Russian). *Problems of Economic Conditions* 3: 1; English trans. in *Econometrica* 5: 105.

(19) Bartlett, M. S. (1946). On the theoretical specification of sampling properties of autocorrelated time series. *J. Roy. Stat. Soc., Ser. B* 8: 27.

(20) Durbin, J. (1960). The fitting of time series models. *Rev. Int. Inst. Stat*. 28: 233.

(21) Jenkins, G. M. (1954, 1956). Tests of hypotheses in the linear autoregressive model. I: *Biometrika* 41: 405; II: *Biometrika* 43: 186.

(22) Wold, H. O. (1938). *A Study in the Analysis of Stationary Time Series*. Almquist and Wicksell,

Uppsala (2nd ed., 1954).

(23) Whittle, P. (1953). Estimation and information in stationary time series. Arkiv. fur Mathematik 2: 423.

(24) Hannan, E. J. (1960). Time Series Analysis. Methuen, London.

(25) Grenander, U., and M. Rosenblatt (1957). Statistical Analysis of Stationary Time Series. John Wiley & Sons, New York.

(26) Barnard, G. A., G. M. Jenkins, and C. B. Winsten (1962). Likelihood inference and time series. J. Roy. Stat. Soc. A125, 321.

(27) Jenkins, G. M. (1978). Practical Experiences with Modelling and Forecasting Time Series. Gwilym Jenkins and Partners, Lancaster, England.

(28) Jenkins, G. M., and Gordon McLeod (1983). Case Studies in Time Series Analysis. Gwilym Jenkins and Partners, Lancaster, England.

(29) Jenkins, G. M., and P. V. Youle (1971). Systems Engineering. Everyman's Library, London.

(30) Finney, D. J. (1945). Fractional replication of factorial arrangements. Ann. Eugen. 12: 291-301.

(31) Plackett, R. L., and J. P. Burman (1946). The design of optimum multifactorial experiments. Biometrika 33: 305-325.

(32) Rao, C. R. (1947). Factorial experiments derivable from combinatorial arrangements of arrays. J. Roy. Stat. Soc., Supplement 9: 128-139.

(33) Fisher, R. A. (1935). The Design of Experiments. Oliver and Boyd, Edinburgh.

(34) Box, G. E. P. (1980). Sampling and Bayes' inference in scientific modelling and robustness. J. Roy. Stat. Soc., Ser. A, 143: 383-430.

4

Distribution Theory, Transformation of Variables, and Nonlinear Estimation

Contents

4.0
Introduction

IRWIN GUTTMAN
University of Toronto

Distribution Theory

George Box's 1949 paper "A General Distribution Theory for a Class of Likelihood Criteria" (4.1), is the classic example of the "good theory-good practice" tenet held dear by Box, and it must have been a thrill, standing in 1949, for the profession to see this work.

The discussion of this paper centers around the following: Suppose that the moments of a statistic W used in testing some hypothesis (see below) are given by

$$E(W^h) = c \left\{ \frac{\prod_{j=1}^{k} (y_j^{y_j})}{\prod_{i=1}^{m} (x_i^{x_i})} \right\}^h \frac{\prod_{i=1}^{m} \{\Gamma[x_i(1+h)+\xi_i]\}}{\prod_{j=1}^{k} \{\Gamma[y_j(1+h)+\xi_j]\}} \tag{1}$$

where

$$\prod_{i=1}^{m} x_i = \prod_{j=1}^{k} y_j.$$

It turns out that for a number of hypothesis in multivariate analysis, the likelihood ratio or some power of it has moments of the form (1), examples of which are the following:

(i) equality of variances and covariances from k samples;
(ii) independence of k groups of variates;
(iii) equality of means, variances, and covariances from k samples.

The quantities x_i, y_j depend on the number of variates, number of samples, and number of degrees of freedom in each sample. Now suppose we let

$$M = -2 \log W \tag{2}$$

and let ρ be any constant not exceeding 1.

The characteristic function, $\Phi(t)$ say, of ρM is found using (1), and then the paper proceeds to consider the cumulant function

$$\Psi(t) = \log \Phi(t) , \tag{3}$$

with the aim of developing an asymptotic expansion for $\Psi(t)$ and hence, by taking antilogs, an asymptotic expansion for $\Phi(t)$. To do this, Box develops an asymptotic expansion for $\log \Gamma(x+h)$. Once the expansion for $\Phi(t)$ is found, the Fourier transform yields an expansion for the cumulative distribution of ρM in terms of χ^2-distributions of successively greater degrees of freedom. The constant ρ can be chosen so as to make the convergence of the expansion more rapid.

Other approximations for the distribution of M are then found, and it is shown that ignoring terms of order x_i^{-2}, y_j^{-2}, M is distributed as $C\chi^2$, for suitable C. Also, ignoring terms of order x_i^{-3}, y_j^{-3}, a function of M is distributed as the variance ratio.

With all these results, he then considers the tests for (i) and (ii) in great detail. In summary, "good theory" developed a usable form for the distributions needed for "good practice," for example, tests for (i) and (ii). A truly remarkable and important paper.

The next remarkable entries in the area of distribution theory are two companion papers (4.2, 4.3) on "Some theorems on quadratic forms applied in the study of analysis of variance problems, I and II," that appeared in 1954. Paper I deals with the effect of the inequality of variances in one-way analysis of variance, and paper II concerns itself with the two-way situation and discusses the effect of inequality of variances and the effect of correlation between errors for this case.

To accomplish this, Box, in paper I (4.2), first derives the exact distribution of a weighted sum of independent chi-squares with *even* degrees of freedom. Using this, the exact distribution of the ratio of independent sums of this type is given. These results are then used to determine the effect of variance-heterogeneity on the distribution of the usual F in the one-way classification. The largest deviations are found to occur when group sample sizes are unequal.

In paper II (4.3), the distributional results are used to determine the effects of inequality of variance and first order serial correlation of errors in the two-way analysis of variance. The results are interesting today, and they must have been very illuminating to the profession in 1954. For example, when the appropriate null hypothesis is true, inequality of variance from column to column results in an *increased* chance of exceeding the significance point for the test of equal column means, and a *decreased* chance for the corresponding test on row means. But for moderate differences in variances, *neither* effect is large. Also, first order serial correlation *within rows* produces a large effect on the "between rows" comparisons, but little effect on the "between columns" comparisons.

These two papers, then, are another example of "good theory–good practice" and are also examples of several papers of Box on effects of "lack of usual assumptions," a continuing interest to this day.

In fact, as recently as 1980, Box and Ljung (4.12) have examined the effect of autocorrelated errors in analysis of variance models, using a *Bayesian* approach. An autoregressive process of order p — $AR(p)$ — is the model for the errors, and the authors discuss the cases of a one-way classification, with $AR(p)$ structure for within groups, and the two-way clasification, with circular autoregressive processes within rows.

Transformation of Variables

The centerpiece of Box's work in this area is the landmark paper written jointly with D. R. Cox (4.9) entitled "An Analysis of Transformations," which appeared in 1964.

The main thrust of this paper follows. We are often in the situation where observations $y_1,\ldots,y_n$ are assumed to be *independently, normally* distributed with *constant variance*, say σ^2. An important example arises in the analysis of variance (e.g., in a two-way table) where the usual assumptions dictate that the observations have constant error variance over cells, have normal distributions, and are independent. (See the preceding section, Distribution Theory, for a discussion of the effect of lack of these assumptions on standard analyses.) More generally, regression analysis requires that the conditional distribution of the dependent variable is normal with constant variance and that the conditional expectation of the dependent variable can be expressed as a linear function of certain explanatory variables. When there is concern that these conditions are not satisfied by the dependent variable (say y) in its original metric, Box and Cox proceed by supposing that these conditions may be satisfied by some transformed metric (say $y^{(\lambda)}$) of the dependent variable. Here it is assumed that for each λ, $y^{(\lambda)}$ is a monotonic function of y.

More specifically, let $\underset{\sim}{y} = (y_1,\ldots,y_n)'$ be the vector of original observations and $\underset{\sim}{y}^{(\lambda)} = (y_1^{(\lambda)},\ldots, y_n^{(\lambda)})'$ the vector of transformed observations. Box and Cox take the bold step of assuming that for some unknown λ, the transformed observations $y_i^{(\lambda)}$, $i = 1,\ldots,n$, satisfy the full normal theory assumptions; that is, they are independent and normal, with constant variance σ^2 and with expectations given by

$$E(\underset{\sim}{y}^{(\lambda)}) = A\underset{\sim}{\theta} \tag{4}$$

with A a known matrix and $\underset{\sim}{\theta}$ a vector of unknown parameters associated with the *transformed* observations.

Hence, the likelihood in relation to the original observations $\underset{\sim}{y}$ is

$$\frac{1}{[2\Pi\sigma^2]^{n/2}} \exp\left\{-\frac{(\underset{\sim}{y}^{(\lambda)} - A\underset{\sim}{\theta})'(\underset{\sim}{y}^{(\lambda)} - A\underset{\sim}{\theta})}{2\sigma^2}\right\} J(\lambda;\underset{\sim}{y}) \tag{5}$$

where

$$J(\lambda;\underset{\sim}{y}) = \prod_{i=1}^{n} \left|\frac{dy_i^{(\lambda)}}{dy_i}\right|$$

is the Jacobian of the transformation. These authors then consider the family of transformations

$$y^{(\lambda)} = \begin{cases} (y^{\lambda} - 1)/\lambda & (\lambda \neq 0) \\ \log y & \lambda = 0 \end{cases} , \tag{6}$$

which is continuous in λ and continuous at $\lambda = 0$. A slightly more general form of (6) that is also considered is

$$y^{(\underset{\sim}{\lambda})} = \begin{cases} [(\underset{\sim}{y} + \lambda_2)^{\lambda_1} - 1]/\lambda & \lambda_1 \neq 0 \\ \log(y_2 + \lambda_2) & \lambda_1 = 0 \end{cases} . \tag{7}$$

Assuming either (6) or (7) in (5), the discussion proceeds in two directions: (i) maximum-likelihood theory is applied to the likelihood of (5); (ii) assuming a non-informative prior for $\underset{\sim}{\theta}$ and σ^2, a Bayesian approach based on the use of (5) is discussed. (The paper ends with two illuminating examples involving a 3×4 factorial design [with replication] and a single replicate of a 3^3 design.) Both these methods result in point and interval estimates of λ (or λ_1, λ_2) and of $\underset{\sim}{\theta}$, and so on. The main thrust is to estimate the appropriate λ (or $\underset{\sim}{\lambda} = (\lambda_1, \lambda_2)'$) and also the accompanying $\underset{\sim}{\theta}$ and σ^2 by either likelihood or Bayesian methods.

The emphasis of this paper, then, is on transformation of the dependent variable, but some comments are given on transformations of the independent variables that would arise in a natural way when transforming y. Box and Cox's idea is very natural and has served as the basis of many subsequent papers. For example, Draper and Guttman (1968) have used their technique in transforming life test data so that the usual assumptions made in that field would hold for the transformed data, and so on. Finally, the Box-Cox paper concludes with an illuminating discussion of the connection between the Box and Cox methods and the techniques of Anscombe and Tukey (1961) based on analysis of residuals.

This paper has not been without its controversies. A recent exchange is of interest, and we refer the reader to Bickel and Doksum (1981) and a reply to this paper by Box and Cox (4.13).

Also, in the area of transformation, the paper by Box and Tidwell (4.8) that appeared in 1962 should be mentioned. As the title "Transformation of the Independent Variables" suggests, the discussion in this paper is devoted to the case where it is "possible to work with a simple functional form in the original variables," and this paper "describes and illustrates a procedure to estimate appropriate transformations (of the independent variable[s]) in this context."

The reader should note, interestingly, that this paper appeared two years before the Box-Cox paper of 1964, but it did give a preview in that, similar to (6), the family of transformations that plays an important role in this paper is

$$\xi = x^{\alpha} \tag{8}$$

where x is an independent variable and the model envisaged is

$$E(y) = f(\underset{\sim}{x}, \underset{\sim}{\theta}) \; . \tag{9}$$

A prominent part of this paper is given over to the help and use of orthogonalization in the resulting "X" matrix found after transformation of the type (8), and after Taylor series expansions are made. The paper concludes with a description of a "lack of fit test" that helps answer the question of whether the transformation(s) on the independent variable(s) is(are) adequate.

Finally, for this section on transformations, mention should be made of Box's paper written jointly with William J. Hill (4.11), entitled "Correcting Inhomogeneity of Variance with Power Transformation Weighting." The main ingredient in this paper is the methodology of the Box-Cox paper mentioned above, in particular, transformation of the type given by (6).

In brief, the problem discussed is the following: Suppose

$$y_u = \eta_u + e_u \; , \; u = 1,\ldots,n \tag{10}$$

where

$$V(e_u) = C_u \sigma^2, \; \mathrm{cov}(e_u, e_u') = 0 \; , \; u \neq u' .$$

Suppose also that for some λ, a transformation (6) is performed. Then it can be shown that

$$V(\sqrt{w_u}\, y_u) \simeq \sigma^2 \tag{11}$$

if

$$w_u \simeq \eta_u^{2\lambda-2}. \tag{11a}$$

The use of w_u's ($u = 1,\ldots,n$) as weights in (an iterative) weighted least squares is proposed. Initial estimates can be obtained, if λ is known, by using

$$w_u \simeq \hat{y}_u^{2\lambda-2} \tag{12}$$

where $\hat{y}_{\hat{u}}$ is the estimated value of η_u in a least squares fit. The problem is, however, that λ is not known. Using the techniques of the Box-Cox paper, Box and Hill outline a Bayesian procedure for obtaining the posterior of λ, and the point of its maximum, say $\hat{\lambda}$. The procedure is illustrated by an interesting and informative example.

Nonlinear Estimation

One of George Box's most important contributions to statistics is his early recognition of the importance of the high-speed computer, which is exemplified in his fundamental and compelling work in the area of nonlinear estimation. In the series of papers (4.4, 4.5, 4.6, 4.7) that started, incidentally, in the year 1956, Box elucidated the problem of nonlinear estimation and the importance of the computer to help with the analysis of any such problem, and he also commented on the problem of design of experiments for this situation. Problems of this sort invariable touch on other aspects, such as transformations of the data, determination of optimum conditions, and so on.

The discussion in the aforementioned papers centers around the following:
Suppose there is interest in exploring a functional relationship

$$\eta = f(\underset{\sim}{\xi},\underset{\sim}{\theta}) \tag{13}$$

connecting a response η (e.g., a yield of a product) with a number of variables $(\xi_1,\ldots,\xi_k)' = \underset{\sim}{\xi}$ (e.g., ξ_1 = time, ξ_2 = temperature, ξ_3 = pressure, etc.) where $\underset{\sim}{\theta} = (\theta_1,\ldots,\theta_p)'$ are unknown parameters. Now, as discussed by Box and his coauthors, two commonly met situations are:

(i) No knowledge of the form of the function $f(\underset{\sim}{\xi},\underset{\sim}{\theta})$ is available, but it is assumed to possess a unique maximum and to be "smooth." One object of experimentation in this case may be the determination based on experimental data of the optimal region of the function f.

(ii) The form of $f(\underset{\sim}{\theta},\underset{\sim}{\xi})$ is not explicitly known, but from theoretical considerations it would be expected that a certain set of simultaneous differential equations would have $f(\underset{\sim}{\theta},\underset{\sim}{\xi})$ as their solution. Again, on the basis of experimental data, an object of experimentation may be to estimate the constants $\underset{\sim}{\theta}$ and their experimental errors, and to ascertain whether the assumed system of differential equations is adequate to explain the data; that is, is the tentatively assumed model f adequate?

For case (i), Box's work centers on the assumption that over a limited region of interest, say R, in the space of variables $\underset{\sim}{\xi}$, the response function f is capable of being adequately represented by some simple, smooth graduating function, such as a polynomial of low degree. This, of course, means that we are dealing with a linear model, and linear least squares can then be brought to bear in estimating the coefficients of the polynomial model and arriving at a fitted polynomial, that is, an estimated graduating function. The problem of how to use this to determine the optimal region of the function f is discussed in this volume in Part 2, Experimental Design and Response Surface Methodology; the question of how to design the experiment so that a so-called lack of fit test may examine whether the polynomial is an adequate graduating function is also discussed there.

For case (ii), Box and his coworkers envisage the general situation where s known differential equations are given, say,

$$\frac{d\eta_u}{dt} = \phi_u(\eta_1, \ldots, \eta_s ; \xi_1, \ldots, \xi_k ; \theta_1, \ldots, \theta_p) \quad u = 1, \ldots, s \tag{14}$$

where the $_u$ are known functions and the ϕ_u may be observed. An example of such a system (see 4.5 for discussion) is

$$\frac{d\eta_1}{d\xi} = \theta_1\eta_1 ,$$
$$\frac{d\eta_2}{d\xi} = \theta_1\eta_1 - \theta_2\eta_2 \tag{15}$$
$$\frac{d\eta_3}{d\xi} = \theta_2\eta_2$$

with boundary conditions $\eta_1 = 1, \eta_2 = 0, \eta_3 = 0$ at time $\xi = 0$. This is an example where an explicit solution is available. for example, η_2 at time ξ is given by

$$\eta_2 = \frac{\theta_1}{\theta_1 - \theta_2}(e^{-\theta_2\xi} - e^{-\theta_1\xi}) . \tag{16}$$

Here $\underset{\sim}{\theta}' = (\theta_1, \theta_2)$ and ξ is a scaler (time). Suppose we observe η_2 at times ξ_u, $u = 1, \ldots, N$, and let y_u be the observed value of η_2 at ξ_u. We are assuming $E(y_u) = n_2$; then Box et al. consider the use of

$$S(\underset{\sim}{\theta}) = \sum_{u=1}^{N} [y_u - f(\xi_u ; \underset{\sim}{\theta})]^2 \tag{17}$$

where f is given in (16). A method (which is a combination of least squares and steepest descent) is now described in papers 4.4, 4.5, 4.6, and 4.7, and asks for the place in $\underset{\sim}{\theta}$-space at which (17) is minimized. Specifically, beginning with an estimated $\underset{\sim}{\theta}$, say $\underset{\sim}{\theta}^{(0)}$, we use a first order Taylor series to find

$$f(\xi_u ; \underset{\sim}{\theta}^*) \simeq f(\xi_u , \underset{\sim}{\theta}^{(0)}) + \sum_{r=1}^{2} (\theta_r^{(0)} - \theta_r^*)\xi_u \tag{18}$$

where $\underset{\sim}{\theta}^{(0)} = (\theta_1^{(0)}, \theta_2^{(0)})'$, and $\underset{\sim}{\theta}^*$ is the true value of $\underset{\sim}{\theta}$. Let $z_u^{(0)} = y_u - f(\xi_u ; \underset{\sim}{\theta}^{(0)})$ so that

$$E(z^{(0)}) \simeq \sum_{r=1}^{2} (\theta_r^{(0)} - \theta_r^*)\xi_{\hat{u}}. \tag{19}$$

The above equation is, of course, a linear model; using the observed $z_u^{(0)}$s, we find, using linear least squares, an estimate of $\theta_r^{(0)} - \theta_r^*$, say $\Delta\theta_r^{(0)}$. We then use as our improved estimate

$$\underset{\sim}{\theta}^{(1)} = \underset{\sim}{\theta}^{(0)} + \Delta\underset{\sim}{\theta}^{(0)} \tag{20}$$

and calculate $S(\underset{\sim}{\theta}^{(1)})$, recording its value. We continue iterating in this way and, provided f is well behaved and $\underset{\sim}{\theta}^{(0)}$ is not too far away from the final value, the adjusted values will converge to the least squares estimate $\hat{\underset{\sim}{\theta}}$. Under normal theory, we have if

$$y_u = f(\xi_u ; \underset{\sim}{\theta}) + \epsilon_u \tag{21}$$

where ϵ_us are IID $N(0,\sigma^2)$s, then $\hat{\underset{\sim}{\theta}}$ is also the maximum likelihood estimate.

The generalization to the case, using usual notation, of $f(\underset{\sim}{\xi},\underset{\sim}{\theta})$, where $\underset{\sim}{\xi}$ is (kxl) and $\underset{\sim}{\theta}$ is (pxl), is immediate. In fact, interestingly, under Box's direction, the above process for the general case was programmed circa 1955 in England at the National Physical Laboratory (Teddington), and at Princeton University on its computer called the MANIAC and also for the IBM 704 machine by Doctors Booth and Peterson of IBM, in cooperation with the Statistical Techniques and Research Group at Princeton, George Box, Director.

The case where explicit fs are not obtainable from (14) may also be similarly treated as discussed in the previously cited papers, and may be handled by the program(s) mentioned above. To my knowledge, these programs are still in use and have been adapted for use on recently appearing computers.

The problem of confidence regions for $\underset{\sim}{\theta}$ obtained by the above process and the question of choice of design are also touched upon by Box et al. in 4.4, 4.5, 4.6, and 4.7. The question of model building is dealt with in 4.7, but it is also discussed in the other three papers. Other related readings are the papers 4.10 and 4.11 — the latter is mentioned in the previous section on transformations.

References Other Than Those of G. E. P. Box

Anscombe, F. J., and Tukey, J. W. 1961. The examination and analysis of residuals. *Technometrics* 5:141–160.

Bickel, P. J., and Doksum, K. A. 1981. An analysis of transformations revisited. *J. Amer. Stat. Assoc.* 76:296–311.

Draper, N. R., and Guttman, Irwin. 1968. Transformations on life test data. *Canadian Mathematical Bulletin* 11:475–478.

4.1
A General Distribution Theory for a Class of Likelihood Criteria

G. E. P. BOX
Imperial Chemical Industries Ltd.

1. Introduction

The likelihood ratio method of Neyman & Pearson (1928) has been used by many different workers for the derivation of criteria appropriate for the testing of a large variety of hypotheses. Plackett (1946), in a recent survey of literature on testing the equality of variances and covariances, lists, on this problem alone, criteria for the testing of no less than thirty-one hypotheses investigated at different times by workers in this field. Most of the criteria either have been or can be arrived at by the likelihood ratio method. In the preface to his survey Plackett says: 'Generally speaking the difficulties in testing such hypotheses lie not so much in deriving criteria—but in finding their exact distributions when the hypotheses are true and determining the best critical region to adopt.'

Although in many cases the exact distributions cannot be obtained in a form which is of practical use, it is usually possible to obtain the moments, and these may be used to obtain approximations. In some cases, for instance, a suitable power of the likelihood statistic has been found to be distributed approximately in the type I form, and good approximations have been obtained by equating the moments of the likelihood statistic to this curve. For example, in the original paper on the L_1 test for homogeneity of variances, Neyman & Pearson (1931) suggested that the distribution could be approximately represented in this way, and later Bishop & Nair (1939) showed that the significance points obtained by Nayer (1936), using this method, were in excellent agreement with the true values. The fitting of the type I curve is simple once the moments are obtained, but these moments, being the products of Γ-functions, are usually rather troublesome to calculate. To overcome this difficulty, Bishop (1939), working on the distribution of the multivariate equivalent of the L_1 test (the test for constancy of variances and covariances in k p-variate samples), derived empirical expressions for the parameters of the appropriate type I curve, thus avoiding the troublesome intermediate step of calculating moments. Bishop mentions that Nair succeeded in finding similar expressions on a theoretical basis, and Tukey & Wilks (1946) give a more general theoretical method to find approximations of this kind.

A different line of approach was adopted by Bartlett (1937). Neyman & Pearson had pointed out in their original paper that, if N' is the total sample size, $-N' \log_e L_1$ would be asymptotically distributed as χ^2. From considerations of sufficiency Bartlett obtained what was in effect a modified form (which, following Hartley & Pearson (1946), we shall call M) of this logarithmic statistic. From the moments of the modified likelihood statistic he was able to develop a scale factor C, which was related to the effective sizes of samples and which approached the value unity as the sample sizes became large. The distribution of M/C was then very well represented by χ^2 even when the samples were small. Bartlett later (1938) used the same method to obtain an approximation for the test of significance in multivariate analysis. In 1940 Hartley, starting from the moments of the modified likelihood statistic,

The Collected Works of George E. P. Box, 1984, Wadsworth, Inc., Belmont, CA 94002.
Originally published in *Biometrika,* vol. XXXVI, parts III and IV (1949), pp. 317–346.

obtained an asymptotic series of χ^2 integrals for the logarithmic statistic M which agreed very closely with the exact distribution. In 1941 Wald & Brookner, investigating an entirely different problem, the distribution of Wilks's statistic for testing independence of k sets of variates, again starting from the moments of the likelihood statistic λ, eventually obtained an expression for the distribution of a logarithmic statistic (a negative multiple of $\log_e \lambda$) in the form of an asymptotic χ^2 series. This was later modified by Rao (1948) in the important special case of two groups, when it corresponds to the test of significance in multivariate analysis previously referred to. Neither Wald & Brookner nor Rao investigated the accuracy of these series.

It is possible therefore to distinguish two definite lines of approach, which have been used in certain cases where the moments of the likelihood criteria are known but the exact distributions are not. On the one hand the moments have been used to fit the Pearson-type curve. This usually gives an adequate approximation, but owing to the amount of labour involved in the calculation of the moments it would not be attractive for routine significance testing unless methods such as Bishop's could be used to obtain the parameters of the fitted curve directly, or the results from the method could be tabled. On the other hand, the general expression for the moments of the likelihood statistic has been used in certain cases to obtain for the distribution of the logarithmic statistic M, a χ^2 approximation and an asymptotic χ^2 series. It will be the object of the present paper to investigate in some detail this second line of attack.

The method will be investigated in particular for two general criteria:

(1) The test of constancy of variance and covariance of k sets of p-variate samples. This includes, as an important special case when $p = 1$, the test for constancy of variance in k samples.

(2) Wilks's test for the independence of k sets of residuals, the lth set having p_l variates. When $k = 2$ this corresponds to the test of significance used in multivariate regression and analysis of variance and covariance, and when $k = 2$ and p_1 or p_2 is unity, it gives the corresponding well-known univariate tests. In the latter case, of course, the exact distributions are known.

We shall refer to these two criteria as generalized tests for homoscedasticity and independence, respectively. The assumption of normality or multinormality for the distributions of the original observations will be made throughout this paper.

Taking for our test function M, a negative multiple of the natural logarithm of the likelihood statistic (or some modification of it), we shall obtain in each case,

(*a*) a series solution which, we shall demonstrate, agrees very closely with the exact distribution,

(*b*) an approximate solution using a single χ^2 distribution,

(*c*) a rather better approximation using a single F distribution.

The accuracy of the various methods and the relation of the results to those of other workers will be discussed.

2. The generalized test of homoscedasticity

The univariate statistic. The L_1 statistic of Neyman & Pearson for testing the homogeneity of a set of variances, takes the form of the ratio of a weighted geometric mean of variances to a weighted arithmetic mean, where the weights are the sample numbers. Welch (1935,

1936) generalized the test to cover the case when residuals from a fitted regression equation were tested for homoscedasticity, and derived the moments for a modified criterion in which the weights could have any values whatever. In 1936 Nayer tabled the approximate significance points for L_1 in the cases of equal sample numbers by fitting type I curves to the distributions by the method of moments, as suggested in the original memoir by Neyman & Pearson.

The statistic proposed by Bartlett (1937) which we shall call M is given by

$$M = N \log_e s - \sum_l \nu_l \log_e s_l,$$

where

$$s = (\Sigma \nu_l s_l)/N,$$

and s_l is the usual unbiased estimate of the variance in the lth group, $l = 1, 2, \ldots, k$, based on sums of squares having ν_l degrees of freedom, and $N = \Sigma \nu_l$. It was later shown (Brown, 1939; Pitman, 1939; Bishop & Nair, 1939) that this criterion is unbiased in the sense used by Neyman & Pearson (1936, 1938). Nair (1939) derived a series solution for the distribution of the likelihood statistic in the case of equal sample numbers; his solution is very involved, but has been used as a standard to check approximations. Bishop & Nair (1939) used this series to check the accuracy of the type I approximation used in Nayer's table. They also checked the Bartlett (1937) approximation and found that both methods were fairly good except when the degrees of freedom were small. In the case of unequal samples, however, Nayer's tables were not available, and in view of the labour involved in the type I fit, the χ^2 method of Bartlett's was preferred.

Hartley's (1940) asymptotic series depended to the degree of approximation used, on two parameters c_1 and c_3 which varied with the effective sample size and relative composition of the groups

$$c_1 = \Sigma \frac{1}{\nu_l} - \frac{1}{N}, \quad c_3 = \Sigma \frac{1}{\nu_l^3} - \frac{1}{N^3}.$$

The first is related to Bartlett's scale factor C, in fact

$$C = 1 + \frac{c_1}{3(k-1)}.$$

Tables were afterwards computed by Thompson & Merrington (1946) from Hartley's formula, and comparisons were made with the values calculated by Bishop & Nair.

The multivariate statistic. In the multivariate case Wilks (1932) derived the likelihood ratio test and obtained the moments of the criterion, which is a generalized form of that used in the univariate test, the determinants of estimated variances and covariances replacing the variances. Bishop (1939) took as his criterion l_1, the $1/N'$th power of the likelihood statistic, N' being the total number of observations. He gave reasons for believing that this criterion could be approximately represented by a type I curve

$$p(l_1) = \text{constant}\, l_1^{m_1-1}(1-l_1)^{m_2-1}, \tag{1}$$

by choosing the value of m_1 and m_2 so that the first two moments of the Pearson curve agreed with those of the criterion. His arguments were supported by the agreement found in a number of trials between the higher moments of the fitted type I curve and those of the criterion. Only in the case of two groups and either one or two variates was it possible to obtain a check against the exact distribution, but in these cases the agreement was very good. Unfortunately the labour involved in the calculation of the first two moments of the criterion was too great to allow this method to be recommended for routine use. Bishop therefore proceeded as follows:

(*a*) For the case of equal sample sizes he obtained, empirically, expressions for m_1 and m_2 in terms of the number of observations n in each group, the number of variates p, and the number of groups k

$$\left.\begin{aligned} m_1 &= k(n-p)-0\cdot 01(k-1)(90-39p+9p^2),\\ m_2 &= 0\cdot 25(k-1)p(p+1). \end{aligned}\right\} \tag{2}$$

(*b*) For unequal sample sizes he proposed approximating to $-2N'\log_e l_1$ by means of a χ^2 distribution using a scale factor G in a similar way to that adopted by Bartlett in the univariate case.

He showed that $-2N'\log_e l_1$ is approximately distributed as $G\chi^2$, where

$$G = 1+\frac{1}{f}\Bigg[\sum_{l=1}^{k}\sum_{i=1}^{p}\{i^2/2n_l+i/(n_l-i)+n_l/3(n_l-i)^2\} \\ -\sum_{i=1}^{p}\{(k+i-1)^2/2N'+(k+i-1)/(N'-k+1-i)+N'/3(N'-k+1-i)^2\}\Bigg]. \tag{3}$$

n_l is the number of observations in the lth group, $\sum_l n_l = N'$ and χ^2 is distributed with $f = \frac{1}{2}(k-1)p(p+1)$ degrees of freedom.

We shall refer to these methods as Bishop's methods (*a*) and (*b*). Bishop remarks that the scale factor G is rather troublesome to calculate unless $n_l = n$. George (1945) was able to evaluate the exact distribution in a number of simple cases. She used her results to test the accuracy of Bishop's approximations and found, in the cases she considered, that (*b*) was superior to (*a*).

Plackett (1947) suggested that in view of the unsatisfactory position with regard to the distribution of this criterion that it might be better to abandon it in favour of an alternative test derived by him which had the advantage that at least when $p = 1$ or 2, and for certain other special cases, the exact distribution was known. Plackett's test, however, has the disadvantages that the results depend on the particular arrangement of observations chosen, and that the samples must be of equal sizes.

2·1. *The present approach*

Suppose s_{ijl} is the usual unbiased estimate of the variance or covariance A^{ijl} between the ith and jth variable in the lth sample based on sums of squares and products having ν_l degrees of freedom, and suppose there are k such samples and s_{ij} is the average variance or covariance $\left(\sum_l \nu_l s_{ijl}\right)\Big/N$, where $N = \sum_l \nu_l$. We take as our criterion a generalized form of Bartlett's criterion

$$M = N\log_e|s_{ij}| - \sum_l(\nu_l\log_e|s_{ijl}|) \tag{4}$$

$$= -N\log_e L_1', \quad \text{where} \quad L_1' = \prod_{l=1}^{k}\left\{\frac{|s_{ijl}|}{|s_{ij}|}\right\}^{\nu_l/N}. \tag{5}$$

When the degrees of freedom are equal, M and L_1' are related with Bishop's l_1 as follows:

$$M = -2N\log_e l_1 \quad \text{and} \quad L_1' = l_1^2. \tag{6}$$

When the degrees of freedom are unequal, L_1' will differ from the likelihood statistic in weighting. When $p = 1$, M is the criterion derived by Bartlett and later used by Hartley.

We proceed to obtain the moments of L_1' when the null hypothesis is true. If c_{ijl} are the sums of squares and products based on ν_l degrees of freedom corresponding to the s_{ijl}, we have $c_{ijl} = \nu_l s_{ijl}$, $c_{ij} = Ns_{ij}$, so that $c_{ij} = \sum_l c_{ijl}$. The joint probability density of the c_{ijl} for the lth sample is given by the distribution discovered by Wishart (1928):

$$p(c_{11l}, c_{12l}, \ldots, c_{ppl}) = K(\nu_l)\,|\,c_{ijl}\,|^{\frac{1}{2}(\nu_l-p-1)} \exp\{-\tfrac{1}{2}\sum_{ij} A_{ijl}c_{ijl}\}, \tag{7}$$

where
$$\{K(\nu_l)\}^{-1} = 2^{\frac{1}{2}(\nu_l p)}\, \pi^{\frac{1}{4}p(p-1)} \prod_{j=0}^{p-1} \Gamma\left(\frac{\nu_l-j}{2}\right) |\,A_l\,|^{-\frac{1}{2}\nu_l}, \tag{8}$$

and A_l is the matrix of the A_{ijl}, the inverse of the matrix of the A^{ijl}.

When the null hypothesis is true, A_l is the same for each of the samples, $A_l = A$, $l = 1, 2, \ldots, k$, and the gth moment of $|\,c_{ij}\,|$ is

$$\int \prod_{l=1}^{k} \left\{K(\nu_l)\,|\,c_{ijl}\,|^{\frac{1}{2}(\nu_l-p-1)}\,|\,c_{ij}\,|^{g/k} \exp\left(-\tfrac{1}{2}\sum_{ij} A_{ij}c_{ijl}\right)\right\} dc_{111}\,dc_{121} \ldots dc_{ppk}; \tag{9}$$

it is also given by

$$\int K(N)\,|\,c_{ij}\,|^{\frac{1}{2}(N-p-1)}\,|\,c_{ij}\,|^{g} \exp\left\{-\tfrac{1}{2}\sum_{ij} A_{ij}c_{ij}\right\} dc_{11}\,dc_{22} \ldots dc_{pp}. \tag{10}$$

Writing $\nu_l(1+2h)$ for ν_l on both sides of the identity and then taking $g = -Nh$ and integrating over the whole space for which the matrices of the c_{ij}, c_{ijl} are positive definite, we have

$$\mathscr{E}\left[\prod_{l=1}^{k}\left\{\frac{|\,c_{ijl}\,|}{|\,c_{ij}\,|}\right\}^{\nu_l}\right]^h \prod_{l=1}^{k}\left[\frac{K\{\nu_l(1+2h)\}}{K(\nu_l)}\right] = \frac{K\{N(1+2h)\}}{K(N)}. \tag{11}$$

That is
$$\mathscr{E}(L_1')^{Nh} = \mathscr{E}(e^{-M})^h = \frac{K\{N(1+2h)\}}{K(N)} \prod_{l=1}^{k}\left[\frac{K(\nu_l)}{K\{\nu_l(1+2h)\}}\left(\frac{N}{\nu_l}\right)^{p\nu_l h}\right] \tag{12}$$

$$= \prod_{l=1}^{k}\left(\frac{N}{\nu_l}\right)^{h\nu_l p} \prod_{j=0}^{p-1}\left[\frac{\Gamma\{\frac{1}{2}(N-j)\}}{\Gamma\{\frac{1}{2}(N(1+2h)-j)\}}\prod_{l=1}^{k}\frac{\Gamma\{\frac{1}{2}(\nu_l(1+2h)-j)\}}{\Gamma\{\frac{1}{2}(\nu_l-j)\}}\right]. \tag{13}$$

We have first proved equation (13) as an analytic identity for real h; it will, however, be generally valid in the range where the functions are analytic. We can thus obtain an expression for the characteristic function of ρM, where ρ is a constant $\leqslant 1$ at our choice, by replacing h by $-it\rho$ in the above expression. The reason for introducing the constant ρ will appear later. Further, if we write $N = \nu k$ (i.e. ν is the average of the degrees of freedom) and define new quantities μ, μ_l, β, β_l by the relations

$$\mu_l = \rho\nu_l, \quad \mu = \rho\nu, \quad \nu = \mu+\beta, \quad \nu_l = \mu_l+\beta_l, \tag{14}$$

we obtain the characteristic function of ρM in the form

$$\Phi(t) = \prod_{l=1}^{k}\left(\frac{k\mu}{\mu_l}\right)^{-it\rho\mu_l p} \prod_{j=0}^{p-1}\left[\frac{\Gamma[\frac{1}{2}\{k\mu+k\beta-j\}]}{\Gamma[\frac{1}{2}\{k\mu(1-2it)+k\beta-j\}]}\prod_{l=1}^{k}\frac{\Gamma[\frac{1}{2}\{\mu_l(1-2it)+\beta_l-j\}]}{\Gamma[\frac{1}{2}(\mu_l+\beta_l-j)]}\right], \tag{15}$$

and taking logarithms we have the cumulant generating function in the form

$$\Psi(t) = g(t)-g(0), \tag{16}$$

where
$$g(t) = -\sum_{l=1}^{k} it\mu_l p \log\left(\frac{k\mu}{\mu_l}\right) + \sum_{j=0}^{p-1}\left[\sum_{l=1}^{k}\{\log\Gamma[\tfrac{1}{2}\{\mu_l(1-2it)+\beta_l-j\}]\} - \log\Gamma[\tfrac{1}{2}\{k\mu(1-2it)+k\beta-j\}]\right], \tag{17}$$

and $g(0)$ is a constant independent of t obtained by putting $t = 0$ in the above expression. Now Barnes (1899) was able to generalize Stirling's theorem, and he showed that for all x, real or complex, $\log \Gamma(x+h)$ may be expanded in an asymptotic series:

$$\log \Gamma(x+h) = \log \sqrt{(2\pi)} + (x+h-\tfrac{1}{2}) \log x - x - \sum_{r=1}^{n} (-1)^r \frac{B_{r+1}(h)}{r(r+1)\,x^r} + R_{n+1}(x), \tag{18}$$

where $R_m(x)$ is a remainder term such that $|R_m(x)| \leqslant \frac{\theta}{|x^m|}$, θ is some constant independent of x and $B_r(h)$ is the Bernoulli polynomial of degree r and order unity defined by

$$\frac{\tau e^{h\tau}}{e^\tau - 1} = \sum_{r=0}^{\infty} \frac{\tau^r}{r!} B_r(h). \tag{19}$$

Expanding each of the Γ-functions in this manner we obtain

$$\Psi(t) = Q - g(0) - \tfrac{1}{4}(k-1)\,p(p+1) \log(1-2it) + \sum_{r=1}^{n} \frac{\alpha_r}{\mu^r}(1-2it)^{-r} + R_{n+1}(\mu, t), \tag{20}$$

where Q does not contain t and is given by

$$Q = \frac{p(k-1)}{2} \log 2\pi + \frac{p}{2}\left[\sum_{l=1}^{k}\left\{\nu_l - \frac{p+1}{2}\right\} \log \frac{\mu_l}{2} - \left\{k\nu - \frac{p+1}{2}\right\} \log \frac{k\mu}{2}\right], \tag{21}$$

$$\alpha_r = \frac{(-1)^{r+1}(2\mu)^r}{r(r+1)} \sum_{j=0}^{p-1}\left[\sum_{l=1}^{k} \frac{B_{r+1}\left(\frac{\beta_l - j}{2}\right)}{\mu_l^r} - \frac{B_{r+1}\left(\frac{k\beta - j}{2}\right)}{\mu^r k^r}\right], \tag{22}$$

and $R_{n+1}(\mu, t)$ is defined by (17) and (18).

From (20) we have

$$\Phi(t) = K(1-2it)^{-\frac{1}{2}f} \sum_{v=0}^{\infty} \frac{a_v}{\mu^v}(1-2it)^{-v} + R'_{n+1}(\mu, t) \tag{23}$$

where $K = \exp\{Q - g(0)\}$, $f = \tfrac{1}{2}(k-1)\,p(p+1)$, and a_v is the coefficient of μ^{-v} in the expansion of $\exp\left\{\sum_{r=1}^{n} \alpha_r/\mu^r\right\}$.

The probability density function of ρM is then given by

$$p(\rho M) = \frac{1}{2\pi}\int_{-\infty}^{+\infty} e^{-it\rho M}\,\Phi(t)\,dt \tag{24}$$

$$= K \sum_{v=0}^{\infty} \frac{a_v}{\mu^v} p(\chi^2_{f+2v}) + R''_{n+1}(\mu, t). \tag{25}$$

The probability that a given value M_0 of the criterion is exceeded is therefore

$$\Pr.\{M \geqslant M_0\} = K \sum_{v=0}^{\infty} \frac{a_v}{\mu^v} P_{f+2v} + R'''_{n+1}(\mu, t), \tag{26}$$

where

$$P_{f+2v} = \int_{\rho M_0}^{\infty} p(\chi^2_{f+2v})\,d\chi^2, \tag{27}$$

$$R'''_{n+1}(\mu, t) = \frac{K}{2\pi}\int_{\rho M_0}^{\infty}\int_{-\infty}^{\infty} e^{-it\rho M} \sum_{v=0}^{\infty} \frac{a_v}{\mu^v}(1-2it)^{-\frac{1}{2}(f+2v)}\left(e^{R_{n+1}(\mu, t)} - 1\right) dt\, d(\rho M)$$

and $p(\chi^2_{f+2v})$ is the probability density function of the χ^2 distribution with $f+2v$ degrees of freedom. For all sufficiently large values of μ, $R'''_{n+1}(\mu, t)$ tends to zero and the required probability will be given with sufficient accuracy by taking a suitable number of terms of the series in (26). Putting $t = 0$ in (20) we have

$$Q - g(0) = -\left\{\sum_{r=1}^{n} (\alpha_r/\mu^r) + R_{n+1}(\mu, 0)\right\}. \tag{28}$$

It is found in practice that by taking a few terms of the series (even in difficult cases usually not more than six), $\exp(-\Sigma\alpha_r/\mu_r)$ is so close to $\exp\{Q-g(0)\} = K$ that direct calculation of that constant is unnecessary.

If we expand $Q-g(0)$ as well as $g(t)$ we obtain instead of (20)

$$\Psi(t) = -\tfrac{1}{2}f\log(1-2it) + \sum_{r=1}^{n}\left\{\frac{\alpha_r}{\mu^r}(1-2it)^{-r}-1\right\} + R_{n+1}(\mu, t) - R_{n+1}(\mu, 0). \tag{29}$$

Proceeding as before we obtain

$$\begin{aligned}\Pr.\{M \geqslant M_0\} = P_f + \alpha_1(P_{f+2}-P_f)\,1/\mu \\ + \left\{\alpha_2(P_{f+4}-P_f) + \frac{\alpha_1^2}{2!}(P_{f+4}-2P_{f+2}+P_f)\right\}1/\mu^2 + \text{etc.}\end{aligned} \tag{30}$$

Thus we may use a suitable number of terms of either of the series given in (26) or (30) to obtain the probability of the criterion exceeding a given value. Formula (26) has been used in this paper, (30) being rather unwieldy if a large number of terms have to be taken.

It should be noted that in the derivation we have used two series, first the asymptotic series for the expansion of the Γ-functions, and then the exponential series. In any particular case we have to decide how many terms we need in the asymptotic series to give a sufficiently close representation of the function and then how many terms we shall use in the exponential series. In those cases investigated here, six terms of the exponential series have nearly always proved adequate as judged by the closeness of agreement between $\Sigma\alpha_r/\mu^r$ and $g(0)-Q$ independently calculated; often fewer terms were necessary, terms in higher powers of $1/\mu$ having negligible effect. In the case of the exponential series the number of terms necessary to represent adequately $\exp\sum_r^n \alpha_r/\mu^r$ is usually not more than eight, but has sometimes been as many as fourteen. It is mainly in order to keep the number of terms required at this stage within manageable limits that the scale factor ρ is introduced, since by suitable choice of this constant, the values of the α's can be kept small and the number of terms required in the exponential series is consequently less. We see that in effect we are fitting a χ^2 series to the statistic M by arranging that, to the order of accuracy chosen in the asymptotic series, the series will have *all* its cumulants identical with those of M. Before we consider the problem of choosing a suitable value for ρ, we shall derive an expression for the α's and hence the a's in a form which is more suitable for computation.

2·2. *Determination of the α's in a form suitable for calculation*

From the well-known properties of Bernoulli polynomials (see, for example, Milne-Thomson, 1933), we may write the symbolic equality

$$B_r(x+y) \doteqdot (B+x+y)^r, \tag{31}$$

where, after expansion, each index of B is to be replaced by the corresponding suffix. Whence

$$B_{r+1}\left(\frac{\beta_i - j}{2}\right) = \sum_{s=0}^{r+1}\binom{r+1}{s}\left(\frac{\beta_i}{2}\right)^{r+1-s} B_s\left(\frac{-j}{2}\right). \tag{32}$$

Also if $P(x)$ is a polynomial in x and $P'(x)$ is the differential coefficient of $P(x)$ with respect to x,

$$\sum_{x=0}^{p-1} P'(x) \doteqdot P(B+p) - P(B). \tag{33}$$

Thus if

$$P'(j) = B_s\left(\frac{-j}{2}\right),$$

$$P(j) = -\frac{2}{s+1} B_{s+1}\left(\frac{-j}{2}\right) + \text{constant}, \tag{34}$$

and

$$\sum_{j=0}^{p-1} B_s\left(\frac{-j}{2}\right) \doteqdot \frac{-2}{s+1}\left[B_{s+1}\left\{-\frac{B+p}{2}\right\} - B_{s+1}\left\{-\frac{B}{2}\right\}\right]. \tag{35}$$

If we denote the expression in square brackets by δ_s, we obtain from (32) and (35)

$$\sum_{j=0}^{p-1}\sum_{l=1}^{k} B_{r+1}\left(\frac{\beta_l - j}{2}\right) \doteqdot -2\sum_{l=1}^{k}\sum_{s=0}^{r+1}\binom{r+1}{s}\frac{1}{s+1}\left(\frac{\beta_l}{2}\right)^{r+1-s}\delta_s, \tag{36}$$

and in a similar way we find

$$\sum_{j=0}^{p-1} B_{r+1}\left(\frac{k\beta - j}{2}\right) \doteqdot -2\sum_{s=0}^{r+1}\binom{r+1}{s}\frac{k^{r+1-s}}{s+1}\left(\frac{\beta}{2}\right)^{r+1-s}\delta_s. \tag{37}$$

Whence from (22) we obtain α_r as a polynomial of degree r in β

$$\alpha_r = \frac{(-1)^r k}{r(r+1)(r+2)}\sum_{s=1}^{r+1}\binom{r+2}{s+1} 2^s D_s \beta^{r+1-s}, \tag{38}$$

where

$$D_s = \delta_s \gamma_s,$$

$$\delta_s \doteqdot B_{s+1}\left\{-\frac{B+p}{2}\right\} - B_{s+1}\left\{-\frac{B}{2}\right\}, \tag{39}$$

and

$$\gamma_s = \frac{1}{k}\sum_{l=1}^{k}\left(\frac{\nu}{\nu_l}\right)^{s-1} - \frac{1}{k^s}. \tag{40}$$

It is interesting to note the relation between these quantities and the c's defined by Hartley; if

$$c_s = \sum_l \frac{1}{\nu_l^s} - \frac{1}{N^s}, \quad \gamma_s = k^{-1}\nu^{s-1}c_{s-1}. \tag{41}$$

In the special case when the samples have equal degrees of freedom

$$\gamma_s = 1 - 1/k^s.$$

The values for δ_s for $s = 0, 1, \ldots, 7$, found from equation (39), are given below:

$$\left.\begin{array}{ll}
s & \delta_s \\
0 & -\frac{1}{2}p, \\
1 & \frac{1}{4}p(p+1), \\
2 & -\frac{1}{16}p(2p^2+3p-1), \\
3 & \frac{1}{16}p(p-1)(p+1)(p+2), \\
4 & -\frac{1}{192}p(6p^4+15p^3-10p^2-30p+3), \\
5 & \frac{1}{128}p(p-1)(p+1)(p+2)(2p^2+2p-7), \\
6 & -\frac{1}{768}p(6p^6+21p^5-21p^4-105p^3+21p^2+147p-5), \\
7 & \frac{1}{768}p(p-1)(p+1)(p+2)(3p^4+6p^3-23p^2-26p+62),
\end{array}\right\} \tag{42}$$

and the values for the first six α's from equation (38) are

$$
\left.\begin{aligned}
\alpha_1 &= -\tfrac{1}{3}k\{3D_1\beta+2D_2\},\\
\alpha_2 &= \tfrac{1}{6}k\{3D_1\beta^2+4D_2\beta+2D_3\},\\
\alpha_3 &= -\tfrac{1}{15}k\{5D_1\beta^3+10D_2\beta^2+10D_3\beta+4D_4\},\\
\alpha_4 &= \tfrac{1}{60}k\{15D_1\beta^4+40D_2\beta^3+60D_3\beta^2+48D_4\beta+16D_5\},\\
\alpha_5 &= -\tfrac{1}{105}k\{21D_1\beta^5+70D_2\beta^4+140D_3\beta^3+168D_4\beta^2+112D_5\beta+32D_6\},\\
\alpha_6 &= \tfrac{1}{42}k\{7D_1\beta^6+28D_2\beta^5+70D_3\beta^4+112D_4\beta^3+112D_5\beta^2+64D_6\beta+16D_7\}.
\end{aligned}\right\} \quad (43)
$$

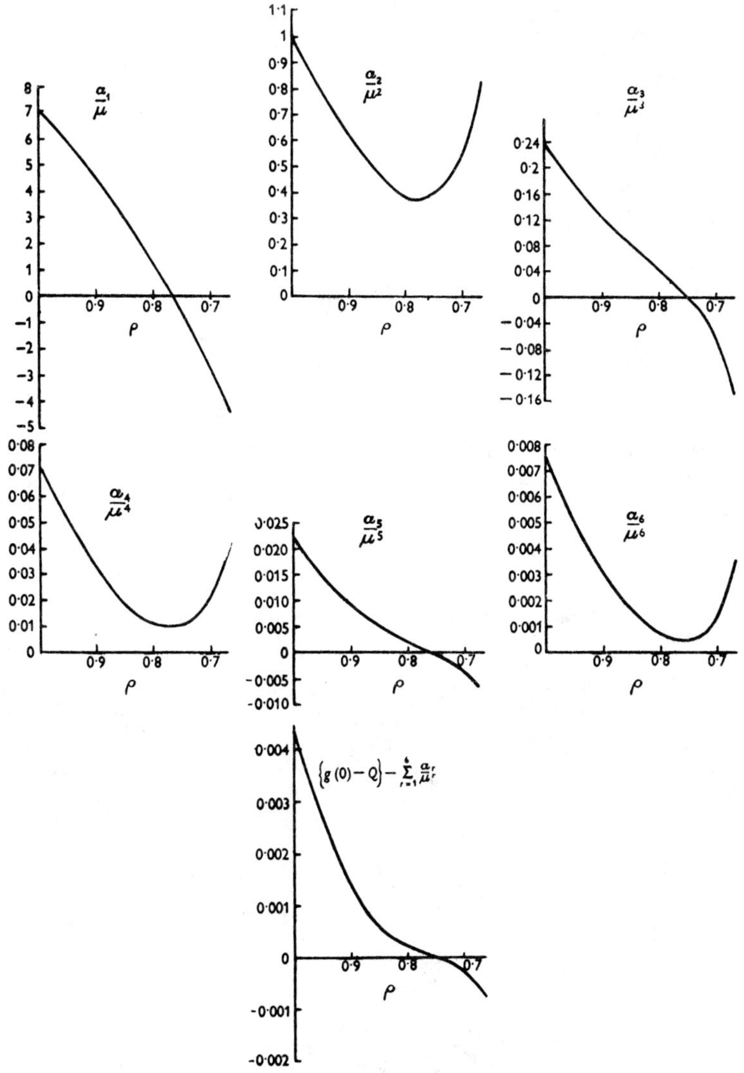

Fig. 1

2·3. *Choice of the value of ρ*

In order that the series should be of practical utility, it must be possible to represent $\exp\{\Sigma\alpha_r\mu^{-r}(1-2it)^{-r}\}$ adequately by a reasonable number of terms of the exponential series; this can be done only if the coefficients α are fairly small. In the *univariate* case, these coefficients will be small even if $\rho = 1$, and in fact if we put $p = 1$ and $\rho = 1$ in equation (26) the series we obtain corresponds exactly with that found by Hartley (1940) using rather a different method of approach. The accuracy of Hartley's series, using only three terms in the asymptotic series, was demonstrated (Hartley & Pearson, 1946) by comparison with the significance levels obtained from Nair's exact expansion; the agreement obtained was good even when the degrees of freedom were as low as three. In the multivariate case, a much more satisfactory series can be obtained if ρ is less than unity.

A typical set of curves showing the values of α_1/μ, α_2/μ^2, ..., α_6/μ^6, and the closeness of agreement between $Q-g(0)$ and $-\sum_{r=1}^{6}\alpha_r/\mu^r$ for varying values of ρ, are plotted in the figure for the case $p = 5$, $k = 5$, $\nu = 9$. The curves all have minima or cross the zero line between $\rho = 0{\cdot}7$ and $\rho = 0{\cdot}8$. The value of ρ which makes α_1 zero is $\rho = 0{\cdot}76296$.

In the calculations carried out here, ρ was chosen so that $\alpha_1 = 0$, since this not only resulted in the other coefficients being small, but the absence of α_1 made the calculation of the a's much easier. Putting $\alpha_1 = 0$ we obtain

$$\rho = 1-\frac{(2p^2+3p-1)}{6(p+1)(k-1)}\left(\Sigma\frac{1}{\nu_l}-\frac{1}{N}\right). \tag{44}$$

2·4. *Example of a calculation using the series*

To check his two working approximations (a) and (b), Bishop used as a standard of reference the values obtained by exact fitting of type I curves to the first two moments of the criterion l_1. In the case $p = 4$, $k = 5$, $\nu = 9$ Bishop found for the 5 % point a value corresponding to $M = 70{\cdot}281$.

To obtain from the series the probability associated with this value, we calculate

$$f = 40,\quad \rho = 0{\cdot}808{,}889,\quad \rho M = 56{\cdot}849{,}5,\quad \mu = \rho\nu = 7{\cdot}28.$$

$r\ (=v)$	α_r/μ^r	a_v/μ^v	P_{f+2v}
0	—	1·000,000	0·040,742
1	0·000,000	0·000,000	—
2	0·143,702	0·143,702	0·092,597
3	0·003,675	0·003,675	0·131,138
4	0·001,793	0·012,118	0·178,763
5	0·000,094	0·000,622	0·235,161
6	0·000,032	0·000,791	0·304,909
7	—	0·000,059	0·369,563
8	—	0·000,044	0·446,178

$\sum_1^6 \alpha_r/\mu^r$	0·149,296	$\sum_1^8 a_v/\mu^v$	1·161,011
$g(0)-Q$	0·149,305	$\exp\{\sum_1^6 \alpha_r/\mu^r\}$	1·161,016
Difference	0·000,009	Difference	0·000,005

$$K = \exp\{Q-g(0)\} = 0{\cdot}861{,}306,\quad \exp\left\{-\sum_1^6\alpha_r\Big/\mu^r\right\} = 0{\cdot}861{,}314$$

$$\Pr.\{M > 70{\cdot}281\} = K\Sigma\{a_v/\mu^v\}P_{f+2v} = 0{\cdot}0492.$$

To illustrate the accuracy with which the asymptotic series represents the function, independent calculations of $g(0)-Q$ have been made. As has already been indicated, however, in practice this rather laborious calculation would not be necessary, K being taken as

$$\exp\{-\Sigma\alpha_r/\mu^r\}.$$

2·5. *Some comparisons between the series and the exact distribution*

For the cases $p = 1, k = 2$ and $p = 2, k = 2$, the exact distribution is known for all values of ν; for $p = 1$ the criterion will simply be a function of the variance ratio, and when $p = 2$ the exact distribution has been found by Pearson & Wilks (1933). Table 1 enables the probabilities obtained from these exact distributions to be compared with those found using the series with scale factor ρ and up to four terms in the asymptotic and exponential series, higher terms having negligible effect. The table shows the values of M corresponding to the 5% and 1% points obtained by Bishop by fitting a type I curve to the first two moments of l_1. The exact probabilities corresponding to these points and those obtained using the series are shown below the values of M.

Table 1. *Comparison of the series with the exact distribution*

		$\nu=9$	$\nu=27$	$\nu=79$
$p=1$, $k=2$	5% point (type I)	4·0499	3·9042	3·8794
	Probability: exact	0·05005	0·05009	0·05002
	series	0·05005	0·05009	0·05003
	1% point (type I)	6·9902	6·7461	6·6991
	Probability: exact	0·00998	0·01001	0·01002
	series	0·00998	0·01001	0·01002
$p=2$, $k=2$	5% point (type I)	8·8801	8·1191	8·0018
	Probability: exact	0·05005	0·04997	0·04979
	series	0·05005	0·04997	0·04979
	1% point (type I)	12·8969	11·7844	11·6074
	Probability: exact	0·00999	0·01000	0·00997
	series	0·00999	0·01000	0·00997

The agreement between the series and the exact values is remarkably good, the series giving five-decimal accuracy in almost every case tested. The more difficult cases, however, are those where p and k are larger, especially when ν is small. For these, the closeness with which $\Sigma\alpha_r/\mu^r$ approaches $g(0)-Q$ and the adequacy, when ρ is suitably chosen, of the exponential series as judged by the comparison of $\exp\{\Sigma(\alpha_r/\mu^r)\}$ and $\Sigma(a_v/\mu^v)$, support belief in the accuracy of this solution. For example, the case $p = 4, k = 5$ which we have used to illustrate the calculation of probabilities from the series, is not a particularly favourable one. It appears, however, that six terms of the asymptotic series and eight of the exponential series will be adequate; in less severe cases of course fewer terms are necessary. Further evidence is supplied later for the accuracy of this type of solution, for in tests of independence to be discussed in § 6, exact distributions are available for comparison, in cases where the series is not favoured, and excellent agreement is found.

APPROXIMATIONS

The series we have found is of rather too complicated a character for routine use; as an alternative, approximations were sought which were relatively simple.

3. Approximations using a single χ^2 distribution

We have for the cumulant generating function of M (putting $\rho = 1$ in equation (20))

$$\Psi(t) = Q - g(0) - \frac{f}{2}\log(1-2it) + \sum_{r=1}^{\infty} \frac{\alpha'_r}{\nu^r}(1-2it)^{-r}, \tag{45}$$

where $f = \frac{1}{2}p(p+1)(k-1)$ and α'_r is obtained by putting $\rho = 1$ (i.e. $\beta = 0$) in equation (43).

Expanding this expression in powers of t we obtain

$$\Psi(t) = \sum_{j=1}^{\infty} \frac{(it)^j}{j!} 2^{j-1}(j-1)!\,f\left\{1 + \sum_{r=1}^{\infty} \binom{j+r-1}{r} \frac{2r\alpha'_r}{\nu^r f}\right\}. \tag{46}$$

The jth cumulant of M is then given by

$$\kappa_j = 2^{j-1}(j-1)!\,f\left\{1 + jA_1 + \frac{j(j+1)}{2}A_2 + \dots\right\}, \tag{47}$$

where

$$A_r = \frac{2r\alpha'_r}{\nu^r f}, \tag{48}$$

and in particular for the generalized test for homoscedasticity which we are considering,

$$\left.\begin{aligned} A_1 &= \frac{2p^2+3p-1}{6(k-1)(p+1)}\left(\sum_{l=1}^{k}\frac{1}{\nu_l} - \frac{1}{N}\right), \\ A_2 &= \frac{(p-1)(p+2)}{6(k-1)}\left(\sum_{l=1}^{k}\frac{1}{\nu_l^2} - \frac{1}{N^2}\right). \end{aligned}\right\} \tag{49}$$

3·1. *The choice of a scale factor in the χ^2 approximation*

Now $2^{j-1}(j-1)!f$ is the jth cumulant of χ^2 with f degrees of freedom. Thus, to order ν^{-1}, (47) is identical with the jth cumulant of $C\chi^2$, where C is either $1+A_1$ or $(1-A_1)^{-1}$. If A_2 were zero then $C = 1+A_1$ would give the first cumulant κ_1 to order ν^{-2} and the remaining cumulants would clearly be less in error than if C were taken as $(1-A_1)^{-1}$. However, if $A_2 = A_1^2$, it would be preferable to put $C = (1-A_1)^{-1}$, since here this form would give agreement to order ν^{-2}.

Clearly this would also be the better form to use if A_2 were near to or greater than A_1^2. In the univariate case $A_2 = 0$ and C should therefore be taken as

$$C = 1 + A_1 = 1 + \frac{1}{3(k-1)}\left(\Sigma\frac{1}{\nu_l} - \frac{1}{N}\right)$$

as has been shown by Bartlett (1937).

For the generalized test for homoscedasticity we find

$$A_2 - A_1^2 = \left(\frac{k}{k-1}\right)^2 \frac{\gamma_2^2}{36(p+1)^2\nu^2}\left\{6(p-1)(p+1)^2(p+2)\left(\frac{\gamma_3}{\gamma_2^2}\frac{k-1}{k}\right) - (2p^2+3p-1)\right\}, \tag{50}$$

where γ_s is defined by equation (40). For $p = 1$, $A_2 = 0$, and consequently this quantity is negative for all values of k. When $p > 1$ it is positive, except in the particular case when $p = 2$ and $k = 2$ and the ν's are equal, when the quantity in curled brackets is equal to -1, and A_1^2 is almost exactly equal to A_2; if the ν's are not equal, this quantity is greater than -1, and it is positive for all larger values of p and k.

For the multivariate statistic, $p > 1$; we therefore take M/C to be approximately distributed as χ^2 with $f = \frac{1}{2}(k-1)p(p+1)$ degrees of freedom and

$$\frac{1}{C} = (1-A_1) = 1 - \frac{(2p^2+3p-1)}{6(p+1)(k-1)}\left(\sum_{l=1}^{k} \frac{1}{\nu_l} - \frac{1}{N}\right);$$

if the degrees of freedom are equal this becomes

$$\frac{1}{C} = 1 - \frac{(2p^2+3p-1)(k+1)}{6(p+1)k\nu}. \tag{51}$$

We note that $1/C$ is the same as the value ρ chosen as scale factor (44) in the series solution. In the case of samples with equal degrees of freedom, the statistic M is equivalent to Bishop's criterion l_1, so that the multivariate scale factor C proposed here is comparable with the scale factor G proposed by Bishop and given in equation (3). Table 2 shows a number of comparisons for the significance levels, together with the values for the probabilities given by the series.

Table 2. *χ^2 approximation; comparisons of scale factors. Significance points for M with probability given by series*

			$p=2$		$p=4$		$p=6$	
$k=5$, $\nu=9$	5%	Bishop (*b*)	23·06	0·0531	67·38	0·0742	142·19	0·1633
		Box	23·27	0·0503	68·93	0·0597	148·30	0·1041
	1%	Bishop (*b*)	28·82	0·0107	77·01	0·0173	156·35	0·0533
		Box	29·01	0·0101	78·74	0·0135	163·16	0·0286
$k=5$, $\nu=19$	5%	Bishop (*b*)	22·13	0·0486	61·36	0·0511	121·29	0·0660
		Box	22·03	0·0501	61·31	0·0515	122·82	0·0556
	1%	Bishop (*b*)	27·34	0·0104	69·95	0·0106	133·60	0·0144
		Box	27·47	0·0100	70·03	0·0105	135·13	0·0116

It appears that, not only is the factor suggested here very much simpler than Bishop's, but that it also gives a better approximation. However, it appears that even with the scale factor C this approximation fails when p is large and ν is small.

4. Approximations using the F distribution

The χ^2 approximation becomes less and less satisfactory as p and k are made larger and ν is made smaller. We know, however, that for all finite p and k, M/C will tend to a type III curve as ν becomes large. When ν is not large we might expect the point corresponding to the distribution of M in the β_1, β_2 plane to lie near the type III line, in either the type I or type VI regions. We shall see that the use of these curves rather than the type III will enable us to absorb a further term in the cumulant series, corresponding to the extra adjustable parameter available with type I and type VI curves, and thus ensure agreement in the cumulants to order ν^{-2}. Although percentage points of the B-function have been tabled (Thompson, 1941), tables of the function F are usually more readily available. For this reason results which occur in the B-function form will be inverted, so that only tables of the F distribution will be required in using these approximations, and they will be referred to as F approximations.

4·1. *Choice of relevant type of curve*

The 'start' of the probability density function for M is at zero. For the Pearson system of frequency curves in which the restriction is made that the start of the curve is at zero, the relation between the cumulants

$$\frac{\kappa_3 \kappa_1}{\kappa_2^2} = 2\tau \tag{52}$$

corresponds with Pearson's type III curve when $\tau = 1$. If τ slightly exceeds unity the curve falls in the type VI region, if it is slightly less than unity it falls in the type I region. Substituting the values for the cumulants of the criterion M, using equation (47), we obtain, ignoring terms of order ν^{-3},

$$\tau = \frac{1+4A_1+7A_2+3A_1^2}{1+4A_1+6A_2+4A_1^2}. \tag{53}$$

Thus for all sufficiently large values of ν the region into which the curve will fall is given by

$$\left.\begin{array}{ccc} A_2 > A_1^2 & A_2 = A_1^2 & A_2 < A_1^2 \\ \tau > 1 & \tau = 1 & \tau < 1 \\ \text{Type VI} & \text{Type III} & \text{Type I} \end{array}\right\} \tag{54}$$

For example, from equation (50) obtained in the case of the generalized test for homoscedasticity, it is clear that for $p = 1$ the curve will be in the type I region, and for nearly all other cases, when p is greater than 1, it will be in the type VI region.

4·2. *Type VI*

The F distribution with $2P$ and $2Q$ degrees of freedom is defined by

$$p(F) = \text{constant } F^{P-1}(PF+Q)^{-(P+Q)}. \tag{55}$$

The rth moment of a quantity bF, where b is a constant, is given by

$$\mu_r'(bF) = \left(b\frac{Q}{P}\right)^r \frac{\Gamma(P+r)\,\Gamma(Q-r)}{\Gamma(P)\Gamma(Q)}, \tag{56}$$

from which, after some algebraic reduction, we obtain the first four cumulants of bF as

$$\kappa_1(bF) = P(b/P)(1-1/Q)^{-1},$$

$$\kappa_2(bF) = P(b/P)^2(1+(P-1)/Q)(1-1/Q)^{-2}(1-2/Q)^{-1},$$

$$\kappa_3(bF) = 2P(b/P)^3(1+(P-1)/Q)(1+(2P-1)/Q)(1-1/Q)^{-3}(1-2/Q)^{-1}(1-3/Q)^{-1},$$

$$\kappa_4(bF) = 6P(b/P)^4\{(P/Q)^2(5/Q-11/Q^2)+(1-1/Q)^2(1-3/Q+2/Q^2+6P/Q-13P/Q^2)\}$$
$$\times(1-1/Q)^{-4}(1-2/Q)^{-2}(1-3/Q)^{-1}(1-4/Q)^{-1}. \tag{57}$$

Now we have seen that M is approximately distributed as $C\chi^2$, so that if τ is greater than unity we would expect to be able to find values b, P and Q, so that bF would be an even better approximation. Since we already know that the distribution is close to type III, we would further expect that Q will be large compared with P since this will be so for type VI curves close to the type III line.

If then we ignore terms of order $(P/Q)^2$, we find

$$\left.\begin{aligned} \kappa_1(bF) &= P(b/P)\,\{1 + 1/Q\}, \\ \kappa_2(bF) &= P(b/P)^2\{1+P/Q+3/Q\}, \\ \kappa_3(bF) &= 2P(b/P)^3\{1+3P/Q+6/Q\}, \\ \kappa_4(bF) &= 6P(b/P)^4\{1+6P/Q+10/Q\}. \end{aligned}\right\} \tag{58}$$

Now put $\quad 2P = f_1 = f, \quad 2Q = f_2 = \dfrac{f_1+2}{A_2 - A_1^2} \quad$ and $\quad b = \dfrac{f_1}{1-A_1-f_1/f_2},$

then we obtain approximately

$$\left.\begin{aligned} \kappa_1(bF) &= f\{1+A_1+A_2\}, \\ \kappa_2(bF) &= 2f\{1+2A_1+3A_2\}, \\ \kappa_3(bF) &= 8f\{1+3A_1+6A_2\}, \\ \kappa_4(bF) &= 48f\{1+4A_1+10A_2\}, \end{aligned}\right\} \tag{59}$$

which are identical to order ν^{-2} with the cumulants of M given by equation (47). Thus M/b will be distributed approximately as F with f_1 and f_2 degrees of freedom, where

$$f_1 = f, \quad f_2 = \frac{f_1+2}{A_2-A_1^2}, \quad b = \frac{f_1}{1-A_1-f_1/f_2}. \tag{60}$$

4·3. *Type I*

We define a quantity X distributed in a type I form with parameters P and Q,

$$p(X) = \text{constant}\, X^{P-1}(1-X)^{Q-1}. \tag{61}$$

The rth moment of bX, where b is a constant, is given by

$$\mu_r'(bX) = b^r \frac{\Gamma(P+r)\,\Gamma(P+Q)}{\Gamma(P)\,\Gamma(P+Q+r)}, \tag{62}$$

from which we find the first four cumulants of bX to be

$$\left.\begin{aligned} \kappa_1(bX) &= P(b/Q)\,(1+P/Q)^{-1}, \\ \kappa_2(bX) &= P(b/Q)^2\,(1+P/Q)^{-2}\,(1+(P+1)/Q)^{-1}, \\ \kappa_3(bX) &= 2P(b/Q)^3\,(1-P/Q)\,(1+P/Q)^{-3}\,(1+(P+1)/Q)^{-1}\,(1+(P+2)/Q)^{-1}, \\ \kappa_4(bX) &= 6P(b/Q)^4\{1+1/Q-2P/Q-4P/Q^2-2P^2/Q^2+P^2/Q^3+P^3/Q^3\} \\ &\quad \times(1+P/Q)^{-4}\,(1+(P+1)/Q)^{-2}\,(1+(P+2)/Q)^{-1}\,(1+(P+3)/Q)^{-1}. \end{aligned}\right\} \tag{63}$$

As before if Q is large compared with P, so that terms of order $(P/Q)^2$ may be ignored, we obtain

$$\left.\begin{aligned} \kappa_1(bX) &= P(b/Q)\,\{1-P/Q\}, \\ \kappa_2(bX) &= P(b/Q)^2\{1-3P/Q+1/Q\}, \\ \kappa_3(bX) &= 2P(b/Q)^3\{1-6P/Q-3/Q\}, \\ \kappa_4(bX) &= 6P(b/Q)^4\{1-10P/Q-6/Q\}, \end{aligned}\right\} \tag{64}$$

and putting $2P = f_1 = f,\quad 2Q = f_2 = \dfrac{f_1+2}{A_1^2-A_2}\quad$ and $\quad b = \dfrac{f_2}{1-A_1+2/f_2}$

we again obtain approximately the values given in (59) which to the order of approximation ν^{-2} are the cumulants of M.

Thus M/b will be distributed as X in expression (61) with $2P = f_1$ and $2Q = f_2$ and

$$f_1 = f,\quad f_2 = \frac{f_1+2}{A_1^2-A_2},\quad b = \frac{f_2}{1-A_1+2/f_2}. \tag{65}$$

Alternatively, $\dfrac{f_2 M}{f_1(b-M)}$ will be distributed as F with f_1 and f_2 degrees of freedom.

We note that although M can vary from 0 to ∞, bX can vary only between the limits 0 and b, so that we are fitting a curve with limited range to one with infinite range. In practice, however, this presents no difficulty (see, for example, the comparisons of Tables 3, 4 and 5), for since the distribution of M will be near to type III, f_2 will be large compared with f_1; consequently b will be large compared with f_1. The mean for such curves will be approximately equal to f_1, so that the range will be large compared with the mean, and the part of the curve ignored by the truncation will be negligible.

4·4. *Application of the F approximation in tests of homoscedasticity*

From (50) we know that when $p = 1$, $A_2-A_1^2$ is negative, and hence the type I form of the approximation is appropriate. When $p \geqslant 2$ we have seen that, except for the case $p = 2, k = 2$, when to this degree of approximation the curve is almost type III, $A_2-A_1^2$ is positive and the type VI form is appropriate.

4·41. *Univariate test* ($p = 1$)

When $p = 1$, A_2 is zero, so that to carry out the test we calculate in turn

$$A_1 = \frac{1}{3(k-1)}\left(\Sigma\frac{1}{\nu_l}-\frac{1}{N}\right),\quad f_1 = (k-1),\quad f_2 = \frac{k+1}{A_1^2},\quad b = \frac{f_2}{1-A_1+2/f_2}, \tag{66}$$

and refer $\dfrac{f_2 M}{f_1(b-M)}$ to tables of the F distribution with f_1 and f_2 degrees of freedom.

In the special case when the degrees of freedom are equal

$$A_1 = \frac{k+1}{3\nu k}. \tag{67}$$

To test the accuracy of the approximation we will compare the values it gives for the 5% and 1% points of M, with those obtained from (1) Bartlett's approximation, (2) Bishop & Nair's (1939) values and (3) the χ^2 series given by Hartley and corresponding to equation (26) with $p = 1$ and $\rho = 1$. Tables 3 and 4 are adapted from those given by Pearson & Hartley (1946) with the value of Bartlett's approximation and the present approximation added. In Table 3 a number of comparisons are made for the special case where the degrees of freedom are equal, and Table 4 shows a few comparisons for the case of five estimates of variance with unequal degrees of freedom.

If the accuracy is judged by the closeness of agreement with the values obtained by Bishop & Nair, it appears that the F approximation is an improvement upon that suggested by

Table 3. *Comparison of approximations. Significance points for M (equal degrees of freedom, $p = 1$)*

k	ν	5%				1%			
		Bartlett (χ^2)	Box (F)	Hartley (series)	Bishop & Nair	Bartlett (χ^2)	Box (F)	Hartley (series)	Bishop & Nair
3	2	7·32	7·20	7·05	7·11*	11·26	10·85	10·57	10·74*
	3	6·88	6·83	6·79	6·80†	10·57	10·41	10·32	10·43†
	4	6·66	6·63	6·61	6·62*	10·23	10·14	10·10	10·13*
	9	6·29	6·29	6·28	6·30†	9·67	9·64	9·64	9·67†
5	2	11·39	11·23	11·01	11·09*	15·93	15·52	15·15	15·32*
	3	10·75	10·69	10·62	10·67†	15·05	14·88	14·76	14·91†
	4	10·44	10·39	10·37	10·38*	14·60	14·42	14·46	14·47*
	9	9·91	9·89	9·90	9·93†	13·87	13·85	13·84	13·86†
10	2	20·02	19·68	19·45	19·62*	25·68	25·22	24·65	24·90*
	3	18·99	18·91	18·79	18·82†	24·31	24·12	23·97	24·09†
	4	18·47	18·41	18·38	18·42*	23·65	23·54	23·49	23·34*
	9	17·61	17·61	17·60	17·64†	22·69	22·58	22·53	22·48†

* Calculated from Nair's exact distribution. † Calculated by fitting type I curve to L_1.

Table 4. *Comparison of approximations. Significance points for M (unequal degrees of freedom, $p = 1$, $k = 5$)*

N	ν_1	ν_2	ν_3	ν_4	ν_5	5%				1%			
						Bartlett (χ^2)	Box (F)	Hartley (series)	Bishop & Nair	Bartlett (χ^2)	Box (F)	Hartley (series)	Bishop & Nair
20	6	6	4	2	2	10·70	10·65	10·54	10·59	14·97	14·82	14·62	14·80
45	16	16	9	2	2	10·45	10·40	10·30	10·35	14·62	14·51	14·31	14·46
20	5	5	4	3	3	10·49	10·46	10·41	10·43	14·68	14·58	14·51	14·59
45	14	14	9	4	4	10·07	10·05	10·04	10·05	14·09	14·05	14·03	14·05

Bartlett and is about as accurate as Hartley's series, whilst it requires no special tables and involves only simple calculations.

Since the approximations proposed by Bartlett, Hartley and the present author are essentially asymptotic, it is to be expected that for small values of ν, and particularly when $\nu = 1$, the approximations will break down. This does in fact happen to a certain extent with all of them, but it seems least serious with the present F approximation; for example, when $k = 4$, $\nu = 1$, we have

Approximation	5% point	1% point
Bartlett (χ^2)	11·1	16·1
Hartley (series)	9·0	11·8
Box (F)	10·3	14·6
Nair's expansion	10·0	14·1

For the case $\nu = 1$, Table 5 compares, for a number of values of k, the 5% and 1 % levels given by Bartlett's approximation and by the present method with values obtained by Bishop & Nair (1939) using Nair's expansion.

Table 5. *Comparison of the approximations when* $\nu = 1$. *Significance points for* M

Value of k		2	3	4	5	6	7	8	9	10
5 % point	Bartlett (χ^2)	5·8	8·7	11·1	13·3	15·4	17·4	19·3	21·3	23·1
	Box (F)	5·1	7·9	10·3	12·6	14·6	16·7	18·6	20·5	22·4
	Nair's expansion	5·1	7·7	10·0	12·0	14·1	15·9	17·9	19·6	21·3
1 % point	Bartlett (χ^2)	10·0	13·3	16·1	18·6	21·0	23·2	25·4	27·5	29·6
	Box (F)	7·9	11·3	14·6	17·1	19·2	21·5	23·7	25·8	27·9
	Nair's expansion	8·3	11·5	14·0	16·5	18·9	21·0	23·1	25·2	27·2

4·42. *Multivariate test* $p \geqslant 2$

To carry out the test we calculate the quantities

$$\left.\begin{aligned} A_1 &= \frac{2p^2+3p-1}{6(k-1)(p+1)}\left(\Sigma\frac{1}{\nu_l}-\frac{1}{N}\right), \quad A_2 = \frac{(p-1)(p+2)}{6(k-1)}\left(\Sigma\frac{1}{\nu_l^2}-\frac{1}{N^2}\right), \\ f_1 &= \tfrac{1}{2}(k-1)p(p+1), \quad f_2 = \frac{f_1+2}{A_2-A_1^2}, \quad b = \frac{f_1}{1-A_1-f_1/f_2}, \end{aligned}\right\} \tag{68}$$

and refer M/b to the tables of the F distribution with f_1 and f_2 degrees of freedom.

When the degrees of freedom are equal

$$A_1 = \frac{(p^2+3p-1)(k+1)}{6(p+1)k\nu}, \quad A_2 = (p-1)(p+2)\frac{(k^2+k+1)}{6k^2\nu^2}. \tag{69}$$

George (1945) was able to evaluate the exact distribution of the generalized L_1 statistic in simple cases, although, when the value of p and k are not very small, the method becomes unmanageable. She used her exact distribution to check Bishop's approximations. Table 6 is taken for George's Table 1 and shows the equivalent value of M obtained by Bishop's empirical formula, method (a), for the 5 % point, together with the exact value of the probability obtained by George by direct integration. The probability corresponding to this value of M has also been calculated by the χ^2 and F approximations suggested here. Thus the closeness with which exact probability approaches 0·0500 indicates the accuracy of Bishop's method, and the closeness with which the probabilities for the χ^2 and F approximations coincide with the exact probability measures the accuracy of these approximations.

We see that the values given by the F approximation are in excellent agreement with the exact probabilities, and even the χ^2 approximation is considerably better than Bishop's method. Unfortunately, no exact values are available in the cases where p and k are larger, when approximation to the curve is more difficult. For these distributions the series given by formula (26), using in most cases up to six* terms in the asymptotic series and up to eight*

* When $\nu = 9$, and $p = 5$ and 6, the coefficients α are rather large, and ten and fourteen terms respectively had to be used in the exponential series. When $p = 6$ there is evidence that further terms in the asymptotic series would give closer agreement.

Table 6. 5 % *points for M given by Bishop's empirical approximation with their associated probabilities calculated by:* (1) *George's exact method,* (2) *the F approximation,* (3) *the* χ^2 *approximation*

	ν	M	Probabilities: Exact (George)	F (Box)	χ^2 (Box)		ν	M	Probabilities: Exact (George)	F (Box)	χ^2 (Box)
$p=2$, $k=2$	9	8·924	0·0492	0·0492	0·0492	$p=3$, $k=2$	9	15·740	0·0461	0·0458	0·0446
	14	7·835	0·0495*	0·0494	0·0496		14	14·434	0·0475	0·0475	0·0470
	19	7·831	0·0496*	0·0496	0·0496		24	13·598	0·0485	0·0486	0·0485
	24	7·828	0·0498*	0·0497	0·0497		29	13·416	0·0488	0·0489	0·0487
							39	13·211	0·0488	0·0489	0·0488
$p=2$, $k=3$	9	14·164	0·0491	0·0491	0·0490						
	19	13·285	0·0496	0·0497	0·0496	$p=3$, $k=3$	14	23·661	0·0481	0·0479	0·0473
	29	13·031	0·0497	0·0499	0·0499		29	22·288	0·0484	0·0480	0·0478
						$p=4$, $k=2$	19	20·989	0·0461	0·0461	0·0455
							29	19·946	0·0477	0·0478	0·0476

* These values have been recalculated and do not agree with the values given in George's table.

Table 7. *Comparisons of approximations. Significance points for M obtained by various methods, with probabilities given by series* (26)

			$p=2$		$p=3$		$p=4$		$p=5$		$p=6$	
$k=5$, $\nu=9$	5 %	Bishop (*a*)	23·40	0·0485	—	—	71·07	0·0434	—	—	173·17	0·0105
		Bishop (*b*)	23·06	0·0531	—	—	67·38	0·0742	—	—	142·19	0·1633
		Box (χ^2)	23·27	0·0503	42·56	0·0532	68·93	0·0597	103·65	0·0673	148·30	0·1041
		Box (F)	23·30	0·0500	42·83	0·0506	69·84	0·0524	106·40	0·0545	153·36	0·0692
		Type I	23·26	0·0504	42·88	0·0502	70·28	0·0492	107·15	0·0500	157·38	0·0488
	1 %	Bishop (*a*)	29·19	0·0096	—	—	81·35	0·0082	—	—	192·34	0·0010
		Bishop (*b*)	28·82	0·0107	—	—	77·01	0·0173	—	—	156·35	0·0533
		Box (χ^2)	29·01	0·0101	50·24	0·0111	78·74	0·0135	113·54	0·0225	163·16	0·0286
		Box (F)	29·05	0·0100	50·58	0·0102	79·84	0·0105	118·56	0·0122	168·94	0·0165
		Type I	29·07	0·0099	50·59	0·0102	80·45	0·0097	120·20	0·0098	173·78	0·0097
$k=5$, $\nu=19$	5 %	Bishop (*a*)	22·14	0·0486	—	—	61·63	0·0489	—	—	124·25	0·0469
		Bishop (*b*)	22·13	0·0486	—	—	61·36	0·0511	—	—	121·29	0·0660
		Box (χ^2)	22·03	0·0501	39·09	0·0506	61·31	0·0515	89·08	0·0532	122·82	0·0556
		Box (F)	22·04	0·0500	39·14	0·0501	61·47	0·0502	89·47	0·0505	123·61	0·0508
		Type I	21·92	0·0516	39·10	0·0505	61·36	0·0511	89·59	0·0496	123·70	0·0503
	1 %	Bishop (*a*)	27·55	0·0098	—	—	70·50	0·0095	—	—	137·09	0·0088
		Bishop (*b*)	27·34	0·0104	—	—	69·95	0·0106	—	—	133·60	0·0144
		Box (χ^2)	27·47	0·0100	46·14	0·0102	70·03	0·0105	99·56	0·0109	135·13	0·0116
		Box (F)	27·48	0·0100	46·20	0·0100	70·23	0·0101	100·01	0·0101	136·03	0·0103
		Type I	27·55	0·0098	46·24	0·0099	70·27	0·0100	100·24	0·0098	136·32	0·0099

terms in the exponential series, may be used as a standard for comparison. Table 7 shows the significance points for M obtained by five different methods together with the probabilities calculated from the series. The methods are: Bishop's empirical approximation (a), Bishop's approximation (b), the χ^2 and F approximations suggested in this paper, and the fitting of a type I curve by exact calculation of the first two moments. The values for M for Bishop's approximations and the type I approximation have been calculated from Bishop's significance points for l_1 given in his Tables 9 and 10.

If we take the series solution as supplying essentially accurate values, we confirm Bishop's suggestion that the type I curve, fitted exactly to the first two moments of l_1, provides an exceedingly good approximation. Of the working approximations, the F approximation suggested here appears to be the best and the χ^2 approximation with the generalized scale factor C will be fairly satisfactory if p and k are not greater than five and ν is not less than, say, twenty.

Table 8 supplies a few comparisons with equal and unequal degrees of freedom.

Table 8. *Significance points for M from χ^2 and F approximations for some equal and unequal groupings, when $p = 4$ and $k = 5$, with associated probability given by series* (26)

N	ν_1	ν_2	ν_3	ν_4	ν_5		5%		1%	
95	19	19	19	19	19	χ^2	61·31	0·0515	70·03	0·0105
						F	61·47	0·0502	70·23	0·0101
95	9	9	19	29	29	χ^2	63·22	0·0578	72·33	0·0124
						F	63·99	0·0521	73·14	0·0107
95	9	9	9	9	59	χ^2	66·32	0·0627	75·76	0·0139
						F	67·39	0·0535	77·07	0·0110
45	9	9	9	9	9	χ^2	68·93	0·0597	78·74	0·0135
						F	69·84	0·0524	79·84	0·0105

It appears, at least for unequal samples with none of the degrees of freedom less than 9, that the F approximation will be fairly satisfactory.

5. Generalization of the procedure

The method we have developed has so far been illustrated in the case of the univariate and multivariate tests of homoscedasticity; its application is, however, more general. In fact, the method can be used whenever, by choosing a suitable power of the original criterion, we can obtain a statistic W which has its hth moment of the form

$$\mathscr{E}(W)^h = \text{constant} \times \left[\frac{\prod_{j=1}^{k} (y_j^{y_j})}{\prod_{i=1}^{m} (x_i^{x_i})}\right]^h \frac{\prod_{i=1}^{m} [\Gamma\{x_i(1+h)+\xi_i\}]}{\prod_{j=1}^{k} [\Gamma\{y_j(1+h)+\eta_j\}]}, \tag{70}$$

where $$\sum_{i=1}^{m} x_i = \sum_{j=1}^{k} y_j.$$

(The constant will be obtained of course by putting $h = 0$ in (70) and taking the reciprocal.) Many of the tests in Plackett's review, referred to in the introduction to this paper, fall into this category. We have already seen that the generalized L_1 statistic is of this type; others are Wilks's test for the independence of k groups of variates (which has some important special cases; and will be considered in detail in the next section); the generalized test for constancy of means, variances and covariances for k samples given by Wilks (1932); and the tests for 'compound symmetry' of variance-covariance matrices discussed by Votaw (1948).

Another group of criteria, which has been studied by Mauchly (1940) and Wilks (1946), arises from tests made on a single sample of n p-variate observations. Mauchly's criterion tests the hypothesis that the variances of the variates are all equal and that the covariances between the variates are all zero. Wilks considered criteria for testing three further hypotheses:

(*a*) That the p means, p variances, and $\frac{1}{2}p(p-1)$ covariances for the variates have respectively the same unknown values.

(*b*) That the variances are the same and the covariances are the same irrespective of what values the means have.

(*c*) That the means are the same (assuming (*b*) true).

It is hoped to consider some of these tests rather more closely in a later paper. Here we shall merely note that, except for Wilks's third criterion (which is always distributed exactly in type I form), the exact distribution of the test function is, in general, not exactly known. The expression for the hth moment, however, is in each case of the form of equation (70) and, as is shown below, our previous approach will provide approximations in all these cases. Tukey & Wilks (1946) have considered this class of statistics and have pointed out that they all possess in common the property that, when the null hypothesis is true, they are distributed as a product of independent components, each component being distributed in type I form.

Consider the expression (70) for the hth moment of any statistic W of this type. If we take $M = -2\log W$ as our working statistic, and write $(1-\rho)x_i = \beta_i$, $(1-\rho)y_j = \epsilon_j$, where ρ is a constant $\leqslant 1$ at our choice, we find for the cumulant generating function of ρM

$$\Psi(t) = g(t) - g(0),$$

where $$g(t) = 2it\rho\left[\sum_{i=1}^{m} x_i \log x_i - \sum_{j=1}^{k} y_j \log y_j\right] + \sum_{i=1}^{m} \log \Gamma\{\rho x_i(1-2it) + \beta_i + \xi_i\} - \sum_{j=1}^{k} \log \Gamma\{\rho y_j(1-2it) + \epsilon_j + \eta_j\}, \quad (71)$$

and $g(0)$ is independent of t and is obtained by writing $t = 0$ in (71). Expanding the logarithms of the Γ-functions by (18), we obtain the cumulant generating function of ρM in the form

$$\Psi(t) = Q - g(0) - \frac{f}{2}\log(1-2it) + \sum_{r=1}^{\infty} \omega_r (1-2it)^{-r}, \quad (72)$$

where
$$f = -2\left\{\sum_{i=1}^{m}\xi_i - \sum_{j=1}^{k}\eta_j - \tfrac{1}{2}(m-k)\right\}, \tag{73}$$

$$\omega_r = \frac{(-1)^{r+1}}{r(r+1)}\left\{\sum_{i=1}^{m}\frac{B_{r+1}(\beta_i+\xi_i)}{(\rho x_i)^r} - \sum_{j=1}^{k}\frac{B_{r+1}(\epsilon_j+\eta_j)}{(\rho y_j)^r}\right\}, \tag{74}$$

$$Q = \tfrac{1}{2}(m-k)\log 2\pi - \frac{f}{2}\log\rho + \sum_{i=1}^{m}(x_i+\xi_i-\tfrac{1}{2})\log x_i - \sum_{j=1}^{k}(y_j+\eta_j-\tfrac{1}{2})\log y_j. \tag{75}$$

From the cumulant generating function (72), the asymptotic χ^2 series corresponding to (26) and (30) are immediately obtainable. Alternatively, we may obtain approximations in the manner given in § 3; the method outlined there is clearly perfectly general for this whole class of statistics. We need the quantities $A_1 = 2\omega_1'/f$ and $A_2 = 4\omega_2'/f$, where ω_r' is the value taken on by ω_r when $\rho = 1$. The scale factor C for the χ^2 approximation will be $1+A_1$ or $(1-A_1)^{-1}$; a decision between the two alternative forms can be reached by the considerations set out in § 3·1. Then, to this order of approximation, M/C will be distributed as χ^2. If greater accuracy is required we may use the F type approximations described in § 4. The particular form is decided by the sign of the quantity $A_2 - A_1^2$. If this quantity is positive, the curve of best fit will be type VI. Putting

$$f_1 = f, \quad f_2 = \frac{f_1+2}{A_2-A_1^2}, \quad b = \frac{f_1}{1-A_1-f_1/f_2}, \tag{76}$$

M/b is distributed approximately as the variance ratio F with f_1 and f_2 degrees of freedom. Alternatively, if $A_2 - A_1^2$ is negative the best fitting curve will be type I, and if we put

$$f_1 = f, \quad f_2 = \frac{f_1+2}{A_1^2-A_2}, \quad b = \frac{f_2}{1-A_1+2/f_2}, \tag{77}$$

then approximately $\dfrac{f_2 M}{f_1(b-M)}$ will be distributed as F with f_1 and f_2 degrees of freedom.

There are thus a number of possible levels of approximation as measured by the order of agreement between the cumulants of the statistic and those of the fitted curve.

(1) Ignoring terms of order x_i^{-1}, y_j^{-1}, M is distributed as χ^2.

(2) Ignoring terms of order x_i^{-2}, y_j^{-2}; by a technique originally used by Bartlett and here generalized, a quantity C can be found such that M is distributed as $C\chi^2$.

(3) Ignoring terms of order x_i^{-3}, y_j^{-3}, a function of M can be obtained which is distributed as the variance ratio F.

(4) Finally, for very precise work and for checking other approximations, a χ^2 series solution may be used and here agreement with the cumulants (as represented by their asymptotic expansions) of the statistic can be obtained to as great an order as seems profitable.

In practice method (4) is sometimes rather long, although it has been found very accurate, but (3) involves very little labour and will often be sufficiently precise.

As a second example of the application of this technique we consider Wilks's generalized test of independence.

6. The generalized test for independence

Wilks (1935) considered the following problem: suppose we have a sample of $\nu+u$ observations for a kp variate normal population and we have some *a priori* reason for dividing the variates into k groups containing $p_1, \ldots, p_n, \ldots, p_l, \ldots, p_k$ variates (where $\sum_l p_l = kp$ and p is thus the average size of the groups and is not necessarily integer). It is required to test the hypothesis that the k groups of residuals, obtained after fitting u independent constants to each of the variates, are mutually independent.

If $|c_{ij}|$ is the $kp \times kp$ determinant of sums of squares and products of residuals for the kp variates and $|c_{ij}|_l$ is the $p_l \times p_l$ determinant of sums of squares and products of residuals of the lth group, then the likelihood ratio criterion obtained by Wilks is

$$\lambda = \frac{|c_{ij}|}{\prod_{l=1}^{k} |c_{ij}|_l} = \frac{|r_{ij}|}{\prod_{l=1}^{k} |r_{ij}|_l}, \tag{78}$$

where $|r_{ij}|$ and $|r_{ij}|_l$ are the corresponding determinants of sample correlation coefficients having ν degrees of freedom. Wilks obtained the moments, and also, for special sets of values of k and p_l, the exact distribution of his criterion which generalizes a very large class of statistical tests. Problems in which there are more than two groups of variates, i.e. where $k > 2$, occur for example in educational research; we may have some prior reason for believing that a battery of, say, ten different tests applied to pupils may be divided up into a number of groups, each group concerned with some distinct ability, and may wish therefore to test the hypothesis that, when the means are eliminated, the selected groups are independent of each other.

When $k = 2$, we consider only two groups of variates containing p_1 in the first and p_2 in the second. Since the criterion and its distribution will be unaffected if the set of p_2 variates are fixed independent variables and the set of p_1 variates 'dependent' variables distributed in a p_1-variate normal distribution, the function is then appropriate for testing the general multivariate linear hypothesis (see, for example, Bartlett, 1934, 1938, 1947). If, in addition, $p_1 = 1$, then the likelihood criterion is $\lambda = 1 - R^2$, where R is the coefficient of multiple correlation between the single dependent variate and the p_2 independent variates. A second special case of Wilks's statistic which is of some interest, and is considered more fully later in this section, occurs when there is only one variate in each of the k groups. The statistic then supplies an overall test for independence between k variates. For the general statistic Wald & Brookner (1941), using a rather different technique from that of Wilks, were able to extend the catalogue of values of k and p_l for which the distribution of λ is exactly known in terms of elementary functions, to include all cases where at most one group contains an odd number of variates. These distributions, although exact, are rather complicated in character. As an alternative and to cover the remaining cases, these authors obtained a series solution and Rao (1948) modified this series in the important case where $k = 2$ to provide an improved test in problems of multivariate analysis. These series will later appear as special cases of that which we are now investigating.

6·1. *Derivation of the series*

The hth moment of λ is given by Wilks as

$$\prod_{l=1}^{k} \prod_{j=0}^{p_l-1} \left\{ \frac{\Gamma\left(\frac{\nu-j}{2}\right)}{\Gamma\left(\frac{\nu-j}{2}+h\right)} \right\} \prod_{m=0}^{kp-1} \left\{ \frac{\Gamma\left(\frac{\nu-m}{2}+h\right)}{\Gamma\left(\frac{\nu-m}{2}\right)} \right\}. \tag{79}$$

So that if we write $W = \lambda^{\frac{1}{2}\nu}$, the hth moment of W will be in the form given in equation (70); taking as our logarithmic statistic $M = -2\log W$ we obtain

$$M = -\nu \log \lambda. \tag{80}$$

To obtain the series, we begin as before by defining the relationship between a quantity ρ (less than unity) and quantities μ and β by the equations $\rho = \mu/\nu$, $\nu = \mu+\beta$. It is also convenient to define a set of quantities

$$\Sigma_s = \left(\sum_l p_l\right)^s - \sum_l p_l^s, \tag{81}$$

which appear in the solution in much the same way as the quantities $\Sigma \frac{1}{\nu_l^s} - \frac{1}{N^s}$ appear in the tests for homoscedasticity. Then, as before, we obtain equation (72) for the cumulant generating function of the ρM, and the constants are available by direct substitution in (73), (74) and (75),

$$f = \tfrac{1}{2}\Sigma_2, \tag{82}$$

$$\omega_r = \frac{\alpha_r}{\mu^r} = \frac{(-2)^r}{r(r+1)\mu^r} \sum_{l=2}^{k} \sum_{j=0}^{p_l-1} \left\{ B_{r+1}\left(\frac{\beta-j}{2}\right) - B_{r+1}\left(\frac{\beta - \sum_{n=1}^{l-1} p_n - j}{2}\right) \right\}, \tag{83}$$

$$Q = -\frac{f}{2}\log\frac{\mu}{2}. \tag{84}$$

The calculation of α_r from formula (83) would clearly be extremely laborious for all but small numbers of variates; we therefore seek an alternative simpler form. Using relations (32) and (35), we find

$$\alpha_r = \frac{(-1)^{r+1}}{r(r+1)(r+2)} \sum_{s=0}^{r+1} \binom{r+2}{s+1} 2^s \sum_{l=2}^{k} \left\{ \beta^{r+1-s} - \left(\beta - \sum_{n=1}^{l-1} p_n\right)^{r+1-s} \right\} \delta_s(p_l), \tag{85}$$

where

$$\delta_s(p_l) \doteqdot B_{s+1}\left(-\frac{B+p_l}{2}\right) - B_{s+1}\left(-\frac{B}{2}\right), \tag{86}$$

and the values taken by (86) when $p = 1, 2, \ldots, 7$, are given by putting $p = p_l$ in equation (42). Writing α_r' for the value which α_r has when $\rho = 1$, i.e. when $\beta = 0$, then substituting for $\delta_s(p_l)$ in (85) and summing, we obtain for the first six values of α_r':

$$\left.\begin{aligned}
\alpha_1' &= \tfrac{1}{24}\{2\Sigma_3 + 3\Sigma_2\},\\
\alpha_2' &= \tfrac{1}{48}\{\Sigma_4 + 2\Sigma_3 - \Sigma_2\},\\
\alpha_3' &= \tfrac{1}{720}\{6\Sigma_5 + 15\Sigma_4 - 10\Sigma_3 - 30\Sigma_2\},\\
\alpha_4' &= \tfrac{1}{480}\{2\Sigma_6 + 6\Sigma_5 - 5\Sigma_4 - 20\Sigma_3 + 3\Sigma_2\},\\
\alpha_5' &= \tfrac{1}{840}\{2\Sigma_7 + 7\Sigma_6 - 7\Sigma_5 - 35\Sigma_4 + 7\Sigma_3 + 49\Sigma_2\},\\
\alpha_6' &= \tfrac{1}{2016}\{3\Sigma_8 + 12\Sigma_7 - 14\Sigma_6 - 84\Sigma_5 + 21\Sigma_4 + 196\Sigma_3 - 10\Sigma_2\},
\end{aligned}\right\} \tag{87}$$

where Σ_s is defined by equation (81), whence we have for the α's

$$\left.\begin{aligned}
\alpha_1 &= \alpha_1' - (f/2)\beta,\\
\alpha_2 &= \alpha_2' - \alpha_1'\beta + (f/4)\beta^2,\\
\alpha_3 &= \alpha_3' - 2\alpha_2'\beta + \alpha_1'\beta^2 - (f/6)\beta^3,\\
\alpha_4 &= \alpha_4' - 3\alpha_3'\beta + 3\alpha_2'\beta^2 - \alpha_1'\beta^3 + (f/8)\beta^4,\\
\alpha_5 &= \alpha_5' - 4\alpha_4'\beta + 6\alpha_3'\beta^2 - 4\alpha_2'\beta^3 + \alpha_1'\beta^4 - (f/10)\beta^5,\\
\alpha_6 &= \alpha_6' - 5\alpha_5'\beta + 10\alpha_4'\beta^2 - 10\alpha_3'\beta^3 + 5\alpha_2'\beta^4 - \alpha_1'\beta^5 + (f/12)\beta^6.
\end{aligned}\right\} \tag{88}$$

As before, from the cumulant generating function we obtain the series corresponding to (26) and (30), and if ρ is chosen so that $\alpha_1 = 0$, we have

$$\beta = 2\alpha_1'/f, \quad \rho = 1 - \frac{1}{12\nu f}(2\Sigma_3 + 3\Sigma_2). \tag{89}$$

Wald & Brookner (1941) derived a χ^2 series for this statistic by a different method from that used here; it is not difficult to show, however, that the series they obtained is equivalent to our series, but with $\rho = 1$. In this form the series is of little practical use for small, or even moderate values of ν because of the difficulty we have noted before of adequately approximating to $\exp \Sigma \alpha_r \mu^{-r}(1-2it)^{-r}$ by means of a series, unless α_r is small or μ is large. By introducing the factor ρ, the size of the coefficients α can be greatly reduced and the series be used even for fairly small values of ν. As an example, consider the case of three groups of variates with two variates in each grouping, $k = 3$, $p_1 = 2$, $p_2 = 2$, $p_3 = 2$, and suppose $\nu = 10$. The values of the coefficients α_r/μ^r are shown below when $\rho = 1$ and also when ρ takes on a value making α_1 zero. When $\rho = 1$, μ is of course equal to ν.

Values of α_r/μ^r

r	$\rho=1$	$\rho=0{\cdot}68\dot{3}$
1	1·900,00	0·000,00
2	0·335,00	0·073,17
3	0·086,33	0·003,71
4	0·026,88	0·001,78
5	0·009,40	0·000,23
6	0·003,55	0·000,07
Total	2·361,16	0·078,96
$g(0)-Q$	2·363,61	0·078,98

For the Wald & Brookner series, if ν is small, the coefficients are so large that in practice it would be impossible to represent the exponent adequately by a reasonably small number of terms of the exponential series; by suitably choosing ρ, however, the size of the coefficients are greatly reduced while the agreement between the sum of the terms and $g(0)-Q$ is improved. In the particular example quoted, the exact distribution is known (Wilks, 1935). It appears in rather a complicated form, but has been used here to check the series and the approximations. Table 9 shows the 5% and 1% significance points for the criterion M

Table 9. *Some comparisons for Wilks's statistic*

			χ^2 approximation			F approximation		
			M	Probability		M	Probability	
				Exact	Series		Exact	Series
$k=3$, $p_1=2$, $p_2=2$, $p_3=2$	$\nu=10$	5%	30·770	0·0612	0·0612	31·357	0·0549	0·0548
		1%	38·366	0·0139	0·0139	39·180	0·0117	0·0117
	$\nu=20$	5%	24·982	0·0516	0·0516	25·083	0·0504	0·0504
		1%	31·149	0·0105	0·0105	31·292	0·0101	0·0101

obtained by using the χ^2 and F approximations which are derived in the next section, together with the exact probabilities and the probabilities calculated from the series, using $\rho = 0{\cdot}68\dot{3}$, and six terms in the asymptotic and eight in the exponential series. Agreement to four places of decimals is usually obtained between the series and the exact value for the probability.

χ^2 approximation. Following the previous procedure, we find that M/C is distributed approximately as χ^2 with f degrees of freedom, where

$$\frac{1}{C} = 1 - \frac{1}{12\nu f}(2\Sigma_3 + 3\Sigma_2) \quad \text{and} \quad f = \tfrac{1}{2}\Sigma_2.$$

F approximation. We have

$$f = \tfrac{1}{2}\Sigma_2, \quad A_1 = \frac{1}{12\nu f}(2\Sigma_3 + 3\Sigma_2), \quad A_2 = \frac{1}{12\nu^2 f}(\Sigma_4 + 2\Sigma_3 - \Sigma_2),$$

from which, using equations (76) and (77), the F type approximation can be easily computed.

The quantities Σ_2, Σ_3 and Σ_4 required in these approximations are given by (81), the calculations of Table 9 give some indication of the accuracy to be expected.

6·2. *Special cases*

We consider two important special cases of the statistic, that in which there are only two groups of variates and that in which there is only one variate in each of the k groups.

6·21. *Case $k = 2$*

In this case the expressions for the coefficients in the series simplify considerably. Writing $p_1 = p$, $p_2 = q$, we obtain

$$f = pq, \quad \rho = 1 - \frac{p+q+1}{2\nu}, \quad \beta = \tfrac{1}{2}(p+q+1),$$

$$\alpha_1 = 0, \quad \alpha_2 = \frac{pq}{48}(p^2+q^2-5),$$

$$\alpha_3 = 0, \quad \alpha_4 = \frac{pq}{1920}\{3p^4+3q^4+10p^2q^2-50(p^2+q^2)+159\},$$

$$\alpha_5 = 0, \quad \alpha_6 = \frac{pq}{16{,}128}\{3(p^6+q^6)-105(p^4+q^4)+1{,}113(p^2+q^2) + (21p^2-350+21q^2)p^2q^2-2{,}995\}.$$

Putting these values in (26) and (30) we confirm* the series given to terms in μ^{-4} by Rao (1948) for this case, $k = 2$. Bartlett had already (1938) obtained the χ^2 approximation using the scale factor $\dfrac{1}{C} = 1 - \dfrac{p+q+1}{2\nu}$ (which is of course the factor given by the present procedure). Rao introduced this scale factor into the Wald & Brookner series, so as to obtain a χ^2 series with Bartlett's χ^2 approximation as the leading term, equivalent to (30). As we have seen, this choice of factor results in this particular case in α_1 and the α's of odd order being zero, so that the calculation of the series is correspondingly simpler.

* There appears to be a misprint in Rao's paper in the expression which corresponds with α_4, where the constant 159 is wrongly given as 150.

χ^2 (*Bartlett*) *approximation.* $\dfrac{M}{C} = \left(1 - \dfrac{p+q+1}{2\nu}\right) M$ is approximately distributed as χ^2 with $f = pq$ degrees of freedom.

F approximation. We find $A_2 - A_1^2 = (p^2+q^2-5)/12\nu^2$; thus for p and $q \geqslant 2$, $A_2 - A_1^2 > 0$, and the type VI form will be appropriate. M/b will be approximately distributed as F with f_1 and f_2 degrees of freedom, where

$$f_1 = pq, \quad f_2 = \frac{12\nu^2(pq+2)}{p^2+q^2-5}, \quad b = \frac{pq}{1 - \dfrac{p+q+1}{2\nu} - \dfrac{f_1}{f_2}}.$$

For p or q equal to 1 and 2, the exact distributions are known and provide simple tests (Wilks, 1932, 1935). For these cases λ and $\sqrt{\lambda}$ respectively are distributed in a type I distribution, and the significance test can be made, either by directly entering Thompson's tables of percentage points of the incomplete B-function, or by inversion of the statistic to its equivalent 'variance ratio' form and using tables of F or of Fisher's z (Bartlett, 1934; Rao, 1948). As has been pointed out by Bartlett (1938) if $p = 1$ and $q = 2$ (or $p = 2$ and $q = 1$) $\dfrac{M}{C} = \dfrac{\nu-2}{\nu} M$ is distributed *exactly* as χ^2, and substituting these values for p and q in the expressions for α_r we find that in this case all these coefficients are zero, so providing a useful check. If p and q were both unity, $A_2 - A_1^2$ would be negative and the type I form be appropriate for the F approximation. Of course we shall not need to use the method here because the criterion $\sqrt{(1-\lambda)}$ is the sample correlation coefficient r and the exact distribution is known. The exact distributions are also known in certain other cases (Wilks, 1935; Wald & Brookner, 1941); the form which these take, however, is rather complicated, but they are useful to check approximations. In Table 10 are shown the 5 % significance points of M for a number of combinations of p and q as given by the χ^2 and F methods of approximation. In the cases chosen, the exact distribution is known, and this has been used to calculate the exact probability associated with each of these points. For comparison, the probability given by the series, using terms up to α_6 in the asymptotic series and, for most values, up to a_8 in the exponential series, is also shown.

We see that, providing ν is sufficiently large, Bartlett's approximation is in good agreement with the exact values, and the F approximation, since it involves very little more labour, provides a worth-while improvement. If ν is not large and one is doubtful whether these approximations will be sufficient, a rough but useful indication is provided by comparing the values obtained by the χ^2 and F approximations (in calculating the F approximation one will have already calculated the quantities needed for the χ^2 approximation). If these two approximations give substantially the same value, it may generally be taken as an indication that the approximation is adequate. If they differ markedly, a more accurate value should be calculated from the series.

6·22. *Case* $p_l = 1$, $l = 1, 2, \ldots, k$

If the k groups each contain only one variate, the hypothesis tested is that each of the variates is independent of all the others. The λ criterion then becomes the determinant of the sample correlation matrix, e.g., if $k = 3$,

$$\lambda = \begin{vmatrix} 1 & r_{12} & r_{13} \\ r_{12} & 1 & r_{23} \\ r_{13} & r_{23} & 1 \end{vmatrix},$$

Table 10. 5 % *significance points for M*

p	q	ν	χ^2 (Bartlett)			F (Box)		
			M	Probability		M	Probability	
				Exact	Series		Exact	Series
1	1	9	4·610	0·0494	0·0494	4·592	0·0499	0·0499
1	5	10	17·032	0·0624	0·0624	17·542	0·0555	0·0555
		20	13·419	0·0518	0·0518	13·504	0·0504	0·0504
1	10	20	26·153	0·0666	0·0666	27·022	0·0562	0·0562
		40	21·538	0·0525	0·0525	21·690	0·0505	0·0505
2	2	9	13·137	0·0515	0·0515	13·200	0·0506	0·0506
2	5	10	30·512	0·0737	0·0737	31·654	0·0614	0·0614
		20	22·884	0·0529	0·0529	23·053	0·0507	0·0507
2	10	20	46·534	0·0753	0·0753	48·164	0·0595	0·0595
		40	37·505	0·0535	0·0535	37·775	0·0507	0·0507
4	4	10	47·811	0·0945	0·0940	49·996	0·0735	0·0731*
		20	33·931	0·0542	0·0542	34·216	0·0512	0·0512

* With $\nu = 10$ and $p = q = 4$, six terms were taken in the asymptotic series and twelve in the exponential series; greater accuracy can be obtained by taking more terms in the asymptotic series.

where r_{ij} is the usual sample product moment correlation coefficient between the ith and jth variates. When $k = 2$ the criterion is simply $1 - r_{12}^2$.

The statistic is useful in supplying an overall test of independence between the k variates. For example, when $k = 5$ there will be ten individual correlation coefficients. Even when the null hypothesis, that all the variates are uncorrelated, is true, we shall expect often to come across individual coefficients which are 'significant'. For such a case it will be appropriate to apply the overall test before testing individual correlations. Again, the expressions for the coefficients simplify and we find, choosing ρ so that $\alpha_1 = 0$,

$$f = \tfrac{1}{2}k(k-1), \quad \rho = 1 - \frac{2k+5}{6\nu}, \quad \beta = \frac{2k+5}{6},$$

$$\alpha_1 = 0, \quad \alpha_2 = \frac{k(k-1)}{288}(2k^2 - 2k - 13),$$

$$\alpha_3 = \frac{k(k-1)}{3{,}240}(k-2)(2k-1)(k+1),$$

$$\alpha_4 = \frac{k(k-1)}{34{,}560}(16k^4 - 32k^3 - 252k^2 + 268k + 1147),$$

$$\alpha_5 = \frac{k(k-1)}{136{,}080}(k-2)(k+1)(2k-1)(8k^2 - 8k - 97),$$

$$\alpha_6 = \frac{k(k-1)}{7{,}838{,}208}(496k^6 - 1{,}488k^5 - 12{,}576k^4 + 27{,}632k^3 + 137{,}490k^2 - 151{,}554k - 562{,}103).$$

For the χ^2 approximation we find, from the argument of § 3·1, that we should take $\frac{1}{C} = 1 - \frac{2k+5}{6\nu}$. Thus $\left\{1 - \frac{2k+5}{6\nu}\right\} M$ will be approximately distributed as χ^2 with $\frac{1}{2}k(k-1)$ degrees of freedom.

For the F approximation we have

$$f = \tfrac{1}{2}k(k-1), \quad A_1 = \frac{2k+5}{6\nu}, \quad A_2 = \frac{k^2+3k+2}{6\nu^2}.$$

For $k = 2$ and 3 we use the type I form and for $k \geqslant 4$ the type VI, since $A_2 - A_1^2 = \frac{2k^2-2k-13}{36\nu^2}$ is negative when $k = 2$ or 3 and positive for larger values of k. We then calculate f_1, f_2 and b, required in this approximation, by formulae (77) and (76) respectively.

7. Summary and conclusions

For a particular class of likelihood criteria, whose moments appear as the product of Γ-functions, a general method is described for obtaining probability levels when the null hypothesis is true. A number of statistics whose moments appear in this form are referred to, and a general method developed to obtain:

(*a*) A series which is in close agreement with the exact distribution.

(*b*) An approximate solution, using a single χ^2 distribution, which is sufficiently accurate for moderate or large samples.

(*c*) A rather better approximation, using a single F distribution, giving close agreement even when the samples are rather small.

The method is illustrated for the following two general statistics:

(1) *Tests for constancy of variance and covariance*

(*a*) *Univariate case.* The F approximation is of the same order of accuracy as Hartley's (1940) series solution although it requires very much less calculation, and significance may be judged by consulting tables of the significance points of the variance ratio F alone.

(*b*) *Multivariate case.* The series solution shows remarkably close agreement with the exact distribution when this is known, and is used in other cases to compare approximations. The χ^2 approximation does not correspond with that found by Bishop (1939), but is, in fact, simpler and more accurate.

The series confirms the accuracy of significance points found by fitting a type I curve to the first two moments of l_1. The calculation of the moments involved in this method renders it too laborious for routine use, and Bishop suggested two working approximations; the F approximation developed here is more accurate than these approximations, whilst it involves no more labour and can be used when the sample sizes are unequal.

(2) *Wilks's test for independence of k groups of variates*

The asymptotic series, and χ^2 and F approximations are derived for this case, and the relation of the results with those of Wald & Brookner, Bartlett, and Rao is discussed. The exact distribution is used to assess the accuracy of the proposed methods in a number of cases. The probabilities given by the series are found to be in excellent agreement with the true values, even for fairly small samples. Providing the sample sizes are not too small, the χ^2 and F approximations will be sufficiently accurate, the latter providing the better approximation, and allowing the sample size to be rather smaller than is possible with the χ^2 approximation. When the number of variates in each group is one, we have a test criterion for the hypothesis that k variates are mutually independent, and the same procedure provides the series solution and simple approximations for tests of significance.

In conclusion, I wish gratefully to acknowledge the help and guidance I have received from Dr H. O. Hartley throughout this investigation.

REFERENCES

Barnes, E. W. (1899). *Mess. Math.* p. 64.
Bartlett, M. S. (1934). *Proc. Camb. Phil. Soc.* **30**, 327.
Bartlett, M. S. (1937). *Proc. Roy. Soc.* A, **160**, 268.
Bartlett, M. S. (1938). *Proc. Camb. Phil. Soc.* **34**, 33.
Bartlett, M. S. (1947). *J.R. Statist. Soc. Suppl.* **9**, 176.
Bishop, D. J. (1939). *Biometrika*, **31**, 31.
Bishop, D. J. & Nair, U. S. (1939). *J.R. Statist. Soc. Suppl.* **6**, 89.
Brown, G. W. (1939). *Ann. Math. Statist.* **10**, 119.
George, A. (1945). *Sankhyā*, **1**, 20.
Hartley, H. O. (1940). *Biometrika*, **31**, 249.
Hartley, H. O. & Pearson, E. S. (prefatory note) (1946). *Biometrika*, **33**, 296.
Mauchly, J. W. (1940). *Ann. Math. Statist.* **11**, 204.
Milne-Thomson, L. M. (1933). *The calculus of finite differences.* Macmillan.
Nair, U. S. (1939). *Biometrika*, **30**, 274.
Nayer, P. P. N. (1936). *Statist. Res. Mem.* **1**, 38.
Neyman, J. & Pearson, E. S. (1928). *Biometrika*, **20** A, 175 and 263.
Neyman, J. & Pearson, E. S. (1931). *Bull. int. Acad. Cracovie*, A, p. 460.
Neyman, J. & Pearson, E. S. (1936). *Statist. Res. Mem.* **1**, 1.
Neyman, J. & Pearson, E. S. (1938). *Statist. Res. Mem.* **2**, 25.
Pearson, E. S. & Wilks, S. S. (1933). *Biometrika*, **25**, 353.
Pitman, E. J. G. (1939). *Biometrika*, **31**, 200.
Plackett, R. L. (1946). *J.R. Statist. Soc.* **109**, 457.
Plackett, R. L. (1947). *Biometrika*, **34**, 311.
Rao, C. R. (1948). *Biometrika*, **35**, 71.
Thompson, C. M. (1941). *Biometrika*, **32**, 151.
Thompson, C. M. & Merrington, M. (1946). *Biometrika*, **33**, 296.
Tukey, J. W. & Wilks, S. S. (1946). *Ann. Math. Statist.* **17**, 318.
Votaw, D. F. (1948). *Ann. Math. Statist.* **19**, 447.
Wald, A. & Brookner, R. J. (1941). *Ann. Math. Statist.* **12**, 137.
Welch, B. L. (1935). *Biometrika*, **27**, 145.
Welch, B. L. (1936). *Statist. Res. Mem.* **1**, 52.
Wilks, S. S. (1932). *Biometrika*, **24**, 471.
Wilks, S. S. (1935). *Econometrica*, **3**, 309.
Wilks, S. S. (1946). *Ann. Math. Statist.* **17**, 257.
Wishart, J. (1928). *Biometrika*, **20** A, 32.

4.2

Some Theorems on Quadratic Forms Applied in the Study of Analysis of Variance Problems, I: Effect of Inequality of Variance in the One-Way Classification

G. E. P. BOX
Imperial Chemical Industries Ltd. and North Carolina State College

1. Summary and introduction. This is the first of two papers describing a study of the effect of departures from assumptions, other than normality, on the null-distribution of the F-statistic in the analysis of variance. In this paper, certain theorems required in the study and concerning the distribution of quadratic forms in multi-normally distributed variables are first enunciated and simple approximations tested numerically. The results are then applied to determine the effect of group-to-group inequality of variance in the one-way classification. It appears that if the groups are equal, moderate inequality of variance does not seriously affect the test. However, with unequal groups, much larger discrepancies appear. In a second paper, similar methods are used to determine the effect of inequality of variance and serial correlation between errors in the two-way classification.

2. Distribution of quadratic forms in multi-normal variates. In what follows we write $\chi^2(\nu)$ to denote a quantity distributed as χ^2 with ν degrees of freedom and $F(\nu_1, \nu_2)$ to denote a quantity distributed as the Fisher-Snedecor F with ν_1 and ν_2 degrees of freedom.

By obvious extension of a theorem due to Cochran [1] we have

THEOREM 2.1. *If z denotes a column vector of p random variables $z_1, z_2, \cdots, z_p$ having expectation zero and distributed in a multi-normal distribution with $p \times p$ variance-covariance matrix V, and if $Q = z'Mz$ in any real quadratic form of rank $r \leqq p$, then Q is distributed like a quantity*

$$X = \sum_{j=1}^{r} \lambda_j \chi^2(1) \tag{2.1}$$

where each χ^2 variate is distributed independently of every other and the λ's are the r real nonzero latent roots of the matrix

$$U = VM. \tag{2.2}$$

It readily follows that

THEOREM 2.2. *The* sth *cumulant* $\text{K}_s(Q)$ *is given by*

$$\text{K}_s(Q) = 2^{s-1}(s-1)! \sum_{j=1}^{r} \lambda_j^s \tag{2.3}$$

The Collected Works of George E. P. Box, 1984, Wadsworth, Inc., Belmont, CA 94002. Originally published in *Ann. Math. Stat.,* vol. 25, no. 2 (1954), pp. 290–302.

Calculation of this quantity is often facilitated by using the relation

$$(2.4)\qquad \sum_{j=1}^{r} \lambda_j^s = \operatorname{tr}(VM)^s = \sum_{a=1}^{p}\sum_{b=1}^{p}\cdots\sum_{s=1}^{p} u_{ab}\,u_{bc}\cdots u_{sa}$$

whence the first few cumulants of Q may readily be derived without actually determining the λ's. In particular

$$(2.5)\qquad \mathrm{K}_1(Q) = \sum_{a=1}^{p} u_{aa}$$

$$(2.6)\qquad \mathrm{K}_2(Q) = 2\sum_{a=1}^{p}\sum_{b=1}^{p} u_{ab}\,u_{ba}\,.$$

When the λ_j are all positive, the following χ^2 series due to Robbins and Pitman [2] may be used to obtain the distribution of $X = \sum_{j=1}^{r} \lambda_j \chi^2(\nu_j)$.

THEOREM 2.3. *If X_0 is some constant greater than zero then*

$$(2.7)\qquad P_1 \leqq \Pr\{X > X_0\} \leqq P_2$$

where

$$(2.8)\qquad P_1 = \sum_{l=0}^{n} c_l \Pr\left\{\chi^2(\nu + 2l) > \frac{X_0}{a_1}\right\} + \left(1 - \sum_{l=0}^{n} c_l\right)\Pr\left\{\chi^2(\nu + 2n + 2) > \frac{X_0}{a_1}\right\}$$

$$(2.9)\qquad P_2 = \sum_{l=0}^{n} c_l \Pr\left\{\chi^2(\nu + 2l) > \frac{X_0}{a_1}\right\} + \left(1 - \sum_{l=0}^{n} c_l\right)$$

and the constants c_l are such that $\sum_{l=0}^{\infty} c_l = 1$ and are defined by the identity

$$(2.10)\qquad \prod_{j=2}^{r} \{a_j^{-\nu_j/2}[1 - (1 - a_j^{-1})w]^{-\nu_j/2}\} = \sum_{l=0}^{\infty} c_l w^l$$

$a_1 = \lambda_1$ being the smallest of the λ_j, $a_j = \lambda_j/\lambda_1$ $(j \neq 1)$, and $\sum_{j=1}^{r} (\nu_j) = \nu$.

If the component degrees of freedom ν_j are even, a finite χ^2 series, derived below, may be used whether the λ_j are positive or not.

THEOREM 2.4. *The exact distribution of $X = \sum_{j=1}^{r} \lambda_j \chi^2(\nu_j)$, where the $\nu_j = 2g_j$ are even integers, is a weighted finite sum of χ^2 distributions,*

$$(2.11)\qquad \Pr(X > X_0) = \sum_{j=1}^{r}\sum_{s=1}^{g_j} \alpha_{js} \Pr\{\chi^2(2s) > X_0/\lambda_j\}$$

and α_{js} is a constant involving only the λ's and is given by

$$(2.12)\qquad \alpha_{js} = f_j^{(g_j - s)}(0)/(g_j - s)!$$

where $f_j^{(h)}(0)$ is obtained by differentiating $f_j(y)$ h times with respect to y and then putting $y = 0$ and

$$(2.13)\qquad f_j(y) = \prod_{i \neq j}^{r} \left[\frac{\lambda_j - \lambda_i}{\lambda_j} + y\frac{\lambda_i}{\lambda_j}\right]^{-\nu_i/2}.$$

In the special case $r = 2$, a series of this type has been used by Satterthwaite [3]. The general theorem is conveniently proved as follows.

PROOF. Since $g_j = \nu_j/2$ is an integer, the characteristic function

$$\varphi(t) = \prod_{j=1}^{r} (1 - 2it\lambda_j)^{-g_j} \tag{2.14}$$

can be resolved into partial fractions

$$\prod_{j=1}^{r} (1 - 2it\lambda_j)^{-g_j} = \sum_{j=1}^{r} \sum_{s=1}^{g_j} \alpha_{js}(1 - 2it\lambda_j)^{-s} \tag{2.15}$$

where the α_{js} are constants not containing t. We recognise this expression as the characteristic function of a quantity v whose probability density function is

$$p(v) = \sum_{j=1}^{r} \sum_{s=1}^{g_j} \alpha_{js}\, p\{\lambda_j \chi^2(2s)\}. \tag{2.16}$$

Hence X is distributed like v and equation (2.11) follows at once.

To find the values for the constants, (2.15) is written in the form

$$\begin{aligned}(1 - 2it\lambda_i)^{-g_i} \prod_{j \neq i}^{r} (1 - 2it\lambda_j)^{-g_j} &= \sum_{s=1}^{g_i} \alpha_{is}(1 - 2it\lambda_i)^{-s} \\ &\quad + \sum_{j \neq i}^{r} \sum_{w=1}^{g_j} \alpha_{jw}(1 - 2it\lambda_j)^{-w}.\end{aligned} \tag{2.17}$$

Putting $y = 1 - 2it\lambda_i$ and multiplying both sides of the identity by y^{g_i} we have

$$\begin{aligned}\prod_{j \neq i}^{r} \left\{\frac{\lambda_i - \lambda_j}{\lambda_i} + y\,\frac{\lambda_j}{\lambda_i}\right\}^{-g_j} &= \sum_{s=1}^{g_i} \alpha_{is}\, y^{g_i - s} \\ &\quad + \sum_{j \neq i}^{r} \sum_{w=1}^{g_i} \alpha_{jw} \left\{\frac{\lambda_i - \lambda_j}{\lambda_i} + y\,\frac{\lambda_j}{\lambda_i}\right\}^{-w} y^{g_i}.\end{aligned} \tag{2.18}$$

If y is put equal to 0 we have

$$\alpha_{ig_i} = \prod_{j \neq i}^{r} \left\{\frac{\lambda_i}{\lambda_i - \lambda_j}\right\}^{g_j}. \tag{2.19}$$

To obtain the remaining constants we differentiate both sides of identity h times and then put $y = 0$. There will be no contribution from the second member on the right-hand side of (2.18) and the term $\sum_{s=1}^{g_i} \alpha_{is} y^{g_i - s}$ will contribute $h!\alpha_{ig_i - h}$. Thus

$$\alpha_{ig_i - h} = \frac{f_i^{(h)}(0)}{h!} \quad \text{where} \quad f_i(y) = \prod_{j \neq i}^{r} \left\{\frac{\lambda_i - \lambda_j}{\lambda_i} + y\,\frac{\lambda_j}{\lambda_i}\right\}^{-g_j}. \tag{2.20}$$

In practice the constants can be most easily found as follows.

$$f_i(y) = \prod_{j \neq i}^{r} \left\{\frac{\lambda_i}{\lambda_i - \lambda_j}\right\}^{g_j} \prod_{j \neq i}^{r} \left\{1 + y\,\frac{\lambda_j}{\lambda_i - \lambda_j}\right\}^{-g_j} \tag{2.21}$$

$$= \prod_{j \neq i}^{r} \left\{\frac{\lambda_i}{\lambda_i - \lambda_j}\right\}^{g_j} \exp\left\{-\sum_{j \neq i}^{r} g_j \log\left[1 + y\,\frac{\lambda_j}{\lambda_i - \lambda_j}\right]\right\}. \tag{2.22}$$

Since t can always be chosen so that $|y| < 1$, we can expand each side of the equation and equate coefficients.

$$(2.23)\qquad \begin{aligned} \sum_{h=0}^{\infty} \alpha_{ig_i-h}\, y^h &= \sum_{h=0}^{\infty} \frac{f_i^{(h)}(0)}{h!}\, y^h \\ &= \prod_{j\neq i}^{r} \left\{\frac{\lambda_i}{\lambda_i - \lambda_j}\right\}^{g_j} \exp\left\{\sum_{h=1}^{\infty} \frac{y^h}{h} \sum_{j\neq i}^{r} g_j \left[\frac{-\lambda_j}{\lambda_i - \lambda_j}\right]^h\right\}. \end{aligned}$$

The relation between α_{ig_i-h} and the coefficient of y^h on the right-hand side of (2.23) is the same as that between the hth moment about the origin μ_h' and the hth cumulant K_h. The well known equalities expressing the moments in terms of the cumulants may be used, therefore, in calculating the coefficients required above. For if we write

$$(2.24)\qquad K_{ih} = (h-1)! \sum_{j\neq i}^{r} \left\{g_j \left[\frac{-\lambda_j}{\lambda_i - \lambda_j}\right]^h\right\}$$

then

$$(2.25)\qquad \alpha_{ig_i-h} = (\mu_{1h}'/h!)\alpha_{ig_i}$$

$$(2.26)\qquad \alpha_{ig_i} = \prod_{j\neq i}^{r} \left\{\frac{\lambda_i}{\lambda_i - \lambda_j}\right\}^{g_j}.$$

3. Investigation of the accuracy of a simple approximation to the distribution of a nonnegative quadratic form. Welch [4], [5] and Fairfield Smith [6] have represented the distribution of a particular nonnegative quadratic form, when $r = 2$, by that of $Z = g\chi^2(h)$, the constants g and h being chosen so that the distribution has the same first two moments as Q. Satterthwaite [3] has suggested its use in the general case ($r \geqq 2$) when we have

THEOREM 3.1.

$$Q = z'Mz = \sum_{j=1}^{r} \lambda_j \chi^2(\nu_j)$$

is distributed approximately as $g\chi^2(h)$ *where*

$$(3.1)\qquad g = \frac{1}{2}\frac{K_2(Q)}{K_1(Q)} = \frac{\sum \nu_j \lambda_j^2}{\sum \nu_j \lambda_j} \quad \text{and} \quad h = \frac{2\{K_1(Q)\}^2}{K_2(Q)} = \frac{(\sum \nu_j \lambda_j)^2}{\sum \nu_j \lambda_j^2}.$$

It should be noted that the effective degrees of freedom, h, are necessarily *less* than the number appropriate if the λ_j were all equal. For if

$$\nu_1, \cdots, \nu_j, \cdots, \nu_r; \quad \lambda_1, \cdots, \lambda_j, \cdots, \lambda_r;$$

and μ are any positive real numbers, then

$$(3.2)\qquad \sum_{j=i}^{r} \nu_i(\lambda_j - \mu)^2 \geqq 0,$$

and if μ is the weighted mean of the λ's, that is if

TABLE 1

Comparison of Approximate and Exact Distributions of Some Quadratic Forms

Values of λ_j					Degrees of freedom					Exact probability (%) of exceeding approx. $100\alpha\%$ significance point					
λ_1	λ_2	λ_3	λ_4	λ_5	ν_1	ν_2	ν_3	ν_4	ν_5	$100\alpha\% = 5.00$	10.00	25.00	50.00	75.00	95.00
1	2	3			2	4	6			5.04	9.94	24.56	49.74	75.07	95.96
3	2	1			2	4	6			5.08	9.87	24.41	49.41	75.27	95.69
10	5	1			2	4	6			5.10	9.75	23.91	49.95	75.63	96.54
10	2	1			2	4	6			5.20	9.54	22.66	47.61	77.25	98.16
10	2	1			2	2	2			5.08	9.68	23.56	48.68	76.95	97.82
10	5	1			4	8	12			5.16	9.92	24.22	49.06	75.16	98.22
5	4	3	2	1	2	2	4	6	6	5.15	10.15	24.40	49.32	75.14	95.72

$$\mu = \left(\sum_{j=1}^{r} \nu_j \lambda_j\right) \Big/ \sum_{j=1}^{r} \nu_j, \tag{3.3}$$

then (3.2) is equivalent to

$$\sum_{j=1}^{r} \nu_j \lambda_j^2 - \left(\sum_{j}^{r} \nu_j \lambda_j\right)^2 \Big/ \sum_{j=1}^{r} \nu_j \geqq 0, \tag{3.4}$$

that is

$$\left(\sum_{j=1}^{r} \nu_j \lambda_j\right)^2 \Big/ \sum_{j=1}^{r} \nu_j \lambda_j^2 \leqq \sum_{j=1}^{r} \nu_j . \tag{3.5}$$

Although an approximation of this kind has often been used (see for example Patnaik [7]), investigations of its accuracy seem limited to the case $k = 2$ studied by Satterthwaite [3].

Table 1 shows the exact probabilities (calculated from the finite series of Theorem 2.4) of exceeding the significance points obtained from the approximation for a number of particular quadratic forms. This brief investigation suggests that the simple approximation is fairly good over a wide range of values of ν and λ. However, when small differences in probability were to be examined, it would be necessary to apply the method with caution and make checks by the exact methods.

4. Distribution of the ratio of independently distributed quadratic forms in multi-normal varieties. By canonical reduction of numerator and denominator, the ratio $Y = Q_1/Q_2$ is seen to be distributed like the quantity

$$X_1/X_2 = \left\{\sum_{j'=1}^{r'} \lambda'_{j'} \chi^2(\nu'_{j'})\right\} \Big/ \left\{\sum_{j=1}^{r} \lambda_j \chi^2(\nu_j)\right\}. \tag{4.1}$$

By representing numerator and denominator by infinite χ^2 series, Pitman and Robbins [2] have obtained the distribution of Y, when the λ's are all positive, as an infinite series in which each term contains a probability calculated from the F distribution (or more conveniently from the incomplete B-function). In our application it will always be possible to arrange that component χ^2's (at least of the denominator) have even degrees of freedom.

Employing the Robbins-Pitman [2] infinite series in the numerator and the finite series in the denominator, we readily obtain Theorem 4.1 (for example by using Cramer's theorem [8] concerning the characteristic function of a ratio).

THEOREM 4.1. *If the* λ'*'s of the numerator are all positive and if* $\lambda_1' = a'$ *is the smallest of the* λ'*'s and* $\sum_{j'=1}^{r'} \nu'_{j'} = \nu'$, *then*

$$P_1 \leqq P_r\,(Y > Y_0) \leqq P_2 \tag{4.2}$$

where

$$\begin{aligned} P_1 = {} & \sum_{l'=0}^{n'} \sum_{j=1}^{r} \sum_{s=1}^{g_j} c_{l'}'\, \alpha_{js}\{I_{x_j}(s, \tfrac{1}{2}\nu' + l')\} \\ & + \left\{1 - \sum_{l'=0}^{n'} c_{l'}'\right\} \sum_{j=1}^{r} \sum_{s=1}^{g_j} \alpha_{js}\{I_{x_j}(s, \tfrac{1}{2}\nu' + n' + 1)\}, \end{aligned} \tag{4.3}$$

$$P_2 = \sum_{l'=0}^{n'} \sum_{j=1}^{r} \sum_{s=1}^{g_j} c_{l'}'\, \alpha_{js}\{I_{x_j}(s, \tfrac{1}{2}\nu' + l')\} + \left\{1 - \sum_{l'=0}^{n'} c_{l'}'\right\}. \tag{4.4}$$

The c'*'s are obtained from* (2.10), *the* α_{js} *from* (2.24), (2.25) *and* (2.26), $I_x(pq)$ *is the incomplete Beta function integral, and* $x_j = \{1 + (\lambda_j/a_1')Y_0\}^{-1}$.

If both numerator and denominator of (4.1) have even degrees we may use the finite series in both numerator and denominator and obtain

THEOREM 4.2. *If* $\nu_j = 2g_j$ *and* $\nu_{j'}' = 2g_{j'}'$ *are even, then*

$$\Pr\,(Y \leqq Y_0) = \sum_{j'=1}^{r'} \sum_{s'=1}^{g'_{j'}} \sum_{j=1}^{r} \sum_{s=1}^{g_j} \alpha_{j's'}'\, \alpha_{js}\{I_{x_{jj'}}(s, s')\} \tag{4.5}$$

where the $\alpha_{j's'}'$ *and* α_{js} *are obtained from* (2.24), (2.25) *and* (2.26), *and* $x_{jj'} = \{1 + \lambda_j Y_0/\lambda_{j'}'\}^{-1}$.

Alternatively, if the forms are nonnegative it is usually simpler to use the following.

THEOREM 4.3. *If* $\lambda_1', \lambda_2', \cdots, \lambda_r'$ *and* $\lambda_1, \lambda_2, \cdots, \lambda_r$ *are all positive and the* ν_j *and* ν_j' *are all even then*

$$\Pr\left[\left\{\sum_{j'=1}^{r'} \lambda_{j'}'\, \chi^2(\nu_{j'}')\right\} \Big/ \left\{\sum_{j=1}^{r} \lambda_j\, \chi^2(\nu_j)\right\} > Y_0\right] = \sum_{i=1}^{r'} \sum_{s=1}^{g_i} \alpha_{is} \tag{4.6}$$

where the α_{is} $(i = 1, 2, \cdots r'; s = 1, \cdots g_j)$ *are constants calculated from* (2.24), (2.25) *and* (2.26) *for the form* $\sum_{i=1}^{r'+r} \zeta_i \chi^2(\nu_i)$ *in which* $\zeta_1 = \lambda_1', \zeta_2 = \lambda_2', \cdots, \zeta_{r'} = \lambda_{r'}'; \zeta_{r'+1} = -Y_0\lambda_1, \zeta_{r'+2} = -Y_0\lambda_2, \cdots, \zeta_{r'+r} = -Y_0\lambda_r$.

PROOF. Since the quadratic forms are nonnegative, the left-hand side of (4.6) may be written

$$P = \Pr\left[\left\{\sum_{j'=1}^{r'} \lambda_{j'}'\, \chi^2(\nu_{j'}') - \sum_{j=1}^{r} Y_0 \lambda_j\, \chi^2(\nu_j)\right\} > 0\right] \tag{4.7}$$

which is of the form

TABLE 2

Comparison of Approximate and Exact Distributions of Ratios of Quadratic Forms

Numerator								Denominator										Exact probability (%) of exceeding approx. $100\alpha\%$ significance point			
Degrees of freedom				Values of λ's				Degrees of freedom					Values of λ's								
ν'_1	ν'_2	ν'_3	ν'_4	λ'_1	λ'_2	λ'_3	λ'_4	ν_1	ν_2	ν_3	ν_4	ν_5	λ_1	λ_2	λ_3	λ_4	λ_5	$100\alpha\% = 5.00$	10.00	25.00	50.00
2	2			1	2			2	4	6			3	2	1			4.71	9.63	24.60	49.96
2	2			1	2			2	4	6			10	5	1			4.24	9.15	24.36	50.00
2	2			1	3			2	4	6			10	5	1			4.26	9.12	24.19	49.93
2	2			1	2			4	8	12			3	2	1			4.88	9.76	24.59	49.91
2	2			1	3			4	8	12			10	5	1			4.72	9.55	24.31	49.80
2	2	2	2	1	2	3	4	2	2	4	6	6	5	4	3	2	1	5.03	9.95	24.86	50.05

$$\Pr\left[\sum_{i=1}^{r'+r} \zeta_i \chi^2(\nu_i) > 0\right]. \tag{4.8}$$

Using (2.11)

$$P = \sum_{i=1}^{r'} \sum_{s=1}^{g_i} \alpha_{is} \Pr\{\chi^2(2s) > 0\} + \sum_{j=r+1}^{r'+r} \sum_{w=1}^{g_j} \alpha_{jw} \Pr\{\chi^2(2w) < 0\}. \tag{4.9}$$

But

$$\Pr\{\chi^2(2s) > 0\} = 1 \quad \text{and} \quad \Pr\{\chi^2(2w) < 0\} = 0. \tag{4.10}$$

Therefore,

$$P = \sum_{i=1}^{r'} \sum_{s=1}^{g_i} \alpha_{is}. \tag{4.11}$$

It will be noted that when this series is applicable, the required probability may be obtained directly without the rather tedious interpolation in the B-function tables required by the method of Theorem 4.2.

5. Use of Theorem 4.3 with quadratic forms that are not independent. This method may be used in suitable cases even if the quadratic forms Q_1 and Q_2 are not distributed independently. For if Q_1 and Q_2 are nonnegative, we may write

$$\begin{aligned} P &= \text{Ps}\{Q_1/Q_2 > \varphi\} = \Pr\{Q_1 - \varphi Q_2 > 0\} = \Pr(z'M_1 z - \varphi z'M_2 z) > 0 \\ &= \Pr\{z'Mz > 0\} = \Pr\left\{\sum_{i=1}^{r'} \zeta_i \chi^2(\nu_i) + \sum_{j=1}^{r} \zeta_j \chi^2(\nu_j) > 0\right\} \end{aligned} \tag{5.1}$$

where ζ_i $(i = 1, \cdots, r')$ is a positive latent root, repeated ν_i times, of the matrix $M = M_1 - M_2$ and ζ_j $(j = 1, \cdots, r)$ is a negative latent root of the same matrix repeated ν_j times. In certain investigations (for example in the study of the two-way classification of the analysis of variance table) it is possible to ensure that the ν_i and ν_j are all even, whence we may apply Theorem 4.3 and obtain

$$P = \sum_{i=1}^{r'} \sum_{s=1}^{\nu i/2} \alpha_{is}\,. \tag{5.2}$$

6. The accuracy of a simple approximation to the distribution of the ratio of independent nonnegative quadratic forms. Since approximation to the distribution of a positive quadratic form Q by $g\chi^2(h)$ is fairly satisfactory, we may attempt to approximate the ratio of two independent quadratic forms Q' and Q by fitting χ^2 distributions in both numerator and denominator, in the manner described.

THEOREM 6.1. *If Q' is distributed approximately as $g'\chi^2(h')$ and Q as $g\chi^2(h)$, a quantity whose distribution approximates to that of the ratio Q'/Q is $bF(h', h)$, where $b = (g'h')/(gh)$ and the g's and h's are found from* (3.1). *In fact*

$$b = \mathrm{K}_1(Q')/\mathrm{K}_1(Q), \qquad h' = 2\{\mathrm{K}_1(Q')\}^2/\mathrm{K}_2(Q'), \qquad h = 2\{\mathrm{K}_1(Q)\}^2/\mathrm{K}_2(Q). \tag{6.1}$$

In Table 2 are shown the exact probabilities (calculated from the finite series of Theorem 4.3) of exceeding the significance points obtained from the approximation. The approximation is not of great accuracy, but may be usefully employed to supplement the accurate (but less suggestive) exact methods.

7. The one-way analysis of variance classification. Data, which it is desired to test for group to group homogeneity of mean value, often are obtained in circumstances where group-to-group homogeneity of variance is not to be expected. To quote one of many examples; in biological work where each observation is the response observed with a particular animal and the subject of enquiry is the comparison of the effects of treatments applied to the animal, the application of the treatment itself would often be expected to cause extra variability, and the extent of this extra variability would vary with the type and manner of treatment applied.

The problem of the effect of unequal group variances was considered in the case of the t test by Welch [5]. He obtained approximate probabilities from which it appeared that the effect was small when the groups were of equal size, but larger when they were different in size. Later some exact probabilities for this case were found by Hsu [9] and another investigation by a different approximate method was made by Grunow [10]. Both these investigations confirmed Welch's results. Quensel [11] considered the one-way analysis of variance classification more generally and obtained an approximate expression for the variance of the F criterion when the group variances differed. He concluded that the test would not be greatly affected if the group sizes were equal.

David and Johnson [12], [13], [14] have discussed the general problem of the power function of analysis of variance criteria when the observations are distributed independently but do not necessarily follow the normal distribution or have constant variance. As a special case they consider the one-way classifi-

TABLE 3

Analysis of Variance. Group Variances Unequal

Source	Deg. fr.	Sum of squares Q	Expectation of Q	Null distribution of Q
Between groups	$k-1$	$Q_B = \sum_{t=1}^{k} n_t(\bar{y}_{t.} - \bar{y}_{..})^2$	$\sum_{t=1}^{k} n_t\gamma_t^2 + \sum_{t=1}^{k}\left(1 - \frac{n_t}{N}\right)\sigma_t^2$	$\sum_{t=1}^{k-1} \lambda_t\, \chi^2(1)$
Within groups	$N-k$	$Q_W = \sum_{t=1}^{k}\sum_{i=1}^{n_t} (y_{ti} - \bar{y}_{t.})^2$	$\sum_{t=1}^{k} (n_t - 1)\,\sigma_t^2$	$\sum_{t=1}^{k} \sigma_t^2\, \chi^2(n_t - 1)$

cation in which the observations are normally distributed but the variances differ from group to group. Their method is different from that given here and is an approximate one. At the time of writing they have published few numerical results and these [14] are confined to the case in which the sizes of the groups are all equal. Confirming the results of Quensel, only slight changes in probability, from those expected if the assumptions were true, have been found.

Using the theorems on quadratic forms discussed above, the required probabilities may be found exactly, while the simple F approximation enables the nature of the effects found to be presented in a readily appreciated form.

Suppose we have $N = \sum_{t=1}^{k} n_t$ observations classified into k groups. The ith observation in the tth group is y_{ti}, the mean of the tth column $\bar{y}_{t.}$, and the grand mean $\bar{y}_{..}$, and there are n_t observations in the tth group. Then in the usual analysis of variance, the quantities Q_B and Q_W shown in the third column of Table 3 are calculated and are associated with degrees of freedom shown in the second column of the table.

It is usually assumed that

$$y_{ti} = \eta_t + z_{ti} \tag{7.1}$$

where $\eta_t = \alpha + \gamma_t$ is the population mean for the tth group, $\sum_{t=1}^{k} n_t\gamma_t = 0$ and the z_{ti} are errors distributed normally and independently about zero with the same variance σ^2. We retain all the assumptions except the last, and, instead of supposing the variance constant, we postulate that $\mathcal{E}(z_{ti}^2) = \sigma_t^2$ where $\sigma_1^2, \sigma_2^2, \cdots, \sigma_t^2, \cdots, \sigma_k^2$ are not necessarily all equal. Then

$$Q_B = \sum_{t=1}^{k} n_t(\gamma_t + \bar{z}_{t.} - \bar{z}_{t..})^2, \tag{7.2}$$

$$\mathcal{E}(Q_B) = \sum_{t=1}^{k} n_t\gamma_t^2 + \mathcal{E}\sum_{t=1}^{k} n_t(\bar{z}_{t.} - \bar{z}_{..})^2. \tag{7.3}$$

We notice that when the null hypothesis is true, Q_B is a quadratic form in the variables$_1$ $\bar{z}, \cdots, \bar{z}_{t.}, \cdots, \bar{z}_k$. The matrix $M = \{m_{ts}\}$ of this form is evidently $N^{-1}\{\delta_{ts}n_tN - n_tn_s\}$, where δ_{ts} is the Kronecker delta. Also the variables follow the multi-normal distribution with diagonal variance-covariance matrix V whose tth element is σ_t^2/n_t. It follows that the matrix U of Theorem 2.1 is

$$U = VM = N^{-1}\{\delta_{ts}\sigma_t^2N - \sigma_t^2n_s\}. \tag{7.4}$$

Using (7.3) and Theorems 2.1 and 2.2, the expectations and null-distribution of Q_B are those shown in the last two columns of Table 3, $\lambda_1, \cdots, \lambda_{k-1}$ being the latent roots of the matrix U. Again, from (7.1),

$$Q_E = \sum_{t=1}^{k} \sum_{i=1}^{n_t} (\bar{y}_{ti} - \bar{y}_{t.})^2 = \sum_{t=1}^{k} \sum_{i=1}^{n_t} (z_{ti} - \bar{z}_{t.})^2. \tag{7.5}$$

Also, since $\sum_{i=1}^{n_t} (z_{ti} - \bar{z}_{t.})^2$ is distributed independently of $\bar{z}_{t.}$ like $\sigma_t^2 \chi^2 (n_t - 1)$, it follows that Q_W is distributed like $\sum_{t=1}^{k} \sigma_t^2 \chi^2\ (n_t - 1)$ independently of Q_B. We may now employ Theorem 4.1 to obtain the exact probability that the ratio of mean squares will exceed the significance points of the tabled F distribution, for any chosen set of group variances. In addition, an approximate value of this probability may be obtained using Theorem 6.1 with equations (2.5) and (2.6). We find that the ratio of mean squares is distributed approximately as $bF(h', h)$ where

$$b = \frac{N - k}{N(k - 1)} \sum_t (N - n_t)\sigma_t^2 / \sum_t (n_t - 1)\sigma_t^2, \tag{7.6}$$

$$h' = \{\sum_t (N - n_t)\sigma_t^2\}^2 / \{\sum_t n_t \sigma_t^2\}^2 + N \sum_t (N - 2n_t)\sigma_t^4\}, \tag{7.7}$$

$$h = \{\sum_t (n_t - 1)\sigma_t^2\}^2 / \{\sum_t (n_t - 1)\sigma_t^4\}. \tag{7.8}$$

A number of calculations of both exact and approximate probabilities are shown in Table 4. It is seen that in the cases studied the approximation, although not

TABLE 4

Probabilities of Exceeding 5% Point when Variances are Unequal in the One-Way Analysis of Variance Table

	Group variances					Number of Observations						Probability (%) of exceeding 5% point		Values in approximating distribution, b F (h′, h),		
						in Groups					Total					
	σ_1^2	σ_2^2	σ_3^2	σ_4^2	σ_5^2	n_1	n_2	n_3	n_4	n_5	N	Exact	Approx.	b	h′	h
(1)	1	1	1			Any					15	5.00		1	2	12
(2)	1	2	3			5	5	5			15	5.58	5.78	1	1.85	10.29
(3)	1	2	3			3	9	3			15	5.55	5.72	1	1.74	11.08
(4)	1	2	3			7	5	3			15	9.25	9.57	1.28	1.86	10.00
(5)	1	2	3			3	5	7			15	4.03	3.35	0.80	1.82	10.89
(6)	1	1	3			5	5	5			15	5.87	5.82	1	1.72	9.09
(7)	1	1	3			7	5	3			15	10.70	9.78	1.35	1.67	9.14
(8)	1	1	3			9	5	1			15	17.41	18.04	1.93	1.62	12.00
(9)	1	1	3			1	5	9			15	1.31	1.73	0.60	1.85	10.32
(10)	1	1	1	1	1	Any					25	5.00		1	4	20
(11)	1	1	1	1	3	5	5	5	5	5	25	7.42	6.86	1	3.21	15.08
(12)	1	1	1	1	3	9	5	5	5	1	25	14.64	15.56	1.48	3.04	20
(13)	1	1	1	1	3	1	5	5	5	9	25	2.49	2.60	0.73	3.40	15.43

of great accuracy, faithfully indicates the order and direction of the effects and enables a clear idea to be gained of the general effects to be expected.

8. Equal groups. For equal groups the comparison of mean squares is unbiased in the sense that the expectations of the mean squares are equal when the null hypothesis is true. In fact

$$\mathcal{E}(Q_B)/(k-1) = \mathcal{E}(Q_W)/(N-k) = \bar{\sigma}^2 \qquad \text{where } \bar{\sigma}^2 = \left(\sum \sigma_t^2\right)/k.$$

The mean squares are distributed independently but their ratio does not follow the distribution $F(k-1, N-k)$. Instead, applying the approximation, we find after a little reduction of (7.7) and (7.8) that the ratio of mean squares is distributed approximately as $F\{(k-1)\epsilon', (N-k)\epsilon\}$ where ϵ' and ϵ, the factors by which the degrees of freedom are reduced, are given by

$$\epsilon' = \left\{1 + \frac{k-2}{k-1}c^2\right\}^{-1}, \qquad \epsilon = (1+c^2)^{-1}, \tag{8.1}$$

and c is the coefficient of variation of the variances. That is to say, c^2 is the variance of the variances divided by the square of the mean variance $\bar{\sigma}^2$:

$$c^2 = \frac{1}{k}\sum_{t=1}^{k} (\sigma_t^2 - \bar{\sigma}^2)^2/(\bar{\sigma}^2)^2. \tag{8.2}$$

Since, when the variances are unequal, ϵ' and ϵ are less than unity, one would expect that the significance of effects would be somewhat overestimated. Comparison in Table 4, of rows (2) and (6) with (1), and of row (11) with row (10), confirms this, and shows that for the differences in variance considered, only moderate discrepancies in probability occur.

Now the σ^2's are essentially positive and it is easily seen that

$$0 \leqq c^2 \leqq k-1, \tag{8.3}$$

and if the variances range from a lower value σ^2 to an upper value $a\sigma^2$, then the largest possible value for c is attained when $k-1$ of the variances are equal to σ^2 and the remaining one is equal to $a\sigma^2$, and that in this case

$$c^2 = (k-1)(a-1)^2/(a-1+k)^2. \tag{8.4}$$

TABLE 5

Values of c^2, ϵ' and ϵ

Largest Variance is a Times Larger than Each of the Remaining Variances

k	3 groups			6 groups			10 groups		
a	3	6	10	3	6	10	3	6	10
c^2	0.32	0.78	1.12	0.31	1.03	1.80	0.25	1.00	2.02
ϵ'	0.86	0.72	0.64	0.80	0.55	0.41	0.82	0.52	0.36
ϵ	0.76	0.56	0.47	0.76	0.50	0.36	0.80	0.50	0.33

Values for c^2 greater than one or at most two probably would be extremely rare in practice, although from the inequality (8.3) it is seen that values of c^2 as great as $k - 1$, and hence values of ϵ as small as $1/k$, *could* occur. Some idea of the discrepancy arising in a very unfavorable case is obtained by considering the example $k = 7$, $n_1 = n_2 = \cdots = n_7 = 3$, $a = 10$; the exact probability of exceeding the assumed 5 per cent point is then 12.0 per cent.

9. Unequal groups. It will be observed that the more serious discrepancies in Table 4 arise when the groups are unequal. This is because with unequal groups the comparison of mean squares is usually biased. If we write $\dot{\sigma}^2$ for the *weighted* mean variance $\sum \nu_t \sigma_t^2 / \sum \nu_t$, and $\bar{\sigma}^2$ for the unweighted mean variance $\sum \sigma_t^2/k$, where the weight $\nu_t = n_t - 1$ is the number of degrees of freedom in the tth group then the expression (7.6) for the bias coefficient reduces to the form

$$b = 1 + \frac{1 - 1/N}{1 - 1/k}\left\{\frac{\bar{\sigma}^2}{\dot{\sigma}^2} - 1\right\}. \tag{9.1}$$

The bias is seen to depend upon the ratio of the unweighted and weighted means of the variances.

In this connection it is of interest to consider the examples of rows (2), (3), (4) and (5) in Table 4. In each case the total number of observations is 15 and the unweighted mean variance is 2. In (2) the numbers in the groups are equal, the weighted and unweighted means agree, and there is no bias, the discrepancy in probability being small. In (3) the numbers are unequal but the distribution is symmetrical, the weighted and unweighted means again agree, and again the discrepancy is small and of the same order as that found before. In (4) the weighted mean variance of 1.67 is lower than the unweighted mean variance of 2, causing an upward bias and a marked discrepancy in the direction of over-estimation of significance. In (5) on the other hand, the weighted mean of 2.33 exceeds 2, causing a downward bias corresponding with a discrepancy in probability resulting in underestimation of significance.

We have seen that in the case of equal groups, the discrepancy, as measured by a reduction in the degrees of freedom of the approximating F distribution, is dependent on the *spread* of the distribution of variances as measured by the coefficient of variation. The feature of the distribution of variances which affects the bias, on the other hand, is related to the "skewness" of that distribution as measured by the ratio of weighted and unweighted means.

The factors which multiply the degrees of freedom in the approximation may be written in this case of unequal groups

$$\epsilon' = [1 + \{c(\lambda)\}^2]^{-1}, \qquad \epsilon = [1 + \dot{c}^2]^{-1}, \tag{9.2}$$

where $c(\lambda)$ is the coefficient of variation of the λ's and $\dot{c}$ is the weighted coefficient of variation of the variances. That is to say, $\dot{c}^2$ is the weighted variance of the variances divided by the weighted mean variance $\dot{\sigma}^2$:

$$\dot{c}^2 = (N - k)^{-1} \sum_{t=1}^{k} \nu_t(\sigma_t^2 - \dot{\sigma}^2)^2/(\dot{\sigma}^2)^2. \tag{9.3}$$

The variation among the λ's will be similar although somewhat less in extent than the variation among the σ_i^2's so that, as before, the coefficients ϵ' and ϵ will depend upon the *spread* of the σ_i^2's.

Study of Table 4, particularly rows (4), (7), (8), (9), (12), and (13), shows that quite large discrepancies can occur when the groups are unequal for even moderate variations of variance. Furthermore, it is clear that these discrepancies will persist in larger samples, for as the sample sizes are increased proportionately the bias coefficient will tend to the fixed limit

$$1 + (k/k - 1)\,\{\bar{\sigma}^2/\dot{\sigma}^2 - 1\}. \tag{9.4}$$

Acknowledgement. I am indebted to Mrs. Margaret Edmondson for valuable assistance with the computations.

REFERENCES

[1] W. G. COCHRAN, "The distribution of quadratic forms in a normal system, with applications to the analysis of covariance," *Proc. Cambridge Philos. Soc.*, Vol. 30 (1934), pp. 178–191.

[2] H. ROBBINS AND E. J. G. PITMAN, "Application of the method of mixtures to quadratic forms in normal variates," *Ann. Math. Stat.*, Vol. 20 (1949), pp. 552–560.

[3] F. E. SATTERTHWAITE, "Synthesis of Variance," *Psychometrika*, Vol. 6 (1941), pp. 309–316.

[4] B. L. WELCH, "The specification of rules for rejecting too variable a product, with particular reference to an electric lamp problem," *J. Roy. Stat. Soc., Suppl.*, Vol. 3 (1936), pp. 29–48.

[5] B. L. WELCH, "The significance of the difference between two means when the population variances are unequal," *Biometrika*, Vol. 29 (1937), pp. 350–362.

[6] H. FAIRFIELD SMITH, "The problem of comparing the results of two experiments with unequal errors," *Journal of the Council for Scientific and Industrial Research* (Australia), Vol. 9, pp. 211–212.

[7] P. B. PATNAIK, "The non-central χ^2 and F-distributions and their applications," *Biometrika*, Vol. 36 (1949), pp. 202–232.

[8] H. CRAMÉR, *Random Variables and Probability Distributions*, Cambridge University Press, 1937, p. 46.

[9] P. L. HSU, "Contributions to the theory of 'Students' t-test as applied to the problem of two samples," *Statistical Research Memoirs*, Vol. 2 (1938), pp. 1–24.

[10] D. G. C. GRUNOW, "Test for the significance of the difference between means in two normal populations having unequal variances," *Biometrika*, Vol. 38 (1951), pp. 252–256.

[11] C. E. QUENSEL, "The validity of the z-criterion when the variates are taken from different normal populations," *Skand. Aktuarietids.*, Vol. 30 (1947), pp. 44–55.

[12] F. N. DAVID AND N. L. JOHNSON, "The effect of non-normality on the power function of the F-test in the analysis of variance," *Biometrika*, Vol. 38 (1951), pp. 43–57.

[13] F. N. DAVID AND N. L. JOHNSON, "A method of investigating the effect of non-normality and heterogeneity of variance on tests of the general linear hypothesis," *Ann. Math. Stat.*, Vol. 22 (1951), pp. 382–392.

[14] F. N. DAVID AND N. L. JOHNSON, "The sensitivity of analysis of variance tests with respect to random between groups variation," *Trabajos de Estadistica*, Vol. 2 (1951) pp. 179–188.

4.3

Some Theorems on Quadratic Forms Applied in the Study of Analysis of Variance Problems, II: Effects of Inequality of Variance and of Correlation between Errors in the Two-Way Classification

G. E. P. BOX
Imperial Chemical Industries Ltd. and North Carolina State College

1. Summary and Introduction. Theorems already enunciated in a previous paper on quadratic forms are used to determine the effects of inequality of variance and first order serial correlation of errors in the two-way classification on the analysis of variance. It is found that when the appropriate null hypothesis is true, inequality of variance from column to column results in an increased chance of exceeding the significance point for the test on homogeneity of column means, and a decreased chance for the corresponding test on row means. For moderate differences in variance neither effect is large. First order serial correlation within rows produces a large effect on the "between rows" comparisons, but little effect on the "between columns" comparisons.

2. The two-way analysis of variance classification. Consider the analysis of variance for a two-way table with k columns and n rows, with one observation in each cell. Experiments in which k treatments are tested in n blocks are an important source of data classified in this way. In such tables the variance might change from treatment to treatment due to the influence of the treatments themselves. Changes in variance might also occur from block to block, for in some circumstances where experimental material was inhomogeneous in mean from block to block it might well be inhomogeneous in variance also.

A further source of departure from the assumptions usually made in the analysis of variance concerns possible lack of independence between the "error" components of the observations. In many types of experiments this difficulty is met by the introduction of randomisation. Data occur, however, in circumstances where there is no possibility of using this device, usually because the factor which is to be studied is the effect of time or position, which itself gives rise to the correlation.

For instance, the first example of analysis of variance of a two-way table in R. A. Fisher's *Statistical Methods for Research Workers* [1] concerns data quoted from Shaw [2] on the frequency of rainfall classified by hour of the day and month of the year. As Fisher himself points out, strong serial correlation between errors within months occurs because showers of rain which last more than one hour are recorded in successive hours. No question of randomisation arises in

Received 6/15/53.

The Collected Works of George E. P. Box, 1984, Wadsworth, Inc., Belmont, CA 94002.
Originally published in *Ann. Math. Stat.,* vol. 25, no. 3 (1954), pp. 484–498.

this example. In discussing the analysis of variance table, Fisher remarks that the serial correlation between hours within months entirely invalidates the "between months" comparisons, but that the "between hours" comparisons may still be made (as an approximate test). The truth of the latter part of this statement is perhaps not immediately obvious, and it is of interest to make a closer study of such examples.

Other instances of two-way tables in which serial correlation between errors might be expected are quoted by Daniels [3] in experiments in wool research where, for example, the variation in weight of slubbing coming from adjacent positions on the wool card is considered. Daniels recognised that correlation effects might invalidate the analysis of variance procedure and carried out some theoretical investigation of the problem [4]. He considered the effects of small inequalities of variance and small correlations between errors, using an approximate method. Tests for the existence of departures from assumptions in the two-way table were discussed by Box in 1950 [5], when reference was given to the results now published.

In what follows we retain the assumption of normality, but allow the variance to differ from column to column and correlation to occur within rows. By substituting columns for rows we can also study the effect of differences in variance from row to row and the effect of correlation within columns.

3. Distribution of items in the analysis of variance table. We need to refer to theorems, equations and sections of a previous paper [6] with the same general title. We indicate such reference by the addition of a prime to the number of the theorem, etc. Thus Theorem 2.1′ and Section 5′ refer to Theorem 2.1 and Section 5, respectively, of the previous paper.

Suppose we have a two-way classification of observations with k columns and n rows and y_{ti} is the observation in the tth column and i^{th} row. Then we can perform an analysis of variance corresponding to the entries in the first three columns of Table 1. We make the usual assumptions that y_{ti} may be represented by a linear model

$$y_{ti} = \alpha + \beta_i + \gamma_t + z_{ti}, \qquad \sum_{i=1}^{n} \beta_i = 0; \qquad \sum_{t=1}^{k} \gamma_t = 0. \tag{3.1}$$

Alternatively we can denote the model for all the elements of the tth column of the table ($t = 1, 2, \cdots, k$) by

$$\mathbf{y}_{t.} = \alpha \mathbf{1}_n + \boldsymbol{\beta} + \gamma_t \mathbf{1}_n + \mathbf{z}_{t.} \tag{3.2}$$

where $\mathbf{y}_{t.}$ is the $n \times 1$ vector of entries in the tth column, $\mathbf{z}_{t.}$ is the corresponding vector of errors, $\mathbf{1}_n$ is an $n \times 1$ vector of unit elements, and $\boldsymbol{\beta}$ an $n \times 1$ vector of row constants $\beta_1, \beta_2, \cdots, \beta_n$. We shall also need the notation $\mathbf{y}_{.i}$ $\mathbf{z}_{.i}$ to denote $k \times 1$ vectors of observations and errors in the ith row of the table.

We do not make the usual assumption that the z_{ti} have the same variance and are uncorrelated. Instead we assume that $\mathbf{z}_{.i}$ follows the normal multivariate

TABLE 1

Analysis of variance for a two-way table with column variances unequal and correlation of errors within rows

Source	D/F	Sums of Squares, Q	*Expectation of Q	*Null distribution of Q
Rows	$n-1$	$Q_R = k\sum_{i=1}^{n}(\bar{y}_{.i}-\bar{y}_{..})^2$	$k\sum_{i=1}^{n}\beta_i^2+(n-1)(\bar{v}_{tt}+(k-1)\bar{v}_{ts})$ $k\sum_{i=1}^{n}\beta_i^2+(n-1)\bar{\sigma}^2$	† $(\bar{v}_{tt}+(k-1)\bar{v}_{ts})\chi^2(n-1)$ † $\bar{\sigma}^2\chi^2(n-1)$
Columns	$k-1$	$Q_C = n\sum_{t=1}^{k}(\bar{y}_{t.}-\bar{y}_{..})^2$	$n\sum_{t=1}^{k}\gamma_t^2+(k-1)(\bar{v}_{tt}-\bar{v}_{ts})$ $n\sum_{t=1}^{k}\gamma_t^2+(k-1)\bar{\sigma}^2$	‡ $\sum_{j=1}^{k-1}\lambda_j\chi^2(1)$
Residual (Error)	$(n-1)\times(k-1)$	$Q_E = \sum_{t=1}^{k}\sum_{i=1}^{n}(y_{ti}-\bar{y}_{t.}-\bar{y}_{.i}+\bar{y}_{..})^2$	$(n-1)(k-1)(\bar{v}_{tt}-\bar{v}_{ts})$ $(n-1)(k-1)\bar{\sigma}^2$	$\sum_{j=1}^{k-1}\lambda_j\chi^2(n-1)$

* The upper expressions hold in the general case, the lower ones when the correlations are zero, that is, when only differences in variance from column to column occur. In this case $\{\underline{u}_{ts}\} = \{(\delta_{ts}-k^{-1})\sigma_t^2\}$. In both cases, $\bar{v}_{tt}=\bar{\sigma}^2$ is the average variance $\Sigma_t\, v_{tt}/k$, and $\bar{v}_{ts}$ is the average covariance $\Sigma\,\Sigma_{t\neq s}\, v_{ts}/\{k(k-1)\}$, while $\lambda_1, \lambda_2, \cdots, \lambda_{k-1}$ are the $k-1$ nonzero latent roots of $\{\underline{u}_{ts}\} = \{v_{ts}-\bar{v}_{t.}\}$, where $\bar{v}_{t.}$ is the average value of the elements in the tth row or column of v.

† Distribution not independent of Q_E.

‡ Distribution independent of Q_E.

law with variance-covariance matrix $\mathcal{E}(\mathbf{z}_{.i}\mathbf{z}'_{.i}) = \mathbf{v} = \{v_{ts}\}$. We further assume that $\mathbf{z}_{.j}$ $(j = 1, 2, \cdots, i-1, i+1, \cdots, n)$ follows the same law independently of $\mathbf{z}_{.i}$. Thus $v_{11}, \cdots, v_{tt}, \cdots, v_{kk}$ are the k variances and $v_{12}, v_{13}, \cdots, v_{ts}, \cdots, v_{k-1\,k}$ the $\frac{1}{2}k(k-1)$ covariances, the same for every row. This enables us to study the effects of column to column heterogeneity of variance and/or "within rows" correlation of errors. The expected values and null distributions of the sums of squares, when the observations are so represented, are shown in Table 1. They are derived below.

Let $\mathbf{Y}_{t.}$ be an $n \times 1$ vector of elements $Y_{t1}, Y_{t2}, \cdots, Y_{tn}$ obtained from $\mathbf{y}_{t.}$ by orthogonal transformation $\mathbf{Y}_{t.} = \mathbf{p}\mathbf{y}_{t.}$, and let the $n \times n$ orthogonal matrix $\mathbf{p}$ have all the elements of its last row equal to $n^{-1/2}$, thus ensuring that $Y_{tn} = n^{1/2}\bar{y}_{t.}$. Then

$$\mathbf{Y}_{t.} = \mathbf{p}\mathbf{y}_{t.} = \alpha\boldsymbol{\delta} + \mathbf{B} + \gamma_t\boldsymbol{\delta} + \mathbf{Z}_{t.} \tag{3.3}$$

where $\boldsymbol{\delta} = \mathbf{p}\mathbf{1}_n$, $\mathbf{B} = \mathbf{p}\boldsymbol{\beta}$ and $\mathbf{Z}_{t.} = \mathbf{p}\mathbf{z}_{t.}$.

Due to the nature of $\mathbf{p}$, in the vector $\boldsymbol{\delta}$ the last element is $n^{1/2}$ and the remaining elements are zeros, and in $\mathbf{B}$ the last element is zero, since $\sum_1^n \beta_i = 0$. The transformed columns of the original two-way table and the transformed column of row means may now be written out as follows:

					Row Means
$B_1 + Z_{11}$	$\cdots$	$B_1 + Z_{t1}$	$\cdots$	$B_1 + Z_{k1}$	$B_1 + \bar{Z}_{.1}$
$\cdots$	$\cdots$	$\cdots$	$\cdots$	$\cdots$	$\cdots$
$B_i + Z_{1i}$	$\cdots$	$B_i + Z_{ti}$	$\cdots$	$B_i + Z_{ki}$	$B_i + \bar{Z}_{.i}$
$\cdots$	$\cdots$	$\cdots$	$\cdots$	$\cdots$	$\cdots$
$B_{n-1} + Z_{1n-1}$	$\cdots$	$B_{n-1} + Z_{tn-1}$	$\cdots$	$B_n + Z_{kn-1}$	$B_{n-1} + \bar{Z}_{.n-1}$
$n^{1/2}(\alpha + \gamma_1) + Z_{1n}$	$\cdots$	$n^{1/2}(\alpha + \gamma_t) + Z_{tn}$	$\cdots$	$n^{1/2}(\alpha + \gamma_k) + Z_{kn}$	$n^{1/2}\alpha + \bar{Z}_{.n}$

Now consider the $nk \times 1$ partitioned vector $\mathbf{z}$ and the $nk \times nk$ partitioned matrices $\mathbf{P}$ and $\mathbf{V}$ defined by

$$\mathbf{z}' = [\mathbf{z}'_{1.} \mid \mathbf{z}'_{2.} \mid \cdots \mid \mathbf{z}'_{t.} \mid \cdots \mid \mathbf{z}'_{k.}]$$

$$\mathbf{P} = \begin{bmatrix} \mathbf{p} & \mathbf{0}_n & \cdots & \mathbf{0}_n \\ \mathbf{0}_n & \mathbf{p} & \cdots & \mathbf{0}_n \\ \cdots & \cdots & \cdots & \cdots \\ \mathbf{0}_n & \mathbf{0}_n & \cdots & \mathbf{p} \end{bmatrix} \qquad \mathbf{V} = \begin{bmatrix} v_{11}\mathbf{I}_n & v_{12}\mathbf{I}_n & \cdots & v_{1k}\mathbf{I}_n \\ v_{12}\mathbf{I}_n & v_{22}\mathbf{I}_n & \cdots & v_{2k}\mathbf{I}_n \\ \cdots & \cdots & \cdots & \cdots \\ v_{1k}\mathbf{I}_n & v_{2k}\mathbf{I}_n & \cdots & v_{kk}\mathbf{I}_n \end{bmatrix} \tag{3.4}$$

where $\mathbf{I}_n$ is the $n \times n$ unit matrix and $\mathbf{0}_n$ is the $n \times n$ null matrix. For $\mathbf{Z}$, denoting the fector $\mathbf{Pz}$ of transformed variables, the matrix of variances and covariances is

$$\mathcal{E}(\mathbf{ZZ}') = \mathcal{E}(\mathbf{Pzz}'\mathbf{P}') = \mathbf{P}\mathcal{E}(\mathbf{zz}')\mathbf{P}' = \mathbf{P}'\mathbf{VP} = \mathbf{V}. \tag{3.5}$$

Since we are concerned with normal variates, it follows that the Z_{ti} are distributed in precisely the same manner as are the z_{ti}, that is the vector $\mathbf{Z}_{.i}$ of transformed errors in the ith row follows the normal multivariate law, with variance-covariance matrix $\mathbf{v}$, independently of the errors in the other rows.

Between columns sum of squares.

$$(3.6) \qquad Q_C = n \sum_{t=1}^{k} (\bar{y}_{t.} - \bar{y}_{..})^2 = \sum_{t=1}^{k} (n^{1/2}\gamma_t + Z_{tn} - \bar{Z}_{.n})^2$$

and the matrix of the quadratic form $q_n = \sum_{t=1}^{k} (Z_{tn} - \bar{Z}_{.n})^2$ is

$$(3.7) \qquad \mathbf{m} = \mathbf{I}_k - k^{-1}\mathbf{1}_k\mathbf{1}_k'$$

while the variance-covariance matrix for the vector of errors $\mathbf{Z}_{.n}$ is $\mathbf{v}$. We have therefore

$$(3.8) \qquad \mathbf{u} = \{u_{ts}\} = \mathbf{vm} = \{v_{ts} - \bar{v}_{t.}\}$$

where u_{ts} is the element of the tth row and s^{th} column of the matrix $\mathbf{u}$ and $\bar{v}_{t.}$ is the arithmetic mean of the entries in the tth row (or column) of $\mathbf{v}$. It follows from equation (2.5′) that the expectation and null distribution of Q_C are those shown in Table 1 where the λ's are the latent roots of $\mathbf{u}$.

Residual (error) sum of squares.

$$(3.9) \qquad Q_E = \sum_{i=1}^{n} \sum_{t=1}^{k} (y_{ti} - \bar{y}_{t.} - \bar{y}_{.i} + \bar{y}_{..})^2$$

$$= \sum_{i=1}^{n-1} \sum_{t=1}^{k} (Y_{ti} - \bar{Y}_{.i})^2 = \sum_{i=1}^{n-1} \sum_{t=1}^{k} (Z_{ti} - \bar{Z}_{.i})^2.$$

Denote $\sum_{t=1}^{k} (Z_{ti} - \bar{Z}_{.i})^2$ by q_i $(i = 1, 2, \cdots, n)$. Then q_i follows the same distribution as q_j $(j = 1, 2, \cdots, n)$ independently of q_j. In particular it follows the same distribution as q_n discussed above. Also, $\sum_{i=1}^{n-1} q_i = Q_E$ is distributed independently of Q_C, in the form indicated in Table 1.

Between rows sum of squares.

$$(3.10) \qquad Q_R = k \sum_{i=1}^{n} (\bar{y}_{.i} - \bar{y}_{..})^2 = k \sum_{i=1}^{n-1} Y_{.i}^2 = k \sum_{i=1}^{n-1} (B_i + \bar{Z}_{.i})^2.$$

Remembering that $\sum_{i=1}^{n-1} B_i^2 = \sum_{i=1}^{n} \beta_i^2$ we have

$$(3.11) \qquad \mathcal{E}(Q_R) = k \sum_{i=1}^{n} \beta_i^2 + (n - 1)\{\bar{v}_{tt} + (k - 1)\bar{v}_{ts}\},$$

where $\bar{v}_{tt}$ is the average variance $\sum_t v_{tt}/k$ and $\bar{v}_{ts}$ is the average covariance $\sum\sum_{t \neq s} v_{ts}/\{k(k - 1)\}$. Now $\bar{Z}_{.i}$ is distributed normally and independently of $\bar{Z}_{.j}$ $(i \neq j = 1, 2, \cdots, n)$. Hence, when the null hypothesis that $\sum_1^n \beta_1^2 = 0$ is true, Q_R is distributed like $X_R = \{\bar{v}_{tt} + (k - 1)\bar{v}_{ts}\}\chi^2(n - 1)$.

Since $\mathbf{Z}_{.n}$ is distributed independently of $\mathbf{Z}_{.i}$ $(i = 1, \cdots, n - 1)$, Q_R and Q_C

are distributed independently. Usually Q_R will not be distributed independently of Q_E, however, as will now be shown.

Dependence of Q_R and Q_E. To investigate the dependence of Q_R and Q_E we transform the $k \times 1$ vector $\mathbf{Z}_{.i}$ of the transformed variates in the ith row of the two-way table to the vector $\mathbf{W}_{.i}$ by means of the orthogonal transformation $\mathbf{W}_{.i} = \mathbf{RZ}_{.i}$, where the elements of the last (k^{th}) row of $\mathbf{R} = \{r_{ts}\}$ are all equal to $k^{1/2}$ so that $W_{ki} = k^{1/2}\bar{Z}_{.i}$. The variance-covariance matrix for the new variates is now given by

$$\mathcal{E}(\mathbf{W}_{.i}\mathbf{W}_{.i}') = \mathcal{E}(\mathbf{RZ}_{.i}\mathbf{Z}_{.i}'\mathbf{R}') = \mathbf{R}\mathcal{E}(\mathbf{Z}_{.i}\mathbf{Z}_{.i}')\mathbf{R}' = \mathbf{RvR}' \tag{3.12}$$

and therefore $\mathcal{E}(W_{ti}W_{ki}) = k^{1/2} \sum_{s=1}^{k} \bar{v}_{s.}r_{ts}$. Now $\sum_{s=1}^{k} r_{ts} = 0$ for $t = 1, 2, \cdots k - 1$. The covariances between W_{ki} and $W_{1i}, W_{2i}, \cdots, W_{k-1i}$ cannot therefore *all* be zero unless $\bar{v}_{s.}$, the mean of entries in the sth row or column of $\mathbf{v}$, is constant for all s, since in a k-space only one vector can be simultaneously at right angles to $k - 1$ other linearly independent vectors. In particular the condition that $\bar{v}_{s.}$ is constant for all s is satisfied when the observations are independent and the variances are equal (when $\mathbf{v} = \sigma^2\mathbf{I}_k$) and also when the observations are circularly correlated. This condition usually will not be satisfied, however. In particular it will not be satisfied when the observations are independent but the variances are unequal, or when the variances are equal and the observations are serially but not circularly correlated.

If W_{ki} is not distributed independently of W_{ti} ($t = 1, 2, \cdots, k - 1$), then W_{ki}^2 will not be distributed independently of $\sum_{t=1}^{k-1} W_{ti}^2$ and $Q_R = \sum_{i=1}^{n-1} W_{ki}^2$ will not be distributed independently of $Q_E = \sum_{i=1}^{n-1} \sum_{t=1}^{k-1} W_{ti}^2$.

4. Distribution of test criteria.

Between columns test. When the appropriate null-hypothesis is true, the ratio of mean squares $(n - 1)Q_C/Q_E$ is distributed like

$$X_C/X_E = \left\{\sum_{j=1}^{k-1} \lambda_j \chi^2(1)\right\} \Big/ \left\{\sum_{j=1}^{k-1} \lambda_j \chi^2(n-1)\right\}, \tag{4.1}$$

where the λ's, which are the same for both numerator and denominator, are the $k - 1$ nonzero latent roots of the matrix $\mathbf{u} = \{v_{ts} - \bar{v}_{t.}\}$, and the numerator and denominator are distributed independently. We may use the exact series of Theorem 4.1′ to find the value of Pr $(Q_C/Q_E > Y_0)$ and so provide a check on the F approximation, provided we choose examples in which n is odd so that $n - 1$ is even.

To use the approximation of Theorem 6.1′ we require the first two cumulants of Q_C and Q_E when the null hypothesis is true. Using equations (2.5′) and (2.6′) we have

$$K_1(Q_C) = k(\bar{v}_{tt} - \bar{v}_{..}) = (k - 1)(\bar{v}_{tt} - \bar{v}_{ts}) \tag{4.2}$$

$$K_2(Q_C) = 2 \sum_{t=1}^{k} \sum_{s=1}^{k} (v_{ts} - \bar{v}_{t.})(v_{ts} - \bar{v}_{s.})$$

$$(4.3)\qquad = 2\left\{\sum_{t=1}^{k}\sum_{s=1}^{k} v_{ts}^2 - 2k\sum_{t=1}^{k}\bar{v}_{t.}^2 + k^2\bar{v}_{..}^2\right\}$$

$$= 2\sum_{t=1}^{k}\sum_{s=1}^{k}(v_{ts} - \bar{v}_{t.} - \bar{v}_{s.} + \bar{v}_{..})^2$$

where $\bar{v}_{tt} = \sum_{t=1}^{k} v_{tt}/k$, while $\bar{v}_{..} = \sum_{t=1}^{k}\sum_{s=1}^{k} v_{ts}/k^2$ and $\bar{v}_{t.} = \sum_{s=1}^{k} v_{ts}/k$. Now $K(Q_E) = (n-1)K_1(Q_c)$ and $K_2(Q_E) = (n-1)K_2(Q_C)$. Hence the null distribution of the ratio of mean squares $(n-1)\,Q_C/Q_E$ is approximately that of $F\{(k-1)\epsilon, (k-1)(n-1)\epsilon\}$ where

$$(4.4)\qquad \epsilon = k^2(\bar{v}_{tt} - \bar{v}_{..})^2/(k-1)\left\{\sum_{t=1}^{k}\sum_{s=1}^{k} v_{ts}^2 - 2k\sum_{t=1}^{k}\bar{v}_{t.}^2 - k^2\bar{v}_{..}^2\right\}.$$

We notice that the comparison of column and residual mean squares is without bias, whatever the nature of the matrix $\mathbf{v}$. The discrepancy that arises is represented in the approximation as a reduction by the same fraction ϵ of both degrees of freedom in the F ratio.

Between rows test. For testing row means the appropriate ratio of mean squares is $(k-1)Q_R/Q_E$. As we have seen, Q_R and Q_E are not distributed independently and the comparison is biassed unless the average covariance $\bar{v}_{ts}$ is zero.

To obtain under the null hypothesis the exact probability

$$P = \Pr\{Q_R/Q_E > \phi_\alpha\}$$

where $(k-1)\phi_\alpha = F_\alpha\{n-1, (n-1)(k-1)\}$ is the α probability point of the F distribution with $n-1$ and $(n-1)(k-1)$ degrees of freedom, we rewrite the probability in the form $\Pr\{(Q_R - \phi Q_E) > 0\}$ and employ Theorem 4.3′ as explained in Section 5′.

Let $\dot{\mathbf{Z}}$ be a $k(n-1)\times 1$ vector of the Z_{ti} arranged in the order $Z_{11}, \cdots, Z_{k1}$; $Z_{12}, \cdots, Z_{k2}$; $Z_{1(n-1)}, \cdots, Z_{k(n-1)}$. Let $\dot{\mathbf{V}}$ be the variance-covariance matrix for the Z_{ti} arranged in this order; thus $\dot{\mathbf{V}} = \mathcal{E}(\dot{\mathbf{Z}}\dot{\mathbf{Z}}')$. Then under the null hypothesis $\sum_{i=1}^{n}\beta_i^2 = \sum_{i=1}^{n-1}B_i^2 = 0$, the quadratic forms Q_R and Q_E are each functions of $\dot{\mathbf{Z}}$,

$$(4.5)\qquad Q_R = k\sum_{i=1}^{n-1}\bar{Z}_{.i}^2 = \dot{\mathbf{Z}}'\mathbf{M}_R\dot{\mathbf{Z}}$$

$$(4.6)\qquad Q_E = \sum_{i=1}^{n-1}\sum_{t=1}^{k}(Z_{ti} - \bar{Z}_{.i})^2 = \dot{\mathbf{Z}}'\mathbf{M}_E\dot{\mathbf{Z}}.$$

We require the probability that $\dot{\mathbf{Z}}'\mathbf{M}\dot{\mathbf{Z}}$ exceeds zero, where $\mathbf{M} = (\mathbf{M}_R - \phi\mathbf{M}_E)$. Now $\mathbf{M}_R$ is a $k(n-1)\times k(n-1)$ matrix partitioned after every k^{th} row and column, with each of its $n-1$ diagonal positions occupied by a $k\times k$ matrix $\mathbf{m}_R = k^{-1}\mathbf{1}_k\mathbf{1}_k'$ and zeros elsewhere:

$$\mathbf{M}_R = \begin{bmatrix} \mathbf{m}_R & \mathbf{0}_k & \cdots & \mathbf{0}_k \\ \mathbf{0}_k & \mathbf{m}_R & \cdots & \mathbf{0}_k \\ \cdots & \cdots & \cdots & \cdots \\ \mathbf{0}_k & \mathbf{0}_k & \cdots & \mathbf{m}_R \end{bmatrix}. \tag{4.7}$$

Also, $\mathbf{M}_E$ and $\dot{\mathbf{V}}$, and hence $\dot{\mathbf{V}}(\mathbf{M}_R - \phi\mathbf{M}_E)$, are of this same form with the $k \times k$ matrices in the diagonal positions equal respectively to $\mathbf{m}_E = \mathbf{I}_k - k^{-1}\mathbf{1}_k\mathbf{1}_k'$, to $\mathbf{v}$, and to $\mathbf{v}(m_R - \phi\mathbf{m}_E)$. Hence the $(n - 1)k$ roots of the determinental equation

$$| \dot{\mathbf{V}}(\mathbf{M}_R - \phi\mathbf{M}_E) - \lambda\mathbf{I}_{k(n-1)} | = 0 \tag{4.8}$$

are the k roots of the equation

$$\Delta_k = | \mathbf{v}(\mathbf{m}_R - \phi\mathbf{m}_E) - \lambda\mathbf{I}_k | = | \{\bar{v}_{t.} - \phi(v_{ts} - \bar{v}_{t.}) - \lambda\delta_{ts}\} | = 0, \tag{4.9}$$

each repeated $n - 1$ times where δ_{ts} is the Kronecker delta. Thus

$$\Pr\{Q_R/Q_E > \phi\} = \Pr\left\{\sum_{i=1}^{r'} \lambda_i\chi^2(n-1) + \sum_{j=r'+1}^{k} \lambda_j\chi^2(n-1) > 0\right\}, \tag{4.10}$$

where λ_i and λ_j are respectively positive and negative roots of equation (4.9). No serious lack of generality in conclusions will be introduced if, in the examples we consider, we make the number of rows n odd so that $n - 1$ is even. Then we can apply Theorem 4.3′ and the required probability is

$$\Pr\{Q_R/Q_E > \phi\} = \sum_{i=1}^{r'} \sum_{s=1}^{(n-1)/2} \alpha_{is}, \tag{4.11}$$

where the α's are obtained from equations (2.24′) (2.25′), and (2.26′).

The theory above may be used to study the distributions of the test criteria for any matrix $\mathbf{v}$. We use it here to consider the effect upon the significance test when

(i) the errors are independent but inequality of variance from column to column occurs,

(ii) the errors have equal variance but are serially correlated within rows.

5. Effect of inequality of column variances in two-way table. If we assume that the variance-covariance matrix $\mathbf{v}$ of "errors within rows" is diagonal, with elements $v_{11} = \sigma_1^2$, $v_{22} = \sigma_2^2$, $\cdots$, $v_{kk} = \sigma_k^2$, we have the case in which the variance changes from column to column but the errors are distributed independently.

Between columns test. The matrix $\mathbf{u}$ of equation 3.8 reduces to

$$\mathbf{u} = \{(\delta_{ts} - k^{-1})\sigma_t^2\}. \tag{5.1}$$

Taking $n - 1$ even we can obtain the exact distribution of Q_c/Q_E under the null hypothesis, using Theorem 4.1′.

On simplification of equation (4.4) we find that $(n - 1)Q_c/Q_E$ is distributed approximately as $F\{(k - 1)\epsilon, (k - 1)(n - 1)\epsilon\}$ where

$$\epsilon = \{1 + c^2(k - 2)/(k - 1)\}^{-1}$$

and c is the coefficient of variation of the variances, given in equation (8.2′). The calculated values in Table 2 indicate that, as would be expected, the divergencies are similar to those for equal groups with the one-way classification.

Between rows test. Since the covariances v_{ts} are all zero, the comparison of row and error mean squares is not biassed. However, the row and column mean squares are not distributed independently. After substituting σ_t^2 for v_{tt} and zero for $v_{ts}(t \neq s)$ in Δ_k of equation (4.9), the resulting determinant may be simplified still further.

Here and in what follows we shall refer to the columns of a determinant, counting from left to right, as $c_1, c_2, \cdots$, etc., and the rows, counting from top to bottom, as $r_1, r_2, \cdots$, etc. By adding $c_2, c_3, \cdots, c_k$ to c_1, then subtracting $r_1 \times \sigma_j^2/\sigma_1^2$ from r_j $(j = 2, 3, \cdots, k)$, and finally dividing each row by k and changing signs in the last $k - 1$ rows, we find that the required k values of λ are the solutions of the determinantal equation

$$(5.2)\qquad (-1)^{k-1}\Delta_k = \begin{vmatrix} \sigma_1^2 - \lambda & \sigma_1^2(1+\phi)/k & \sigma_1^2(1+\phi)/k & \cdots & \sigma_1^2(1+\phi)/k \\ \lambda(1 - \sigma_2^2/\sigma_1^2) & \phi\sigma_2^2 + \lambda & 0 & \cdots & 0 \\ \lambda(1 - \sigma_3^2/\sigma_1^2) & 0 & \phi\sigma_3^2 + \lambda & \cdots & 0 \\ \cdots & \cdots & \cdots & \cdots & \cdots \\ \lambda(1 - \sigma_k^2/\sigma_1^2) & 0 & 0 & \cdots & \phi\sigma_k^2 + \lambda \end{vmatrix} = 0$$

In the one-way classification it appeared that for a given range of variances the greatest discrepancies might be expected when $k - 1$ of the variances were equal, while the kth was larger (say a times as large as the others). Suppose the variances are $1, 1, \cdots, 1, a$. Then (5.2) reduces to

$$(5.3)\qquad (\phi + \lambda)^{(k-2)}[k\lambda^2 - \{(k - 1)(1 - \phi a) + (a - \phi)\}\lambda - ka\phi] = 0$$

from which all the λ's are readily obtained.

The results of a number of calculations using methods described above are shown in Table 2. It appears that the discrepancies in probability both for the test on rows as well as for the test on columns are not very large.

As was the case for the one-way classification, the effect of column-to-column differences in variance is to cause the significance of column differences in mean to be overestimated, although the differences in variance would have to be large for the effect to become serious. In the row comparisons, discrepancies of similar order but in the opposite direction occur, leading to underestimation of significance. Comparison of the first and third lines with the second and fourth

TABLE 2

Probabilities of exceeding 5% point when column variances are unequal in the two-way analysis of variance table

Number of Rows n	Number of Columns k	Column Variances	True Chance (per cent) of Exceeding 5% point		Values in approximating distribution $F(h', h)$ of ratio of mean squares*	
			Row Test, Exact	Columns Test, Approx.	h'	h
11	3	1 2 3	4.25	5.49	1.85 (2)	18.46 (20)
5	3	1 2 3	4.27	5.59	1.85 (2)	7.38 (8)
11	3	1 1 3	3.76	5.93	1.72 (2)	17.24 (20)
5	3	1 1 3	3.91	6.12†	1.72 (2)	6.90 (8)
3	5	1 1 1 1 3	4.47	6.92‡	3.21 (4)	6.43 (8)
3	11	1 1 ··· 1 3	4.86	7.09	7.90 (10)	15.79 (20)

* Bracketted values show appropriate degrees of freedom when variances are equal.

† 5.98 by the exact method.

‡ 6.75 by the exact method.

lines in Table 2 shows that the effects are worst when all the variances but one are at the lower end of the range. Comparison of the last four lines in the table suggests that the between-rows discrepancy is worse when the number of rows exceeds the number of columns, while the between-columns discrepancy is worse when the number of columns exceeds the number of rows.

6. Effect of serial correlation of errors within rows. Suppose that the normally distributed errors $z_{1i}, z_{2i}, \cdots, z_{ki}$ in the ith row of the analysis of variance table all have equal variance σ^2 but are not distributed independently. Thus $\mathbf{V} = \sigma^2 \boldsymbol{\varrho}$, where $\boldsymbol{\varrho} = \{\rho_{ts}\}$ is a $k \times k$ matrix with diagonal elements all unity and the element ρ_{ts} of the tth column and s^{th} row is the coefficient of correlation between z_{ti} and z_{si}, the same for all i. The theory described above enables us to examine the effect of any such correlation we choose.

A type of correlation of particular interest in practice is serial correlation which might be expected to arise when the observations within rows or columns were made at equally spaced intervals of time or space. This occurs when the rows of the two-way table are associated with a time factor, as in Fisher's example [1] and in the growth and wear curve examples of [5], or with a space factor as in Daniel's examples [3].

Normally the first order coefficient ρ_1, or ρ as we shall denote it, will be the largest of the serial correlations. We shall study the case where this first order serial correlation is taken into account but the effect of other correlations is ignored. Thus we shall assume

$$\boldsymbol{\varrho} = \begin{bmatrix} 1 & \rho & 0 & \cdots & 0 & 0 \\ \rho & 1 & \rho & \cdots & 0 & 0 \\ 0 & \rho & 1 & \cdots & 0 & 0 \\ \cdots & \cdots & \cdots & \cdots & \cdots & \cdots \\ 0 & 0 & 0 & \cdots & 1 & \rho \\ 0 & 0 & 0 & \cdots & \rho & 1 \end{bmatrix}. \tag{6.1}$$

To ensure positive definiteness we also assume

$$| \rho | < [2 \cos \{\pi/(k + 1)\}]^{-1}. \tag{6.2}$$

"Between columns" test. In order to determine the exact probability $\Pr \{Q_C/Q_E > \phi\}$, we require the latent roots of $\mathbf{u}$ of equation (3.8). Making the substitution $\mathbf{v} = \boldsymbol{\varrho}\sigma^2$, where $\boldsymbol{\varrho}$ is defined in equation (6.1), and writing $\lambda = \lambda'\sigma^2$, the determinantal equation multiplied by k/σ^2 is

$$| k\rho\mathbf{m} - k\lambda'\mathbf{I}_k |$$

$$= \begin{vmatrix} k - (1+\rho) - k\lambda' & k\rho - (1+\rho) & -(1+\rho) & \cdots & -(1+\rho) \\ k\rho - (1+2\rho) & k - (1+2\rho) - k\lambda' & k\rho - (1+2\rho) & \cdots & -(1+2\rho) \\ -(1+2\rho) & k\rho - (1+2\rho) & k - (1+2\rho) - k\lambda' & \cdots & -(1+2\rho) \\ \cdots & \cdots & \cdots & \cdots & \cdots \\ -(1+2\rho) & -(1+2\rho) & -(1+2\rho) & \cdots & k\rho - (1+2\rho) \\ -(1+\rho) & -(1+\rho) & -(1+\rho) & \cdots & k - (1+\rho) - k\lambda' \end{vmatrix} = 0. \tag{6.3}$$

To solve the equation, the determinant is first reduced to a more tractable form by a series of elementary transformations as follows:

(i) Add $c_2 + c_3 + \cdots + c_k$ to c_1.

(ii) Divide c_1 by $-k\lambda'$, and add $(2\rho + 1) \times c_1$ to $c_2, c_3, \cdots, c_k$, in turn.

(iii) Substitute $\lambda' = 1 + \rho\vartheta$ and divide $c_2, \cdots, c_k$ by $k\rho$.

(iv) Add c_3 to c_2, c_4 to c_3, $\cdots$, c_k to c_{k-1}.

(v) Add r_1 to r_2, $r_1 + r_2$ to r_3, $\cdots$, $r_1 + r_2 \cdots + r_{k-1}$ to r_k, and multiply c_k by k.

(vi) Add $(\vartheta - 2)c_1 + 2c_2 + 3c_3 \cdots + (k - 1)c_{k-1}$ to c_k, change the sign of c_k, and interchange c_2 and c_k.

(vii) Add a new first row $r_0 = 10100 \cdots 0$ and a new first column $c_0 = 100 \cdots 0$, which leaves the value of the determinant unchanged.

(viii) Add r_0 to $r_1, r_2, \cdots, r_k$ in turn, and interchange rows and columns.

We now have the equation in the more manageable form

$$\begin{vmatrix} 1 & 1 & 1 & 1 & \cdots & 1 & 1 & 1 \\ 0 & 1 & 2 & 3 & \cdots & k-2 & k-1 & k \\ 1 & -\vartheta & 1 & 0 & \cdots & 0 & 0 & 0 \\ 0 & 1 & -\vartheta & 1 & \cdots & 0 & 0 & 0 \\ \cdots & \cdots & \cdots & \cdots & \cdots & \cdots & \cdots & \cdots \\ 0 & 0 & 0 & 0 & \cdots & -\vartheta & 1 & 0 \\ 0 & 0 & 0 & 0 & \cdots & 1 & -\vartheta & 1 \end{vmatrix} = 0, \tag{6.4}$$

the nonzero solutions of which give the required λ's via the relation $\lambda_t = (1 + \rho\vartheta_t)\sigma^2$. Denote the matrix of the determinant of (6.4) by $\mathbf{L}$ and a column vector of real numbers $(x_0, x_1, \cdots, x_k)$ by $\mathbf{x}$. Then the necessary and sufficient condition that a nontrivial solution exists for the equations $\mathbf{Lx} = \mathbf{0}$ is that $|\mathbf{L}| = 0$. Thus corresponding to each value of ϑ satisfying (6.4) there exists a set of solutions $x_0, x_1, \cdots, x_k$.

The equations $\mathbf{Lx} = \mathbf{0}$ may be written as

$$\sum_{t=0}^{k} x_t = 0, \qquad \sum_{t=0}^{k} t x_t = 0, \tag{6.5}$$

$$x_t - \vartheta x_{t+1} + x_{t+2} = 0, \qquad t = 0, 1, \cdots, k-2. \tag{6.6}$$

The difference equation (6.6) with boundary conditions given by (6.5) is readily solved by standard methods. With $\vartheta = 2\cos\phi$, a set of solutions

$$x_t = e^{it\phi} - e^{i(k-t)\phi} \tag{6.7}$$

is obtained if $\phi = 2s\pi/k + 1$ or if $(k+2)\sin(\frac{1}{2}k\phi) = k \sin \frac{1}{2}(k+2)\phi$. In the first case,

$$\vartheta = 2\cos\left(\frac{2s\pi}{k+1}\right), \qquad s = \begin{cases} 1, \cdots, \frac{1}{2}k; & k \text{ even}, \\ 1, \cdots, \frac{1}{2}(k-1); & k \text{ odd}. \end{cases} \tag{6.8}$$

In the second case, the remaining solutions for ϑ are most readily obtained by putting $t = \tan \frac{1}{2}\phi$, yielding

$$\begin{cases} \displaystyle\sum_{s=1}^{k/2} (-1)^{s-1}(k - 2s - 1)\binom{k+2}{2s-1}(t^2)^{\frac{1}{2}k-s} = 0 & k \text{ even}, \\ \displaystyle\sum_{s=0}^{(k-1)/2} (-1)^{s}(k - 2s - 1)\binom{k+2}{2s}(t^2)^{\frac{1}{2}(k-1)-s} = 0 & k \text{ odd}. \end{cases} \tag{6.9}$$

TABLE 3

Values of ϑ for first order serial correlation

$k = 2$	3	4	5	6	7	8	9	10
−1.0000	0.0000	0.6180	1.0000	1.2470	1.4142	1.5321	1.6180	1.6825
	−1.3333	−0.5000	0.1165	0.5486	0.8544	1.0760	1.2405	1.3655
		−1.6180	−1.0000	−0.4450	0.0000	0.3473	0.6180	0.8308
			−1.7165	−1.2153	−0.7258	−0.3057	0.0407	0.3229
				−1.8019	−1.4142	−1.0000	−0.6180	−0.2846
					−1.8429	−1.5203	−1.1588	−0.8113
						−1.8794	−1.6180	−1.3097
							−1.9001	−1.6772
								−1.9190

TABLE 4

Between columns test: Values of ϵ

ρ	$k = 3$	$k = 5$	$k = 10$
−0.4	0.9576	0.8862	0.8233
−0.2	0.9863	0.9640	0.9453
+0.2	0.9769	0.9507	0.9222
+0.4	0.8832	0.8033	0.7718

Since $\vartheta = 2(1 - t^2)/(1 + t^2)$ is a single valued function of t^2, the polynomial equations (6.9) in t^2 supply the $\frac{1}{2}k - 1$ and $\frac{1}{2}(k - 1)$ values of ϑ required when k is even and odd, respectively, to give with (6.8) the total $k - 1$ solutions.

Values for $k = 2, 3, \cdots, 10$ are shown in Table 3, whence values of the λ's may be obtained for any chosen values of ρ and σ^2 from the relation $\lambda_t = (1 + \rho\vartheta_t)\sigma^2$. Using these values we may obtain the required probabilities from the exact series of Theorem 4.1′.

If we use the F approximation we consider that the ratio of mean squares $(n - 1)Q_C/Q_E$ is distributed approximately as $F\{(k - 1)\epsilon, (n - 1)(k - 1)\epsilon\}$, where

$$\epsilon = \{1 + \rho^2 2(k + 1)(k - 2)^2/(k - 1)(k - 2\rho)^2\}^{-1}. \tag{6.10}$$

Values of the constant ϵ for various values of k and ρ are shown in Table 4. Since there is no bias and the effect of moderate correlation does not greatly reduce the degrees of freedom in the approximation, no large discrepancies will be expected in the between-columns comparison of the analysis of variance. Some calculated values are given in Table 6.

"*Between rows*" test. As we have seen already, the expectations of the row and error mean squares are equal only if the average covariance $\bar{v}_{ts} = 0$. For

TABLE 5

Between rows test: values of bias B

ρ	$k = 3$	$k = 5$	$k = 10$
−0.4	0.3684	0.3103	0.2593
−0.2	0.6471	0.6296	0.6154
+0.2	1.4615	1.4348	1.4167
+0.4	2.0909	1.9524	1.8696

TABLE 6

Probability (per cent) of exceeding 5 per cent point when first order serial correlation between errors within rows is ρ, for analysis of variance table with 5 rows and 5 columns

First order serial correlation, ρ	−0.4	−0.2	0.0	0.2	0.4
Exact per cent probability for test on rows	0.03	1.01	5.00	13.05	24.70
Approximate per cent probability for test on columns	5.90	5.27	5.00	5.37	*6.68

* By exact method, per cent probability is 6.43.

the case of first order serial correlation considered above, the expectations, under the null hypothesis, of the row and error mean squares are, respectively,

$$(1 + 2\rho(k - 1)/k)\sigma^2 \quad \text{and} \quad (1 - 2\rho/k)\sigma^2.$$

The ratio B of these expectations is

$$B = 1 + 2\rho k/(k - 2\rho). \tag{6.11}$$

Values of B for a number of values of k and ρ are shown in Table 5. This bias coefficient can be large even with only moderate correlation, and we shall therefore expect discrepancies to arise in the between-rows comparisons. Using equations (4.9), (4.10), and (4.11), exact probabilities for the between-rows test are obtained.

The results of a number of calculations for the case of the two-way table with five rows and columns are shown in Table 6. These confirm that very large discrepancies in the between-rows test in the directions expected do in fact occur, but that the between-columns comparisons are much less seriously affected. In particular, the remarks of R. A. Fisher concerning the analysis of the rainfall data are seen to be justified.

Acknowledgement. I am indebted to Mrs. Margaret Edmondson for valuable assistance with the calculations.

REFERENCES

[1] R. A. FISHER, *Statistical Methods for Research Workers*, 8th ed., Oliver and Boyd, Edinburgh, 1941, p. 226.
[2] N. SHAW, *The Air and its Ways*, Cambridge University Press, 1922, p. 211.
[3] H. E. DANIELS "Some problems of statistical interest in wool research," *J. Roy. Stat. Soc., Suppl.*, Vol. 5 (1938), pp. 89–128.
[4] H. E. DANIELS, "The effect of departures from ideal conditions other than non-normality on the t and z tests of significance," *Proc. Cambridge Philos. Soc.*, Vol. 34 (1938), pp. 321–328.
[5] G. E. P. BOX, "Problems in the analysis of growth and wear curves," *Biometrics*, Vol. 6 (1950), pp. 362–389.
[6] G. E. P. BOX, "Some theorems on quadratic forms applied in the study of analysis of variance problems, I. Effect of inequality of variance in the one-way classification," *Ann. Math. Stat.*, Vol. 25 (1954), pp. 290–302.

4.4

Application of Digital Computers in the Exploration of Functional Relationships

G. E. P. BOX and G. A. COUTIE

(*The paper was first received 6th February, and in revised form 2nd May*, 1956. *It was read at the* CONVENTION ON DIGITAL-COMPUTER TECHNIQUES, 10*th April*, 1956.)

SUMMARY

In order to discover the optimum process conditions and to supply basic data for process control, it is necessary to determine, at least approximately, the process law

$$\eta = f(\xi_1 \ldots \xi_k; \theta_1 \ldots \theta_p) \quad . \quad . \quad . \quad . \quad (1)$$

connecting a response η (such as yield of product) with levels of the variables $\xi_1 \ldots \xi_k$ (such as temperature, time, pressure, etc.), $\theta_1 \ldots \theta_p$ being unknown parameters.

The problem is discussed of using an electronic digital computer for fitting the function eqn. (1), to a set of observed data,

(*a*) when the functional form is unknown but can be locally represented by a multivariate polynomial, and

(*b*) when the functional form is not known explicitly, but is thought to be the solution of *s* simultaneous differential equations.

In case (*b*), starting with any guessed values of the parameters $\theta_1 \ldots \theta_p$ the electronic computer is caused to follow a series of trial values which result in progressively smaller discrepancies between observed and calculated values of η. The procedure provides the least-squares estimates of the parameters and their standard errors; it also gives a criterion from which the adequacy of the form of the assumed sets of differential equations may be judged.

An example of the application of the method is described.

(1) INTRODUCTION

In the chemical industry the problem often arises of exploring experimentally a functional relationship

$$\eta = f(\xi_1, \xi_2, \ldots, \xi_k; \theta_1, \theta_2, \ldots, \theta_p) \quad . \quad . \quad . \quad (1)$$
$$= f(\boldsymbol{\xi}; \boldsymbol{\theta})$$

connecting a response η (such as yield of product) with a number of variables $\xi_1, \xi_2, \ldots, \xi_k$ (such as time, temperature, pressure and concentration of the various reactants) in which $\theta_1, \theta_2, \ldots, \theta_p$ are unknown parameters.

Two reasons for studying such problems are:

(i) It may be required to find those conditions of the variables $\xi_1, \xi_2, \ldots, \xi_k$ which in some sense give η an optimum value. Thus if η was the yield of product* it might be required to find the values of $\xi_1, \xi_2, \ldots, \xi_k$ which maximized η;

(ii) It may be necessary to know the characteristics of $f(\boldsymbol{\xi}; \boldsymbol{\theta})$, at least in the neighbourhood of the optimum operating conditions, so as to determine the best way to control the process.

There will, in practice, be a greater or lesser extent of ignorance concerning the function for $f(\boldsymbol{\xi}; \boldsymbol{\theta})$ which is appropriate in eqn. (1). Two common situations are:

(*a*) No knowledge of the form of the function $f(\boldsymbol{\xi}; \boldsymbol{\theta})$ is available, but it is assumed to possess a unique maximum and to be 'smooth'. From a limited use of experimental facilities it is required that the optimal region of the function should be determined and mapped.

* More generally a function of η that measured efficiency or profit would be sought.

Dr. Box and Mr. Coutie are with Imperial Chemical Industries Limited.

(*b*) The form of $f(\boldsymbol{\xi}; \boldsymbol{\theta})$ is not explicitly known but from theoretical considerations it would be expected that a certain set of simultaneous differential equations would have $f(\boldsymbol{\xi}; \boldsymbol{\theta})$ as their solution. Some data are available or can be generated, and from this it is required to estimate the constants $\boldsymbol{\theta}$ and their experimental errors, and to ascertain whether the assumed system of differential equations is adequate to explain the data.

In the paper, consideration is given first to an iterative experimental method for determining the position of a maximum or minimum which was specifically developed[1] for use in situation (*a*) above, and it is shown how this procedure is expedited by the use of the electronic computer. It is then shown how, by substituting computer calculations for experiments, an analogous method may be adopted to deal with the second situation (*b*).

(2) AN ITERATIVE EXPERIMENTAL PROCEDURE TO ATTAIN A MAXIMUM OR MINIMUM

For clarity, consider the situation where there are only two variables (e.g. temperature ξ_1 and concentration ξ_2). Then eqn. (1) defines an unknown 'response surface' that can be conveniently represented by the contours of η in the space of the variables ξ_1 and ξ_2, as in Fig. 1.

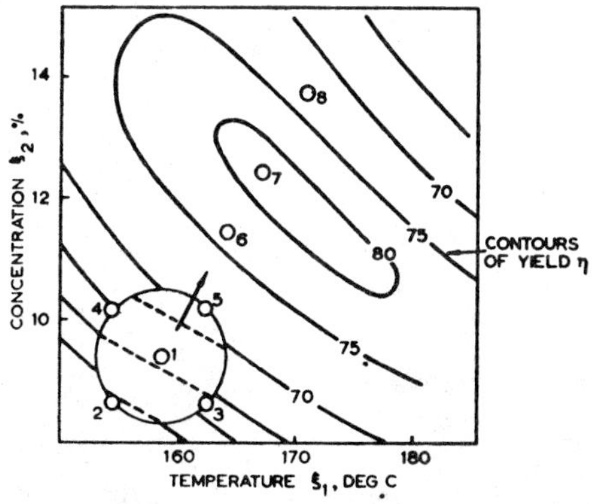

Fig. 1.—Contour representation of a response surface and first-order experimental design.

To carry out a close grid of experiments sufficiently widespread to cover the whole region of possible operating conditions would usually be a policy too prodigal to contemplate. Such a complete exploration is, however, not usually required, since conditions giving an unsatisfactory level of the response are of no real interest. What is needed is to determine the region of optimal response and to map this region only. An iterative procedure for doing this which has proved effective in a number of applications is now described.

Over a sufficiently small region of the space in the vicinity of

The Collected Works of George E. P. Box, 1984, Wadsworth, Inc., Belmont, CA 94002.
Originally published in *Proc. Inst. Elec. Engrs.,* vol. 103, part B, supplement no. 1 (1956), pp. 100–107.

the point (ξ_{10}, ξ_{20}) the function could usually be represented by the lower-order terms of its Taylor series expansion

$$\eta = \beta_0 x_0 + (\beta_1 x_1 + \beta_2 x_2) + (\beta_{11} x_1^2 + \beta_{22} x_2^2 + \beta_{12} x_1 x_2) + (\beta_{111} x_1^3 + \beta_{222} x_2^3 + \beta_{112} x_1^2 x_2 + \beta_{122} x_1 x_2^2) +, \text{etc.} \quad (2)$$

where $x_0 = 1$; $x_1 = (\xi_1 - \xi_{10})/S_1$; $x_2 = (\xi_2 - \xi_{20})/S_2$; S_1 and S_2 are suitable scaling units, and the β's are suitable multiples of the partial derivatives at (ξ_{10}, ξ_{20}) of the function $f(\boldsymbol{\xi}; \boldsymbol{\theta})$ with respect to the variables. An expansion of this sort which includes all terms of order d and less will be called a polynomial approximation of order d.

The iterative procedure is begun at the best conditions known. In this neighbourhood enough experiments are performed to allow, by the method of least squares, a polynomial approximation to be fitted of sufficient order to provide a local representation of the surface. Knowledge of the 'local geography' thus provided normally makes it possible to choose a further region at which higher responses are to be expected. Further experiments are now performed in this region and the process is repeated until no further gain is achieved.

Now, at least n observations are needed to fit an approximating function containing n constants. Consequently the greatest economy in experimentation as well as the greatest simplicity will normally be attained if at each stage the polynomial of lowest order needed to make further progress possible is used. A beginning is therefore made by assuming that a first-order approximation is to be used. This assumption would be abandoned, and a second-order approximation adopted, only when the first-order approximating function had proved inadequate.

To allow such a technique to be used efficiently, special patterns of experimental points (called experimental designs) are needed. Where possible an experimental design of order d is chosen so that

(*a*) it allows an approximating polynomial of order d (tentatively assumed to be representationally adequate) to be fitted most accurately;

(*b*) it allows some check to be made on the adequacy of the assumed function by allowing the size of certain combinations of higher-order terms to be examined;

(*c*) in the event that the assumed order of polynomial proves inadequate the already completed design of order d forms a nucleus from which a design of order $(d + 1)$ may be built up;

(*d*) it contains as small a number of experimental points as possible.

It frequently happens in practice that only first- and second-order polynomial approximations need be used, although at the final stage a third-order approximation[2,3] is sometimes of value. A number of examples in which the method has been successfully applied to attain a maximum in as many as five variables using only first- and second-order approximations have been given[1,4,10] but the process is best illustrated in terms of two variables with the help of Fig. 1. The generalization to more variables introduces no new principles.

(2.1) First-Order Procedure

Suppose that the best-known conditions are at the point labelled '1' in Fig. 1. Then this point is treated as the temporary origin (ξ_{10}, ξ_{20}) and the scaling units S_1 and S_2 are chosen so that appreciable and comparable effects in the response might be expected by changing ξ_1 and ξ_2 by the amounts S_1 and S_2 respectively. A suitable first-order design then consists of the original point 1, together with the four further points 2, 3, 4 and 5, whose co-ordinates are $(\xi_{10} \pm S_1, \xi_{20} \pm S_2)$. For the variables $x_1 = (\xi_1 - \xi_{10})/S_1$, $x_2 = (\xi_2 - \xi_{20})/S_2$ scaled, in working units, therefore, the first-order design consists of the point (0, 0) together with the four points $(\pm 1, \pm 1)$. It should be noted that the designs given here are chosen partly to attain simplicity in exposition. A fuller discussion of the design problem will be found elsewhere.[5,6]

Denoting observations of the response η at the labelled points by y_1, y_2, y_3, y_4, and y_5, and least-squares estimates of the coefficients β_0, β_1, etc., by b_0, b_1, etc., it is easily shown

(*a*) that the approximating polynomial of first order which fits the observations best (in the sense of least squares) is

$$y = b_0 + b_1 x_1 + b_2 x_2 \quad . \quad . \quad . \quad . \quad (3)$$

where $b_0 = \bar{y}$, $b_1 = (y_3 + y_5 - y_2 - y_4)/4$ and $b_2 = (y_5 + y_4 - y_2 - y_3)/4$;

(*b*) that, if an approximation of second order is needed, estimates $b_{12} = (y_5 + y_2 - y_4 - y_3)/4$ and $b_{11} + b_{22} = \frac{1}{4}[y_2 + y_3 + y_4 + y_5 - 4y_1]$ of second-order terms are available from which the adequacy of the first-order equation can be checked.

Eqn. (3) represents the plane which best fits the experimental points. The response contours of this best-fitting plane might have the appearance shown by the dotted lines in Fig. 1. These differ somewhat from the contours of the true, but unknown, response surface, owing first to the inability of this simple approximation to represent accurately the surface, and secondly to experimental error. In some examples the large size of the experimental error would make it essential to replicate the experiments. If the size of experimental error is not known it is best to proceed sequentially, performing further experiments if the standard errors of the coefficients estimated from the first set are too large. In the example illustrated the representation is quite adequate to allow progress to be made.

To proceed to a region where the response is greater, it is natural to follow the direction indicated by the arrow at right angles to the contours of the fitted surface. This is the direction of steepest ascent of the fitted surface with the particular choice of units adopted. Experiment 6 would indicate that an improvement in this direction could indeed be attained, while further experiments 7 and 8 would indicate that attention should now be focused on a region about 7.

(2.2) Second-Order Procedure

Following the strategy already outlined, a first-order design would now be placed about the point 7. In the example illustrated, the observations would indicate that further terms were needed in the approximating polynomial to obtain a useful representation of the local surface. Further points would therefore be added to form a second-order design which would now become of the octagonal type illustrated in Fig. 2. Just as with the first-order approximating polynomial, the coefficients of the best-fitting approximating polynomial of second order

$$y = b_0 + b_1 x_1 + b_2 x_2 + b_{11} x_1^2 + b_{22} x_2^2 + b_{12} x_1 x_2 \quad . \quad (4)$$

would be determined by the method of least squares as simple linear functions of the observations. The type of approximation that would be achieved by the fitted second-order surface is indicated by the dotted contour line in Fig. 2.

The information contained in the second-degree approximating equation is most readily comprehended by rewriting the equation in the canonical form

$$y - y_S = B_{11} Z_1^2 + B_{22} Z_2^2 \quad . \quad . \quad . \quad . \quad (5)$$

where y_S is the response predicted at the centre S of the system whose co-ordinates (x_{1S}, x_{2S}) are the solutions of the equations

$$\left.\begin{aligned} b_1 + 2b_{11} x_1 + b_{12} x_2 &= 0 \\ b_2 + b_{12} x_1 + 2b_{22} x_2 &= 0 \end{aligned}\right\} \quad . \quad . \quad . \quad . \quad (6)$$

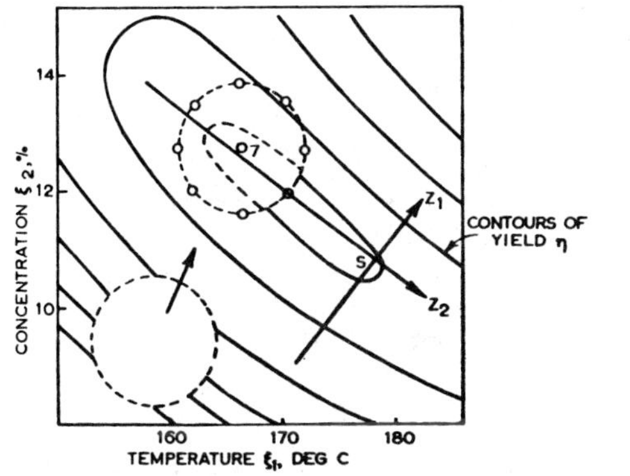

Fig. 2.—Contour representation of a response surface and second-order experimental design.

B_{11} and B_{22} are the latent roots of the matrix

$$\begin{bmatrix} b_{11} & \frac{1}{2}b_{12} \\ \frac{1}{2}b_{12} & b_{22} \end{bmatrix} \quad . \quad . \quad . \quad . \quad . \quad . \quad . \quad (7)$$

Z_1 and Z_2 are new 'canonical' variables of the form

$$\left.\begin{aligned} Z_1 &= m_{11}(x_1 - x_{1S}) + m_{12}(x_2 - x_{2S}) \\ Z_2 &= m_{21}(x_1 - x_{1S}) + m_{22}(x_2 - x_{2S}) \end{aligned}\right\} \quad . \quad . \quad (8)$$

and $\begin{bmatrix} m_{11} \\ m_{12} \end{bmatrix}, \begin{bmatrix} m_{21} \\ m_{22} \end{bmatrix}$ are the normalized latent vectors of the matrix (7).

In summarizing the information and indicating what further action should be taken, the canonical form performs the same function for the second-degree approximating polynomial as does the direction of steepest ascent for the first-degree approximating polynomial. Its use is described in detail elsewhere.[4, 10] In the present example the coefficients B_{11} and B_{22}, which measure curvature in the surface in the directions Z_1 and Z_2 respectively, would both be negative and B_{11} would be considerably greater numerically than B_{22}. Since B_{11} and B_{22} are both negative, the fitted surface has a maximum at the centre S, but because this point is outside the immediate region of experimentation it cannot be assumed that this was even approximately true for the underlying response surface. In fact, S should be regarded only as a useful construction point.

The analysis would show that, because of the small size of B_{22} relative to B_{11}, the fitted equations represented a ridge-like system, the crest of the ridge being the axis Z_2. From the nature of the coefficients m_{11} and m_{12} it would be apparent that the ridge was oblique to the directions of the original variables x_1 and x_2. It would therefore be difficult to study this oblique ridge system in terms of the original variables x_1 and x_2, which in this situation are said to 'interact' or to show 'dependence' in their effects on the response, the response surface expressed in terms of these variables being said to be 'poorly conditioned'. Further exploration could best be conducted in terms of a new co-ordinate system with axes Z_1 and Z_2.

New variables $z_1 = \sqrt{(-B_{11})}Z_1$ and $z_2 = \sqrt{(-B_{22})}Z_2$ would therefore be introduced. In terms of these variables the surface could be expressed in a much more symmetrical form, as will be seen from Fig. 3, which shows the surface plotted in terms of z_1 and z_2. It will be seen that a final second-order design, set out in terms of the new variables z_1 and z_2 about the best point attained along the Z_2 axis, would almost certainly straddle the true maximum and make it possible to determine its position fairly accurately.

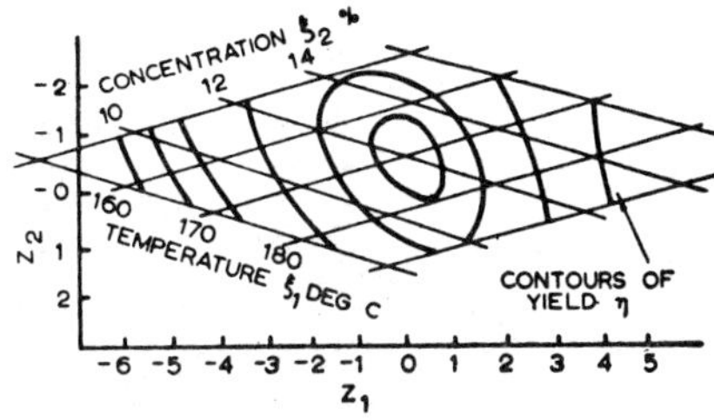

Fig. 3.—Response surface plotted in terms of the transformed variables z_1 and z_2.

(2.3) Use of the Computer

The computer is of particular value at this second-order stage of experimentation. For k variables there are $(k + 1)(k + 2)/2$ constants to be estimated in the second-order approximating polynomial, i.e. 10 constants for 3 variables, 15 constants for 4 variables, and 21 constants for 5 variables. Application of the method of least squares thus generates $(k + 1)(k + 2)/2$ linear equations in the same number of unknowns, the solutions of which are the required estimates b_0, b_1, etc. Now it so happens that, for data generated by efficient experimental designs (such as those described), these equations are highly specialized and easily solved. For this reason the above process, to the extent that it includes only data generated by these special designs, can be readily carried out without undue labour with only a desk calculating machine. However, towards the later stages of experimentation the siting of experiments tends more and more to be dictated by the features of the surface being explored rather than by consideration of efficient experimental design, and it would often be desired to include information from extra points not included in a special design, or to combine information from designs. The use of the computer makes such calculations feasible.

The calculations are readily summarized in matrix notation. The representation of the true response η at the n sets of experimental conditions by an approximating polynomial containing m terms is expressed by

$$\boldsymbol{\eta} = X\boldsymbol{\beta} \quad . \quad . \quad . \quad . \quad . \quad . \quad . \quad . \quad (9)$$

where $\boldsymbol{\eta}$ is an $n \times 1$ vector of the true responses, $\boldsymbol{\beta}$ is an $m \times 1$ vector of the coefficients β_0, β_1 etc., and X is an $n \times m$ matrix whose columns contain the values of $x_0, x_1, x_2, \ldots, x_1^2, x_2^2, \ldots, x_1x_2, \ldots$, etc., at the n sets of conditions. The matrix X is calculated from the $n \times k$ matrix D which gives the levels of the x's at the n sets of experimental conditions and is called the design matrix.

If Y is the $n \times 1$ vector of observations recorded at the n sets of experimental conditions, the $m \times 1$ vector B of least-squares estimates, whose elements are $b_0, b_1, \ldots$, etc., is

$$B = (X'X)^{-1}X'Y \quad . \quad . \quad . \quad . \quad . \quad (10)$$

If the errors in the observations are uncorrelated and have the same variance σ^2, the matrix of variances and covariances of these estimates is given by

$$E(B - \boldsymbol{\beta})(B - \boldsymbol{\beta})' = (X'X)^{-1}\sigma^2 \quad . \quad . \quad . \quad (11)$$

where E denotes mathematical expectation. On the assumption that the approximating polynomial, eqn. (9), adequately represents the function in the locality of the experiments,

$$(Y - XB)'(Y - XB) = Y'Y - Y'XB$$

supplies an unbiased estimate of $(n - m)\sigma^2$.

In performing these calculations on the Deuce computer it was found convenient to punch the original data in decimal on Hollerith cards so that each card corresponded to one experiment. A single card recorded the number of the experiment, the experimental conditions in working units and the response observed. From a set of n such cards, providing the design matrix $\mathbf{D}$ and the vector $\mathbf{Y}$, the machine has been programmed to produce in binary numbers the matrix $\mathbf{X}$ and the vector $\mathbf{Y}$. Standard matrix programmes produced by the Mathematics Division of the National Physical Laboratory have then been used

(a) to form the matrix of least-squares estimates

$$\mathbf{B} = (\mathbf{X}'\mathbf{X})^{-1}\mathbf{X}'\mathbf{Y}$$

(b) to form the estimate $s^2 = (\mathbf{Y}'\mathbf{Y} - \mathbf{B}'\mathbf{X}'\mathbf{Y})/(n - m)$ of the experimental error variance;

(c) to form the matrix $(\mathbf{X}'\mathbf{X})^{-1}s^2$ of estimated variances and covariances of the b's;

(d) to calculate the latent roots and vectors of the matrix of second-order estimates required in the canonical analysis.

The provision that, when the original data are fed into the machine, one experiment is represented by one card facilitates the sequential employment of the computer. Thus the effect of adding or omitting the results from a particular experiment or group of experiments is readily discovered by running the programme with the appropriate cards added or omitted.

(3) FITTING DIFFERENTIAL EQUATIONS TO DATA

So far, it has been supposed that nothing is known of the form of the relationship $\eta = f(\boldsymbol{\xi}, \boldsymbol{\theta})$ connecting the value of the response η and the levels of the variables $\boldsymbol{\xi}$. In chemical problems it is rare for this functional relationship to be known explicitly, but not infrequently a set of simultaneous differential equations derived from consideration of the reaction kinetics and containing the constants could be postulated. Given the form of these differential equations and some data on the values of $\boldsymbol{\xi}$ at various experimental conditions, it is required to estimate the constants $\boldsymbol{\theta}$, to determine the precision of these estimates and to check the adequacy of the equations to describe the system.

As an example, consider a chemical system in which substances M_1 and M_2 react to form M_3, which then further reacts with more M_1 to form M_4 in accordance with the scheme

$$M_1 + M_2 \rightarrow M_3, \; M_1 + M_3 \rightarrow M_4$$

If η_1, η_2, η_3 and η_4 are the concentrations of M_1, M_2, M_3 and M_4 relative to the initial concentration c_{20} of M_2, α, β and ρ are unknown constants and T is the absolute temperature, on assumptions fully set out elsewhere,[8] the progress of the reaction is represented by the following set of simultaneous differential equations:

$$\left.\begin{aligned}
\frac{d\eta_1}{dt} &= -\,2c_{20}\alpha\eta_1(\rho\eta_2 + \eta_3)\exp(-\beta/T)\\
\frac{d\eta_2}{dt} &= -\,\rho c_{20}\alpha\eta_1\eta_2\exp(-\beta/T)\\
\frac{d\eta_3}{dt} &= c_{20}\alpha\eta_1(\rho\eta_2 - \eta_3)\exp(-\beta/T)\\
\frac{d\eta_4}{dt} &= c_{20}\alpha\eta_1(\rho\eta_2 + \eta_3)\exp(\beta/T)
\end{aligned}\right\} \quad (12)$$

with the boundary conditions

$$\eta_1 = \eta_{10},\ \eta_2 = 1,\ \eta_3 = 0,\ \eta_4 = 0 \text{ at time } t = 0$$

In this particular example these equations can be reduced to a somewhat simpler form. To preserve generality in the method of attack they are here discussed in the form in which they naturally arise.

Experiments may be performed in which the temperature, T, the ratio of the initial concentrations of the starting materials, η_{10}, the time elapsing before the reaction is stopped, t, and the initial concentration of M_2, c_{20}, are varied. It is required to estimate from the results the values of the constants α, β and ρ. Thus, in the notation previously used T, η_{10}, t and c_{20} would correspond with the experimental variables ξ_1, ξ_2, ξ_3, ξ_4, and α, β and ρ with the constants θ_1, θ_2 and θ_3 or with suitable functions of them. In the general chemical problem there will be s simultaneous differential equations of the form

$$\frac{d\eta_u}{dt} = \phi_u(\eta_1, \ldots, \eta_s; \xi_1, \ldots, \xi_k; \theta_1, \ldots, \theta_p) \qquad (u = 1, 2, \ldots, s) \quad . \quad (13)$$

describing the kinetic mechanism of some chemical system where the time t itself is usually one of the ξ's and they would have to be 'fitted' to a set of n observations in one or more of the η's. In the paper consideration will be limited to the case where only one response η is observed, although the procedure may be extended to cover the more general case.

If $\eta = f(\boldsymbol{\xi}, \boldsymbol{\theta})$ is the solution of the differential equations for the observed response at some particular set of experimental conditions $\boldsymbol{\xi}$, and if y is the observed value of η at these conditions, then least-squares estimates of the k constants $\boldsymbol{\theta}$ would be obtained by minimizing the quantity

$$S(\boldsymbol{\theta}) = \Sigma[y - f(\boldsymbol{\xi}, \boldsymbol{\theta})]^2 \quad . \quad . \quad . \quad . \quad (14)$$

where the summation is over the n sets of experimental conditions. On the assumption that the experimental errors are normally distributed, the estimates $\hat{\boldsymbol{\theta}}$ so obtained are also maximum likelihood estimates. If the function $f(\boldsymbol{\xi}, \boldsymbol{\theta})$ were known explicitly the values $\hat{\boldsymbol{\theta}}$ making $S(\boldsymbol{\theta})$ a minimum could be found by differentiating in the usual way. In some cases numerical methods would be needed to solve the resulting equations, but this would usually present no particular difficulty. In the problem in question, however, it is supposed that (as is most often the case in practice) no explicit solution of the differential equations is available, so that $S(\boldsymbol{\theta})$ cannot be minimized by differentiation, and an alternative method must be followed.

(3.1) Method for Determining and Minimizing $S(\boldsymbol{\theta})$

A sub-routine using the Runge–Kutta technique as modified by Gill[7] has been developed for the Deuce by the Mathematics Division of the National Physical Laboratory. This produces numerically the values η_1, η_2, . . ., η_s for up to $s = 10$ simultaneous differential equations. To use the sub-routine for a specific problem it is only necessary to arrange that the machine shall calculate the functions ϕ_u in eqn. (13), which are completely unspecialized. Given a set of data $\{y\}$ representing n observations of η at the n sets of experimental conditions, then for any specified set of simultaneous differential equations and any chosen set of constants $\boldsymbol{\theta}$ the Runge–Kutta programme can be used to produce a calculated set of values $[\eta(\boldsymbol{\theta})]$ and from these the value $S(\boldsymbol{\theta}) = \Sigma[y - \eta(\boldsymbol{\theta})]^2$ can be produced. The problem now becomes analogous to that discussed in Section 2. The quantity S corresponds to the response to be minimized, the parameters θ_1, θ_2, . . . , θ_p to the variables, and calculations on the machine take the place of experiments. Now, even with a high-speed computer, the calculation of S takes an appreciable time. For example, with only three simultaneous equations and nineteen observations the calculation of each value of S took a minute and a half on the Deuce. Calculations over an exhaustive multi-dimensional grid of points would therefore be unreasonably expensive in computer time and the more economical technique for minimizing S discussed in Section 2 is adopted.

Starting with a trial set of values $\boldsymbol{\theta}_1$ a first-order design* of parameter points is calculated by the computer. From these the direction of steepest descent is deduced. This direction is then followed until an increase in S is found in successive calculations. A new first-order pattern is then performed about the best point reached and the process is repeated. As soon as this process becomes ineffective a second-order design pattern† is used. Usually this second-order pattern would not lead at once to the final minimum but would reveal a state of poor conditioning in the S surface. This could arise from either or both of the following causes

(*a*) Dependence between the θ's in their effects on the quantity $S(\boldsymbol{\theta})$;

(*b*) An unsymmetrical choice of the steps by which the parameters had been varied.

Adoption at this stage of the canonical variables corresponding to z_1, z_2 in Fig. 3 as new working variables would greatly improve the conditioning of the surface and would usually allow the minimum to be determined fairly readily in one or two sets of second-order experiments scaled in the new variables.

(3.2) Testing the Adequacy of the Assumed Form of the Equations

If the set of differential equations is adequate to represent the system studied it may be shown that the mathematical expectation of $s_1^2 = S(\hat{\boldsymbol{\theta}})/(n - p)$ is approximately equal to σ^2 and that if the data are normally distributed $(n - p)s_1^2$ is approximately distributed[11] as $\chi^2_{(n-p)}\sigma^2$. Thus if the truth of the hypothesis concerning the adequacy of the equations can be assumed, s_1^2 provides an estimate of the experimental error variance having $n - p$ degrees of freedom.

In practice the truth of this hypothesis is rarely certain and it is important to be able to test it. To make such a test an independent estimate s_2^2 of the experimental error variance is necessary. It can be obtained, for example, by duplicating the experiments, or an estimate may be available from previous experimentation. Suppose this estimate has ν_2 degrees of freedom. If the hypothesis of the adequacy of the equations is untrue $S(\hat{\boldsymbol{\theta}})$ will tend to be increased. An appropriate test criterion therefore is $s_1^2/s_2^2 = S(\hat{\boldsymbol{\theta}})/(n - p)s_2^2$, which may be referred to a table of the F distribution[11] with $n - p$ and ν_2 degrees of freedom. This will show the probability of obtaining an apparent discrepancy from hypothesis as large as or larger than that observed when the hypothesis is really true. If this probability is very small (e.g. less than 0·05) the hypothesis should be rejected and the equations be regarded as inadequate to describe the data. A useful lead on the nature of the inadequacy is often provided by examining the size of the n individual residual quantities $[y - f(\boldsymbol{\xi}, \boldsymbol{\theta})]$. When the estimates s_1^2 and s_2^2 are of comparable magnitudes they may be used to give a combined estimate of experimental error

$$s^2 = [(n - p)s_1^2 + \nu_2 s_2^2]/(n - p + \nu_2) \quad . \quad . \quad (15)$$

with $\nu = n - p + \nu_2$ degrees of freedom.

(3.3) Accuracy of the Estimates

It can be shown that the variances and covariances of the least-squares estimates $\hat{\boldsymbol{\theta}}$ are given approximately by the elements of the $p \times p$ matrix $V^{-1}\sigma^2$, where $2V$ is the matrix of the partial second derivatives

$$S^{(ij)} = \left[\frac{\partial^2 S(\boldsymbol{\theta})}{\partial\theta_i \partial\theta_j}\right]_{\boldsymbol{\theta}=\hat{\boldsymbol{\theta}}} \quad . \quad . \quad . \quad . \quad . \quad (16)$$

$i, j = 1, 2, \ldots, p$, and σ^2 is the experimental error variance.

Assuming, as would normally be the case, that the minimization process has terminated with the calculation of a second-degree equation which closely fits in the neighbourhood of the minimum, then the derivative $S^{(ii)}$ is given approximately by twice the coefficient of θ_i^2, and the derivative $S^{(ij)}$ by the coefficient of $\theta_i\theta_j$ in the fitted equation. Thus, given an estimate s^2 of the experimental error, an estimate of the standard error of θ_i is given by $s\sqrt{V^{ii}}$, where V^{ii} is the ith diagonal element in the matrix V^{-1}.

When the S surface is poorly conditioned, the errors in the estimates will be highly correlated; for example, the data may be equally consistent with the occurrence of a high value of θ_1 in conjunction with a low value of θ_2, or with a low value of θ_1 in conjunction with a high value of θ_2, but not with high or low values of both constants in combination. In this circumstance the standard errors of individual constants are inadequate, and in order to obtain an appreciation of the zone of error of $\boldsymbol{\theta}$ it is necessary to determine a confidence region. This is the region which includes the true value with any given degree of probability $1 - \alpha$. In practice α is often put equal to 0·05 and we thus obtain the 95% confidence region which includes the true value 19 times out of 20. Such a region is given approximately by the inequality

$$\tfrac{1}{2}\sum_{i=1}^{p}\sum_{j=1}^{p} S^{ij}(\theta_i - \hat{\theta}_i)(\theta_j - \hat{\theta}_j) \leqslant ps^2 F_\alpha(p, \nu) \quad . \quad . \quad (17)$$

where s^2 is an independent estimate of experimental error having ν degrees of freedom, and $F_\alpha(p, \nu)$ is the α probability point of the F distribution with p and ν degrees of freedom. It will be noted that, to the extent that the second-degree approximation is adequate, the above inequality simply defines a contour of the S surface surrounding the point $\hat{\boldsymbol{\theta}}$. This contour denotes the boundary along which the parameter points are equally discrepant with the data as measured by the criterion S at the probability level α.

(3.4) The Approximation Used

It has been mentioned that the results in Sections 3.2 and 3.3 are approximations. The closeness of the approximation differs between one example and another, but can be checked in any particular case by methods which will be discussed more fully elsewhere.

A necessary condition for the adequacy of the approximation is that $S(\boldsymbol{\theta})$ should be nearly a quadratic function of θ in the region of interest, which can be roughly defined as being the 95% confidence region. There is some advantage, therefore, in arranging that the last-fitted second-order design is so scaled as roughly to cover this region.

(4) EXAMPLE

A consecutive reaction in which

$$M_1 \to M_2, \; M_2 \to M_3$$

can be represented by the simultaneous differential equations

$$\left.\begin{aligned} d\eta_1/dt &= -k_1\eta_1 \\ d\eta_2/dt &= k_1\eta_1 - k_2\eta_2 \\ d\eta_3/dt &= k_2\eta_2 \end{aligned}\right\} \quad . \quad . \quad . \quad . \quad (18)$$

* The design problem here is somewhat different from that in the experimental situation discussed in Section 2. In the present computational problem there is no experimental error and requirement (*a*) of Section 2 does not apply. Simple designs in which the variables are changed one at a time [e.g. those containing, for two variables, the points (0, 0), (0, 1) and (1, 0)] can be used to estimate the differentials, but these designs have the disadvantage that they provide no check on the assumptions.

† A calculating procedure for determining a maximum based on the fitting of a second-degree equation was used by Koshal.[9]

with boundary conditions $\eta_1 = 100\%$, $\eta_2 = 0$, $\eta_3 = 0$ at time $t = 0$.

In this particular example, although the differential equations can be solved explicitly, the general method is followed in order to illustrate the technique. Take η_1, η_2, η_3 to be the percentages of the reactant M_1 left unchanged, and in the form of M_2 and M_3 respectively. Six duplicate experiments were performed in random order, in which the observations y_2 of η_2, the percentage of the intermediate product, shown in Table 1, were made at the times specified.

Table 1

Time (t)	Yield of M_2 (y_2)	Mean result ($\bar{y}_2$)
min	%	%
10	19·2 14·0	16·6
20	14·4 24·0	19·2
40	42·3 30·8	36·55
80	42·1 40·5	41·3
160	40·7 46·4	43·55
320	27·1 22·3	24·7

It was required

(*a*) to ascertain whether the form of the differential equations is consistent with the data;

(*b*) to estimate the parameters k_1 and k_2;

(*c*) to estimate the accuracy with which these values are determined.

In the least-squares calculation the means $\bar{y}$ of the paired results have been used throughout and, following usual practice, k_1 and k_2 were expressed as logarithms to base 10. The constants θ_1 and θ_2 to be determined were defined as $\theta_1 = 3 + \log k_1$, $\theta_2 = 3 + \log k_2$.

To obtain starting values it was noted that at time zero $d\eta_2/dt = 100k_1$. Since after 10 minutes $\bar{y}_2$ was 16·6, very roughly the initial rate was 1·66 and $k_1 = 0{\cdot}0166t^{-1}$. The constant k_2 was expected to be of the same order of magnitude as k_1, so values of θ_1 and θ_2 each in the neighbourhood of $3 + \log(0{\cdot}0166) = 1{\cdot}2$ were used to start the process.

The course of the calculation is shown in Table 2 and illustrated in Fig. 4. The initial first-order design was centred at the point (1·19, 1·19), the scaling or working unit being chosen as 0·01

Table 2

COURSE OF LEAST-SQUARES CALCULATION

	(θ_1, θ_2)	Values of $S(\theta_1, \theta_2)$	Working units	Coefficients in working units
First-order design (1)	(1·19 ± 0·01, 1·19 ± 0·01)	1069·89 1078·41 1030·93 984·85 995·07	θ_1 0·01 θ_2 0·01	$b_0 = 1030{\cdot}93$ $b_1 = 4{\cdot}69$ $b_2 = 42{\cdot}10$ $\lambda = 20{\cdot}7$ $b_{12} = -0{\cdot}43$ $b_{11} + b_{22} = 1{\cdot}13$
Path of steepest descent (1)	(1·1677, 0·99) (1·1455, 0·79) (1·1232, 0·59)	333·24 Minimum → (1·1483, 0·8150) 139·81 408·00		Reduce θ_1 in steps of 0·02226 Reduce θ_2 in steps of 0·20
First-order design (2)	(1·1483 ± 0·01, 0·8150 ± 0·01)	110·92 138·69 127·24 118·69 148·58	θ_1 0·01 θ_2 0·01	$b_0 = 127{\cdot}24$ $b_1 = 14{\cdot}41$ $b_2 = -4{\cdot}42$ $\lambda = 4{\cdot}8$ $b_{12} = -0{\cdot}54$ $b_{11} + b_{22} = 1{\cdot}98$
Path of steepest descent (2)	(1·0983, 0·8303) (1·0483, 0·8457)	79·26 Minimum → (1·0745, 0·8361) 90·23		Reduce θ_1 in steps of 0·05 Increase θ_2 in steps of 0·01534
First-order design (3)	(1·0795 ± 0·01, 0·8361 ± 0·01)	81·88 81·20 76·75 74·84 76·16	θ_1 0·01 θ_2 0·01	$b_0 = 76{\cdot}75$ $b_1 = 0{\cdot}16$ $b_2 = 3{\cdot}02$ $\lambda = 0{\cdot}9$ $b_{12} = -0{\cdot}50$ $b_{11} + b_{22} = 1{\cdot}77$
Path of steepest descent (3)	(1·0790, 0·8261) (1·0785, 0·8161) (1·0779, 0·8061)	74·50 Minimum → (1·0785, 0·8168) 73·82 75·50		Reduce θ_1 in steps of 0·0005 Reduce θ_2 in steps of 0·01
Second-order design (4)	(1·0785 ± 0·0707, 0·8168 ± 0·0707) (1·0785 ± 0·1000, 0·8168) (1·0785, 0·8168 ± 0·1000)	158·01 175·92 145·03 152·59 73·79 179·00 135·75 205·06 171·78	θ_1 0·0707 θ_2 0·0707	$b_0 = 74{\cdot}10$ $b_1 = 9{\cdot}47$ $b_2 = -4{\cdot}92$ $b_{12} = -25{\cdot}05$ $b_{11} = 45{\cdot}87$ $b_{22} = 45{\cdot}42$ Sum of squares of discrepancies = 0·19

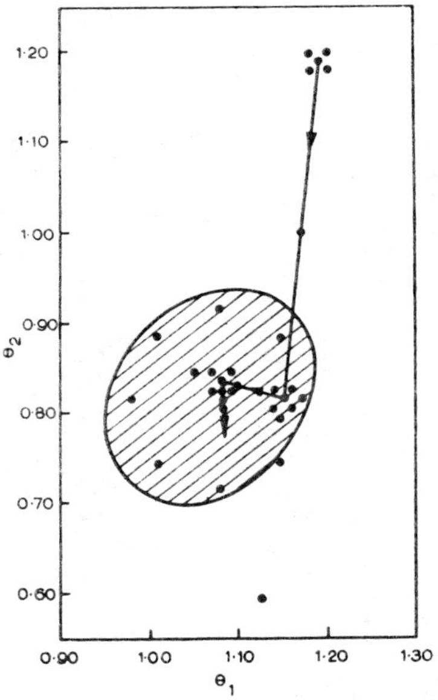

Fig. 4.—Course of least-squares calculation and confidence region.

for both θ_1 and θ_2. The coefficients calculated in terms of these working units are shown in the right-hand column of the table. It will be observed that effects of first order calculated from this design are dominant in the region explored. It would therefore be expected that rapid progress would be possible using the steepest descent procedure. As a very rough guide to the size of step that should be employed in exploring the path of descent the criterion

$$\lambda = \tilde{b}_i/2\tilde{b}_{ii} \quad . \quad . \quad . \quad . \quad . \quad . \quad (19)$$

was used, where $\tilde{b}_i$ is the mean of the moduli of the first-order coefficients and $\tilde{b}_{ii}$ is the mean of the quadratic coefficients b_{ii}. For this first design λ was 20·7. Thus a distance of about 20 working units, i.e. 0·20 in the θ_1, θ_2 space at each step, was indicated. To achieve this approximately, whilst following the steepest descent path, θ_2 was diminished in steps of 0·20 and θ_1 in steps of $0\cdot 20 \times 4\cdot 69/42\cdot 10 = 0\cdot 022\,26$. In two steps the sum of squares was reduced from 1 030·93 to 139·81, but at the third step it rose to 408·00. The minimum on this path was then found by interpolation and the second first-order design carried out about this point.

From the third first-order design it was clear that first-order effects no longer predominated and that a minimum was being closely approached. In a region surrounding the lowest point on the third path of steepest descent a second-order design was therefore applied. For reasons given in Section 3.4 it is desirable that this should roughly cover the confidence region. Now, an estimate of s_2^2, the experimental error variance appropriate for the means of pairs of observations, may be calculated from the differences in the duplicate pairs. In fact, we have

$$V(\bar{y}) = s_2^2 = \tfrac{1}{24}\,[(19\cdot 2 - 14\cdot 0)^2 + (14\cdot 4 - 24\cdot 0)^2 + \ldots + (27\cdot 1 - 22\cdot 3)^2] = 12\cdot 9$$

It follows that if the set of differential equations provides an adequate representation of the surface the minimum value of $S = S(\hat{\boldsymbol{\theta}})$ would be expected to be about $4 \times 12\cdot 9 = 51\cdot 6$. Approximately, therefore, the confidence region for $\boldsymbol{\theta}$ will follow the locus

$$S(\boldsymbol{\theta}) = 51\cdot 6 + ps^2\mathrm{F}(p, \nu)$$

and putting $p = 2$, $s^2 = 12\cdot 9$ and $\mathrm{F} = 4\cdot 10$ we obtain $S(\boldsymbol{\theta}) = 157\cdot 38$.

It will be understood that this is a very rough estimate. By considering the values of S on the path which had been followed, it appeared (again very roughly) that the octagonal design should be given a radius of about 0·10 so that values of the order of 157 should be near its periphery. The working units for θ_1 and θ_2 defined as before were therefore 0·0707. The second-degree equation proved to be an excellent fit over this region, the sum of squares of discrepancies between the observed values and the calculated values at the nine points of the design being only 0·19.

The desired estimates could now be readily calculated.

(4.1) **Estimation of θ_1 and θ_2**

On differentiating the expression

$$S = 74\cdot 10 + 9\cdot 47x_1 - 4\cdot 92x_2 - 25\cdot 05x_1x_2 + 45\cdot 87x_1^2 + 45\cdot 42x_2^2 \quad . \quad (20)$$

we obtain for the minimum point

$$x_{1S} = -0\cdot 096 \quad x_{2S} = 0\cdot 027$$

Using the relations $x_1 = (\theta_1 - 1\cdot 078\,5)/0\cdot 070\,7$ and

$$x_2 = (\theta_2 - 0\cdot 816\,8)/0\cdot 070\,7 \quad . \quad . \quad . \quad (21)$$

we obtain for the least-squares estimates

$$\hat{\theta}_1 = 1\cdot 072 \quad \hat{\theta}_2 = 0\cdot 819$$

(4.2) **Test of Adequacy of the Assumed Form of the Equations and Estimation of Experimental Error**

By substituting $x_1 = x_{1S}$, $x_2 = x_{2S}$ in eqn. (20) it is found that at the minimum point $S(\hat{\boldsymbol{\theta}}) = 73\cdot 58$. Hence the 'residual' variance s_1^2, having four degrees of freedom and calculated from the means of pairs of observations, is

$$s_1^2 = S(\hat{\boldsymbol{\theta}})/(n - p) = 73\cdot 58/4 = 18\cdot 4$$

Comparing this with the estimated experimental-error variance $s_2^2 = 12\cdot 9$, appropriate for means of pairs of observations, we have $s_1^2/s_2^2 = 18\cdot 4/12\cdot 9 = 1\cdot 43$. From a table of the distribution of the variance ratio F we find that such a value of F (or a larger one) would be expected to occur more often than once in three times by pure chance. There is therefore no reason to question the adequacy of the assumed form of differential equations. A combined estimate of the variance of means of pairs of observations is given by $s^2 = (4 \times 18\cdot 4 + 6 \times 12\cdot 9)/10 = 15\cdot 1$. The variance σ^2 of individual observations is thus estimated as $\hat{\sigma}^2 = 15\cdot 1 \times 2 = 30\cdot 2$, and the experimental-error standard deviation $\hat{\sigma}$ as 5·5%.

(4.3) **Accuracy of the Estimates**

Using the relations eqn. (21) we have, to a close approximation,

$$(\partial^2 S/\partial\theta_1^2)_{\boldsymbol{\theta}=\hat{\boldsymbol{\theta}}} = 2(0\cdot 070\,7)^{-2}b_{11} = 18\,348$$

$$(\partial^2 S/\partial\theta_2^2)_{\boldsymbol{\theta}=\hat{\boldsymbol{\theta}}} = 2(0\cdot 070\,7)^{-2}b_{22} = 18\,168$$

$$(\partial^2 S/\partial\theta_1\partial\theta_2)_{\boldsymbol{\theta}=\hat{\boldsymbol{\theta}}} = (0\cdot 070\,7)^{-2}b_{12} = -5010$$

The matrix $\boldsymbol{V}$ defined in Section 3.3 is therefore given by

$$\boldsymbol{V} = \begin{bmatrix} 9174 & -2505 \\ -2505 & 9084 \end{bmatrix}, \text{ whence } \boldsymbol{V}^{-1} = 10^{-6} \times \begin{bmatrix} 117\cdot 9 & 32\cdot 5 \\ 32\cdot 5 & 119\cdot 0 \end{bmatrix}$$

The estimate of experimental-error variance $s^2 = 15\cdot 1$ gives

$$\text{Standard error of } \hat{\theta}_1 = 0\cdot 042, \text{ Standard error of } \hat{\theta}_2 = 0\cdot 042$$

(4.4) Confidence Region

Using eqn. (17) a confidence region which includes the true values θ_1, θ_2 with 95% probability is defined by the inequality

$$9\,174(\theta_1 - 1\cdot 072)^2 - 5\,010(\theta_1 - 1\cdot 072)(\theta_2 - 0\cdot 819) + 9\,084(\theta_2 - 0\cdot 819)^2 \leqslant 123\cdot 82$$

where the right-hand side of the inequality is $ps^2F_{0\cdot 05}(2, 10) = 2 \times 15\cdot 1 \times 4\cdot 10 = 123\cdot 82$. This confidence region is indicated by the shaded zone in Fig. 4.

(4.5) Discussion

It seems certain that the details of the minimization process illustrated above (e.g. the choice of design and the time spent on the first-order process) could be improved. For this reason the use of the computer has so far been limited to the calculation of S, although it is clear that the steepest-descent procedure at least could easily be made automatic.

By following through a number of calculations in detail it is hoped to gain sufficient experience to judge which features are essential to achieve generality and what simplifications can safely be introduced. In the above example, for instance, it is apparent that a simpler first-order design (involving only 3 or possibly 4 points) could have been used if it were agreed to sacrifice knowledge of the magnitude of second-order coefficients. It appears that only one, or at most two, applications of steepest descent were really necessary before the use of the second-order design.

(5) ACKNOWLEDGMENT

The authors are indebted to the staff of the Mathematics Division of the National Physical Laboratory for their generous help and advice in the use of the Deuce computer.

(6) REFERENCES

(1) BOX, G. E. P., and WILSON, K. B.: 'On the Experimental Attainment of Optimum Conditions', *Journal of the Royal Statistical Society*, Series B, 1951, **13**, p. 1.

(2) ARNOLD, D. S., BOX, G. E. P., ERICKSON, E. E., HUNTER, J. S., NELLI, J. R., and PIKE, F. P.: 'The Application of Statistical Procedures to the Flooding Capacity of a Pulse Column', North Carolina State College, Department of Chemical Engineering, Progress Report No. 3 under Contract AT(40–1)–1320.

(3) ERICKSON, E. E., HUNTER, J. S., NELLI, J. R., and PIKE, F. P.: 'The Flooding Capacity of a Pulse Column on the Benzene-Water System', *ibid.*, Progress Report No. 4.

(4) BOX, G. E. P.: 'The Exploration and Exploitation of Response Surfaces: Some General Considerations and Examples', *Biometrics*, 1954, **10**, p. 16.

(5) BOX, G. E. P.: 'Multifactor Designs of First Order', *Biometrika*, 1952, **39**, p. 49.

(6) BOX, G. E. P., and HUNTER, J. S.: 'Multifactor Experimental Designs for Exploring Response Surfaces', *Annals of Mathematical Statistics*. (In the press.)

(7) GILL, S.: 'A Process for the Step-by-Step Integration of Differential Equations in an Automatic Digital Computing Machine', *Proceedings of the Cambridge Philosophical Society*, **47**, p. 96.

(8) BOX, G. E. P., and YOULE, P. V.: 'The Exploration and Exploitation of Response Surfaces: An Example of the Link between the Fitted Surface and the Basic Mechanism of the System', *Biometrics*, 1955, **11**, p. 287.

(9) KOSHAL, R. S.: 'Application of the Method of Maximum Likelihood to the Improvement of Curves fitted by the Method of Moments', *Journal of the Royal Statistical Society*, 1933, **96**, p. 303.

(10) BOX, G. E. P., CONNOR, L. R., COUSINS, W. R., DAVIES, O. L. (Editor), HIMSWORTH, F. R., and SILLITO, G. P.: 'The Design and Analysis of Industrial Experiments' (Oliver and Boyd, 1954).

(11) FISHER, R. A., and YATES, F.: 'Statistical Tables for Biological, Agricultural and Medical Research' (Oliver and Boyd, 4th Edition, 1953).

4.5

Use of Statistical Methods in the Elucidation of Basic Mechanisms

G. E. P. BOX
Princeton University

INTRODUCTION

This paper summarises some of the work covered by the above title which is currently in progress at Princeton University, Statistical Techniques Research Group.

The problem often arises of exploring experimentally a possible functional relationship

$$\eta = f(\xi_1, \xi_2, \ldots, \xi_k; \theta_1, \theta_2, \ldots, \theta_p) = f(\xi; \theta). \tag{1}$$

In the chemical industry, which is the field of application with which I am most familiar, η may be a yield of some chemical product and $\xi_1, \xi_2, \ldots, \xi_k$ may be such variables as time of reaction, temperature and concentration.

We can distinguish three component phases in such investigations each of which can involve the use of appropriate statistical methods. These are:

(i) Screening studies aimed at delineating the important variables;

(ii) Empirical studies aimed at describing certain important features of the relationship between the response and the variables (e.g. describing "effects" of variables in some particular region, or more ambitiously of graduating approximately the principal features of the multi-dimensional graph or "response surface" relating the response and the variables in some important region of the variables); and

(iii) Theoretical studies aimed at discovering plausible and predictive mechanisms for the phenomena studied and estimating to satisfactory accuracy the relevant parameters.

A fourth important phase is the transition from (ii) to (iii). As has been illustrated, for example in reference [1], careful consideration of the results of an empirical study can produce ideas tending to the development of a theoretical model.

We are here concerned with the problems of analysis and design which arise in phase (iii).

The Collected Works of George E. P. Box, 1984, Wadsworth, Inc., Belmont, CA 94002. Originally published in *Bull. Int. Inst. of Stat.,* Stockholm (1957), pp. 215–225.

THEORETICAL MECHANISMS

Either after a number of clues have been provided by results of empirical experimentation or sometimes without such a preliminary investigation, it may be possible for the experimenter to conceive a mechanism which might be responsible for the phenomena which are being observed. Thus, in chemistry, kinetic theory may provide such a mechanism. From a mathematical analysis of this hypothetical mechanism it will then usually be possible to produce some theoretical functional form which should relate the observations to the variables.

Consider for example a simple chemical system in which reactant A decomposes to form B which in turn decomposes to form C in accordance with the scheme

$$A \xrightarrow{k_1} B \xrightarrow{k_2} C, \tag{2}$$

where k_1 and k_2 are the rate constants for the two reactions.

With η_1, η_2 and η_3 representing the yields of A, B and C at time t we can, under certain well defined assumptions, apply the law of mass action to obtain the set of differential equations

$$\begin{aligned} \frac{d\eta_1}{dt} &= -k_1\eta_1, \\ \frac{d\eta_2}{dt} &= k_1\eta_1 - k_2\eta_2, \\ \frac{d\eta_3}{dt} &= k_2\eta_2, \end{aligned} \tag{3}$$

with boundary conditions $\eta_1 = 1$, $\eta_2 = 0$, $\eta_3 = 0$ at time $t = 0$. In this particular example (although not usually) an explicit solution of the differential equations is available. For example, the yield η_2 of the intermediate product B at any given time t is given by

$$\eta_2 = \frac{k_1}{k_1 - k_2}(e^{-k_2 t} - e^{-k_1 t}). \tag{4}$$

Suppose that the results of a number of experiments are available. For the system described by equations (3) and (4) for example, the data shown in Table 1 are given in reference [2].

The problems of *analysis* of the results are then (i) to discover whether the assumed functional form is adequate to describe the data, and if it is not in what way it is not; (ii) if the functional form is adequate to estimate values of the unknown parameters (i.e. the constants k_1 and k_2 in our example) and to determine their precision as measured by their standard errors and confidence region.

A number of problems in statistical *design* arise, some of which are discussed briefly at the end of this paper.

TABLE 1

Time min.	Yield of M_2 (y_2) %	Mean result ($\bar{y}_2$) %
10	19.2	16.6
	14.0	
20	14.4	19.2
	24.0	
40	42.3	36.55
	30.8	
80	42.1	41.3
	40.5	
160	40.7	43.55
	46.4	
320	27.1	24.7
	22.3	

SOME PROBLEMS OF ANALYSIS

If $\eta_u = f(\xi_u; \theta)$ is the true value for the response at some particular set of experimental conditions ξ_u, and if y_u is the observed value of η_u at these conditions, then least squares estimates of the k constants θ would be obtained by minimizing the quantity

$$S(\theta) = \Sigma [y_u - f(\xi_u; \theta)]^2, \tag{5}$$

where the summation is over the n sets of experimental conditions. On the assumption that the experimental errors are normally distributed, the estimates $\hat{\theta}$ so obtained are maximum likelihood estimates. If the function $f(\xi; \theta)$ were known explicitly, the values $\hat{\theta}$ making $S(\theta)$ a minimum could be found by differentiating in the usual way. In some cases numerical methods would be needed to solve the resulting equations but this would usually present no particular difficulty. Usually, however, no explicit expression for $f(\xi; \theta)$ is available, the relationship being most often defined in some less direct way, in particular in terms of differential equations, partial differential equations or integral equations.

A method of finding the least squares estimates $\hat{\theta}$ is therefore needed which can be employed in these circumstances. Now it will generally be possible to calculate values of $f(\xi; \theta)$ for given ξ and θ by using some suitable numerical procedure. If such values can be calculated we can follow some iterative process, each stage of which leads from one set of "guessed" values θ' to a new set of θ'' which should be closer to the least square estimates.

One way of doing this, based on the minimization of $S(\theta)$ by steepest descents and the representation $S(\theta)$ near the minimum by an equation of second degree in θ

was given in reference [2]. (See also [4].) Experience and theoretical work have shown that this method is considerably less effective in the present context than one originated by Gauss [9]. (See also Fisher [3] and Deming [5].) With the number of parameters being equal to p, the amount of computation necessary using the Gauss method is approximately a fraction $2/(p+2)$ of that necessary using the other procedure.

The amount of computation needed with either method is considerable, especially when the values of the function must be calculated at each set of conditions by numerical methods. However, with the coming of electronic digital computing machines, large computations no longer constitute a serious obstacle. In fact a *general* non-linear least squares sub-routine can be devised. This is used in conjunction with appropriate numerical sub-routines for the calculation of the function when this is not given explicitly.

A least squares programme of this sort worked out in cooperation with G. A. Coutie was run last year using the DEUCE computer at the National Physical Laboratory at Teddington, in conjunction with a Runge-Kutta subroutine for the solution of as many as ten simultaneous differential equations. Similar programmes have been worked out for the Princeton MANIAC computer by Bucher and for the IBM 704 machine by Booth and Peterson of International Business Machines Corp. in cooperation with the Princeton group.

The procedure is as follows. Suppose the elements of the vector θ are the true values of the parameters and those of the vector θ' are preliminary estimates. Then if θ' were close to θ we could write as an approximation

$$f(\xi_u;\ \theta) = f(\xi;\ \theta') + \sum_{r=1}^{p} \varphi_r\, x_{ru}, \tag{6}$$

with

$$\begin{aligned} x_{ru} &= \Delta_r f(\xi_u;\ \theta') \\ &= f\{\xi_u;\ \theta_1', \ldots, \theta_r' + \Delta\theta_r', \ldots, \theta_p'\} - f\{\xi_u;\ \theta_1', \ldots, \theta_r', \ldots, \theta_p'\} \end{aligned}$$

and

$$\varphi_r = (\theta_r - \theta_r')/\Delta\,\theta_r'. \tag{7}$$

As is well known, since we have observations y_u of $f(\xi_u, \theta)$ and we can readily calculate values of $\Delta_r f(\xi_u, \theta')$, we can estimate the φ_r and hence the corrections which must be applied to the preliminary estimates to give a second set of estimates θ''. The improved estimates may now be made the basis of a new iteration and the process repeated until the estimates become stable. The approximate variance-covariance matrix for the φ_r, and hence of the final estimates, is obtained from the reciprocal of the matrix of sums of squares and products of the x_r. Furthermore, an approximate test of goodness of fit is provided by comparing the residual sum of squares with some estimate of pure error, provided for example by random replication.

As an example the data of Table 1 on which the "steepest ascent" method was illustrated in reference [2] may be reworked. As before the parameters to be estimated were taken to be

$$\theta_1 = 3 + \log k_1 \quad \theta_2 = 3 + \log k_2 \tag{8}$$

and the starting values

$$\theta_1^* = \theta_2^* = 1.19. \tag{9}$$

The means $\bar{y}_2$ of Table 1 were treated as the "observations". Convergence to virtually exact values is then obtained in two iterations. The second of these was as follows.

Using the values from first iteration of $\theta_1' = 1.04$, $\theta_2' = 0.84$ and the intervals $\Delta\theta_1' = \Delta\theta_2' = 0.02$, we obtain

t_u (min)	$y_u - f_u(\xi_u, \theta)$	x_{1u}	x_{2u}
10	6.5797	0.4530	−0.0172
20	0.8696	0.7802	−0.0617
40	5.8665	1.1470	−0.2066
80	−1.7576	1.1871	−0.5892
160	0.8796	0.4267	−1.2126
320	3.2082	−0.3802	−1.3074

The normal equations are

$$\begin{aligned} 3.865\,\varphi_1 - 1.013\,\varphi_2 &= 7.457 \\ -1.013\,\varphi_1 + 3.574\,\varphi_2 &= -5.604 \end{aligned} \tag{10}$$

giving

$$\varphi_1 = 1.640 \quad \varphi_2 = 1.103 \tag{11}$$

and new estimates

$$\theta_1'' = 1.04 + (1.640 \times 0.020) = 1.073$$

$$\theta_2'' = 0.84 + (1.103 \times 0.020) = 0.818$$

at which values the sum of squares is 73.6.

The approximate test of goodness of fit is conducted by comparing the residual mean square

$$S(\hat{\theta})/(n-p) = 73.6/4 = 18.4$$

with the estimated experimental error variance (from within replicates) $s_2^2 = 12.9$ appropriate for the *means* of pairs of observations.

An approximate F test on 4 and 6 degrees of freedom indicates that there is in this case no reason to doubt the model (3) and hence the mechanism implied by (2). A combined estimate of experimental error is 15.1 based on 10 degrees of freedom.

Using the elements in equations (10) we can now write down the approximate confidence interval for the θ:s as

$$3.865(\theta_1 - 1.073)^2 - 2.025(\theta_1 - 1.073)(\theta_2 - 0.818) + \\ + 3.574(\theta_2 - 0.818)^2 \leqslant (0.02)^2\, 123.82, \qquad (12)$$

where the right hand side of the inequality is

$$(\Delta\,\theta)^2\, p s^2\, F_{0.05}(2,\ 10).$$

In this particular example it is easy to show by actually plotting the boundary of the interval given by (12) that this closely follows the true sum of squares contour.

The rapid convergence experienced in this example is not necessarily found in other cases. It can be shown that the expected value of the error in the adjustment is proportional to the *square* of the discrepancy between starting value and true value. This is evidenced in examples in which the starting values are poorly chosen when violent oscillations in the estimates of the θ:s can occur from one iteration to the next.

Although no example has so far been encountered in which eventual convergence did not occur, the precaution has been taken of building a special routine into the computer programmes. When this sub-routine is switched in, the computer does not necessarily take the vector φ as the appropriate adjustment but instead the indicated direction of modification is explored by automatically making a sequence of trial calculations at a series of points equally spaced along a logarithmic scale, starting with the adjustments $\frac{1}{2}\varphi$, φ, 2φ. The interpolated value along this path with smallest sum of squares is taken as the new starting point.

As a check, at the end of the computation just described, the computer calculates the *actual* sums of squares at $2k$ points surrounding the supposed minimum. In the present programme these points are at the ends of the principal axes of a 95 % confidence ellipsoid calculated assuming linearity. The values for the θ:s at each of these points and the corresponding calculated sums of squares are printed out at the completion of the work.

The computation serves three purposes:

(i) It provides a rough check that a true minimum has been found (the $2k$ values of these sums of squares should all of course be *larger* than the value at what is claimed is the minimum).

(ii) It provides an indication of the degree of *local* non-linearity of the problem. (To ensure that the test of goodness of fit, the confidence region, standard errors, and analysis of variance test are good approximations we require approximate linearity of the parameters not over the whole space of the response but only in a region around the solution.)

If the $2k$ calculated sums of squares are all nearly equal to each other and to the value expected on the 95 % sum of squares contour, then procedures based on local linearity may be applied with some assurance. When marked departures from linearity are indicated the sizes of the calculated values can help to indicate transformations of the parameters in terms of which approximate linearity might be achieved.

(iii) It provides an indication of the nature and extent of the confidence interval.

In addition to the method based on the linearization of equation (6), one might in some circumstances use a quadratic approximation. One such procedure has been studied by M. B. Wilk, and others are being developed.

Computer programmes of these sorts provide a valuable and rapid means of assessing the plausibility of a number of alternative mechanisms and of separating them into two classes; those which are contradicted by the data and those which are not.

SOME PROBLEMS OF DESIGN

A number of design problems now occur. One concerns the situation where we have two or more mechanisms which are plausible on present data and we wish to decide where experiments ought to be performed which will most effectively discriminate between the rival possibilities. Another problem arises because the original data may be totally unsuitable to the purpose of estimating separately the parameters θ and we wish now to generate data which will enable the best possible estimates to be obtained. Important work connected with this problem has been carried out by Elfving [6] and by Chernoff [7].

It is not infrequently found when attempts are made to estimate parameters θ from some initial set of data that estimates of the θ will be very highly correlated. The sums of squares surface then involves marked valleys near minima and a number (say r) of the latent roots of the correlation matrix approach zero. This indicates that only $p-r$ linear combinations of the parameters can be effectively estimated separately. Such a situation may arise because of the unsuitability of the data (i.e. unsuitable location of the ξ:s) or because of the nature of the model.

If the difficulty in separating the θ:s arises from unsuitable data, improvement can be obtained by proper choice of the design. Dr. Henry L. Lucas and the present writer [8] have studied this design problem by proceeding in the following way.

In general an experimental design may be defined by an $n \times k$ matrix $D = (\xi_{iu})$ called the design matrix. The uth row ξ'_u with elements $\xi_{1u}, \ldots, \xi_{iu}, \ldots, \xi_{ku}$ provides the levels of the variables at which the response is to be observed in the uth trial.

In the model of equation (9) t would correspond to ξ_1, and θ_1 and θ_2 to k_1 and k_2. The problem would be to choose a set of times $t_u = \xi_{1u} (u = 1, 2, \ldots, n)$ at which to observe the yield so that from these observations $\theta_1 = k_1$ and $\theta_2 = k_2$ would be estimated as accurately as possible. In this example the design matrix D would simply consist of a single column whose n elements were the n times at which the yield was to be observed.

In practice when variables are quantities such as time, temperature, dosages, concentrations, etc., the levels at which these variables can be set are usually restricted. For example, in the study of the chemical reaction above, the feasible range for the variable time (ξ_1) would be from zero to some value ξ_1(max), the longest time contemplated in the experiment. In general, there would exist some *region*, R, in the ξ-space which would delimit the area within which experiments could actually be conducted. In some cases, though not in all, this region could be defined by a series of inequalities in the ξ:s of the type

$$\xi_i(\min) \leqslant \xi_i \leqslant \xi_i(\max) \quad (i = 1, 2, \ldots, k). \tag{13}$$

In given cases the boundaries will not be parallel to the axes of the ξ:s and may even be curved. For example hazards may occur for certain combinations of the variables (explosive hazards with certain combinations of temperature and pressure in a chemical reaction; toxicity hazards from certain combinations of nutrients in a feeding trial).

CRITERION FOR SELECTION OF THE DESIGN

We denote the response observed at the uth set of experimental conditions by y_u and we suppose that

$$E(y_u) = \eta_u = f(\xi_u, \theta) \tag{14}$$

and

$$E(y_u - \eta_u)(y_v - \eta_v) = \begin{cases} \sigma^2 & u = v \\ 0 & u \neq v \end{cases} \quad (u, v = 1, 2, \ldots, n), \tag{15}$$

where in general σ^2 is unknown.

Now suppose the true values of the parameters are $\theta_{10}, \theta_{20}, \ldots, \theta_{p0}$, denoted by the vector θ_0. Denote the partial derivative of the response function with respect to the rth parameter θ_r for the uth set of experimental conditions taken at the point θ_0 by

$$f_{ru} = \left[\frac{\partial f(\xi_u, \theta)}{\partial \theta_r}\right]_{\theta=\theta_0}, \tag{16}$$

and the $r \times p$ matrix of these derivatives by

$$F = (f_{ru}). \tag{17}$$

Then the least squares estimates $\hat{\theta}$ of θ have a variance-covariance matrix which is approximated by $(F'F)^{-1}\sigma^2$. We proceed by attempting to choose D so that the determinant $|(F'F)^{-1}|$ is made as small as possible.

To the degree of approximation assumed, this choice minimizes Wilks' generalized variance for the estimates. The logic of this choice can be seen by noting that if the estimates are approximately normally distributed the determinant $|(F'F)^{-1}|$ is approximately proportional to the volume, in the space of the parameters, contained within any specific ellipsoidal probability contour for $\hat{\theta}$. The criterion chosen ensures that any such probability contour includes the smallest possible volume.

NECESSITY FOR PRELIMINARY ESTIMATES

We see that the efficiency of different possible designs depends upon the matrix F whose elements are the values of the derivatives of the response function with respect to the θ:s at the point $\theta = \theta_0$. Now the derivatives $\partial f/\partial \theta_r$ can only be independent of the values actually taken by the parameter θ_r if the response function f is linear in θ_r.

For non-linear response functions the values of the derivatives and hence the efficiency of any particular design depends upon the actual values of the p parameters. If we are to suppose design is possible at all, we must assume that *something* is known about the values of the parameters in advance. In practical problems it will almost universally be the case that some such information is available; when this is so this information will be made the basis of a first design.

We proceed by assuming that preliminary values θ^* of the parameters are available, and for the purpose of design we will treat these quantities as if they were the true values. Suppose then that we are given the functional form

$$\eta = f(\xi_u, \theta) \tag{18}$$

and a set of preliminary values, θ^*, then we shall choose the design D such that $|F^{*\prime} F^*|$ is a maximum within the experimental region R in the ξ-space, where

$$F^* = (f^*_{ru}) \tag{19}$$

and

$$f^*_{ru} = \left[\frac{\partial f(\xi_u, \theta_u)}{\partial \theta_r}\right]_{\theta=\theta^*}. \tag{20}$$

In order to ensure uniqueness we have so far assumed that the number of distinct trials (that is the number of distinct combinations of levels of the variables) is equal to the number p of parameters to be estimated. Each of the p trials could of course be replicated to attain any desired degree of precision.

Experiments in which the number of parameters estimated is equal to the number of distinct trials are perhaps not intuitively appealing. This is because one is usually not completely assured that the model is correct; so that one ordinarily has in fact a two-fold objective: (i) to check the model and (ii) to estimate the parameters in it when it proves to be representationally adequate. However, in order for a design to be efficient in "checking" the model we must be somewhat specific about what sorts of departure from the model we suspect. Insofar as we can be completely specific, we will be able to write an "alternative" wider model which makes allowance for these possible departures in terms of extra parameters to be estimated. In this way we may be able to achieve both the objectives mentioned above within the framework which we have discussed. As a simple illustration consider the chemical reaction discussed earlier. Here A was decomposed to form B which in turn decomposed to C. Suppose the second decomposition was not in fact expected to occur. Then the primary model in mind for the production of B would be

$$\eta = 1 - e^{-k_1 t}.$$

However in the light of possible decomposition of B to C it would be appropriate to plan the experiment so as to be most effective in estimating both k_1 and k_2 in model (4).

A geometrical formulation of the problem has been worked out and designs obtained for simple models such as the exponential decay law, the Mitscherlich equation, consecutive decay laws such as (2), and many others.

One point reemphasized by the investigation is that for some models, for example the model

$$\eta = \frac{1 + \theta_1 t}{\theta_2 + \theta_3 t},$$

no design is effective in separating the θ:s.

The problem is now being considered of working out a general computer programme which will be able to determine numerically the best design in more complicated cases.

REFERENCES

1. Box, G. E. P. & Youle, P. V., The exploration and exploitation of response surfaces: an example of the link between the fitted surface and the basic mechanism of the system. *Biometrics*, **11**, No. 3, 1955.
2. Box, G. E. P. & Coutie, G. A., Application of digital computers in the exploration of functional relationships. *Proceedings of the Institution of Electrical Engineers*, **103**, part B, Suppl. 1, 1956.
3. Fisher, R. A., Theory of statistical estimation. *Proc. Camb. Phil. Soc.*, **22**, part 5.
4. Koshal, R. S., Application of the method of maximum likelihood to the improvement of curves fitted by the method of moments. *J. Royal Statistical Society*, **96**, 1933.
5. Deming, W. E. *Statistical Adjustment of Data*. John Wiley and Sons, New York, 1943.
6. Elfving, G., Optimum allocation in linear regression theory. *Annals of Math. Stat.*, **23**, 1952.
7. Chernoff, H., Locally optimal designs for estimating parameters. *Annals of Math. Stat.*, **24**, 1953.
8. Box, G. E. P. & Lucas, H. L., *Design of Experiments in Non-linear Situations*. Statistical Techniques Research Group, Princeton University (technical report in preparation).
9. Gauss, K. F., *Theoria Motus Corporum Coelestium*, Werke, Vol. 7, p. 240–254, 1809.

RÉSUMÉ

Emploi des méthodes statistiques dans l'élucidation des mécanismes basiques

Cette étude expose les recherches indiquées par le titre et poursuivies actuellement à l'Université de Princeton.

En explorant expérimentalement une relation fonctionelle

$$\eta = f(\xi, \theta), \tag{1}$$

où il s'agit d'une série de paramètres inconnus θ et qui donne la relation entre une réponse η et des variables ξ, on peut distinguer trois phases composantes :

i) études criblures pour tracer les variables importantes;
ii) études empiriques visant à décrire par une fonction graduante, tel qu'un polynôme, certains traits importants de la relation;
iii) études théoriques visant à décrire des mécanismes plausibles et révisibles pour les phénomènes étudiés.

Cette étude traite de la partie du travail comprise dans la phase (iii).

Après qu'un certain nombre des indices ait été fourni par expérimentation empirique, ou quelquefois au commencement, il est possible pour l'expérimentateur de concevoir un mécanisme qui peut expliquer la phénomène sous observation et de déduire la forme fonctionelle correspondant à l'équation [1]. Dans ce cas, cette fonction pourrait s'exprimer directement mais, plus souvent, elle ne peut être définie facilement que par le moyen des équations différentielles, des équations partielles différentielles, ou d'une autre façon indirecte.

En avançant dans la compréhension du système étudié, deux types de problèmes se présentent, problèmes *d'analyse* et problèmes *de dessin*. Dans le problème d'analyse nous est donné un ensemble de valeurs de la réponse η à divers « niveaux » des variables ξ et on veut :

i) essayer si la forme fonctionelle supposée suffit à décrire les données,
ii) dans le cas que la forme fonctionelle n'est pas discréditée, apprécier les valeurs des paramètres inconnus et
iii) fournir des erreurs modèles et une région de confiance pour les éléments de θ.

Des procédures sont décrites, permettant d'apprécier une fonction arbitraire non-linéaire par la méthode des moindres carrés et d'évaluer approximativement le « goodness of fit ». Les erreurs modèles et une région de confiance par les θ peuvent être calculés aussi. Les calculations sont un peu longues et sont exécutées, en effet, par une machine à calculer automatique. Le programme de la machine demande seulement la possibilité de calculer les valeurs de la fonction pour valeurs arbitraires de ξ et θ, et peut-être employé, par exemple, en combinaison avec des sous-programmes pour l'évaluation numérique des solutions d'équations différentielles, quand la fonction n'est pas donnée explicitement.

Très souvent il n'est possible d'obtenir que des évaluations fortement corrélatives de θ. Ceci peut arriver, soit lorsque les données sont insuffisantes, c'est à dire que les points expérimentaux ξ, où les observations étaient effectueés, n'étaient pas convenablement choisis, soit encore comme une propriété intrinsèque du modèle.

Bien des problèmes de dessin expérimental se présentent. Premièrement, il est nécessaire de découvrir des dessins, c'est à dire des séries de « niveaux » des variables ξ, qui donneront à toutes les estimations des paramètres θ toute l'exactitude possible. Le second problème est de dessiner des expériments qui soient le plus plein de force dans la distinction entre deux modèles qui sont regardés initialement comme plausibles. Le premier problème a été étudié par l'autcur et le Dr. H. L. Lucas et un certain nombre de dessins ont été complétés par des modèles décrivant les lois de « decay » simple, l'équation de Mitscherlich, les lois de « decay » consécutif et beaucoup d'autres. Un travail plus étendu a été entrepris particulièrement pour établir un programme de machine à calculer, pour les dessins de cas plus compliqués.

4.6

Fitting Empirical Data*

G. E. P. BOX
U.S. Army Mathematics Research Center, Madison, Wisconsin

1. Formulation

Suppose we have an observable response in which level η is functionally related to the levels $\xi_1, \xi_2, \ldots, \xi_k$ of k variables, and suppose that this relationship involves p parameters $\theta_1, \theta_2, \ldots, \theta_p$ whose true but unknown values are $\theta_1^*, \theta_2^*, \ldots, \theta_p^*$. Then the *response function* is defined by

$$\eta = \eta(\xi_1, \ldots, \xi_i, \ldots, \xi_k; \quad \theta_1^*, \ldots, \theta_r^*, \ldots, \theta_p^*) = \eta(\boldsymbol{\xi},\boldsymbol{\theta}^*) \quad (1)$$

Suppose that N experiments or trials are performed and that in the uth trial ($u = 1, 2, \ldots, N$) the variables are held at the known levels $\boldsymbol{\xi}_u' = (\xi_{1u}, \xi_{2u}, \ldots, \xi_{ku})$ and the actual response observed is y_u. We then suppose that $\eta_u = \eta(\boldsymbol{\xi}_u, \boldsymbol{\theta}^*)$ is the expected value $E(y_u)$ of y_u.

We denote by $\mathbf{y}$ an $N \times 1$ vector of observations, by $\boldsymbol{\eta} = E(\mathbf{y})$ an $N \times 1$ vector of their expected values, and by $\boldsymbol{\varepsilon} = \mathbf{y} - \boldsymbol{\eta}$ an $N \times 1$ vector of errors. The levels of the k variables in the complete set of N trials can be represented by an $N \times k$ matrix $\mathbf{D}$ called the *design matrix*, whose uth row is the vector $\boldsymbol{\xi}_u'$. The rows of $\mathbf{D}$ can be regarded conveniently as the coordinates of a set of *experimental points* in the k-dimensional space of the variables.

Now suppose that $\boldsymbol{\varepsilon}$, the vector of errors, has a probability density function of known form that involves a set of q unknown parameters $\boldsymbol{\pi}$ whose true but unknown values are given by $\boldsymbol{\pi}^*$:

$$\begin{aligned} p(\boldsymbol{\varepsilon},\boldsymbol{\pi}^*) = p\{&y_1 - \eta(\boldsymbol{\xi}_1,\boldsymbol{\theta}^*), \quad y_2 - \eta(\boldsymbol{\xi}_2,\boldsymbol{\theta}^*), \ldots, \\ &y_N - \eta(\boldsymbol{\xi}_N,\boldsymbol{\theta}^*); \quad \pi_1^*, \pi_2^*, \ldots, \pi_q^*\} \end{aligned} \quad (2)$$

then

$$\begin{aligned} p(\boldsymbol{\varepsilon},\boldsymbol{\pi}^*) &= F\{\mathbf{y},\mathbf{D},\boldsymbol{\theta}^*,\boldsymbol{\pi}^*\} \\ &= F\{\mathbf{y},\mathbf{D},\boldsymbol{\phi}^*\} \end{aligned} \quad (3)$$

where $\boldsymbol{\phi}^*$ is the set of $p + q$ parameters $\boldsymbol{\theta}^*$ and $\boldsymbol{\pi}^*$. When a set of trials is completed, the observations $\mathbf{y}$ and the design matrix $\mathbf{D}$ are known; regarding these as fixed but the parameters $\boldsymbol{\phi}$ (that is, $\boldsymbol{\theta}$ and $\boldsymbol{\pi}$) as variable we may rewrite $F\{\mathbf{y},\mathbf{D},\boldsymbol{\theta},\boldsymbol{\pi}\} = F\{\mathbf{y},\mathbf{D},\boldsymbol{\phi}\}$ as $L(\boldsymbol{\theta},\boldsymbol{\pi}) = L(\boldsymbol{\phi})$, a function of the unknown parameters only called the *likelihood function.*[1] The value this function takes for any set of values of the parameters gives the *likelihood* for this set of values of the parameters.

In many situations in which empirical data are analyzed, some knowledge of the values of the parameters $\boldsymbol{\theta}$ and $\boldsymbol{\pi}$ prior to availability of the data

* Part of the work reported in this paper was performed under Office of Ordnance Contract No. DA 36-034-ORD-2297, at the Statistical Techniques Research Group, Princeton University, Princeton, N. J.

The Collected Works of George E. P. Box, 1984, Wadsworth, Inc., Belmont, CA 94002.
Originally published in *Ann. N.Y. Acad. of Sciences,* vol. 86, no. 3 (1960), pp. 792–816.

does exist. Furthermore, it is such that in the relevant parameterization it may be represented by a (subjective) prior probability distribution for $\boldsymbol{\theta}$ and $\boldsymbol{\pi}$ which is likely to be nearly constant over the region where the likelihood has an appreciable value and is unlikely to be many orders of magnitude greater in some neighboring region. In these circumstances, as emphasized by G. A. Barnard (personal communication) and by L. J. Savage,[2] the likelihood function can be regarded, by a straightforward application of Bayes' theorem, as providing a close approximation to the posterior distribution of the unknown parameters $\boldsymbol{\theta}$ and $\boldsymbol{\pi}$.

2. The Dual Problems of Design and Analysis

Considering the situation before the experiments are performed, we see that the information about the parameters that would be contained in the likelihood function depends upon the choice of **D**, that is, upon the program of experiments to be performed. In particular, if we take data at points chosen haphazardly in the space of the variables, then the estimates may be imprecise or we may not be able to obtain separate estimates of the individual parameters at all, even though a different choice of **D** would have given separate and precise estimates.

We thus have two problems. These concern:

(1) The manner in which we choose to *generate* the data. This is the problem of the statistical design of experiments involving the choice of **D**. Many different situations can arise, depending on the particular objectives we wish to achieve.

(2) The statistical analysis of the experimental data once it is generated. The objective at this stage is to extract all the information contained in the data relevant to the questions of interest.

It will be convenient to consider the second problem first; before we do so, however, let us consider some properties of the likelihood function and some special and important simplifications.

3. The Analysis of Experimental Data

In some problems we may be interested in the parameters directly; in others we may be concerned with certain functions of them, in particular with the response function $\eta(\boldsymbol{\xi},\boldsymbol{\theta}^*)$ itself. In any case, the likelihood function contains all the information available from the data about the parameters or functions of them; moreover, most questions relevant to the fitting of data once they have been obtained are answered by appropriate study of this function.

Where there are only a few parameters it is instructive and perfectly feasible to plot the likelihood function over the whole range of parameter values likely to be of interest. For example, likelihood contour diagrams in the parameter space for two and three parameters provide us with a visual representation of what experimental evidence entitles us to believe about the parameters. The calculations required are laborious, but are repetitive and ideally suited to electronic computers. With many parameters we cannot proceed by visual representation, and a summary of the information contained in the likelihood function in a readily appreciated form is desirable. Such a summary is obtainable as long as the likelihood function itself is sufficiently simple to be simply described. Description is often facilitated by use of the logarithm of the likelihood, designated $l(\boldsymbol{\phi})$, rather than the likelihood $L(\boldsymbol{\phi})$ itself.

One feature of interest is the set of values $\hat{\boldsymbol{\phi}}$ of $\boldsymbol{\phi}$ which make $L(\boldsymbol{\phi})$ and hence $l(\boldsymbol{\phi})$ a maximum. These are the so-called maximum likelihood estimates. As in most practical problems where maxima are of importance, knowledge merely of the position of the maximum is inadeqı ate and often misleading. It is almost always necessary to study at least the local behavior of the function in the neighborhood of the maximum. Often $l(\boldsymbol{\phi})$ may be closely aproximated in the parameter space over a region of sufficient size in the neighborhood of $\hat{\boldsymbol{\phi}}$ by an expression quadratic in $\boldsymbol{\phi}$:

$$l(\boldsymbol{\phi}) = l(\hat{\boldsymbol{\phi}}) + \tfrac{1}{2}\sum_{rs} (\phi_r - \hat{\phi}_r)(\phi_s - \hat{\phi}_s)l_{rs} \tag{4}$$

with

$$l_{rs} = \left\{\frac{\partial^2[l(\boldsymbol{\phi})]}{\partial\phi_r\partial\phi_s}\right\}_{\boldsymbol{\phi}=\hat{\boldsymbol{\phi}}}$$

When this representation is adequate, virtually all the information of interest is contained in $\hat{\boldsymbol{\phi}}$ and in the matrix of second derivatives $\{l_{rs}\}$.

When such representation is not adequate, of course, some more elaborate approximation is necessary: in addition to $\hat{\boldsymbol{\phi}}$ and $\{l_{rs}\}$ other quantities will be needed to describe the likelihood function and hence to summarize the information contained in the data.

Often a sufficiently close representation is obtained by assuming that the errors ϵ are independently distributed in normal distributions each having the same variance σ^2. In this case the contours of probability density in the N-dimensional space of the observations are hyperspheres; hence we call the joint distribution of errors a spherical normal distribution. The single parameter of the error distribution is σ^2, and

$$-2l(\boldsymbol{\theta},\sigma^2) = \text{const.} + S(\boldsymbol{\theta})/\sigma^2 \tag{5}$$

The function

$$S(\boldsymbol{\theta}) = \sum_{u=1}^{N} \{y_u - \eta(\boldsymbol{\xi}_u, \boldsymbol{\theta})\}^2 \tag{6}$$

which measures the sum of squares of discrepancies between observed and postulated responses for any value $\boldsymbol{\theta}$, is called the *sum of squares function.*

Thus, when normal distributions are assumed, the contours of the likelihood function are also contours of the sum of squares function, and the maximum likelihood estimates are least-squares estimates.

The assumption of normally distributed errors is usually not a burdensome one. It is known[3] for response functions linear in the parameters, that with many choices of design serious discrepancies from normality would not greatly affect conclusions based on the assumption of normality. Furthermore, in the case of such linear response functions, many properties of the least-squares estimates (their unbiased nature, the sizes of their variances and covariances, and the fact that of all linear estimates of the parameters those given by the method of least squares have smallest variance) are independent of the normality assumption.

4. Linear Least Squares

Suppose errors are spherically normally distributed and in addition that the response function *is linear in the parameters*, so that

$$\eta(\xi_u, \boldsymbol{\theta}) = \sum_{r=1}^{p} \theta_r^* x_{ru} \qquad (7)$$

where the $x_{ru} = \partial\eta_u/\partial\theta_r$ are known or calculable constants, functions only of the elements of ξ_u. Then the expected values of the response at the chosen set of N experimental conditions can be written in matrix notation as

$$\mathbf{n} = \mathbf{X}\boldsymbol{\theta}^* \qquad (8)$$

where $\mathbf{X} = \{x_{ru}\}$ is an $N \times p$ matrix of constants x_{ru} and the elements of $\boldsymbol{\theta}^*$ are the true values of the parameters. We then may write

$$-2l(\boldsymbol{\theta}) - \text{const.} = S(\boldsymbol{\theta})/\sigma^2 = \sum_{u=1}^{N} (y_u - \sum_{r=1}^{p} \theta_r x_{ru})^2/\sigma^2 \qquad (9)$$

and

$$S(\boldsymbol{\theta}) = S(\hat{\boldsymbol{\theta}}) + (\boldsymbol{\theta} - \hat{\boldsymbol{\theta}})'\mathbf{X}'\mathbf{X}(\boldsymbol{\theta} - \hat{\boldsymbol{\theta}}) \qquad (10)$$

where $\hat{\boldsymbol{\theta}}$ is the set of parameter values that make $S(\boldsymbol{\theta})$ a minimum.

We see that in this case considerable simplification arises because of the following consequences of our assumptions:

(1) The sum of squares function, and hence $l(\boldsymbol{\theta}, \sigma^2)$, are exactly quadratic functions of the θ's.

(2) It is easily demonstrated that $\hat{\boldsymbol{\theta}}$ is simply the solution of the *linear* equations $(\mathbf{X}'\mathbf{X})\hat{\boldsymbol{\theta}} = \mathbf{X}'\mathbf{y}$, and that $E(\hat{\boldsymbol{\theta}}) = \boldsymbol{\theta}^*$.

(3) The variance-covariance matrix for $\hat{\boldsymbol{\theta}}$ is provided exactly by $(\mathbf{X}'\mathbf{X})^{-1}\sigma^2$.

(4) On the assumption that the model is adequate, $S(\hat{\boldsymbol{\theta}})/(n - p)$ provides an unbiased estimate s_r^2 of σ^2.

Objectives of the Experimenter

Commonly the experimenter will desire to obtain from his data some indications of the possible adequacy or inadequacy of the assumed mathematical model, estimates of the various parameters, and measures of the precision of these estimates. Let us consider how this can be done in this special situation in which the response function is linear in the parameters.

Adequacy of the model. If $\boldsymbol{\eta} = \mathbf{X}\boldsymbol{\theta}^*$, the model assumed, is inadequate, we suppose that an adequate model would be of the form

$$\boldsymbol{\eta} = \mathbf{X}\boldsymbol{\theta}^* + \mathbf{X}_1\boldsymbol{\theta}_1^* \tag{11}$$

where $\boldsymbol{\theta}_1^*$ contains p_1 further parameters and $\mathbf{X}_1 = \{x_{su}\}$ $(s = 1, 2, \ldots, p_1)$ is the corresponding $N \times p_1$ matrix of constants. Now suppose the experimenter, believing the model $\boldsymbol{\eta} = \mathbf{X}\boldsymbol{\theta}^*$ to be adequate, estimates the parameters by $\hat{\boldsymbol{\theta}} = (\mathbf{X}'\mathbf{X})^{-1}\mathbf{X}'\mathbf{y}$. The expected values $E(\hat{\boldsymbol{\theta}})$ now contain elements of $\boldsymbol{\theta}_1^*$ as well as $\boldsymbol{\theta}^*$, for

$$E(\hat{\boldsymbol{\theta}}) = \boldsymbol{\theta}^* + (\mathbf{X}'\mathbf{X})^{-1}\mathbf{X}'\mathbf{X}_1\boldsymbol{\theta}_1^* \tag{12}$$

Information on the adequacy of a postulated model can be obtained as follows: if the values $\hat{\boldsymbol{\theta}}$ are substituted in the response function, we can calculate the vector $\hat{\mathbf{y}}$ of estimated values of the response at the N experimental points from

$$\hat{\mathbf{y}} = \mathbf{X}\hat{\boldsymbol{\theta}} = \mathbf{X}(\mathbf{X}'\mathbf{X})^{-1}\mathbf{X}'\mathbf{y} \tag{13}$$

The discrepancies $\mathbf{y} - \hat{\mathbf{y}}$ between observed and estimated responses are called the residuals of $\mathbf{y}$ on $\mathbf{X}$. Now, for any set of parameter values $\boldsymbol{\theta}$, the decomposition of the sum of squares function in EQUATION 10 can be written

$$(\mathbf{y} - \mathbf{X}\boldsymbol{\theta})'(\mathbf{y} - \mathbf{X}\boldsymbol{\theta}) = (\mathbf{y} - \hat{\mathbf{y}})'(\mathbf{y} - \hat{\mathbf{y}}) + (\hat{\mathbf{y}} - \mathbf{X}\boldsymbol{\theta})'(\hat{\mathbf{y}} - \mathbf{X}\boldsymbol{\theta}) \tag{14}$$

This is simply a statement of Pythagoras's theorem for the orthogonal vectors in the equality

$$\mathbf{y} - \mathbf{X}\boldsymbol{\theta} = (\mathbf{y} - \hat{\mathbf{y}}) + (\hat{\mathbf{y}} - \mathbf{X}\boldsymbol{\theta}) \tag{15}$$

Now, since $\hat{\mathbf{y}} - \mathbf{X}\boldsymbol{\theta}$ is the component of $\mathbf{y} - \mathbf{X}\boldsymbol{\theta}$ in the space spanned by the r N-dimensional vectors that are columns of $\mathbf{X}$, it can tell us nothing of the adequacy or inadequacy of a model that assumes adequate representation in terms of $\mathbf{X}$. All information on the adequacy of the postulated model is contained in the vector of residuals $\mathbf{y} - \hat{\mathbf{y}}$.

Both the length and the direction of the vector $\mathbf{y} - \hat{\mathbf{y}}$ may be used to throw light on particular aspects of the adequacy of fit. An *over-all* test of the adequacy of the model is provided by considering the length l of the residual vector. We have

$$E(l^2) = E(\mathbf{y} - \hat{\mathbf{y}})'(\mathbf{y} - \hat{\mathbf{y}}) = E\{S(\hat{\boldsymbol{\theta}})\} = (N - p)\sigma^2 + \Lambda^2 \tag{16}$$

where

$$\Lambda^2 = \boldsymbol{\theta}_1'\mathbf{X}_1'\{\mathbf{I} - \mathbf{X}(\mathbf{X}'\mathbf{X})^{-1}\mathbf{X}'\}\mathbf{X}_1\boldsymbol{\theta}_1 \tag{17}$$

When the original model is adequate, $\boldsymbol{\theta}_1$ is null and the mean value of $S(\hat{\boldsymbol{\theta}})$ is $(N - p)\sigma^2$; otherwise this mean value is inflated by Λ^2, a nonnegative definite quadratic form in $\boldsymbol{\theta}_1$. Suppose certain trials have been replicated, from which an estimate s_e^2 of σ_e^2 having ν_e degrees of freedom can be calculated; then

$$E\{S(\hat{\boldsymbol{\theta}}) - \nu_e s_e^2\} = (N - p - \nu_e)\sigma^2 + \Lambda^2$$

A judgment as to the over-all adequacy of the model is provided by comparing $\{S(\hat{\boldsymbol{\theta}}) - \nu_e s_e^2\}/(N - p - \nu_e)$ and s_e^2. This judgment is assisted by the consideration that with an adequate model and the usual assumptions, the ratio of these two quantities follows an F distribution with $N - p - \nu_e$ and ν_e degrees of freedom.

Possible causes for inadequacy are often suggested by study of the direction of the residual vector. Such studies are usually called tests on residuals. In this connection the vector of expected values of the residuals is

$$E(\mathbf{y} - \hat{\mathbf{y}}) = \{\mathbf{I} - \mathbf{X}(\mathbf{X}'\mathbf{X})^{-1}\mathbf{X}'\}\mathbf{X}_1\boldsymbol{\theta}_1 \tag{18}$$

Now $(\mathbf{X}'\mathbf{X})^{-1}\mathbf{X}'\mathbf{X}_1 = \mathbf{A} = \{a_{rs}\}$ is a $p \times p_1$ matrix of the regression coefficients of the p_1 columns of $\mathbf{X}_1$ on the p columns of $\mathbf{X}$. Whence

$$\{\mathbf{I} - \mathbf{X}(\mathbf{X}'\mathbf{X})^{-1}\mathbf{X}'\}\mathbf{X}_1 = \mathbf{X}_1 - \mathbf{X}\mathbf{A} = \mathbf{X}_1 - \hat{\mathbf{X}}_1 = \mathbf{X}_{1\cdot 0} \tag{19}$$

is an $N \times p_1$ matrix of the residuals of $\mathbf{X}_1$ on $\mathbf{X}$.

Thus EQUATION 18 can be written

$$E(\mathbf{y} - \hat{\mathbf{y}}) = (\mathbf{X}_1 - \hat{\mathbf{X}}_1)\boldsymbol{\theta}_1 = \mathbf{X}_{1\cdot 0}\boldsymbol{\theta}_1 \tag{20}$$

The vector $E(\mathbf{y} - \hat{\mathbf{y}})$ is null when $\boldsymbol{\theta}_1$ is null. Thus when the assumed model is adequate, $\mathbf{y} - \hat{\mathbf{y}}$ contains no systematic component from $\mathbf{X}_1$. If the assumed model is inadequate, however, it contains a component in the space of the p_1 column vectors of $\mathbf{X}_{1\cdot 0}$ which are components of the column vectors of $\mathbf{X}_1$ orthogonal to those of $\mathbf{X}$. Most procedures for analyzing residuals are concerned essentially with searching for vectors $\mathbf{X}_1$ that could produce the observed systematic component. Suppose, for example, that in application of a particular linear model $\hat{\mathbf{y}} = \sum_{r=1}^{p} \hat{\theta}_r \mathbf{x}_r$ a lack of fit was found, and that a linear time trend that might have occurred during the course of the experiment was suspected as a possible cause. We could proceed by plotting the residuals $\mathbf{y} - \hat{\mathbf{y}}$ against the residuals $\mathbf{t} - \hat{\mathbf{t}}$ from the fitted function $\hat{\mathbf{t}} = \sum_{r=1}^{p} \delta_r \mathbf{x}_r$. The existence of a linear relationship between $\mathbf{y} - \hat{\mathbf{y}}$ and $\mathbf{t} - \hat{\mathbf{t}}$ would confirm our suspicions.

Again, suppose that lack of fit has been found, its cause being unknown; suppose it is postulated that the lack of fit is associated with the addition of a *single* extra term $\theta_{p+1}\mathbf{x}_{p+1}$ to the model. Then clearly $\mathbf{x}_{p+1}$, the vector responsible, is estimated by an expression of the form

$$\mathbf{x}_{p+1} = c_0(\mathbf{y} - \hat{\mathbf{y}}) + \sum_{r=1}^{p} c_r \mathbf{x}_r \tag{21}$$

where the c's are unknown constants. The possible identity of the unknown vector is frequently suggested by experiments with possible choices of the c's defining various linear combinations of $\mathbf{y} - \hat{\mathbf{y}}$ and the vectors $\mathbf{x}_r$.

Estimates of the parameters. These are supplied by the least-squares estimates

$$\hat{\boldsymbol{\theta}} = (\mathbf{X}'\mathbf{X})^{-1}\mathbf{X}'\mathbf{y} \tag{22}$$

Precision of the estimates. In the space of the parameters $\boldsymbol{\theta}$, the sum of squares contours and hence the likelihood contours are a series of concentric ellipsoids (or hyperellipsoids) centered about $\hat{\boldsymbol{\theta}}$. These contours show the way in which the likelihood falls off as we move away from $\hat{\boldsymbol{\theta}}$. This in turn shows the precision of the estimates. If a dichotomy of likely and unlikely values is desired, then we can choose a particular contour at a suitable level of likelihood and regard all those parameter points within the contour as likely and those outside as unlikely.

If we proceed using the theory of confidence intervals, then on the assumption that the model is adequate with s^2 an estimate of σ^2 having ν degrees of freedom, we can write the exact probability statement

$$\Pr\left\{\frac{(\boldsymbol{\theta} - \hat{\boldsymbol{\theta}})'\mathbf{X}\mathbf{X}'(\boldsymbol{\theta} - \hat{\boldsymbol{\theta}})}{ps^2}\right\} \leqq F_\alpha(p,\nu) = 1 - \alpha \tag{23}$$

where $F_\alpha(p,\nu_1)$ is the α significance point of the F distribution with p and ν degrees of freedom. The interior of the hyperellipsoid defined by this inequality is called the confidence region for $\boldsymbol{\theta}$. The chance that this region includes the true parameter point $\boldsymbol{\theta}^*$ is exactly $1 - \alpha$.

The boundary of the region together with $\hat{\boldsymbol{\theta}}$ supplies a useful summary of what is known about the parameter. An alternative and in many respects more satisfactory way of regarding these elements of information is that together they serve roughly to indicate the nature of the likelihood surface itself. Better still, by the interpretation of Section 1, they indicate the approximate posterior distribution of the parameters $\boldsymbol{\theta}$. The boundary has then the property that the posterior probability density for any $\boldsymbol{\theta}$ inside it is greater than for any $\boldsymbol{\theta}$ outside it, as well as the property that only a portion α of the distribution lies outside it.

However one interprets the meaning of such a boundary, there is real difficulty in appreciating its actual nature when we have more than 2 or 3 parameters to consider jointly. One can outline a hyperellipsoid with clarity and considerable economy by listing the coordinates of the $2p$ points that are the ends of its principal axes. Unfortunately, such a procedure is not invariant with changes in scale of the parameters.

A useful although somewhat arbitrary convention I have used[4] is to scale each parameter in units of its standard deviation. In this case, the rth coordinate of the points at the ends of the tth principal axis is given by

$$\theta_r = \hat{\theta}_r \pm u_{tr}\{c^{rr}\lambda_t p F_\alpha(p,\nu_1)s^2\}^{1/2} \qquad (24)$$

where

$$(u_{t1}, u_{t2}, \ldots, u_{tp}) = \mathbf{u}_t'$$

$$\mathbf{u}_t'\mathbf{R} = \lambda_t \mathbf{u}_t'$$

$$\mathbf{R} = \mathbf{G}^{-1/2}\mathbf{C}^{-1}\mathbf{G}^{-1/2}$$

$$\mathbf{C}^{-1} = \{c^{tr}\} = (\mathbf{X}'\mathbf{X})^{-1}$$

and $\mathbf{G}$ is a diagonal matrix whose tth diagonal element is c^{tt} so that $\mathbf{R}$ is the correlation matrix for the estimates $\hat{\boldsymbol{\theta}}$. A listing of the λ_t and of the associated $2p$ coordinates of the ends of the principal axes makes it possible to appreciate fairly readily the nature of the confidence region; experience has shown that an analysis of this kind is essential to any proper understanding of the situation. In particular, the relative magnitudes of the characteristic roots λ_t indicate the state of conditioning of the likelihood surface or, equivalently, the state of correlation between the estimates.

Since

$$\sum_{t=1}^{p} \lambda_t = p$$

if the estimates are completely uncorrelated, each λ_t is unity. With the scaling of this convention, the likelihood contours are then spheres or hyperspheres. Frequently the ellipsoidal contours of the likelihood surface are extremely attenuated along one or more of the obliquely oriented principal axes. In these circumstances, the point estimates $\hat{\boldsymbol{\theta}}$ are highly correlated and may be extremely misleading without ancillary information. Practical situations frequently arise where the attenuation is so great that there is a line, plane, or space where the likelihood is very little different from its maximum value. This implies that with the available data it is virtually impossible to obtain separate estimates of the parameters. This situation is easily detected with the above type of analysis. When it occurs, one or more of the latent roots λ_t are large compared with the others. Each latent vector $\mathbf{u}_t$ corresponding to a large latent root indicates a linear combination of parameters estimable only with large variance. When there are $q > 1$ large roots, the q-dimensional space spanned by the latent vectors associated with the large roots defines the totality of poorly estimated combinations of parameters. These linear combinations are not uniquely defined, of course. Any q independent linear combinations of the selected latent vectors also define poorly estimated combinations of parameters. The space of poorly estimated parameter combinations is to some extent dependent on the original scaling convention; however, it is easy to show that as the magnitudes of the large latent roots becomes greater relative to the remainder, this dependence becomes smaller. In cases of interest, therefore, the dependence on the scaling convention is not of much practical importance.

5. The Importance of Linear Least Squares

Situations where the response is truly a linear function of the unknown parameters occur in nature apparently rather rarely. However, the linear situation is important for two reasons:

(1) Not infrequently, although the true form of the response function is unknown, nevertheless it can be assumed that over a limited region of interest R in the space of the variables, the response function is capable of being adequately represented by some simple smooth graduating function

$$g(\xi_1, \xi_2, \ldots, \xi_k) = g(\boldsymbol{\xi})$$

To be of value, the graduating function should be flexible and easy to fit. A polynomial of sufficiently high degree does contain a great deal of useful flexibility and, since the polynomials are linear in their coefficients, they may be fitted with use of linear least-squares theory. A great deal of useful information about the general characteristics of the response relationship over the region R can be obtained by this form of graduation. The use of linear least squares for the fitting of a graduating function such as a polynomial is familiarly known as regression analysis when unplanned data are being fitted. It plays an important part in response surface methodology when the data are generated from a set of experimental runs specially chosen to allow efficient fitting.

(2) In some cases the supposedly true functional form of the response relationship can be postulated, usually from consideration of the mechanism of the phenomenon under study. Frequently the mechanism is described most directly by a set of ordinary or partial differential equations; the response function is then the solution of these differential equations. Whether or not the response function is thus defined, it is exceptional for the parameters to appear linearly. Nevertheless, as was originally pointed out by Gauss,[5] fitting of nonlinear functions by least squares often can be achieved by an iterative method involving a series of *linear approximations*. At each stage of the iteration, linear least-squares theory is used to obtain the next approximation. This second use of linear least squares in fitting nonlinear functions is perhaps less familiar, although it has been discussed in detail by Deming[6] and others. With the advance of electronic computers

and the consequent removal of much burdensome calculation, considerable attention recently has been directed toward it.

6. Nonlinear Least Squares

Suppose that within a region in the parameter space in the neighborhood of the true parameter point $\boldsymbol{\theta}^*$ fair accuracy is given by the linear approximation

$$\eta\{\xi_u, \boldsymbol{\theta}^*\} \approx \eta\{\xi_u, \boldsymbol{\theta}^0\} + \sum_{r=1}^{p} (\theta_r^0 - \theta_r^*) x_{ru}^0 \tag{25}$$

where

$$x_{ru}^0 = \left\{\frac{\partial \eta(\xi_u, \boldsymbol{\theta})}{\partial \theta_r}\right\}_{\boldsymbol{\theta}=\boldsymbol{\theta}^0} \tag{26}$$

and $\boldsymbol{\theta}^0$ is some point within the region. Suppose $\boldsymbol{\theta}^0$ is a set of guessed starting values. In substituting these starting values in the response function, let $\mathbf{z}^0$ be the vector of discrepancies $z_u^0 = y_u - \eta(\xi_u, \boldsymbol{\theta}^0)$ between the observations and the calculated responses. We then can write

$$E(z_u^0) \approx \sum_{r=1}^{p} (\theta_r^0 - \theta_r^*) x_{ru}^0 \tag{27}$$

which is a linear model of the same form as EQUATION 7. Treating the z's as observations, we can by means of ordinary least squares obtain an estimate $\Delta\boldsymbol{\theta}$ of $\boldsymbol{\theta}^0 - \boldsymbol{\theta}^*$. If the function were truly linear in the parameters, then the adjusted values $\boldsymbol{\theta}_0 + \Delta\boldsymbol{\theta}$ for the θ's would be the elements of $\hat{\boldsymbol{\theta}}$, the least-squares values. Because of nonlinearity, however, $\boldsymbol{\theta}_0 + \Delta\boldsymbol{\theta}$ does not give $\hat{\boldsymbol{\theta}}$ at once, but a vector of "improved values" which now may be substituted for $\boldsymbol{\theta}^0$ to provide the starting point for a second iteration, and so on. In this way, provided the functions are well behaved and the starting values $\boldsymbol{\theta}^0$ are not too far from the final values, the adjusted values will converge to the least-squares estimates $\hat{\boldsymbol{\theta}}$. The linear theory described in Section 4 can be used to judge adequacy of fit, to obtain an approximate representation of the sum of squares function in the neighborhood of its minimum, using EQUATION 10, and to obtain an approximate confidence region, using EQUATION 23. The elements x_{ru} of the matrix $\mathbf{X}$ which appears in EQUATIONS 10 and 23 are simply the derivatives at the maximum likelihood estimates

$$\left\{\frac{\partial \eta(\xi_u, \boldsymbol{\theta})}{\partial \theta_r}\right\}_{\boldsymbol{\theta}=\hat{\boldsymbol{\theta}}}$$

As we have explained, a supposedly true functional form is usually derived from consideration of the mechanism of the phenomenon under study, and can frequently be written most directly in terms of a set of ordinary or partial differential equations. In some cases such equations can be integrated to provide an explicit form for the response function; more frequently, however, this is not possible. In practice, therefore, we may not assume that the derivatives x_{ru} are obtainable by direct differentiation; instead, small changes can be made in each of the parameters in turn and the derivatives calculated from the differences. With this device it is necessary only to be able to compute the *value* of the function for any given values of $\boldsymbol{\xi}_u$ and $\boldsymbol{\theta}$ in order to carry out the iterative process.

Not infrequently it is found that the above linearizing method, which we have called the Gauss method, results at least in the early stages in overshoot. This means that, when the full adjustment $\Delta\boldsymbol{\theta}$ is added, we may actually obtain a sum of squares for the adjusted values larger than that for the initial values. Nevertheless, it is possible to show for the fairly general conditions stated recently by H. O. Hartley[7] that a change from $\boldsymbol{\theta}_0$ along the vector $\Delta\boldsymbol{\theta}$ must *initially* reduce the sum of squares. For this reason it is sometimes useful to use for adjusted values the quantities $\boldsymbol{\theta}_0 - c\Delta\boldsymbol{\theta}$ where c is some suitable constant found by trial.[4] In many cases the best value of c is less than unity, but in some it is greater than unity. A useful method incorporated in the Princeton-IBM program is based on sequential halving and doubling of the interval and interpolation between the values that give smallest sums of squares.

The Princeton-IBM 704 Program

During 1957 and 1958 a cooperative project between the Princeton Statistical Techniques Research Group and the International Business Machines Corporation resulted in the Princeton-IBM nonlinear least-squares 704 program,[8] now available under SHARE. The program uses the modified Gauss method described above. In practice it was found that the time taken in calculating intermediate trial values along the vector $\Delta\boldsymbol{\theta}$ in the manner described above was sometimes greater than the time saved by preventing overshoot. For this reason it is arranged that this feature is optional and may be switched in when the behavior of the iteration suggests that this is desirable. The program requires only that it shall be possible to calculate the function, provision being made to insert the necessary subroutines to do this for each specific problem. The program thus may be used for the fitting of differential equations, partial differential equations, integral equations and, in principle, any calculable function, the derivatives x_{ru} being calculated automatically from differences. A latent root–latent vector routine can be switched in at any stage to analyze the state of conditioning of the sum of squares surface. When $\hat{\boldsymbol{\theta}}$ has been found, the $2p$ values of $\boldsymbol{\theta}$ at the ends of the principal axes of the (linear theory) confidence ellipsoid are automatically calculated with use of EQUATION 24. The actual sums of squares at these $2p$ points are also calculated. These calculations serve to confirm that a minimum sum of squares indeed has been reached, to outline the confidence region and to measure the degree of nonlinearity which exists in the particular problem. These calculations are illustrated by an example in the next section.

Since there is an infinite number of ways in which functions may be nonlinear, it is probable that no iterative program can ever be devised that will never give trouble for certain examples. Experience over the past four years with many examples programmed in nonlinear least squares on the DEUCE computer at Teddington in England and, more recently, on the IBM 704 supports the following general conclusions.

Time will be saved in the long run if careful preliminary analysis is made for each problem. In particular, preliminary calculations by makeshift methods often can provide starting values for the parameters that are not too far from the final values. Such provision greatly enhances the speed and certainty of convergence of the iterative process. Where the sum of squares function is poorly conditioned, convergence may be extremely slow. Typically in this situation the sum of squares surface contains an extremely attenuated oblique curved ridge and the iterative process tends to oscillate. Often this trouble can be cured by working with suitable, less highly correlated transformations of the parameters. Study of a particular problem will often suggest the type of transformation desirable. For example, suppose two parameters θ_0 and θ_1 always appear together in the form $\theta_0 \exp \{\theta_1 x\}$, and that the range of variation in the x's is not large compared with their mean value. Then, typically, the likelihood surface in θ_0 and θ_1 will be an attenuated, curved ridge: for by increasing θ_0 from its maximum likelihood value at the same time that θ_1 is suitably decreased from *its* maximum likelihood value, a series of values can be found for pairs (θ_0, θ_1) such that the corresponding $\theta_0 \exp \{\theta_1 x\}$ are very nearly equal to $\hat{\theta}_0 \exp \{\hat{\theta}_1 x\}$. A natural transformation that will eliminate this effect is

$$\phi_0 = \theta_0 \exp \{\theta_1 \bar{x}_1\}$$

$$\phi_1 = \theta_1$$

Typically, the likelihood surface in terms of ϕ_0 and ϕ_1 will not demonstrate the characteristic oblique ridge found for θ_0 and θ_1.

7. Fitting a Set of Simultaneous Differential Equations by Least Squares

To illustrate a typical problem in nonlinear least squares for which the IBM 704 program was used, we summarize an analysis of some data given by Box and P. V. Youle.[9] This analysis will be given in more detail elsewhere.

Formulation of the Problem

These authors describe a chemical experiment in which two reactants M_1 and M_2 combined to form reactants M_3 and M_4. The concentrations of these reactants at a particular time t are denoted by η_1, η_2, η_3, and η_4, respectively, and a kinetic model is postulated that leads to the following set of simultaneous differential equations to describe the progress of the reaction:

$$\begin{aligned} -\frac{d\eta_1}{dt} &= 2\eta_1\eta_2\theta_1^{-\theta_2/T} + 2\eta_1\eta_3\theta_3^{-\theta_4/T} \\ -\frac{d\eta_2}{dt} &= \eta_1\eta_2\theta_1^{-\theta_2/T} \\ \frac{d\eta_3}{dt} &= \eta_1\eta_2\theta_1^{-\theta_2/T} - \eta_1\eta_3\theta_3^{-\theta_4/T} \\ \frac{d\eta_4}{dt} &= \eta_1\eta_3\theta_3^{-\theta_4/T} \end{aligned} \tag{25}$$

where T is the absolute temperature and at time zero

$$\eta_1 = \eta_{10}, \qquad \eta_2 = \eta_{20}, \qquad \eta_3 = 0, \qquad \eta_4 = 0$$

TABLE 1 shows observations y on the fractional yield $\eta = \eta_3/\eta_{20}$ of the product M_3 in 19 sets of independent experimental runs. The experimental conditions changed from run to run were $\xi_1 = T$, the absolute temperature, $\xi_2 = \eta_{10}/\eta_{20}$, the ratio of the starting concentrations of M_1 and M_2, and $\xi_3 = t$, the length of time the reaction was allowed to proceed.

TABLE 1

OBSERVED VALUES OF FRACTIONAL YIELD OF PRODUCT IN 19 SETS OF EXPERIMENTAL CONDITIONS

Run	Experimental conditions			y = Observed fractional yield of M_3
	$\xi_1 = T$ (°K.)	$\xi_2 = \eta_{10}/\eta_{20}$	$\xi_3 = t$ (hr.)	
1	435	35.30	5	0.459
2	435	35.30	8	0.533
3	435	42.36	5	0.575
4	435	42.36	8	0.588
5	445	35.30	5	0.606
6	445	35.30	8	0.580
7	445	42.30	5	0.586
8	445	42.30	8	0.524
9	440	38.83	6.5	0.569
10	450	38.83	6.5	0.554
11	430	38.83	6.5	0.469
12	440	45.89	6.5	0.575
13	440	31.77	6.5	0.550
14	440	38.83	9.5	0.589
15	440	38.83	3.5	0.503
16	450	28.29	6.5	0.611
17	450	28.29	6.5	0.629
18	433	48.01	7.5	0.600
19	433	48.01	7.5	0.606

Our problems are the following: given the observations, (1) is there reason to suspect that the proposed model is inadequate to represent the data? (2) if not, what are the best estimates of the parameters θ_1, θ_2, θ_3, and θ_4? and (3) how accurately are they known?

The above simultaneous differential equations cannot be integrated to give, in terms of tabulated functions, an explicit expression for $\eta = \eta_3/\eta_{20}$. Therefore, the values required for η for the desired values of ξ_1, ξ_2, and ξ_3 and θ_1, θ_2, θ_3, and θ_4 were computed with the use of a suitable numerical subroutine.

Reparametrization of the Model

In view of the fact that the levels of the absolute temperature T cover a range that is small compared with the mean value, attenuated ridges associated with the pairs (θ_1, θ_2) and (θ_3, θ_4) were to be expected. Therefore the following transformations were made:

$$\alpha_1 = \ln \theta_1 - \theta_2 \overline{T^{-1}}$$

$$\alpha_2 = \ln \theta_3 - \theta_4 \overline{T^{-1}}$$

where $\overline{T^{-1}}$ is the average reciprocal absolute temperature. The quantities θ_2 and θ_4, which measure the activation energies, are expected to be roughly equal, of about equal variance, and correlated. Such correlation should be reduced by considering instead of θ_2 and θ_4 their mean and their difference. Similarly, α_1 and α_2, which measure the logarithms of the rates of two successive chemical reactions at the average absolute temperature, may be expected to be of the same order of magnitude and to have roughly equal variances. Correlation may be reduced by reparametrization in terms of their mean and difference or, alternatively, by using the value of one and the difference. The latter parametrization was used, in fact, for this example. Hence the calculations were performed in terms of the following transformed parameters:

$$\phi_1 = \alpha_1 - \alpha_2 = \ln \theta_1 - \ln \theta_2 - (\theta_2 - \theta_4)\overline{T^{-1}}$$

$$\phi_2 = \alpha_2 = \ln \theta_3 - \theta_4 \overline{T^{-1}}$$

$$\phi_3 = \tfrac{1}{2}(\theta_2 + \theta_4)$$

$$\phi_4 = \theta_2 - \theta_4$$

In the rough fit made by Youle and myself, approximate values $\phi_1 = -6.06$, $\phi_2 = 1.22$, $\phi_3 = 10{,}090$, and $\phi_4 = 0$ were obtained. Starting from these values the iteration proceeded as shown in TABLE 2. After the fourth iteration the values remained stable.

An impression of the adequacy of the model can be formed by comparing the residual sum of squares which, by analogy with linear theory, has $19 - 4 = 15$ degrees of freedom, with an independent estimate of the experimental error variance σ^2. There was some evidence that the true value of σ^2 was about 3 or possibly less. The residual sum of squares, 93.5, divided by 15 equals 6.23; examination of the appropriate tables confirms the fact that there is considerable doubt as to the adequacy of this model.

Since inadequacy of the model was now suspected, analysis of the residuals was undertaken in an attempt to discover a possible reason. It was found that the residuals were correlated with the level of the variable $\xi_2 = \eta_{10}/\eta_{20}$ measuring the ratio of the starting concentrations. This suggested that an incorrect assumption had been made concerning the order of the reaction. With a five-parameter model in which the fifth parameter measured the order of the reactions with respect to the substituent M_1, an excellent fit was obtained.

For the moment we shall proceed as if there were no evidence of lack of fit and consider the precision of these estimates. Application of the linear

TABLE 2
CONVERGENCE OF ITERATION ONTO THE LEAST-SQUARES VALUES

	Iteration				
	Start	1	2	3	4
ϕ_1	−6.06	−6.20	−6.27	−6.27	−6.27
ϕ_2	1.22	1.11	1.20	1.20	1.20
ϕ_3	10,090	10,180	10,930	10,930	10,840
ϕ_4	0	2050	1730	1710	1700
Residual sum of squares	450.4	134.1	93.5	93.5	93.5

theory of Section 4 and in particular EQUATION 24 yields the least-squares estimates and hence an outline of the approximate confidence region as shown in TABLE 3. The outline presents the coordinates of the ends of the principal axes of the linear 95 per cent confidence elipsoid with the scaling convention explained in Section 4. The individual confidence limits for the parameters obtained from the t tables by direct application of linear theory are also shown.

A feature of the table is the last column in which the sums of squares associated with the various parameter points are shown. Also shown is the value of the 95 per cent critical sum of squares that would be obtained on the boundary of the confidence region if the model were exactly linear in the parameters. This table serves a number of purposes.

(1) It provides some confirmation that a minimum has in fact been found. It will be seen in the above example that the sums of squares recorded at the 8 parameter points about the minimum are all *larger* than the minimum sum of squares, 94.

(2) It checks the linearity. If the function had been exactly linear over the confidence region, each one of the 8 sums of squares would have equaled 170. That they do in fact differ from this value somewhat indicates non-linearity that is moderate but probably not so great that linear theory will not yield a useful approximation both for the lack of fit assessment and for the confidence region.

(3) The table shows the state of conditioning of the likelihood surface with respect to the parameters ϕ_1, ϕ_2, ϕ_3, and ϕ_4. In some examples, consideration of which combinations of parameters have been poorly determined can be most instructive as to more adequate design of later experimental runs. A study of the statistical properties of confidence regions of the kind discussed here has been published recently by E. M. L. Beale.[10]

TABLE 3
TABLE FOR CHECK ON MINIMIZATION, OUTLINE OF CONFIDENCE REGION, AND ASSESSMENT OF NONLINEARITY

	λ	ϕ_1*	ϕ_2	ϕ_3	ϕ_4	Sum of squares
Minimum		−6.27	1.20	10,840	1,700	94
Outline of (linear theory) confidence region	2.43	−6.08	1.08	6,890	4,870	216
		−6.46	1.32	14,790	−1,470	161
	0.97	−6.21	1.26	13,140	5,290	162
		−6.33	1.14	8,540	−1,890	201
	0.41	−6.33	1.25	8,540	2,700	158
		−6.21	1.15	13,140	700	189
	0.19	−6.33	1.16	11,340	2,670	174
		−6.21	1.24	10,340	730	166
	Theoretical value of critical sum of squares on 95% contour:					170
Individual (linear theory) confidence limits		6.40 6.14	1.29 1.11	13,980 7,700	4,740 −1,340	

* Values of λ are given to the left of the brace points.

8. CHOICE IN GENERATION OF EXPERIMENTAL DATA: DESIGN OF EXPERIMENTS

The amount of information about the parameters that will be contained in the likelihood function often can be greatly increased by careful choice of the levels at which the variables are to be run, that is, by careful choice of the design matrix **D**. Concerning this choice, different situations can occur appropriate to the different questions at issue and to the type of model postulated. As we have seen already, two situations are of importance:

(1) The true functional form is unknown, but it is assumed that in some region R in the space of the variables a simple function such as a polynomial

may be used to graduate it. In this situation there is never any real basis for supposing that the form of function fitted could exactly represent the true function.

(2) The functional form, having been deduced from some mechanistic theory of the phenomenon studied, is known. In this case, it can be logically assumed, provisionally, that the function is capable of exactly representing the phenomenon studied. The functional forms usually involved are nonlinear in the parameters and often are not explicit in the sense explained in Section 6.

9. Designs for Fitting Graduating Polynomials

When, as is often the case, the true functional form $\eta(\boldsymbol{\xi},\boldsymbol{\theta}^*)$ of the response relationship is unknown, we may proceed by employing a flexible graduating function $g(\boldsymbol{\xi})$, such as a polynomial, which it is hoped may represent the principal features of the actual relationship over some region of interest R in the space of the variables. It should be noted that usually such a function has no theoretical basis and we usually can be certain that however the coefficients of the polynomial are chosen, $E(y) = \eta(\boldsymbol{\xi},\boldsymbol{\theta}^*)$ will not correspond exactly with $g(\boldsymbol{\xi})$ except at a few isolated points. In this situation therefore the discrepancy between the fitted function and the true function can arise from two sources: (1) experimental error, a useful measure of the magnitude of which is the variance, whence our term *variance error*; (2) the inadequacy of the fitted graduating function to represent the basic relationship. We shall call this kind of error *bias error*.

Until recently, in studies of experimental design the influence of bias error has been largely ignored, on the assumption that exact representation of the response relationship was possible by means of an empirical mathematical model (for example, a polynomial function). A notable exception is H. Hotelling's study[11] in which designs intended specifically to eliminate or reduce certain biases were investigated. We first review some of this important work in which the approach was to minimize in some way the consequences of variance error only, and bias has been either ignored or only incidentally considered.

10. Designs for Minimizing Variance Error

Fitting a Polynomial of First Degree

Suppose we are to generate data under circumstances in which it is expected that over the region of interest R a polynomial of first degree will supply adequate graduation. A suitable arrangement of experiments will be called a *first-order design*. The appropriate model can be written as

$$\eta = \beta_0 + \sum_{i=1}^{k} \beta_i x_{iu} \tag{26}$$

where

$$x_{iu} = (\xi_{iu} - \xi_i^0)/S_i \tag{27}$$

and the location constants ξ_i^0 and scale constants S_i are chosen such that restriction of the experimental points to the particular region of interest is expressed by

$$\sum_{u=1}^{N} x_{iu} = 0$$
$$\sum_{u=1}^{N} x_{iu}^2 = N \qquad (i = 1, 2, \ldots, k) \tag{28}$$

This model is linear not only in the coefficients of the polynomial β_i, but also in the elements ξ_{iu} of the design matrix and, as was first formally shown by Hotelling,[12] the smallest variance for each of the estimates $b_i = \hat{\beta}_i$ of the coefficients of the polynomial is achieved when the $N \times k$ matrix $\{x_{iu}\}$, which defines **D** via EQUATION 27 has orthogonal columns. The requirement that the joint confidence region for the β_i have smallest volume leads to this same conclusion.

Examples of designs having the orthogonal property are the factorial designs originally proposed by R. A. Fisher.[13] These, particularly the two-level factorial and fractional factorial designs, have been employed for many years with great success. A desirable feature of the orthogonal designs, in addition to the first-order minimum variance property mentioned above, is that they give rise to easily calculated and uncorrelated estimates.

An interesting and somewhat remarkable property of the first-order orthogonal designs is that the variance of the estimated response $\hat{y}_x$ is constant on spheres in the space of the x's. That is,

$$V(\hat{y}_x) = f\left\{\sum_{i=1}^{k} x_i^2\right\} \tag{29}$$

We can regard the reciprocal of the variance of $\hat{y}_x$ as a measure of the amount of information about the response the design yields at the point x. Thus it is evident that the orthogonal first-order designs have the property of providing a highly symmetrical generation of information in the space of the x's. This property and all the others mentioned above are retained on rotation. Such an arrangement may therefore be called a *rotatable* design.

It should be noted that for first-order designs the properties of orthogonality and rotatability are such that one implies the other. Therefore, an alternative approach to the problem of experimental design would be to make symmetrical generation of information (as provided by the property of rotatability) the basic requirement. This would lead automatically to orthogonality for first-order designs.

Geometrical Properties

It is of interest to consider certain geometrical properties of the optimal first-order designs. Suppose that with k variables $k + 1$ experiments are run; let us consider the positions of the $k + 1$ points in the k-dimensional $\mathbf{x}$ space. It is easy to show that for an orthogonal design these points must be at the vertices of a regular simplex in the k space.

Where N, the number of experimental points, is greater than $k + 1$, the design can be regarded as a projection of an $N - 1$ dimensional regular simplex onto the k space. The wide choice of experimental designs then available makes it possible,[14] while retaining the orthogonal property, to arrange that columns of the design matrix are all orthogonal to a set of $r \leqq N - 1 - k$ further x's that may represent systematic effects such as time trends, "blocks," and other extraneous disturbances that can be foreseen, but whose exact nature are unknown. Such disturbances then can be eliminated without loss of efficiency.

Fitting Polynomials of Higher Degree

The property of orthogonality that gives best first-order designs in the sense I have defined cannot be generalized directly when it is desired to fit polynomials of degree d higher than the first. The property of rotatability, however, which for first-order designs is equivalent to orthogonality, can be so generalized. It is readily shown[15] that for a polynomial of degree d EQUATION 29 is satisfied by a certain choice of moments of order $2d$ and less of the design points in the k space of the variables. A wide variety of arrangements of experimental points can provide higher-order rotatable designs. Important and useful examples of rotatable designs in which the number of experimental points is not excessive are obtained from points at the vertices of regular figures. For example, a valuable design called a *composite* rotatable design, which is suitable for fitting a second-degree equation in $k = 3$ variables, may be built up from the eight vertices of a cube combined with the six vertices of a symmetrically placed octahedron and one or more center points. Once more, of course, any rotation of this design provides another equally good arrangement. Generalizations of the composite designs into higher space are easily obtained and have been used with considerable success in fitting empirical data to functions in the neighborhood of a stationary point. Details of first- and second-order designs, of fitting procedures, and of the analysis of the fitted equations are given elsewhere.[15-17] The nature of the analysis of equations of first and second degree is such that the principal geometric characteristics of the fitted response surface are readily appreciated even when many variables are studied simultaneously. No such ready appreciation seems to be possible for polynomials of degree higher than the second. Since situations can easily arise in which the first- and second-degree polynomials supply an inadequate representation, the usefulness and flexibility of the empirical method are considerably increased by allowing the possibility of fitting simple polynomials on other graduating functions in a suitably chosen transformation of the variables rather than in the variables themselves. A method has been worked out recently[18] by means of which the estimation of such transformations can easily be carried out.

11. Designs for Minimizing Bias as well as Variance Error

The justification for the experimental arrangements thus far described has come from minimization of the variance of the regression coefficients b_i in the case of first-order designs and behavior of the variance function $V(\hat{y})$, more generally. As we have seen already, polynomials and similar functions are usually employed to graduate some unknown and quite unrelated function over a specific region. Even if there existed no experimental error, the fitted graduating function and the true function would necessarily be discrepant to a greater or lesser extent except at a few isolated points. The realization that to ignore this bias might lead to arrangements not necessarily very good in practical circumstances has resulted recently in considerable work[19-21] whose object has been to find arrangements that minimize discrepancies due to bias as well as those arising from variance error.

One approach[19] is as follows: suppose that $\hat{y}_x$ is the estimated response at the point **x** obtained after fitting by least squares the response function to a particular set of N design points. Suppose also that the value η_x is the true response at x:

$$\hat{y}_x - \eta_x = \{\hat{y}_x - E(y_x)\} + \{E(\hat{y}_x) - \eta_x\} \tag{30}$$

where we have

$$E\{\hat{y}_x - \eta_x\}^2 = E\{\hat{y}_x - E(\hat{y}_x)\}^2 + \{E(\hat{y}_x) - \eta_x\}^2 \tag{31}$$

which we can write as

$$M(\hat{y}_x) = V(\hat{y}_x) + B^2(\hat{y}_x) \tag{32}$$

The quantity on the left-hand side measures the mean square error at the point **x**, while the right-hand side shows this resolved into two parts, one the variance of $\hat{y}$ at **x** and the other the squared bias at **x**.

Now suppose that in the k space of the x's we have a region of immediate interest R over which we desire to graduate the response function. Then, if Av_R refers to the average value over the region R, we may employ as an over-all measure of the representational adequacy of the fitted expression the average mean square error

$$\frac{N}{\sigma^2} Av_R M(\hat{y}_x) = \frac{N}{\sigma^2} Av_R V(\hat{y}_x) + \frac{N}{\sigma^2} Av_R B^2(\hat{y}_x) \tag{33}$$

or

$$M = V + B^2 \tag{34}$$

where M is average mean square error, V the average variance, B^2 the average squared bias over the region R, and the multiplier N/σ^2 is a normalizing factor included to allow fair comparison between situations in which there are unequal numbers of observations and different experimental error variances. It seems reasonable to attempt to select the design matrix $\mathbf{D}$ so that M is as small as possible.

In application, the region R over which we desire to graduate the function would usually be only a small part of what may be called the *operability region* O, that is, the region in the space of the x's over which it is possible to perform experiments. It is supposed therefore that R is contained within O, which is sufficiently large that the design points never reach its boundary. In particular, with this approach the design points will not necessarily lie within R; some or all of the points will lie outside R if smaller values of M may thus be obtained.

To visualize an application of this approach, suppose that the true function η_x can be exactly represented throughout O by a polynomial of degree d_2 and that, unaware of this, an experimenter attempts to graduate the function over the limited region R by fitting by least squares a polynomial of degree d_1. We can illustrate the method and the type of conclusions to which this approach leads us by considering the simple case of a single variable x, $d_2 = 2$, and $d_1 = 1$. Thus the true situation is that, over the whole region of operability (which we regard as large), the exact response function is the quadratic

$$\eta_x = \beta_0 + \beta_1 x + \beta_{11} x^2 \tag{35}$$

and an attempt is being made to graduate the response function over a more limited region (which we shall suppose extends from $x = -1$ to $x = 1$) by the linear function

$$\hat{y}_x = b_0 + b_1 x \tag{36}$$

to be fitted by the method of least squares.

We suppose that

$$\sum_{u=1}^{N} x_u = 0$$

and we denote the second and third moments of the design as follows:

$$N^{-1} \sum_{u=1}^{N} x_u^2 = [11] = \sigma_x^2 \tag{37}$$

$$N^{-1} \sum_{u=1}^{N} x_u^3 = [111] \tag{38}$$

It is readily shown that, whatever the values of σ_x and β_{11} for the minimization of M, the third moments [111] must be set equal to zero. Making this substitution, we obtain the average mean square

$$M = V + B^2 = \left\{1 + \frac{1}{3\sigma_x^2}\right\} + \alpha_{11}^2 \left\{\left(\sigma_x^2 - \frac{1}{3}\right)^2 + \frac{4}{45}\right\} \qquad (39)$$

where $\alpha_{11}^2 = N\beta_{11}^2\sigma^{-2}$ is a standardized measure of the quadratic curvature showing the magnitude of this curvature relative to the sampling error. The only design characteristic in the above expression is σ_x which measures the spread of the design points relative to the region R. The value of σ_x that minimizes M clearly depends on the value of α_{11}.

When α_{11} is very small, corresponding to the situation where the quadratic curvature is very small compared with the sampling error, the optimal σ_x tends to infinity. This is the familiar result that if a linear model could be assumed over the whole range O, then the spread of points should be as large as possible.

When α_{11} tends to infinity, corresponding to the situation where the experimental error is negligible and all the discrepancy arises from the inadequacy of the model to represent the quadratic curvature, M is minimized by choosing the x's such that their standard deviation σ_x approaches its minimal value of $1/\sqrt{3} = 0.58$. Bearing in mind that the region R extends from -1 to $+1$, we see that for many distributions of x's this would lead to a scatter of x's covering very much the same range as the region R.

At first sight it appears that this approach is fruitless, since in practice we shall not know α_{11} and we see that, depending on the size of this constant, the optimal design will cover at one extreme an *infinite* range and at the other extreme only a range *comparable with the size of R*. Some idea of the situation likely to occur in practice can be gained, however, from the following argument. If a rational experimenter were graduating a function by a polynomial over a limited region R, logically he should employ a polynomial of sufficient but no higher degree than he expected would be necessary to bring out those perturbations in the true response surface that would not inevitably be engulfed in the sampling error of the estimated response. It seems therefore that optimal designs for use by rational experimenters should be such that B^2, the average size of error in $\hat{y}$ arising from bias, is about as large as V, the average size of error in $\hat{y}$ arising from variance. We therefore should be particularly interested in ratios V/B^2 of about 1. TABLE 4 shows the optimal values of σ_x for the extreme cases previously considered—completely dominant variance ($V/B^2 = \infty$) and completely dominant bias ($V/B^2 = 0$)—together with certain intermediate cases.

We see that for V/B^2 of 1 the optimal value of σ_x^2 is 0.62, a value very little different from 0.58 which is appropriate when the bias is dominant and variance is completely ignored. Even with $V/B^2 = 4$, the optimal value of σ_x is only 0.72. If, as seems inescapable, we accept the fact that these two intermediate cases represent the situations likely to be met in practice, we are led to the somewhat remarkable conclusion that design characteristics should approximate those appropriate when the bias contribution is completely dominant and the variance contribution is entirely ignored. This conclusion is particularly striking when it is remembered that most work on experimental design has been based tacitly on the other extreme assumption. Similar results are obtained for the case where a linear expression is fitted in not one but k variables over a spherical region R, the true function being quadratic.

The importance of designs that minimize the contribution of squared bias renders of particular note the following general theorem.[19] If a polynomial of degree d_1 in k variables is fitted by the method of least squares when the true function is a polynomial of degree $d_2 > d_1$, then the squared bias, averaged over *any* region R, is minimized for all values of the coefficients of the neglected terms by making the moments of order $d_1 + d_2$ and less of the design points equal to the corresponding moments of a uniform distribution over R.

TABLE 4
OPTIMAL VALUE OF σ_x FOR VARIOUS VALUES OF V/B^2

V/B^2	0	1	4	∞
Optimal value of σ_x	0.58	0.62	0.72	∞

12. Designs for Fitting "Theoretical Functions" Nonlinear in the Parameters

The problem of designing experiments in this situation has received comparatively little attention by statisticians. One approach[22] is to attempt to choose $\mathbf{D}$ so that the volume of the confidence region for $\boldsymbol{\theta}$ is minimized. By the linear approximation, the volume of this region is proportional to the reciprocal of the determinant $\Delta = \| \mathbf{X}'\mathbf{X} \|$, where $\mathbf{X} = \{x_{ru}\}$ and

$$x_{ru} = \left\{ \frac{\partial f\{\boldsymbol{\xi}_u, \boldsymbol{\theta}\}}{\partial \theta_r} \right\}_{\theta = \hat{\theta}} \quad (40)$$

Unfortunately, since we do not know the values $\hat{\boldsymbol{\theta}}$ of the parameters (indeed, to obtain these estimates is the object of carrying out the experiment), we cannot know the derivatives x_{ru}. In most cases, however, some knowledge of the size of the θ's will be available. Suppose preliminary guessed values $\boldsymbol{\theta} = \boldsymbol{\theta}^0$ are available and that x_{ru}^0 is the derivative obtained when we substitute $\boldsymbol{\theta}^0$ for $\hat{\boldsymbol{\theta}}$ in EQUATION 40. The elements of $\mathbf{X}^0 = \{x_{ru}\}$ and of the determinant $\Delta^0 = \| \mathbf{X}^{0\prime}\mathbf{X}^0 \|$ are then explicit functions of the $\boldsymbol{\xi}_u$. The design $\mathbf{D}$ may therefore be chosen so as to minimize Δ^0.

The need for some preliminary knowledge of the parameters is not peculiar to this method of procedure, but appears to be a logical necessity; it is an example of the more general fact that, if nothing is known about the experimental situation then, properly speaking, no experiment can be designed. The statistical design of experiments is, in fact, the art of determining *on current evidence* which additional trials will furnish most knowledge of the situation.

H. L. Lucas and I have presented a number of specific examples of designs that result from application of the above theory.[22] One matter of importance brought out concerns the choice of design to obtain estimates that are not heavily associated. We have mentioned already that not infrequently the contours of the likelihood surface are attenuated and obliquely oriented to the coordinates of the parameters, so that a whole surface of almost equally acceptable but widely different values of $\boldsymbol{\theta}$ may exist. If none of the parameters are known and we are interested in simultaneous estimation of all of them, this presents a very unsatisfactory situation. If, on the analysis of a particular set of data arising from N experiments, this phenomenon occurred, it might be due to a poor choice of design, and another set of N experiments with $\mathbf{D}$ suitably chosen might produce estimates very much less dependent. However, situations can occur in which association of the parameters arises to a large extent from the nature of the response function itself, and no design can yield estimates that are not heavily associated. An example of such a response function is

$$\eta = \frac{1 + \theta_1\xi_1}{\theta_2 + \theta_3\xi_1}$$

For certain values of the θ's and of N, almost any choice of the ξ's gives rise to estimates highly associated one with another.

References

1. Fisher, R. A. 1925. Theory of statistical estimation. Proc. Camb. Phil. Soc. **22:** 700.
2. Savage, L. J. 1959. Subjective probability and statistical practice. Technical note 59-1161. Univ. of Chicago. Chicago, Ill.
3. Box, G. E. P. & G. S. Watson. On the robustness of linear least squares. Princeton Univ. S.T.R.G. Tech. Report. In preparation.
4. Box, G. E. P. 1957. Use of statistical methods in the elucidation of basic mechanisms. Bull. inst. international statist. **36**(3): 215.
5. Gauss, C. F. Theory of Least Squares. English translation by Hale F. Trotter. Princeton Univ. S.T.R.G. Tech. Report **No. 5**.
6. Deming, W. E. 1943. Statistical Adjustment of Data. Wiley. New York, N.Y.
7. Hartley, H. O. 1959. The modified Gauss-Newton method for the fitting of nonlinear regression functions by least squares. Technical Report No. 2, 3. Statistical Laboratory, Iowa State Univ. Iowa City, Iowa.
8. Nonlinear Estimation. 1959. (Princeton-IBM) 704 Program WLNLI. Mathematics & Applications Dept., I. B. M. Corp., New York, N. Y.
9. Box, G. E. P. & P. V. Youle. 1955. The exploration and exploitation of response surfaces. An example of the link between the fitted surface and the basic mechanism of the system. Biometrics. **11:** 287.

10. BEALE, E. M. L. Confidence regions in nonlinear estimation. J. Roy. Stat. Soc. **B**. In press.
11. HOTELLING, H. 1941. The experimental determination of the maximum of a function. Ann. Math. Stat. **12:** 20.
12. HOTELLING, H. 1943. Some improvements in weighing and other experimental techniques. Ann. Math. Stat. **15:** 297.
13. FISHER, R. A. 1935. The Design of Experiments. Oliver & Boyd. Edinburgh, Scotland.
14. BOX, G. E. P. 1952. Multifactor designs of first order. Biometrika. **39:** 49.
15. BOX, G. E. P. & J. S. HUNTER. 1956. Multi-factor experimental designs for exploring response surfaces. Ann. Math. Stat. **28:** 195.
16. BOX, G. E. P. & K. B. WILSON. 1951. On the experimental attainment of optimum conditions. J. Roy. Stat. Soc. **B13:** 1.
17. BOX, G. E. P. 1954. The exploration and exploitation of response surfaces: some general considerations and examples. Biometrics. **10:** 16.
18. BOX, G. E. P. 1959. Transformations on the independent variable. Paper presented at the Chemical Division Conference of the American Society for Quality Control, September 1959, Houston, Texas.
19. BOX, G. E. P. & N. R. DRAPER. 1959. A basis for the selection of a response surface design. J. Am. Stat. Assoc. **59:** 622.
20. DAVID, H. A. & B. E. ARENS. 1959. Optimal spacing in regression analysis. Ann. Math. Stat. **30:** 1072.
21. FOLKS, J. L. 1958. Comparison of designs for exploration of response relationships. Paper read at the 18th Annual Meeting Am. Stat. Assoc., December 1958, Chicago, Ill.
22. BOX, G. E. P. & H. L. LUCAS. 1959. Design of experiments in nonlinear situations. Biometrika. **46:** 77.

4.7
A Useful Method for Model-Building

G. E. P. BOX and WILLIAM G. HUNTER†
University of Wisconsin

A simple technique useful in iterative model-building is described. In this a statistical analysis is applied to the estimated parameters rather than to the observations directly. This mode of procedure pinpoints the source and nature of possible inadequacies and these, interacting with the experimenter's special knowledge of the problem, can suggest what modifications are necessary for improvement. The cycle may be repeated till an adequate model, which should of course be checked by additional critical experiments, is found.

1. Introduction

The object of much experimentation is to build or discover a suitable model. This is done by an iterative procedure in which a particular model is tentatively entertained, strained in various ways over the region of application, and its defects found. The nature of the defects interacting with the experimenter's technical knowledge can suggest changes and remedies leading to a new model which, in turn, is tentatively entertained, and submitted to a similar straining process.

In this paper a simple method is presented for checking and modifying a model in which the estimated constants of the system are treated as the observations. By definition parameters should not be affected when the settings of the variables are changed; in short, "constants" should stay constant. In such an analysis, therefore, the usual effects and interactions have expected value zero if the model is correct. By a slight extension of this idea examples occur where we should not expect the "constants" to stay constant but to change in a prescribed way (e.g., linearly) as the variables are changed. In such cases the non-existence of interactions provides a check on the adequacy of the model while the sizes of the linear effects estimate important parameters. This extension is illustrated by means of an example.

If significant departures from expectation are found, the model is shown to be inadequate. The precise nature of the inadequacy is pinpointed and valuable clues about appropriate modifications of the model are provided. When a modification has been conjectured the cycle is repeated.

Figure 1 emphasizes the adaptive features of this method. In the iterative process of model-building the experimenter, who can call upon pertinent technical knowledge that the statistician will in general lack, plays a vital role in modifying

† This research was supported by the United States Navy through the Office of Naval Research, under Contract Nonr-1202(17), Project NR 062 222. Reproduction in whole or in part is permitted for any purpose of the United States Government.

The Collected Works of George E. P. Box, 1984, Wadsworth, Inc., Belmont, CA 94002.
Originally published in *Technometrics,* vol. 4, no. 3 (1962), pp. 301–318.

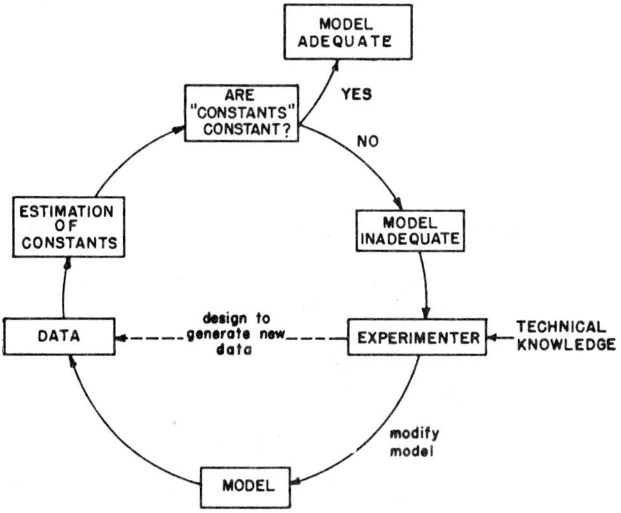

FIGURE 1—Adaptive model-building.

the model as the nature of its defects are revealed. Depending on the nature of the modification the calculation may go forward using the same data or further appropriate data may be required. We have no reason to suppose that this method is new but we feel that it is of sufficient general interest to be formally recorded.

For purposes of model-building, the factorials seem particularly suitable because using these designs the straining process is applied with great efficiency. In some examples, however, it would be advantageous to use for the "factors" not the natural variables of the process but appropriate functions of them deemed by the experimenter to be of more fundamental interest.

2. Empirical and Theoretical Models

In some cases so little is known about a physical, chemical, or biological process that in studying it we are forced to adopt a heavily empirical approach which assumes little more than the process function can be represented as locally smooth. In these circumstances we may be able to do no more than represent the relation by a graduating function such as a polynomial. Such a graduating function acts as a mathematical French curve and makes no claim of course to represent the relationship functionally. At the other extreme, cases may occur in which a theoretical model based on an intimate knowledge of the physical mechanism is available. In such a case the functional relation is often defined by time- or space-dependent differential equations. If a model adequately represents a given situation, we can estimate the behavior of the system by estimating the constants and their precision [1, 2, 4, 7, 8]. In the first case we know practically nothing about the functional form; in the second, we know everything about the functional form.

Most frequently, the situation is somewhere in between the two extremes. The theory is usually inadequately understood, but by a series of planned experiments using an iterative process like that already described we hope to arrive at a working model which takes account of the main theoretical aspects of the problem. Strictly speaking, no model can ever give a precise description of what happens. A working theoretical model, however, supplies information on the system under study over important ranges of the variables by means of equations which reflect at least the major features of the mechanism. Quite possibly it fails to hold over more extensive ranges where effects which previously could be safely ignored become important.

It is common to find too timid an attitude toward model-building. There is a tendency to think that unless the functional form is almost exactly known, the theoretical approach is unjustified. This results in many purely empirical studies where a semi-theoretical approach in which knowledge of the functional form was built up step by step would have been more effective and more rewarding.

The functional form derived from a theoretical development is usually non-linear in the unknown parameters. The empirical approach is sometimes adopted merely to avoid the complication of fitting these non-linear equations. The *fitting* of non-linear equations is chiefly a task in numerical analysis, and various devices can be used to make it reasonably simple. In particular the present availability of electronic computers has largely removed the computational labor which would otherwise make this approach unattractive.

The effect of adopting the empirical approach, when a theoretical one would have been possible, is often indirectly to invite trouble. In a purely empirical formulation non-linearities in the form of interactions and complex effects may arise which can be largely avoided by a semi-theoretical approach. Not only are such non-linearities difficult to interpret but it may be necessary to employ an excessively large number of experiments to estimate the proliferation of constants which result from the use of an empirical functional form.

In this paper we show how by working with the estimated parameters of the problem and treating these rather than the original responses as the "observations," this introduction of unnecessary non-linearity is avoided, often-times with great simplification in the experimentation required. The method can best be illustrated by means of an example. Since no data of the required sort is at present available to the authors, this example has been constructed.

3. A Chemical Example with Four Variables

In Table 1 are yields which might have been obtained from a 2^4 factorial experiment on a chemical process in which the variables studied were $[A]_0$ and $[B]_0$ the initial concentrations of reactants A and B, the concentration $[C]$ of catalyst C, and the temperature D. We denote these variables in the factorial experiment by A, B, C, and D. During each run, samples were taken at five times (80, 160, 320, 640 and 1280 minutes), and the yield (the concentration of the product F) was determined.

3.1 Conventional Analysis

The usual factorial analysis of these results is shown in Table 2. We suppose that the between-run standard deviation in the units of concentration is known

Table 1

Experimental Results

Run	Design $A\ B\ C\ D$	Concentration of F(moles × 10^2 per liter) t = 80 min	160 min	320 min	640 min	1280 min
1	− − − −	3.17	5.39	8.66	15.9	22.6
2	+ − − −	14.7	23.4	34.3	34.0	20.3
3	− + − −	4.80	10.8	22.5	34.6	42.0
4	+ + − −	23.2	39.0	55.6	63.4	41.6
5	− − + −	3.72	3.81	17.2	20.0	23.9
6	+ − + −	17.9	28.3	40.5	34.2	21.6
7	− + + −	8.60	13.3	25.9	39.8	50.8
8	+ + + −	30.9	51.4	72.2	76.4	38.9
9	− − − +	7.48	9.93	20.0	30.9	24.9
10	+ − − +	25.3	35.3	39.1	28.4	7.50
11	− + − +	13.3	27.1	43.0	58.0	49.4
12	+ + − +	50.8	75.6	84.2	57.0	11.5
13	− − + +	9.15	15.8	27.5	33.9	23.0
14	+ − + +	30.8	44.4	46.7	24.9	2.94
15	− + + +	22.8	37.2	57.9	69.1	53.9
16	+ + + +	62.6	88.0	89.5	43.4	5.80

	A (moles/liter)	B(moles/liter)	C(millimoles/liter)	D(°C)
+	40	2	1.0	175
−	20	1	0.5	165

to be of the order of 2 so that the standard deviation associated with the effects in Table 2 is about 1. We therefore appear to be dealing with an exceedingly complex phenomenon in which two-factor, three-factor, and four-factor interactions as well as main effects are of importance. Furthermore, the main effects and interactions change markedly and in a complex fashion with regard to time. (Note, for example, the A effect, the AD and the ABD interactions.) If we regarded time as a further factor in a "split-plot" design we would be faced with the interpretation of a five-factor interaction.

3.2 Alternative Approach

Suppose in this investigation the experimenter was able to conceive of a theoretical model which might act at least as a starting point in explaining the time dependency of the data. The reaction studied was of the type

$$A + B \rightarrow F$$

$$A + F \rightarrow G$$

where F was the desired product. The concentration of $[A]_0$ employed was large compared with that of $[B]_0$, roughly in the ratio of 20 to 1. Thus, the percentage of A used up during the reaction was small and consequently, as an approximation, its concentration could be treated as constant for any given run.

TABLE 2

Factorial Analysis of the Yields

Effects (moles per liter × 10^2)	t(min) 80	160	320	640	1280
A	21.45	30.32	26.46	3.55	−20.15
B	11.65	19.57	23.63	23.40	15.80
C	6.92	9.40	12.23	6.33	2.73
D	15.86	22.18	19.86	7.30	−7.74
AB	5.15	6.20	4.65	−1.80	−9.63
AC	3.04	5.18	3.65	0.48	−0.45
BC	4.18	4.83	4.77	5.52	3.70
AD	7.74	8.00	1.32	−13.10	−10.72
BD	7.54	11.05	11.70	3.95	−0.23
CD	2.02	−0.31	−3.40	−7.08	−4.64
ABC	1.42	2.75	4.20	3.22	0.22
ABD	4.30	5.13	3.98	−2.00	−2.50
ACD	−1.50	−3.79	−6.02	−8.28	−2.77
BCD	−0.65	−2.94	−3.50	−6.02	−2.38
$ABCD$	−1.80	−2.98	−6.63	−7.78	−2.10
Mean	19.85	30.58	41.06	39.55	26.24

Standard deviation of effects is about 1.

Tentative Model

The simplest possible kinetic model describing the behaviour of the system would be one which assumed that the reactions were first order with respect to the concentrations of B and F. Then for a *given* initial concentration of the reactant A, a *given* concentration of the catalyst C and a *given* temperature D, the course of the reaction would be described by the simultaneous differential equations:

$$-\frac{d[A]}{dt} = k_1[B] + k_2[F]$$

$$-\frac{d[B]}{dt} = k_1[B]$$

$$\frac{d[F]}{dt} = k_1[B] - k_2[F]$$

$$\frac{d[G]}{dt} = k_2[F]$$

where $[A]$, $[B]$, $[F]$, $[G]$ refer to concentrations at time t.

The rate equations can be integrated to give for the concentration of F,

$$[F] = \frac{[B]_0 k_1}{k_1 - k_2}(e^{-k_2 t} - e^{-k_1 t}) \tag{1}$$

In this expression $k_1 = k_1([A]_0, [C], D)$ and $k_2 = k_2([A]_0, [C], D)$ would be expected to be functions of the experimental variables $[A]_0$, $[C]$, and D but not of $[B]_0$ if the model is adequate.

For a given experimental run we have five observations of the concentration of F at different times (Table 1) and from these we can estimate k_1 and k_2 by the method of least squares. To avoid digression from the main theme at this stage, we discuss this estimation procedure in the Appendix, where we also illustrate the use of the likelihood surface in the appreciation of the estimation situation.

The fitted curves for each of the 16 runs are shown in Figure 2. Careful inspection of the data points in relation to these curves and of the residuals shown in Table 4 would give no reason to doubt the adequacy of the assumptions so far.

Figure 3 exemplifies the estimation situation. In each case the cross indicates the least squares estimates in the k_1, k_2 space and the shaded region is that enclosed by a particular sum of squares contour. (On the usual Normal assump-

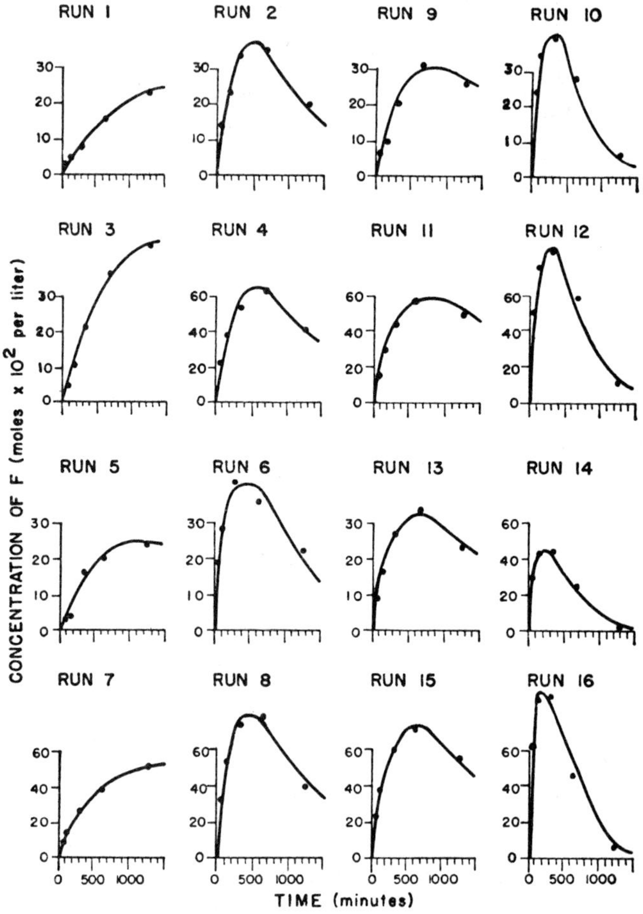

FIGURE 2—Fitted curves.

tions this corresponds to a likelihood contour.) For each run this "critical contour" has been taken so that

$$S_c = S_{\min} + s^2 p F_{0.05}(2, 48)$$

where S_c is the value of the "critical contour," $S_{\min}$ is the minimum value of the sum of squares (i.e., the value when the least squares estimates are substituted), s^2 is the pooled estimate of within run variance assumed to be based on 48 degrees of freedom, $p = 2$ the number of parameters fitted, and $F_{0.05}(2, 48)$ is the usual 5% value of the F criterion. If the model were linear the shaded region would thus provide a 95% confidence region and in the present circumstances provides a useful indication of the uncertainty associated with each

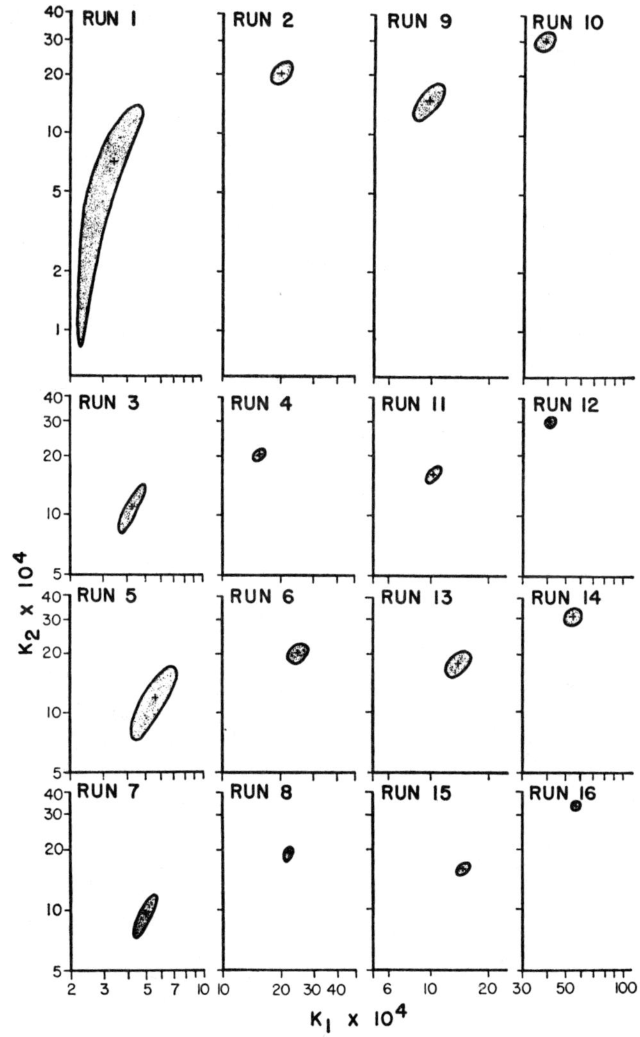

FIGURE 3—Regions of uncertainty for k_1 and k_2 based on the sum of squares contours.

pair of estimates. By comparing each of these figures with its counterpart in Figure 2 we see how these regions accurately reflect the adequacies and inadequacies of the data in various cases. For example, in Runs 3, 4, 7, 8, 11, 12, 15 and 16 shown in the second and fourth rows of the Figures the concentration of B was at its high level and the height of all these graphs is about double that in the first and third rows in Figure 2. (Note the changes in scale in the ordinate.) The percentage accuracy associated with the points, therefore, is higher when the concentration of B is high and this is reflected in the smaller regions of uncertainty in Figure 3. Furthermore, we see when the curve is more completely "covered" by the data as in Run 14 a small near circular region of uncertainty is obtained. When, however, as in Run 1, the data only cover the early portion of the curve a large and ridge-like region is obtained reflecting the fact that many different curves having similar beginnings but different endings could be drawn through the points. Some further discussion will be found in the Appendix.

Analysis of the Rate "Constants"

There is some advantage in working with the logarithms rather than the rate "constants" themselves. In Table 3 we show $-10 \ln \hat{k}_1$ and $-10 \ln \hat{k}_2$ for each of the 16 runs.* The factorial analysis applied to these values is shown in Table 5.

We notice first that if our assumptions were true that the reactions were first order with respect to the concentration of reactant B, then no main effects or interactions containing B should be present. Inspection of the effects in Table 5 provides no reason to doubt this assumption.

We have now to consider the probable nature of the dependence of k_1 and k_2

TABLE 3

Least Squares Estimates of "Constants"

Run	$-10 \ln \hat{k}_1$	$-10 \ln \hat{k}_2$
1	79.74	72.68
2	61.99	62.09
3	77.49	68.03
4	64.56	62.04
5	74.80	67.25
6	59.91	62.19
7	75.99	69.35
8	61.01	62.40
9	69.48	65.09
10	55.70	58.05
11	68.78	64.31
12	54.82	58.19
13	65.71	63.42
14	52.04	57.86
15	65.22	64.44
16	51.76	57.01

* The notation ln is used for natural logarithms and log for the logarithms to the base 10.

TABLE 4

Residuals

Run	$[y_t - \eta_t(\hat{k}_{1j}, \hat{k}_{2j})] \times 100 \begin{cases} t = 80, \cdots, 1280 \\ j = 1, 2, \cdots, 16 \end{cases}$ $t = 80$ min	160 min	320 min	640 min	1280 min
1	0.45	0.32	−0.68	0.08	−0.23
2	0.90	−0.08	0.31	−1.01	0.76
3	−1.68	−1.38	0.84	0.72	−0.36
4	1.38	1.20	−1.12	−0.58	0.70
5	−0.49	−4.03	3.57	−0.66	−0.07
6	1.21	0.40	1.47	−4.04	3.13
7	1.04	−0.96	0.64	−0.38	0.10
8	0.48	0.12	−1.18	1.26	−0.68
9	0.52	−2.70	−0.78	2.71	−1.22
10	2.12	−0.01	−1.91	0.64	1.04
11	−1.58	0.42	−0.40	0.88	−0.44
12	0.78	0.28	−1.40	1.22	−0.76
13	−0.72	−1.60	0.48	1.27	−0.87
14	−0.45	−0.18	0.92	−0.12	−1.31
15	1.92	0.28	0.22	−1.56	0.92
16	−0.56	−0.56	1.84	−1.12	−0.64

on the remaining three factors A, C, and D—the concentration of A, the catalyst concentration and the temperature. In accordance with simple kinetic theory, it might be expected that the temperature dependence of the rate constants will be given by the Arrhenius law, that is to say that $\ln k_i$ $(i = 1, 2)$ will be

TABLE 5

Factorial Analysis of $\ln \hat{k}_1$ *and* $\ln \hat{k}_2$

Effects	on $10 \ln \hat{k}_1$	on $10 \ln \hat{k}_2$
A	14.43	6.84
B	−0.03	0.36
C	3.26	0.82
D	9.00	4.71
AB	−0.60	−0.22
AC	−0.18	−0.59
BC	−0.35	−0.98
AD	−0.71	−0.30
BD	0.64	−0.24
CD	0.25	−0.09
ABC	0.56	1.16
ABD	0.59	0.46
ACD	0.02	0.55
BCD	0.14	0.78
$ABCD$	−0.67	−0.46
MEAN	−64.94	−63.40

a linear function of the reciprocal absolute temperature. Also, experimental evidence supported at least in part by kinetic theory suggests that the constants will be proportional to some power of the concentration of A and to some power of the catalyst concentration. Thus, the equations

$$k_1 = k_1([A]_0, [C], T) = [A]_0^{p_1}[C]^{q_1}\alpha_1 e^{-\beta_1/T}$$
$$k_2 = k_2([A]_0, [C], T) = [A]_0^{p_2}[C]^{q_2}\alpha_2 e^{-\beta_2/T}$$

are now tentatively entertained as describing the dependence of the rate constants on the variables. Equivalently,

$$\begin{aligned} \ln k_1 &= p_1 \ln [A]_0 + q_1 \ln [C] + \ln \alpha_1 - \beta_1/T \\ \ln k_2 &= p_2 \ln [A]_0 + q_2 \ln [C] + \ln \alpha_2 - \beta_2/T \end{aligned} \tag{2}$$

Since these expressions are linear in $\ln [A]_0$, $\ln [C]$, and T^{-1}, all interactions involving $[A]_0$, $[C]$, and T would be zero if our tentative assumptions were correct. Again, consideration of Table 5 lends support to our hypotheses. Inspection of these interactions and plotting on probability paper [6] show no abnormalities.

We shall proceed, therefore, on the assumption that Equations (2) provide an adequate representation of the dependence of the constants on the factors A, C, and D. With this same assumption, the pooled interactions supply an estimate of the standard errors of the effects. (One would not expect in this example that the variances of the estimated constants would necessarily remain the same from run to run. It may be shown, however, that for the case of a two-level factorial design, which is here employed, unbiassed estimates of the effects *and* their standard errors are obtained even so [3].) Denoting by s_1 and s_2 the standard errors of the effects calculated for $\ln \hat{k}_1$ and $\ln \hat{k}_2$ respectively we have, for example,†

$$s_1^2 = \frac{(-.060)^2 + (-.018)^2 + \cdots + (-.067)^2}{11} = 0.00236, \quad \text{or} \quad s_1 = 0.049$$

The 95% confidence intervals for the true effects are obtained by adding to and subtracting from the estimated effects 2.201 s_1 and 2.201 s_2 respectively, where 2.201 is the 5% level of Student's t with eleven degrees of freedom.

Having checked our assumptions, it now remains to estimate the unknowns in Equations (2). Denoting averages by bars, we have for the "average log rate constants"

$$\overline{\ln k_i} = p_i \overline{\ln [A]_0} + q_i \overline{\ln [C]} + \ln \alpha_i - \beta_i \overline{T^{-1}} \qquad (i = 1, 2)$$

On subtraction from Equations (2) we obtain

$$\begin{aligned} \ln k_1 &= \overline{\ln k_1} + p_1(\ln [A]_0 - \overline{\ln [A]_0}) + q_1(\ln [C] - \overline{\ln [C]}) - \beta_1(T^{-1} - \overline{T^{-1}}) \\ \ln k_2 &= \overline{\ln k_2} + p_2(\ln [A]_0 - \overline{\ln [A]_0}) + q_2(\ln [C] - \overline{\ln [C]}) - \beta_2(T^{-1} - \overline{T^{-1}}) \end{aligned} \tag{3}$$

Estimates of the unknowns are thus provided by

† The entries in Table 5 are divided by 10 to provide "effects" approximate to $\ln k_1$ and $\ln k_2$.

$$\hat{p}_i = \frac{A_i \text{ effect}}{\ln [A]_0^+ - \ln [A]_0^-}, \hat{q}_i = \frac{C_i \text{ effect}}{\ln [C]^+ - \ln [C]^-}, \hat{\beta}_i = \frac{D_i \text{ effect}}{\dfrac{1}{T^-} - \dfrac{1}{T^+}} \quad (i = 1, 2)$$

where the A_1 effect refers to the "A effect" calculated from $\ln \hat{k}_1$, the A_2 effect that calculated from $\ln \hat{k}_2$.

The 95% confidence intervals for these unknowns are

$$\hat{p}_i \pm \frac{2.201 s_i}{\ln [A]_0^+ - \ln [A]_0^-}, \hat{q}_i \pm \frac{2.201 s_i}{\ln [C]^+ - \ln [C]^-}, \hat{\beta}_i \pm \frac{2.201 s_i}{\dfrac{1}{T^-} - \dfrac{1}{T^+}},$$

Thus finally we obtain

$$\begin{aligned} \hat{p}_1 &= 2.08 \pm 0.16; & \hat{\beta}_1 &= (17.7 \pm 2.1) \times 10^3 \\ \hat{p}_2 &= 0.99 \pm 0.20; & \hat{\beta}_2 &= (9.2 \pm 2.7) \times 10^3 \\ \hat{q}_1 &= 0.47 \pm 0.16; & \ln \alpha_1 &= 26.59 \pm 4.70 \\ \hat{q}_2 &= 0.12 \pm 0.20; & \ln \alpha_2 &= 11.26 \pm 0.65 \end{aligned}$$

The activation energies E of the two reactions are given by

$$E = R\beta$$

where R is the gas constant 1.987 cal. deg.$^{-1}$ mole^{-1}. Thus

$$E_1 = 35.2 \pm 4.2 \text{ kcal}$$

$$E_2 = 18.5 \pm 5.4 \text{ kcal}.$$

Kinetic theory renders most plausible the existence of reactions of orders zero, one, two, or three for the concentrations of A and B. The situation with regard to catalyst concentration is less well understood; however, in some circumstances half-order reactions are plausible on theoretical grounds.

We see that the data are readily explained in terms of a kinetic model which behaves as if the first reaction ($A + B \rightarrow F$) is second-order with respect to $[A]$, first-order with respect to $[B]$, and half-order with respect to the catalyst concentration. The data support the hypothesis that the second reaction is first order with respect to $[A]$ and $[B]$ while the effect of catalyst concentration appears to be slight. Actual trial shows that the function

$$[F] = \frac{[B]_0[A]_0^2[C]^{0.5}k_1'}{[A]_0^2[C]^{0.5}k_1' - [A]_0 k_2'} (e^{-[A]_0 k_2' t} - e^{-[A]_0^2[C]^{0.5}k_1' t})$$

where

$$k_1' = 3.518 \times 10^{11} e^{-35200/RT}$$

$$k_2' = 7.795 \times 10^4 e^{-18300/RT}$$

closely reproduces the yields of F at the various times and the various experimental conditions. The data were in fact generated by using a mathematical model of the form here recovered with $p_1 = 2.0$, $p_2 = 1.0$, $q_1 = 0.5$, $q_2 = 0.0$, $k_1' = 9.28 \times 10^{11} e^{-35580/RT}$, and $k_2' = 3.99 \times 10^3 e^{-15740/RT}$. The "experimental" values were obtained by adding random normal deviates to the calculated values. The "within run" error had a standard deviation of 0.020 and the "between run" error a standard deviation of 0.015. In real examples not one but a number of iterative cycles would usually be necessary before a satisfactory model would be obtained. After such an iterative process it is of course essential to check the model by further critical experiments. We can with some confidence use it for prediction purposes at least over the range of variables actually tested.

5. Discussion

The device adopted is of rather general application. Many problems in chemistry, engineering, physics, and biology appear to be of a kind which could usefully be tackled in this manner. In those instances where the functional form is defined by differential equations which are not integrable to explicit solutions, numerical integration may be employed. The Princeton-IBM 704 program [9] is one example of a general computer program which uses a suitable subroutine for numerically solving the equations and automatically applies an iterative procedure to obtain the least squares solution. The analog computer provides an alternative means for fitting the constants although care should be taken when using these machines to explore the estimation situation by obtaining at least a rough plot of the sum of squares surface. If this is not done, excessive ill-conditioning of the sum of squares surface may go unnoticed and lead to serious trouble.

The advantage of an analysis of the "constants" of the system is that unnecessary non-linearities are removed and inadequacies in the model can be shown up in readily understood form. Where these inadequacies are of minor character and not such as to cast doubt on the general usefulness of the basic model, allowances are readily made. We are often in one of two simple situations. Either (1) the effects of the variables have already been taken into account in the model so that no effects will occur (this is the case in the example with respect to the initial concentration of reactant B); or, alternatively, (2) the only effects that occur will be linear (this is the case for the variables A, C and D).

Our aim, therefore, should be to subject our model to the largest possible number of "strains" by varying each of the variables which should theoretically have no effect or possibly only a *linear* effect. The existence of non-zero effects or interactions indicates the necessity for modification. It is seen that we are here in an ideal situation for the application of fractional factorial designs. Such designs are particularly valuable where experiments are expensive and the number of runs necessarily limited. When maximum economy is essential, fractional factorial designs which are of particular value are those of resolution IV [5]. For example using one such design, the behavior of the model with respect to as many as three principal variables and eight minor variables can be checked in only 16 runs. The principal variables are those which it is feared might give trouble by showing not only main effects but also interactions; and the minor variables are those to be checked only in so far as their main effects are concerned. In this context in the above example the minor variables would include the initial concentration of B and any other variables the effects of which it is believed are already taken into account in the model and which would, therefore, be expected to have zero effect.

In our example we have illustrated the method using a factorial design which supplies a very efficient way of exerting appropriate strains on the model from which the data are readily analyzed. It will be noticed that this same method can be applied, although less efficiently, when the levels of the variables do not follow a prescribed design. The constants will then appear as the dependent variables in a multiple regression equation.

Appendix

Least Squares Estimation of k_1 and k_2

The constants are found by an iterative least squares method using an approximate linearization of the function (1) [2, 3, 7, 8]. It is often possible to obtain a closer linear approximation by expanding in terms of suitable *functions* of the parameters rather than the untransformed variables themselves [4]. For the particular functional form here considered it is readily shown by considering higher order terms that expansion in terms of ln k_1 and ln k_2 provides a closer approximation than the more direct expansion in k_1 and k_2.

If we let $\theta_1 = 4 + \log k_1$ and $\theta_2 = 4 + \log k_2$ and denote the function by $\eta(\theta_1, \theta_2)$, we have

$$\eta(\theta_1, \theta_2) = \eta(\theta_1^*, \theta_2^*) + (\theta_1 - \theta_1^*)\left[\frac{\partial \eta}{\partial \theta_1}\right]_{\theta=\theta^*} + (\theta_2 - \theta_2^*)\left[\frac{\partial \eta}{\partial \theta_2}\right]_{\theta=\theta^*} + \epsilon \qquad (4)$$

ignoring terms of degree higher than the first in $(\theta - \theta^*)$. In the common case in which the tentative function is not known explicitly but is for example given only in terms of differential equations which can be solved numerically, the derivative can be replaced by a difference approximation. The difference method is usually most convenient even when direct differentiation is available and we shall use it here. Letting θ_i^* denote our first guess of the true value θ_i $(i = 1, 2)$, and writing

$$x_1 = \eta(\theta_1^* + \delta_1, \theta_2^*) - \eta(\theta_1^*, \theta_2^*) \qquad x_2 = \eta(\theta_1^*, \theta_2^* + \delta_2) - \eta(\theta_1^*, \theta_2^*)$$

$$Y = y - \eta(\theta_1^*, \theta_2^*)$$

equation (4) takes on the simple "regression" form

$$Y = \beta_1 x_1 + \beta_2 x_2 + \epsilon$$

where $\beta_i = (\theta_i - \theta_i^*)/\delta_i$. Estimates b_1 and b_2 for β_1 and β_2 can be obtained by the usual linear least squares and may be used to obtain new estimates θ_1^{**} and θ_2^{**} of θ_1 and θ_2 using

$$\theta_1^{**} = \theta_1^* + \delta_1 b_1$$

$$\theta_2^{**} = \theta_2^* + \delta_2 b_2 .$$

If the function is well-behaved, these new estimates will be closer to the least squares values and may in turn be used as starting values for further iteration. The process can be repeated until there is no material improvement.

For our present purpose we shall show the estimation of k_1 and k_2 for the first run. The rapidity of convergence of this non-linear estimation method depends markedly on the closeness of the original guesses k_1^* and k_2^* . It is always worthwhile to take a little trouble in obtaining reasonably good preliminary estimates therefore.

Obtaining Preliminary Estimates

For the present example, the preliminary estimates were found as follows. Using Equation (1) we have at time zero

$$\left(\frac{d[F]}{dt}\right)_{t=0} = k_1[B]_0$$

whence

$$k_1 = \left(\frac{d[F]}{dt}\right)_{t=0} \bigg/ [B]_0 \tag{5}$$

Also if t_{max} represents the time at which $[F]$ attains its maximum value, then it is readily shown by direct differentiation that

$$t_{max} = \frac{\ln k_1 - \ln k_2}{k_1 - k_2}. \tag{6}$$

We can therefore proceed by making a rough sketch of the curve of the yield of F against time for a particular run, estimating $(d[F]/dt)_{t=0}$ and t_{max} , and substituting these estimates in Equations (5) and (6) to obtain starting values k_1^* and k_2^* for k_1 and k_2 . For illustration, a plot of the curve for the first run is shown in Figure 4.

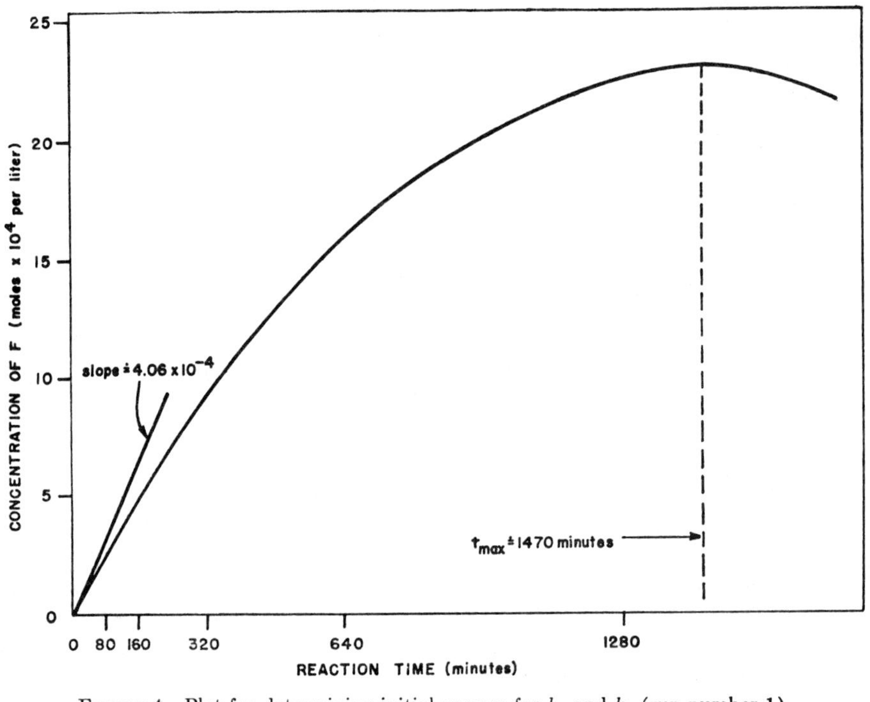

FIGURE 4—Plot for determining initial guesses for k_1 and k_2 (run number 1).

The values from this curve are $(dF/dt)_{t=0} = 4.06 \times 10^{-4}$ moles per liter/min. and $t_{max} = 1470$ min. yielding the preliminary estimates $k_1^* = 4.06 \times 10^{-4}$ min^{-1} and $k_2^* = 10.52 \times 10^{-4}$ min^{-1}. Using these estimates as starting values we can now carry through the more precise estimation of k_1 and k_2 by least squares.

Newton-Gauss Iteration

For the present example we have $\theta_1^* = 0.6090$ and $\theta_2^* = 1.0220$. Taking $\delta_1 = .05$ and $\delta_2 = .02$ we readily obtain

	{1}	{2}	{3}	{4}
t(min)	observed value y	$\eta(0.61, 1.02)$	$\eta(0.66, 1.02)$	$\eta(0.61, 1.04)$
80	.0317	.0307	.0345	.0306
160	.0539	.0579	.0648	.0577
320	.0866	.1032	.1149	.1024
640	.1590	.1643	.1815	.1618
1280	.2260	.2104	.2281	.2047

Thus	{1}−{2}	{3}−{2}	{4}−{2}
	Y	x_1	x_2
	.0010	.0038	−.0001
	−.0040	.0069	−.0002
	−.0166	.0117	−.0008
	−.0053	.0172	−.0025
	.0156	.0177	−.0057

And

$$\sum x_1^2 = 80.807 \times 10^{-5} \quad \sum x_2^2 = 3.943 \times 10^{-5} \quad \sum x_1x_2 = -15.501 \times 10^{-5}$$

$$\sum Yx_1 = -3.306 \times 10^{-5} \quad \sum Yx_2 = -6.169 \times 10^{-5}$$

In this case we obtain the Normal equations

$$80.807\, b_1 - 15.501\, b_2 = -3.306$$

$$-15.501\, b_1 + 3.943\, b_2 = -6.169$$

whence

$$b_1 = -1.388 \quad \text{an} \quad b_2 = -7.021$$

and

$$\theta_1^{**} = 0.6090 + (.05)(-1.388) = 0.5396$$

$$\theta_2^{**} = 1.0220 + (.02)(-7.021) = 0.8816$$

These new estimates θ_1^{**} and θ_2^{**} are then used as the starting values for further iteration. The course of the iteration up to the fourth stage is shown below.

Iteration	θ_1	θ_2
1st	0.6090	1.0220
2nd	0.5396	0.8816
3rd	0.5340	0.8485
4th	0.5369	0.8437

A similar process can be used to find estimates of the constants for each of the sixteen runs. The final values taken to the base e instead of to the base 10 are shown in Table 3.

When, as in this example, the function is not markedly non-linear in the parameters over the region in which the iteration takes place, it may be unnecessary to calculate $\sum x_1^2$, $\sum x_2^2$, and $\sum x_1x_2$ at each iteration. In the present example, for instance, it is found that these quantities remain essentially the same from one trial to the next. In general if hand calculation is employed, considerable saving is achieved when these values can be reused from one iteration to the next.

Checking the Assumptions

Before proceeding further with the analysis, it is of course essential to confirm that the assumptions underlying the calculation of these preliminary estimates are reasonable. In a non-linear situation it is necessary to make a careful and critical study.

For any given run, the estimation situation concerning k_1 and k_2 is completely revealed by a study of the sum of squares surface. On the assumption that the experimental errors are normally distributed with the same variance, the contours of the sum of squares $S(\theta_1, \theta_2) = \sum \{y_i - \eta_i(\theta_1, \theta_2)\}^2$ in the θ_1, θ_2 space are simply the contours of constant likelihood. In these expressions y_i is the observed yield for the time t_i and $\eta(\theta_1, \theta_2)$ is the calculated value for the yield with k_1 and k_2 the values of the constants. The sum of squares contours for the data from the first run are shown in Figure 5. The contours of the sum of squares surface which would have been appropriate if the situation had been truly linear are shown as dotted lines.

It is seen that the estimation situation does not contain serious abnormalities. In particular the sum of squares surface is reasonably well-conditioned. That is to say it has a clearly defined unique minimum and the contours in the region of this minimum are not seriously attenuated in oblique ridges. The latter situation would indicate serious dependence between the estimates such as might make separate estimation of the constants difficult. Inspection of Figure 3 shows that Run 1 is in fact that for which the most serious ill-conditioning and non-linearity occurs. In all other cases the approximate local linearity, which is assumed for example when the within run variance is computed, is obtained.

As a general rule, a check of this kind is essential to make sure that the routine calculation not concealing difficulties which may later cause trouble. In the case where there are several parameters, such a study is particularly important. In this case, a complete plot of the likelihood function may be out of the question; however, a survey can be made, by using the method described in [4]. Here the confidence region is "outlined" by determining the points which are the principal axes of the linear theory confidence ellipsoid in a conventional scaling. The theoretical sums of squares and the actual sums of squares are then compared. The objects are:

(a) to provide confirmation that a minimum has indeed been found,
(b) to check linearity (this would be of especial importance in cases where, for example, estimates of the variances and covariances are used in approximate linear theory tests), and
(c) to show the state of conditioning of the likelihood surface. When there

are a number of estimates, it may reveal that certain combinations of the parameters are virtually confounded and therefore it is not really possible to obtain separate estimates.

In Table 4 the residuals from the fitted curves are presented. Relying on the local linearity of the fitted function, the sum of squares of the five residuals from each curve can be used to calculate an estimate of variance based on three degrees of freedom. Pooling the estimates from all the curves, we obtain an estimate of the "within runs" variance of $s^2 = 2.86 \times 10^{-4}$ based on 48 degrees of freedom.

The regions shown in Figure 3 are based on this within run variance and thus take account of course only of the within run error and not of errors persisting

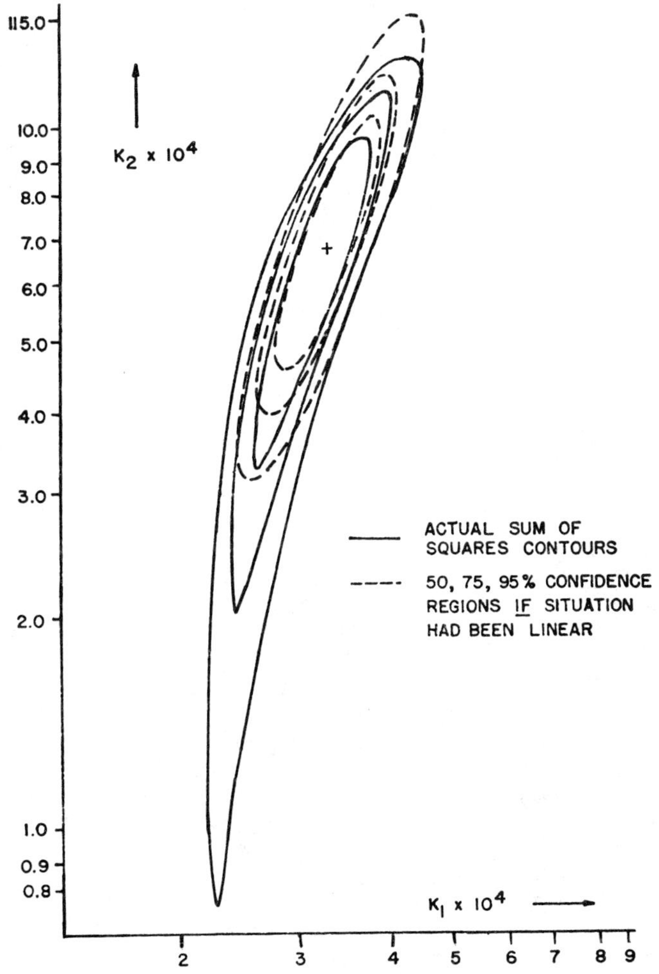

FIGURE 5—Contours of actual sum of squares surface and of corresponding linear theory surface for run number 1.

throughout a run. The latter errors which may be of even greater magnitude than the within run errors arise, for example, from slight misadjustments of the apparatus, discrepancies in the concentrations of solutions, deviations from intended temperature settins, etc. In our analysis of the effects we take these errors into account. By employing an estimate of error from the higher order interactions we take account of both the errors arising within runs and between runs.

References

1. Box, G. E. P., and Coutie, G. A. (1956), "Application of digital computers in the exploration of functional relationships," *Proceedings of the Institution of Electrical Engineers, 103, Part B*, Supplement No. 1, 100–107.
2. Box, G. E. P. (1957), "Use of statistical methods in the elucidation of basic mechanisms," *Bull. Inst. International Statistics, 36*, 215.
3. Box, G. E. P. (1960), "The effect of errors in the factor levels and experimental design," *Bull. Inst. International Statistics, 38*, 133.
4. Box, G. E. P. (1960), "Fitting empirical data," *Proceedings N. Y. Academy of Sciences, 86*, 792.
5. Box, G. E. P., and Hunter, J. S. (1961), "The 2^{k-p} fractional factorial designs," *Technometrics, 3*, 311.
6. Daniel, Cuthbert (1959), "Use of half-normal plots in interpreting factorial two level experiments," *Technometrics, 1*, 311.
7. Gauss, K. F., *Theoria Motus Corporum Coelestium, Werke, Vol. 7*, p. 240–254, 1809.
8. Hartley, H. O. (1961), "The modified Gauss-Newton method for the fitting of non-linear regression functions by least squares," *Technometrics, 3*, 269.
9. Non-linear Estimation (1959), Princeton-IBM 704 Program WLNL 1, Mathematics and Applications Department, IBM Corporation, New York, New York.

4.8
Transformation of the Independent Variables

G. E. P. BOX and PAUL W. TIDWELL
University of Wisconsin

In representing a realationship between a response and a number of independent variables, it is preferable when possible to work with a simple functional form in transformed variables rather than with a more complicated form in the original variables. This paper describes and illustrates a procedure to estimate appropriate transformations in this context.

1. Introduction

Much experimental work is concerned with the study of a response function $\eta = E(y) = f(\mathbf{x}, \boldsymbol{\theta})$ where $\mathbf{x}$ are the levels of k variables $x_1, x_2, \cdots, x_k$; $\boldsymbol{\theta} = \theta_1, \theta_2, \cdots \theta_p$ are a set of p parameters. The objective is usually (a) to check the adequacy of the assumed functional form, (b) to estimate the values of the parameters $\boldsymbol{\theta}$ and hence the response $E(y_{\mathbf{x}_0})$ at any chosen $\mathbf{x}$, and (c) to obtain measures of precision for the estimates.

Suppose that n observations have been made. The uth observation y_u at known levels of the variables $\mathbf{x}_u = x_{u1}, x_{u2}, \cdots, x_{uk}$ is then given by

$$y_u = f(\mathbf{x}_u, \boldsymbol{\theta}) + \epsilon_u . \tag{1}$$

Then the convenient and well known least squares method of analysis may be appropriately applied if the errors ϵ_u can be regarded as independently and normally distributed with constant variance σ^2. The majority of the published work on transformations is concerned with transforming the y's to achieve this simplicity when the necessary assumptions could not otherwise be realistically made (2, 12, 14). Most emphasis has been placed therefore on transformations of $E(y)$ which may be expected to stabilize the variance or reduce $f(\mathbf{x}, \boldsymbol{\theta})$ to linearity in the θ's. Frequently the assumption of independence between the errors is not in question, or alternately randomization has been introduced so it is legitimate to analyze as if the errors are independent (11, 12).

Possible reasons and methods for transforming the θ's have recently received attention in connection with the fitting of non-linear models (3, 4, 5).

1.1 *Transformations on the x's.*

In the present investigation we suppose that the errors in the y's are at least approximately normally and independently distributed with constant variance, and we shall concentrate on finding transformations in the x's to reduce the

This research was supported by the United States Navy through the Office of Naval Research, under Contract Nonr-1202(17), Project NR 042 222. Reproduction in whole or in part is permitted for any purpose of the United States Government.

The Collected Works of George E. P. Box, 1984, Wadsworth, Inc., Belmont, CA 94002.
Originally published in *Technometrics,* vol. 4, no. 4 (1962), pp. 531–550.

function in these transformed variables to as simple a form as possible. Although not limited to this application, the method is perhaps of most general use when the true form $f(\mathbf{x}, \boldsymbol{\theta})$ of the response function is unknown and we represent it by a "graduating function".

1.2 *Graduating Functions*

When the true functional relationship $f(\mathbf{x}, \boldsymbol{\theta})$, arising out of the mechanism of the process is unknown, progress often can be made by adopting some suitable graduating function $g(\mathbf{x}, \boldsymbol{\beta})$ which, over a limited region in the x space, can be expected to represent the true relationship reasonably well. The elements of $\boldsymbol{\beta}$ are a set of h empirical constants, and h is not necessarily equal to p, the number of parameters $\boldsymbol{\theta}$.

In particular $g(\mathbf{x}, \boldsymbol{\beta})$ is often chosen to be a polynomial of first or second degree in the x's, and special designs may be employed to ensure efficient estimation. This technique for empirical study of the local characteristics of the response function has come to be called "Response Surface Methodology", and has been rather widely applied in experimental work, particularly in the chemical industry. The number of experimental runs required to allow estimation of the polynomial and the difficulty of appreciating the implications of the polynomial that has been fitted rapidly becomes greater as its degree increases.

Our aim then is to work with a polynomial of low degree in the transformed variables rather than with a polynomial of higher degree in the original variables.

1.3 *Approximation by Polynomials of First Degree.*

It is not uncommon to find that over a particular region of interest of the x's, the function is strictly monotonic although not entirely linear. For example, figure 1a shows contours of a particular function $f(x_1, x_2, x_3)$ over a cubical region in the x space. Figure 1b shows contours of the same function plotted in terms of the transformed variables $\xi_1 = \ln x_1$, $\xi_2 = x_2^{-1}$, and $\xi_3 = x_3^{\frac{1}{2}}$. If a polynomial were directly fitted to the x's, an equation of at least 2nd degree might be needed to obtain a suitable fit, but in terms of the transformed variables $\boldsymbol{\xi}$ a linear equation would be sufficient. The diagrams in figure 1 have actually been drawn using the function

$$\eta = 25 + 5 \ln x_1 + 100x_2^{-1} + 6x_3^{\frac{1}{2}} .$$

The advantages of approximating the expression by a linear form are that (1) its implications can be more readily appreciated and (2) it simplifies subsequent calculations for which this representation might be needed. A commonly occurring use to which response surface analysis is put is to find a set of levels $x_{10}, x_{20}, \cdots, x_{k0}$ in the operating variables such that a series of j responses dependent on $x_1, x_2, \cdots, x_k$ (such as quality, purity and yield of some material) shall be within certain specified limits. When approximate linear graduating functions in suitably transformed variables can be found, the problem of simultaneously satisfying these j inequalities is a standard linear programming problem, and is readily solved using standard techniques. The corresponding problem when the functions $g(\mathbf{x}, \boldsymbol{\beta})$ are non-linear is more difficult.

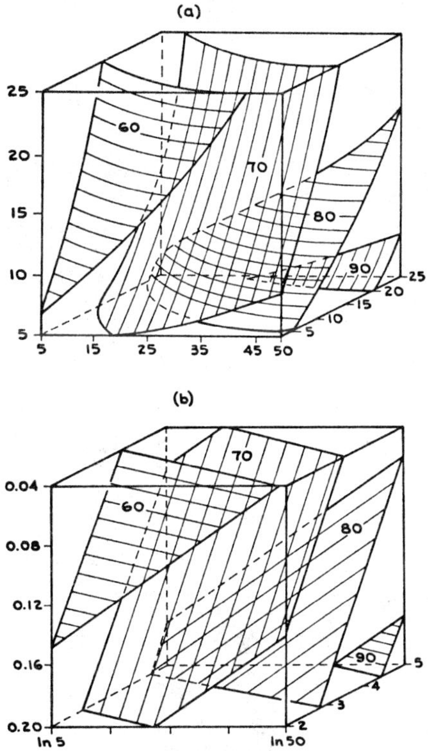

FIGURE 1
A Three Variable Response Function before and after Transformation

1.4 *Approximation by Polynomials of Second Degree.*

In some important cases, we study $\eta = f(\mathbf{x}, \boldsymbol{\theta})$ in a region where η has an extremum. The graduating polynomial will then have to be at least of 2nd degree. Now the manipulation and appreciation of fitted 2nd degree polynomials is a relatively easy matter (7), but higher order polynomials are much more difficult to deal with. In cases where the untransformed 2nd degree polynomial is not adequate, it is preferable to attempt to obtain adequate representation in terms of a 2nd degree polynomial in the transformed variables rather than to fit higher order polynomials in the untransformed variables. In this way it is frequently possible to obtain suitable representation of a maximum, minimum, ridge system or a minimax by an equation of only 2nd degree.

1.5 *More General Transformations.*

The situations envisaged above and throughout the rest of this paper are those in which suitable graduation can be obtained by having each transformed variable a function of only one original variable. Most generally each transformed variable would be a function of all the original variables. For example, with two variables x_1 and x_2, we would need to consider the transformations

$\xi_1 = \xi_1(x_1, x_2)$, $\xi_2 = \xi_2(x_1, x_2)$. We shall not investigate this more complicated case here.

2. The Transformation Procedure

Suppose observations $y_1, y_2, \cdots, y_u, \cdots, y_n$ are available at n sets of conditions $\mathbf{x}_1, \mathbf{x}_2, \cdots, \mathbf{x}_u, \cdots, \mathbf{x}_n$, where $\mathbf{x}_u$ is a $k \times 1$ vector giving the levels of the x's for the uth observation. Suppose further that

$$E(y_u) = \eta_u \tag{2}$$

and

$$E(y_u - \eta_u)(y_v - \eta_v) = \begin{cases} \sigma^2 & u = v \\ 0 & u \neq v \end{cases} \tag{3}$$

where σ^2 is unknown.

Suppose that the response η can be closely represented over the region of interest by some simple function

$$\eta = f(\boldsymbol{\xi}, \boldsymbol{\beta}) \tag{4}$$

where the elements $\xi_1, \xi_2, \cdots, \xi_k$ of the vector $\boldsymbol{\xi}$ are the x's transformed in some suitable way so that

$$\xi_i = \xi_i(x_i\,;\boldsymbol{\alpha}_i)$$

with $\boldsymbol{\alpha}_i$ a p_i-dimensional vector with elements $\alpha_{i1}, \alpha_{i2}, \cdots, \alpha_{ij}, \cdots, \alpha_{ip_i}$ the unknown constants of the transformation. The functional relationship also involves l unknown constants whose values $\beta_1, \beta_2, \cdots, \beta_t, \cdots, \beta_l$, elements of the vector $\boldsymbol{\beta}$ depend upon the particular transformations of the ξ's employed (that is, they depend on the choice of the α's).

Now suppose $\boldsymbol{\alpha}_i^{(0)} = (\alpha_{i1}^{(0)}, \alpha_{i2}^{(0)}, \cdots, \alpha_{ip_i}^{(0)})$ are the first guesses for the constants of the ith transformation. At the first iteration, these first guesses would often be chosen such that $\xi_i = x_i$, that is such that no transformation at all was applied. Then write $\boldsymbol{\xi}_u^{(0)}$ for the vector whose ith element is $\xi_i(x_{iu}\,;\boldsymbol{\alpha}_i^{(0)})$. Expanding about these guessed values and ignoring terms of higher than first order in $(\alpha_{ij} - \alpha_{ij}^{(0)})$ we have approximately

$$\eta_u = f(\boldsymbol{\xi}_u^{(0)}, \boldsymbol{\beta}) + \sum_{i=1}^{k} \sum_{j=1}^{p_i} (\alpha_{ij} - \alpha_{ij}^{(0)}) \left\{ \frac{\partial f(\boldsymbol{\xi}, \boldsymbol{\beta})}{\partial \alpha_{ij}} \right\}_{\substack{\boldsymbol{\xi}=\boldsymbol{\xi}_u^{(0)} \\ \boldsymbol{\alpha}_i=\boldsymbol{\alpha}_i^{(0)}}} \tag{6}$$

Now

$$\left\{ \frac{\partial f(\boldsymbol{\xi}, \boldsymbol{\beta})}{\partial \alpha_{ij}} \right\}_{\substack{\boldsymbol{\xi}=\boldsymbol{\xi}_u^{(0)} \\ \boldsymbol{\alpha}_i=\boldsymbol{\alpha}_i^{(0)}}} = \left\{ \frac{\partial f(\boldsymbol{\xi}, \boldsymbol{\beta})}{\partial \xi_i} \right\}_{\boldsymbol{\xi}=\boldsymbol{\xi}_u^{(0)}} \left\{ \frac{\partial \xi_{iu}}{\partial \alpha_{ij}} \right\}_{\boldsymbol{\alpha}_i=\boldsymbol{\alpha}_i^{(0)}}.$$

The quantities $\{\partial \xi_{iu}/\partial \alpha_{ij}\}_{\boldsymbol{\alpha}=\boldsymbol{\alpha}_i^{(0)}}$ can be calculated at once from knowledge of the particular form of the transformations (5) which it has been decided to use, but the quantities $\{\partial f(\boldsymbol{\xi}, \boldsymbol{\beta})/\partial \xi_i\}_{\boldsymbol{\xi}=\boldsymbol{\xi}_u^{(0)}}$ must be estimated in some way. This can be done conveniently by making a preliminary fitting of the observations to $\eta_u = f(\boldsymbol{\xi}_u^{(0)}, \boldsymbol{\beta})$ by the method of least squares or any other convenient method. The fitted expression $y_u = f(\boldsymbol{\xi}_u^{(0)}, \mathbf{b})$ where $\mathbf{b}$ are the least squares estimates of $\boldsymbol{\beta}$ may be differentiated to obtain approximate values of the required quantities $\{\partial f(\boldsymbol{\xi}, \boldsymbol{\beta})/\partial \xi_i\}_{\boldsymbol{\xi}=\boldsymbol{\xi}_u^{(0)}}$.

The $n \sum_{i=1}^{k} p_i$ approximate values of

$$\left\{\frac{\partial f(\boldsymbol{\xi}, \boldsymbol{\beta})}{\partial \alpha_{ij}}\right\}_{\substack{\boldsymbol{\xi}=\boldsymbol{\xi}_u^{(0)} \\ \boldsymbol{\alpha}_i=\boldsymbol{\alpha}_i^{(0)}}},$$

which can now be calculated, provide a set of "independent variables" from which we can obtain the adjustments to the constants of the transformations by refitting the observations to the whole expression on the right hand side of equation (6) by the method of least squares. These adjusted constants can now take the place of the "first guesses" in the above calculation and the whole cycle repeated.

3. Some Simple Applications Using Power Transformations

One of the simplest types of transformation we can employ is the power transformation

$$\xi_i = \begin{cases} x_i^{\alpha_i} & \alpha_i \neq 0 \\ \ln x_i & \alpha_i = 0. \end{cases} \tag{7}$$

This includes many of the forms commonly found useful, for example the reciprocal transformation, and the square root transformation. In practice, if we make guesses different from 1 for the values of α_i in the first iteration, we can always regard these $x_i^{\alpha_i}$ as being the basic variables x_i so that we need only consider the case where $\alpha_i = \alpha_j = 1$. Having obtained new values from the set of starting values we can treat the resulting transformed x's as new x's to perform a second iteration and so on. We then have the model

$$\eta = f(\boldsymbol{\xi}, \boldsymbol{\beta}) \quad \text{with} \quad \xi_i = x_i^{\alpha_i}$$

and $\alpha_i^{(0)} = 1$ ($i = 1, 2, \cdots, k$) for our first guesses.

Then approximately

$$\eta_u = f(\mathbf{x}_u, \boldsymbol{\beta}) + \sum_{i=1}^{k} (\alpha_i - 1)\left\{\frac{\partial f(\boldsymbol{\xi}, \boldsymbol{\beta})}{\partial \alpha_i}\right\}_{\substack{\boldsymbol{\xi}=\mathbf{x}_u \\ \alpha_i=1}}, \tag{8}$$

that is

$$\eta_u = f(\mathbf{x}_u, \boldsymbol{\beta}) + \sum_{i=1}^{k} (\alpha_i - 1)\{\partial f(\mathbf{x}_u, \boldsymbol{\beta})/\partial x_{iu}\}x_{iu} \ln x_{iu}. \tag{9}$$

We proceed by first fitting

$$\hat{y}_u = f(\mathbf{x}_u, \mathbf{b}).$$

We then differentiate this expression to obtain the quantities $\{\partial f(\mathbf{x}_u, \mathbf{b})/\partial x_{iu}\}$ which estimate the quantities in the braces in (9). We then multiply by $x_{iu} \ln x_{iu}$ to provide newly constructed independent variables

$$z_{iu} = \{\partial f(\mathbf{x}_u, \mathbf{b})/\partial x_{iu}\}x_{iu} \ln x_{iu}.$$

On refitting by least squares the quantities

$$\eta_u = f(\mathbf{x}_u, \boldsymbol{\beta}) + \sum_{i=1}^{k} (\alpha_i - 1)z_{iu}$$

we obtain estimates $a_i - 1$ of $\alpha_i - 1$. The variables $x_i^{a_i}$ may now be treated as the x_i's in a further iteration.

3.1 *Fitting a Function Linear in the ξ_i's.*

If

$$\eta_u = f(\xi_u , \beta) = \beta_0 + \sum_{i=1}^{k} \beta_i \xi_{iu} \quad \text{where} \quad \xi_{iu} = x_{iu}^{\alpha_i} \tag{10}$$

then following the above procedure we first fit

$$\hat{y}_u = b_0 + \sum_{i=1}^{k} b_i x_{iu} \, . \tag{11}$$

We can now refit

$$\hat{y}'_u = b'_0 + \sum_{i=1}^{k} b'_i x_{iu} + \sum_{i=1}^{k} c_i x_{iu} \ln x_{iu} \tag{12}$$

and the first estimates of the α's are provided by

$$a_i = (c_i/b_i) + 1. \tag{13}$$

The process can then be repeated with the new variables $x'_i = x_i^{a_i}$. This particular application has been investigated by Turner (13).

In some cases it is convenient to make the calculation by first computing the orthogonal functions

$$[w_{iu}] = x_{iu} \ln x_{iu} - \sum_{h=0}^{k} \delta_h x_{hu}$$

where

$$x_{0u} = 1 \qquad (u = 1, 2, \cdots , n)$$

and

$$\sum_{u=1}^{n} [w_{iu}] x_{hu} = 0 \qquad (i = 1, 2, \cdots , k; h = 0, 1, 2, \cdots , k).$$

We can then perform the single fit

$$\hat{y}_u = b_0 + \sum_{i=1}^{k} b_i x_{iu} + \sum_{i=1}^{k} c_i [w_{iu}]$$

to obtain the estimates $a_i = (c_i/b_i) + 1$ as before.

3.2 *Numerical Examples for $k = 1$*

A somewhat remarkable feature of the procedure is its rapidity of convergence as will be seen from the following examples in which the data (without error) were generated by the models $\eta_1 = 10 + \sqrt{x}$, $\eta_2 = 10 + \ln x$, $\eta_3 = 10 + x^{-1}$, $\eta_4 = 10 + x^{-2}$ for values of x covering the range 1, 2, 3, 4, 5.

x	$x - \bar{x}$	$[w]$	η_1	η_2	η_3	η_4
1	−2	0.396	11.00	10.00	11.00	11.0
2	−1	−0.243	11.41	10.69	10.50	10.25
3	0	−0.359	11.73	11.09	10.33	10.11
4	1	−0.135	12.00	11.38	10.25	10.06
5	2	0.342	12.24	11.60	10.20	10.04
		$c =$	−0.1593	−0.4180	0.3940	0.6260
		$b =$	0.3058	0.3912	−0.1850	−0.2108
		$a = (c/b) + 1 =$	0.48	−0.07	−1.13	−1.97

A second iteration produces the values

$$c = 0.022 \quad -1.103 \quad -0.119 \quad -0.024$$

$$b = 1.061 \quad -15.547 \quad 0.950 \quad 1.003$$

leading to the estimates

$$a = 0.50 \quad 0.00 \quad -1.01 \quad -1.99$$

3.3 *Fitting a Function Quadratic in the* ξ_i*'s*

If

$$\eta_u = f(\boldsymbol{\xi}_u, \boldsymbol{\beta}) = \sum_{i=0}^{k} \sum_{j \geq i}^{k} \beta_{ij} \xi_{iu} \xi_{ju}$$

where

$$\xi_{iu} = x_{iu}^{\alpha_i}, \qquad \beta_{0j} = \beta_j, \qquad x_{0u} = 1, \qquad \beta_{00} = \beta_0$$

then, following the above procedure, we first fit a quadratic expression to the untransformed variables

$$\hat{y} = \sum_{i=0}^{k} \sum_{j \geq i}^{k} b_{iu} x_{iu} x_{ju}.$$

We now refit

$$\hat{y}_u = \sum_{i=0}^{k} \sum_{j \geq i}^{k} b'_{ij} x_{iu} x_{ju} + \sum_{i=1}^{k} (a_i - 1) z_{iu}$$

where

$$z_{iu} = (b_i + b_{i1} x_{1u} + b_{i2} x_{2u} + \cdots + 2b_{ii} x_{iu} + \cdots + b_{ik} x_{ku}) x_i \ln x_i$$

to obtain estimates a_i of the α_i .

3.4 *Examples for Two Independent Variables*

The data (without error) in the first three columns of Table 1 are generated from the model

$$\eta = 80 - 25(\ln x_1 - 0.8)^2 - 45(\ln x_1 - 0.8)(x_2^{-1} - 0.6) - 40(x_2^{-1} - 0.6)^2$$

which is quadratic in $\ln x_1$ and x_2^{-1}. It was supposed that the experiment was

TABLE 1

Generated Data Using a 4 × 4 Factorial Showing Values of $[z_i]$

x_1	x_2	η	$[z_1]$	$[z_2]$
1	1	72.0	−0.81	−1.42
2	1	75.3	−0.71	−0.33
3	1	70.0	3.84	0.76
4	1	54.8	−2.33	1.85
1	2	60.0	0.28	1.12
2	2	78.8	−1.89	−0.06
3	2	78.7	2.93	−1.23
4	2	73.4	−1.33	−2.41
1	3	51.4	1.37	2.01
2	3	75.4	−3.07	1.10
3	3	78.5	2.02	0.19
4	3	75.7	−0.33	−0.72
1	4	46.5	2.46	−1.71
2	4	73.1	−4.24	−0.71
3	4	77.6	1.11	0.28
4	4	75.8	0.67	1.28

laid out as a 4 × 4 factorial with x_1 and x_2 each at levels 1, 2, 3, and 4. The best fitting quadratic model fitted to the untransformed x's is

$$\hat{y} = 79.55 + 3.79\dot{x}_1 - 0.18\dot{x}_2 - 6.11\dot{x}_1^2 - 1.68\dot{x}_2^2 + 4.84\dot{x}_1\dot{x}_2$$

where $\dot{x}_1 = x_1 - \bar{x}_1$ and $\dot{x}_2 = x_2 - \bar{x}_2$.

The quantities z_{1u} , z_{2u} are thus

$$z_{1u} = (3.79 - 12.22\dot{x}_{1u} + 4.84\dot{x}_{2u})x_{1u} \ln x_{1u}$$

$$z_{2u} = (-\ 0.18 + 4.84\dot{x}_{1u} - 3.36\dot{x}_{2u})x_{2u} \ln x_{2u}$$

These values were calculated for each of the experimental points.

It is convenient to proceed in the case of this somewhat specialized design by next determining the values $[z_{1u}]$ and $[z_{2u}]$ which are the parts of the z's orthogonal to the independent variables 1, $\dot{x}_1$, $\dot{x}_2$, $\dot{x}_1^2$, $\dot{x}_2^2$, $\dot{x}_1\dot{x}_2$. These orthogonal parts are simply the residuals from the second degree equation in $\dot{x}_1$ and $\dot{x}_2$ fitted to z_1 and z_2 respectively. These orthogonal parts $[z_1]$ and $[z_2]$ are shown in the fourth and fifth columns of the table. Because of the particular choice of the design $[z_1]$ and $[z_2]$ are themselves orthogonal. The first iteration produces the estimates

$$a_1 = (\sum \eta[\dot{z}_1]/\sum [z_2]^2) + 1 = 0.30$$

$$a_2 = (\sum \eta[z_2]/\sum [z_2]^2) + 1 = -0.94$$

as compared with the correct values $\alpha_1 = 0$ and $\alpha_2 = -1$.

3.41 *An example in which the observations are subject to error*

In the study of the relationship of the yield in a chemical process to the two variables time and concentration of a reactant, an experimenter used the oc-

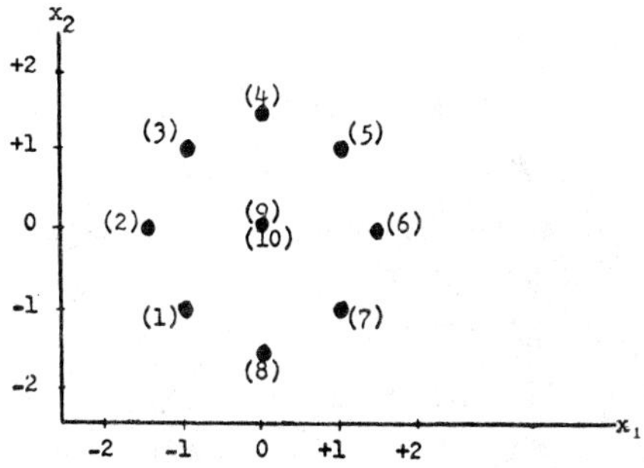

FIGURE 2
A Two Variable Rotable Design

tagonal design (6) shown in figure (2), with

$$\dot{x}_1 = \frac{\text{Time} - 35.0}{25/\sqrt{2}} \qquad \dot{x}_2 = \frac{\text{Concentration} - 40.00}{20/\sqrt{2}}$$

Data obtained from duplicate runs are shown in Table 2. The best fitting

TABLE 2
Data From a Two Variable Rotable Design

x_1	x_2	Time	Concentration	Yield	
-1	-1	17.3	25.86	56.30	53.66
$-\sqrt{2}$	0	10.0	40.00	55.73	48.61
-1	$+1$	17.3	54.14	64.26	69.62
0	$+\sqrt{2}$	35.0	60.00	42.04	45.08
$+1$	$+1$	52.7	54.14	53.90	52.10
$+\sqrt{2}$	0	60.0	40.00	68.35	70.63
$+1$	-1	52.7	25.86	74.09	75.71
0	$-\sqrt{2}$	35.0	20.00	65.84	64.90
0	0	35.0	40.00	82.58	85.60
0	0	35.0	40.00	78.90	77.30

quadratic equation in the standardized variables is

$$\hat{y} = 81.10 + 3.77\dot{x}_1 - 5.13\dot{x}_2 - 8.91\dot{x}_1^2 - 12.09\dot{x}_2^2 - 8.39\dot{x}_1\dot{x}_2$$

The quantities z_{1u} and z_{2u} are thus*

$$z_{1u} = \left(\frac{\partial \hat{y}_u}{\partial T}\right) T_u \ln T_u = \frac{\sqrt{2}}{25}(3.77 - 17.82\dot{x}_{1u} - 8.39\dot{x}_{2u})T_u \ln T_u$$

$$z_{2u} = \left(\frac{\partial \hat{y}_u}{\partial C_u}\right) C_u \ln C_u = \frac{\sqrt{2}}{20}(-5.13 - 8.39\dot{x}_{1u} - 24.18\dot{x}_{2u})C_u \ln C_u.$$

These quantities were calculated for each of the experimental points and are

* Although calculations are usually simplified by working with the standarized variables $\dot{x}_1$ and $\dot{x}_2$, appropriate scale factors must be introduced in calculating the z's to ensure that these have the values they would have if computed from the untransformed variables T and C.

tabulated in Table 4. The orthogonal parts of z_{1u} and z_{2u} , $[z_{1u}]$ and $[z_{2u}]$, are shown in the fifth and sixth columns of this table.

For the data in Table 2, the first estimates a_i of α_i are

$$a_1 = 0.105, \qquad a_2 = 1.515.$$

The values to which these coefficients finally converge after repetition of the process using an electronic computer are

$$a_1 = 0.186 \pm 0.699, \qquad a_2 = 1.823 \pm 1.577.$$

The numbers following the plus and minus signs are the asymptotic standard errors calculated by the method of maximum likelihood in a manner described later. It will be seen that even with this experimental data which is subject to quite large experimental error (the estimated standard error obtained from the observations is $s = 2.974$), the first iteration is reasonably close to the final values. It will also be noted from the sizes of the standard errors that there is considerable uncertainty in the transformation and no strong evidence that any transformation is required, since twice the standard errors would certainly include $\alpha = 1$.

4. Tabled Orthogonal Functions

An important application of this technique is to the fitting of a straight line to a transformed independent variable. Since it appears that the first iteration produces a useful estimate of the required power transformation, and since the original data would often be taken at equally spaced values of the untransformed variable, it is worthwhile to table the orthogonal functions associated with $x \ln x$.

Suppose that n equally spaced values of the untransformed independent variable x have mean m and range r. It is convenient to define the quantity

$$q = \frac{\tfrac{1}{2} \text{ range of the independent variable}}{\text{mean of the independent variable}} = \frac{\tfrac{1}{2}r}{m}.$$

In the accompanying Table 3 the quantity W is tabulated where

$$W = [x \ln x]/mq^2$$

and $[x \ln x]$ indicates the orthogonal part of $x \ln x$. Use of the table is best illustrated by an example.

4.1 *Example of the Use of the Tables of the Orthogonal Functions.*

An experimenter observes the quantity of polymer formed in one hour under steady-state conditions at various equally spaced charges of a catalyst as follows:

Grams of catalyst charged to reactor x	Grams of Polymer formed y	Value of W for $q = 0.8$ W
0.5	28.45	0.321 041
1.5	47.78	−0.207 594
2.5	64.35	−0.281 630
3.5	74.67	−0.098 120
4.5	83.65	0.266 304

Divisor = Δ = 0.195 854

The observed response y is plotted against x in Figure 3a.

If we fit $\hat{y} = b_0 + b_1(x - m)$, a simple regression of y on x, we obtain the equation $\hat{y} = 59.78 + 13.729\ (x - 2.5)$. For this example $m = 2.5$ grams and $r = 4$ grams, hence $q = \frac{1}{2}r/m = 0.8$. From the table of the orthogonal functions for $q = 0.8$ and $n = 5$ we have the values for W as shown above, whence

$$c = \sum_u W_u y_u / m\Delta = -8.0844.$$

Then a, the estimate of α, is

$$\frac{-8.0844}{13.729} + 1 = 0.411.$$

Additional iterations can be made, using as the independent variable $x_u' = x_u^a$. For this example further iterations proceed as follows:

	a	b
2nd iteration	0.458	44.306
3rd iteration	0.474	42.443

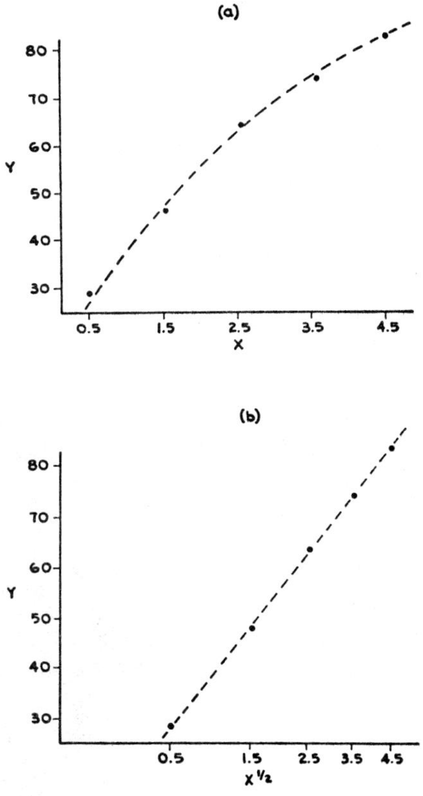

FIGURE 3
A One Variable Response Function before and after Transformation

TABLE 3

The Orthogonal Function W

n	$q = 0.1$	$q = 0.2$	$q = 0.3$	$q = 0.4$	$q = 0.5$
	0.164 188	0.167 630	0.169 236	0.171 699	0.174 426
3	−0.328 372	−0.335 258	−0.338 473	−0.343 395	−0.348 851
	0.164 186	0.167 628	0.169 238	0.171 696	0.174 425
△	0.001 617	0.006 743	0.015 466	0.028 300	0.045 636
	0.223 211	0.227 277	0.230 371	0.235 360	0.242 019
	−0.226 206	−0.233 572	−0.239 360	−0.248 022	−0.258 367
4	−0.217 217	−0.214 687	−0.212 395	−0.210 034	−0.209 322
	0.220 213	0.220 982	0.221 384	0.222 696	0.225 671
△	0.001 996	0.008 045	0.018 403	0.033 698	0.055 017
	0.252 898	0.255 899	0.262 961	0.268 583	0.277 135
	−0.130 178	−0.135 714	−0.143 711	−0.151 149	−0.161 301
5	−0.250 328	−0.249 611	−0.254 121	−0.255 869	−0.259 486
	−0.120 391	−0.117 229	−0.112 467	−0.109 146	−0.105 663
	0.248 007	0.246 656	0.247 338	0.247 582	0.249 316
△	0.002 195	0.008 831	0.020 538	0.037 385	0.060 869
	0.271 279	0.277 440	0.281 119	0.289 587	0.298 854
	−0.058 219	−0.064 560	−0.069 628	−0.075 965	−0.084 156
6	−0.216 925	−0.221 357	−0.223 704	−0.229 815	−0.235 597
	−0.211 782	−0.210 092	−0.207 228	−0.208 449	−0.208 227
	−0.049 203	−0.044 702	−0.041 514	−0.037 732	−0.034 404
	0.264 845	0.263 271	0.260 955	0.262 375	0.263 530
△	0.002 414	0.009 823	0.022 201	0.040 985	0.066 472
	0.282 555	0.286 768	0.293 724	0.303 261	0.313 798
	−0.003 918	−0.008 251	−0.012 613	−0.018 351	−0.024 341
	−0.170 997	−0.174 834	−0.180 021	−0.187 409	−0.194 474
7	−0.222 861	−0.222 596	−0.224 000	−0.227 183	−0.230 504
	−0.163 418	−0.159 647	−0.158 521	−0.155 314	−0.154 531
	0.003 670	0.007 065	0.009 847	0.013 999	0.017 388
	0.274 969	0.271 496	0.271 584	0.270 996	0.272 665
△	0.002 611	0.010 466	0.024 120	0.044 287	0.072 135
	0.290 020	0.295 276	0.304 083	0.312 805	0.324 454
	0.037 713	0.034 609	0.031 638	0.026 806	0.022 548
	−0.126 674	−0.131 162	−0.137 114	−0.143 951	−0.151 635
	−0.205 822	−0.207 994	−0.212 938	−0.215 563	−0.221 210
8	−0.202 252	−0.201 066	−0.202 056	−0.200 824	−0.202 133
	−0.118 340	−0.114 946	−0.112 345	−0.109 036	−0.106 501
	0.043 675	0.046 294	0.049 772	0.052 080	0.055 950
	0.281 681	0.278 990	0.278 960	0.277 683	0.278 527
△	0.002 801	0.011 298	0.026 222	0.047 647	0.077 653
	0.295 707	0.302 588	0.311 436	0.320 148	0.332 448
	0.070 473	0.068 566	0.066 226	0.063 172	0.060 111
	−0.087 358	−0.092 168	−0.098 555	−0.103 854	−0.111 649
	−0.179 584	−0.183 608	−0.188 723	−0.192 732	−0.199 407
9	−0.207 905	−0.209 318	−0.210 930	−0.212 182	−0.215 205
	−0.173 930	−0.172 504	−0.170 619	−0.170 062	−0.168 486
	−0.079 185	−0.076 076	−0.072 323	−0.070 066	−0.066 224
	0.074 885	0.077 309	0.080 113	0.082 449	0.085 296
	0.286 899	0.285 210	0.283 378	0.283 128	0.283 116
△	0.002 999	0.012 206	0.028 103	0.051 236	0.083 220

TABLE 3 (*continued*)
The orthogonal function W

n	$q = 0.6$	$q = 0.7$	$q = 0.8$	$q = 0.9$
	0.178 335	0.183 899	0.191 687	0.203 567
3	−0.356 671	−0.367 800	−0.383 375	−0.407 135
	0.178 336	0.183 901	0.191 688	0.203 568
△	0.068 695	0.099 428	0.141 096	0.201 397
	0.250 037	0.260 556	0.275 140	0.297 059
	−0.270 373	−0.285 711	−0.306 441	−0.336 866
4	−0.209 364	−0.210 247	−0.212 538	−0.217 445
	0.229 700	0.235 402	0.243 839	0.257 252
△	0.083 597	0.122 077	0.175 512	0.255 298
	0.287 974	0.301 956	0.321 041	0.349 844
	−0.173 116	−0.187 834	−0.207 594	−0.236 974
5	−0.264 880	−0.243 421	−0.281 630	−0.296 189
	−0.102 790	−0.100 336	−0.098 120	−0.096 078
	0.252 812	0.258 207	0.266 304	0.279 396
△	0.092 714	0.135 816	0.195 854	0.286 390
	0.311 428	0.327 596	0.349 781	0.383 380
	−0.094 444	−0.106 971	−0.123 977	−0.149 842
	−0.243 614	−0.253 812	−0.267 139	−0.286 300
6	−0.208 948	−0.210 970	−0.214 564	−0.220 682
	−0.030 625	−0.026 717	−0.022 461	−0.017 269
	0.266 203	0.270 875	0.278 358	0.290 713
△	0.101 057	0.147 870	0.213 188	0.311 780
	0.327 388	0.345 133	0.369 468	0.406 476
	−0.032 501	−0.042 656	−0.056 492	−0.078 177
	−0.203 361	−0.214 735	−0.229 491	−0.250 424
7	−0.234 217	−0.239 548	−0.247 154	−0.258 528
	−0.153 321	−0.152 535	−0.152 571	−0.153 806
	0.020 912	0.025 138	0.030 221	0.036 920
	0.275 100	0.279 204	0.286 019	0.297 539
△	0.109 467	0.159 878	0.230 046	0.335 689
	0.338 901	0.357 906	0.383 816	0.423 341
	0.016 667	0.008 877	−0.001 855	−0.019 355
	−0.160 507	−0.171 957	−0.186 860	−0.207 999
	−0.227 103	−0.234 645	−0.244 791	−0.259 405
8	−0.203 837	−0.206 102	−0.209 723	−0.215 765
	−0.104 752	−0.102 839	−0.100 984	−0.099 352
	0.059 650	0.063 746	0.069 019	0.076 358
	0.280 980	0.285 014	0.291 377	0.302 177
△	0.117 899	0.172 064	0.247 041	0.359 407
	0.347 522	0.367 475	0.394 686	0.436 178
	0.056 274	0.050 766	0.042 946	0.029 413
	−0.120 077	−0.130 956	−0.145 270	−0.165 731
	−0.206 641	−0.215 507	−0.227 158	−0.243 669
9	−0.219 140	−0.223 877	−0.230 293	−0.239 845
	−0.168 623	−0.169 284	−0.170 568	−0.173 150
	−0.063 569	−0.061 065	−0.058 341	−0.055 092
	0.089 150	0.093 594	0.098 861	0.106 433
	0.285 106	0.288 853	0.295 137	0.305 464
△	0.126 283	0.184 195	0.264 151	0.383 239

Hence the final fitted equation is

$$\hat{y} = 59.78 + 42.44(x^{0.474} - \overline{x^{0.474}}).$$

In this case then, the data suggest that a convenient transformation would be $x^{\frac{1}{2}}$. The observed response y plotted against $x^{\frac{1}{2}}$ is shown in figure 3b.

The orthogonal tables cannot be used after the first iteration since the independent variable is then not equally spaced. For many applications this "first iteration" should be sufficient. We shall see later that the confidence region appropriate for α is usually rather large.

4.2 *Construction of the Table*

The levels of any equally spaced ordered independent variables can be written

$$x_u = m\left(1 + \frac{2u - (n+1)}{n-1}\,q\right) \qquad (u = 1, 2, \cdots, n)$$

Consequently

$$\begin{aligned} x_u \ln x_u &= m\left(1 + \frac{2u-(n+1)}{n-1}\,q\right) \ln\left\{m\left(1 + \frac{2u-(n+1)}{n-1}\,q\right)\right\} \\ &= m\left(1 + \frac{2u-(n+1)}{n-1}\,q\right) \ln\left(1 + \frac{2u-(n+1)}{n-1}\,q\right) \\ &\qquad + m\left(1 + \frac{2u-(n+1)}{n-1}\,q\right) \ln m. \end{aligned}$$

If

$$w_u = \left(1 + \frac{2u-(n+1)}{n-1}\,q\right) \ln\left(1 + \frac{2u-(n+1)}{n-1}\,q\right),$$

then

$$x_u \ln x_u = m\left(1 + \frac{2u-(n+1)}{n-1}\,q\right) \ln m + m w_u$$

If we write $[v_u]$ for $(v_u - \hat{v}_u)$ where $\hat{v}_u = d_0 + d_1 x_u$, and d_0 and d_1 are the usual regression coefficients, we have $[x_u \ln x_u] = m[w_u]$. It is clear that we can tabulate $[w_u]$ once for all for any desired n and q and calculate c from

$$c = \sum [w_u]y_u/m \sum [w_u]^2.$$

In practice it is found that the quantity $W_u = [w_u]/q^2$ changes only slowly with changing q. It is this quantity $W_u = [w_u]/q^2$ which is tabulated since interpolation is greatly facilitated. Now

$$c = \frac{\sum [x_u \ln x_u]y_u}{\sum [x_u \ln x_u]^2} = \frac{\sum [w_u]y_u}{m \sum [w_u]^2} = \frac{\sum W_u y_u}{mq^2 \sum W_u^2}$$

The divisor Δ shown in the table is $q^2 \sum W_u^2$. Thus to calculate c we simply compute

$$c = \Sigma W_u y_u / m\,\Delta.$$

5. Circumstances Where the Method May Be Appropriately Applied

Experience with this method has shown that while it has always given good results when appropriately used, its unthinking application can sometimes give rise to difficulties. It is clear that when there is no real evidence of curvature then equally there is no evidence that a transformation is needed. The method can still be applied in suitable cases to determine what range of transformations are not contradicted by the data. These ranges are calculated from the confidence limits given in Section 7 or, where this method is not available, from the asymptotic standard errors.

Examples can arise in which in successive iterations the estimate varies wildly. This can occur in particular when q is so small that with widely different values of α curves which agree with the limits of error can be obtained. In many such cases there is no evidence that any such transformation is needed. Reasonably good estimates of the required transformation can only be expected when q is not too small or σ too large.

Particular attention should be paid to rounding error. In some examples it has been found that violent oscillation in successive iterations is caused simply by failing to carry a sufficient number of decimal places.

6. A General Method for Calculating the Required Orthogonal Functions

We now describe a convenient method for obtaining orthogonal functions appropriate to important specific statistical designs.

Suppose we have a $n \times k$ matrix $\mathbf{X} = \{x_{ui}\}$ and an $n \times 1$ column vector $\mathbf{w} = \{w_u\}$. Then the vector $[\mathbf{w}]$ which is the orthogonal part of the vector $\mathbf{w}$ with respect to the matrix $\mathbf{X}$ is

$$[\mathbf{w}] = (\mathbf{I} - \mathbf{X}(\mathbf{X}'\mathbf{X})^{-1}\mathbf{X}')\mathbf{w} = (\mathbf{I} - \mathbf{R})\mathbf{w}.$$

It is worthwhile computing the matrix $(\mathbf{I} - \mathbf{R})$ for specific designs so that the required orthogonal vectors can be easily calculated. Several of these matrices have been computed by Draper (8).

As a further example we consider the fitting of a second degree polynomial in two variables such as

$$\eta = \beta_0 + \beta_1 x_1 + \beta_2 x_2 + \beta_{11} x_1^2 + \beta_{22} x_2^2 + \beta_{12} x_1 x_2$$

using an octagonal rotatable design already discussed in Section 3.4.

For this design the matrix $\mathbf{X}$ is as follows:

Run Number	x_0	x_1	x_2	x_1^2	x_2^2	x_1x_2
1	1	-1	-1	1	1	1
2	1	$-\sqrt{2}$	0	2	0	0
3	1	-1	1	1	1	-1
4	1	0	$\sqrt{2}$	0	2	0
$\mathbf{X} =$ 5	1	1	1	1	1	1
6	1	$\sqrt{2}$	0	2	0	0
7	1	1	-1	1	1	-1
8	1	0	$-\sqrt{2}$	0	2	0
9	1	0	0	0	0	0
10	1	0	0	0	0	0

TABLE 4

Calculation of z_1 and z_2 for two variable rotable design

$\frac{\partial \hat{y}}{\partial T}$	$\frac{\partial \hat{y}}{\partial C}$	z_1	z_2	$[z_1]$	$[z_2]$
1.6959	1.9401	84.291	168.267	−1.973	−1.694
1.6388	0.4759	37.735	70.213	4.208	−0.815
0.7476	−1.4800	37.158	−319.834	−4.187	2.684
−0.4576	−2.7810	−56.942	−683.181	1.927	−2.818
−1.2694	2.6665	−264.592	−576.242	1.250	1.131
−1.2123	−1.2014	−297.812	−177.273	−3.484	1.379
−0.3202	0.7543	−66.742	63.465	3.464	−3.248
0.8842	2.0555	110.027	123.154	−1.203	3.382
0.2133	−0.3627	26.542	−53.518	0	0
0.2133	−0.3627	26.542	−53.518	0	0

Whence for (**I** − **R**) we obtain

$$(\mathbf{I}-\mathbf{R}) = \frac{1}{8}\left[\begin{array}{cccccccc|cc} a & b & c & d & -c & d & c & b & 0 & 0 \\ b & a & b & c & d & -c & d & c & 0 & 0 \\ c & b & a & b & c & d & -c & d & 0 & 0 \\ d & c & b & a & b & c & d & -c & 0 & 0 \\ -c & d & c & b & a & b & c & d & 0 & 0 \\ d & -c & d & c & b & a & b & c & 0 & 0 \\ c & d & -c & d & c & b & a & b & 0 & 0 \\ b & c & d & -c & d & c & b & a & 0 & 0 \\ \hline 0 & 0 & 0 & 0 & 0 & 0 & 0 & 0 & 4 & -4 \\ 0 & 0 & 0 & 0 & 0 & 0 & 0 & 0 & -4 & 4 \end{array}\right]$$

$$a = 3 \qquad c = 1$$
$$b = -(\sqrt{2}+1) \qquad d = \sqrt{2}-1$$

As might be expected the matrix formed by the first eight rows and columns of (**I** − **R**) is a circulant (1) and consequently the portion of any variable **z** orthogonal to the columns of the corresponding **X** matrix is readily obtained by a simple repetitive routine.

The method is illustrated with the data shown in Table 4, where the constructed variables $\mathbf{z}_1$ and $\mathbf{z}_2$ are tabulated. The vectors $[\mathbf{z}_1]$ and $[\mathbf{z}_2]$ are calculated using the coefficients in (**I** − **R**). For example

$$\begin{aligned}[z_{11}] = \tfrac{1}{8}\{&3(84.291) - (\sqrt{2}+1)(37.735) + 1(37.158) \\ &+ (\sqrt{2}-1)(-56.942) - 1(-264.592) \\ &+ (\sqrt{2}-1)(-297.812) + 1(-66.742) \\ &- (\sqrt{2}+1)(110.027) + 0(26.542) + 0(26.542)\} \\ = -1.973&\end{aligned}$$

$$\begin{aligned}[z_{12}] = \tfrac{1}{8}\{&-(\sqrt{2}+1)(84.291) + 3(37.735) - (\sqrt{2}+1)(37.158) \\ &+ 1(-56.942) + (\sqrt{2}-1)(-264.592) \\ &- 1(-297.812) + (\sqrt{2}-1)(-66.742) \\ &+ 1(110.027) + 0(26.542) + 0(26.542)\} \\ = 4.208&\end{aligned}$$

7. Confidence Intervals for the Parameter of the Transformation

In practice we require measures of precision for the estimates of the parameters of the transformations. Suppose that in a problem which involves p variables, the constructed orthogonal variables $[z_1], [z_2], \cdots, [z_p]$ have been obtained where

$$z_i = \frac{\partial \hat{y}}{\partial x_i} x_i \ln x_i$$

and the squared brackets refer as usual to the appropriate orthogonal functions. Denote by $\mathbf{Z}$ the $n \times p$ matrix which has the elements $[z_i]$ for its ith column. Consider the matrix $(\mathbf{Z}'\mathbf{Z})^{-1}\sigma^2$. Adopting the usual argument associated with the method of maximum likelihood (10), the elements of this matrix will tend, i.e. for large n, to the variances and covariances for the $\hat{\alpha}$'s. However for moderate n these estimates will be imprecise, since they ignore the fact that the z_i's themselves contain estimated regression coefficients.

In the particular case of the fitting of the transformed linear function, this difficulty may be avoided by the use of Fieller's theorem (9), which makes an appropriate allowance for the uncertainty arising from the inclusion of estimates in the variable z_i.

7.1 *The Linear Case*

Consider the linear case in which there is only one variable. Consider the estimates

$$b = \frac{\sum (x - \bar{x})y}{\sum (x - \bar{x})^2} \quad \text{and} \quad c = \frac{\sum [w]y}{m \sum [w]^2}$$

where

$$\sigma_b^2 = \frac{\sigma^2}{\sum (x - \bar{x})^2} \quad \text{and} \quad \sigma_c^2 = \frac{\sigma^2}{m \sum [w]^2}.$$

At a given stage in the iteration let

$$E(b) = \beta \quad \text{and} \quad E(c) = \gamma \quad \text{and let} \quad \frac{\gamma}{\beta} = \phi.$$

Then $E(\phi b - c) = 0$, and $V(\phi b - c) = \phi^2\sigma_b^2 + \sigma_c^2$. Then using Fieller's theorem, the $100(1 - \pi)\%$ confidence limits for ϕ are given by

$$\left\{\frac{bc}{b^2 - \chi_\pi^2\sigma_b^2} \pm \chi_\pi \sqrt{\frac{\sigma_c^2}{b^2 - \chi_\pi^2\sigma_b^2} + \frac{c^2\sigma_b^2}{(b^2 - \chi_\pi^2\sigma_b^2)^2}}\right\}$$

If we iterate until the process converges the $c = 0$ and the confidence limits for ϕ are given by

$$\pm\chi_\pi \sqrt{\frac{\sigma_c^2}{b^2 - \chi_\pi^2\sigma_b^2}}.$$

This expression is to be compared with that given by the direct likelihood approach, which provides the limits $\chi_\pi\sigma_c/b$. We notice that the limits given by Fieller's theorem are widened by an amount which depends on σ_b^2.

Alternatively if an estimate s^2 of σ^2 based on ν degrees of freedom and distributed independently of b and c is available, then $\nu s^2/\sigma^2$ has a χ^2 distribution with ν degrees of freedom, and we obtain for the confidence limits of ϕ

$$\left\{\frac{bc}{b^2 - t_{\nu\pi}^2 s_b^2} \pm t_{\nu\pi}\sqrt{\frac{s_c^2}{b^2 - t_{\nu\pi}^2 s_b^2}\frac{c^2 s_b^2}{(b^2 - t_{\nu\pi}^2 s_b^2)^2}}\right\}$$

When the process converges, the limits are therefore

$$\pm t_{\nu\pi}\sqrt{\frac{s_c^2}{b^2 - t_{\nu\pi}^2 s_b^2}}$$

Hence at the last iteration, the above limits are the $100(1 - \pi)\%$ confidence limits for $\alpha - 1$.

We illustrate the use of the limits for the example in Section 5.1 above.

Suppose the five observations in the example were in fact averages of two observations as follow.

x	$\bar{y}$	y_1	y_2
0.5	28.45	27.86	29.04
1.5	47.78	47.70	47.86
2.5	64.35	65.78	62.92
3.5	74.67	74.18	75.16
4.5	83.65	82.83	84.47

$$s_y^2 = 1.324, \qquad s_{\bar{y}}^2 = 0.662$$

Then the $100(1 - \pi)\%$ confidence limits for $\phi = \alpha - 1$ on the last iteration are

$$\pm t_{\nu\pi}\sqrt{\frac{\dfrac{0.662}{0.023945}}{(42.4)^2 - t_{\nu\pi}^2\,\dfrac{0.662}{1.073824}}}.$$

Choosing $\pi = 0.05$, then $t_{\nu\pi} = 2.571$ and the 95% confidence limits for α are 0.474 ± 0.319.

8. Adequacy of the Transformation

Suppose experiments have been performed at p distinct sets of conditions, certain of which have been replicated so that in all there are $n = p + r$ observations, and suppose that s transformation parameters have been estimated. Suppose finally that there are f constants in the untransformed model. Then after the final iteration we can calculate the value of $\hat{y}$ for each experimental run, where $\hat{y}$ represents the estimated value using the finally transformed response function. We may now calculate a residual sum of squares $S_r = \sum (y - \hat{y})^2$. If the model fits then approximately this residual sum of squares would be distributed as $\chi^2\sigma^2$ with $p + r - f - s$ degrees of freedom. Furthermore, a sum of squares S_e can be isolated having r degrees of freedom which measures the sum of squares of deviations of the observations from the p means at the distinct experimental runs, and which is distributed on the usual assumptions as $\chi^2\sigma^2$ with r degrees of freedom.

We may now obtain by subtraction a sum of squares $S_f = S_r - S_e$ which is, adopting the usual linear approximation, distributed as $\chi^2\sigma^2$ with $p - f - s$ degrees of freedom independently of S_e. When there is lack of fit, S_f and hence

S_f, will become inflated, but S_e will remain unaffected. Consequently an approximate test statistic for the inadequacy of our fitted transformed function is supplied by the ratio

$$\frac{r}{(p - f - s)} \frac{S_f}{S_e} = \frac{r(S_r - S_e)}{(p - f - s)S_e}$$

which, when the fit is perfect, is approximately distributed as F with $p - f - s$ and r degrees of freedom.

For example, in the example given in Section 5.1 and 7.1 above, the analysis of variance after the transformation technique is applied is the following.

Source	Sum of Squares	Degrees of Freedom	Mean Squares	
Total	39 618.9934	$p + r = 10$		
Regression	39 605.4044	$f + s = 3$		
Lack of Fit	6.9652	$p - f - s = 2$	3.4876	
Error	6.6238	$r = 5$	1.3248	$F = 2.63$

The analysis of variance before the transformation technique was applied is as follows

Source	Sum of Squares	Degrees of Freedom	Mean Squares	
Total	39 618.9934	$p + r = 10$		
Regression	39 506.1928	$f = 2$		
Lack of Fit	106.1768	$p - f = 3$	35.3923	
Error	6.6238	$r = 5$	1.3248	$F = 26.8$

The 20%, 5% and 1% F ratios for 2 and 5 degrees of freedom are respectively 2.26, 5.79 and 13.27, so that whereas there is strong reason to doubt the adequacy of the linear fit, after transformation the agreement is satisfactory.

Bibliography

(1) Aitken, A. A. (1959) *Determinants and Matrices.* Interscience Publishers, Incorporated, New York.

(2) Bartlett, M. S. (1947) The Use of Transformations. *Biometrics 3* 39–57.

(3) Beale, E. M. L. (1960) Confidence Regions in Non-Linear Estimation. *Jour. Roy. Stat. Soc. B 22* 41–76.

(4) Box, G. E. P. (1957) Some Notes on Non-Linear Estimation (S.T.R.A. Report Princeton University).

(5) Box, G. E. P. and Coutie, G. A. (1956) Application of Digital Computers in the Exploration of Functional Relationships. *Proceedings of the Institution of Electrical Engineers* Vol. 103 Part B Supplement No. 1 100–107.

(6) Box, G. E. P. and Hunter, J. S. (1957) Multifactor Experimental Designs for Exploring Response Surfaces. *Annals of Mathematical Statistics 28* 195–241.

(7) Box, G. E. P. and Wilson, K. B. (1951) On the Attainment of Optimum Conditions. *Jour. Roy. Stat. Soc. B 13* 1–38.

(8) Draper, N. R. (1960) *Missing Value Formulae for Certain Three Factor, Second Order Response Surface Designs.* Mathematics Research Center, U. S. Army, The University of Wisconsin, Technical Report No. 201.

(9) Fieller, E. C. (1940) The Biological Standardisation of Insulin. *Jour. Roy. Stat. Soc. Supplement 7* 1–53.

(10) Fisher, R. A. (1922) On the Mathematical Foundations of Theoretical Statistics. *Phil-*

osophical Transactions of the Royal Society of London, Series A, Vol. 222 309–368 (reprinted by John Wiley and Sons, New York, 1950).

(11) FISHER, R. A. (1960) *The Design of Experiments*. Hafner Publishing Company, Incorporated, New York.

(12) KEMPTHORNE, O. (1952) *The Design and Analysis of Experiments*. John Wiley and Sons, Incorporated, New York.

(13) TURNER, M. E., MONROE, R. J. and LUCAS, H. L. (1961) Generalized Asymptotic Regression and Non-Linear Path Analysis. *Biometrics 17* 120–143.

(14) TUKEY, J. W. (1949) One Degree of Freedom for Non-Additivity. *Biometrics 5* 232–242.

4.9
An Analysis of Transformations

G. E. P. BOX
University of Wisconsin

D. R. COX
Birkbeck College, University of London

[Read at a RESEARCH METHODS MEETING of the SOCIETY, April 8th, 1964, Professor D. V. LINDLEY in the Chair]

SUMMARY

In the analysis of data it is often assumed that observations $y_1, y_2, \ldots, y_n$ are independently normally distributed with constant variance and with expectations specified by a model linear in a set of parameters $\boldsymbol{\theta}$. In this paper we make the less restrictive assumption that such a normal, homoscedastic, linear model is appropriate after some suitable transformation has been applied to the y's. Inferences about the transformation and about the parameters of the linear model are made by computing the likelihood function and the relevant posterior distribution. The contributions of normality, homoscedasticity and additivity to the transformation are separated. The relation of the present methods to earlier procedures for finding transformations is discussed. The methods are illustrated with examples.

1. INTRODUCTION

THE usual techniques for the analysis of linear models as exemplified by the analysis of variance and by multiple regression analysis are usually justified by assuming

(i) simplicity of structure for $E(y)$;
(ii) constancy of error variance;
(iii) normality of distributions;
(iv) independence of observations.

In analysis of variance applications a very important example of (i) is the assumption of additivity, i.e. absence of interaction. For example, in a two-way table it may be possible to represent $E(y)$ by additive constants associated with rows and columns.

If the assumptions (i)–(iii) are not satisfied in terms of the original observations, y, a non-linear transformation of y may improve matters. With this in mind, numerous special transformations for use in the analysis of variance have been examined in the literature; see, in particular, Bartlett (1947). The main emphasis in these studies has tended to be on obtaining a constant error variance, especially when the variance of y is a known function of the mean, as with binomial and Poisson variates.

In multiple regression problems, and in particular in the analysis of response surfaces, assumption (i) might be that $E(y)$ is adequately represented by a rather simple empirical function of the independent variables $x_1, x_2, \ldots, x_l$ and we would want to transform so that this assumption, together with assumptions (ii) and (iii), is approximately satisfied. In some cases transformation of independent as well as of dependent variables might be desirable to produce the simplest possible regression model in the transformed variables. In all cases we are concerned not merely to find a transformation which will justify assumptions but rather to find, where possible, a metric in terms of which the findings may be succinctly expressed.

The Collected Works of George E. P. Box, 1984, Wadsworth, Inc., Belmont, CA 94002.
Originally published in *J. Roy. Stat. Soc.,* Series B, vol. 26, no. 2 (1964), pp. 211–252.

Each of the considerations (i)–(iii) can, and has been, used separately to select a suitable candidate from a parametric family of transformations. For example, to achieve additivity in the analysis of variance, selection might be based on

(a) minimization of the F value for the degree of freedom for non-additivity (Tukey, 1949); or

(b) minimization of the F ratio for interaction versus error; or

(c) maximization of the F ratio for treatments versus error (Tukey, 1950).

Tukey and Moore (1954) used method (a) in a numerical example, plotting contours of F against (λ_1, λ_2) for transformations in the family $(y+\lambda_2)^{\lambda_1}$. They found that in their particular example the minimizing values were very imprecisely determined.

In both (a) and (b) the general object is to look for a scale on which effects are additive, i.e. to see whether an apparent interaction is removable by a transformation. Of course, only a particular type of interaction is so removable. Whereas (a) can be applied, for example, to a two-way classification without replication, method (b) requires the availability of an error term separated from the interaction term. Thus, if applied to a two-way classification, method (b) could only be used when there was some replication within cells. Finally, method (c) can be used even in a one-way analysis to find the scale on which treatment effects are in some sense most sensitively expressed. In particular, Tukey (1950) suggested multivariate canonical analysis of (y, y^2) to find the linear combination $y+\lambda y^2$ most sensitive to treatment effects. Incidentally, care is necessary in using $y+\lambda y^2$ over the wide ranges commonly encountered with data being considered for transformation, for such a transformation is sensible only so long as the value of λ and the values of y are such that the transformation is monotonic.

For transformation to stabilize variance, the usual method (Bartlett, 1947) is to determine empirically or theoretically the relation between variance and mean. An adequate empirical relation may often be found by plotting log of the within-cell variance against log of the cell mean. Another method would be to choose a transformation, within a restricted family, to minimize some measure of the heterogeneity of variance, such as Bartlett's criterion. We are grateful to a referee for pointing out also the paper of Kleczkowski (1949) in which, in particular, approximate fiducial limits for the parameter λ in the transformation of y to $\log(y+\lambda)$ are obtained. The method is to compute fiducial limits for the parameters in the linear relation observed to hold when the within-cell standard deviation is regressed on the cell mean.

Finally, while there is much work on transforming a single distribution to normality, constructive methods of finding transformations to produce normality in analysis of variance problems do not seem to have been considered.

While Anscombe (1961) and Anscombe and Tukey (1963) have employed the analysis of residuals as a means of detecting departures from the standard assumptions, they have also indicated how transformations might be constructed from certain functions of the residuals.

In regression problems, where both dependent and independent variables can be transformed, there are more possibilities to be considered. Transformation of the independent variables (Box and Tidwell, 1962) can be applied without affecting the constancy of variance and normality of error distributions. An important application is to convert a monotonic non-linear regression relation into a linear one. Obviously it is useless to try to linearize a relation which is not monotonic, but a transformation is sometimes useful in such cases, for example, to make a regression relation more nearly quadratic around its maximum.

2. GENERAL REMARKS ON TRANSFORMATIONS

The main emphasis in this paper is on transformations of the dependent variable. The general idea is to restrict attention to transformations indexed by unknown parameters λ, and then to estimate λ and the other parameters of the model by standard methods of inference. Usually λ will be a one-, or at most two-, dimensional parameter, although there is no restriction in principle. Our procedure then leads to an interesting synthesis of the procedures reviewed in Section 1. It is convenient to make first a few general points about transformations.

First, we can distinguish between analyses in which either (a) the particular transformation, λ, is of direct interest, the detailed study of the factor effects, etc., being of secondary concern; or (b) the main interest is in the factor effects, the choice of λ being only a preliminary step. Type (b) is likely to be much the more common. Nevertheless, (a) can arise, for example, in the analysis of a preliminary set of data. Or, again, we may have two factors, A and B, whose main effects are broadly understood, it being required to study the λ, if any, for which there is no interaction between the factors. Here the primary interest is in λ. In case (b), however, we shall need to fix one, or possibly a small number, of λ's and go ahead with the detailed estimation and interpretation of the factor effects on this particular transformed scale. We shall choose λ partly in the light of the information provided by the data and partly from general considerations of simplicity, ease of interpretation, etc. For instance, it would be quite possible for the formal analysis to show that say $\sqrt{y}$ is the best scale for normality and constancy of variance, but for us to decide that there are compelling arguments of ease of interpretation for working say with $\log y$. The formal analysis will warn us, however, that changes of variance and non-normality may need attention in a refined and efficient analysis of $\log y$. That is, the method developed below for finding a transformation is useful as a guide, but is, of course, not to be followed blindly. In Section 7 we discuss briefly some of the consequences of interpreting factor effects on a scale chosen in the light of the data.

In regression studies, it is sometimes necessary to take an entirely empirical approach to the choice of a relation. In other cases, physical laws, dimensional analysis, etc., may suggest a particular functional form. Thus, in a study of a chemical system one would expect reaction rate to be proportional to some power of the concentration and to the antilog of the reciprocal of absolute temperature. Again, in many fields of technology relationships of the form

$$y \propto x_1^{\beta 1} \ldots x_l^{\beta l}$$

are very common, suggesting a log transformation of all variables. In such cases the reasonable thing will often be first to apply the transformations suggested by the prior reasoning, and after that consider what further modifications, if any, are needed. Finally, we may know the behaviour of y when the independent variables x_i tend to zero or infinity, and certainly, if we are hopeful that the model might apply over a wide range, we should consider models that are consistent with such limiting properties of the system.

We can distinguish broadly two types of dependent variable, extensive and non-extensive. The former have a relevant property of physical additivity, the latter not. Thus yield of product per batch is extensive. The failure time of a component would be considered extensive if components are replaced on failure, the main thing of interest being the number of components used in a long time. Properties like temperature, viscosity, quality of product, etc., are not extensive. In the absence of the sort of prior consideration mentioned in the previous paragraph there is no reason to prefer the initial form of a non-extensive variable to any monotonic function of it. Hence, transformations can be applied freely to non-extensive variables. For extensive variables, however, the population mean of y is the parameter determining the long-run behaviour of the system. Thus in the two examples mentioned above, the total yield of product in a long period and the total number of components used in a very long time are determined respectively by the population mean of yield per batch and the mean failure time per component, irrespective of distributional form.

In a narrowly technological sense, therefore, we are interested in the population mean of y, not of some function of y. Hence we either analyse linearly the untransformed data or, if we do apply a transformation in order to make a more efficient and valid analysis, we convert the conclusions back to the original scale. Even in circumstances where, for immediate application, the original scale y is required, it may be better to think in terms of transformed values in which, say, interactions have been removed.

In general, we can regard the usual formal linear models as doing two things:

(a) specifying the questions to be asked, by defining explicitly the parameters which it is the main object of the analysis to estimate;

(b) specifying assumptions under which the above parameters can be simply and effectively estimated.

If there should be conflict between the requirements for (a) and for (b), it is best to pay most attention to (a), since approximate inference about the most meaningful parameters is clearly preferable to formally "exact" inference about parameters whose definition is in some way artificial. Therefore in selecting a transformation we might often give first attention to simplicity of the model structure, for example to additivity in the analysis of variance. This allows simplicity of description and also the main effect of a factor A, measured on a scale for which there appears to be no interaction with a factor B, often has a reasonable possibility of being valid for levels of B outside those of the initial experiment.

3. Transformation of the Dependent Variable

We work with a parametric family of transformations from y to $y^{(\lambda)}$, the parameter λ, possibly a vector, defining a particular transformation. Two important examples considered here are

$$y^{(\lambda)} = \begin{cases} \dfrac{y^{\lambda}-1}{\lambda} & (\lambda \neq 0), \\ \log y & (\lambda = 0), \end{cases} \tag{1}$$

and

$$y^{(\lambda)} = \begin{cases} \dfrac{(y+\lambda_2)^{\lambda_1}-1}{\lambda_1} & (\lambda_1 \neq 0), \\ \log(y+\lambda_2) & (\lambda_1 = 0). \end{cases} \tag{2}$$

The transformations (1) hold for $y > 0$ and (2) for $y > -\lambda_2$. Note that since an analysis of variance is unchanged by a linear transformation (1) is equivalent to

$$y^{(\lambda)} = \begin{cases} y^{\lambda} & (\lambda \neq 0), \\ \log y & (\lambda = 0); \end{cases} \tag{3}$$

the form (1) is slightly preferable for theoretical analysis because it is continuous at $\lambda = 0$. In general, it is assumed that for each λ, $y^{(\lambda)}$ is a monotonic function of y over the admissible range. Suppose that we observe an $n \times 1$ vector of observations $\mathbf{y} = \{y_1, \ldots, y_n\}$, and that the appropriate linear model for the problem is specified by

$$E\{\mathbf{y}^{(\lambda)}\} = \mathbf{a\theta}, \tag{4}$$

where $\mathbf{y}^{(\lambda)}$ is the column vector of *transformed* observations, $\mathbf{a}$ is a known matrix and $\boldsymbol{\theta}$ a vector of unknown parameters associated with the transformed observations.

We now assume that for some unknown λ, the transformed observations $y_i^{(\lambda)}$ $(i = 1, \ldots, n)$ satisfy the full normal theory assumptions, i.e. are independently normally distributed with constant variance σ^2, and with expectations (4). The probability density for the untransformed observations, and hence the likelihood *in relation to these original observations*, is obtained by multiplying the normal density by the Jacobian of the transformation.

The likelihood in relation to the original observations $\mathbf{y}$ is thus

$$\frac{1}{(2\pi)^{\frac{1}{2}n}\,\sigma^n} \exp\left\{-\frac{(\mathbf{y}^{(\lambda)} - \mathbf{a\theta})'(\mathbf{y}^{(\lambda)} - \mathbf{a\theta})}{2\sigma^2}\right\} J(\lambda; \mathbf{y}), \tag{5}$$

where

$$J(\lambda; \mathbf{y}) = \prod_{i=1}^{n} \left| \frac{dy_i^{(\lambda)}}{dy_i} \right|.$$

We shall examine two ways in which inferences about the parameters in (5) can be made. In the first, we apply "orthodox" large-sample maximum-likelihood theory to (5). This approach leads directly to point estimates of the parameters and to approximate tests and confidence intervals based on the chi-squared distribution. In the second approach, via Bayes's theorem, we assume that the prior distributions of the θ's and $\log \sigma$ can be taken as essentially uniform over the region in which the likelihood is appreciable and we integrate over the parameters to obtain a posterior distribution for λ; for general discussion of this approach, see, in particular, Jeffreys (1961).

We find the maximum-likelihood estimates in two steps. First, for given λ, (5) is, except for a constant factor, the likelihood for a standard least-squares problem. Hence the maximum-likelihood estimates of the θ's are the least-squares estimates for the dependent variable $y^{(\lambda)}$ and the estimate of σ^2, denoted for fixed λ by $\hat{\sigma}^2(\lambda)$, is

$$\hat{\sigma}^2(\lambda) = \mathbf{y}^{(\lambda)\prime}\mathbf{a}_r\,\mathbf{y}^{(\lambda)}/n = S(\lambda)/n \tag{6}$$

where, when $\mathbf{a}$ is of full rank,

$$\mathbf{a}_r = \mathbf{I} - \mathbf{a}(\mathbf{a}'\mathbf{a})^{-1}\mathbf{a}', \tag{7}$$

and $S(\lambda)$ is the residual sum of squares in the analysis of variance of $y^{(\lambda)}$.

Thus for fixed λ, the maximized log likelihood is, except for a constant,

$$L_{\max}(\lambda) = -\tfrac{1}{2}n \log \hat{\sigma}^2(\lambda) + \log J(\lambda; \mathbf{y}). \tag{8}$$

In the important special case (1) of the simple power transformation, the second term in (8) is

$$(\lambda - 1)\Sigma \log y_i. \tag{9}$$

In (2), when an unknown origin λ_2 is included, the term becomes

$$(\lambda_1 - 1)\Sigma \log(y_i + \lambda_2). \tag{10}$$

It will now be informative to plot the maximized log likelihood $L_{\max}(\lambda)$ against λ for a trial series of values. From this plot the maximizing value $\hat{\lambda}$ may be read off and we can obtain an approximate $100(1-\alpha)$ per cent confidence region from

$$L_{\max}(\hat{\lambda})-L_{\max}(\lambda)<\tfrac{1}{2}\chi^2_{\nu_\lambda}(\alpha), \tag{11}$$

where ν_λ is the number of independent components in λ. The main arithmetic consists in doing the analysis of variance of $\mathbf{y}^{(\lambda)}$ for each chosen λ.

If it were ever desired to determine $\hat{\lambda}$ more precisely this could be done by determining numerically the value $\hat{\lambda}$ for which the derivatives with respect to λ are all zero. In the special case of the one parameter power transformation $y^{(\lambda)}=(y^\lambda-1)/\lambda$,

$$\frac{d}{d\lambda}L_{\max}(\lambda)=-n\frac{\mathbf{y}^{(\lambda)\prime}\mathbf{a}_r\mathbf{u}^{(\lambda)}}{\mathbf{y}^{(\lambda)\prime}\mathbf{a}_r\mathbf{y}^{(\lambda)}}+\frac{n}{\lambda}+\Sigma\log y_i, \tag{12}$$

where $\mathbf{u}^{(\lambda)}$ is the vector of components $\{\lambda^{-1}y_i^\lambda\log y_i\}$. The numerator in (12) is the residual sum of products in the analysis of covariance of $\mathbf{y}^{(\lambda)}$ and $\mathbf{u}^{(\lambda)}$.

The above results can be expressed very simply if we work with the normalized transformation

$$\mathbf{z}^{(\lambda)}=\mathbf{y}^{(\lambda)}/J^{1/n},$$

where $J=J(\lambda;\mathbf{y})$. Then

$$L_{\max}(\lambda)=-\tfrac{1}{2}n\log\hat{\sigma}^2(\lambda;\mathbf{z}),$$

where

$$\hat{\sigma}^2(\lambda;\mathbf{z})=\frac{\mathbf{z}^{(\lambda)\prime}\mathbf{a}_r\mathbf{z}^{(\lambda)}}{n}=\frac{S(\lambda;\mathbf{z})}{n},$$

where $S(\lambda;\mathbf{z})$ is the residual sum of squares of $\mathbf{z}^{(\lambda)}$. The maximized likelihood is thus proportional to $\{S(\lambda;\mathbf{z})\}^{-n}$ and the maximum-likelihood estimate is obtained by minimizing $S(\lambda;\mathbf{z})$ with respect to λ.

For the simple power transformation

$$z^{(\lambda)}=\frac{y^\lambda-1}{\lambda\dot{y}^{\lambda-1}},$$

where $\dot{y}$ is the geometric mean of the observations.

For the power transformation with shifted location

$$z^{(\lambda)}=\frac{(y+\lambda_2)^{\lambda_1}-1}{\lambda_1\{\mathrm{gm}\,(y+\lambda_2)\}^{\lambda_1-1}},$$

where $\mathrm{gm}\,(y+\lambda_2)$ is the sample geometric mean of the $(y+\lambda_2)$'s.

Consider now the corresponding Bayesian analysis. Let the degrees of freedom for residual be $\nu_r=n-\mathrm{rank}\,(\mathbf{a})$, and let

$$s^2(\lambda)=\frac{\mathbf{y}^{(\lambda)\prime}\mathbf{a}_r\mathbf{y}^{(\lambda)}}{\nu_r}=\frac{S(\lambda)}{\nu_r} \tag{13}$$

be the residual mean square in the analysis of variance of $\mathbf{y}^{(\lambda)}$; note the distinction between $\hat{\sigma}^2(\lambda)$, the maximum-likelihood estimate with divisor n, and $s^2(\lambda)$ the "usual" estimate, with divisor the degrees of freedom ν_r. We first rewrite the likelihood (5), i.e. the conditional probability density function of the y's given $\boldsymbol{\theta}$, σ^2, λ, in the form

$$p(\mathbf{y}\,|\,\boldsymbol{\theta},\sigma^2,\lambda) = \frac{1}{(2\pi)^{\frac{1}{2}n}\,\sigma^n}\exp\left\{-\frac{\nu_r s^2(\lambda)+(\boldsymbol{\theta}-\hat{\boldsymbol{\theta}}_\lambda)'\mathbf{a}'\mathbf{a}(\boldsymbol{\theta}-\hat{\boldsymbol{\theta}}_\lambda)}{2\sigma^2}\right\}J(\lambda;\,\mathbf{y}), \qquad (14)$$

where $\hat{\boldsymbol{\theta}}_\lambda$ is the least-squares estimate of $\boldsymbol{\theta}$ for given λ.

Now consider the choice of the joint prior distribution for the unknown parameters. We first parametrize so that the θ's are linearly independent and hence $n-\nu_r$ in number. Let $p_0(\lambda)$ denote the marginal prior density of λ. We assume that it is reasonable, when making inferences about λ, to take the conditional prior distribution of the θ's and $\log\sigma$, given λ, to be effectively uniform over the range for which the likelihood is appreciable. That is, the conditional prior element given λ is

$$g(\lambda)\,d\boldsymbol{\theta}_\lambda\,d(\log\sigma_\lambda), \qquad (15)$$

where, for definiteness, we for the moment denote the effects and variance measured in terms of $y^{(\lambda)}$ by a suffix λ. The factor $g(\lambda)$ is included because the general size and range of the transformed observations $y^{(\lambda)}$ may depend strongly on λ. If the conditional prior distribution (15) were assumed independent of λ, nonsensical results would be obtained.

To determine $g(\lambda)$ we argue as follows. Fix a standard reference value of λ, say λ_1. Suppose provisionally that, for fixed λ, the relation between $y^{(\lambda)}$ and $y^{(\lambda_1)}$ over the range of the observations is effectively linear, say

$$y^{(\lambda)} = \text{const}+l_\lambda\,y^{(\lambda_1)}. \qquad (16)$$

We can then choose $g(\lambda)$ so that when (16) holds, the conditional prior distributions (15) are consistent with one another for different values of λ. In fact, we shall need to apply the answer when the transformations are appreciably non-linear, so that (16) does not hold. There may be a better approach to the choice of a prior distribution than the present one.

It follows from (16) that

$$\log\sigma_\lambda^2 = \text{const}+\log\sigma_{\lambda_1}^2 \qquad (17)$$

and hence, to this order, the prior density of σ_λ^2 is independent of λ. However, the θ_λ's are linear combinations of the expected values of the $y^{(\lambda)}$'s, so that

$$\frac{d\theta_\lambda}{d\theta_{\lambda_1}} = l_\lambda.$$

Since there are $n-\nu_r$ independent components to $\boldsymbol{\theta}$, it follows that $g(\lambda)$ is proportional to $1/l_\lambda^{n-\nu_r}$.

Finally we need to choose l_λ. In passing from λ_1 to λ, a small element of volume of the n dimensional sample space is multiplied by $J(\lambda;\,\mathbf{y})/J(\lambda_1;\,\mathbf{y})$. An average scale change for a single y component is the nth root of this and, since λ_1 is only a standard reference value, we have approximately

$$l_\lambda = \{J(\lambda;\,\mathbf{y})\}^{1/n}. \qquad (18)$$

Thus, approximately, the conditional prior density (15) is

$$\frac{d\boldsymbol{\theta}_\lambda\,d(\log\sigma_\lambda)}{\{J(\lambda;\,\mathbf{y})\}^{(n-\nu_r)/n}}.$$

The combined prior element of probability is thus

$$\frac{d\boldsymbol{\theta}\, d(\log\sigma)}{\{J(\lambda;\, \mathbf{y})\}^{(n-\nu_r)/n}} p_0(\lambda)\, d\lambda, \tag{19}$$

where we now suppress the suffix λ on $\boldsymbol{\theta}$ and σ.

This is only an approximate result. In particular, the choice of (18) is somewhat arbitrary. However, when a useful amount of information is actually available from the data about the transformation, the likelihood will dominate and the exact choice of (19) is not critical. The prior distribution (19) is interesting in that the observations enter the approximate standardizing coefficient $J(\lambda;\, \mathbf{y})$.

We now have the likelihood (14) and the prior density (19) and can apply Bayes's theorem to obtain the marginal posterior distribution of λ in the form

$$K'_y \frac{I(\lambda \mid y)\, p_0(\lambda)}{\{J(\lambda;\, y)\}^{(n-\nu_r)/n}}, \tag{20}$$

where K'_y is a normalizing constant independent of λ, chosen so that (20) integrates to one with respect to λ, and

$$I(\lambda \mid y) = \int_{-\infty}^{\infty} d(\log\sigma) \int_{-\infty}^{\infty} d\boldsymbol{\theta}\, p(\mathbf{y} \mid \boldsymbol{\theta}, \sigma^2, \lambda). \tag{21}$$

The integral (21) can be evaluated to give

$$I(\lambda \mid y) = \frac{|\mathbf{a}'\mathbf{a}|^{-\frac{1}{2}}\, 2^{\frac{1}{2}\nu_r}\, \Gamma(\frac{1}{2}\nu_r)}{(2\pi)^{\frac{1}{2}\nu_r} \{s^2(\lambda)\}^{\frac{1}{2}\nu_r}\, \nu_r^{\frac{1}{2}\nu_r}} J(\lambda;\, y).$$

Substituting into (20), we have that the posterior distribution of λ is

$$K_y \frac{\{J(\lambda;\, \mathbf{y})\}^{\nu_r/n}}{\{s^2(\lambda)\}^{\frac{1}{2}\nu_r}} p_0(\lambda),$$

where K_y is a normalizing constant independent of λ.

Thus the contribution of the observations to the posterior distribution of λ is represented by the factor

$$\{J(\lambda;\, y)\}^{\nu_r/n} / \{s^2(\lambda)\}^{\frac{1}{2}\nu_r}$$

or, on a log scale, by the addition of a term

$$L_b(\lambda) = -\tfrac{1}{2}\nu_r \log s^2(\lambda) + (\nu_r/n) \log J(\lambda;\, y) \tag{22}$$

to $\log p_0(\lambda)$.

Once again if we work with the normalized transformation $z^{(\lambda)} = y^{(\lambda)}/J^{1/n}$, the result is expressed with great simplicity, for

$$L_b(\lambda) = -\tfrac{1}{2}\nu_r \log s^2(\lambda;\, \mathbf{z}) \tag{23}$$

and the posterior density is

$$p(\lambda) = \text{const} \times p_0(\lambda) \times \{S(\lambda;\, \mathbf{z})\}^{-\frac{1}{2}\nu_r}.$$

In practice we can plot $\{S(\lambda; \mathbf{z})\}^{-\frac{1}{2}\nu_r}$ against λ, combining it with any prior information about λ. When the prior density of λ can be taken as locally uniform, the posterior distribution is obtained directly by plotting

$$p_u(\lambda) = k\{S(\lambda; \mathbf{z})\}^{-\frac{1}{2}\nu_r}, \tag{24}$$

where k is chosen to make the total area under the curve unity.

We normally end by selecting a value of λ in the light both of this plot and of other relevant considerations discussed in Section 2. We then proceed to a standard analysis using the indicated transformation.

The maximized log likelihood and the log of the contribution to the posterior distribution of λ may be written respectively as

$$L_{\max}(\lambda) = -\tfrac{1}{2}n\log\{S(\lambda; \mathbf{z})/n\}, \quad L_b(\lambda) = -\tfrac{1}{2}\nu_r\log\{S(\lambda; \mathbf{z})/\nu_r\}.$$

They differ only by substitution of ν_r for n. They are both monotonic functions of $S(\lambda; \mathbf{z})$ and their maxima both occur when the sum of squares $S(\lambda; \mathbf{z})$ is minimized. For general description, $L_{\max}(\lambda)$ and $L_b(\lambda)$ are substantially equivalent. However, it can easily happen that ν_r/n is appreciably less than one, even when n is quite large. Therefore, in applications, the difference cannot always be ignored, especially when a number of models are simultaneously considered.

There are some reasons for thinking $L_b(\lambda)$ preferable to $L_{\max}(\lambda)$ from a non-Bayesian as well as from a Bayesian point of view; see, for example, the introduction by Bartlett (1937) of degrees of freedom into his test for the homogeneity of variance. The general large-sample theorems about the sampling distributions of maximum-likelihood estimates, and the maximum-likelihood ratio chi-squared test, apply just as much to $L_b(\lambda)$ as to $L_{\max}(\lambda)$.

4. Two Examples

We have supposed that after suitable transformation from y to $y^{(\lambda)}$, (a) the expected values of the transformed observations are described by a model of simple structure; (b) the error variance is constant; (c) the observations are normally distributed. Then we have shown that the maximized likelihood for λ, and also the approximate contribution to the posterior distribution of λ, are each proportional to a negative power of the residual sum of squares for the variate $z^{(\lambda)} = y^{(\lambda)}/J^{1/n}$.

The "overall" procedure seeks a set of transformation parameters λ for which (a), (b) and (c) are simultaneously satisfied, and sample information on all three aspects goes into the choice. In this Section we now apply this overall procedure to two examples. In Section 5 we shall show how further analysis can show the separate contributions of (a), (b) and (c) in the choice of the transformation. We shall then illustrate this separation using the same two examples.

The above procedure depends on specific assumptions, but it would be quite wrong for fruitful application to regard the assumptions as final. The proper attitude of sceptical optimism is accurately expressed by saying that we tentatively entertain the basis for analysis, rather than that we assume it. The checking of the plausibility of the present procedure will be discussed in Section 5.

A Biological Experiment using a 3×4 Factorial Design with Replication

Table 1 gives the survival times of animals in a 3×4 factorial experiment, the factors being (a) three poisons and (b) four treatments. Each combination of the two factors is used for four animals, the allocation to animals being completely randomized.

We consider the application of a simple power transformation $y^{(\lambda)} = (y^{\lambda}-1)/\lambda$. Equivalently we shall actually analyse the standardized variate $z^{(\lambda)} = (y^{\lambda}-1)/(\lambda\dot{y}^{\lambda-1})$.

TABLE 1

Survival times (unit, 10 hr) of animals in a 3 × 4 factorial experiment

Poison	*Treatment* A	B	C	D
I	0·31	0·82	0·43	0·45
	0·45	1·10	0·45	0·71
	0·46	0·88	0·63	0·66
	0·43	0·72	0·76	0·62
II	0·36	0·92	0·44	0·56
	0·29	0·61	0·35	1·02
	0·40	0·49	0·31	0·71
	0·23	1·24	0·40	0·38
III	0·22	0·30	0·23	0·30
	0·21	0·37	0·25	0·36
	0·18	0·38	0·24	0·31
	0·23	0·29	0·22	0·33

We are tentatively entertaining the model that after such transformation

(a) the expected value of the transformed variate in any cell can be represented by additive row and column constants, i.e. that no interaction terms are needed,
(b) the error variance is constant,
(c) the observations are normally distributed.

The maximized likelihood and the posterior distribution are functions of the residual sum of squares for $\mathbf{z}^{(\lambda)}$ after eliminating row and column effects. This sum of squares is denoted $S(\lambda; \mathbf{z})$. It has 42 degrees of freedom and is the result of pooling the "within groups" and the "interaction" sums of squares.

Table 2 gives $S(\lambda; \mathbf{z})$ together with $L_{\max}(\lambda)$ and $p_u(\lambda)$ over the interesting ranges. The constant k in $k\,e^{L_b(\lambda)} = p_u(\lambda)$ is the reciprocal of the area under the curve $Y = e^{L_b(\lambda)}$ determined by numerical integration. Graphs of $L_{\max}(\lambda)$ and of $p_u(\lambda)$ are shown in Fig. 1. This analysis points to an optimal value of about $\hat{\lambda} = -0{\cdot}75$. Using (11) the curve of maximized likelihood gives an approximate 95 per cent confidence interval for λ extending from about $-1{\cdot}13$ to $-0{\cdot}37$.

The posterior distribution $p_u(\lambda)$ is approximately normal with mean $-0{\cdot}75$ and standard deviation 0·22. About 95 per cent of this posterior distribution is included within the limits $-1{\cdot}18$ and $-0{\cdot}32$.

The reciprocal transformation has a natural appeal for the analysis of survival times since it is open to the simple interpretation that it is the *rate of dying* which is to be considered. Our analysis shows that it would, in fact, embody most of the advantages obtainable. The complete analysis of variance for the untransformed data and for the reciprocal transformation (taken in the z form) is shown in Table 3.

Whereas no great change occurs on transformation in the mean squares associated with poisons and treatments, the within groups mean square has shrunk to a third of

TABLE 2

Biological data. Calculations based on an additive, homoscedastic, normal model in the transformed observations

λ	$S(\lambda; \mathbf{z})$	$L_{max}(\lambda)$	λ	$S(\lambda; \mathbf{z})$	$L_{max}(\lambda)$
1·0	1·0509	91·72	−1·0	0·3331	119·29
0·5	0·6345	103·83	−1·2	0·3586	117·52
0·0	0·4239	113·51	−1·4	0·4007	114·86
−0·2	0·3752	116·44	−1·6	0·4625	111·43
−0·4	0·3431	118·58	−2·0	0·6639	102·74
−0·6	0·3258	119·82	−2·5	1·1331	89·91
−0·8	0·3225	120·07	−3·0	2·0489	75·69

λ	$p_u(\lambda)$	λ	$p_u(\lambda)$
0·0	0·01	−0·8	1·82
−0·1	0·02	−0·9	1·42
−0·2	0·08	−1·0	0·92
−0·3	0·26	−1·1	0·47
−0·4	0·49	−1·2	0·19
−0·5	0·94	−1·3	0·07
−0·6	1·46	−1·5	0·01
−0·7	1·82		

$L_{max}(\lambda) = -24 \log \hat{\sigma}^2(\lambda; \mathbf{z}) = \log\{S(\lambda; \mathbf{z})\}^{-24} + 92{\cdot}91$; $p_u(\lambda) = k\, e^{L_b(\lambda)} = 0{\cdot}866 \times 10^{-10}\{S(\lambda; \mathbf{z})\}^{-21}$.

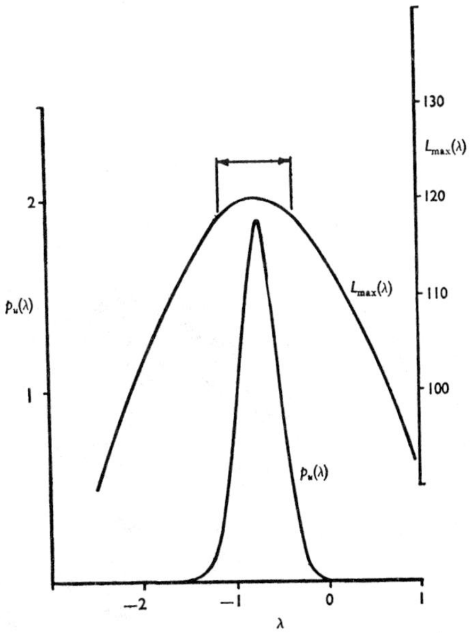

FIG. 1. Biological data. Functions $L_{max}(\lambda)$ and $p_u(\lambda)$. Arrows show approximate 95 per cent. confidence interval for λ.

its value and the interaction mean square is now much closer in size to that within groups. Thus, in the transformed metric, not only is greater simplicity of interpretation possible but also the sensitivity of the experiment, as measured by the ratios

TABLE 3

Analyses of variance of biological data

		Mean squares × 1000	
	Degrees of freedom	*Untransformed*	*Reciprocal transformation (z form)*
Poisons	2	516·5	568·7
Treatments	3	307·1	221·9
$P\times T$	6	41·7	8·5
Within groups	36	22·2	7·8

of the poisons and the treatments mean squares to the residual square, has been increased almost threefold. We shall not here consider the detailed interpretation of the factor effects.

A Textile Experiment using a Single Replicate of a 3^3 *Design*

In an unpublished report to the Technical Committee, International Wool Textile Organization, Drs A. Barella and A. Sust described some experiments on the behaviour of worsted yarn under cycles of repeated loading. Table 4 gives the numbers of cycles to failure, y, obtained in a single replicate of a 3^3 experiment in which the factors are

x_1: length of test specimen (250, 300, 350 mm.),
x_2: amplitude of loading cycle (8, 9, 10 mm.),
x_3: load (40, 45, 50 gm.).

In Table 4 the levels of the x's are denoted conventionally by $-1, 0, 1$.

It is useful to describe first the results of a rather informal analysis of Table 4. Barella and Sust fitted a full equation of second degree in x_1, x_2 and x_3, but the conclusions were very complicated and messy. In view of the wide relative range of variation of y, it is natural to try analysing instead $\log y$, and there results a great simplification. All linear regression terms are very highly significant and all second-degree terms are small. Further, it is natural to take logs also for the independent variables, i.e. to think in terms of relationships like

$$y \propto x_1^{\beta_1} x_2^{\beta_2} x_3^{\beta_3}. \tag{25}$$

The estimates of the β's, from the linear regression coefficients of $\log y$ on the $\log x$'s, are, with their estimated standard errors,

$$\hat{\beta}_1 = 4{\cdot}96 \pm 0{\cdot}20, \quad \hat{\beta}_2 = -5{\cdot}27 \pm 0{\cdot}30, \quad \hat{\beta}_3 = -3{\cdot}15 \pm 0{\cdot}30.$$

Since $\hat{\beta}_1 \simeq -\hat{\beta}_2$, the combination $\log x_1 - \log x_2 = \log(x_1/x_2)$ is suggested by the data as of possible importance. In fact, x_2/x_1 is just the fractional amplitude of the loading cycle; indeed, naïve dimensional considerations suggest this as a possible factor, although there are in fact other relevant lengths, so that dependence on x_1

and x_2 separately is not inconsistent with dimensional considerations. If, however, we write $x_2/x_1 = x_4$ and round the regression coefficients, we have the simple formula

$$y \propto x_4^{-5} x_3^{-3}$$

which fits the data remarkably well.

TABLE 4

Cycles to failure of worsted yarn: 3^3 *factorial experiment*

Factor levels x_1	x_2	x_3	*Cycles to failure, y*
−1	−1	−1	674
−1	−1	0	370
−1	−1	+1	292
−1	0	−1	338
−1	0	0	266
−1	0	+1	210
−1	+1	−1	170
−1	+1	0	118
−1	+1	+1	90
0	−1	−1	1,414
0	−1	0	1,198
0	−1	+1	634
0	0	−1	1,022
0	0	0	620
0	0	+1	438
0	+1	−1	442
0	+1	0	332
0	+1	+1	220
+1	−1	−1	3,636
+1	−1	0	3,184
+1	−1	+1	2,000
+1	0	−1	1,568
+1	0	0	1,070
+1	0	+1	566
+1	+1	−1	1,140
+1	+1	0	884
+1	+1	+1	360

In this case, there seem strong general arguments for starting with a log transformation of all variables. Power laws are frequently effective in the physical sciences; also, provided that the signs of the β's are right, (25) has sensible limiting behaviour for $x_2, x_3 \rightarrow 0, \infty$; finally, the obvious normal theory model based on transforming (25) gives distributions over positive values of y only.

Nevertheless, it is interesting to see whether the method of the present paper applied directly to the data of Table 4 produces the log transformation. In this paper, transformations of the dependent variable alone are considered; in fact, since the relative range of the x's is not very great, transformation of the x's does not have a big effect on the linearity of the regression.

We first consider the application of a simple power transformation in terms, as before, of the standardized variate $z^{(\lambda)} = (y^{\lambda}-1)/(\lambda \dot{y}^{\lambda-1})$. We tentatively suppose that after such transformation

(a) the expected value of the transformed response can be represented merely by a model *linear* in the x's,
(b) the error variance is constant,
(c) the observations are normally distributed.

The maximized likelihood and the posterior distribution are functions of the residual sum of squares for $z^{(\lambda)}$ after fitting only a linear model to the x's. Since there are four constants in the linear regression model this residual sum of squares has $27-4=23$ degrees of freedom; we denote it by $S(\lambda; \mathbf{z})$.

Table 5 shows $S(\lambda; \mathbf{z})$ together with $L_{\max}(\lambda)$ and $p_u(\lambda)$ over the interesting ranges and the results are plotted in Fig. 2. The optimal value for the transformation parameter is $\hat{\lambda} = -0{\cdot}06$. The transformation is determined remarkably closely in this

TABLE 5

Textile data. Calculations based on normal linear model in the transformed observations

λ	$S(\lambda; \mathbf{z})$	$L_{\max}(\lambda)$	λ	$S(\lambda; \mathbf{z})$	$L_{\max}(\lambda)$
1·00	5·4810	21·52	−0·20	0·2920	61·11
0·80	2·9978	29·67	−0·40	0·5478	52·61
0·60	1·5968	38·17	−0·60	1·1035	43·16
0·40	0·8178	47·21	−0·80	2·1396	34·22
0·20	0·4115	56·48	−1·00	3·9955	25·79
0·00	0·2519	63·10			

λ	$p_u(\lambda)$	λ	$p_u(\lambda)$
0·20	0·02	−0·10	4·66
0·15	0·09	−0·15	2·36
0·10	0·42	−0·20	0·77
0·05	1·58	−0·25	0·19
0·00	4·18	−0·30	0·04
−0·05	5·64	−0·35	0·01

$L_{\max}(\lambda) = -13{\cdot}5 \log \hat{\sigma}^2(\lambda; \mathbf{z}) = \{S(\lambda; \mathbf{z})\}^{-13\cdot 5} + 44{\cdot}49.$
$p_u(\lambda) = k\, e^{L_b(\lambda)} = 0{\cdot}540 \times 10^{-6} \{S(\lambda; \mathbf{z})\}^{-11\cdot 5}.$

example, the approximate 95 per cent confidence range extending only from $-0{\cdot}18$ to $+0{\cdot}06$. The posterior distribution $p_u(\lambda)$ has its mean at $-0{\cdot}06$. About 95 per cent of the distribution is included between $-0{\cdot}20$ and $+0{\cdot}08$. As we have mentioned, the advantages of a log transformation corresponding to the choice $\lambda = 0$ are very great and such a choice is now seen to be strongly supported by the data.

The complete analysis of variance for the untransformed and the log transformation, taken in the z form, is shown in Table 6.

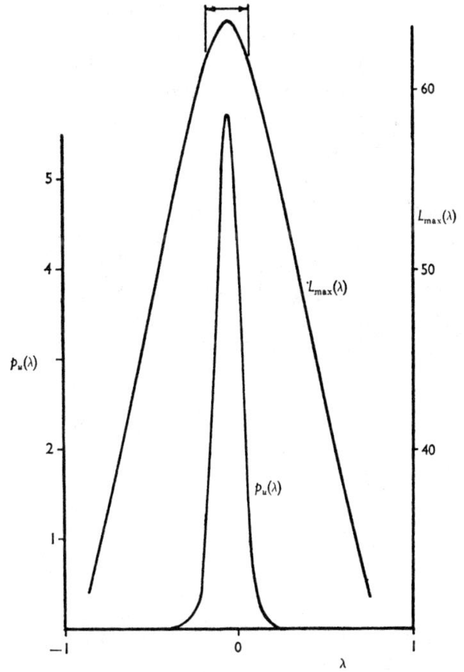

FIG. 2. Textile data. Functions $L_{max}(\lambda)$ and $p_u(\lambda)$. Arrows show approximate 95 per cent confidence interval for λ.

TABLE 6

Analyses of variance of textile data

		Mean squares × 1000	
	Degrees of freedom	*Untransformed*	*Logarithmic transformation (z form)*
Linear . .	3	4,916·2	2,374·4
Quadratic . .	6	704·1	8·1
Residual . .	17	73·9	11·9

The transformation eliminates the need for second-order terms in the regression equation while at the same time increasing the sensitivity of the analysis by about three, as judged by the ratio of linear and residual mean squares.

For this example we have also tried out the procedures we have discussed using the two parameter transformation $y^{(\lambda)} = \{(y+\lambda_2)^{\lambda_1} - 1\}/\lambda_1$ or in the z form actually used here $z^{(\lambda)} = \{(y+\lambda_2)^{\lambda_1}-1\}/\{\lambda_1 \operatorname{gm}(y+\lambda_2)\}^{\lambda_1-1}$. Incidentally the calculation and print out of 77 analysis of variance tables, involving in each case the fitting of a general equation of second degree, and calculation of residuals and fitted values took 2 min. 6 sec. on the C.D.C. 1604 electronic computer. The full numerical results can be obtained from the authors, but are not given here. Instead approximate contours of $-11{\cdot}5 \log S(\lambda; \mathbf{z})$, and hence of $S(\lambda; \mathbf{z})$ itself, of the maximized likelihood and of $p_u(\lambda_1, \lambda_2)$, are shown in Fig. 3. If the joint posterior distribution $p_u(\lambda_1, \lambda_2)$ were normal then a region which excluded 100α per cent of the total posterior probability could be given by

$$L_b(\hat{\lambda}_1, \hat{\lambda}_2) - L_b(\lambda_1, \lambda_2) = \chi_2^2(\alpha). \tag{26}$$

The shape of the contours indicates that the normal assumption is not very exact. Nevertheless, the quantity 100α obtained from (26) has been used to label the contours in Fig. 3 which thus roughly indicates the posterior probability distribution. For this example no appreciable improvement results from the addition of the further transformation parameter λ_2.

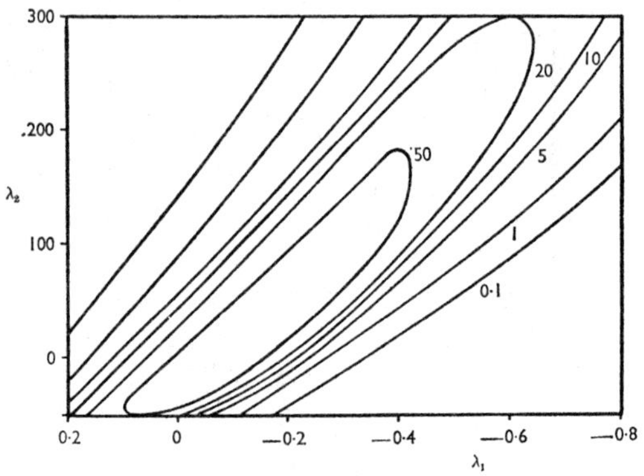

FIG. 3. Textile data. Transformation to $(y+\lambda_2)^{\lambda_1}$. Contours of $p_u(\lambda_1, \lambda_2)$ labelled with approximate percentage of posterior distribution excluded.

5. FURTHER ANALYSIS OF THE TRANSFORMATION

5.1. *General Procedure for Further Analysis*

The general procedure discussed above seeks to achieve simultaneously a model with (a) simple structure for the expectations, (b) constant variance and (c) normal distributions. Further analysis is sometimes profitable to see the separate contributions of these three elements to the transformation. Such analysis may indicate

(i) how simple a model we are justified in using;
(ii) what weight is given to the considerations (a) – (c) in choosing λ;
(iii) whether different transformations are really needed to achieve the different objectives and hence whether or not the value of λ chosen using the overall procedure is a compatible compromise.

Of course, quite often careful inspection of the data will answer (i)–(iii) adequately for practical purposes. Nevertheless, a further analysis is of interest.

We aim at simplicity both to achieve ease of understanding and to allow an efficient analysis. Validity of the formal tests associated with analysis of variance may, in virtue of the robustness of these tests, often hold to a good enough approximation even with the untransformed data. We stress, however, that such approximate validity is not by itself enough to justify an analysis; sensitivity must be considered as well as robustness. Thus in the biological example we have about one-third the sensitivity on the original scale as on the transformed scale. The approximate validity of significance tests on the original scale would be very poor consolation for the substantial loss of information involved in using the untransformed analysis. In any case even such validity is usually only preserved under the null hypothesis that all treatment effects are zero.

For the further analysis we again explore two approaches, one via maximum likelihood and the other via Bayes's theorem. Consider a general model to which a constraint C can be applied or relaxed, so that the relative merits of the simple and of the more complex model can be assessed. For example, the general model may include interaction terms, the constraint C being that the interaction terms are zero.

If $L_{\max}(\lambda)$ and $L_{\max}(\lambda | C)$ denote maximized log likelihoods for the general model and for the constrained model, then

$$L_{\max}(\lambda | C) = L_{\max}(\lambda) + \{L_{\max}(\lambda | C) - L_{\max}(\lambda)\}. \tag{27}$$

Here the second term on the right-hand side is a statistic for testing for the presence of the constraint.

More generally, with a succession of constraints, we have

$$\begin{aligned} L_{\max}(\lambda | C_1, C_2) = L_{\max}(\lambda) &+ \{L_{\max}(\lambda | C_1) - L_{\max}(\lambda)\} \\ &+ \{L_{\max}(\lambda | C_1, C_2) - L_{\max}(\lambda | C_1)\}, \end{aligned} \tag{28}$$

and the three terms on the right of (28) can be examined separately. The detailed procedure should be clear from the examples to follow.

To apply the Bayesian approach, we write the posterior density of λ

$$p(\lambda | C) = p(\lambda) \times \frac{p(C | \lambda)}{p(C)}, \tag{29}$$

where $p(C) = E_\lambda\{p(C | \lambda)\}$ is a constant independent of λ. That is, the posterior density of λ under the constrained model is the posterior density under the general model multiplied by a factor proportional to the conditional probability of the constraint given λ. Successive factorization can be applied when there is a series of successively applied constraints, giving, for example,

$$p(\lambda | C_1, C_2) = p(\lambda) \times \frac{p(C_1 | \lambda)}{p(C_1)} \times \frac{p(C_2 | \lambda, C_1)}{p(C_2 | C_1)}, \tag{30}$$

where $p(C_2 | C_1) = E_\lambda\{p(C_2 | \lambda, C_1)\}$ is a further constant independent of λ. Note that we are concerned here not with the probabilities that the constraints are true, but with the contributions of the constraints to the final function $p(\lambda | C_1, C_2)$.

5.2. *Structure of the Expectation*

Now very often the most important question is: how simple a form can we use for $E\{y^{(\lambda)}\}$? Thus in the analysis of the biological example in Section 4, we assumed, among other things, that additivity can be achieved by transformation. In fact, interaction terms may or may not be needed. Similarly, in our analysis of the textile example we took a linear model with four parameters; the full second-degree model with ten parameters may or may not be necessary.

Now let A, H and N denote respectively the constraints to the simpler linear model (without interaction or second-degree terms), to a heteroscedastic model and to a model with normal distributions. Then,

$$L_{\max}(\lambda \mid A, H, N) = L_{\max}(\lambda \mid H, N) + \{L_{\max}(\lambda \mid A, H, N) - L_{\max}(\lambda \mid H, N)\}. \qquad (31)$$

Let the parameter $\boldsymbol{\theta}$ in the expectation under the general linear model be partitioned $(\boldsymbol{\theta}_1, \boldsymbol{\theta}_2)$ where $\boldsymbol{\theta}_2 = 0$ is the constraint A. Denote the degrees of freedom associated with $\boldsymbol{\theta}_1$ and $\boldsymbol{\theta}_2$ by ν_1 and ν_2. If ν_r is the number of degrees of freedom for residual in the complex model, the number in the simpler model is thus $\nu_r + \nu_2$.

As before, we work with the standardized variable $z^{(\lambda)} = y^{(\lambda)}/J^{1/n}$. If we identify residual sums of squares by their degrees of freedom, we have

$$L_{\max}(\lambda \mid \boldsymbol{\theta}_2 = 0, H, N) = -\tfrac{1}{2} n \log\{S_{\nu_r+\nu_2}(\lambda; \mathbf{z})/n\}, \qquad (32)$$

whereas

$$L_{\max}(\lambda \mid H, N) = -\tfrac{1}{2} n \log\{S_{\nu_r}(\lambda; \mathbf{z})/n\}. \qquad (33)$$

Thus, in the textile example, S_{ν_r} refers to the residual sum of squares from a second-degree model and $S_{\nu_r+\nu_2}$ refers to the residual sum of squares from a first-degree model. Quite generally

$$S_{\nu_r+\nu_2}(\lambda; \mathbf{z}) = S_{\nu_r}(\lambda; \mathbf{z}) + S_{\nu_2.\nu_1}(\lambda; \mathbf{z}),$$

where $S_{\nu_2.\nu_1}(\lambda; \mathbf{z})$ denotes the extra sum of squares of $\mathbf{z}^{(\lambda)}$ for fitting $\boldsymbol{\theta}_2$, adjusting for $\boldsymbol{\theta}_1$, and has ν_2 degrees of freedom.

Thus with (32) and (33) the decomposition (31) becomes

$$L_{\max}(\lambda \mid \boldsymbol{\theta}_2 = 0, H, N) = L_{\max}(\lambda \mid H, N) - \tfrac{1}{2} n \log\left\{1 + \frac{\nu_2}{\nu_r} F(\lambda; \mathbf{z})\right\}, \qquad (34)$$

where

$$F(\lambda; \mathbf{z}) = \frac{S_{\nu_2.\nu_1}(\lambda; \mathbf{z})/\nu_2}{S_{\nu_r}(\lambda; \mathbf{z})/\nu_r} \qquad (35)$$

is the standard F ratio, in the analysis of variance of $\mathbf{z}^{(\lambda)}$, for testing the restriction to the simpler model.

Equation (34) thus provides an analysis of the overall criterion into a part taking account only of homoscedasticity (H) and normality (N) plus a part representing the additional requirement of a simple linear model, given that H and N have been achieved.

In the corresponding Bayesian analysis (30) gives

$$p(\lambda \mid \boldsymbol{\theta}_2 = 0, H, N) = p(\lambda \mid H, N) \times k_A \, p(\boldsymbol{\theta}_2 = 0 \mid \lambda, H, N), \qquad (36)$$

where

$$1/k_A = E_{\lambda \mid H, N}\{p(\boldsymbol{\theta}_2 = 0 \mid \lambda, H, N)\},$$

the expectation being taken over the distribution $p(\lambda \mid H, N)$.

Note that since the condition $\boldsymbol{\theta}_2 = 0$ is given, there is no component for these parameters in the prior distribution, so that the left-hand side of (36) is the posterior density obtained previously assuming A. Thus, in terms of the standardized variable $\mathbf{z}^{(\lambda)}$, the left-hand side is

$$p_0(\lambda)\, C_{\nu_r+\nu_2} \{S_{\nu_r+\nu_2}(\lambda; \mathbf{z})\}^{-\frac{1}{2}(\nu_r+\nu_2)}, \qquad (37)$$

where the normalizing constant is given by

$$C_{\nu_r+\nu_2}^{-1} = \int p_0(\lambda)\{S_{\nu_r+\nu_2}(\lambda;\, \mathbf{z})\}^{-\frac{1}{2}(\nu_r+\nu_2)}\, d\lambda.$$

Similarly, in the general model with $\boldsymbol{\theta}_1$ and $\boldsymbol{\theta}_2$ both free to vary, we obtain the first factor on the right-hand side of (36) as

$$p(\lambda \mid H, N) = p_0(\lambda)\, C_{\nu_r}\{S_{\nu_r}(\lambda;\, \mathbf{z})\}^{-\frac{1}{2}\nu_r}, \tag{38}$$

with

$$C_{\nu_r}^{-1} = \int p_0(\lambda)\{S_{\nu_r}(\lambda;\, \mathbf{z})\}^{-\frac{1}{2}\nu_r}\, d\lambda.$$

Thus, from (37) and (38), the second factor on the right-hand side of (36) must be

$$\frac{C_{\nu_r+\nu_2}}{C_{\nu_r}} \frac{\{S_{\nu_r}(\lambda;\, \mathbf{z})\}^{\frac{1}{2}\nu_r}}{\{S_{\nu_r+\nu_2}(\lambda;\, \mathbf{z})\}^{\frac{1}{2}(\nu_r+\nu_2)}}. \tag{39}$$

Now the general equation (36) shows that this last expression must be proportional to $p(\boldsymbol{\theta}_2 = 0 \mid \lambda, H, N)$. It is worth proving this directly. To do this, consider a transformed scale on which constant variance and normality have been attained and the standard estimates $\hat{\boldsymbol{\theta}}_2$ and s^2 calculated. For the moment, we need not indicate explicitly the dependence on λ and z. We denote the matrix of the reduced least-squares equations for $\boldsymbol{\theta}_2$, eliminating $\boldsymbol{\theta}_1$, by $\mathbf{b}$, so that the covariance matrix of $\boldsymbol{\theta}_2$ is $\sigma^2 \mathbf{b}^{-1}$. The elements of $\mathbf{b}$ and $\mathbf{b}^{-1}$ are denoted b_{ij} and b^{ij}. Also we write $\rho_{ij} = b^{ij}/\sqrt{(b^{ii} b^{jj})}$ and $\{\rho^{ij}\}$ for the matrix inverse to $\{\rho_{ij}\}$. Then the joint distribution of

$$t_i = \frac{\theta_{2i} - \hat{\theta}_{2i}}{s\sqrt{b^{ii}}}$$

is (Cornish, 1954; Dunnett and Sobel, 1954)

$$\text{const} \times \left(1 + \frac{\Sigma \rho^{ij} t_i t_j}{\nu_r}\right)^{-\frac{1}{2}(\nu_r+\nu_2)}$$

where here and later the constant involves neither the parameters nor the observations. With uniform prior distributions for the θ's and for $\log \sigma$, this is also the posterior distribution of the quantities $(\theta_{2i} - \hat{\theta}_{2i})/(s/\sqrt{b^{ii}})$, where now the θ_{2i} are the random variables. Transforming from the t_i's to the θ_{2i}'s we have that

$$p(\boldsymbol{\theta}_2 \mid \lambda, H, N) = \text{const} \times (s_{\nu_r}^2)^{-\frac{1}{2}\nu_2} \left\{1 + \frac{(\boldsymbol{\theta}_2 - \hat{\boldsymbol{\theta}}_2)' \mathbf{b} (\boldsymbol{\theta}_2 - \hat{\boldsymbol{\theta}}_2)}{\nu_r s_{\nu_r}^2}\right\}^{-\frac{1}{2}(\nu_r+\nu_2)}$$

whence

$$p(\boldsymbol{\theta}_2 = 0 \mid \lambda, H, N) = \text{const} \times (S_{\nu_r})^{-\frac{1}{2}\nu_2} \left\{1 + \frac{\hat{\boldsymbol{\theta}}_2' \mathbf{b} \hat{\boldsymbol{\theta}}_2}{S_{\nu_r}}\right\}^{-\frac{1}{2}(\nu_r+\nu_2)}$$

$$= \text{const} \times \frac{S_{\nu_r}^{\frac{1}{2}\nu_r}}{(S_{\nu_r+\nu_2})^{\frac{1}{2}(\nu_r+\nu_2)}}. \tag{40}$$

If now we restore in our notation the dependence on λ, comparison of (40) with (39) proves the required result; the appropriateness of the constant is easily checked.

Thus (36) provides an analysis of the overall density into a part $p(\lambda \mid H, N)$ taking account only of homoscedasticity and normality, and a second part, (39), in which the influence of the simplifying constraint is measured.

Equation (39) can be rewritten

$$\text{const} \times \{S_{\nu_r}(\lambda;\, \mathbf{z})\}^{-\frac{1}{2}\nu_2} \left\{1 + \frac{\nu_2}{\nu_r} F(\lambda;\, \mathbf{z})\right\}^{-\frac{1}{2}(\nu_r+\nu_2)} \tag{41}$$

Now, by (34), the corresponding expression in the maximum-likelihood approach is given, in a logarithmic version, by

$$-\tfrac{1}{2}n \log \left\{1 + \frac{\nu_2}{\nu_r} F(\lambda;\, \mathbf{z})\right\}. \tag{42}$$

The essential difference between (41) and (42) is the occurrence of the term in $S_{\nu_r}(\lambda;\, \mathbf{z})$ in (41). In conventional large sample theory, ν_r is supposed large compared with ν_2 and then in the limit the variation with λ of the additional term is negligible, and the effect of both terms can be represented by plotting the standard F ratio as a function of λ. In applications, however, ν_2/ν_r may well be appreciable; thus in the textile example $\nu_2/\nu_r = 6/17$.

Hence (41) and (42) could lead to appreciably different conclusions, for example, if we found a particular value of λ giving a low value of $F(\lambda;\, z)$ but a relatively high value of $S_{\nu_r}(\lambda;\, \mathbf{z})$.

The distinction between (41) and (42) from a Bayesian point of view can be expressed as follows. In (41) there occurs the *ordinate* of the posterior distribution of $\boldsymbol{\theta}_2$ at $\boldsymbol{\theta}_2 = 0$. On the other hand, the F ratio, which determines (42), is a monotonic function of the *probability mass* outside the contour of the posterior distribution passing through $\boldsymbol{\theta}_2 = 0$. Alternatively, a calculation of the posterior probability of a small region near $\boldsymbol{\theta}_2 = 0$ having a length proportional to σ_z in each of the ν_2 component directions gives an expression equivalent to (42). The difference between (41) and (42) will be most pronounced if there exists an extreme transformation producing a low value of $F(\lambda;\, z)$ but a large value of $S_{\nu_r}(\lambda;\, z)$, corresponding to a large spread of the posterior distribution of $\boldsymbol{\theta}_2$. Expression (42) would give an answer tending to favour this transformation, whereas (41) would not.

5.3. *Application to Textile Example*

We now illustrate the above analysis using the textile data. The calculations are set out in Table 7 and displayed in Figs. 4 and 5. We discuss the conclusions in some detail here. In practice, however, the most useful aspect of this approach is the opportunity for graphical assessment.

Fig. 4 shows that the curvature of $L_{\max}(\lambda \mid H, N)$ is much less than that of $L_{\max}(\lambda \mid A, H, N)$ previously given in Fig. 2, the constraint A here being that the second-degree terms are supposed zero. The inequality

$$L_{\max}(\hat{\lambda} \mid H, N) - L_{\max}(\lambda \mid H, N) < \tfrac{1}{2}\chi_1^2(\alpha) \tag{43}$$

thus gives the much wider approximate 95 per cent confidence interval $(-0{\cdot}48,\, 0{\cdot}13)$ for λ indicated by *HN* in Fig. 4 and compared with the previous interval, marked *AHN*. Since the constraint has 6 degrees of freedom the sampling distribution of

$$-2\{L_{\max}(\lambda \mid A, H, N) - L_{\max}(\lambda \mid H, N)\} \tag{44}$$

for fixed normalizing λ is asymptotically χ_6^2. Alternatively, (44), being a monotonic function of F, can be tested exactly. Thus we can decide for which λ's, if any, the inclusion of the constraint is compatible with the data. In Fig. 5, $F(\lambda;\, \mathbf{z})$ is close to unity over the interesting range of λ close to zero, so that we can use the simpler model in this neighbourhood. The range indicated by C in Fig. 4 is that for which F is less than 2·70, the 5 per cent significance point.

TABLE 7

Textile data. Calculations for the analysis of the transformation

λ	$L_{max}(\lambda \mid A, H, N)$	$L_{max}(\lambda \mid H, N)$	*Difference* $= -13{\cdot}5 \times$ $\log(1+\frac{6}{17}F(\lambda; z))$	$F(\lambda; z)$
1·00	21·52	41·41	−19·89	9·52
0·80	29·67	49·14	−19·47	9·15
0·60	38·17	55·65	−17·48	7·50
0·40	47·21	60·59	−13·38	4·80
0·20	56·48	63·99	− 7·51	2·09
0·00	63·10	66·02	− 2·92	0·68
−0·20	61·11	66·89	− 5·78	1·51
−0·40	52·61	66·07	−13·46	4·84
−0·60	43·16	62·68	−19·52	9·19
−0·80	34·22	56·44	−22·22	11·85
−1·00	25·79	48·18	−22·39	12·03

λ	$p_u(\lambda \mid A, H, N)$	$p_u(\lambda \mid H, N)$	$k_A\, p_u(A \mid \lambda, H, N)$
0·20	0·02	0·32	0·05
0·15	0·09	0·49	0·18
0·10	0·42	0·69	0·62
0·05	1·58	0·93	1·71
0·00	4·18	1·19	3·51
−0·05	5·64	1·47	3·84
−0·10	4·66	1·76	2·65
−0·15	2·36	1·96	1·20
−0·20	0·77	2·06	0·37
−0·25	0·19	2·03	0·09
−0·30	0·04	1·88	0·02
−0·35	0·01	1·59	0·01

The Bayesian analysis follows parallel lines. In Fig. 4, $p_u(\lambda \mid H, N)$ has a much greater spread than $p_u(\lambda \mid A, H, N)$. Fig. 5 shows $p_u(\lambda \mid H, N)$ with the component $k_A\, p(A \mid \lambda, H, N)$ from the constraint. When multiplied together they give the overall density $p_u(\lambda \mid A, H, N)$. A value of λ near zero maximizes the posterior density assuming the constraint and is consistent with the information in $p_u(\lambda \mid H, N)$.

There is, however, nothing in our Bayesian analysis itself to tell us whether the simplified model with the constraint is compatible with the data, even for the best possible λ. There is an important general point here. All probability calculations in statistical inference are conditional in one way or another. In particular, Bayesian posterior distributions such as $p_u(\lambda \mid A, H, N)$ are conditional on the model, in particular here on assumption A. It could easily happen that there is no value of λ for which A is at all reasonable, but to check on this we need to supplement the

Bayesian argument (Anscombe, 1961). Here we can do this by a significance test based on the sampling distribution of a suitable function of the observations, namely $F(\lambda; \mathbf{z})$. For λ around zero the value of $F(\lambda; z)$ is, in fact, well within the significance limits, so that we can reasonably use the posterior distribution of λ in question.

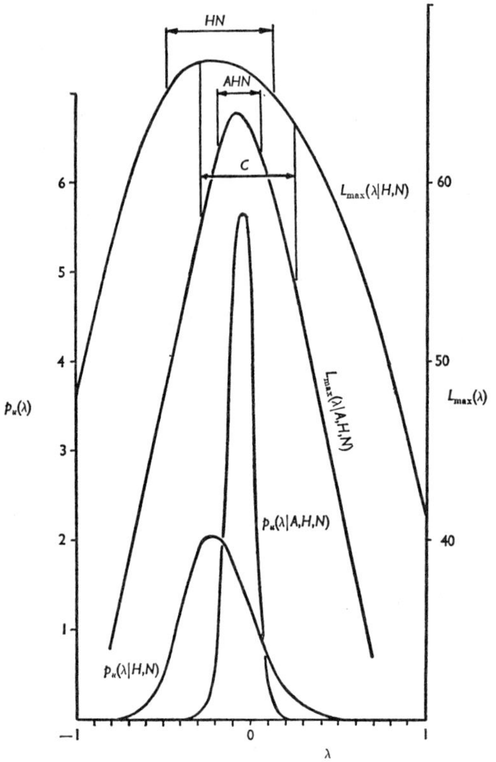

FIG. 4. Textile data. Functions $L_{\max}(\lambda)$ and $p_u(\lambda)$ under different models. A: additivity. H: homogeneity of variance. N: normality. Arrows HN, AHN show approximate 95 per cent confidence intervals for λ. Arrows C show range for which F for second-degree terms is not significant at 5 per cent level.

5.4. *Homogeneity of Variance*

Suppose that we have k groups of data, the expectation and variance being constant within each group. In the lth group, let the variance be σ_l^2 and let $S^{(l)}$ denote the sum of squares of deviations, having $\nu_l = n_l - 1$ degrees of freedom. Write $\Sigma n_l = n$, $\Sigma \nu_l = n - k$. Thus in our biological example, $k = 12$, $\nu_1 = \ldots = \nu_{12} = 3$, $n_1 = \ldots = n_{12} = 4$ and $\nu = 36$, $n = 48$.

Now suppose that a transformation to $y^{(\lambda)}$ exists which induces normality simultaneously in all groups. Then in terms of the standardized variable $z^{(\lambda)}$, the maximized log likelihood is

$$L_{\max}(\lambda \mid N) = -\tfrac{1}{2}\Sigma n_l \log\{S^{(l)}(\lambda; \mathbf{z})/n_l\}, \tag{45}$$

where $S^{(l)}(\lambda; \mathbf{z})$ is the sum of squares $S^{(l)}$, considered as a function of λ and calculated from the standardized variable $z^{(\lambda)}$.

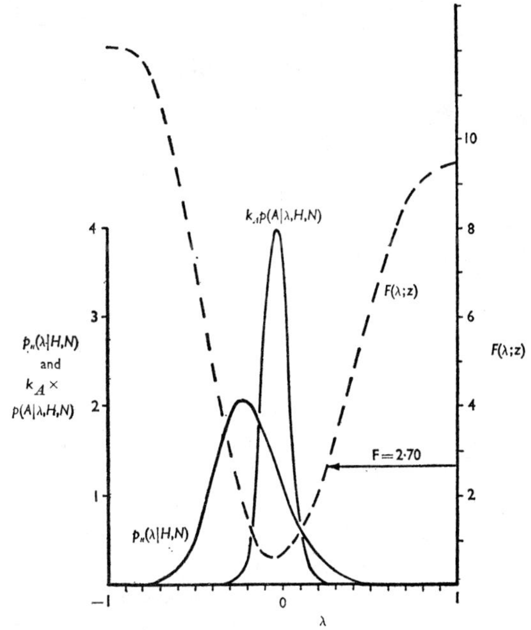

FIG. 5. Textile data. ——— Components of posterior distribution. - - - - - Variance ratio, $F(\lambda; \mathbf{z})$. Arrow gives 5 per cent significance level.

We now consider the constraint $H, \sigma_1^2 = \ldots = \sigma_k^2$, i.e. look at the possibility that a transformation exists simultaneously achieving normality and constant variance. Then if $S_\nu = \Sigma S^{(l)}$ is the pooled sum of squares within groups

$$L_{\max}(\lambda | H, N) = -\tfrac{1}{2}n \log\{S_\nu(\lambda; \mathbf{z})/n\}. \tag{46}$$

Therefore

$$L_{\max}(\lambda | H, N) = L_{\max}(\lambda | N) + \log\left[\frac{\Pi\{S^{(l)}(\lambda; \mathbf{z})/n_l\}^{\frac{1}{2}n_l}}{\{S_\nu(\lambda; \mathbf{z})/n\}^{\frac{1}{2}n}}\right]$$

$$= L_{\max}(\lambda | N) + \log L_1(\lambda; \mathbf{z}), \tag{47}$$

say. Here the second factor is the log of the Neyman–Pearson L_1 criterion for testing the hypothesis $\sigma_1^2 = \ldots = \sigma_k^2$.

In the corresponding Bayesian analysis, (29) gives

$$p(\lambda | H, N) = p(\lambda | N) \times k_H \, p(\sigma_1^2 = \ldots = \sigma_k^2 | \lambda, N), \tag{48}$$

where

$$k_H^{-1} = E_{\lambda|N}\{p(\sigma_1^2 = \ldots = \sigma_k^2 | \lambda, N)\}.$$

For the general model in which $\sigma_1^2, \ldots, \sigma_k^2$ may be different, the prior distribution is

$$p_0(\lambda)(\Pi d\theta_l)(\Pi d \log \sigma_l) J^{-\nu/n}$$

and

$$p(\lambda \mid N) = p_0(\lambda)\, c\Pi\{S^{(l)}(\lambda;\, \mathbf{z})\}^{-\frac{1}{2}\nu_l}, \tag{49}$$

with

$$c^{-1} = \int p_0(\lambda)\, \pi\{S^{(l)}(\lambda;\, \mathbf{z})\}^{-\frac{1}{2}\nu_l} d\lambda.$$

For the restricted model in which the variances are all equal to σ^2, the appropriate prior distribution is

$$p_0(\lambda)\,(\Pi d\theta_l)\,(d \log \sigma)\, J^{-\nu/n}$$

and

$$p(\lambda \mid H, N) = \{p_0(\lambda)\, c_\nu(\lambda;\, z)\}^{-\frac{1}{2}\nu}. \tag{50}$$

Hence, on dividing (50) by (49), we have that the second factor in (48) is

$$\frac{c_\nu}{c} \frac{\Pi\{S^{(l)}(\lambda;\, \mathbf{z})\}^{-\frac{1}{2}\nu_l}}{\{S_\nu(\lambda;\, \mathbf{z})\}^{-\frac{1}{2}\nu}} = \frac{c_\nu}{c} \frac{\Pi \nu_l^{\frac{1}{2}\nu_l}}{\nu^{\frac{1}{2}\nu}} e^{-\frac{1}{2}M(\lambda;\, \mathbf{z})}, \tag{51}$$

where (Bartlett, 1937)

$$M(\lambda;\, \mathbf{z}) = \nu \log \left\{\frac{S_\nu(\lambda;\, \mathbf{z})}{\nu}\right\} - \Sigma \nu_l \log \left\{\frac{S^{(l)}(\lambda;\, \mathbf{z})}{\nu_l}\right\}$$

is the modification of the L_1 statistic for testing homogeneity of variance, replacing sample sizes by degrees of freedom.

From our general argument, (51) must be proportional to $p(\sigma_1^2 = \ldots = \sigma_k^2 \mid \lambda, N)$. This can be verified directly by finding the joint posterior distribution of $\sigma_1^2, \ldots, \sigma_k^2$, transforming to new variables $\sigma_1^2, \sigma_2^2/\sigma_1^2, \ldots, \sigma_k^2/\sigma_1^2$, integrating out σ_1^2, and then taking unit values of the remaining arguments.

5.5. *Application to Biological Example*

In the biological example, we can now factorize the overall criterion into three parts. These correspond to the possibilities that in addition to normality within each group, we may be able to get constant variance and that it may be unnecessary to include interaction terms in the model, i.e. that additivity is achievable.

In terms of maximized likelihoods,

$$L_{\max}(\lambda \mid A, H, N) = L_{\max}(\lambda \mid N) + \log L_1(\lambda;\, \mathbf{z}) - \tfrac{1}{2} n \log \left\{1 + \frac{\nu_2}{\nu_r} F(\lambda;\, \mathbf{z})\right\}, \tag{52}$$

where $L_1(\lambda;\, \mathbf{z})$ is the criterion for testing constancy of variance given normality and $F(\lambda;\, \mathbf{z})$ is the criterion for absence of interaction given normality and constancy of variance.

The corresponding Bayesian analysis is

$$p(\lambda \mid A, H, N) = p(\lambda \mid N) \times k_H\, p(\sigma_1^2 = \ldots = \sigma_k^2 \mid \lambda, N) \times k_A\, p(\boldsymbol{\theta}_2 = 0 \mid \lambda, N, H). \tag{53}$$

The results are set out in Table 8 and in Figs. 6–8. The graphs of $L_{\max}(\lambda \mid N)$ and $p_u(\lambda \mid N)$ in Fig. 6 show that the information about λ coming from within group normality is very slight, values of λ as far apart as -1 and 2 being acceptable on this basis. The requirement of constant variance, however, has a major effect on the choice of λ; further, some information is contributed by the requirement of additivity.

TABLE 8

Biological data. Calculations for analysis of the transformation

λ	$L_{max}(\lambda \mid A, H, N)$	$L_{max}(\lambda \mid H, N)$	$L_{max}(\lambda \mid N)$	$M(\lambda; z)$	$F(\lambda; z)$
4·0			125·33		1·17
3·0			128·50		1·48
2·0	62·97	69·36	130·78	92·13	1·83
1·0	91·72	98·24	131·93	50·54	1·88
0·5	103·83	109·55	132·15	33·90	1·62
0·0	113·51	117·96	131·9[illegible]	20·99	1·22
−0·2	116·44	120·37	131·79	17·13	1·07
−0·4	118·58	122·13	131·59	14·19	0·95
−0·6	119·82	123·21	131·35	12·21	0·90
−0·8	120·07	123·60	131·04	11·16	0·94
−1·0	119·29	123·30	130·69	11·09	1·09
−1·2	117·52	122·35	130·29	11·91	1·33
−1·4	114·86	120·76	129·85	13·64	1·67
−1·6	111·43	118·55	129·37	16·23	2·08
−2·0	102·74	112·50	128·27	23·66	3·01
−2·5	89·91	102·46	126·68	36·33	4·12
−3·0	75·69	90·10	124·84	52·11	4·93

λ	$p_u(\lambda \mid A, H, N)$	$p_u(\lambda \mid H, N)$	$p_u(\lambda \mid N)$	$k_H\, p(H \mid \lambda, N)$	$k_A\, p(A \mid \lambda, H, N)$
1·0			0·335		
0·5		0·006	0·398		0·03
0·0	0·006	0·021	0·342	0·06	0·28
−0·1	0·023	0·055	0·324	0·17	0·39
−0·2	0·076	0·127	0·304	0·42	0·60
−0·3	0·257	0·261	0·283	0·92	0·98
−0·4	0·492	0·471	0·261	1·80	1·04
−0·5	0·942	0·754	0·240	3·14	1·25
−0·6	1·462	1·059	0·218	4·85	1·38
−0·7	1·823	1·320	0·196	6·73	1·38
−0·8	1·823	1·430	0·173	8·27	1·27
−0·9	1·419	1·360	0·153	8·88	1·04
−1·0	0·923	1·136	0·134	8·47	0·81
−1·1	0·468	0·850	0·116	7·33	0·55
−1·2	0·194	0·558	0·099	5·64	0·35
−1·3	0·067	0·329	0·083	3·96	0·20
−1·4	0·019	0·170	0·069	2·46	0·11
−1·5	0·005	0·078	0·058	1·34	0·06
−1·6	0·001	0·032	0·050	0·64	0·03
−1·7		0·009			

From Fig. 7, which shows the detailed separation of the maximum-likelihood and Bayesian components, any transformation in the region y^{-1} to $y^{-\frac{1}{2}}$ gives a compatible compromise.

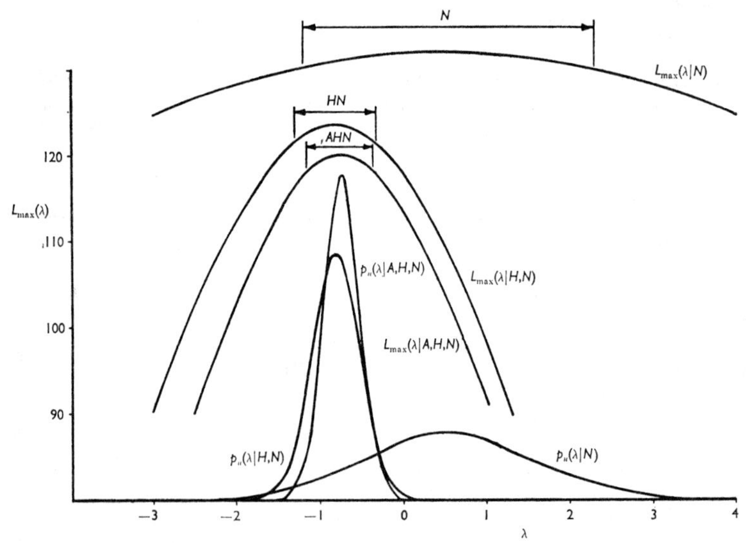

FIG. 6. Biological data. Functions $L_{max}(\lambda)$ and $p_u(\lambda)$ under different models. A: additivity. H: homogeneity of variance. N: normality. Arrows N, HN, AHN show approximate 95 per cent confidence intervals for λ.

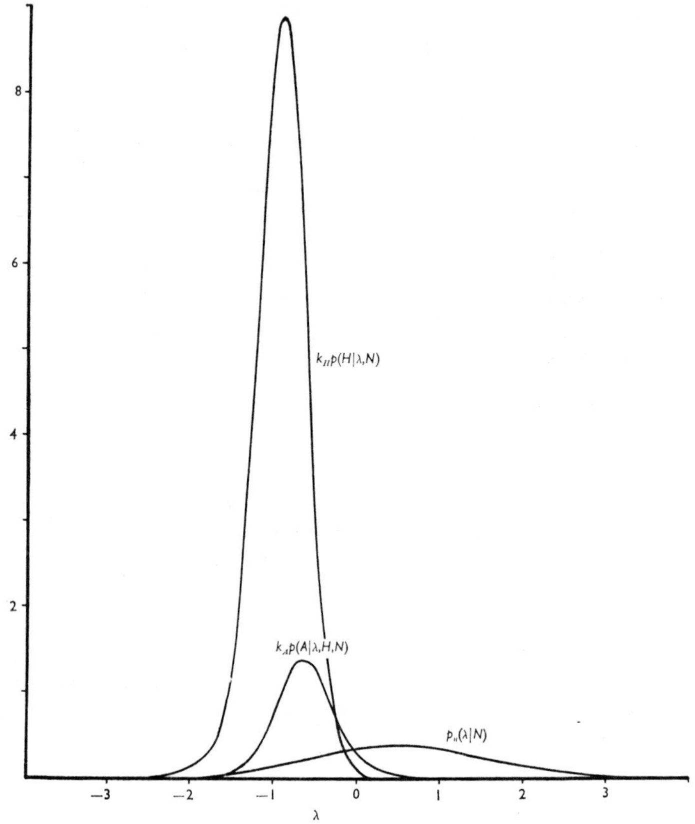

FIG. 7. Biological data. Components of posterior distribution.

Since the groups all contain four observations

$$-2\log L_1(\lambda;\, \mathbf{z}) = \tfrac{4}{3} M(\lambda;\, \mathbf{z})$$

and the graph of $M(\lambda;\, \mathbf{z})$ in Fig. 8 is equivalent to one of $L_1(\lambda;\, \mathbf{z})$. Since on the null hypothesis the distribution of $M(\lambda;\, \mathbf{z})$ is approximately χ^2_{11}, we can use Fig. 8 to

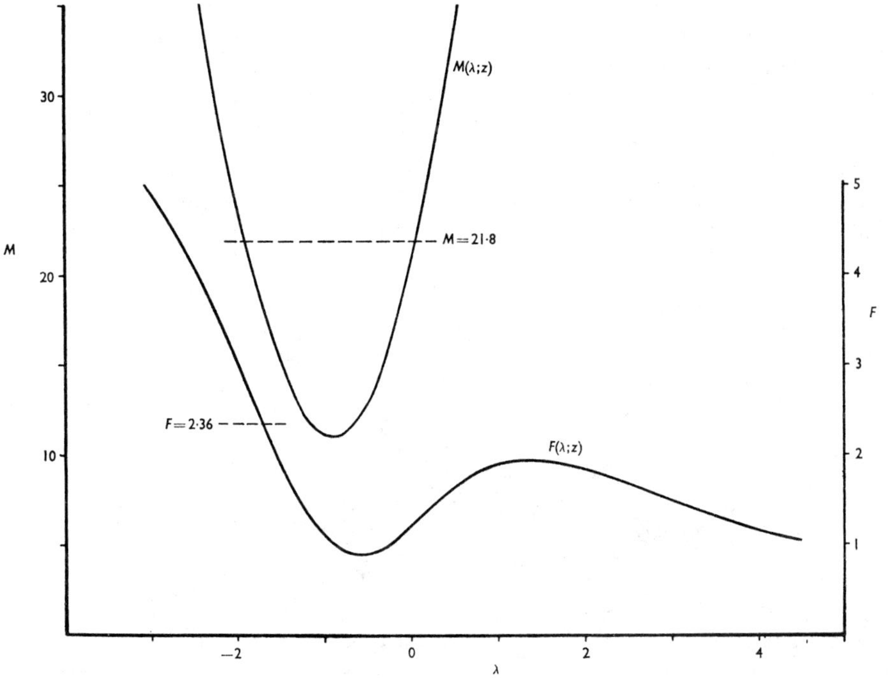

FIG. 8. Biological data. Variance ratio, $F(\lambda;\, \mathbf{z})$, for interaction against error as a function of λ. Bartlett's criterion, $M(\lambda;\, \mathbf{z})$, for equality of cell variances as a function of λ. Dotted lines give 5 per cent significance limits.

find the range in which the data are consistent with homoscedasticity. Similarly the graph of $F(\lambda;\, \mathbf{z})$ indicates the range within which the data are consistent with additivity. The dotted lines indicate the 5 per cent significance levels of M and of F.

The minimum of $M(\lambda;\, \mathbf{z})$ is very near $\lambda = -1$. It is of interest that the regression coefficient of log(sample variance) on log(sample mean) is nearly 4, so that the reciprocal transformation is suggested also by the usual approximate argument for stabilizing variance.

6. ANALYSIS OF RESIDUALS†

We now examine briefly a connection between the methods of the present paper and those based on the analysis of residuals. The analysis of residuals is intended

† We are greatly indebted to Professor F. J. Anscombe for pointing out an error in the approximation for α as we originally gave it. In the present modified version terms originally neglected in this Section have been included to correct the discrepancy.

primarily to examine what happens on one particular scale, although its use to indicate a transformation has been suggested (Anscombe and Tukey, 1963). Corresponding to an observation y, let Y be the deviation $\hat{y}-\bar{y}$ of the fitted value $\hat{y}$ from the sample mean and let $r = y-\hat{y}$ be the residual. If the ideal assumptions are satisfied r and Y will be distributed independently. Different sorts of departures from ideal assumptions can be measured, therefore, by studying the deviations of the statistics $T_{ij} = \Sigma r^i Y^j$ from $nE(r^i)E(Y^j)$. In addition to graphical analysis, a number of such functions have indeed been proposed for particular study (Anscombe, 1961; Anscombe and Tukey, 1963).

Specifically, the statistics

$$T_{30} = \Sigma r^3, \quad T_{40} = \Sigma r^4, \quad T_{21} = \Sigma r^2 Y, \quad T_{12} = \Sigma r Y^2 \tag{54}$$

were put forward as measures respectively of skewness, kurtosis, heterogeneity of variance and non-additivity. Tukey's degree of freedom for non-additivity (Tukey, 1949) involves the sum of squares corresponding to T_{12} considered as a contrast of residuals with "fixed" coefficients Y^2.

Suppose now that we consider the family of power transformations and, writing $z = y/\dot{y}$, and $w = z-1$, make the expansion

$$\begin{aligned} z^{(\lambda)} = \frac{z^\lambda - 1}{\lambda} &= w + \tfrac{1}{2}(\lambda-1)\,w^2 + \tfrac{1}{6}(\lambda-1)(\lambda-2)\,w^3 + O(w^4) \\ &= w - \alpha w_2 + \tfrac{2}{3}\alpha(\alpha+\tfrac{1}{2})\,w_3 + O(w^4), \end{aligned} \tag{55}$$

where $w_2 = w^2$, $w_3 = w^3$ and $\alpha = 1-\lambda$.

Now, $L_{\max}(\lambda)$ and $L_b(\lambda)$ are determined by the residual sum of squares of $z^{(\lambda)}$, which is approximately

$$\{\mathbf{w} - \alpha\mathbf{w}_2 + \tfrac{2}{3}\alpha(\alpha+\tfrac{1}{2})\,\mathbf{w}_3\}'\mathbf{a}_r\{\mathbf{w} - \alpha\mathbf{w}_2 + \tfrac{2}{3}\alpha(\alpha+\tfrac{1}{2})\,\mathbf{w}_3\}. \tag{56}$$

If we take terms up to the fourth degree in w and then differentiate with respect to α, we have that the maximum-likelihood estimate of α is approximately

$$\hat{\alpha} = \frac{3\mathbf{w}'\mathbf{a}_r\mathbf{w}_2 - \mathbf{w}'\mathbf{a}_r\mathbf{w}_3}{3\mathbf{w}_2'\mathbf{a}_r\mathbf{w}_2 + 4\mathbf{w}'\mathbf{a}_r\mathbf{w}_3}. \tag{57}$$

If we write $y_1 = y-\dot{y}$, $y_2 = (y-\dot{y})^2$, $y_3 = (y-\dot{y})^3$ and denote by $\hat{y}_1, \hat{y}_2, \hat{y}_3$ the values obtained by fitting y_1, y_2 and y_3 to the model, the above approximation may be expressed in terms of the original observations as

$$\hat{\alpha} = \frac{3\dot{y}(\mathbf{y}_1'\mathbf{a}_r\mathbf{y}_2) - \mathbf{y}_1'\mathbf{a}_r\mathbf{y}_3}{3\mathbf{y}_2'\mathbf{a}_r\mathbf{y}_2 + 4\mathbf{y}_1'\mathbf{a}_r\mathbf{y}_3} = \frac{3\dot{y}\Sigma(y_1-\hat{y}_1)(y_2-\hat{y}_2) - \Sigma(y_1-\hat{y}_1)(y_3-\hat{y}_3)}{3\Sigma(y_2-\hat{y}_2)^2 + 4\Sigma(y_1-\hat{y}_1)(y_3-\hat{y}_3)}. \tag{58}$$

To see the relation between this expression and the T statistics, write $d = \bar{y}-\dot{y}$. Then $y_1 = y-\dot{y} = r+Y+d$. Bearing in mind that $\mathbf{a}_r\mathbf{Y} = 0$, $\mathbf{a}_r\mathbf{r} = \mathbf{r}$, $\mathbf{Y}'\mathbf{r} = 0$, $\mathbf{a}_r\mathbf{1} = 0$, $\mathbf{1}'\mathbf{r} = 0$, where $\mathbf{1}$ denotes a vector of ones, terms such as $\mathbf{y}_1'\mathbf{a}_r\mathbf{y}_2$ can easily be expressed in terms of sums of powers and products of r, Y and d. In particular, on writing S for Σr^2, we find the numerator of (58) to be

$$3(\bar{y}-3d)(T_{30}+2T_{21}+T_{12}) - (T_{40}+3T_{31}+3T_{22}+T_{13}) + 3d(2\bar{y}-3d)S. \tag{59}$$

To this order of approximation the maximum-likelihood estimate of α thus involves all the T statistics of orders 3 and 4.

As a very special case, for data assumed to form a single random sample

$$\hat{\alpha} = \frac{3\bar{y}\Sigma(y-\bar{y})(y_2-\bar{y}_2)+\Sigma(y-\bar{y})(y_3-\bar{y}_3)}{3\Sigma(y_2-\bar{y}_2)^2+4\Sigma(y-\bar{y})(y_3-\bar{y}_3)}.$$

Here questions such as non-additivity and non-constancy of variance do not arise and the transformation is attempting only to produce normality. Correspondingly in (59), $T_{21} = T_{12} = T_{31} = T_{22} = T_{13} = 0$, since $Y = \hat{y}-\bar{y} = 0$. In fact if we write $m_1 = \bar{y}$, $m_p = n^{-1}\Sigma(y-\bar{y})^p$ $(p = 2, 3, \ldots)$ and make the approximation $d = \frac{1}{2}m_2/m_1$, we have that

$$\hat{\alpha} = \frac{m_1 m_3 - \frac{1}{3}\left\{(m_4-3m_2^2)+\frac{3m_2 m_3}{m_1}+\frac{9}{4}\frac{m_2^3}{m_1^2}\right\}}{6m_2^2+\frac{1}{3}\left\{7(m_4-3m_2^2)+12\frac{m_2 m_3}{m_1}+6\frac{m_2^3}{m_1^2}\right\}}. \tag{60}$$

For distributions in which m_1, m_2, m_3 and $m_4-3m_2^2$ are of the same order of magnitude, the terms in curly brackets are of one order higher in $1/m_1$ than are the other terms of the numerator and denominator. If we ignore the higher-order terms, we have

$$\hat{\alpha} \simeq \frac{m_1 m_3}{6m_2^2}.$$

A useful check suggested by Anscombe is to consider the χ^2 distribution for moderate degrees of freedom and the Poisson distribution for not too small a mean. For χ^2 we find $\alpha \simeq \frac{1}{6}$, whence $\lambda \simeq \frac{1}{3}$, corresponding to the well-known Wilson–Hilferty transformation. For the Poisson distribution, $\alpha \simeq \frac{1}{3}$, whence $\lambda \simeq \frac{2}{3}$.

7. Analysis of Effects after Transformation

In Section 2 we suggested that, having chosen a suitable λ, we should make the usual detailed estimation and interpretation of effects on this transformed scale. Thus in our two examples we recommended that the detailed interpretation should be in terms of a standard analysis of respectively $1/y$ and $\log y$. Since the value of λ used is selected at least partly in the light of the data, the question arises of a possible need to allow for this selection when interpreting the factor effects.

To investigate an appropriate allowance, we regard λ as an unknown parameter with "true" value λ_0, say, and suppose the true factor effects to be measured in terms of the scale λ_0. If we were, for instance, to analyse the factor effects on the scale corresponding to the maximum-likelihood estimate $\hat{\lambda}$, we might expect some additional error arising from the difference between $\hat{\lambda}$ and λ_0. We now investigate this matter, although the present formulation of the problem is not always completely realistic. For example, in our biological example, having decided to work with $1/y$, we shall probably be interested in factor effects measured on this scale and not those measured in some unknown scale corresponding to an unknown "true" λ_0. On the other hand, if we are interested in whether there is interaction between two factors, it is possibly dangerous to answer this by testing for interaction on the scale $\hat{\lambda}$, since $\hat{\lambda}$ may be selected at least in part to minimize the sample interaction. A more reasonable formulation here may often be: on some unknown "true" scale λ_0, are interaction terms necessary in the model?

From the maximum-likelihood approach, the most useful result is that significance tests for null hypotheses, such as that just mentioned about the absence of interaction, can be obtained in a straightforward way in terms of the usual large-sample chi-squared test. Thus, in the textile example, we could test the null hypothesis that second-degree terms are absent for some unknown "true" λ_0, by testing twice the difference of the maxima of the two curves of $L_{\max}(\lambda)$ in Fig. 4 as χ_6^2. Note that the maxima occur at different values of λ. In this particular example, such a test is hardly necessary.

It would be possible to obtain more detailed results by evaluating the usual large-sample information matrix for the joint estimation of λ, σ^2 and $\boldsymbol{\theta}$. Since, however, more specific results can be obtained from the Bayesian analysis, we shall present only those. The general conclusion will be that to allow for the effect of analysing in terms of $\hat{\lambda}$ rather than λ_0, the residual degrees of freedom need only be reduced by ν_λ, the number of component parameters in λ. This result applies provided that the population and sample effects are measured in terms of the normalized variables $\mathbf{z}^{(\lambda)}$.

Consider locally uniform prior densities for $\boldsymbol{\theta}$, $\log \sigma$ and λ. Then the posterior density for $\boldsymbol{\theta}$ is

$$\frac{\int \{(\mathbf{z}^{(\lambda)} - \mathbf{a}\boldsymbol{\theta})' (\mathbf{z}^{(\lambda)} - \mathbf{a}\boldsymbol{\theta})\}^{-\frac{1}{2}n}\, d\lambda}{\int \{\nu_r s^2(\lambda;\, \mathbf{z})\}^{-\frac{1}{2}\nu_r}\, d\lambda}. \tag{61}$$

Approximate evaluation of the integral in (61) is done by expansion around the maxima of the integrands. The maximum of the integrand in the denominator is at the maximum-likelihood estimate $\hat{\lambda}$, and that of the numerator is near $\hat{\lambda}$, so long as $\boldsymbol{\theta}$ is near its maximum-likelihood value. The answer is that (61) is approximately

$$\frac{\{(\mathbf{z}^{(\hat{\lambda})} - \mathbf{a}\boldsymbol{\theta})' (\mathbf{z}^{(\hat{\lambda})} - \mathbf{a}\boldsymbol{\theta})\}^{-\frac{1}{2}(n-\nu_\lambda)}}{\{\nu_r s^2(\hat{\lambda};\, \mathbf{z})\}^{-\frac{1}{2}(\nu_r-\nu_\lambda)}}. \tag{62}$$

This is exactly the posterior density of $\boldsymbol{\theta}$ for some known fixed λ with the degrees of freedom reduced by ν_λ.

To derive (62) from (61), we need to evaluate integrals of the form

$$I = \int \{q(\lambda)\}^{-\frac{1}{2}\nu}\, d\lambda, \tag{63}$$

where ν is large, and $q(\lambda)$ is assumed positive and to have a unique minimum at $\lambda = \hat{\lambda}$, with a finite Hessian determinant Δ_q at the minimum. We can then make a Laplace expansion, writing

$$\begin{aligned} I &= \int \exp\left[-\frac{\nu}{2}\log q(\hat{\lambda}) - \frac{\nu}{2}\log\left\{1 + \frac{q(\lambda)-q(\hat{\lambda})}{q(\hat{\lambda})}\right\}\right] d\lambda \\ &\simeq \frac{\{q(\hat{\lambda})\}^{-\frac{1}{2}\nu-\frac{1}{2}\nu_\lambda}}{\Delta_q^{\frac{1}{2}}} \times \text{const}; \end{aligned} \tag{64}$$

for this we expand the second logarithmic term as far as the quadratic terms and then integrate over the whole ν_λ-dimensional space of λ. In our application the terms $\Delta_q^{\frac{1}{2}}$ in numerator and denominator are equal to the first order.

Finally, we can obtain an approximation to the posterior distribution $p_u(\boldsymbol{\lambda})$ of $\boldsymbol{\lambda}$ that is better than the usual type of asymptotic normal approximation. For an expansion about $\boldsymbol{\lambda}$ gives that

$$p_u(\boldsymbol{\lambda}) = \frac{\{s^2(\boldsymbol{\lambda};\mathbf{z})\}^{-\frac{1}{2}\nu_r}}{\int\{s^2(\boldsymbol{\lambda};\mathbf{z})\}^{-\frac{1}{2}\nu_r}\,d\boldsymbol{\lambda}}$$

$$\simeq \frac{\text{const}}{\left\{1+\dfrac{(\boldsymbol{\lambda}-\hat{\boldsymbol{\lambda}})'\mathbf{b}(\boldsymbol{\lambda}-\hat{\boldsymbol{\lambda}})}{\nu_r s^2(\hat{\boldsymbol{\lambda}};\mathbf{z})}\right\}^{\frac{1}{2}\nu_\lambda}}. \tag{65}$$

Here

$$\mathbf{b} = \mathbf{d}'(\hat{\boldsymbol{\lambda}})\,\mathbf{a}_r\,d(\hat{\boldsymbol{\lambda}}), \tag{66}$$

with $\mathbf{d}(\boldsymbol{\lambda})$ being the $n\times\nu_\lambda$ matrix with elements

$$\frac{\partial z_i^{(\lambda)}}{\partial\lambda_j} \quad (i=1,\ldots,n;\ j=1,\ldots,\nu_\lambda).$$

The matrix $\mathbf{b}$ determines the quadratic terms in the expansion of $s^2(\boldsymbol{\lambda};\mathbf{z})$ around $\hat{\boldsymbol{\lambda}}$.

Thus the quantities $(\lambda_j-\hat{\lambda}_j)/\{s(\hat{\boldsymbol{\lambda}};\mathbf{z})\sqrt{b^{ii}}\}$ have approximately a posterior multivariate t distribution and

$$\frac{(\boldsymbol{\lambda}-\hat{\boldsymbol{\lambda}})'\mathbf{b}(\boldsymbol{\lambda}-\hat{\boldsymbol{\lambda}})}{\nu_r s^2(\hat{\boldsymbol{\lambda}};\mathbf{z})}$$

a posterior F distribution. In fact, however, it will usually be better to examine the posterior distribution of $\boldsymbol{\lambda}$ directly, as we have done in the numerical examples.

8. FURTHER DEVELOPMENTS

We now consider in much less detail a number of possible developments of the methods proposed in this paper. Of these, the most important is probably the simultaneous transformation of independent and dependent variables in a regression problem. Some general remarks on this have been made in Section 1.

Denote the dependent variable by y and the independent variables by $x_1,\ldots,x_l$. Consider a family of transformations from y into $y^{(\lambda)}$ and $x_1,\ldots,x_l$ into $x_1^{(\kappa_1)},\ldots,x_l^{(\kappa_l)}$, the whole transformation being thus indexed by the parameters $(\lambda;\kappa_1,\ldots,\kappa_l)$. It is not necessary that the family of transformations of say x_1 into $x_1^{(\kappa_1)}$ and x_2 into $x_2^{(\kappa_2)}$ should be the same, although this would often be the case.

We now assume that for some unknown $(\lambda;\kappa_1,\ldots,\kappa_l)$ the usual normal theory assumptions of linear regression theory hold. We can then compute say the maximized log likelihood for given $(\lambda;\kappa_1,\ldots,\kappa_l)$, obtaining exactly as in (8)

$$L_{\max}(\lambda;\kappa_1,\ldots,\kappa_l) = -\tfrac{1}{2}\log\hat{\sigma}^2(\lambda;\kappa_1,\ldots,\kappa_l)+\log J(\lambda;\mathbf{y}), \tag{67}$$

where $\hat{\sigma}^2(\lambda;\kappa_1,\ldots,\kappa_l)$ is the maximum-likelihood estimate of residual variance in the standard multiple regression analysis of the transformed variable. The corresponding expression from the Bayesian approach is

$$L_b(\lambda;\kappa_1,\ldots,\kappa_l) = -\tfrac{1}{2}\nu_r\log s^2(\lambda;\kappa_1,\ldots,\kappa_l)+\frac{\nu_r}{n}\log J(\lambda;\mathbf{y}). \tag{68}$$

The straightforward extension of the procedure of Section 3 is to compute (67) or (68) for a suitable set of $(\lambda; \kappa_1, \ldots, \kappa_l)$ and to examine the resulting surface especially near its maximum. This is, however, a tedious procedure, except perhaps for $l = 1$. Further, graphical presentation of the conclusions will not be easy if $l > 1$; for $l = 1$ we can plot contours of the functions (67) and (68).

When λ is fixed, i.e. transformations of the independent variables only are involved, Box and Tidwell (1962) developed an iterative procedure for the corresponding non-linear least-squares problem. In this the independent variables are, if necessary, first transformed to near the optimum form. Then two terms of the Taylor expansion of $x_1^{(\kappa_1)}, \ldots, x_l^{(\kappa_l)}$ are taken. For example if $x_1^{(\kappa_1)} = x^{\kappa_1}$ and the best value for κ_1 is thought to be near 1, we write

$$x_1^{\kappa_1} = x_1 + (\kappa_1 - 1)\, x_1 \log x_1. \tag{69}$$

A linear regression term $\beta_1 x_1^{\kappa_1}$ can then be written approximately

$$\beta_1 x_1 + \beta_1(\kappa_1 - 1)\, x_1 \log x_1 = \beta_1 x_1 + \gamma_1 x_1 \log x_1,$$

say. If the linear model involves linear regression on $x_1, \ldots, x_l$ and if all the transformations of the independent variable are to powers, we can therefore take the linear regression on $x_1, \ldots, x_l$, $x_1 \log x_1, \ldots, x_l \log x_l$ in order to estimate the β's and γ's and hence also the κ's. The procedure can then be iterated. Transformation of the dependent variable will usually be the more critical. Therefore, a reasonable practical procedure will often be to combine straightforward investigation of transformation of the dependent variable with Box and Tidwell's method applied to the independent variables.

It is possible also to consider simplifications of the procedure for determining a transformation of the dependent variable. The main labour in straightforward application of the method of Section 3 is in applying the transformation for various values of λ and then computing the standard analysis of variance for each set of transformed data. Such a sequence of similar calculations is straightforward on an electronic computer. It is perfectly practicable also for occasional desk calculation, although probably not for routine use. There are a number of possible simplifications based, for example, on expansions like (69) or even (55), but they have to be used very cautiously.

In the present paper we have concentrated largely on transformations for those standard "fixed-effects" analysis of variance situations where the response can be treated as a continuous variable. The same general approach could be adopted in dealing with "random-effects" models, and with various problems in multivariate analysis and in the analysis of time series. We shall not go into these applications here.

An important omission from our discussion concerns transformations specifically for data suspected of following the Poisson or binomial distributions. There are two difficulties here. One is purely computational. Suppose we assume that our observations, y, follow, for example, Poisson distributions with means that obey an additive law on an unknown transformed scale. Thus, in a row–column arrangement, it might be assumed that the Poisson mean in row i and column j has the form

$$(\mu + \alpha_i + \beta_j)^{1/\lambda} \quad (\lambda \neq 0),$$

$$\mu \alpha_i \beta_j \quad (\lambda = 0),$$

where λ is unknown. Then λ and the other parameters of the model can be estimated by maximum likelihood (Cochran, 1940). It would probably be possible to develop reasonable approximations to this procedure although we have not investigated this matter.

An essential distinction between this situation and the one considered in Section 3 is that here the untransformed observations y have known distributional properties. The analogous normal theory situation would involve observations y normally distributed with constant variance on the untransformed scale, but for which the population means are additive on a transformed scale. The maximum-likelihood solution in this case would involve, at least in principle, a straightforward non-linear least-squares problem. However, this situation does not seem likely to arise often; certainly, it is inappropriate in our examples.

An important possible complication of the analysis of data connected with Poisson and binomial distributions has been particularly stressed by Bartlett (1947). This is the presence of an additional component of variance of unknown form on top of the Poisson or binomial variation. If inspection of the data shows that such additional variation is substantial, it may be adequate to apply the methods of Section 3. For integer data with range (0, 1, ...) it will often be reasonable to consider power transformations. For data in the form of proportions of "successes" in which "successes" and "failures" are to be treated symmetrically, Professor J. W. Tukey has, in an unpublished paper, suggested the family of transformations from y to

$$y^{\lambda}-(1-y)^{\lambda}.$$

For suitable λ's this approximates closely to the standard transforms of proportions, the probit, logistic and angular transformations. The methods of the present paper could be applied with this family of transformations.

ACKNOWLEDGEMENT

We thank many friends for remarks leading to the writing of this paper.

REFERENCES

ANSCOMBE, F. J. (1961), "Examination of residuals", *Proc. Fourth Berkeley Symp. Math. Statist. and Prob.*, **1**, 1–36.

—— and TUKEY, J. W. (1963), "The examination and analysis of residuals", *Technometrics*, **5**, 141–160.

BARTLETT, M. S. (1937), "Properties of sufficiency and statistical tests", *Proc. Roy. Soc.* A, **160**, 268–282.

—— (1947), "The use of transformations", *Biometrics*, 3, 39–52.

BOX, G. E. P. and TIDWELL, P. W. (1962), "Transformation of the independent variables", *Technometrics*, **4**, 531–550.

COCHRAN, W. G. (1940), "The analysis of variance when experimental errors follow the Poisson or binomial laws", *Ann. math. Statist.*, **11**, 335–347.

CORNISH, E. A. (1954), "The multivariate t distribution associated with a set of normal sample deviates", *Austral. J. Physics*, **7**, 531–542.

DUNNETT, C. W. and SOBEL, M. (1954), "A bivariate generalization of Student's t distribution", *Biometrika*, **41**, 153–169.

JEFFREYS, H. (1961), *Theory of Probability*, 3rd ed. Oxford University Press.

KLECZKOWSKI, A. (1949), "The transformation of local lesion counts for statistical analysis", *Ann. appl. Biol.*, **36**, 139–152.

TUKEY, J. W. (1949), "One degree of freedom for non-additivity", *Biometrics*, **5**, 232–242.

—— (1950), "Dyadic anova, an analysis of variance for vectors", *Human Biology*, **21**, 65–110.

—— and MOORE, P. G. (1954), "Answer to query 112", *Biometrics*, **10**, 562–568.

DISCUSSION ON PAPER BY PROFESSOR BOX AND PROFESSOR COX

Mr J. A. NELDER: May I begin with a definition (from the Concise Oxford Dictionary): "Box and Cox—two persons who take turns in sustaining a part." I must admit to having spent some time in trying to deduce which person was sustaining which part of this most interesting paper. I do not think the exercise was very successful, and this testifies to some sound collaboration on the part of the authors.

It seems to me that there are two basic problems besetting all conscientious data analysts (to borrow Professor Tukey's term). One is how to check that the data are not contaminated with rogue observations and what action to take if they are. The other is how to check that the model being used to analyse the data is substantially the right one. Looking through the corpus of statistical writings one must be struck, I think, by how relatively little effort has been devoted to these problems. The overwhelming preponderance of the literature consists of deductive exercises from *a priori* starting points. Now, of course, there must always be some assumptions made *a priori*; in data analysis the important thing is that they should not be much stronger than previous evidence justifies. The first of the two problems, that of gross errors or rogue observations, we are not directly concerned with now, but the question of scale for analysis, which is discussed here, is fundamental to the second. One sees not infrequently remarks to the effect that the design of an experiment determines the analysis. Life would be easier if this were true. To the information from the design we must add the analyst's prior judgements, preconceptions or prejudices (call them what you will) about questions of additivity, homoscedasticity and the like. Frequently these prior assumptions are unjustifiably strong, and amount to an assertion that the scale adopted will give the required additivity, etc. The great virtue of this paper lies in its showing us how to weaken these prior assumptions and allow the data to speak for themselves in these matters. The data analyst's two problems are closely intertwined, however; for if rogue observations are present their residuals tend to dominate the residual sum of squares, and may thus seriously affect the estimation of λ.

The two approaches, via likelihood and via Bayes theorem, run side by side, and give results which will often be very similar. I am not entirely happy about the derivation of equation (19) and wonder whether the appearance of the observations in the prior probability is not only "interesting", as the authors state, but also illegal. They remark (on p. 219) that, "There are some reasons for thinking $L_b(\lambda)$ preferable to $L_{\max}(\lambda)$ from a non-Bayesian as well as from a Bayesian point of view." I agree and, furthermore, I believe that a suitable modification of the likelihood approach may be found to produce just this result. The starting point is that fixed effects are unrealistic in a model. If we measure a treatment effect in an experiment, it is common experience that a further experiment will give us a further estimate of the effect which often differs from the original estimate by more than the internal standard errors of the experiments would lead us to expect. If we construct a model with this in mind, then for a single normal sample of n we might obtain

$$y_i = m + e_i$$

where $m = N(\mu, \sigma'^2)$ and $e_i = N(0, \sigma^2)$. If we now do an orthogonal transformation of the data $\mathbf{z} = \mathbf{Hy}$ where $\mathbf{H}$ is an orthogonal matrix of known coefficients having its first row with elements $n^{-\frac{1}{2}}$, then the log likelihood is given by

$$L = \text{const} - \tfrac{1}{2} \ln V - (z_1 - \mu\sqrt{n})^2/2V - \tfrac{1}{2}(n-1)\log\sigma^2 - \sum_2^n z_i^2/2\sigma^2,$$

where $V = \sigma^2 + n\sigma'^2$. Clearly we cannot estimate V unless μ is known, which in general it is not. However, for any fixed *but unknown* V, we have L maximized by taking

$$\hat{\mu} = \bar{y}, \quad \text{and} \quad \hat{\sigma}^2 = \Sigma(y-\bar{y})^2/(n-1).$$

Thus $L_{max}(\lambda)$ following equation (24) is replaced (apart from an unknown constant) by $L_b(\lambda)$. By extensions of this argument we obtain Bartlett's criterion for testing the homogeneity of variances instead of the L_1 criterion, and the likelihood criterion for a restricted hypothesis on the means (equation (35)) becomes the same (apart from an unknown constant factor) as the Bayesian one. Thus some of the apparent differences between the two approaches may result from the restrictions implied by fixed effects in a model, these being equivalent to assertions of zero variance in repetitions of the experiment.

Taken with the work of Tukey, Daniel and others, on the detection of rogue observations, the results of this paper should lead before long to substantial improvements in computer programmes for the analysis of experiments. "First generation" programmes, which largely behave as though the design *did* wholly define the analysis, will be replaced by new second-generation programmes capable of checking the additional assumptions and taking appropriate action. It is hardly necessary to stress what an advance this would be.

I suppose that the converse of "two persons who take turns in sustaining a part" would be "one person who takes turns in sustaining two parts". Such a person is often the proposer of the vote of thanks, the parts being those of congratulator and critic; the latter has been known to overwhelm the former, but not, I hope, today. We must all be grateful for the clear exposition of an important problem, for the practical value of the results obtained and for the possibilities opened up for future investigations. It is a real pleasure, therefore, for me to propose the vote of thanks today.

Dr J. HARTIGAN: I would like to suggest a non-parametric approach to Box and Cox's problem. Suppose in the ith experiment we observe y_i under conditions x_i and that it is desired to find the probability distribution of y given x for various x. The only general principle that seems to apply is a similarity principle—"What will happen under present circumstances will probably be similar to what happened under similar circumstances in the past" or more simply "like equals likely". The Meteorological Office does seem to be acting according to this principle in its long-range forecasts, where the procedure is to look at this month's weather, look in the records for a similar month, see what happened the following month then, and predict the same thing will happen next month, now—they would say, to predict what y_0 will be under conditions x_0, look among the (y_i, x_i) for an x_i close to x_0, then predict $y_0 = y_i$.

It does seem possible to offer a non-parametric method for predicting a new y at x_0; in least squares theory this would be the fitted value Y_0. The general procedure is to smooth from the various readings (y, x) in the neighbourhood of x_0, values of y being given greater or less weight according to x's "similarity" to x_0; just how the weights are to be chosen, or how the y's are to be combined is an open question; the least squares answer is $Y_0 = \Sigma\alpha_i y_i$, where the weights α_i (possibly negative, but not very, and nearly always adding to one) are calculated from the linear model.

Box and Cox are assuming that for some transformed set of observations $f(y_i)$, the model is valid, and their smoothed value would be given by

$$f(Y_0) = \Sigma\alpha_i f(y_i).$$

A "non-parametric" approach would be to order the observations $y_{(1)}, \ldots, y_{(n)}$ and select Y_0 such that

$$\sum_{y_{(i)}<Y_0} \alpha_i = \tfrac{1}{2}\Sigma\alpha_i.$$

Essentially, Y_0 is the median of the distribution consisting of points $y_{(i)}$ with probability α_i (possible negative values confuse this interpretation). The justification of this procedure is that Y_0 should not be too far from the value obtained by Box and Cox's procedure, since the median of the $f(y_i)$'s will be approximately equal to the mean of the $f(y_i)$'s; but this procedure is invariant under *any* monotonic transformation of the observations.

I have tried this with Box and Cox's 3^3 experiment, when x_0 is at the centre of the cube (0, 0, 0). The weights α_i will depend on the linear model; for a complete factorial model $\alpha_i = 1$ at (0, 0, 0) and 0 elsewhere so that no smoothing takes place; for the second-degree polynomial model $\alpha_i = 7$ at the centre, 4 at the midpoint of a face, 1 at the midpoint of an edge and -2 at a vertex; for the first- and zero-degree polynomials, $\alpha_i = 1$ everywhere and the smoothing is excessive.

The smoothed values with various similarity coefficients (we may regard α_i as the relevance of the ith observation to Y_0) and various methods of combination are

Degree of Polynomial	*Mean*	*Mean* log	*Median*
0,1	861	564	566
2	724	610	604
CF	620	620	620

Negative weights are a nuisance, and, also, we would like the similarity coefficients to decrease with distance. However, least squares is the only general way of generating the coefficients at present.

I wonder if the interquartile range of the distribution over the y_i with weights α_i would be a reasonable (transformation invariant) measure of dispersion of a new observation y about Y_0. In general this would tend to be large if y_i's which were observed under highly similar conditions were a long way from the predicted Y_0 at x_0.

A preliminary analysis of the above type based on the order statistics would be invariant under monotonic transformation, and so would seem an appropriate method of finding a transformation in which an ordinary "metric" analysis might be performed.

I have found this paper extremely informative and stimulating and it gives me great pleasure to second the vote of thanks to Professors Box and Cox.

The vote of thanks was put to the meeting and carried unanimously.

The following written contribution was read by Professor D. G. Kendall.

Professor J. W. TUKEY: The results reported by Professors Box and Cox clearly represent a substantial step forward; all those concerned with the actual analysis of data should be pleased to know that they do exist, both because of the new and modified techniques which they urge us to try, and because these results were obtained by using almost "all the allowed principles of witchcraft" as of the year 1964: normality assumptions, maximum-likelihood estimation, Bayesian inference and *a priori* distributions invariant under natural, transitive groups. This last fact makes it inevitable that intelligent choice of modes of expression for the observed responses will become both socially acceptable and widely taught and that the long-run consequences for the analysis of data will be very desirable.

While this is a useful step forward, it is, I think, important not to overestimate its conclusiveness. From the point of view of the man who does indeed have data to analyse, these results are merely further guidance about a situation only reasonably close to the one he actually faces. This is, of course, no novelty in statistics, but some aspects of the present discussion make it important to re-emphasize some things that should be familiar to all of us. In the authors' discussion, as in all to nearly all of our presently available theory, all the approaches are at least formally based upon a model involving normality—or, as I would rather say, Gaussianity. I think that this is stressed by the discussion in Section 5 where one is asked to look first at the evidence from assumed Gaussianity, then at the evidence from an additional assumption of constancy of variance in the presence of Gaussianity and, finally, at the evidence from a further assumption of additivity in the presence of both other assumptions. So long as we are going to work with tight specifications, where only a few parameters can be allowed to enter, it is hard to see how things can be done in any other way than this. But from the point of view of the man with the actual data, it would make much more sense to ask—possibly in vain—for an analysis in which one could examine first the evidence derived from assumed additivity in the absence of other assumptions, secondly (in those situations where this was appropriate) the evidence provided by an additional assumption of constant variance in the presence of additivity, and thirdly (in perhaps a few cases) the additional evidence provided by assumed Gaussianity, in the presence of both additivity and constancy of variance. (If additivity —or, more generally, parsimony—is at issue, considerations of constancy of variance and Gaussianity of distribution are usually negligible, at least so far as the choice of a mode of expression is concerned. If additivity is not at issue, constancy of variance usually dominates Gaussianity of distribution.) If all of us can have enough good ideas over a long enough period of time, perhaps we can come, eventually, to a theory which corresponds more directly to what we desire. It may well be that, with the exception of very rare instances, the differences in practice associated with such an approach would be inappreciably different from those suggested by the present approach. The widespread tendency for additivity, constancy of variance and Gaussianity of distribution to come and go as a group offers us such a hope. It would be nice to know whether or not this hope is justified.

We are all used to having maximum-likelihood estimation combine different bits of evidence with quite appropriate weights. Accordingly, we may hope that this is still the case in the present situation, but I must report that the relative weighting of the evidence provided by interaction sums of squares and error sums of squares does not feel as if it were being quite fairly weighted when one merely looks, as in Table 3, at the total of these two sums of squares. Perhaps the decomposition into the three parts mentioned above, and concentration upon the part associated with the additivity assumption, might produce a much heavier weighting of the interaction sums of squares. Again it would be interesting to know whether or not this is true.

In most circumstances one is going to be more interested in reaching additivity than in maximizing the formal sensitivity of the main effects. There will be, however, a few instances where the reverse is true. I am not clear, from the discussion of Table 6, to what extent the results of applying the proposed approach rigorously and without thought will differ from the results obtained by seeking maximum sensitivity. If there should be differences which persist as the amount of data is increased without limit, I think one will have, in the long run, to look more carefully into the choice of criterion, where a decision to look need not imply an ultimate decision to adopt a different criterion.

Clearly Box and Cox have made a major step forward in the succession of approximations which give us better and better answers to an important problem of practice.

The following written contribution was read by the Honorary Secretary.

Professor R. L. PLACKETT: The authors have come up with the interesting ideas we would have expected from them, and deserve our congratulations for a paper which will be widely appreciated. They have made full use of modern computational facilities and the two systems of inference which are currently competing for our attention. An impression left by reading their paper is that the data should be fed into a large and powerful machine which will very quickly draw all the necessary graphs and print out the best analysis of variance available in the circumstances. Those accustomed to the blissful ease of the standard analysis of variance calculations will need to be convinced that such hard work is really necessary, and will ask for assurance that too much responsibility has not been delegated.

So much has recently been said on Bayesian procedures that it is a relief to find that the authors are not really Bayesians at all, but have been very ingenious in using Bayesian arguments without ever becoming fully committed to them. Thus they call for uniform distributions, but only over the region where the likelihood is appreciable, and they justify their preference for a Bayesian procedure on the grounds that the confidence coefficients from asymptotic distribution theory are closer to their nominal values if L_b is used instead of $L_{\max}$. It is true that in the further analysis separating out A and H they suggest that the two procedures may lead to appreciably different conclusions, but the circumstances in which this might occur are not closely defined. Surely it is not the magnitude of either $S_{\nu_r}(\lambda; z)$ or $F(\lambda; z)$ which is relevant, but that of the derivatives of these quantities with respect to λ. In any case, the authors do not tell us what they would do if the conclusions differ markedly; but it accords with the spirit of this long-awaited collaboration that we should be left in doubt as to which method of inference to follow.

Likelihood procedures have also been well publicized and discussed, but there is a practical point which seems not to have been emphasized in the midst of a good deal of mathematical and logical argument. It arises because the likelihood function contains much that is taken for granted in the way of distributional forms, and is no substitute for an inspection of the data. As a simple illustration, consider a large sample of measurements in which half are clustered round the value a and half round the value b $(a \neq b)$. The assumption that this constitutes a sample from a normal distribution with mean μ and standard deviation σ leads to an exactly parabolic log likelihood function for μ, but the inferences that this would suggest conflict with those obtained directly from the data.

It is tempting to contrast the smooth and deceptive character of a likelihood function with the spotty but straightforward nature of Anscombe and Tukey's procedures. They fit a full linear model to the original data and plot residuals against fitted values. Residuals are something which the authors have not calculated, but it would have been interesting to see other methods at work on the same examples. One might consider a modification of the Anscombe–Tukey procedure in which the predicted value Y is plotted against the observed value y. This will lead to a linearizing transformation $Y = f(y)$ (e.g. by Dolby's, 1963, analysis of the simple family); the procedure can be iterated if necessary and should converge under reasonable conditions. It may be objected that the possibility of differing variances is not taken into account, but the usual argument is that the same transformation does for both. If a greatly differing transformation is necessary to equalize the variances, then the experiment is unlikely to be very successful.

In the second part of their paper, the authors separate the contributions of linearity, constant variance and normality, but the place of normality in their analysis is logically different from that occupied by the other two, since normality is not a constraint which they either apply or relax. For that, they would presumably need to carry through the entire analysis with some other distribution.

Professor M. S. BARTLETT: Like Professor Tukey, I think that the authors have made a major step forward in this paper on the theory of transformations. I think also, like Professor Plackett, I was a little uneasy about the extent to which complicated analysis might seem necessary.

Again, like Mr Nelder, I found myself wondering about the Box and Cox nature of the paper and in particular whether this kind of oscillatory character between likelihood and Bayes analysis had any relevance to the Box and Cox aspect! Perhaps Professor Cox may wish to comment on this; on this point of Bayes versus likelihood I would especially welcome his views on whether he is advocating them as equally useful or whether he has reached any conclusions as to whether one is better than the other. In particular I would certainly draw attention to the point made in the paper, and I think Professor Plackett made this point also, that whichever analysis you make, the inference is very conditional on your set of assumptions from which you start.

Now to come to other minor points, I think I have only two to make. One was in the approximation used for the log likelihood, the max log likelihood and the use of χ^2 with this, and I wondered whether Professor Cox, or for that matter, Professor Box, could make any comment on the accuracy in this in other than very large samples. One knows that the distribution is valid up to but not including order $1/n$, and one knows, for example, from Professor Box's work, that if you want to go to order $1/n$ you have to bring in a different multiplying factor to your χ^2 approximation. And it would help to know whether there is any possibility of getting the sort of confidence limits based on the χ^2 analysis a bit more exact, and if not, how misleading they might occasionally be.

I think my last point is one that was raised by Professor Tukey and that is, I did wonder about the uniqueness of this order of taking the various factors, normality, additivity and homogeneity of variances, and whether you would reach anything like the same sort of conclusion if you tried to take them in a different order.

Dr M. R. SAMPFORD: Like Professor Tukey, I am rather nervous about the effect of the assumed normality of the transformed variable on the additivity, in particular, and to a lesser extent on the homogeneity of variance, when in fact no single transformation will achieve all three properties. The relatively small amount of information about λ obtained from the normality assumption in the example (Table 8) seems to be reassuring on this point, but the possible effects when the transformed distribution is rather far from normal might still be serious. Of course, one can sometimes advance a more plausible distributional model, and in this context it may be worth suggesting that, though the title of this paper should more properly be "An Analysis of Transformations to Normality", the ingenious approach on which it is based could perfectly well be applied to other distributions. For example, I have several times encountered response-time distributions—in particular, distributions of time to death—that appear log-normal at the lower end of the scale, but have a secondary mode in the upper tail. This might suggest that some animals die as a direct result of damage caused by the treatment, but that others, having a high tolerance or being, by chance, little damaged, may survive the initial shock, only to die later as a result of physiological disturbance caused by the damage. One might, by making some assumptions about distributions of damage and tolerances, derive a more or less plausible class of distributions for transformed times that might be expected to be consistent with variance homogeneity and at least approximate additivity. The method of this paper could then be applied to determine the most satisfactory transformation leading to a distribution in this class. This is perhaps a rather extreme example, but I hope suggests the potential value of the authors' approach in situations where additivity need not be expected to involve, as it often does, near-normality.

Dr C. A. B. SMITH: I merely wish to draw attention to a recent paper by A. F. Naylor (1964). He applied the arcsine, logit, log-log and normal equivalent deviate transformations to four sets of biological data. He concluded that for all practical purposes they could be considered as equivalent. For example, in most of the entries the expected numbers calculated from the four transformations differ only slightly in the first decimal place.

Mr D. KERRIDGE: I have two comments to make, one general and one particular. The general comment is that it is very pleasant to have a paper in which the idea is obvious. I am not saying this in any derogatory sense. I think all the great ideas were obvious ones. Nothing could be more obvious than the idea of taking a parametric family and estimating the parameter. It is strange that such an obvious idea should take such a long time to be seen, but in many ways, the simpler the idea, the greater the discovery. There is, for example, much more chance that a simple idea will be used in practice. The particular comment concerns the rather strange prior distribution which has the interesting property that it contains the observations. We cannot let the night go without saying something about that. Clearly this is not an expression of belief, so some people would not call it a probability. It is not prior, because it is determined *a posteriori*, and so it is a pseudo-prior pseudo-probability. Now I am not against it because of its strangeness, since obviously the authors have extremely good reasons for using it. They use it because it works. It is very interesting indeed to find a practical example in which you have to use something which clearly is a pseudo-probability. I believe that as we get to use Bayes's theorem instead of talking about it, as I hope we are going to do in the future, we are going to come up against many more of these peculiar things. For example, I think that to get sensible significance tests in Bayesian theory we are going to have to use prior probabilities which depend on the number of observations. These again will be pseudo-probabilities, in a sense pseudo-prior too. So this is a very interesting first example of something which will eventually, I think, shed some light on what probabilities really are. My view is that they do not express beliefs. They are a convenient figment introduced to do something we do not really understand yet, but by examining examples of this sort I hope that one day we will achieve understanding.

Mr E. M. L. BEALE: I should like to add my thanks to Professors Box and Cox for a most valuable paper, and to ask one question. Would the authors ever consider using a transformation of the type (1) when some y's are negative, or one of type (2) where some $y_i+\lambda_2$ is negative? Such a transformation obviously has strange arithmetic properties. It gives a real answer if λ_1 is integral, and I think one can always overcome any problems created by the fact that y may not be uniquely determined by the value of $y^{(\lambda)}$. But would the transformation ever make sense statistically?

The following written contribution was received after the meeting:

Professor F. J. ANSCOMBE: The authors are to be congratulated on a most remarkable paper. The basic idea is highly original, and the tackling of horrendous difficulties is breath-taking. The examples are illuminating, and the preliminary "rather informal" analysis of the textile example is statistry in the grand manner—but, indeed, the whole paper is that.

Because of my own efforts with residuals, I have been particularly interested by Section 6. In my 1961 paper I gave a formula for roughly estimating the power transformation that would remove Tukey's type of removable non-additivity, and also one for estimating the power transformation that would remove an exponential dependence of error variance on the mean. The formulas were based essentially on the statistics denoted by T_{12} and T_{21}, respectively, in this paper. I did *not* also give a formula aimed at removing skewness of the error distribution, based on the statistic here denoted by T_{30}, though I have since used such a formula; in the notation of my 1961 paper the formula goes

$$p = 1-2g_1\,\bar{y}/3(2+g_2)\,s.$$

(My p is Box and Cox's λ, $\bar{y}$ is the overall sample mean, s the residual root mean square, and g_1 and g_2 are analogues of Fisher's g-statistics.) It was my thought that one would calculate one or more of these expressions, and (if more than one) hope they would somewhat agree. No doubt, with factorial data showing pronounced effects for at least two factors, one would attach primary importance to additivity. With only one effective factor, there would be no question of additivity, and one would attach primary importance to constancy of variance. With no effective factors, and in particular with a simple homogeneous sample, there would be nothing to worry about except skewness.

Now Professors Box and Cox have shown that these three separate estimates should (very nearly) be averaged in a certain proportion to yield a best estimate of the power. This result, for the relatively simple calculations based on residuals from a least-squares analysis on one scale, parallels the subtle decomposition of the likelihood function into three parts in Section 5.

Professor Cox replied briefly at the meeting and the authors subsequently replied more fully in writing as follows:

We are very grateful to the speakers for their encouraging and helpful remarks.

One important general issue raised by Professors Tukey, Plackett, Bartlett and Dr Sampford concerns priorities for the criteria of simplicity of the model and specifically of additivity, A, homogeneity of variance, H, and normality, N. We certainly agree on the importance of the first of these, as indeed we indicate in our remarks at the end of Section 2. In the formal analysis of Section 5 we have considered *N*, *HN*, *AHN* as three models in that order. If one is to employ a parametric approach one must, it seems, start from some distributional assumption although, of course, if desired this could be broader than that adopted here. Furthermore, there is no reason in principle why *A* should not have been taken before *H* in discussing the biological example. We would then have to fit an additive model with separate within-cell variances. The rough justification for thinking that the procedure given in the paper genuinely separates out the effects of *N*, *H* and *A* is that $M(\lambda; \mathbf{z})$, on which (47) and (51) depend, is a valid descriptive measure of heterogeneity of variance independently of *N*. Likewise $F(\lambda; \mathbf{z})$ is a descriptive measure of non-additivity independently of *H* and *N*. If we started from a non-normal model, we would get a different measure of heterogeneity of variance, but except in extreme circumstances it is unlikely that it would be minimized by a value of λ very different from that minimizing $M(\lambda; \mathbf{z})$. An analogous remark applies to $F(\lambda; \mathbf{z})$. Under non-normality the weighting of the different requirements will be different, but it is hard to see how a radically different value of λ could emerge from the final analysis.

Concerning Professor Tukey's point about the appropriateness of the weighting given by the likelihood in the biological example, the truth seems to be that in this example non-additivity is not in fact the major contribution in determining λ. The sizes of the mean squares in Table 3 seem rather to bear this out than to contradict it. Concerning Tables 3 and 6 a striking thing is not only the removal of non-additivity, or correspondingly in Table 6 the simplification of the model, but also the large increase in sensitivity of the experiment. The result achieved by transformation is in fact equivalent to threefold increase in experimental effort.

In the paper we were at pains to stress that, where the procedures do seem relevant, we recommend using them in a flexible way, and that the assumptions on which they are based are a tentative working basis for the analysis rather than anything to be adopted irrevocably. In particular, in the discussion of the textile example we deliberately gave first the "common-sense" analysis before the more elaborate one. As Mr Kerridge has very rightly stressed, the basic idea is an extremely simple one; in particular, the absence of iterative calculations is a considerable practical advantage. We hope that this will reassure Professor Plackett that we are not advocating unnecessary elaboration. Mr Nelder has stated extremely clearly the need for a more searching examination of "assumptions".

We have not specifically investigated the point raised by Professor Bartlett concerning the adequacy of the chi-squared approximation for confidence intervals for λ. However, the line we have followed in finding a closer approximation to the posterior density of λ leads to posterior intervals based on the *F* distribution and a similar approximation might be found for confidence intervals. The use of $L_b(\lambda)$ instead of $L_{\max}(\lambda)$ was suggested by analogy with Bartlett's (1937) procedure of applying the likelihood-ratio procedure after suitable contrasts have been removed by transformation. The difficulty when λ is unknown is that the transformations to remove the parameters θ depend on λ, so that the argument is at best approximate. We were most interested in Mr Nelder's remarks on this point and hope that he will develop his ideas further.

The maximum-likelihood approach and the Bayesian approach have deliberately been given as entirely separate but parallel developments. Professor Plackett suggests that we justify the Bayesian approach only because it leads to "better" confidence intervals; this is not so. Several speakers have commented on the special prior distribution (19) which involves the observations. As we remarked in the paper, it is possible that there is an alternative and better approach to this; one way may be to make the prior distributions for the contrasts depend on the general population mean. However, the observations enter (19) only in a mild way in establishing the overall level of the observations, usually the overall geometric mean in our special cases. It is essential that some allowance should be made for the fact that the prior distribution for the magnitude of the contrasts depends on the overall magnitude of the observations.

In answer to Mr Beale's question, we feel that, while it is probably possible to develop the theory for non-monotonic transformations of the dependent variable, we cannot think of any situations where such transformations would be physically allowable.

We are grateful to Dr Smith for his reference to Naylor's work. However, Naylor seems to be considering situations where the transformations are, over the relevant range, practically linear functions of one another. In our examples the relative range of variation of the observations is high, the transformations are very non-linear and this is of course why we are able to obtain fairly sharp discrimination between the different values of λ. In the quantal response case, the transformations in question become essentially different only in the tails of the response curve, and observations there would be required for the differences to be detectable and of practical importance.

We are very interested in Professor Anscombe's remarks on residuals. Further comparisons of the analysis of residuals with the methods of our paper would be of value.

We are interested in Dr Hartigan's problem and formulation. However, this seems essentially different from ours, partly because in our applications we are primarily interested in changes in response, rather than in absolute responses, and partly because one of our primary objectives is to find a scale on which the factor effects are succinctly characterized by a few parameters. Even if the distributional assumptions were to be phrased non-parametrically (which we would in any case not wish to do), we must have parameters in order to describe at all concisely the changes in response in a complex system.

References in the Discussion

Dolby, J. L. (1963), "A quick method for choosing a transformation", *Technometrics*, **5**, 317–326.

Naylor, A. F. (1964), "Comparisons of regression constants fitted by maximum likelihood to four common transformations of binomial data", *Ann. hum. Genet., Lond.*, **27**, 241–246.

4.10

Some Problems Associated with the Analysis of Multiresponse Data

ASQC Chemical Division Technical Conference
1971 Prize Winning Paper

G. E. P. BOX*, W. G. HUNTER, J. F. MacGREGOR**, and J. ERJAVEC***
University of Wisconsin

Experience has shown that unless special care is exercised in analyzing multiresponse data serious mistakes can be made. In this paper some problems associated with fitting multiresponse models are identified and discussed. In particular, three kinds of dependencies are considered: dependence among the errors, linear dependencies among the expected values of the responses, and linear dependencies in the data. Since ignoring such dependencies can lead to difficulties, a method is described for detecting and handling them. The concepts involved are illustrated with a chemical example.

KEY WORDS

Estimation
Multiresponse data
Linear dependencies
Eigenvalue-Eigenvector analysis
Nonlinear models
Chemical kinetics
Multivariate data

1. INTRODUCTION

Engineers and scientists frequently need to analyze multiresponse data. When studying a chemical reaction for instance, for each setting of a group of "input" variables determining the reaction conditions, not one but a number of "output" variables or responses (such as the concentrations of each of the chemical constituents) may be measured. The capability of making such multiple measurements has greatly increased with the advent of better analytical tools such as the gas chromatograph. This capability has, in turn, increased the potential information generated by a particular experimental run, making possible more precise discrimination among models, more adequate checking of models, and more accurate estimation of parameters.

But with the capability to measure with comparative ease all the substituents in the reaction mixture comes the necessity to take account of possible dependencies

* Supported by the Air Force Office of Scientific Research under Grant AF-AFOSR 72-2362.
** Present address: Dept. of Chemical Engineering, McMaster Univ., Hamilton, Ontario
*** Present address: American Cyanamid, Bound Brook, New Jersey
Received Sept. 1970; revised May 1972.

The Collected Works of George E. P. Box, 1984, Wadsworth, Inc., Belmont, CA 94002.
Originally published in *Technometrics,* vol. 15, no. 1 (1973), pp. 33–51.

among them (arising, for example, as a consequence of the law of conservation of matter). If a particular chemical mechanism is given, it may follow, for example, that to maintain the carbon balance a certain linear relationship must exist among the amounts of the substituents in *all* the experimental runs. A second such relationship may exist to maintain the nitrogen balance and so on. Relationships of this kind are called stoichiometric. Now in practice the experimenter is *not* given the mechanism but must learn about it as a result of an iterative conversation between the generated data and the theoretical possibilities sparked off in his mind. He tentatively entertains possibilities and by suitably planned experiments and suitable analysis he allows the data to comment on these.

Various kinds of problems all associated with dependencies of one kind or another then arise:

(i) Whether or not stoichiometric type dependencies exist, to fit a tentative model to the data one must properly take account of *correlations* between the errors in different substituents.

(ii) If the nature of dependencies among the expected values of the various substituents is known, this knowledge might confirm or deny the validity of the model under study or suggest the appropriateness of some new mechanistic form.

(iii) The investigator frequently uses one or more stochiometric type relations to deduce the presumed values of substituents which are difficult to measure, or to adjust measured values to agree with these relations. He may be unaware of the implications of such practices on the estimation process and the data analyst may not always be aware of what the experimenter has done.

Since the advent of new methods of chemical analysis and increasing interest in mechanistic studies using multiresponse data, we have encountered a rash of problems which point to the need for discussing and distinguishing the different types of dependencies and for providing methods for use in practical data analysis whereby mistakes can be avoided, and forgotten or unknown relationships can be made manifest.

2. Three Kinds of Dependencies

Suppose that, in an experimental program, n sets of reaction conditions (not necessarily all different) are run, and at each set of conditions r responses $(y_1, y_2, \cdots, y_i, \cdots, y_r)$ are recorded. Suppose furthermore that we can write a mathematical model for the ith response at the uth set of reaction conditions

$$y_{iu} = \eta_i(\boldsymbol{\xi}_u, \boldsymbol{\theta}) + \epsilon_{iu} \qquad i = 1, \cdots, r \quad u = 1, \cdots, n \tag{1}$$

where ϵ_{iu} is the error in the ith response for the uth run, $\boldsymbol{\theta}$ are unknown parameters and $\boldsymbol{\xi}_u$ are the values of the input variables defining the reaction conditions for the uth run.

Three kinds of dependencies among the responses will be considered in this paper along with the effects of each on the fitting of multiresponse models.

2.1 *Dependence Among the Errors*

Consider the r errors committed in the uth run, $\boldsymbol{\epsilon}_u' = (\epsilon_{1u}, \epsilon_{2u}, \cdots, \epsilon_{ru})$. It will usually be true that these errors are correlated. It is important that the statistical treatment of the data should take account of this correlation, and this has not always been done. For example, one technique for estimating parameters which has been used (for example, see Ball (1966)) is to find those parameters which minimize the overall residual sum of squares from all the responses

$$\text{RSS} = \sum_{i=1}^{r} \sum_{u=1}^{n} [y_{iu} - \eta_i(\xi_u, \boldsymbol{\theta})]^2. \tag{2}$$

However, it is easily seen (Box & Draper (1965), Hunter (1967)) that this criterion is appropriate only if (a) the errors are all uncorrelated and (b) the errors all have equal variances. In practice neither of these circumstances is likely to be true, and analysis of data as if it were true can give incorrect results (Eakman (1969), Erjavec (1969)). A method which overcomes these difficulties was developed by Box and Draper (1965). Assuming that the errors were distributed according to a multivariate Normal distribution with unknown variance-covariance matrix $\boldsymbol{\Sigma} = E(\boldsymbol{\varepsilon}_u \boldsymbol{\varepsilon}_u')$ and using a "non-informative" prior distribution, they showed that, given the data, the posterior distribution for all the parameters, $\boldsymbol{\theta}$, is proportional to

$$|\mathbf{V}_r|^{-n/2} \tag{3}$$

where

$$V = \{v_{ij}\} \quad \text{and} \quad v_{ij} = \sum_{u=1}^{n} (y_{iu} - \eta_{iu})(y_{ju} - \eta_{ju}).$$

Estimates of $\boldsymbol{\theta}$ yielding maximum posterior density are those obtained by minimizing the determinant $|\mathbf{V}_r|$. In the particular case of a single response, this procedure leads to the method of least squares.

2.2 *Linear Dependence Among Expected Values of Responses*

In chemical systems stoichiometry, material and energy balances, or steady-state conditions will usually require that certain linear relationships exist among the expected values of the responses. For example, for every run, stoichiometry may dictate that the sum of the expected values of the number of moles of the r constituents in the system must be the same. That is,

$$\sum_{i=1}^{r} E(y_{iu}) = \sum_{i=1}^{r} \eta_i(\xi_u, \boldsymbol{\theta}) = a_0 \qquad u = 1, \cdots, n \tag{4}$$

More generally, there may be m independent linear relations among the expectations which must be satisfied for each run,

$$\sum_{i=1}^{r} a_{qi} E(y_{iu}) = \sum_{i=1}^{r} a_{qi} \eta_i(\xi_u, \boldsymbol{\theta}) = a_{q0} \qquad u = 1, \cdots, n \quad q = 1, \cdots, m \tag{5}$$

Or, in matrix notation,

$$\mathbf{A}E(\mathbf{y}_u) = \mathbf{a}_0, \qquad u = 1, \cdots, n \tag{6}$$

where $\mathbf{A} = \{a_{qi}\}$ is an $m \times r$ matrix and $\mathbf{y}_u = (y_{1u}, \cdots, y_{ru})'$ is the uth data vector.

2.3 *Linear Dependencies in the Data*

Suppose an experimenter knows that a formula such as (4), for example, expressing some material balance relationship must be true for each of his experimental runs and yet, because of experimental error, it is not exactly satisfied by his observed responses. As is commonly done in practice, the experimenter may force his observations to fit his relationship by some normalizing calculation to give

$$y_1 + y_2 + \cdots + y_r = a_0 \tag{7}$$

This is often done by multiplying each of the originally measured values of the

r observations $y_{1u}^*, y_{2u}^*, \cdots, y_{ru}^*$ for a given run by the factor $a_0/\sum_{u=1}^{r} y_u^*$. In so doing he produces an exact linear dependence among the responses $y_1, y_2, \cdots, y_r$. Another way in which exact linear dependence in the data is introduced is when the investigator measures only $r - 1$ responses independently and calculates the rth response using the relationship (7). More generally, by placing absolute trust in the specific model he favors, the experimenter might make use of $m_1 \leq m$ of the relationships in (6). Thus he might make only $r - m_1$ independent determinations of the substituents and obtain the others by calculation, or, he might complete r independent determinations and then force the m_1 relationships to hold by normalizing or otherwise adjusting the data. Suppose the first $m_1 < m$ relations are used for normalizing, and the matrix $\mathbf{A}$ and the vector $\mathbf{a}_0$ are partitioned after the m_1th row so that (6) may be written

$$\begin{bmatrix} \mathbf{A}_1 \\ -- \\ \mathbf{A}_2 \end{bmatrix} E(\mathbf{y}_u) = \begin{bmatrix} \mathbf{a}_{10} \\ -- \\ \mathbf{a}_{20} \end{bmatrix}$$

If $\mathbf{y}_u$ is an $r \times 1$ vector of "data" values including those obtained by calculation, then there will be the following m_1 exact linear relationships connecting each data vector

$$\mathbf{A}_1 \mathbf{y}_u = \mathbf{a}_{10} \tag{8}$$

Whenever possible, of course, normalizing or adjusting "data" in the manner described above should be avoided. The experimenter should be prepared to go to some trouble to determine each response independently, and having done so he should refrain from forcing observed responses to satisfy theoretical relationships that he believes to be true. He should do this because independent information on each of the m relationships can not only provide better estimates of the parameters but also make possible a more comprehensive check on the model which is currently being entertained. In some situations, however, the avoidance of linear dependencies in the data is not possible. In some instances the analytical procedures or equipment necessarily make use of such relationships. Such is the case with chemical composition data obtained from gas chromatographs, for example, where it is only possible to calculate relative percentages.

When it is impossible to determine all the substituents independently or to avoid "normalized" data, careful note should be made of which observations are independent, which were obtained by calculation, and which of the expectation relationships (6) have been employed in obtaining the data. Ignoring such dependencies can lead to serious mistakes in interpretation. Unfortunately, even if the investigator can identify the precise nature of dependencies in the data, he may be unaware of the importance of ensuring that they are taken account of in the subsequent statistical analysis.

Careful preliminary analysis of the system under study ought in principle to reveal all the dependencies in the expected values of the responses, and adequate inquiry ought to show which of the relationships may have been utilized, either consciously or otherwise, to normalize the results. Unfortunately, we have found from bitter experience that in practice dependencies are frequently overlooked and we therefore now regard it a practical necessity to look for such relationships empirically as a preliminary to the analysis of multiresponse data. Frequently, the results of such an analysis reveals unexpected but highly informative dependencies. Further analysis should not be proceeded with until these dependencies have been satisfactorily explained.

3. Eigenvalue-Eigenvector Analysis

Suppose the data analyst were a priori unaware of the nature of the possible linear relationships existing among the responses. He could then proceed as follows. Ordinarily, the vector of constants $\mathbf{a}_0$ on the right of (6) are unknown, but they can be eliminated by working with the matrix $\mathbf{D} = \{d_{iu}\} = \{y_{iu} - \bar{y}_i\}$ of deviations from the individual averages. The eigenvalues λ_k and the r-dimensional eigenvectors $\mathbf{z}_k$ of $\mathbf{DD}'$ are such that

$$\mathbf{z}_k'\mathbf{DD}' = \lambda_k\mathbf{z}_k' \tag{9}$$

and the eigenvectors can be normalized so that

$$\mathbf{z}_k'\mathbf{z}_k = 1 \qquad k = 1, \cdots, r. \tag{10}$$

If there are m_1 independent exact linear relations in the *data*

$$\mathbf{A}_1\mathbf{y}_u = \mathbf{a}_{10} \qquad u = 1, 2, \cdots, n \tag{11}$$

then

$$\mathbf{A}_1\mathbf{d}_u = \mathbf{0} \qquad u = 1, 2, \cdots, n. \tag{12}$$

In this case therefore there will be m_1 zero eigenvalues, $\lambda_1 = 0, \cdots, \lambda_{m_1} = 0$, and the associated m_1 eigenvectors $\mathbf{z}_1, \mathbf{z}_2, \cdots, \mathbf{z}_{m_1}$ will define the same hyperplane as do the m_1 rows of $\mathbf{A}_1$. Thus, if $\mathbf{Z}_1$ is the $m_1 \times r$ matrix whose rows consist of these eigenvectors, then a non-singular transformation exists such that

$$\mathbf{A}_1 = \mathbf{TZ}_1 \tag{13}$$

In general, in addition to the m_1 exact linear relations in the *data* we will have $m_2 = m - m_1$ further linear relations in the *expected values* so that

$$\mathbf{A}_2\mathbf{d}_u = \mathbf{e}_u \qquad u = 1, \cdots, n \tag{14}$$

where $\mathbf{A}_2$ is an $m_2 \times r$ matrix of coefficients and $\mathbf{e}_u$ is an $m_2 \times 1$ vector of errors all of whose elements have expected value zero. To correspond with these relations one can expect a further m_2 small eigenvalues whose expected size depends on the experimental errors via the following relation derived in the Appendix:

$$E(\lambda_k) = (n - 1)\mathbf{z}_k'\mathbf{\Sigma}\mathbf{z}_k \qquad k = m_1 + 1, \cdots, m_1 + m_2 \tag{15}$$

where $\mathbf{\Sigma} = E(\boldsymbol{\varepsilon}_u\boldsymbol{\varepsilon}_u')$ is the $r \times r$ variance-covariance matrix for the errors in the r responses. The corresponding eigenvectors (which we suppose form the rows of the $m_2 \times r$ matrix $\mathbf{Z}_2$) define a hyperplane which approximately coincides with the hyperplane given by those components of the m_2 vectors in $\mathbf{A}_2$ which are orthogonal to $\mathbf{A}_1$.

In the situation envisaged, then, there would be

(i) m_1 eigenvalues produced by adjustment of data which differ from zero, if at all, only because of rounding error*;

(ii) a further $m_2 = m - m_1$ eigenvalues produced by other relations among the expected values and whose magnitudes are determined by experimental error;

(iii) $r - m$ values which would typically be very much larger since they would be quadratic functions of the changes in response produced by changes in the experimental conditions.

These three different kinds of roots usually differ in size by orders of magnitude. To separate them into appropriate classes all that is usually needed is a rough estimate of the anticipated size of the λ's associated with the expectation rela-

* For a precise measure of how large these near zero eigenvalues could be a covariance matrix $\mathbf{\Sigma}_{re}$ for rounding error could be substituted in (15).

tionships given by (15). When runs have been replicated, then the sample estimate $\hat{\mathbf{\Sigma}}$ may be substituted in (15) to yield estimates of the λ_k's. When no replicated experiments exist sometimes previous experimentation will provide a rough value for $\mathbf{\Sigma}$. Since only a very approximate "order of magnitude" value is really needed, if high correlations between experimental errors in the various responses are not expected it will be enough to approximate $\mathbf{\Sigma}$ by $\mathbf{I}\bar{\sigma}^2$ where $\bar{\sigma}^2$ is an average value for the residual error variance of the r responses. Then

$$E(\lambda_k) \simeq (n - 1)\mathbf{z}_k'\mathbf{I}\mathbf{z}_k\bar{\sigma}^2 = (n - 1)\bar{\sigma}^2 \qquad (16)$$

If the order of magnitude of the r responses is very different, for example because of different measurement units, then it would be better to approximate $\mathbf{\Sigma}$ by a diagonal matrix containing the appropriate estimates of the variances of the individual responses.

After proceeding with the subsequent main analysis in which the parameters $\boldsymbol{\theta}$ are estimated a value for $\mathbf{\Sigma}$ will be available from the residuals so that at this later stage we can return to the preliminary analysis and recheck for agreement.

4. A Simple Illustration

For illustration, suppose that $r = 3$ constituents can be measured, and that

$$\text{(i)}\ E(y_1) + E(y_2) + E(y_3) = 6 \qquad (17)$$

$$\text{(ii)}\ E(y_1) - 2E(y_2) = 3. \qquad (18)$$

Suppose, in fact, the experimenter chemically determines y_1 and y_2 separately but estimates y_3 "by difference" according to

$$y_1 + y_2 + y_3 = 6 \qquad (19)$$

(This same linear dependency (19) would also result if the experimenter measured all three responses for a given run u, y_{1u}^*, y_{2u}^*, and y_{3u}^*, and then "normalized" the data by multiplying each measurement for that run by the factor $6/(y_{1u}^* + y_{2u}^* + y_{3u}^*)$). Then a typical set of data might be as follows:

$$\mathbf{Y}' = \begin{array}{c} \begin{matrix} y_1 & y_2 & y_3 \end{matrix} \\ \begin{bmatrix} 2 & 0 & 4 \\ -1 & -2 & 9 \\ 3 & 0 & 3 \\ 6 & 1 & -1 \\ 5 & 1 & 0 \end{bmatrix} \end{array} \qquad (20)$$

with averages $\bar{y}_1 = 3$, $\bar{y}_2 = 0$, and $\bar{y}_3 = 3$. The relationship (19) is exactly true for every row, but because y_1 and y_2 are subject to error, the relationship (18) yields

$$(y_1 - \epsilon_1) - 2(y_2 - \epsilon_2) = 3 \qquad (21)$$

so that

$$y_1 - 2y_2 = 3 + e \qquad (22)$$

where

$$e = \epsilon_1 - 2\epsilon_2 . \qquad (23)$$

The matrix **D** of deviations from the individual averages is

$$\mathbf{D}' = \begin{bmatrix} -1 & 0 & 1 \\ -4 & -2 & 6 \\ 0 & 0 & 0 \\ 3 & 1 & -4 \\ 2 & 1 & -3 \end{bmatrix} \tag{24}$$

and

$$\mathbf{DD}' = \begin{bmatrix} 30 & 13 & -43 \\ 13 & 6 & -19 \\ -43 & -19 & 62 \end{bmatrix} \tag{25}$$

The eigenvalues and eigenvectors for $\mathbf{DD}'$ are shown in Table 1. Since λ_1 is zero, $\mathbf{z}_1$ corresponds to an exact linear dependence among the responses,

$$0.5774\ y_{1u} + 0.5774\ y_{2u} + 0.5774\ y_{3u} = \text{a constant} \tag{26}$$

Or, after multiplying both sides by 1/0.5774, we have

$$y_{1u} + y_{2u} + y_{3u} = \text{a constant} \tag{27}$$

TABLE 1

The Eigenvalues and Eigenvectors of $\mathbf{DD}'$ *for the Simple Illustration.*

i	λ_i	$\mathbf{z}_i'$
1	0	(0.5774 0.5774 0.5774)
2	0.338	(0.6006 -0.7793 0.1787)
3	97.66	(-0.5531 -0.2436 0.7967)

agreeing with (19). Suppose that the error variance for the substituents y_{iu} is known to be of the order of 0.1. Since λ_2 ($= 0.338$) is of the same order of magnitude as $(n - 1)\bar{\sigma}^2$ ($= 4 \times 0.1 = 0.4$) we anticipate that $\mathbf{z}_2$ corresponds to a linear relationship among the expected values of the responses, or, more specifically, to that part of the relationship which is orthogonal to $\mathbf{z}_1$. To see that this is so, note that the component of $\mathbf{A}_2 = (1, -2, 0)$ which is orthogonal to $\mathbf{A}_1 = (1, 1, 1)$ is $(0.617, -0.772, 0.154)$ when normalized. This vector is very close to the eigenvector $\mathbf{z}_2$ as expected. We see in this simple illustration how eigenvalue-eigenvector analysis helps in uncovering the linear relationships which exist among the observations and among the expected values.

5. IMPLICATIONS IN DATA FITTING

When there are r responses containing m_1 exact linear relations the $r \times r$ matrix $\mathbf{V}_r$ formed as in equation (3) has rank $r - m_1$. Attempts which have occasionally been made to make inferences about the parameters by studying $|\mathbf{V}_r|$ have led to nonsensical results since this determinant is equal to zero for all values of the parameters. Small rounding errors have further confused the picture since, when these are present, $|\mathbf{V}_r|$ will be not quite zero and will change as the parameters are changed. A correct analysis is obtained by studying $|\mathbf{V}_{r-m_1}|$. This $(r - m_1) \times (r - m_1)$ determinant should contain only those responses which have been independently determined (or more generally any $r - m_1$ independent linear combinations of them). In particular, on the assumptions made earlier the posterior distribution of the parameters $\boldsymbol{\theta}$ is proportional to $|\mathbf{V}_{r-m_1}|^{-n/2}$ and the estimates yielding maximum posterior density are obtained by minimizing this determinant.

The existence of m_2 further independent linear relationships in the *expected values* of the r responses should *not* lead to elimination of m_2 further responses. These relationships do not cause singularities in $|\mathbf{V}_{r-m_1}|$ and they contain valuable information allowing us to check the adequacy of the model and to obtain more precise estimates of the parameters $\boldsymbol{\theta}$. An eigenvalue-eigenvector analysis can draw attention to both kinds of relationships. The former kind must be allowed for in subsequent data fitting; the latter are of interest to the experimenter and in some cases may confirm or deny the adequacy of the model which is being fitted, but they will not directly affect the fitting of the data.

Each problem must be considered on its merits. The object of this paper is to point out that a preliminary eigenvalue-eigenvector analysis is a useful tool and, preferably, further analysis should not be proceeded with until the relationships it points to have been satisfactorily explained.

6. Chemical Example

The thermal isomerization of α-pinene to dipentene and allo-ocimene which in turn yields α- and β-pyronene and a dimer was studied by Fuguitt and Hawkins (1947). The proposed reaction scheme for this homogeneous chemical reaction is:

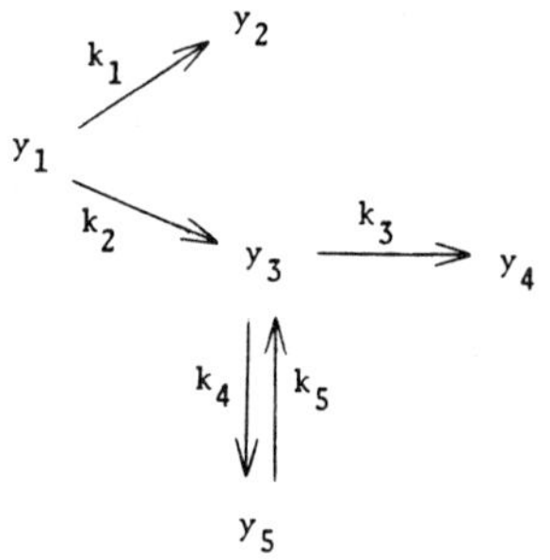

The concentrations of the reactant and the four products were reported by Fuguitt and Hawkins at eight time intervals, and these data are reporduced in Table 2. Mathematical models can be derived which give the concentration of the various species as a function of time if the chemical reaction orders are known. Hunter and MacGregor (1967), assuming first-order kinetics throughout, derived the following equations:

$$\eta_{1u} = y_{10}e^{-\phi t_u}, \qquad \eta_{2u} = \frac{\theta_1 y_{10}}{\phi}(1 - e^{-\phi t_u}), \qquad \eta_{3u} = C_1 e^{-\phi t_u} + C_2 e^{\beta t_u} + C_3 e^{\gamma t_u},$$

$$\eta_{4u} = \theta_3\left(\frac{C_1}{\phi}(1 - e^{-\phi t_u}) + \frac{C_2}{\beta}(e^{\beta t_u} - 1) + \frac{C_3}{\gamma}(e^{\gamma t_u} - 1)\right),$$

and

$$\eta_{5u} = \theta_4\left(\frac{C_1}{(\theta_5 - \phi)} e^{-\phi t_u} + \frac{C_2}{(\theta_5 + \beta)} e^{\beta t_u} + \frac{C_3}{(\theta_5 + \gamma)} e^{\gamma t_u}\right),$$

where y_{10} = the value of y_1 at $t = 0$, $\alpha = \theta_3+\theta_4+\theta_5$, $\beta = (-\alpha + \sqrt{\alpha^2 - 4\theta_3\theta_5})/2$, $\gamma = (-\alpha - \sqrt{\alpha^2 - 4\theta_3\theta_5})/2$, $\phi = \theta_1 + \theta_2$, $C_1 = \theta_2 y_{10}(\theta_5 - \phi)/((\phi + \beta)(\phi + \gamma))$, $C_2 = \theta_2 y_{10}(\theta_5 + \beta)/((\phi + \beta)(\beta - \gamma))$, $C_3 = \theta_2 y_{10}(\theta_5 + \gamma)/((\phi + \gamma)(\gamma - \beta))$.

TABLE 2

Concentration vs. Time Data for the Isomerization of α-pinene at 189.5°C.

Time (min.)	y_1 α-pinene	y_2 dipentene	y_3 allo-ocimene	y_4 pyronene	y_5 dimer
1230	88.35	7.3	2.3	0.4	1.75
3060	76.4	15.6	4.5	0.7	2.8
4920	65.1	23.1	5.3	1.1	5.8
7800	50.4	32.9	6.0	1.5	9.3
10680	37.5	42.7	6.0	1.9	12.0
15030	25.9	49.1	5.9	2.2	17.0
22620	14.0	57.4	5.1	2.6	21.0
36420	4.5	63.1	3.8	2.9	25.7

Assuming these models to be appropriate, we can obtain the posterior distribution of the parameters (the five rate constants) following Box and Draper (1965). In particular, we can find those parameter values which have the highest posterior density by minimizing the determinant criterion. If data dependencies are ignored, however, and parameter values which minimize the determinant criterion $|\mathbf{V}_5|$ (using all five responses) are found, the result is the unsatisfactory data fit shown in Figure 1. This example demonstrates how analysis of multiresponse data that ignores dependencies can lead to meaningless answers if linear dependencies are present.

From an examination of Fuguitt and Hawkins' paper, it can be found that y_4 (α- plus β-pyronene), because of experimental difficulties, was not measured independently but rather was assumed to constitute three percent of the total conversion of α-pinene (y_1). That is, it was assumed that

$$y_4 = 0.03(100 - y_1).$$

Thus, there is the following exact linear relationship among the observations:

$$(0.03)y_1 + (0)y_2 + (0)y_3 + (1)y_4 + (0)y_5 = 3. \tag{28}$$

Furthermore, by reducing the first order differential equations defining the kinetic

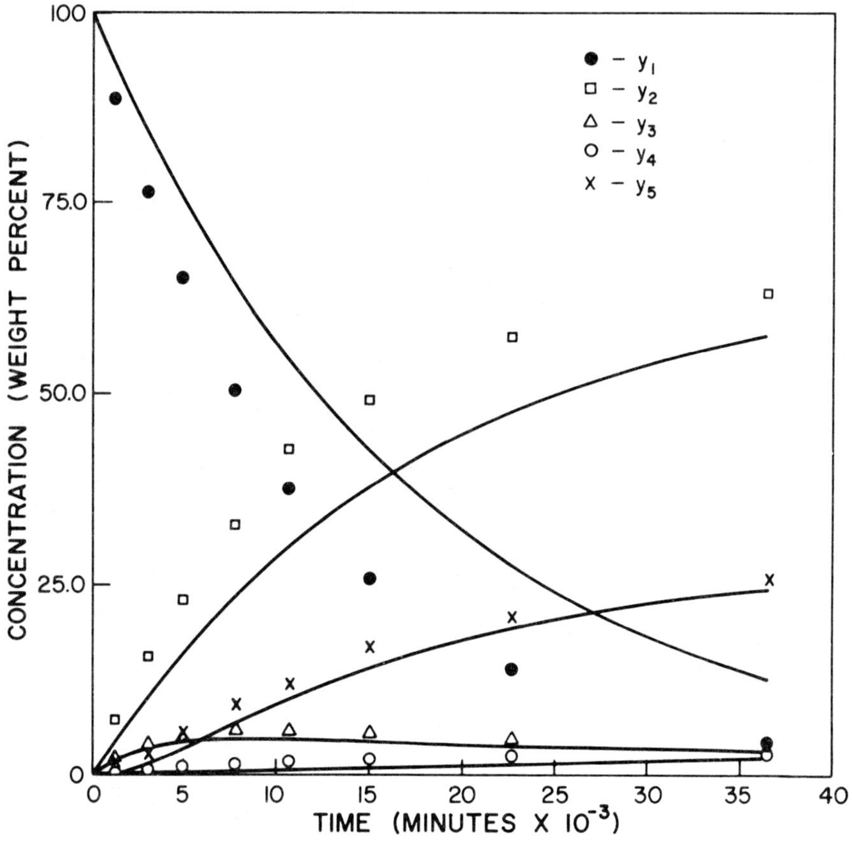

FIGURE 1—Data Fit with Parameter Values Which, Ignoring Dependencies, Minimize $|\mathbf{V}_5|$.

system to a canonical form following Ames (1962), it can be shown that the following two relationships must exist among the expectations of the y's:

$$E(y_1) + E(y_2) + E(y_3) + E(y_4) + E(y_5) = \text{const.} \tag{29}$$

and

$$E(y_1) + (1 + \theta_2/\theta_1)E(y_2) = \text{const.} \tag{30}$$

The first of these relationships (29) simply expresses an overall mass balance for the system while the second (30) results from the fact that the isomerizations of α-pinene (y_1) are assumed irreversible. The experimenters in this study chose to report their data in "normalized" weight percentage form, and so in effect, consciously or otherwise, used their knowledge of the expectation relationship (29) to force the following relationship among the observed responses:

$$y_1 + y_2 + y_3 + y_4 + y_5 = 100. \tag{31}$$

Thus there exist three linear relationships (28), (30), and (31) in this system, two of which, (28) and (31), are exact apart from rounding error. Because of the rounding error the value of the determinant was not exactly zero for all values of the parameters and this made it technically possible to obtain a minimum for $|\mathbf{V}_5|$ which however is meaningless.

6.1 *Eigenvalue-Eigenvector Analysis*

We now illustrate how empirical analysis could reveal these relationships. To implement the eigenvalue-eigenvector analysis, the individual means are first eliminated from the various responses to form a new data matrix, $\mathbf{D} = \{d_{iu}\}$, where $d_{iu} = y_{iu} - \bar{y}_i$. The eigenvalues and the associated eigenvectors of the $\mathbf{DD}'$ matrix are then obtained. These quantities are given in Table 3 for the present example.

TABLE 3

Eigenvalues and Eigenvectors of $\mathbf{DD}'$

Eigenvalues of $\underset{\sim}{DD}'$

λ_1	λ_2	λ_3	λ_4	λ_5
.0013	.0168	1.21	25.8	9660.

Eigenvectors of $\underset{\sim}{DD}'$

$\underset{\sim}{z}_1$	$\underset{\sim}{z}_2$	$\underset{\sim}{z}_3$	$\underset{\sim}{z}_4$	$\underset{\sim}{z}_5$
-.169	.476	-.296	.057	.809
-.211	.490	-.611	-.224	-.540
-.161	.435	.640	-.612	-.013
.931	.364	-.010	.004	-.024
-.185	.459	.360	.756	-.231

By employing the overall residual sum of squares (RSS) obtained from minimizing (2), one can obtain a crude estimate of the average experimental error variance by calculating

$$\overline{\sigma^2} \simeq \frac{\text{RSS}}{\text{d.f.}} = \frac{19.87}{(40 - 5)} = 0.6$$

where d.f. stands for residual degrees of freedom. Hence by equation (16) we can expect those eigenvalues arising from linear relationships among the expected values of the responses to be of the order of $(n - 1)\overline{\sigma^2} = 4.2$. The eigenvalue λ_3 is seen to be of this order of magnitude and will be shown later to correspond to equation (30). The eigenvalues λ_4 and λ_5 are however much larger and we would therefore not expect them to be associated with any linear relationships. It is clear that the capability of estimating the parameters in our models is coming mostly from $\mathbf{z}_5$, the eigenvector associated with λ_5.

To see whether there are any exact linear relationships among the responses, one should look for eigenvalues which are zero. In this case, there are none which are exactly zero but both λ_1 and λ_2 are very small and much smaller than the value 4.2 which one would expect from a linear expectation relationship. Thus one might suspect that eigenvalues $\mathbf{z}_1$ and $\mathbf{z}_2$ correspond to linear relationships among the recorded responses which are exact except for rounding error.

To obtain an estimate of the expected value of an eigenvalue when there is only rounding error present, we can assume that the rounding error is distributed uniformly with range -0.5 to $+0.5$ of the last digit reported. Rounding error variance σ^2_{RE} is then given by the range squared divided by 12, and since in this example all responses have been rounded to the same number of significant figures, the corresponding expected value of an eigenvalue using (3) may be approximated by

$$E(\lambda_k) = (n - 1)\mathbf{z}_k'\mathbf{I}\mathbf{z}_k\sigma^2_{\mathrm{re}} = (n - 1)\sigma^2_{\mathrm{re}} .$$

The concentration data here were reported to the nearest 0.1 percent and therefore the range is from -0.05 to $+0.05$ or 0.10. Thus, the expected value of an eigenvalue is $7(0.10)^2/12$ or approximately 0.006. Both eigenvalues λ_1 and λ_2 are of this order of magnitude, thus helping to confirm that $\mathbf{z}_1$ and $\mathbf{z}_2$ represent exact dependencies among the responses.

The question now is to determine, if possible, what are the true linear dependencies among the responses that are causing this two dimensional singularity plane. As shown in equation (13) the plane defined by $\mathbf{Z}_1 = [\mathbf{z}_1 , \mathbf{z}_2]'$ will only be some non-singular transformation of the true constraint matrix $\mathbf{A}_1$. We previously stated that the linear relationships given by equations (28) and (31) were expected to be present in the data. To test whether these account for the two dimensional singularity region represented by $\mathbf{Z}_1$ we need to test for the coplanarity of the regions defined by $\mathbf{A}_1 = [\mathbf{a}_1 , \mathbf{a}_2]'$ and $\mathbf{Z}_1$ where $\mathbf{a}_1' = (.03, 0, 0, 1.0, 0)$ and $\mathbf{a}_2' = (1., 1., 1., 1., 1.)$. A very simple check on this is to calculate the cosine of the angle made by each of the vectors $\mathbf{a}_1$ and $\mathbf{a}_2$ with the plane $\mathbf{Z}_1$. Doing this yielded cosines of .9999 and .9993 respectively, implying that $\mathbf{a}_1$ and $\mathbf{a}_2$ do indeed lie almost entirely within the plane of $\mathbf{Z}_1$.

If the true underlying relationships are not known it may be possible to use the empirical eigenvectors to provide some indication of what they may be. For instance, if the smallest of the "zero" eigenvalues is considerably smaller than the others, its corresponding eigenvector may correspond fairly closely to the most exact of the linear relationships. However, this probably would not be so if the "zero" eigenvalues happened to be of nearly equal magnitude. Looking at Table 3 in our example the first eigenvector $\mathbf{z}_1'$ corresponds reasonably closely to $\mathbf{a}_1' = (0.03, 0, 0, 1.0, 0)$, and $\mathbf{z}_2'$ to that component of $\mathbf{a}_2' = (1., 1., 1., 1., 1.)$ which is orthogonal to $\mathbf{z}_1'$, namely (.465, .468, .464, .363, .466).

We are also in a position now to check whether the third eigenvalue $\mathbf{z}_3$ corresponds to the relationship among the expected values of the responses given by (30). If we use the current estimates of θ_1 and θ_2 and take that component of the vector corresponding to the relation (31) which is orthogonal to both $\mathbf{z}_1$ and $\mathbf{z}_2$ and normalize it, we get $(-.308, -.665, .482, .008, .482)$ which is indeed very similar to $\mathbf{z}_3' = (-.296, -.611, .640, -.010, .360)$.

The confirmation of an expectation relationship such as (30) provides a valuable check of the tentatively entertained model structure for the system, in this case of the chemical stochiometry. Had either of the isomerization reactions been reversible, then (30) would not have been true.

6.2 *Analysis of the Data:*

Before any meaningful analysis of the data can be conducted, the two-dimensional singularity resulting from the relationships (28) and (31) must be removed. Perhaps the most natural approach to this problem is to think in terms of the responses themselves and ask the question: which responses should be dropped? It must be done in such a way as to leave an independent subset of three responses. In many cases, the structure of the problem and of the dependencies may well dictate a natural way for dropping responses. In our problem, if we look at the two dependencies which we know to exist, it is obvious that at least one of y_1 or y_4 must be eliminated, preferably y_4 since it is known to be the fabricated response. The second relationship (31) contains all five responses and if there were no rounding error, it would make absolutely no difference which of the additional four responses was dropped. Another approach to this problem is to conduct the final analysis of the data on three linearly independent combinations of all five responses. Again if there were no rounding error it would make no difference which three linearly independent combinations were used so long as they adequately defined the 3-dimensional subspace orthogonal to the 2-dimensional singularity plane defined by the vectors $\mathbf{a}_1' = (0.03, 0, 0, 1.0, 0)$ and $\mathbf{a}_2' = (1., 1., 1., 1., 1.)$. From an estimation point of view, it would be convenient to use three orthogonal vectors to define this space. However, since the singularity relationships are not exactly satisfied by the data, and since the likelihood surface for the parameters is poorly conditioned, the final parameter estimates and their variance-covariance matrix will be sensitive to some extent to how the singularities are removed.

6.3 *Use of Empirical Eigenvectors*

Were it not for the presence of roundoff error, the eigenvalues λ_1 and λ_2 would have been zero and their corresponding eigenvectors $\mathbf{z}_1$ and $\mathbf{z}_2$ would define the exact singularity plane defined by $\mathbf{a}_1$ and $\mathbf{a}_2$. A natural set of vectors to use in defining the remaining three dimensions would then be the remaining three empirical eigenvectors $\mathbf{z}_3$, $\mathbf{z}_4$, and $\mathbf{z}_5$ since these satisfy the requirements of being independent, orthogonal vectors, all orthogonal to the singularity plane. Therefore, in practice, when it has not been possible to pinpoint the true singularity relationships ($\mathbf{a}_1$ and $\mathbf{a}_2$) or if the roundoff error is considered to be negligible, then these eigenvectors corresponding to the non-zero eigenvalues can be used to form the three independent linear combinations $f_{3u} = \mathbf{z}_3'\mathbf{y}_u$, $f_{4u} = \mathbf{z}_4'\mathbf{y}_u$, and $f_{5u} = \mathbf{z}_5'\mathbf{y}_u$. By minimizing the determinant $|\mathbf{V}_3|$ where

$$\mathbf{V}_3 = \{(\mathbf{f}_i - E(\mathbf{f}_i))'(\mathbf{f}_j - E(\mathbf{f}_j))\}, \qquad i, j = 3, 4, 5 \tag{32}$$

where $E(f_{iu}) = \mathbf{z}_i'\mathbf{n}_u$, one obtains the point estimates of the rate constants given in the second row of Table 4. The first row contains those rate constant estimates obtained previously by minimizing the overall residual sum of squares of all five responses. Figure 2 shows the fits of the responses $\mathbf{y}_i$ obtained using these parameter estimates.

6.4 *Use of Theoretical Eigenvectors*

In this example, however, the true, or theoretical linear dependencies ($\mathbf{a}_1'y_u$ and $\mathbf{a}_2'y_u$) have in fact been uncovered and it is therefore better to use three independent linear combinations of the five responses which are orthogonal to this true singularity plane rather than the approximate one represented by the eigenvectors $\mathbf{z}_1$ and $\mathbf{z}_2$ (although these will differ only very slightly since the roundoff error is relatively small). For this purpose, we used as the basis for our three linear combinations the vector components ($\mathbf{a}_3$, $\mathbf{a}_4$, and $\mathbf{a}_5$) of the eigenvectors $\mathbf{z}_3$, $\mathbf{z}_4$, and $\mathbf{z}_5$ which are orthogonal to $\mathbf{a}_1$ and $\mathbf{a}_2$. By minimizing the determinant of the form (32) where now $f_{iu} = \mathbf{a}_i'\mathbf{y}_u$ and $E(f_{iu}) = \mathbf{a}_i'\mathbf{n}_u$ we obtained the point estimates of the

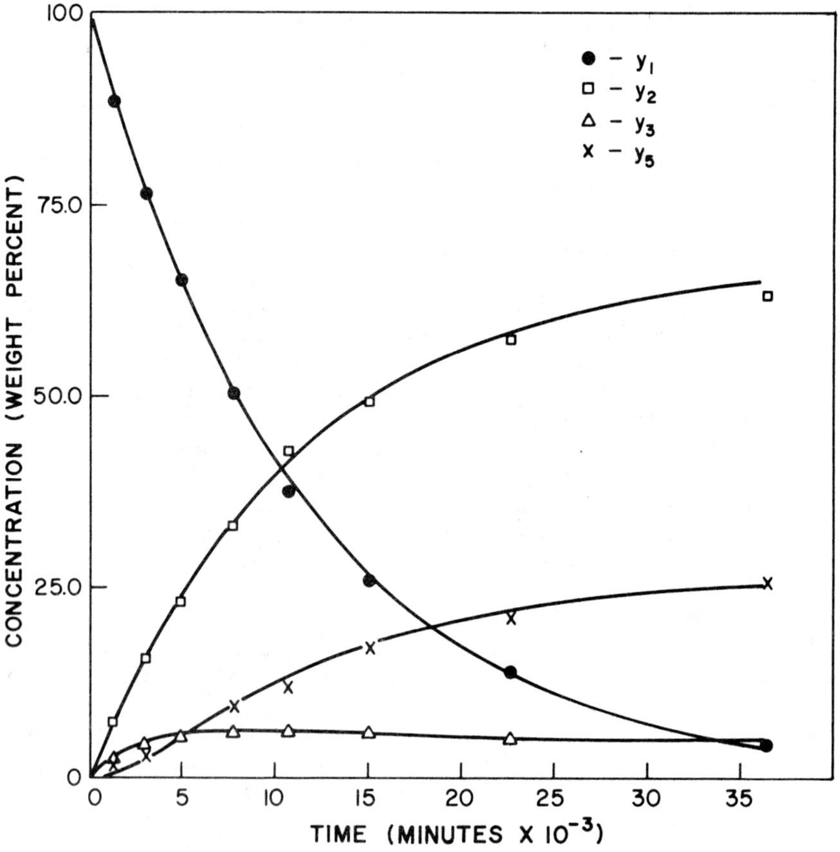

FIGURE 2—Fit of the Responses Using Parameter Values Which Minimize $|\mathbf{V}_3|$ Based on the Three Linear Combinations ($\mathbf{f}_i = \mathbf{z}'\mathbf{y}_u$, $i = 3, 4, 5$).

rate constants shown in the third row of Table 4. These are the values which give the highest posterior probability density. It can be seen that, as expected, they differ very little from those values in the second row. The resulting fit of the data is obviously very much better now that the singularities have been removed (see Figures 1 and 2).

The estimate of θ_3 is imprecise as is reflected by an extremely flat surface of the determinant function in the θ_3 direction. This was to be expected since most of the information on the θ_3 rate constant is contained in the singularity plane of the response space and, in particular, in y_4 .

The confidence region for the parameters can be computed using the general formula (Box and Draper, 1965)

$$|\mathbf{V}|_{(1-\alpha)} \simeq |\mathbf{V}|_{\min} \exp \{\chi_p^2(1 - \alpha)/n\} \tag{33}$$

where p = number of parameters
n = number of observations
$\chi_p^2(1 - \alpha)$ = chi-square value for p degrees of freedom and $(1 - \alpha) \times 100\%$ probability level.

For our example $p = 5$ and $n = 8$.

TABLE 4

Estimates of Rate Constants.

Minimization of	Rate Constants, 10^{-5} Min.$^{-1}$				
	$\hat{\theta}_1$	$\hat{\theta}_2$	$\hat{\theta}_3$	$\hat{\theta}_4$	$\hat{\theta}_5$
RSS	5.93	2.96	2.05	27.5	4.00
$\lvert \underset{\sim}{V}_3 \rvert$ (empirical eigenvectors)	5.95	2.84	0.43	31.3	5.74
$\lvert \underset{\sim}{V}_3 \rvert$ (theoretical eigenvectors)	5.95	2.85	0.50	31.5	5.89

7. SUMMARY

In the process of model-building using multiresponse data, one should always be alert for possible linear relationships among the responses. When there is doubt as to what relationships, if any, are present, an empirical eigenvalue-eigenvector analysis should be used. When dependencies are found, there must be good reason for them and considerable effort should be made to uncover their causes. Such studies can confirm or deny the validity of the model under study or suggest the appropriateness of a new and previously unsuspected mechanistic form. If m_1 exact linear relationships are known to exist, then m_1 dependent responses or m_1 linear combinations of them must be deleted before the data are analyzed; otherwise parameter estimation procedures yield meaningless results. So far as the actual analysis of the data at hand is concerned, however, knowing these true or theoretical dependencies is not absolutely necessary. One can proceed by making use of the empirically determined eigenvectors which do not necessarily represent true dependencies. A simplified version of the proposed procedure is shown in flow diagram form in Figure 3. Sections of this paper in which there is a fuller explanation are indicated.

It is desirable to have a check of some kind for data dependencies (such as the eigenvalue-eigenvector analysis described in this paper) built into general purpose computer programs for multiresponse fitting problems in the same way it is desirable to have a check on possible singularity or near-singularity incorporated in standard regression programs so that the user is given warning that answers produced by the program may not be meaningful. The eigenvalue-eigenvector analysis can be used to distinguish between dependencies in the data $\mathbf{y}$ on the one hand and in the expectations $E(\mathbf{y})$ on the other. In the first case, the eigenvalues will be of the order of magnitude of the mean square rounding error while in the second they are of the order of magnutide of the mean square residual error. Usually the latter is several orders of magnitude larger than the former so that it is possible to distinguish between the two kinds of dependencies so they can be handled accordingly.

We are indebted to a referee for pointing out that one way in which dependencies

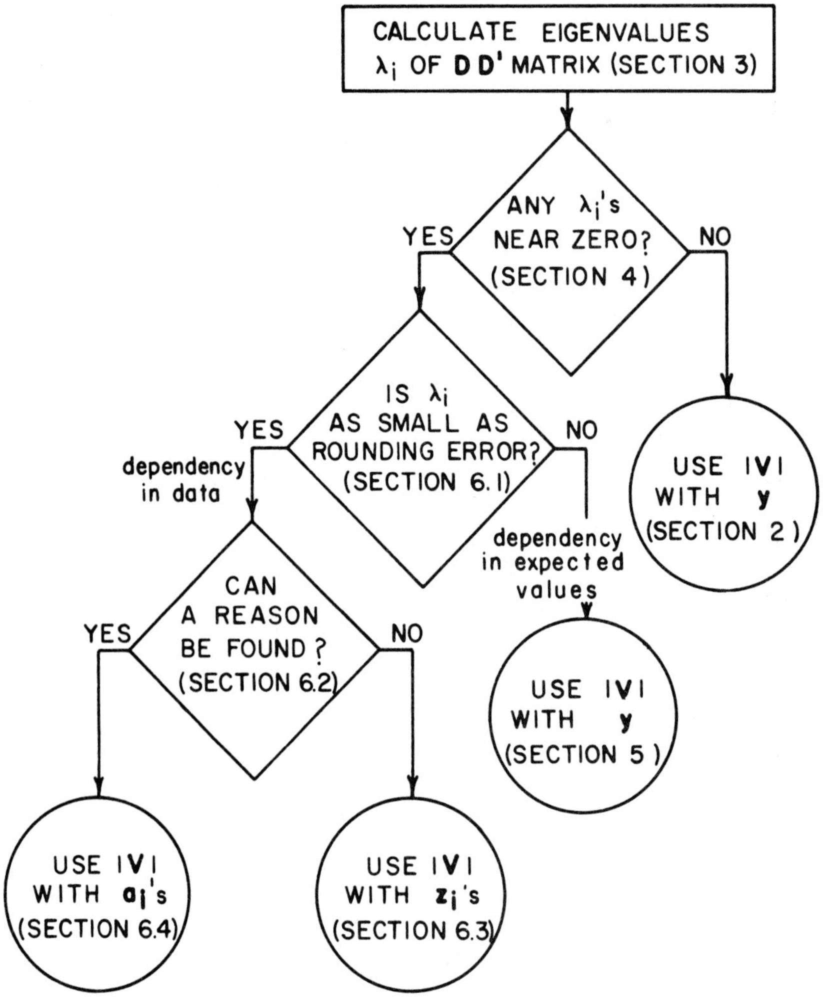

FIGURE 3—Simplified Version of Procedure for Fitting a Multiresponse Models.

among the expected values may enter a problem is through a steady-state relationship. This situation will usually be a local one in that it will depend upon the range of the experimental conditions. For instance, if in our example the data had been collected only over the restricted time range of 5000 to 15000 seconds where the rate of formation of alloocimene (y_3) is relatively constant, the following kinetic relationship would nearly hold:

$$\frac{d\eta_3}{dt} = \theta_2\eta_1 - \theta_3\eta_3 - \theta_4\eta_3 + \theta_5\eta_5 = 0$$

yielding a linear relationship among the expected values of several responses which would need to be appropriately explained.

8. Appendix

Expected Value of an Eigenvalue

The expected value of an eigenvalue, λ, whose associated eigenvector, $\mathbf{z}$, corresponds to a linear relationship in the expected values of the responses, can be found as follows:

Since $\mathbf{z}$ is an eigenvector of $\mathbf{DD}'$

$$\mathbf{DD}'\mathbf{z} = \lambda\mathbf{z} \tag{A.1}$$

Premultiplying equation (A.1) by $\mathbf{z}'$ we get

$$\mathbf{z}'\mathbf{DD}'\mathbf{z} = \mathbf{z}'\lambda\mathbf{z} = \lambda\mathbf{z}'\mathbf{z} \tag{A.2}$$

But since the $\mathbf{z}$ vector is scaled such that $\mathbf{z}'\mathbf{z} = 1$,

$$\lambda = \mathbf{z}'\mathbf{DD}'\mathbf{z} \tag{A.3}$$

By definition, $\mathbf{D} = \{d_{iu}\} = \{y_{iu} - \bar{y}_i\}$. We can further write $y_{iu} = \eta_{iu} + \epsilon_{iu}$ and $\bar{y}_i = \bar{\eta}_i + \bar{\epsilon}_i$ so that

$$\mathbf{D} = \{\eta_{iu} + \epsilon_{iu} - \bar{\eta}_i - \bar{\epsilon}_i\} \tag{A.4}$$

The ijth element of $\mathbf{DD}'$ is then

$$(\mathbf{DD}')_{ij} = \sum_{u=1}^{n} (\eta_{iu} + \epsilon_{iu} - \bar{\eta}_i - \bar{\epsilon}_i)(\eta_{ju} + \epsilon_{ju} - \bar{\eta}_j - \bar{\epsilon}_j) \tag{A.5}$$

which can be expanded to

$$\begin{aligned}(\mathbf{DD}')_{ij} = {} & \sum_{u=1}^{n} (\eta_{iu}\eta_{ju} - \eta_{iu}\bar{\eta}_j - \bar{\eta}_i\eta_{ju} + \bar{\eta}_i\bar{\eta}_j) \\ & + \sum_{u=1}^{n} (\eta_{iu}\epsilon_{ju} - \eta_{iu}\bar{\epsilon}_j - \bar{\eta}_i\epsilon_{ju} + \bar{\eta}_i\bar{\epsilon}_j + \epsilon_{iu}\eta_{ju} - \epsilon_{ju}\bar{\eta}_j - \bar{\epsilon}_i\eta_{ju} + \bar{\epsilon}_i\bar{\eta}_j) \\ & + \sum_{u=1}^{n} (\epsilon_{iu}\epsilon_{ju} - \epsilon_{iu}\bar{\epsilon}_j - \bar{\epsilon}_i\epsilon_{ju} + \bar{\epsilon}_i\bar{\epsilon}_j)\end{aligned} \tag{A.6}$$

The last three terms in the first summation of equation (A.6) can be seen to be equivalent,

$$\sum_{u=1}^{n} \eta_{iu}\bar{\eta}_j = \sum_{u=1}^{n} \bar{\eta}_i\eta_{ju} = \sum_{u=1}^{n} \bar{\eta}_i\bar{\eta}_j = n\bar{\eta}_i\bar{\eta}_j \tag{A.7}$$

Assuming that the models, η_{iu}, are correct so that $E(\epsilon_{iu}) = 0$, it follows that $E(\bar{\epsilon}_i) = 0$ also. Then, when the expected value of $(\mathbf{DD}')_{ij}$ is taken, all the terms in the second summation of (A.6) become zero.

We further assume that the errors in the responses have a variance-covariance matrix $\mathbf{\Sigma} = \{\sigma_{ij}^2\}$ so that $E(\epsilon_{iu}\epsilon_{ju}) = \sigma_{ij}^2$, and that these errors are independent from run to run. That is, $E(\epsilon_{iu}\epsilon_{iv}) = 0$ for all $u \neq v$. Now the expectations of the terms in the third summation of (A.6) can be evaluated.

$$E\left(\sum_{u=1}^{n} \epsilon_{iu}\epsilon_{ju}\right) = \sum_{u=1}^{n} E(\epsilon_{iu}\epsilon_{ju}) = \sum_{u=1}^{n} \sigma_{ij}^2 = n\sigma_{ij}^2 \tag{A.8}$$

$$E\left(\sum_{u=1}^{n} \epsilon_{iu}\bar{\epsilon}_j\right) = nE\left(\epsilon_{iu} \sum_{v=1}^{n} \epsilon_{jv}/n\right) = nE(\epsilon_{iu}\epsilon_{ju}/n) = \sigma_{ij}^2 \tag{A.9}$$

$$E\left(\sum_{u=1}^{n} \bar{\epsilon}_i \epsilon_{ju}\right) = nE\left[\sum_{v=1}^{n} (\epsilon_{iv}/n)\epsilon_{ju}\right] = nE[(\epsilon_{iu}/n)\epsilon_{ju}] = \sigma_{ij}^2 \tag{A.10}$$

$$E\left(\sum_{u=1}^{n} \bar{\epsilon}_i \bar{\epsilon}_j\right) = nE\left(\sum_{u=1}^{n} \frac{\epsilon_{iu}}{n}\right)\left(\sum_{v=1}^{n} \frac{{}_{jv}}{n}\right) = \frac{1}{n} E\left(\sum_{u=1}^{n} \epsilon_{iu}\epsilon_{ju}\right) = \sigma_{ij}^2 \tag{A.11}$$

When the above information is incorporated in equation (A.6) it becomes

$$E(\mathbf{DD}')_{ij} = \sum_{u=1}^{n} \eta_{iu}\eta_{ju} - n\bar{\eta}_i\bar{\eta}_j - n\bar{\eta}_i\bar{\eta}_j + n\bar{\eta}_i\bar{\eta}_j + n\sigma_{ij}^2 - \sigma_{ij}^2 - \sigma_{ij}^2 + \sigma_{ij}^2 \tag{A.12}$$

which, upon simplification, reduces to

$$E(\mathbf{DD}')_{ij} = \sum_{u=1}^{n} \eta_{iu}\eta_{ju} - n\bar{\eta}_i\bar{\eta}_j + (n - 1)\sigma_{ij}^2 \tag{A.13}$$

The expected value of λ can now be found by taking expectations of both sides of equation (A.3).

$$E(\lambda) = E(\mathbf{z}'\mathbf{DD}'\mathbf{z}) = \mathbf{z}'E(\mathbf{DD}')\mathbf{z} = \sum_{i=1}^{r}\sum_{j=1}^{r} z_i z_j E(\mathbf{DD}')_{ij} \tag{A.14}$$

Substituting (A.13) into (A.14) we obtain

$$E(\lambda) = \sum_{i=1}^{r}\sum_{j=1}^{r} z_i z_j \left[\sum_{u=1}^{n} \eta_{iu}\eta_{ju} - n\bar{\eta}_i\bar{\eta}_j + (n - 1)\sigma_{ij}^2\right] \tag{A.15}$$

Since z represents a linear relationship in the expected values of the responses, and the expected value of y_{iu} is η_{iu}, we know that

$$\sum_{i=1}^{r} z_i \eta_{iu} = a_0 \tag{A.16}$$

which is the same constant, a_0, for every run, u. It follows that

$$\sum_{i=1}^{r} z_i \bar{\eta}_i = \sum_{i=1}^{r} z_i \sum_{u=1}^{n} \frac{\eta_{iu}}{n} = \sum_{u=1}^{n}\sum_{i=1}^{r} \frac{z_i\eta_{iu}}{n} = \sum_{u=1}^{n} \frac{a_0}{n} = a_0 \tag{A.17}$$

By expanding equation (A.15) we obtain

$$E(\lambda) = \sum_{u=1}^{n}\sum_{i=1}^{r} z_i\eta_{iu} \sum_{j=1}^{r} z_j\eta_{ju} - n \sum_{i=1}^{r} z_i\bar{\eta}_i \sum_{j=1}^{r} z_j\bar{\eta}_j + (n - 1) \sum_{i=1}^{r}\sum_{j=1}^{r} z_i z_j \sigma_{ij}^2 \tag{A.18}$$

Then, when we insert (A.16) and (A.17) into (A.18) we get

$$E(\lambda) = \sum_{u=1}^{n} (a_0)(a_0) - n(a_0)(a_0) + (n - 1)\mathbf{z}'\mathbf{\Sigma}\mathbf{z} \tag{A.19}$$

which simplifies to

$$E(\lambda) = (n - 1)\mathbf{z}'\mathbf{\Sigma}\mathbf{z} \tag{A.20}$$

If $\mathbf{\Sigma}$ is known, the expected value of λ can be found directly from equation (A.20). And even when $\mathbf{\Sigma}$ is unknown, it may be possible to estimate it, for example, there may be some replication in the data. The best estimate for the expected value of λ in this case, too, is found from equation (A.20).

If $\mathbf{\Sigma}$ is unknown and cannot be estimated, we can get a very crude estimate of $E(\lambda)$ by approximating $\mathbf{\Sigma}$ by $\mathbf{I}\overline{\sigma^2}$. Then $E(\lambda) \simeq (n - 1)\overline{\sigma^2}$, where the average variance, $\overline{\sigma^2}$, may be estimated from the residual sum of squares.

9. Acknowledgment

The authors would like to thank Paul W. Tidwell of the Monsanto Company for valuable and enjoyable discussion on ideas presented in this paper.

References

[1] Ames, W. F. (1962). "Canonical Forms for Nonlinear Kinetic Differential Equations", *Ind. Eng. Chem. Fundamentals, 1,* 214.

[2] Ball, W. E. and Groenweghe, L. C. D. (1966). "Determination of Best-Fit Rate Constants in Chemical Kinetics", *Ind. Eng. Chem. Fundamentals, 5,* 181.

[3] Box, G. E. P. and Draper, N. R. (1965). "Bayesian Estimation of Common Parameters from Several Responses", *Biometrika, 52,* 355.

[4] Eakman, J. M. (1969). "Strategy for Estimation of Rate Constants from Isothermal Reaction Data", *Ind. & Eng. Chem. Fundamentals, 8,* 53.

[5] Erjavec, J. (1970). "Strategy for Estimation of Rate Constants from Isothermal Reaction Data", *Ind. & Eng. Chem. Fundamentals, 9,* 187.

[6] Fuguitt, R. E. and Hawkins, J. E. (1947). "Rate of Thermal Isomerization of α-Pinene in the Liquid Phase", *J.A.C.S., 69,* 319.

[7] Hunter, W. G. (1967). "Estimation of Unknown Constants from Multiresponse Data", *Ind. & Eng. Chem. Fundamentals, 6,* 461.

[8] Hunter, W. G. and McGregor, J. F. (1967). "The Estimation of Common Parameters from Several Responses: Some Actual Examples", Unpublished Report, The Department of Statistics, The University of Wisconsin.

4.11

Correcting Inhomogeneity of Variance with Power Transformation Weighting

G. E. P. BOX*
University of Wisconsin

WILLIAM J. HILL
Allied Chemical Corp.

A method is presented for using power transformation weights in least squares analysis to account for inhomogeneity of variance. The need for this form of weighting is common. In particular, it frequently arises when, as is illustrated with an example, a linearized kinetic rate expression is analyzed.

KEY WORDS

Inhomogeneity of Variance
Weighted Least Squares
Iterative
Linearize
Transformation
Residuals
Bayesian Approach

1. INTRODUCTION

Model fitting by unweighted least squares is efficient if errors besides being Normal and independent have constant variance. While moderate inhomogeneity is unlikely to have serious consequences, large differences in variance can lead to inefficient estimation. Serious inhomogeneity of variance can arise when, for example, a kinetic rate model which involves its parameters nonlinearly, is linearized by transformation. For instance, investigators frequently linearize chemical rate expressions by working with the reciprocal r^{-1} of the reaction rate r. But if, as is dramatically illustrated in section 4 and in [7], r has approximately constant variance over an extensive range then the variance of r^{-1} will be inhomogeneous or nonconstant. More generally, there would usually be no reason to suppose that r itself had constant variance, and a less restrictive assumption would be that the variance of r was some function of its mean value.

A method is presented here for obtaining approximate weights in a weighted least squares analysis when the variance of the fitted dependent variable is a function of its expected value. The method is applicable both for linear and nonlinear least squares analysis, and whether or not inhomogeneity of variance exists initially or is induced by transformation of the data as in the example we present.

* Supported by the Air Force Office of Scientific Research under grant AFOSR 72-2363A.
Received Oct. 1972; revised Nov. 1973.

2. DEVELOPMENT OF A WEIGHTING RELATIONSHIP

Let $\mathbf{y} = (y_1, y_2, \cdots, y_n)'$ represent an $n \times 1$ vector of independent observations. A least squares solution of the p parameters $\boldsymbol{\theta} = (\theta_1, \theta_2, \cdots, \theta_p)'$ will be efficient if

$$y_u = \eta_u + e_u \qquad u = 1, 2, \cdots, n \tag{1}$$

where the η_u's are functions of $\boldsymbol{\theta}$ and the e_u's are independent Normally distributed errors with constant variance. We relax the last assumption and suppose instead that the variance of e_u is finite and is given by

$$V(e_u) = c_u\sigma^2 \tag{2}$$

where c_u is unknown and must be estimated.

Suppose there exists some *unknown* transformation $y_u^{(\phi)}$ that has constant variance, where $y_u^{(\phi)}$ is defined as

$$y_u^{(\phi)} = \begin{cases} \dfrac{y_u^{\phi} - 1}{\phi} & \phi \neq 0 \\ \log y_u & \phi = 0 \end{cases} \tag{3}$$

using the power transformation notation of Box and Cox [4]. The variance of $y_u^{(\phi)}$ is expressed as

$$V(y_u^{(\phi)}) = \sigma^2 \tag{4}$$

An approximate variance expression is now developed for y_u using Bartlett's [3] method for stabilizing variance. That is,

$$V(y_u) \simeq V(y_u^{(\phi)})\left(\frac{dy_u}{dy_u^{(\phi)}}\right)^2_{\eta_u} \tag{5}$$

$$= \sigma^2\eta_u^{2-2\phi}$$

The Collected Works of George E. P. Box, 1984, Wadsworth, Inc., Belmont, CA 94002.
Originally published in *Technometrics,* vol. 16, no. 3 (1974), pp. 385-389.

Combining expressions (2) and (5), we have

$$c_u \simeq \eta_u^{2-2\phi} \tag{6}$$

Therefore, the weight w_u to be assigned to y_u in least squares analysis will be

$$w_u = \frac{1}{c_u} \simeq \eta_u^{2\phi-2} \tag{7}$$

such that

$$V(\sqrt{w_u}\, y_u) \simeq \sigma^2 \tag{8}$$

Since η_u is unknown and must be estimated, a further approximation of the weight for y_u will be

$$w_u \simeq \hat{y}_u^{2\phi-2} \tag{9}$$

where $\hat{y}_u$ is the predicted value of y_u.

3. Estimating ϕ

The unknown weighting parameter ϕ will be estimated here via a Bayesian approach. To initiate this development, the Normality assumption for y_u gives the likelihood function

$$L = \frac{\left(\prod_{u=1}^{n} w_u\right)^{\frac{1}{2}}}{(2\pi\sigma^2)^{n/2}} \exp\left\{-\frac{1}{2\sigma^2}\sum_{u=1}^{n} w_u(y_u - \eta_u)^2\right\} \tag{10}$$

It is assumed that η_u is either a linear function of p parameters $\boldsymbol{\theta} = (\theta_1, \theta_2, \cdots, \theta_p)'$ or that η_u is approximately linear in the region of the least squares estimates $\hat{\boldsymbol{\theta}}$ of $\boldsymbol{\theta}$. In either case, we assume that we can express the model precisely or approximately as

$$\eta_u = \mathbf{X}_u\boldsymbol{\theta} \tag{11}$$

where $\mathbf{X}_u$ is a $1 \times p$ vector $(x_{u1}, x_{u2}, \cdots, x_{up})$ which are functions of m independent variables $\boldsymbol{\xi} = (\xi_1, \xi_2, \cdots, \xi_m)$. If η_u is nonlinear in the parameters whereby (11) is a linear approximation using a Taylor expansion, then

$$x_{ui} = \left.\frac{\partial \eta_u}{\partial \theta_i}\right|_{\boldsymbol{\theta}=\boldsymbol{\theta}_0} \qquad \begin{matrix} i = 1, 2, \cdots, p \\ u = 1, 2, \cdots, n \end{matrix} \tag{12}$$

where $\boldsymbol{\theta}_0$ is an initial set of values for $\boldsymbol{\theta}$.

If locally uniform prior distributions are assumed for $\boldsymbol{\theta}$ and ϕ, and the prior for σ is assumed proportional to σ^{-1} [9], then a prior distribution for the parameters is

$$p(\phi, \boldsymbol{\theta}, \sigma) \propto \sigma^{-1} \tag{13}$$

From Bayes' theorem, the joint posterior density is written as

$$p(\phi, \boldsymbol{\theta}, \sigma \mid \mathbf{y}) = \frac{p(\phi, \boldsymbol{\theta}, \sigma)\cdot L}{\int_\sigma\int_\theta\int_\phi p(\phi, \boldsymbol{\theta}, \sigma)\cdot L\, d\phi\, d\boldsymbol{\theta}\, d\sigma}$$

$$\propto \frac{\left(\prod_{u=1}^{n} w_u\right)^{\frac{1}{2}}}{(2\pi)^{n/2}\sigma^{n+1}} \exp\left\{-\frac{1}{2\sigma^2}\sum_{u=1}^{n} w_u(y_u - \mathbf{X}_u\boldsymbol{\theta})^2\right\} \tag{14}$$

The marginal posterior density for ϕ is written as

$$p(\phi \mid \mathbf{y}) \propto \int_\sigma\int_\theta \frac{\left(\prod_{u=1}^{n} w_u\right)^{\frac{1}{2}}}{(2\pi)^{n/2}\sigma^{n+1}} \cdot \exp\left\{-\frac{1}{2\sigma^2}\sum_{u=1}^{n} w_u(y_u - \mathbf{X}_u\boldsymbol{\theta})^2\right\} d\boldsymbol{\theta}\, d\sigma \tag{15}$$

Upon integration this reduces to

$$p(\phi \mid \mathbf{y}) = K\, |\mathbf{X}'\mathbf{W}\mathbf{X}|^{-\frac{1}{2}} \cdot \left(\prod_{u=1}^{n} w_u\right)^{\frac{1}{2}} \left\{\sum_{u=1}^{n} w_u(y_u - \mathbf{X}_u\hat{\boldsymbol{\theta}})^2\right\}^{-(n-p)/2} \tag{16}$$

where $\mathbf{X}$ is an $n \times p$ matrix consisting of elements x_{ui} $(i = 1, 2, \cdots, p)$, $\mathbf{W}$ is an $n \times n$ matrix with diagonal elements w_u and zero off-diagonal elements, $|\mathbf{M}|$ is the determinant of the matrix $\mathbf{M} = \mathbf{X}'\mathbf{W}\mathbf{X}$, and K is a constant of proportionality. Also, $\hat{\boldsymbol{\theta}}$ is a weighted linear least squares estimate of $\boldsymbol{\theta}$ where

$$\hat{\boldsymbol{\theta}} = (\mathbf{X}'\mathbf{W}\mathbf{X})^{-1}\mathbf{X}'\mathbf{W}\mathbf{y} \tag{17}$$

A section on weighted linear least squares appears in [6].

With the aid of a computer, expression (16) may be used iteratively as follows to estimate ϕ:

a) initially, substitute y_u for $\hat{y}_u$ in w_u;
b) select a ϕ;
c) calculate $\hat{\boldsymbol{\theta}}$ using (17) and hence calculate $p(\phi \mid \mathbf{y})$ using (16);
d) return to (b) and continue until finding a value $\hat{\phi}$ that maximizes $p(\phi \mid \mathbf{y})$;
e) corresponding to maximizing value $\hat{\phi}$, find $\hat{y}_u$ using the model and substitute these into w_u;
f) return to (b) and start new iteration. End operation when iterations converge on a common value for $\hat{\phi}$.

(Note: Steps (b)–(d) can be replaced by a gradient search that will find the maximizing value of ϕ for each iteration.)

4. Example

Carr [5] studied the catalytic isomerization of n-pentane to isopentane. One of the candidate reaction mechanisms that he studied was a single site mechanism where the surface reaction rate was assumed to be the rate controlling step. From Hougen and Watson theory [8] of reactions on solid catalysts, the following mathematical model was hypothesized:

$$r = \frac{k_0 k_2(\xi_2 - \xi_3/1.632)}{1 + k_1\xi_1 + k_2\xi_2 + k_3\xi_3} \tag{18}$$

where r is the reaction rate, k_0 is a constant dependent on catalyst and temperature, k_1, k_2, and k_3 are adsorption equilibrium constants. The variables ξ_1 , ξ_2 , and ξ_3 are partial pressures of hydrogen, *n*-pentane, and isopentane, respectively.

The model given in expression (18) is nonlinear in the reaction parameters k_0 , k_1 , k_2 , and k_3 but can be linearized via an inverse transformation on r so that

$$y = r^{-1} = \theta_1 x_1 + \theta_2 x_2 + \theta_3 x_3 + \theta_4 x_4 \qquad (19)$$

where

$$\theta_1 = \frac{1}{k_0 k_2}, \quad \theta_2 = \frac{k_1}{k_0 k_2}, \quad \theta_3 = \frac{1}{k_0}, \quad \theta_4 = \frac{k_3}{k_0 k_2} \qquad (20)$$

and the new independent variables are

$$x_1 = \frac{1}{\xi_2 - \xi_3/1.632},$$

$$x_2 = \xi_1 x_1, \quad x_3 = \xi_2 x_1, \quad x_4 = \xi_3 x_1 \qquad (21)$$

The original reaction constants can easily be back-calculated once the $\hat{\theta}$'s are found as follows:

$$\hat{k}_0 = \frac{1}{\hat{\theta}_3}, \quad \hat{k}_1 = \frac{\hat{\theta}_2}{\hat{\theta}_1}, \quad \hat{k}_2 = \frac{\hat{\theta}_3}{\hat{\theta}_1}, \quad \hat{k}_3 = \frac{\hat{\theta}_4}{\hat{\theta}_1} \qquad (22)$$

TABLE 1—*Data used in Carr's example* [5]*

Run #	ξ_1	ξ_2	ξ_3	r
1	205.8	90.9	37.1	3.541
2	404.8	92.9	36.3	2.397
3	209.7	174.9	49.4	6.694
4	401.6	187.2	44.9	4.722
5	224.9	92.7	116.3	.593
6	402.6	102.2	128.9	.268
7	212.7	186.9	134.4	2.797
8	406.2	192.6	134.9	2.451
9	133.3	140.8	87.6	3.196
10	470.9	144.2	86.9	2.021
11	300.0	68.3	81.7	.896
12	301.6	214.6	101.7	5.084
13	297.3	142.2	10.5	5.686
14	314.0	146.7	157.1	1.193
15	305.7	142.0	86.0	2.648
16	300.1	143.7	90.2	3.303
17	305.4	141.1	87.4	3.054
18	305.2	141.5	87.0	3.302
19	300.1	83.0	66.4	1.271
20	106.6	209.6	33.0	11.648
21	417.2	83.9	32.9	2.002
22	251.0	294.4	41.5	9.604
23	250.3	148.0	14.7	7.754
24	145.1	291.0	50.2	11.590

* Reprinted from I & EC, *52*, May 1960, p. 393. Copyright 1960 by the American Chemical Society. Reprinted by permission of copyright owner.

TABLE 2—*Least squares results for example*

Parameters	Unweighted*	Weighted
$\hat{\phi}$	1	-.8
$\hat{k}_0$	189.65	40.00
$\hat{k}_1$	-.38	.75
$\hat{k}_2$	-.03	.35
$\hat{k}_3$	-1.04	1.85

* Estimates derived from Carr's analysis [5].

Note: Estimates on $\hat{k}_1$, $\hat{k}_2$ and $\hat{k}_3$ were multiplied by 14.7 to convert units from $(\text{psia})^{-1}$ to $(\text{atm})^{-1}$.

An unweighted linear least squares analysis was performed by Carr on the data in Table 1 leading to the results shown in column 2 of Table 2. The negative estimates for k_1 , k_2 , and k_3 were contrary to the popular physical interpretation, and in the past a model of this type has often been rejected when estimates were produced with the "wrong" sign.

The seriousness of the constant variance assumption can be easily depicted when r^{-1} is plotted versus r as shown in Figure 1. Here a comparison is made of runs 19 and 20 before and after trans-

PLOT OF r^{-1} VERSUS r ILLUSTRATING EFFECT OF INVERSE TRANSFORMATION ON VARIANCE OVER WIDE RANGE IN r

FIGURE I

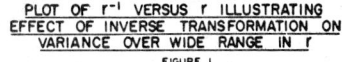

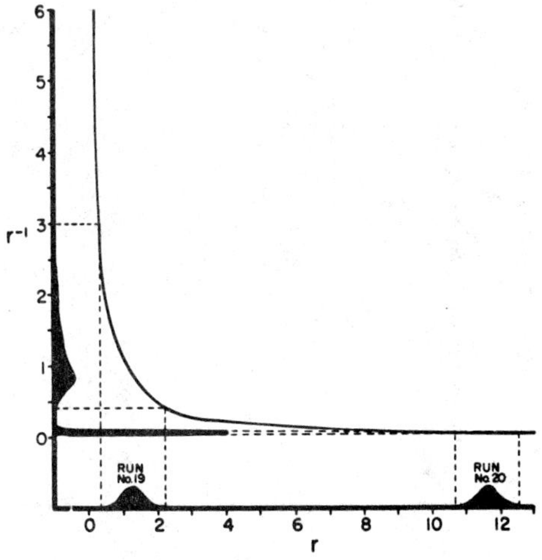

formation assuming that $y^{-1} = r$ has approximately constant variance over the whole range of values (i.e., ϕ is assumed to be approximately -1). From expression (6) with r^{-1} substituted for η, the ratio of the variances of r^{-1} for runs 19 and 20 is approximately 7050 to 1. Differences in variances this large and larger can easily be found in this data set, and it becomes apparent that weighting is needed for efficient least squares estimation.

The results of a weighted linear least squares analysis on $y = r^{-1}$ are shown in column 3 of Table 2. As is illustrated in Figure 2, a converged estimate $\hat{\phi} = -.8$ of the weighting parameter ϕ was found after 3 iterations. This implied that $y^{-.8}$ or $r^{.8}$ (since $y = r^{-1}$) had constant variance. Also, reasonable positive estimates were obtained for the reaction parameters (Table 2). It is seen in this example that a properly weighted least squares analysis avoided spurious negative parameter estimates which could have led the investigator to reject the model.

Residual analysis as described in ([1], [2], and [6]) is a valuable tool for studying the effects of nonhomogeneity of variance. Here, it is especially illustrative of why weighting is necessary when the linearized model (19) is used. A plot of the unweighted residuals $(r^{-1} - \widehat{r^{-1}})$ versus $\widehat{r^{-1}}$ in Figure 3(a) shows the seriousness of the inhomogeneity of variance. A plot of the weighted residuals $\sqrt{w}\,(r^{-1} - \widehat{r^{-1}})$ versus $\widehat{r^{-1}}$ in Figure 3(b) illustrates the results of the corrective remedy offered by the weighting procedure.

In a more recent paper [10], a less flexible version of the method here was used, such that ϕ was assumed known and fixed at -1 (i.e., the variance of r was assumed constant). A weighted linear least squares analysis in [10] led to the conclusion that the single site mechanism compared favorably with other candidate mechanisms under consideration in Carr's study. In a rebuttal note at the end of [10], Carr stated that the isopentane yield exhibited a maximum with respect to total pressure which he concluded did not reflect the nature of the single site mechanism.

Our purpose in studying this example is not to argue in favor of one model over another, especially when we haven't studied the other models. Instead, our purpose is to show that efficient estimation methods, such as weighted least squares in the presence of inhomogeneity of variance, can improve the quality of the investigative process. For the example here, the weighting of the data has prevented the model from being rejected prematurely because of a deficiency in the estimation method and not because of deficiency in the model itself. With efficient estimation, the model is better able to be evaluated on its own merits or weaknesses.

CONVERGENCE ON ESTIMATE OF ϕ IN EXAMPLE

FIGURE 2

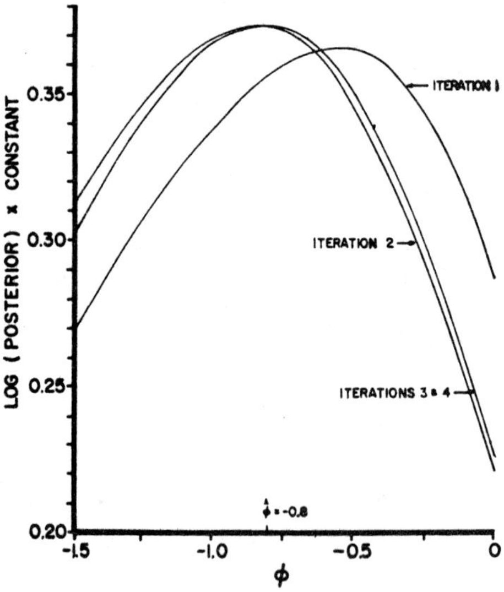

PLOT OF UNWEIGHTED RESIDUALS VS PREDICTED VALUES

FIGURE 3a

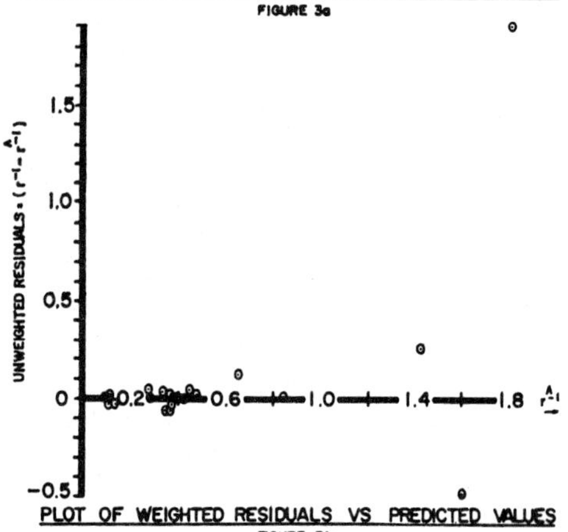

PLOT OF WEIGHTED RESIDUALS VS PREDICTED VALUES

FIGURE 3b

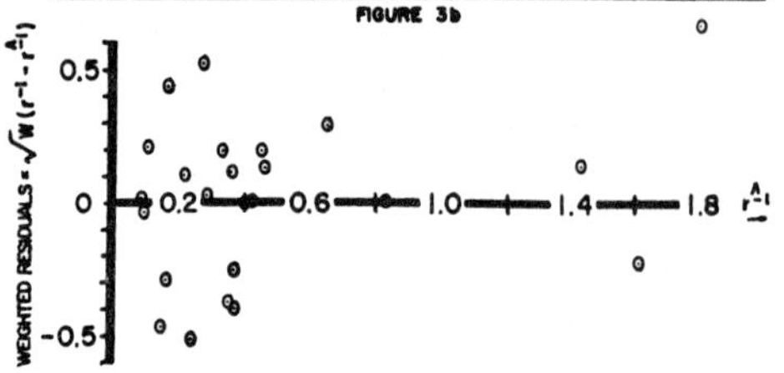

5. Summary

A method has been proposed here for estimating weights in a weighted least squares analysis when the variance of the dependent variable is related to the expected value. The importance of this method has been demonstrated by example where the weighting corrected earlier inefficient estimation.

6. Acknowledgement

The authors wish to acknowledge the valuable assistance and suggestions given by Drs. W. G. Hunter and R. Mezaki. The authors wish to acknowledge the computer programming assistance from Messrs. E. K. Roop and P. N. Sheldon, and are also grateful to the two referees whose suggestions contributed greatly to the clarity and organization of the paper.

This work was supported in part by the Air Force Office of Scientific Research under Grant No. AF-AFOSR 72-2363.

References

[1] Anscombe, F. J. (1961). Examination of residuals. *Proc. Fourth Berkeley Symp. Math. Statist.* and *Prob.*, *1*, 1–36.

[2] Anscombe, F. J., and Tukey, J. W. (1963). The examination and analysis of residuals. *Technometrics*, *5*, 141–160.

[3] Bartlett, M. S. (1947). The use of transformations. *Biometrics*, *3*, 39–52.

[4] Box, G. E. P., and Cox, D. R. (1964). The analysis of transformations. *J. Roy. Statist. Soc.*, Series B, *26*, No. 2, 211–252.

[5] Carr, N. L. (1960). Kinetics of catalytic isomerization of n-pentane. *Indus. Eng. Chem.*, *52*, May, 391–396.

[6] Draper, N. R., and Smith, H. (1966). *Applied Regression Analysis*. John Wiley and Sons, Inc., New York.

[7] Hill, W. J. (1966). Statistical techniques for model-building. Ph.D. thesis, University of Wisconsin, Madison, Wis.

[8] Hougen, O. A., and Watson, K. M. (1947). *Chemical Process Principles*. Part III, John Wiley and Sons, Inc., New York.

[9] Jeffreys, H. (1961). *The Theory of Probability*. Clarendon Press, Oxford.

[10] Mezaki, R., Johnson, R. A., and Standal, N. A. (1968). An examination of Hougen-Watson rate models including a weighted least squares approach. *I & EC Fundamentals*, *7*, No. 1, Feb., 181–183.

4.12

Analysis of Variance with Autocorrelated Observations

GRETA M. LJUNG
Boston University

G. E. P. BOX
University of Wisconsin

ABSTRACT. In this paper we utilize Bayesian methods to make inferences about the parameters in analysis of variance models with autocorrelated error terms. Specifically, we discuss inferences concerning a single mean value, the comparison of two or more mean values and the comparison of row effects and column effects in a two-way classification model with correlation within rows. It is assumed that the correlation structure can be described by an autoregressive process of order p. The sensitivity of the inferences to assumptions about the correlation is illustrated by examples.

Key words: analysis of variance, autocorrelation, autoregressive process, circular autocorrelation, Bayesian inference, robustness

1. Introduction

Consider the linear model

$$\mathbf{y} = \mathbf{X}\boldsymbol{\mu} + \mathbf{u},$$

where $\mathbf{y} = (y_1, ..., y_n)'$ is the $n \times 1$ observation vector, $\mathbf{X}$ is a $n \times k$ matrix of fixed elements, $\boldsymbol{\mu} = (\mu_1, ..., \mu_k)'$ is a $k \times 1$ parameter vector and $\mathbf{u} = (u_1, ..., u_n)'$ is the $n \times 1$ vector of error terms. In applications of this model it is often assumed that the errors are independent random variables with mean zero and constant variance σ^2. When these assumptions are satisfied the best linear unbiased estimates of the parameters are given by the method of least squares. If in addition the u_i's are normally distributed, well known inference procedures based on the t and F distributions are available.

The assumption of independent errors is often violated when the data are collected sequentially in time or space. The resulting effects on the inference procedures have been studied extensively in the literature; see Anderson (1954), Box (1954), Watson (1967) and Anderson (1971, Ch. 10). When randomization can be performed, the usual t and F tests can be justified as approximations to the randomization tests, which do not assume independence. However, randomization is not always possible and methods are needed which allow for correlation among the errors.

Within the sampling theory framework, large sample inference procedures have been studied by Durbin (1960), Pierce (1971) and others. Zellner & Tiao (1964) adopted a Bayesian approach and developed an inference method for regression models with errors generated by a first order autoregressive model. Sredni (1970) generalized the results to cover autoregressive error models of general order. The first order case was studied further by Pallesen (1978) using slightly different prior distributions for the parameters.

In the present paper we utilize Bayesian methods to make inferences about the parameters in analysis of variance models with correlated errors. The special case where all observations have a common mean is considered in Section 2. In Section 3 we discuss the comparison of two or more mean values and finally in Section 4 the comparison of row effects and column effects in a two-way classification model, where the errors are correlated within rows. We assume that the correlation can be described by an autoregressive model of order p. For the two-way classification model, we use Hotelling's circular definition of this process. This model is appropriate for the rainfall data given by Fisher (1925, p. 234) which are analyzed in Section 4. The sensitivity of the inferences to assumptions about the correlation is illustrated by examples.

2. Inferences concerning a single mean

Consider the model

$$y_i = \mu + u_i, \quad i = 1, ..., n,$$

where the errors u_i follow the stationary autoregressive model

$$u_i - \phi_1 u_{i-1} - ... - \phi_p u_{i-p} = a_i$$

with the a_i's i.i.d. $N(0, \sigma^2)$ random variables. Alternatively, defining $y_i' = y_i - \phi_1 y_{i-1} - ... - \phi_p y_{i-p}$ and

The Collected Works of George E. P. Box, 1984, Wadsworth, Inc., Belmont, CA 94002.
Originally published in *Scandinavian Journal of Statistics* 7:4 (1980), pp. 172–180.

$\mu' = \mu(1 - \phi_1 - \ldots - \phi_p)$, we can write

$$y_t' = \mu' + a_t, \tag{2.1}$$

where the errors now are independent.

Under the assumptions associated with the model, the likelihood function for the parameters μ, $\boldsymbol{\phi} = (\phi_1, \ldots, \phi_p)'$ and σ is

$$L(\mu, \boldsymbol{\phi}, \sigma \mid \mathbf{y}) = (2\pi\sigma^2)^{-n/2} |\boldsymbol{\Sigma}|^{-1/2} \exp\left\{ -\frac{1}{2\sigma^2} S(\mu, \boldsymbol{\phi}) \right\}, \tag{2.2}$$

where

$$S(\mu, \boldsymbol{\phi}) = (\mathbf{y} - \mathbf{x}\mu)'\boldsymbol{\Sigma}^{-1}(\mathbf{y} - \mathbf{x}\mu),$$

$\boldsymbol{\Sigma} = \sigma^{-2}$ cov (y) and $\mathbf{x}$ is $n \times 1$ column vector of ones. In an earlier paper the authors showed that

$$\boldsymbol{\Sigma}^{-1} = \mathbf{L}'\mathbf{L} - \mathbf{V}\mathbf{V}' \tag{2.3}$$

(Ljung & Box, 1979). Here $\mathbf{L}$ is a lower triangular bandmatrix with 1's on the main diagonal, $-\phi_1$ on the first subdiagonal, $-\phi_2$ on the second subdiagonal and so on, and

$$\mathbf{V} = \left[\frac{\mathbf{V}_1}{\mathbf{0}}\right],$$

where $\mathbf{V}_1$ is $p \times p$ upper triangular bandmatrix with elements $v_{ij} = \phi_{p+i-j}$, $i \leq j$, $j = 1, \ldots, p$.

Hence,

$$S(\mu, \boldsymbol{\phi}) = \sum_{t=1}^{n} (u_t - \phi_1 u_{t-1} - \ldots - \phi_p u_{t-p})^2 - \phi_p^2 u_1^2 - (\phi_{p-1} u_1 + \phi_p u_2)^2 - \ldots - (\phi_1 u_1 + \phi_2 u_2 + \ldots + \phi_p u_p)^2,$$

with $u_t = y_t - \mu$, for $t = 1, \ldots, n$ and $u_t = 0$, for $t \leq 0$. The determinant $|\boldsymbol{\Sigma}|$ in the likelihood function is equal to

$$|\boldsymbol{\Sigma}_p| = |\mathbf{L}_1'\mathbf{L}_1 - \mathbf{V}_1\mathbf{V}_1'|^{-1}, \tag{2.4}$$

where $\mathbf{L}_1$ is the $p \times p$ submatrix in the upper left corner of $\mathbf{L}$. The matrix $\boldsymbol{\Sigma}_p$ is proportional to the covariance matrix of p successive observations.

The posterior distribution of the parameters is proportional to the likelihood function multiplied by a chosen prior density. An analysis would often be required in cases where little is known about the parameters initially. Thus, following arguments similar to those given by Box & Tiao (1973, Section 1.3) we use the noninformative prior distribution

$$p(\mu, \boldsymbol{\phi}, \sigma) = p(\sigma) \cdot p(\mu, \boldsymbol{\phi}) \propto \sigma^{-1} |\mathbf{I}(\mu, \boldsymbol{\phi})|^{\frac{1}{2}}$$

where

$$\mathbf{I}(\mu, \boldsymbol{\phi}) = n \left[\begin{array}{c|c} \sigma^{-2}\mathbf{x}'\boldsymbol{\Sigma}^{-1}\mathbf{x} & \mathbf{0}' \\ \hline \mathbf{0} & \mathbf{I}(\boldsymbol{\phi}) \end{array}\right]$$

is the information matrix of $(\mu, \boldsymbol{\phi})$.

The quantity $\mathbf{x}'\boldsymbol{\Sigma}^{-1}\mathbf{x}$ is a function of $\boldsymbol{\phi}$ but does not involve μ. Thus, if μ and $\boldsymbol{\phi}$ were assumed independent a priori, $p(\mu, \boldsymbol{\phi})$ would be proportional to $|\mathbf{I}(\boldsymbol{\phi})|^{\frac{1}{2}}$. In the present case, however, we will consider the parameters μ and $\boldsymbol{\phi}$ jointly and choose

$$p(\mu, \boldsymbol{\phi}) = p(\mu \mid \boldsymbol{\phi}) \cdot p(\boldsymbol{\phi}) \propto (\mathbf{x}'\boldsymbol{\Sigma}^{-1}\mathbf{x})^{\frac{1}{2}} |\mathbf{I}(\boldsymbol{\phi})|^{\frac{1}{2}}.$$

Since for large n, $\mathbf{x}'\boldsymbol{\Sigma}^{-1}\mathbf{x} \propto (1 - \phi_1 - \ldots - \phi_p)$, this is nearly equivalent to choosing the prior for the parameter μ' in (2.1) locally uniform and independent of $\boldsymbol{\phi}$. For given $\boldsymbol{\phi}$, μ' may be interpreted as a location parameter in a model with independent error terms. The prior is thus consistent with that obtained by Box & Tiao (1973, p. 52) using the argument of data translated likelihoods.

By combining the joint prior density

$$p(\mu, \boldsymbol{\phi}, \sigma) \propto \sigma^{-1}(\mathbf{x}'\boldsymbol{\Sigma}^{-1}\mathbf{x})^{\frac{1}{2}} |\mathbf{I}(\boldsymbol{\phi})|^{\frac{1}{2}}$$

with the likelihood (2.2) and using the approximation

$$\mathbf{I}(\boldsymbol{\phi}) \simeq n\boldsymbol{\Sigma}_p$$

(Box & Jenkins, 1970, p. 280), we obtain the joint posterior density

$$p(\mu, \boldsymbol{\phi}, \sigma \mid \mathbf{y}) \propto \sigma^{-(n+1)}(\mathbf{x}'\boldsymbol{\Sigma}^{-1}\mathbf{x})^{1/2} \exp\left\{ -\frac{1}{2\sigma^2} S(\mu, \boldsymbol{\phi}) \right\}, \tag{2.5}$$

which integrated over the nuisance parameter σ gives

$$p(\mu, \boldsymbol{\phi} \mid \mathbf{y}) \propto (\mathbf{x}'\boldsymbol{\Sigma}^{-1}\mathbf{x})^{\frac{1}{2}} \{S(\mu, \boldsymbol{\phi})\}^{-(n/2)}. \tag{2.6}$$

For any given value of $\boldsymbol{\phi}$, the quadratic form $S(\mu, \boldsymbol{\phi})$ can be partitioned as

$$S(\mu, \boldsymbol{\phi}) = \nu s^2(\boldsymbol{\phi}) + (\mu - \hat{\mu})(\mathbf{x}'\boldsymbol{\Sigma}^{-1}\mathbf{x})(\mu - \hat{\mu}),$$

where $\hat{\mu}$ is the generalized least squares estimate of μ, i.e.

$$\hat{\mu} = (\mathbf{x}'\boldsymbol{\Sigma}^{-1}\mathbf{x})^{-1}\mathbf{x}'\boldsymbol{\Sigma}^{-1}\mathbf{y},$$

and

$$\nu s^2(\boldsymbol{\phi}) = S(\hat{\mu}, \boldsymbol{\phi}) = (\mathbf{y} - \mathbf{x}\hat{\mu})'\boldsymbol{\Sigma}^{-1}(\mathbf{y} - \mathbf{x}\hat{\mu}). \tag{2.7}$$

Substitution in (2.6) shows that

$$p(\mu|\phi, \mathbf{y}) \propto \left[1 + \frac{(\mu - \hat{\mu})^2}{\nu s^2(\phi)(\mathbf{x}'\boldsymbol{\Sigma}^{-1}\mathbf{x})^{-1}}\right]^{-n/2}. \quad (2.8)$$

Hence, conditional on ϕ, μ has a t distribution with $\nu = n - 1$ degrees of freedom, location parameter $\hat{\mu}$ and scale parameter $s^2(\phi)(\mathbf{x}'\boldsymbol{\Sigma}^{-1}\mathbf{x})^{-1}$. Equivalently,

$$t = \frac{\sqrt{\mathbf{x}'\boldsymbol{\Sigma}^{-1}\mathbf{x}}(\mu - \hat{\mu})}{s(\phi)}$$

has a t distribution with $n - 1$ degrees of freedom. For the special case $p = 1$,

$$\hat{\mu} = \frac{y_1 + y_n + (1 - \phi)\sum_{i=2}^{n-1} y_i}{2 + (n - 2)(1 - \phi)}$$

$$\mathbf{x}'\boldsymbol{\Sigma}^{-1}\mathbf{x} = 2(1 - \phi) + (n - 2)(1 - \phi)^2$$

and

$$\nu s^2(\phi) = (1 - \phi^2)\hat{u}_1^2 + \sum_{i=2}^{n} (\hat{u}_i - \phi\hat{u}_{i-1})^2,$$

where $\hat{u}_i = y_i - \hat{\mu}$. The location parameter $\hat{\mu}$ is thus a weighted average of the observations $\mathbf{y}$ with the first and last observations given weights proportional to one and the remaining observations weights proportional to $1 - \phi$. For $\phi = 0$, $\hat{\mu}$ is the sample average $\bar{y}$. For ϕ close to one, $\hat{\mu}$ is very nearly the average of y_1 and y_n. The factor $\mathbf{x}'\boldsymbol{\Sigma}^{-1}\mathbf{x}$ in the scale parameter is nearly proportional to $(1 - \phi)^2$ and decreases rapidly as ϕ increases. The inferences about μ are therefore sensitive to assumptions about ϕ.

To determine which values of ϕ are plausible in light of the data and the prior assumptions, we obtain the posterior distribution of ϕ. Integration of the joint density function (2.5) with respect to μ and σ yields

$$p(\phi|\mathbf{y}) \propto \{S(\hat{\mu}, \phi)\}^{-(n-1)/2},$$

where $S(\hat{\mu}, \phi)$ is given by (2.7).

The marginal posterior distribution of μ is obtained by integration of $p(\mu, \phi|\mathbf{y})$ in (2.6) with respect to ϕ. Equivalently,

$$p(\mu|\mathbf{y}) = \int_R p(\mu|\phi, \mathbf{y}) \cdot p(\phi|\mathbf{y})\, d\phi, \quad (2.9)$$

where R represents the stationarity region of the autoregressive process. Although the densities on the right hand side of (2.9) have a relatively simple form, it does not seem possible to get a simple closed form expression for $p(\mu|\mathbf{y})$. The density can, however, be evaluated numerically for small values of p.

If the density $p(\phi|\mathbf{y})$ is concentrated and nearly symmetric, the integral can be approximated by

$$p(\mu|\mathbf{y}) \doteq p(\mu|\hat{\phi}, \mathbf{y}),$$

where $\hat{\phi}$ is the mode of $p(\phi|\mathbf{y})$.

In many applications, the error process is adequately approximated by an autoregressive process of low order, very often of order one. A rough visual check to determine if correlation is present in the data is obtained by plotting y_t against y_{t-1}. If the error process is of first order, we can write $y_t = \mu(1 - \phi) + \phi y_{t-1} + a_t$, which shows that ϕ can be regarded as the slope of a regression line of y_t on y_{t-1}, while $\mu(1 - \phi)$ is its intercept with the y_t axes. If the sample size is large, further information about the error process can be obtained by examining the autocorrelation pattern of the residuals $\hat{u}_t = y_t - \hat{\mu}$, where these residuals may be computed assuming no correlation.

The results for the first order error model in this section deviate slightly from those given by Zellner and Tiao (1964) because of differences in the choice of prior distributions. The prior distribution employed in this study agrees with that used by Pallesen (1978) who argued that an assumption of prior independence between regression and autoregressive parameters could lead to inconsistent results.

An example

To illustrate the above results we consider an example of 20 observations generated from the model $y_t = \mu + u_t$, $u_t = \phi u_{t-1} + a_t$ with $\mu = 0$, $\phi = 0.7$ and var $(a_t) = 1$. The observations were 0.44, 0.17, 0.59, 1.90, 1.88, 0.86, 0.21, 1.27, 3.65, 2.67, 1.73, 0.37, −1.55, −1.34, −1.88, 0.51, 0.26, 0.42, 0.90, 1.70. A plot of y_t against y_{t-1} clearly indicates a positive correlation between successive observations.

The conditional density $p(\mu|\phi, \mathbf{y})$ in (2.8) is shown in Fig. 1 for different values of ϕ. The center of the distribution is relatively insensitive to changes in ϕ, but the spread of the distribution increases considerably as ϕ increases. The inferences about μ are thus very sensitive to assumptions about ϕ. This is also shown by Fig. 2, where the probability associated with the region

$$p(\mu|\phi, \mathbf{y}) < p(\mu = 0|\phi, \mathbf{y})$$

is plotted for different values of ϕ.

The posterior distribution of ϕ is shown in Fig. 3. The distribution has its mode at $\hat{\phi} = 0.65$. The density at $\phi = 0$ is very low, showing the correlation in the data. By comparing Fig. 2 and Fig. 3 we see that overall we would not discard the possibility that μ is zero. This is also seen from Fig. 4 where the solid curve shows the distribution of μ in (2.9). We note

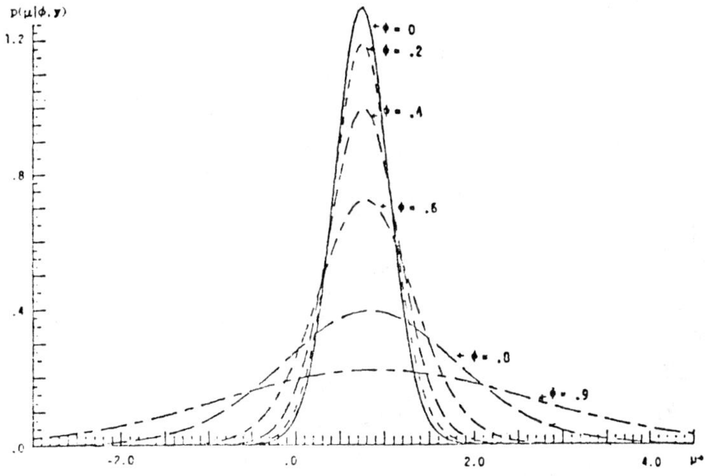

Fig. 1. Posterior distribution of μ for various choices of ϕ: Generated data.

that this distribution is very different from the distribution obtained under the independence assumption ($\phi = 0$) in Fig. 1.

The dotted curve in Fig. 4 shows the distribution $p(\mu \mid \hat{\phi}, \mathbf{y})$, where $\hat{\phi}$ is the mode of $p(\phi \mid \mathbf{y})$. This distribution supplies a surprisingly good approximation to the distribution of μ.

3. Comparison of two or more means

Suppose there are k sets of observations $\mathbf{y}_1 = (y_{11}, \ldots, y_{n_1,1}), \ldots, \mathbf{y}_k = (y_{1k}, \ldots, y_{n_k,k})'$, $k \geq 2$, and we wish to compare the means $\mu_1, \ldots, \mu_k$ of the distributions from which these observations are presumed to be samples. The model for the $n = \Sigma n_j$ observations $\mathbf{y} = (\mathbf{y}_1', \ldots, \mathbf{y}_k')'$ can be written as

$$\mathbf{y} = \mathbf{X}\boldsymbol{\mu} + \mathbf{u},$$

where $\mathbf{X}$ is a $n \times k$ matrix of indicator variables and $\mathbf{u}$ the $n \times 1$ error vector. We assume that the errors within groups are generated by an autoregressive process of order p, identically in all groups, but that there is no correlation between groups.

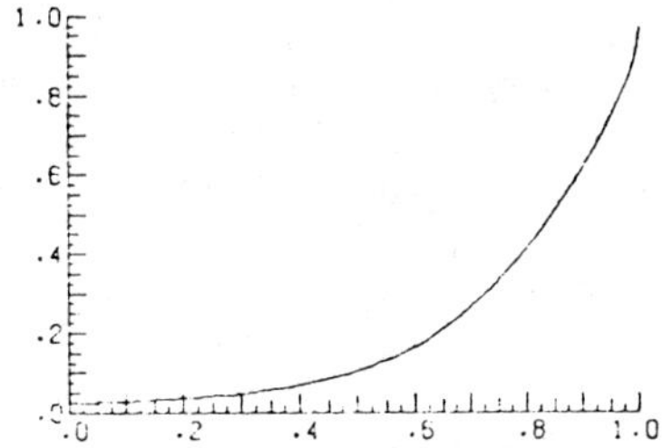

Fig. 2. The posterior probability associated with the region where $p(\mu \mid \phi, \mathbf{y}) < p(0 \mid \phi, \mathbf{y})$ plotted as a function of ϕ.

Defining $\text{cov}(\mathbf{y}) = \sigma^2\boldsymbol{\Sigma}$ and $\text{cov}(\mathbf{y}_j) = \sigma^2\boldsymbol{\Sigma}_j$, the likelihood function is

$$L(\boldsymbol{\mu}, \boldsymbol{\phi}, \sigma \mid \mathbf{y}) = (2\pi\sigma^2)^{-n/2} |\boldsymbol{\Sigma}_p|^{-k/2} \exp\left\{ -\frac{1}{2\sigma^2} S(\boldsymbol{\mu}, \boldsymbol{\phi}) \right\},$$

where

$$S(\boldsymbol{\mu}, \boldsymbol{\phi}) = (\mathbf{y} - \mathbf{X}\boldsymbol{\mu})' \boldsymbol{\Sigma}^{-1} (\mathbf{y} - \mathbf{X}\boldsymbol{\mu})$$
$$= \sum_{j=1}^{k} (\mathbf{y}_j - \mathbf{1}_j \mu_j)' \boldsymbol{\Sigma}_j^{-1} (\mathbf{y}_j - \mathbf{1}_j \mu_j)$$

and $\mathbf{1}_j$ is a $n_j \times 1$ vector of ones.

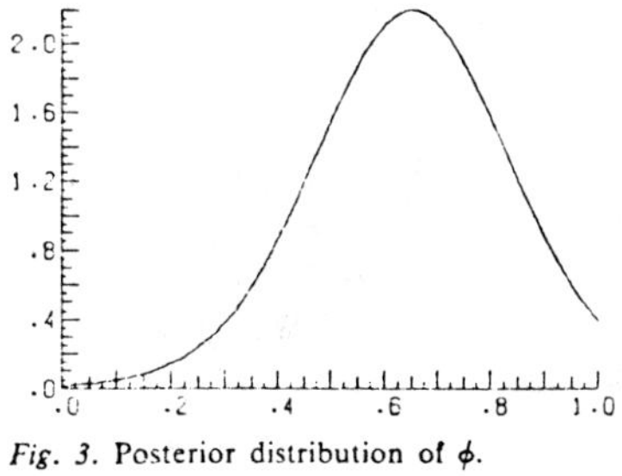

Fig. 3. Posterior distribution of ϕ.

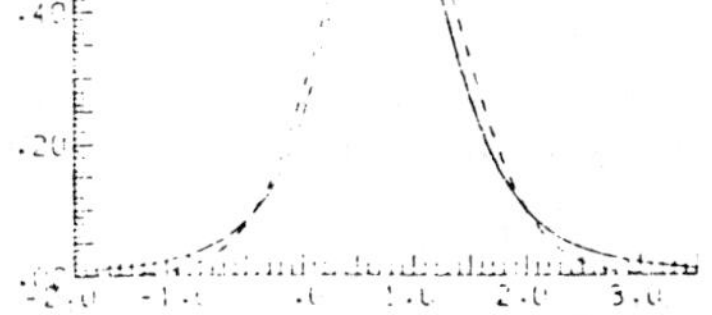

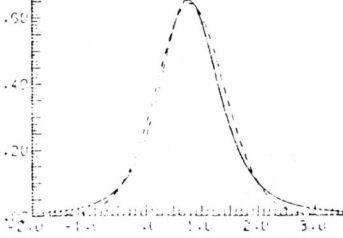

Fig. 4. Posterior distribution of μ.

Combining the likelihood function with a non-informative prior distribution as above we obtain

$$p(\boldsymbol{\mu}, \boldsymbol{\phi}, \sigma \mid \mathbf{y}) \propto \sigma^{-(n+1)} |\mathbf{X}'\boldsymbol{\Sigma}^{-1}\mathbf{X}|^{1/2} |\boldsymbol{\Sigma}_p|^{-(k-1)/2} \times \exp\left\{-\frac{1}{2\sigma^2} S(\boldsymbol{\mu}, \boldsymbol{\phi})\right\}. \quad (3.1)$$

Integration over the nuisance parameter σ gives the joint density function of $(\boldsymbol{\mu}, \boldsymbol{\phi})$ which can be written in the form

$$p(\boldsymbol{\mu}, \boldsymbol{\phi} \mid \mathbf{y}) \propto |\mathbf{X}'\boldsymbol{\Sigma}^{-1}\mathbf{X}|^{1/2} |\boldsymbol{\Sigma}_p|^{-(k-1)/2} \times \{\nu s^2(\boldsymbol{\phi}) + (\boldsymbol{\mu} - \hat{\boldsymbol{\mu}})' \mathbf{X}'\boldsymbol{\Sigma}^{-1}\mathbf{X}(\boldsymbol{\mu} - \hat{\boldsymbol{\mu}})\}^{-n/2},$$

where

$$\hat{\boldsymbol{\mu}} = (\mathbf{X}'\boldsymbol{\Sigma}^{-1}\mathbf{X})^{-1}\mathbf{X}'\boldsymbol{\Sigma}^{-1}\mathbf{y}$$

and

$$\nu s^2(\boldsymbol{\phi}) = S(\hat{\boldsymbol{\mu}}, \boldsymbol{\phi}).$$

For any fixed value of $\boldsymbol{\phi}$, the conditional distribution of $\boldsymbol{\mu}$ is in the form of a k-variate t distribution with $\nu = n - k$ degrees of freedom, mean vector $\hat{\boldsymbol{\mu}}$ and dispersion matrix $s^2(\boldsymbol{\phi})(\mathbf{X}'\boldsymbol{\Sigma}^{-1}\mathbf{X})^{-1}$. The matrix $\mathbf{X}'\boldsymbol{\Sigma}^{-1}\mathbf{X}$ is diagonal with the jth diagonal element given by $n_j' = \mathbf{1}_j'\boldsymbol{\Sigma}_j^{-1}\mathbf{1}_j$, the sum of the elements of $\boldsymbol{\Sigma}_j^{-1}$. Hence,

$$p(\boldsymbol{\mu} \mid \boldsymbol{\phi}, \mathbf{y}) \propto \left[1 + \frac{\sum_{j=1}^{k} n_j'(\mu_j - \hat{\mu}_j)^2}{\nu s^2(\boldsymbol{\phi})}\right]^{-(\nu+k)/2}$$

For the special case when the error process is of order one,

$$n_j' = 2(1-\phi) + (n_j - 2)(1-\phi)^2,$$

$$\hat{\mu}_j = \frac{(1-\phi)}{n_j'}\left\{(y_{1j} + y_{n_j,j}) + (1-\phi)\sum_{t=2}^{n_j-1} y_{tj}\right\}$$

and

$$\nu s^2(\phi) = \sum_{j=1}^{k}\left\{(1-\phi^2)\hat{u}_{1j}^2 + \sum_{t=2}^{n_j}(\hat{u}_{tj} - \phi\hat{u}_{t-1,j})^2\right\}$$

where $\hat{u}_{tj} = y_{tj} - \hat{\mu}_j$.

Conditional on $\boldsymbol{\phi}$, $k-1$ independent linear contrast $\boldsymbol{\delta} = \mathbf{l}'\boldsymbol{\mu}$ of $\boldsymbol{\mu}$ have a $(k-1)$-variate t distribution with $\nu = n - k$ degrees of freedom, location parameter $\mathbf{l}'\hat{\boldsymbol{\mu}}$ and scale parameter $s^2(\boldsymbol{\phi})\mathbf{l}'(\mathbf{X}'\boldsymbol{\Sigma}^{-1}\mathbf{X})^{-1}\mathbf{l}$. In particular, for $k = 2$,

$$\delta = \mu_2 - \mu_1$$

has a univariate t distribution with $\nu = n_1 + n_2 - 2$ degrees of freedom, location parameter $\hat{\mu}_2 - \hat{\mu}_1$ and scale parameter $s^2(\boldsymbol{\phi})(1/n_1' + 1/n_2')$. Equivalently,

$$t = \frac{\delta - (\hat{\mu}_2 - \hat{\mu}_1)}{s(\boldsymbol{\phi})\sqrt{\dfrac{1}{n_1'} + \dfrac{1}{n_2'}}}$$

is distributed as t with ν degrees of freedom. Probability statements about δ can thus be made using a table over the t distribution. In particular, the point $\delta = 0$, corresponding to $\mu_1 = \mu_2$, is included in the $(1-\alpha)$ Highest Posterior Density (H.P.D.) region if and only if

$$\frac{|\hat{\mu}_2 - \hat{\mu}_1|}{s(\boldsymbol{\phi})\sqrt{\dfrac{1}{n_1'} + \dfrac{1}{n_2'}}} < t_{\alpha/2}(\nu).$$

For comparison of more than two means, it is convenient to consider the linear contrasts

$$\delta_j = \mu_j - \frac{1}{n'}\sum_{j=1}^{k} n_j'\mu_j$$

where $n' = \sum_{j=1}^{k} n_j'$. For any given $\boldsymbol{\phi}$, $\boldsymbol{\delta} = (\delta_1, \ldots, \delta_{k-1})'$ is distributed as $(k-1)$ variate t with density function

$$p(\boldsymbol{\delta} \mid \boldsymbol{\phi}, \mathbf{y}) \propto \left[1 + \frac{Q(\boldsymbol{\delta} \mid \boldsymbol{\phi})}{\nu s^2(\boldsymbol{\phi})}\right]^{-\{\nu + (k-1)\}/2}$$

where

$$Q(\boldsymbol{\delta} \mid \boldsymbol{\phi}) = \sum_{j=1}^{k} n_j'(\delta_j - \hat{\delta}_j)^2$$

and

$$\hat{\delta}_j = \hat{\mu}_j - \frac{1}{n'}\sum_{j=1}^{k} n_j'\hat{\mu}_j.$$

The quantity

$$F = \frac{Q(\boldsymbol{\delta} \mid \boldsymbol{\phi})}{(k-1)s^2(\boldsymbol{\phi})}$$

is distributed as F with $(k-1, \nu)$ degrees of freedom. In particular, therefore, the point $\boldsymbol{\delta} = \mathbf{0}$ corresponding to $\mu_1 = \ldots = \mu_k$ is included in the $(1-\alpha)$ H.P.D. region if and only if

$$\frac{Q(\mathbf{0} \mid \boldsymbol{\phi})}{(k-1)s^2(\boldsymbol{\phi})} < F_\alpha(k-1, \nu)$$

where $F_\alpha(k-1, \nu)$ denotes the upper α percentage point of the F distribution. The calculations are set

Table 1. *Analysis of variance table to determine whether for given* ϕ *the point* $\mu_1 = \ldots = \mu_k$ *is included within a given H.P.D. region*

Source of variation	Sum of squares	D.F.	Mean square	F ratio
Inequality of means	$Q(0\mid\phi) = \sum_{j=1}^{k} n_j'\left(\hat\mu_j - \frac{1}{n'}\sum_{i=1}^{k} n_i'\hat\mu_i\right)^2$	$k-1$	$Q(0\mid\phi)/(k-1)$	$\dfrac{Q(0\mid\phi)}{(k-1)s^2(\phi)}$
Residual	$S(\hat{\mu},\phi) = \sum_{j=1}^{k}(\mathbf{y}_j - \mathbf{1}_j\hat\mu_j)'\Sigma_j^{-1}(\mathbf{y}_j - \mathbf{1}_j\hat\mu_j)$	$n-k$	$s^2(\phi)$	

out in the form of analysis of variance table in Table 1.

Integration of (3.1) with respect to μ and σ gives the posterior density of ϕ

$$p(\phi\mid\mathbf{y}) \propto |\Sigma_p|^{-(k-1)/2}\{S(\hat{\mu},\phi)\}^{-(n-k)/2}$$

This density is for moderate or large samples dominated by the term involving $S(\hat{\mu},\phi)$, since the determinant $|\Sigma_p|$ is independent of n.

The marginal posterior density of $\boldsymbol{\delta}$ is of the form

$$p(\boldsymbol{\delta}\mid\mathbf{y}) = \int_R p(\boldsymbol{\delta}\mid\phi,\mathbf{y})\,p(\phi\mid\mathbf{y})\,d\phi.$$

Again, it does not seem possible to express the marginal distribution in a simple closed form. The integral can, however, be evaluated numerically. In situations where $p(\phi\mid\mathbf{y})$ is sharp and nearly symmetric, the distribution might be approximated by $p(\boldsymbol{\delta}\mid\hat\phi,\mathbf{y})$, where $\hat\phi$ is the mode of $p(\phi\mid\mathbf{y})$. Statements about the relative values of the parameters can thus be made using the t and F distributions.

4. Inferences concerning row and column effects in the two-way classification model

One of the earliest examples of a two-way analysis of variance table, given by Fisher (1925, p. 234), is one where one would expect strong serial correlation between successive observations. The data represent frequencies of rainfall over a ten year period classified by hour of the day and month of the year. Correlation is expected between hours within months because showers which last more than one hour are recorded in successive cells. Fisher remarked that this correlation entirely invalidates the "between months" comparisons, but that the "between hours" comparisons can still be made as an approximate test. Fisher's assessment of the effect of the correlation was stated without proof, but was confirmed by Box (1954), who studied the significance levels of the usual F tests for row and column effects for different degrees of autocorrelation within rows.

Assuming no interaction effects, the usual two-way classification model is

$$y_{ij} = \mu + \alpha_i + \beta_j + u_{ij}, \quad i = 1, \ldots, I, \quad j = 1, \ldots, J,$$

$$\sum_{i=1}^{I}\alpha_i = 0, \quad \sum_{j=1}^{J}\beta_j = 0,$$

where μ is a common location parameter, α_i is the ith row effect, and β_j is the jth column effect. We shall assume that the errors u_{ij} are independent between rows but correlated within rows with the correlation structure defined by the circular autoregressive process

$$u_{ij} = \phi_1 u_{ij-1} + \ldots + \phi_p u_{ij-p} + a_{ij}$$

where $u_{is} \equiv u_{i(J+s)}$, $s = 0, -1, \ldots, -(p-1)$. This error model implies that the first observations in each row depend on the last observations in the same row. Dependence of this kind is expected in cases such as the rainfall example, where the observations represent sums over several time periods and the data are recorded without interruption between time periods.

Defining $\boldsymbol{\theta} = (\mu, \alpha_1, \ldots, \alpha_I, \beta_1, \ldots, \beta_J)'$, we can write the model in matrix form as

$$\mathbf{y} = \mathbf{X}\boldsymbol{\theta} + \mathbf{u}.$$

The $IJ \times (I+J+1)$ matrix $\mathbf{X}$ is not of full rank and it is therefore not possible to obtain a nonsingular posterior distribution for $\boldsymbol{\theta}$ using a noninformative prior distribution for the parameters. To avoid this difficulty, the model is reparametrized using the restrictions $\Sigma\alpha_i = 0$ and $\Sigma\beta_j = 0$. Defining $\boldsymbol{\gamma} = (\mu, \alpha_1, \ldots, \alpha_{I-1}, \beta_1, \ldots, \beta_{J-1})'$, we can write

$$\mathbf{y} = \mathbf{Z}\boldsymbol{\gamma} + \mathbf{u},$$

where $\mathbf{Z}$ is a $IJ \times (I+J-1)$ matrix of full rank.

The quadratic form in the exponent of the likelihood function is

$$S(\gamma, \phi) = (\mathbf{y} - \mathbf{Z}\gamma)'\mathbf{\Sigma}^{-1}(\mathbf{y} - \mathbf{Z}\gamma).$$

The matrix $\mathbf{\Sigma}$ is blockdiagonal with each diagonal block equal to $\mathbf{\Sigma}_J = (\mathbf{C}_J'\mathbf{C}_J)^{-1}$, where $\mathbf{C}_J$ is the $J \times J$ circulant matrix

$$\mathbf{C}_J = \begin{bmatrix} 1 & & & & -\phi_p & \cdots & -\phi_1 \\ -\phi_1 & 1 & & & & -\phi_p \cdots & -\phi_2 \\ \vdots & -\phi_1 & \ddots & & & \ddots & \vdots \\ -\phi_p & \vdots & \ddots & & & & -\phi_p \\ & -\phi_p & \ddots & & & & \\ & & \ddots & & & & \\ & & & & -\phi_p & \cdots -\phi_1 & 1 \end{bmatrix}$$

Hence,

$$S(\gamma, \phi) = \sum_{i=1}^{I}\sum_{j=1}^{J} \{y_{ij} - \phi_1 y_{ij-1} - \cdots - \phi_p y_{ij-p}$$
$$- (1 - \phi_1 - \cdots - \phi_p)\mu - (1 - \phi_1 - \cdots - \phi_p)\alpha_i$$
$$- (\beta_j - \phi_1\beta_{j-1} - \cdots - \phi_p\beta_{j-p})\}^2.$$

The determinant is

$$|\mathbf{\Sigma}| = |\mathbf{\Sigma}_J|^I$$
$$= |\mathbf{C}_J|^{-2I}$$

For the special case $p = 1$, $|\mathbf{C}_J| = 1 - \phi^J$. For general p, $|\mathbf{C}_J| = \Pi_{i=1}^{p}(1 - x_i^J)$, where the x_i's are the inverse roots of $1 - \phi_1 B - \ldots - \phi_p B^p = 0$ (Daniels, 1956).

Combination of the likelihood with a noninformative prior distribution gives

$$p(\gamma, \phi, \sigma \,|\, \mathbf{y}) \propto \sigma^{-(IJ+1)} |\mathbf{I}(\phi)|^{1/2} |\mathbf{Z}'\mathbf{\Sigma}^{-1}\mathbf{Z}|^{1/2} |\mathbf{\Sigma}_J|^{-I/2} \times \exp\left\{-\frac{1}{2\sigma^2} S(\gamma, \phi)\right\}$$

Proceeding a before we find that the conditional distribution of γ, given ϕ, is in the form of a $(I + J - 1)$ variate t with $\nu = (I-1)(J-1)$ degrees of freedom

$$p(\gamma \,|\, \phi, \mathbf{y}) \propto \left[1 + \frac{Q(\gamma\,|\,\phi)}{\nu s^2(\phi)}\right]^{-IJ/2},$$

where

$$Q(\gamma \,|\, \phi) = (\gamma - \hat{\gamma})'\mathbf{Z}'\mathbf{\Sigma}^{-1}\mathbf{Z}(\gamma - \hat{\gamma})$$
$$\hat{\gamma} = (\mathbf{Z}'\mathbf{\Sigma}^{-1}\mathbf{Z})^{-1}\mathbf{Z}'\mathbf{\Sigma}^{-1}\mathbf{y}$$

and

$$\nu s^2(\phi) = S(\hat{\gamma}, \phi).$$

The elements of the vector $\hat{\gamma} = (\hat{\mu}, \hat{\alpha}, \ldots, \hat{\alpha}_{I-1}, \hat{\beta}_1, \ldots, \hat{\beta}_{J-1})'$ are given by $\hat{\mu} = \bar{y}_{..}$, $\hat{\alpha}_i = \bar{y}_{i.} - \bar{y}_{..}$, and $\hat{\beta}_j = \bar{y}_{.j} - \bar{y}_{..}$, where $\bar{y}_{..}$ is the overall average, $\bar{y}_{i.}$ is the average for the ith row and $\bar{y}_{.j}$ is the average for the jth column. Because of the correlation structure assumed, these estimates are independent of ϕ.

The quadratic form $Q(\gamma\,|\,\phi)$ can be partitioned as

$$Q(\gamma\,|\,\phi) = Q_0(\mu\,|\,\phi) + Q_1(\alpha\,|\,\phi) + Q_2(\beta\,|\,\phi) \tag{4.1}$$

where

$$Q_0(\mu\,|\,\phi) = IJ(1 - \phi_1 - \ldots - \phi_p)^2(\mu - \hat{\mu})^2,$$
$$Q_1(\alpha\,|\,\phi) = J(1 - \phi_1 - \ldots - \phi_p)^2 \textstyle\sum_{i=1}^{I} (\alpha_i - \hat{\alpha}_i)^2$$

and

$$Q_2(\beta\,|\,\phi) = I\sum_{j=1}^{J} \times \{(\beta_j - \hat{\beta}_j) - \phi_1(\beta_{j-1} - \hat{\beta}_{j-1}) - \cdots - \phi_p(\beta_{j-p} - \hat{\beta}_{j-p})\}^2,$$

with $\beta_s \equiv \beta_{J+s}$ and $\hat{\beta}_s \equiv \hat{\beta}_{J+s}$, $s = 0, -1, \ldots, -(p-1)$.

From the properties of the multivariate t distribution it follows that μ, α, and β are for given ϕ separately distributed as k-variate t with k equal to 1, $I-1$, and $J-1$, respectively. The densities are functions of the quantities Q in (4.1). It is apparent that some of these quantities depend heavily on the parameters ϕ. The inferences about the parameters are therefore expected to be sensitive to assumptions about ϕ.

The conditional density of γ is a monotonic function of

$$F = \frac{Q(\gamma\,|\,\phi)}{ks^2(\phi)},$$

where $k = (I + J - 1)$. This quantity has a F distribution with (k, ν) degrees of freedom. A particular point γ_0 is thus included in an H.P.D. of content $1 - \alpha$ if and only if

$$\frac{Q(\gamma_0\,|\,\phi)}{ks^2(\phi)} < F_\alpha(k, \nu).$$

The posterior distribution of ϕ is

$$p(\phi\,|\,\mathbf{y}) \propto |\mathbf{I}(\phi)|^{1/2} |\mathbf{\Sigma}_J|^{-I/2} \{S(\hat{\gamma}, \phi)\}^{-(I-1)(J-1)/2} \tag{4.2}$$

When the number of observations is moderate or large, the distribution is dominated by the term

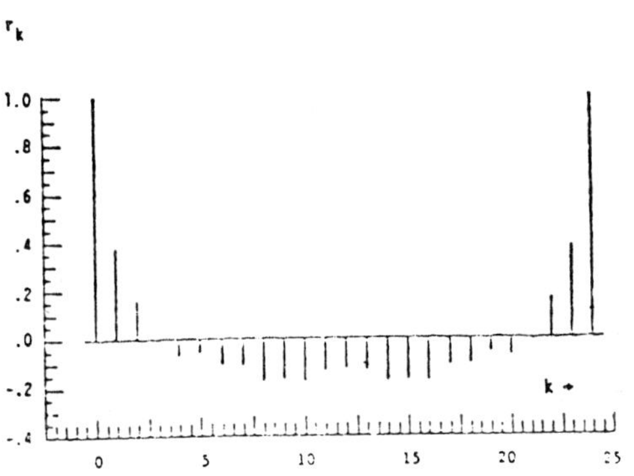

Fig. 5. Autocorrelation function of the residuals $\hat{u}_i$: Rainfall data.

involving $S(\hat{\gamma}, \phi)$, so that

$$p(\phi \mid \mathbf{y}) \,\dot{\propto}\, \{S(\hat{\gamma}, \phi)\}^{-(I-1)(J-1)/2} \qquad (4.3)$$

Application to Fisher's rainfall data

Residual autocorrelation coefficients for the rainfall data, calculated using the circular definition

$$r_k = \frac{\sum_{i=1}^{I} \sum_{j=1}^{J} \hat{u}_{ij}\hat{u}_{ij-k}}{\sum_{i=1}^{I} \sum_{j=1}^{J} \hat{u}_{ij}^2}, \quad k = 0, 1, \ldots, J,$$

where $\hat{u}_{is} \equiv \hat{u}_{i(J+s)}$, $s = 0, -1, \ldots, -(k-1)$, are shown in Fig. 5. The pattern suggests that an autoregressive error model of order 1 or 2 may be appropriate. By examining the posterior distribution of ϕ, one finds that the first order model is sufficient.

The solid curve in Figure 6 shows the posterior distribution of the parameter ϕ in (4.2). The density function is nearly symmetric about the mode $\hat{\phi} \doteq 0.38$. A 95% H.P.D. region includes values between 0.25 and 0.50 approximately. The density at $\phi = 0$ is practically zero, which verifies the suspected dependence between the errors. The density of ϕ is

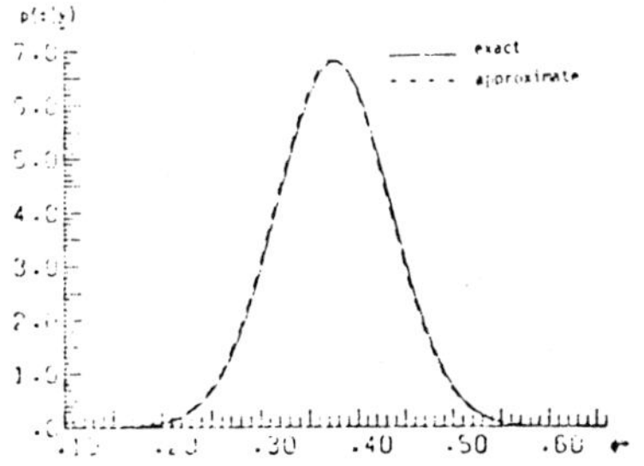

Fig. 6. Posterior distribution of ϕ: Rainfall data.

Table 2. *Sums of squares and F ratios for different values of ϕ: rainfall data*

	Sum of squares			F ratio	
ϕ	Row (months)/(hours) $Q_1(0 \mid \phi)$	Column (months)/(hours) $Q_2(0 \mid \phi)$	Residual $S(\hat{\gamma}, \phi)$	Row	Column
.00	6 568.6	1 539.3	3 819.6	39.55	4.43
.20	4 203.9	1 135.9	3 400.2	28.44	3.67
.40	2 364.7	855.7	3 286.3	16.55	2.86
.60	1 051.0	698.6	3 478.0	6.95	2.21
.80	262.7	664.6	3 975.3	1.52	1.84
.95	16.4	720.0	4 548.8	.08	1.74

closely approximated by (4.3), which is shown by the dotted curve in Fig. 6.

The sums of squares Q and F-ratios corresponding to $\alpha = 0$ and $\beta = 0$ are shown in Table 2 for different values of ϕ. Comparing with values from an F table, we find that the points fall outside the 99% H.P.D. region for all values of ϕ in the interval where $p(\phi \mid \mathbf{y})$ is appreciable. The results therefore indicate an influence of the time of the day and the time of the year on the frequency of rainfall.

Although for this particular data set the same conclusion would be reached assuming $\phi = 0$, the inferences can be very sensitive to assumptions about ϕ. This is particularly illustrated by the 500 fold decrease in the F ratios for rows as ϕ increases from 0 to 0.95. The F-ratios for columns are more stable, indicating that the comparison of column effects are less affected by assumptions about ϕ. This can also be seen by plotting the conditional distributions of $\alpha_r - \alpha_s$ and $\beta_r - \beta_s$ for different values of ϕ.

If the marginal distributions of α and β were needed, close approximations would be obtained by substituting the modal value $\hat{\phi} = 0.38$ for ϕ in the conditional distributions above.

We have assumed in this section that the dependence between the observations can be described by a circularly defined autoregressive process. If the number of row cells J is large, the results can be used as an approximation in cases where a non-circular model would be more appropriate. Alternatively, using the inverse (2.3), one could easily derive expressions for the non-circular case which properly take the unknown initial conditions into account.

Acknowledgement

This work was supported by the U.S. Army Research Office under Grant DAAG29-80-C-0041 and by the Air Force Office of Scientific Research under Grant AFOSR 72-2363D.

References

Anderson, R. L. (1954). The problem of autocorrelation in regression analysis. *J. Amer. Statist. Assoc.* **49**, 113–129.

Anderson, T. W. (1971). *The statistical analysis of time series.* John Wiley, New York.

Box, G. E. P. (1954). Some theorems on quadratic forms applied in the study of analysis of variance problems. II. Effects of inequality of variance and of correlation between errors in the two-way classification. *Ann. Math. Statist.* **25**, 484–498.

Box, G. E. P. & Jenkins, G. M. (1970). *Time series analysis, forecasting and control.* Holden-Day, San Francisco.

Box, G. E. P. & Tiao, G. C. (1973). *Bayesian inference in statistical analysis.* Addison-Wesley, Reading, Mass.

Daniels, H. E. (1956). The approximate distribution of serial correlation coefficients. *Biometrika* **43**, 169–185.

Durbin, J. (1960). Estimation of parameters in time-series regression models. *J. Roy. Statist Soc. Ser. B* **22**, 139–153.

Fisher, R. A. (1925). *Statistical methods for research workers.* Oliver and Boyd, Edinburgh.

Ljung, G. M. & Box, G. E. P. (1979). The likelihood function of stationary autoregressive-moving average models. *Biometrika* 66, 625–633.

Pallesen, L. C. (1978). Studies in the analysis of serially dependent data. University of Wisconsin, MRC Technical Summary Report no. 1837.

Pierce, D. A. (1971). Least squares estimation in the regression model with autoregressive-moving average errors. *Biometrika* **58**, 299–312.

Sredni, J. (1970). Problems of design, estimation and lack of fit in model building. Unpublished Ph.D. thesis, University of Wisconsin, Madison.

Watson, G. S. (1967). Linear least squares regression. *Ann. Math. Statist.* **38**, 1697–1699.

Zellner, A. & Tiao, G. C. (1964). Bayesian analysis of the regression model with autocorrelated errors. *J. Amer. Statist. Assoc.* **59**, 763–778.

Greta Ljung
Boston University
704 Commonwealth Avenue
Boston, MA 02215
USA

4.13

An Analysis of Transformations Revisited, Rebutted

G. E. P. BOX and D. R. COX*

Transformation has long been a powerful tool in developing parsimonious representations and interpretations of data. In 1964 we examined the formal estimation of a suitable transformation. In particular, suppose that a response y is transformed to $y^{(\lambda)}$, where

$$y^{(\lambda)} = (y^{\lambda} - 1)/\lambda \quad (\lambda \neq 0)$$
$$\log y \quad (\lambda = 0),$$

and that we assume provisionally that for some unknown λ, the vector $\mathbf{y}^{(\lambda)} = (y_1^{(\lambda)}, \ldots, y_n^{(\lambda)})$ of n transformed observations satisfies a linear model

$$E(\mathbf{y}^{(\lambda)}) = \mathbf{X\theta},$$

where $\boldsymbol{\theta}$ is unknown, the errors being independently normally distributed with zero mean and constant variance σ^2. Estimation of λ, $\boldsymbol{\theta}$, and σ^2 can be by Bayesian or maximum likelihood methods.

Bickel and Doksum (1981), in a technically impressive paper, studied in particular the joint estimation of λ and $\boldsymbol{\theta}$, examining consistency and asymptotic variances. They report that the cost of not knowing λ and having to estimate it, can be severe; that ". . . the performance of all Box-Cox type procedures is unstable and highly dependent on the parameters of the model in structured models with small to moderate error variances." That is, the estimates $\hat{\lambda}$ and $\hat{\boldsymbol{\theta}}$ can be highly correlated, so that the marginal variances of the $\hat{\theta}$'s can be inflated by large factors over the conditional variances for fixed λ.

It seems to us that this general conclusion is qualitatively obvious and at the same time scientifically irrelevant.

To illustrate first the obviousness, take as a simple example the comparison of two groups of modest size, the observations $\mathbf{y}$ in group one being near 995 and those in group two being near 1005, the scatters within the two groups being roughly normal with standard deviations close to unity. A parameter θ representing the difference between groups on the y scale is quite precisely estimated to be about 10 y-units. Suppose that the possibility of transformation were contemplated. For a very wide range of λ the function $y^{(\lambda)}$ is very nearly linear in y over the span of the data, and, in particular, unless the sample sizes were very large indeed, it would be quite impossible to distinguish from the data whether y or y^{-1} gave better fit to the standard normal assumptions: if the parameter θ were to refer to a difference on the y^{-1} scale it is quite precisely estimated to be near -10^{-5} y^{-1}-units (or 10^{-5} $y^{(-1)}$-units, where $y^{(-1)} = (1/y - 1)/(-1)$). Thus if the target parameter θ is defined in terms of unknown λ in such a case as this, where λ is poorly determined, the numerical value of θ (in units of y^{λ} or $y^{(\lambda)}$) could be virtually anything.

As to the scientific implications of this, how can it be sensible scientifically to state a conclusion as a number measured on an unknown scale? Surely to know that some effect has magnitude 10 units is without content unless one knows the scale and units in which the effect is defined. To say in the above idealized example that θ, defining the difference between groups, is ill determined because the data establish a wide range of functions as virtually equivalent, seems to be very misleading.

There is, of course, no dispute with Bickel and Doksum over mathematics: the issue is one of scientific relevance. As with any procedure it is necessary to use some common sense in estimating transformations, and in particular (see, e.g., Box, Hunter, and Hunter 1978, p. 241) not to expect this to be possible or relevant when for the particular data and class of transformations in mind the transformation is essentially linear.

Of course the gross correlation effects would be avoided if, following our paper, the investigation had been conducted in terms of

$$z^{(\lambda)} = (y^{\lambda} - 1)/(\lambda \dot{y}^{(\lambda - 1)}), \quad (\lambda \neq 0)$$
$$\dot{y} \log y \quad (\lambda = 0),$$

which takes account of the Jacobian of the transformation. (For the above examples the differences in means for both $z^{(1)}$ and $z^{(-1)}$ would then have been very nearly 10 units.) However, some question of scientific relevance would still remain.

There are numerous aspects of transformations that merit further study. These include in particular the further development of simple ways of assessing *transformation potential*; that is, of providing some more formal measure of the ability of particular data to provide useful information about a class of transformations. Further, a referee has made the perceptive comment that the following issue remains unresolved. Suppose that the parameter of

* G. E. P. Box is Professor, Mathematics Research Center, University of Wisconsin, Madison, WI 53706. D. R. Cox is Professor of Statistics, Department of Mathematics, Imperial College of Science and Technology, Queensgate, London, England SW7 2AZ.

The Collected Works of George E. P. Box, 1984, Wadsworth, Inc., Belmont, CA 94002.
Originally published in *J. Amer. Stat. Assoc.,* vol. 77, no. 377 (1982), pp. 209–210.

interest (difference, regression coefficient, etc.) is defined on the data-dependent scale $\hat{\lambda}$; in what circumstances do confidence intervals for these parameters calculated in the "usual" way, as if $\hat{\lambda}$ were preassigned, provide an adequate approximation?

[*Received October 1981. Revised November 1981.*]

REFERENCES

BICKEL, P.J., and DOKSUM, K.A. (1981), "An analysis of transformations revisited," *Journal of the American Statistical Association*, 76, 296–311.

BOX, G.E.P., and COX, D.R. (1964), "An analysis of transformations," *Journal of the Royal Statistical Society*, Ser. B, 26, 211–252.

BOX, G.E.P., HUNTER, W.G., and HUNTER, J.S. (1978), *Statistics for Experimenters*, New York: John Wiley.

5
Application of Statistics

Contents

5.0
Introduction

RONALD D. SNEE
E. I. DuPont de Nemours & Co.

Box has pointed out on many occasions that statistics is more than an intellectual pursuit; it is a set of concepts and tools that we use to solve problems. He has emphasized that in order to develop and use statistics effectively, there must be constant iteration between practice and theory. A study of his applications papers is and enlightening and stimulating experience. We see him working as a practicing statistician solving the problems that confront his employer. This is particularly true of the papers he published while working in the British army during World War II, later at Imperial Chemical Industries (ICI), and more recently while consulting on the analysis of the Los Angeles air pollution data.

It is clear from Box's work that he sees the need to solve practical problems as the stimulus for theoretical advances. If there is to be an iteration between theory and practice, one must begin somewhere. For Box, this iteration began most frequently with a practical problem that needed solving. This is, of coure, the most effective and efficient way to develop statistical methodology, because the result is put to work immediately and we avoid the situation of having a methodology looking for a problem to solve. Further understanding of Box's views and work on applications of statistics can be found in his videotapes, "Importance of Practice in the Development of Statistics" and "Practice and Theory: Some Experiences of George Box," available from the American Statistical Association, 806-15th Street NW, Washington, D.C. 20005.

One cannot help but marvel at Box's clear and interesting style of exposition, which is very important if a statistician is to communicate effectively with scientists. The ability to write clearly and succinctly is characteristic of a deep understanding of the subject. Box is world renowned for this quality, and it is one of the secrets to his success in getting his methodology broadly used in many fields of science and engineering. The fact that he learned statistics mostly by self-study was likely very important in this regard, for it made evident to him what was needed for a nonstatistician to understand and use statistical methodology.

From the beginning, Box's work makes extensive use of graphics. This comes as no surprise when we note his success as a communicator of statistical concepts and methods. While many statisticians have only recently become comfortable with the use of graphical tools, scientists have been using graphics effectively for many years. Box uses graphics both to stimulate his insight into problems and develop understanding, and to present the resulting findings and ideas to others. We see clearly in his papers both Box the scientist and Box the statistician at work. This is not surprising, since he was a scientist before he was a statistician.

One particularly revealing way to study Box's work is to read his papers in the order in which he wrote them and observe the evolution of his thinking and the variety of problems he worked on. His two papers with Cullumbine (5.1, 5.2) in 1947 describe his work during World War II while he was a chemist in the British army. At that time he was studying statistics at night in order to solve the research problems he was working on during the day. These papers describe the use of a wide variety of statistical tools (e.g., probit analysis, analysis of variance, factorial designs, chi-square statistics, regression analysis, t-test, transformations, analysis of censored data), the extent of which would likely require at least a two-year university course in biometry to cover adequately. These papers show us a practicing statistician at work, one who has a clear understanding of the role of statistics in science and who has made an excellent start on a very productive career.

The second paper with Cullumbine (5.2) is important, for it describes the problems Box was working on when he visited R. A. Fisher to discuss the use of transformations. An account of this meeting is contained in his videotape "Practice and Theory — Some Experiences of George Box."

The paper on the analysis of growth and wear curves (5.3) in 1950 is a classic and is referenced by all who work in this field. This paper, which describes work done at ICI, was the stimulus for Box's work on the distribution of quadratic forms (4.2, 4.3). It begins with two examples (tire wear, growth of rats), introduces the concepts of growth rate and curve shape and the use of differences to remove serial correlation, discusses the use of analysis of variances, develops multivariate statistics to test assumptions and compare group differences, and mentions future work on robustness. The paper concludes with comment on practical problems in growth curve analysis, such as data reduction by curve fitting, using the average of several readings per week as the response, and the handling of differences in initial weight.

The paper on pigment strength testing with Hobbs and North (5.4) in 1953 is a good case history of the use of experimental design in problem solving. It is also a revealing discussion of the many problems encountered in making dyes of uniform quality. One thing that characterizes this paper as a good applications paper is the summary of the important findings and conclusions that resulted from this work.

The paper "Mathematical Statistics and Rubber Technology" (5.5) is an excellent example of how to introduce scientists to statistics. The important concepts and techniques are presented by a series of examples. The exposition contains several tables and graphs and no equations. This approach is effective and widely used by practicing statisticians today.

Box's work on evolutionary operation (EVOP) (5.6) is classic. In developing a method for increasing industrial productivity, he pointed out that "a process should be run so as to generate product *plus information on how to improve the product*" (emphasis added). To communicate this idea, he used the analogy of natural selection. EVOP was widely used in industry during the late 1950s and 1960s and resulted in many successful applications. Response to problems encountered in the application of EVOP resulted in later papers, one with J. S. Hunter (1959) on "Condensed Calculations for Evolutionary Operation Programs," one with N. R. Draper (1969) on "Isn't My Process Too Variable for EVOP?", and his 1966 paper, "A Simple System of Evolutionary Operation Subject to Empirical Feedback." The late 1950s and early 1960s were a transition period for Box in which his thinking moved from design of experiments and steady-state process optimization to his work with G. M. Jenkins on time series analysis. This transition came about while he was working on adaptive optimization of continuous processes as discussed at the end of his 1960 paper "Some General Considerations in Process Optimization" and in the papers with Chanmugam (1962), "Adaptive Optimization of Continuous Processes," and Kotnour and Altpeter (1966), "A Discrete Predictor Controller Applied to Sinusoidal Perturbation Adaptive Optimization." Some key elements were the recognition that in many processes there are continuous changes in process variables, dynamic effects, and correlated errors in the observations. In discussions with Jenkins, it became clear that time series modeling concepts were needed to develop a useful solution. These initial investigations led to the broad study of time series modeling and eventually to Box's book with Jenkins, *Time Series Analysis Forecasting and Control.*

Box's papers on statistical computation (5.7) in 1969 and the environment (5.8) in 1974 are the start of his philosophical writings on statistics. In these papers we hear about the interaction between statistics and science and problem solving, and how practice stimulates theoretical advances. In the 1969 paper we read about the need for interaction between statisticians and computer scientists and the need for interaction between the computer and "the human mind and so allow the creative process to proceed." Box also comments on how to use the computer effectively and how statisticians should be trained and subsequently practice their profession.

The 1974 paper (5.8) discusses problem-solving philosophy, with emphasis on environmental problems and the use of statistics in their solution. He emphasizes that learning and problem solving are iterative processes with a feedback loop. In statistical analysis, we iterate between being a model sponsor (the fitting process) and a model critic (study of the residuals, coefficients, etc.). He concludes with more comment on the training of statisticians. He notes that "a proper balance of theory and practice is needed and, most important, statisticians must learn how to be good scientists, a talent which, I think, has to be learned by example." He goes on to discuss how this is accomplished at the University of Wisconsin.

In the mid-1970s, Box got involved in the important study of the Los Angeles air pollution data. The works with Tiao, Hamming, and Phadke (5.8, 5.9, the 1975 paper "A Statistical Analysis of the Los Angeles Ambient Carbon Monoxide Data 1955–1972," and the 1976 paper "Empirical-Mechanistic Modeling of Air Pollution"), describe some first-rate problem-solving efforts that include the important model-building concepts of intervention analyses, parsimony, and the combination of empirical and mechanistic information. Also characteristic of these papers are the lucid explanations and the effective use of graphical displays that we saw thirty years earlier in his work with Cullumbine (5.1, 5.2).

Any comment on Box's application work is incomplete without noting the effect that his work has had on statistical practice in general. Briefly stated, it suffices to say that in my case, I design screening experiments using the methods of Box and Hunter (2.10, 2.11); carry out response surface optimization studies using the designs of Box and Wilson (2.1), Box and Hunter (2.5), and Box and Behnken (2.9); do model fitting using the methods of Box and Cox (4.9) and Box and Tidwell (4.8); analyze time series by the methods of Box and Jenkins; and develop growth curve and object shape analysis procedures using Box's 1950 paper (5.3). All of these techniques have their roots in problems Box encountered in his statistical practice. These techniques are also widely used by other statisticians working in industry and in other areas of application. Statisticians and other scientists and engineers are indeed fortunate that George Box chose statistics as his life's work.

5.1
The Effect of Exposure to Sub-Lethal Doses of Phosgene on the Subsequent L(*Ct*)50 for Rats and Mice

G. E. P. BOX and H. CULLUMBINE
Chemical Defence Experimental Station, Porton

(Received November 15, 1946)

It has been suggested that when dogs and goats survive doses of phosgene gas ($COCl_2$) their subsequent susceptibility is apparently lessened. The effect was explained as being probably due to selection ; since the more susceptible animals were killed, the remainder would be more resistant and their average susceptibility lower. The experiments described in this report were carried out in order to obtain further information about the matter.

Experimental

Preliminary experiments were carried out to ascertain the highest concentration to which rats and mice could be exposed for 10 min. without causing death. It was found that exposure of rats and mice to dosages (*Ct*)* of 800 and 600 mg.min./cu.m. ($t = 10$ min.) respectively did not normally produce any deaths, although the animals showed all the symptoms of severe phosgene poisoning. These dosages were therefore used throughout the work as the preliminary or " pregassing " doses.

The exposure to phosgene was carried out in a small chamber (20 litres capacity) in a constant flow of 200 litres/min. of an air-phosgene mixture of the required concentration. The apparatus is shown in Fig. 1. The atmosphere in the chamber was sampled at 10 litres/min. throughout the whole of the exposure.

*The dosage to which animals were exposed is measured here by multiplying the mean concentration (*C*) measured in mg./cu.m. by the time of exposure *t*. (Hence *Ct* in mg.min. cu.m.) The time of exposure (*t*) was 10 min. in all experiments unless otherwise stated.

The Collected Works of George E. P. Box, 1984, Wadsworth, Inc., Belmont, CA 94002.
Originally published in *British Journal of Pharmacology and Chemotherapy,* vol. 2, no. 1 (1947), pp. 38–55.

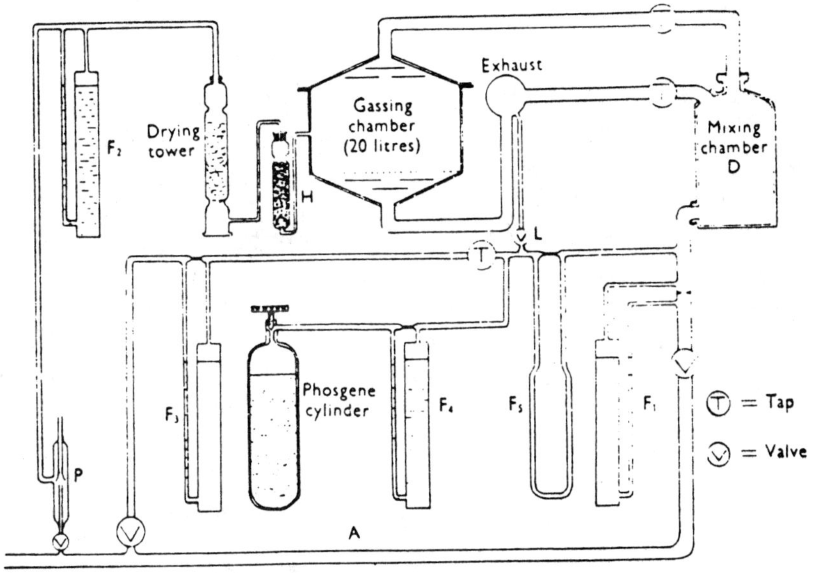

FIG. 1.—The constant-flow phosgene apparatus. 200 litres/min. of air was blown along the main air line A and measured by flowmeter F_1. Phosgene (measured by F_4) was mixed with diluting air (measured by F_3). The mixture regulated by valve L was measured by F_5 and mixed in chamber D with the main air stream. The diluted air–phosgene mixture could then be made to pass through the chamber or run to waste as required. A suction pump (P) drew a continuous sample at 10 litres/min. through the sampling bubbler H.

The effect of phosgene

96 rats (weight 115–125 g.) were taken and divided at random into 4 groups of 24 each. 12 rats in each group were exposed to a dosage (*Ct*) of 800. and the remaining 12 in each group were kept under identical conditions as controls. Five days later each group of 24 was exposed for 10 min. to concentrations of phosgene in the lethal range. The mortalities at the end of 48 hours were:

Dosage (*Ct*) to which rats were exposed (mg.min. cu.m.)	2300	2500	3150	4400	Total
Mortality in controls	8/12	7/11	9/12	11/12	35/47
Mortality in pregassed animals ..	3/12	2/12	3/12	8/12	16/48

Regression lines were fitted between the mortality expressed in probits and the logarithm of the dosage (*Ct*) by the method described by Gaddum (1933), as elaborated by Bliss (1935, 1938) and Fisher and Yates (1943).

Statistical analysis

	Degrees of freedom	χ^2	P
Differences in position of lines	1	17·6	<0·0001
Differences in slope of lines	1	0·2	0·6
Heterogeneity	4	2·2	0·7

L(*Ct*)50 Pregassed 3840 mg.min./cu.m.
" Control 1880 " } $I_{L(Ct)50} = 2{\cdot}0$

Where $I_{L(Ct)50}$ is an index of effectiveness obtained by dividing the L(*Ct*)50 for pregassed by the L(*Ct*)50 for control animals.

There is little doubt that when subsequently exposed to the same concentration there is a lower mortality in the pregassed than in the control animals.

Phosgene and hexamine

Larger doses of phosgene for pregassing might produce greater effects, but would result in deaths in the pregassing. It was therefore decided (i) to pregas with a dosage (*Ct*) of 6,000, but to prevent death by administering oral hexamine* immediately beforehand, and to compare this with normal pregassing ; (ii) to use intervals of both 3 and 7 days between pregassing and lethal gassing so as to compare the effect of altering the periods between gassing ; (iii) to carry out the experiment with mice in order to test whether they behaved similarly to rats ; and (iv) to carry out the whole experiment at two different concentrations of phosgene.

96 mice were divided at random into 8 groups of 8, and 2 of 16. The treatment applied to the groups and the resulting mortalities are shown below. The dose of hexamine used was about 2 g. per kg. (0.2 c.c. of a 20 per cent (w/v) solution orally immediately prior to gassing).

	Pregassed 3 days before 2nd exposure with *Ct*		Pregassed 7 days before 2nd exposure with *Ct*		No pregassing
2nd gassing *Ct* in mg.min. cu.m.	600	6000 and hexamine	600	6000 and hexamine	—
2,450	6/8	5/8	6/8	8/8	16/16
1,500	4/8	3/8	2/8	5/8	14/16

Statistical analysis

By a simple extension of the technique described by Bliss (1935), we can fit regression lines for the 4 treatments and control, and compare these for differ-

*Hexamine has a great chemical affinity for phosgene, and oral administration (2 g. per kg.) *immediately* prior to gassing will prevent phosgene poisoning.

ences in position and slope. The variation between slopes is not significant ($\chi^2 = 1.70$ for 4 degrees of freedom, $P \simeq 0.8$). Comparing the lines for position, we find:

Comparison of methods of pregassing	Degrees of freedom	χ^2	P
Time (3 days *v* 7 days)	1	1·55	0·2
Dosage (600 *v* 6000 + hexamine)..	1	1·55	0·2
Comparison of pregassing (in general) with controls ..	1	11·41	<0·001
L(*Ct*)50 pregassed (average)	1630 mg.min./cu.m.	$\frac{L_{(Ct)50}}{L_{(Ct)50}} = 1{\cdot}6$	
L(*Ct*)50 control	1020 mg.min./cu.m.		

In general pregassing of mice has a highly significant effect. There is no evidence from this experiment that the effect is different after 3 or 7 days. Increase in dosage (*Ct*) of the second gassing from 1,500 to 2,450 mg.min./cu.m. produces a uniform increase of probit in all the groups, and the effect is equally marked at both dosages.

Exposure in the preliminary gassing to a dosage of 6,000 after oral hexamine produces an effect no greater than that of exposure to 600 without hexamine.

Duration of the effect

In order to obtain further information on the duration of the effect, 70 mice were taken and divided at random into 7 groups of 10. One group was kept as a control and the other 6 were exposed to a phosgene dosage of 600 mg.min./cu.m. at 1, 2, 3, 5, 7, and 10 days before the second gassing. The whole 70 mice were then exposed to a dosage (*Ct*) of 5,850 mg.min./cu.m. (i.e., about 3 times the L(*Ct*)50) at the same time. The time of death of the mice was noted. We have shown elsewhere (Box and Cullumbine, 1947) that the reciprocal of survival time is closely correlated with dosage. The survival times were therefore transformed to reciprocals for analysis. The median survival times are given below together with the analysis of variance of the transformed variate. The data are graphed in Fig. 2, using the reciprocal scale.

Second gassing (*Ct.* = 5,850, t = 10) on day Z.

Preliminary gassing on day:	Z-1	Z-2	Z-3	Z-5	Z-7	Z-10	Control
Estimated median survival time (hours):	4·3	5·0	9·1	10·4	7·5	5·3	5·5

ANALYSIS OF VARIANCE OF RECIPROCAL SURVIVAL TIMES ($100 \times$ HR.$^{-1}$)

Source of variation	Degrees of freedom	Sum of squares	Mean square	Variance ratio (F)
Between groups ..	6	1529	255	8·8
Within groups ..	63	1802	29	
Total	69	3331		

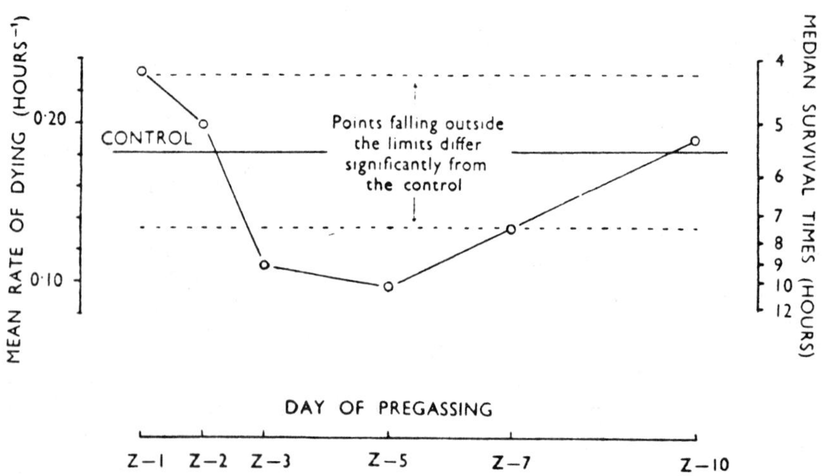

FIG. 2.—The median survival time (plotted as reciprocal) of mice pregassed at different times and exposed to phosgene (dosage Ct = 5,850, t = 10 min.).

It is clear that the variation between groups is highly significant ($P < 0.001$) and that the animals in the group pregassed one day before die significantly faster, whilst those pregassed 3, 5, and 7 days before die significantly slower than the control animals. The index of effectiveness I_T (obtained by dividing survival time in treated group by the survival time in controls) reaches a maximum of 1.9 five days after pregassing.

Pregassing more than once

This experiment was designed to test whether the effect could be increased by exposure more than once to pregassing. The pregassing doses were given 12, 8, and 4 days before the dose in the lethal zone. 96 mice were used and divided at random into 8 groups, A to H, each containing 12 mice.

On day Z−12	groups	C, D, G and H	were exposed to a dosage (Ct) of	600
,, ,, Z−8	,,	D, B, F and H	,, ,, ,,	550
,, ,, Z−4	,,	E, F, G and H	,, ,, ,,	600
,, ,, Z	,,	A, B, C, D, E, F, G, H	,, ,, final dosage of	2,850

MORTALITIES (OUT OF 12) AFTER 48 HOURS

A : 11/12	B : 10/12	E : 7/12	F : 6/12
C : 11/12	D : 12/12	G : 5/12	H : 3/12

The data may be analysed by transforming the proportions to angles using the transformation $x = \sin^{-1}\sqrt{p}$ (where p is the proportion dying). This will have the effect of stabilizing the variance. The technique was exactly as described by Fisher and Yates (1943). The expected values can in this case be obtained by inspection of the data, but in more awkward cases the graphical methods suggested by Richards (1941) have proved helpful.

Analysis of variance of the transformed variate

Effect	Degrees of freedom	Sum of squares	χ^2	P
Z−12 1st Pregassing ..	1	0		
Z−8 2nd "	1	12		
Z−4 3rd "	1	1588	23·4	<0·00001
Interactions				
1 × 2	1	101		
1 × 3	1	0		
2 × 3	1	101		
1 × 2 × 3	1	12		
Total	7	1814	26·5	<0·001

$$\text{Theoretical variance } \frac{820{\cdot}7}{12} = 68$$

Pregassing on Z–12 and Z–8 days had no effect upon mortality, although pregassing at Z–4 produced a highly significant decrease in mortality in all groups.

Lung damage and the effect

To test whether damage to the lungs was necessary to produce the effect, the following experiment was carried out: 48 rats were divided at random into 4 groups of 12, A, B, C, and D.

Group A was given a pregassing dosage on day Z–5,

B was given a pregassing dosage on day Z–5 immediately after oral hexamine,

C was kept as a control,

and D was given oral hexamine *only* on day Z–5.

A, B, C, and D were all exposed on day Z to a dosage of 3,150 mg.min./cu.m. After the first exposure group B showed no signs of distress. (Rats treated in this way showed no signs of pulmonary oedema at autopsy 12 and 24 hours after gassing: thus it appeared that oral hexamine was completely effective in preventing the action of this concentration of phosgene.) The 48-hour mortality after the second exposure was:

	No hexamine	Hexamine
Pregassed	A : 3/12	B : 9/12
Not pregassed	C : 9/12	D : 9/12

Statistical analysis

	Degrees of freedom	χ^2	P
Pregassing with no hexamine	1	4·17	<0·05

Thus pregassing with no hexamine significantly reduces mortality; pregassing *with* hexamine so that no pulmonary damage is produced is ineffective. It would appear necessary therefore to cause lung damage in order to produce the effect. At first sight this experiment appears to contradict our previous experiment in which mice were gassed with a *Ct* of 6,000 after hexamine. However, in that experiment the administration of oral hexamine was not able to neutralize the whole effect of a very high dosage (6,000) of phosgene.

Duration of lung damage

In order to investigate lung damage produced after different time intervals by the pregassing dosage and to try to correlate it with the effect produced, groups of mice were exposed to the pregassing dose and autopsied after 1, 2, 3, 4, 5, 7, 10, and 14 days. The trachea was tied off before the thorax was punctured and paraffin wax sections prepared; these were examined by Professor G. R. Cameron, whose observations are given in Table I.

It would appear that the oedema is present for the first four days and that no damage is visible after the 10th day. When this result is compared with our experiment on the duration of the effect, the return to normality appears to correspond with a cessation of the effect, which is also absent in the first few days when oedema is most severe.

TABLE I

HISTOLOGICAL EXAMINATION OF LUNGS OF MICE AT VARIOUS TIMES AFTER PREGASSING

All mice were exposed to a dose (*Ct*) of 600 mg.min. cu.m. phosgene

No. of days mice killed after pregassing	No. of mice in group	Histological examination of lungs
1 day	5	All show varying degrees of oedema, some with patches of leucocytic infiltration. Bronchial and bronchiolitic epithelia seem intact or in process of being lifted up by oedema. A few leucocytes.
2 days	5	Oedema seems more severe and extensive. Infiltrated with leucocytes and monocytes. Bronchial and bronchiolitic epithelia intact. Collapse.
3 days	5	Varying amounts of oedema in two cases. Severe in others. Very patchy with areas of collapse showing numerous leucocytes. Bronchial and bronchiolitic epithelia intact.
4 days	5	Two still with extensive oedema and collapsed infiltrated alveoli. Remainder with hardly any oedema. Bronchial and bronchiolitic epithelia intact: normal.
5 days	5	Two show extensive patchy oedema and collapse with emphysema. Remainder show slightly patchy oedema resolving. Bronchial and bronchiolitic epithelia intact and normal in all.
7 days	4	Three normal. One shows small patches of collapse and alveolitis. Bronchial and bronchiolitic epithelia normal.
10 days	4	All normal (one shows much recent haemorrhage. Traumatic).
14 days	4	All normal.

Differences in respiration between control and pregassed animals

When toxic substances are injected into animals the variation in response is largely due to the difference in susceptibility between them ; but when animals are exposed to a gas the dose which each animal inhales is partly determined by the respiration of that animal. Experiments were made in order to find out whether the pregassing effect could be explained by differences in respiration in the pregassed and control groups during the terminal gassing. Ideally it would be useful to measure the total air breathed, the oxygen uptake, the CO_2 output, and the respiration rate ; in view of the practical difficulties involved only the last two were measured.

Respiration rate

Two groups of rats were taken at random and one group exposed to the pregassing concentration. Five days later all the rats were exposed in the chamber two at a time, one rat being taken from the pregassed and one from the control group. They were placed in small wire cages close to a glass window in the gassing chamber. and their respirations counted by two observers, possible bias in the observer being eliminated by tossing a coin to decide which observer should count a particular rat.

1. In the first experiment 9 pairs of rats were exposed. Respirations were counted for 15 second periods every two minutes during the 10-min. exposure. Some animals struggled and others held their breath, resulting in great variation between animals and difficulty in counting.

2. In a second experiment the rats were lightly anaesthetized with 0.5 c.c. per kg. of nembutal solution. (This would itself affect the respiration rate, but it was thought that any systematic involuntary difference would still be apparent.) 13 pairs of rats were used and respirations counted for 1/2-min. periods every minute during the 5-min. exposure. The results for these two experiments are plotted in Fig. 3. It is clear that the animals cut

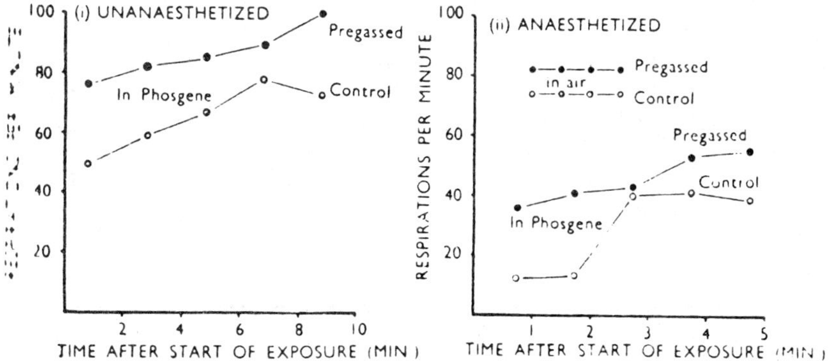

FIG. 3.—The average respiration rates (respiration per min.) of control and pregassed rats during exposure to phosgene. (i) Unanaesthetized. $C = 500$ mg./cu.m. ; $t = 10$ min. (ii) Anaesthetized. $C = 700$ mg./cu.m. ; $t = 5$ min.

down their respiratory rate in phosgene very markedly at first (even when anaesthetized), but breath-holding lessens as the exposure proceeds. (Compare Boyland *et al.*, 1946.)

TABLE II

RESPIRATION RATES PER MINUTE OF ANAESTHETIZED AND UNANAESTHETIZED RATS IN AIR AND AIR-PHOSGENE MIXTURES

	(i) Not anaesthetized $Ct = 5{,}000$ $t = 10$		(ii) Anaesthetized $Ct = 3{,}500$ $t = 5$		(iii) Anaesthetized $Ct = 5{,}900$ $t = 10$	
	Control	Pregassed	Control	Pregassed	Control	Pregassed
Average respiration rate in air	160*	—	74	82	—	—
Average respiration rate in $COCl_2$	66	86	32	46	41	49
Mortality	9/9	9/9	†4/13	2/13	10/10	4/8
Percentage increase in respiratory rate in pregassed group	30		44		19	
t_f = "Students" ratio for f degrees of freedom	$t_{16} = 2{\cdot}10$, $P = 0{\cdot}052$		$t_{24} = 3{\cdot}31$, $P = 0{\cdot}003$		$t_{16} = 1{\cdot}33$, $P = 0{\cdot}200$	
	Combined $P < 0{\cdot}005$					

*Rate obtained from Fig. 54 given by Gaddum (1940).
†In the anaesthetized groups the mortalities were lower than would be expected normally, presumably because of the reduction in breathing rate caused by anaesthesia.

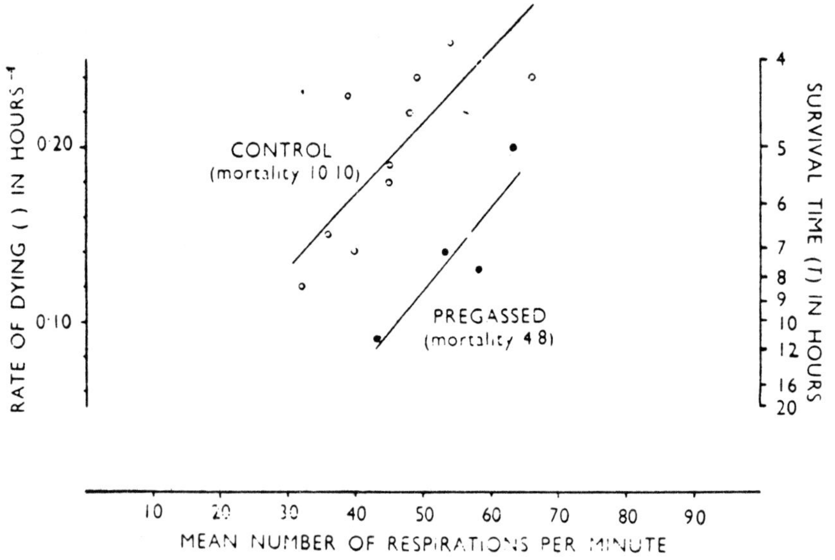

FIG. 4.—A comparison of the respiration rates of anaesthetized rats during exposure to phosgene and their subsequent survival times, in control and pregassed groups. $C = 590$ mg./cu.m.; $t = 10$ min.

3. In a third experiment an exposure time of 10 min. was used; otherwise it was similar in detail to the second experiment. There were originally 10 rats in each group, but 2 died in the pregassing group under anaesthesia.

A summary of the results from these three experiments is given in Table II. The analysis shows that there is little reason for doubt that when exposed to phosgene the pregassed rats breathe faster than the controls.

In the third experiment each rat was marked and its survival time recorded; the results are given in Table III and are plotted together with the lines of best fit in Fig. 4.

TABLE III

AVERAGE NUMBER OF RESPIRATIONS PER MINUTE IN PHOSGENE AND SUBSEQUENT SURVIVAL TIMES OF CONTROL AND PREGASSED RATS

Control group		Pregassed group	
Average No. of respirations per min. in $COCl_2$*	Survival time (hours)	Average No. of respirations per min. in $COCl_2$*	Survival time (hours)
32	8·4	18	Survived
36	6·8	40	Survived
39	4·4	43	11·7
40	7·2	53	7·0
45	5·3	54	Survived
45	5·4	58	7·8
48	4·6	63	5·0
49	4·2	64	Survived
54	3·8		
66	3·9		
Mortality 10/10		Mortality 4/8	

*In order of magnitude to facilitate reference to Fig. 4.

Clearly there is a very strong correlation in the groups between survival time and respiration rate ($r = 0.83$, $P < 0.001$). The regression equation for the control group suggests a relationship of the type $\frac{1}{T} = KX$, where T is the survival time and X the respiration rate. Analysis of the transformed variate ($100 \times \text{hours}^{-1}$) gives:

Analysis of transformed variate ($100 \times \text{hours}^{-1}$)

Source	Degrees of freedom	Sum of squares	Mean square	F
About lines	10	88·4	8·8	} 25
Between lines, slope ..	1	0·4		
Between lines, position ..	1	225·0	225·0	

It is clear that there is a highly significant difference in position between the lines ($P < 0.001$).

CO_2 output during gassing

The method used was similar to that described by Gaddum and Hetherington (1931). The apparatus is sketched in Fig. 5.

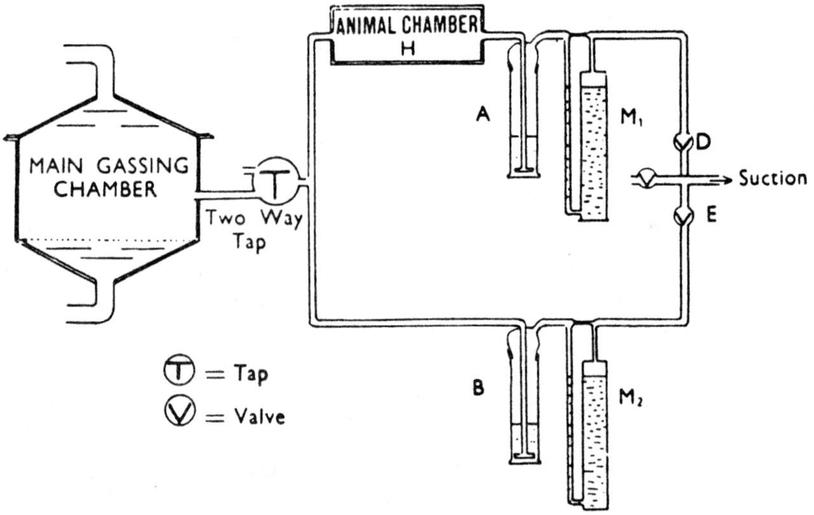

FIG. 5.—Apparatus for measuring CO_2 output of small animals during gassing. The flow of air along tubes E and D was adjusted to 3 litres per min. by means of the valves and the flowmeters M_1 and M_2. H was a chamber sufficiently large to accommodate the animals or animal, without too much dead space. A two-way tap allowed circuit to be opened to outside air or to a main gassing chamber. Bubbler A sampled the atmosphere together with expired air from animals, bubbler B sampled atmosphere only.

The amount of CO_2 produced by the rat was calculated by deducting the CO_2 contained in the air—measured by bubbler B—from that produced by the rat plus the amount contained in the air—measured by bubbler A.

In order to estimate the actual concentrations of phosgene to which the animals were exposed (and not that in the large chamber), a method was devised whereby CO_2 and phosgene could be absorbed and estimated in the same bubbler. Our methods of overcoming certain problems arising in these measurements are discussed in the Appendix.

It was first of all necessary to know whether the CO_2 output during exposure to $COCl_2$ and the dose of phosgene breathed were related. An experiment was therefore carried out in which normal rats were exposed to lethal concentrations of phosgene one at a time. The CO_2 output and phosgene concentrations were estimated and the survival time of each animal recorded. The results are given in Table IV.

It might be expected that most simply the dose breathed would be proportional to the CO_2 output (R). The dose breathed cannot be measured, but we have found (Box and Cullumbine, 1947) that the median mortality time and Ct are related by the type of expression $\frac{1}{T}=K\ (Ct)$ where T is the survival time,

TABLE IV

CO_2 OUTPUT OF RATS IN PHOSGENE AND THEIR SUBSEQUENT SURVIVAL TIMES

(1)	(2)	(3)	(4)	(5)	(6)
Mean concentration *C* in mg./cu.m. phosgene	C.c. of CO_2 produced per kg. body wt. in phosgene in 10 min.* (*R*)	Fractional CO_2 output of average: $R \div 199$	Survival time (hours)	Reciprocal survival time (hours^{-1})	(1) × (3) "*C*" when differences in *R* are allowed for
458	87	0·44	about 24	0·042	202
436	108	0·54	survived	—	235
427	127	0·64	14·5	0·069	273
496	138	0·69	19·0	0·053	342
499	143	0·72	about 24	0·042	359
442	154	0·77	11·0	0·091	340
475	155	0·78	about 24	0·042	370
442	182	0·92	9·3	0·107	407
541	189	0·95	6·6	0·152	514
386	193	0·97	survived	—	374
449	198	0·99	19·5	0·051	445
894	212	1·07	12·2	0·082	957
446	217	1·09	9·7	0·103	486
449	228	1·15	5·3	0·189	516
479	284	1·43	5·2	0·192	685
464	360	1·81	7·0	0·143	840
522	411	2·06	4·4	0·227	1075
	Mean of *R* = 199				

*Rats in order of CO_2 output.

C is the concentration to which the animal is exposed for time *t*, and *K* is constant. Hence the type of expression which might be expected would be:

$$\frac{1}{T} = KCR \qquad \text{(i)}$$

where R = c.c. of CO_2 produced per kg. body weight during the period of gassing.

If this were true, converting to logs we should have:

$$\log\frac{1}{T} = K + b_1 \log C + b_2 \log R \qquad \text{(ii)}$$

and b_1 and b_2 would both be equal to unity.

To test this, the values of $\frac{1}{T}$, C, and R were converted to logarithms. The best fitting equation obtained by the method of least squares was:

$$\log\frac{1}{T} = -5{\cdot}13 + 0{\cdot}56 \log C + 1{\cdot}13 \log R.$$

The values for b_1 and b_2 and their standard errors are $0{\cdot}56 \pm 0{\cdot}58$ and $1{\cdot}13 \pm 0{\cdot}26$ respectively.

The range of values of *C* here is very small and consequently the coefficient of log *C* (0.56) is very unreliable and not significantly different from zero or from unity. But our previous work leads us to expect that this part of equation (ii) is well founded and the true value of the coefficient b_1 will not be far from the expected value, i.e., unity. The coefficient $b_2 = 1.13$ is highly significantly different from zero and not significantly different from one. Hence there are grounds for believing that the relationship is of type (i).

The experiment leaves no doubt as to the correlation between CO_2 output and reciprocal survival time (r = 0.79). As the dosage (*Ct*) was not always constant, the partial correlation coefficient between reciprocal survival time and CO_2 output with the effect of *C* eliminated was worked out; this is still 0.79 correct to two places of decimals, a value which is highly significant.

It is clear, therefore, that the CO_2 output during exposure to phosgene gives a good indication of the amount of gas breathed. The experiment shows that there is an enormous difference in the amount of gas breathed by different animals, presumably owing to the variations in breath-holding and in the degree of activity of the animals in the concentration. In fact, over half the variation ($R^2 = 0.65$) is attributable to factors involved in differences of respiration and concentration of the gas. The remaining variation ($1 - R^2 = 0.35$), or less than half of it, is attributable to other factors such as differences of susceptibility. In order to test the hypothesis that *CR* represented the dose breathed, columns (1) and (3) of Table IV were multiplied to give a product which might be considered as

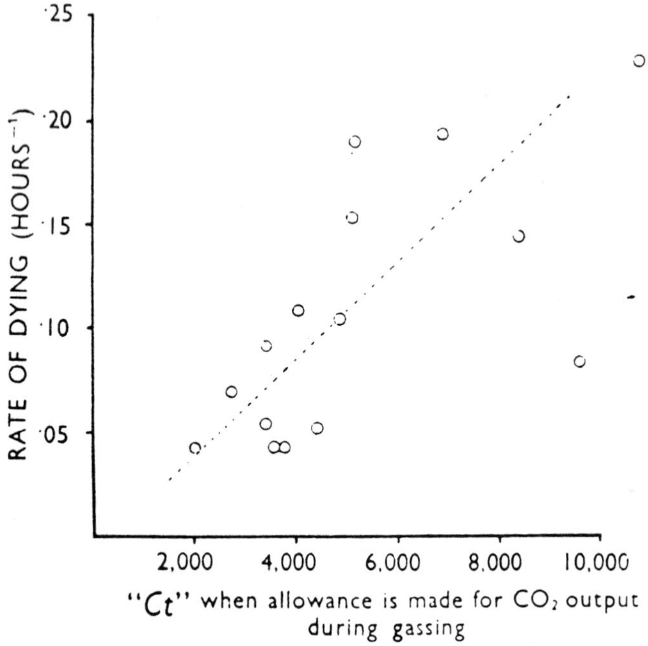

FIG. 6.—The points represent the survival times (plotted as reciprocals) for corresponding values of "*Ct*" for individual animals when allowance is made for differences in CO_2 output during the gassing. The line is that obtained in a previous investigation for median survival time at various dosages.

representing better the "dosage" to which the lungs of animals were exposed. In Fig. 6 reciprocal survival time is plotted against this product multiplied by t (10 min.) and the points are compared with the relationship between Ct and survival time found before (Box and Cullumbine, 1947). It can be seen that the hypothesis is in reasonable agreement with the facts.

CO_2 output of control and pregassed rats during gassing

Twenty rats, which had been submitted to the pregassing concentration 5 days previously, and 20 controls were exposed to phosgene in the apparatus in groups of 4 at a time. The CO_2 production was first determined during a 10-min. period while the animals were breathing air and later during a 10-min. period while they were breathing an air–phosgene mixture of mean $Ct = 1450$ ($t = 10$ min.). The results were as follows:

Control				Pregassed			
No. of rats	Average weight (g.)	Average rate of CO_2 production (c.c./min./kg.)		No. of rats	Average weight (g.)	Average rate of CO_2 production (c.c./min./kg.)	
		In air	In $COCl_2$			In air	In $COCl_2$
4	197	32·4	16·0	4	207	22·6	16·0
4	195	24·3	15·6	4	202	29·5	17·8
4	197	26·8	16·2	4	210	31·0	20·3
4	200	25·4	14·9	4	190	18·3	16·7
4	197	32·4	18·9	4	190	30·9	24·1
	Means	28·2	16·3		Means	26·4	19·0
	Mortality 1/20				Mortality 0/20		

The apparent increase in CO_2 output in phosgene (16.3 to 19.0 c.c. CO_2/min./kg.) is not statistically significant; $t = 1.64$ for 8 degrees of freedom and P is between 0.2 and 0.1.

The dosage used in the previous experiment was rather low, so the experiment was repeated using higher concentrations. Eight pregassed and 8 control rats were used in the experiment and they were exposed in pairs.

Control				Pregassed			
No. of rats	Average wt. (g.)	Ct(mg./min./cu.m.)	c.c. CO_2 min. kg.	No. of rats	Average wt. (g.)	Ct(mg. min./cu.m.)	c.c. CO_2 min. kg.
2	185	1,940	23·4	2	182	2,270	32·4
2	185	3,360	20·4	2	195	3,360	23·1
2	210	2,610	23·4	2	190	2,390	20·5
2	190	3,700	23·3	2	185	3,690	20·3
Means	192	2,900	22·6	Means	188	2,780	24·1
	Mortality 6/8				Mortality 3/8		

As before the apparent increase in CO_2 production (22.6 to 24.1 c.c. CO_2 min. kg.) is not statistically significant; $t = 0.5$ for 6 degrees of freedom and $P = 0.6$.

It must be concluded that there is no significant difference in CO_2 production of pregassed and control rats during subsequent exposure to phosgene.

The respiration experiments suggest that after pregassing and during the period when oedema has subsided but visible lung damage persists (i.e., from about the 4th–7th day), the animals breathe more rapidly but less deeply. It seems likely that this would result in the effect demonstrated, since damage in the second gassing would probably be more superficial with this type of breathing.

Summary and Conclusions

1. When rats and mice are exposed to preliminary non-lethal doses of phosgene a transitory effect (lasting from about the 3rd to the 7th day in mice) is produced, resulting in an apparent increase of resistance to phosgene.

2. In order to produce this effect it is necessary to produce lung damage.

3. Repeated exposures do not produce a cumulative effect.

4. The respiration rate is related to the dose breathed.

5. Rats which have been exposed to a pregassing dose breathe more rapidly in phosgene, but take longer to die.

6. The CO_2 output is related to the dose breathed.

7. There is no significant difference between the CO_2 outputs of pregassed and control rats during exposure to phosgene 5 days later.

It seems likely that the effect can be explained by the more rapid and shallower type of breathing of the pregassed rats in phosgene caused by lung damage in the first exposure.

Appendix

Problems arising in the measurement of CO_2 production of rats and mice during exposure to phosgene (or other toxic agents)

General

The normal amount of CO_2 present in laboratory air is about 0.05 per cent or 0.5 c.c. per litre. The CO_2 output of an average size rat is about 4 c.c. per min. Hence in the type of apparatus described the greater the air flow used the greater will be the proportion of CO_2 absorbed from the air and the less the proportion due to the rat. Hence the air flow must not be too high, otherwise atmospheric CO_2 will tend to swamp the CO_2 expired by the rat and reduce the accuracy of the method. On the other hand, in order that the air already in the rat chamber may be washed out rapidly, the dead space must be cut to a minimum and flow kept as high as possible. Further, the type of bubbler with sintered glass diffuser plates cannot be used at flows much above 3 litres/min. without frothing.

Bearing these considerations in mind, the following design was adopted:

1. Not more than 1 litre of chamber space per rat (dead space about 750 c.c.).

2. Flow of 3 litres/min.

3. Phosgene and CO_2 were sampled in the same bubbler.

With this arrangement the chamber would be completely washed out in a quarter of a minute and the error would be negligible in 10-min. exposures. Further, for every 4 c.c.

of CO_2 produced by the rat only about 1.5 c.c. would pass through the bubbler from the air. The estimation of CO_2 and phosgene in the same bubbler (1) ensures that the actual atmosphere breathed by the rat is sampled, (2) gives double the rate of flow possible with two bubblers using the same rate through the chamber, and (3) avoids duplication of solutions, bubblers, flowmeters, and air lines.

The absorbent was a mixture of 25 c.c. of $N/1$ NaOH and 50 c.c. of 6 per cent (w/v) hexamine. The bubblers were Dreschel bottles with sintered glass diffuser plates.

Method

The contents of the bubbler were washed out into a beaker and 50 c.c. of 66 per cent (w/v) solution of barium nitrate added to precipitate the carbonate. The volume was made up to 170 c.c. and a few drops of phenolphthalein added. The residual alkali was titrated with $N/1$ nitric acid. The endpoint (red-colourless) is extremely sharp, as the precipitated barium carbonate acts as a white background for the colour change. The hydrogen ion concentration was then adjusted to *p*H 2 by adding sufficient $N/1$ HNO_3 to make the total volume added 48.5 c.c.

The solution remaining was colourless and the chloride was estimated by titrating with $M/50$ mercuric nitrate, using 1.5 c.c. of 1 per cent diphenylcarbazone in alcohol as indicator (Roberts, 1936). The large amount of excess acid necessary was due to the buffering action of the hexamine.

It was found by using a *p*H meter that the addition of the amounts of reagents described always resulted in the correct *p*H. Blanks were carried out for both titrations using all the reagents. It was necessary to make a correction for the CO_2 produced by the breakdown of phosgene; this correction was in practice very small, and is described below.

The efficiency of absorption of phosgene

As the procedure described above and the concentration of the reactants were not quite the same as we normally used for sampling and analysing phosgene concentrations, tests were carried out as follows: a phosgene–air mixture in which the $COCl_2$ was about 1,000 mg./cu.m. (which is higher than any concentration used in the experiment) was sucked through two bubblers in series, using the absorbent mixture described; in a set of three experiments no chloride was detected in the second bubbler.

The efficiency of absorption of CO_2

The first experiments were carried out with ordinary bead bubblers, but the efficiency of absorption of CO_2 with 15 c.c. in each bubbler and rates of flow of 3–5 litres min. was in no case greater than 50 per cent. In order to increase the efficiency larger amounts of the reagents and Dreschel bottles with sintered glass diffuser plates were used. With a total of 75 c.c. of reagents, efficiencies of over 80 per cent with a rate of flow of 3 litres/min. were obtained.

It was found that owing to small physical differences the efficiencies of the bubblers varied. Each bubbler had therefore to be standardized and its efficiency measured.

This can be done (1) if the exact concentration of CO_2 bubbled through is known or (2) by using a long train of bubblers so that the leak at the end is negligible. However, the exact determination of the CO_2 without the use of bubblers is not an easy task, and the use of a long train of bubblers is clumsy and involves a large number of titrations: also the back pressure set up by these bubblers is high, and this causes added complications. In view of the above the following device was used.

If the efficiency of absorption is independent of the CO_2 concentration, imagine two bubblers of unequal efficiency, arranged in series.

Let the fractional leak of the first bubbler be L_1 and that of the second L_2. Then if the concentration of CO_2 entering the first bubbler is c, cL_1 leaves the first bubbler and enters the second and cL_1L_2 leaves the second.

∴ Concentration absorbed by the first bubbler $= c(1-L_1)$ which is proportional to T_1
and ,, ,, ,, ,, second ,, $= cL_1(1-L_2)$ which is proportional to T_2
where T_1 and T_2 are the titrations in bubbler 1 and 2.

$$\text{Then } \frac{T_2}{T_1} = L_1\frac{(1-L_2)}{1-L_1} = \text{say } R_1$$

if the order of the bubblers is reversed we have

$$\frac{T_2}{T_1} = L_2\frac{(1-L_1)}{1-L_2} = \text{say } R_2$$

Then by simple algebra

$$L_2 = R_2\frac{(1+R_1)}{1+R_2} \text{ and } E_2 = (1-L_2)$$

$$L_1 = R_1\frac{(1+R_2)}{1+R_1} \text{ and } E_1 = (1-L_1)$$

Hence the efficiencies (E_1 and E_2) of the bubblers may be found without knowing the actual concentration of gas passing. All that is required is the ratio R_1 of the titrations when arranged in a given order and the ratio R_2 when that order is reversed.

Independence of efficiency and CO_2 concentration

The above assumes that the efficiency is independent of the CO_2 concentration. This was tested in the following manner.

Dreschel bottles 1 and 2 were set up in series, each bottle containing 25 c.c. of N/1 NaOH and 50 c.c. of 6 per cent (w/v) hexamine as before. Air was drawn through the chamber containing the rats and then through the bubblers at a rate of 3 litres/min.: four experiments were carried out and the concentration of CO_2 varied by altering the number of rats. Each run lasted 10 min., and in each case a duplicate run was made with the order of the bubblers reversed.

The results were:

Expt.	Order	No. of rats	Blank minus titration		R_1	R_2	Efficiency per cent	
			Bubbler (1)	Bubbler (2)			Bubbler (1)	Bubbler (2)
1	1–2	1	4·00	0·55	0·137		86	
	2–1		4·30	0·80		0·186		82
2	1–2	2	7·25	1·05	0·149		85	
	2–1		7·60	1·40		0·184		82
3	1–2	3	10·40	1·70	0·163		84	
	2–1		10·50	1·95		0·185		82
4	1–2	4	16·65	2·45	0·147		85	
	2–1		11·95	1·95		0·163		83

Mean efficiencies: (1) 85 and (2) 82 per cent.

Hence it would appear that the efficiency of absorption of the bubblers is independent of the CO_2 concentration over the range actually used during the experiment.

The calculation of the CO_2 output of the rats in a concentration of phosgene

The bubblers were standardized by the method described and their efficiencies recorded.

An example is given to show the calculations involved: 4 rats in the chamber. Air flow through each bubbler 3 litres/min. Efficiency of bubblers: (1) 85 and (2) 82 per cent.

Bubbler (1) was connected in position A (see Fig. 5) and absorbed CO_2 from air and rats, bubbler (2) in position B absorbed CO_2 from the air only. Expected phosgene concentration 500 mg./cu.m. Duration of run 10 min.

	Bubbler A (rats and air)			Bubbler B (air only)	
	$COCl_2$ titration		CO_2 titration	$COCl_2$ titration	CO_2 titration
	c.c. of $M/50$ $Hg(NO_3)_2$		c.c. of $N.HNO_3$	c.c. of $M/50$ $Hg(NO_3)_2$	c.c. of $N.HNO_3$
Blank	..	0·10	24·80	0·10	24·80
Test	..	14·50	12·65	15·80	23·50
Difference		14·40	12·15	15·70	1·30

$COCl_2$: 1 ml. of $M/50$ $Hg(NO_3)_2 \equiv 0{\cdot}99$ mg.$COCl_2$

average titration $\frac{A + B}{2} = 15{\cdot}05$. ∴ Concentration $= \frac{15{\cdot}05 \quad 0{\cdot}99 \quad 1000}{30} = 497$ mg. cu.m.

Hence $Ct = 4{,}970$ mg.min./cu.m.

CO_2 correction

In the hydrolysis of $COCl_2$, CO_2 will be formed. Correction is made for this here, but as may be seen the effect of this correction is normally so small that it may be ignored.

If x is the $COCl_2$ titration (c.c. of $M/50$ $Hg(NO_3)_2$) then

correction of CO_2 will be simply $\frac{x}{50}$ c.c. N. acid.

CO_2 : 1 c.c.N.acid $\equiv 11{\cdot}2$ c.c. CO_2

Total CO_2 through bubbler A $\equiv \frac{12{\cdot}15}{0{\cdot}85} = 14{\cdot}30$ c.c.N.acid

Correction for $COCl_2$ $\equiv \frac{14{\cdot}4}{50} = 0{\cdot}28$ c.c.N.acid

Difference (CO_2 from rat and air) $\equiv 14{\cdot}02$ c.c.N.acid

Total CO_2 through bubbler B $\equiv \frac{1{\cdot}30}{0{\cdot}82} = 1{\cdot}58$ c.c.N.acid

Correction for $COCl_2$ $\equiv \frac{15{\cdot}7}{50} = 0{\cdot}31$ c.c.N.acid

Difference (CO_2 from air only) $\equiv 1{\cdot}27$ c.c.N.acid

Hence CO_2 from 4 rats $\equiv 14{\cdot}02 - 1{\cdot}27 = 12{\cdot}75$ c.c.N.acid

and this is equivalent to 143 c.c. CO_2.

Hence CO_2 output per rat was 3·57 c.c./min.

Our thanks are due to the Director-General of Scientific Research (Defence) for permission to publish these results, and to our colleagues at Porton for their advice and criticism. Our special thanks are due to Prof. C. R. Cameron for his help and encouragement.

REFERENCES

Bliss, C. I. (1935). *Ann. appl. Biol.*, **22**, 134, 307.
Bliss, C. I. (1938). *Quart. J. Pharm. and Pharmacol.*, **11 (2)**, 192.
Box, G. E. P., and Cullumbine, H. (1947). *Brit. J. Pharmacol.*, **2**, 27.
Boyland, E., McDonald, F. F., and Rumens, M. J. (1946). *Brit. J. Pharmacol.*, **1**, 81.
Fisher, R. A., and Yates, F. (1943). *Statistical Tables for Biological, Agricultural and Medical Research*, 2nd Ed., p. 11. London and Edinburgh: Oliver and Boyd.
Gaddum, J. H., and Hetherington (1931). *Quart. J. Pharm. and Pharmacol.*, **4**, 183.
Gaddum, J. H. (1933). *Med. Res. Council Spec. Rept.*, No. 183.
Gaddum, J. H. (1940). *Pharmacology*, 1st Ed., p. 253. London: Oxford University Press.
Richards, F. J. (1941). *Ann. Bot. Lond.*, N.S. **5**, 249.
Roberts, I. (1936). *Ind. Eng. Chem. (Anal. Ed.)*, **8**, 365.

5.2

The Relationship between Survival Time and Dosage with Certain Toxic Agents

G. E. P. BOX and H. CULLUMBINE
Chemical Defence Experimental Station, Porton

(Received November 15, 1946)

In toxicity experiments with chemical warfare agents it was frequently noticed that animals which had received the largest doses died more quickly than those receiving smaller doses. The investigations described here were carried out in order to find whether a quantitative relationship existed between dosage and survival time and, if so, how best this relationship could be expressed and used. It has not been possible for the present authors to pursue this investigation very far, but in view of the interesting results so far obtained it is hoped that others may extend this study to other substances.

The substances used in our investigations were mustard gas, $S(CH_2.CH_2Cl)_2$, and phosgene gas, $COCl_2$. When breathed by animals these substances exert their effect in quite different ways: animals dying after exposure to phosgene consistently showed, at autopsy, a picture of pulmonary oedema ; those dying after exposure to mustard gas vapour, however, presented a more varied pathology, pseudo-membranous tracheitis, pulmonary oedema, broncho-pneumonia, and enteritis being the predominant autopsy findings.

The survival times of the animals were quite different: with mustard gas animals may die from the direct effect of the vapour 20 days after exposure, whereas with phosgene nearly all animals die in the first 48 hours.

Mice exposed to mustard gas

Mice were exposed to mustard gas vapour in a constant-flow apparatus. The advantage of this type of apparatus over the static chamber is that the concentration of gas does not tend to fall off during the exposure as it does with the latter owing to adsorption on to the surface of the chamber and the animals' fur. In the constant-flow apparatus a continuous flow of air-gas mixture of the required concentration is drawn through the chamber. In the experiments described here the volume of the chamber was 20 litres and the flow of air-gas mixture was 200 litres per minute. The liquid mustard was evaporated into the air stream at the required rate by means of an electric heating coil. The atmosphere in the chamber was continuously sampled throughout the exposure and accurate determinations (within about 1 or 2%) of the concentrations in the chamber could be made.

240 male albino mice (of weight 25–30 g.) were mixed up together and 8 groups of 30 mice selected at random. The groups were exposed all on the same day to 8 different

The Collected Works of George E. P. Box, 1984, Wadsworth, Inc., Belmont, CA 94002.
Originally published in *British Journal of Pharmacology and Chemotherapy,* vol. 2, no. 1 (1947), pp. 27–37.

concentrations of mustard vapour. The exposure time in each case was 10 minutes. The exposures were carried out in wire cages which had separate compartments for each mouse in order to avoid the possibility of the mice huddling together and breathing through each other's fur.

The mortalities at the end of successive 24-hour periods are shown in Table I.

TABLE I

DISTRIBUTION OF SURVIVAL TIMES IN GROUPS OF 30 MICE EXPOSED TO VARIOUS CONCENTRATIONS OF MUSTARD VAPOUR FOR 10 MIN. IN A CONSTANT-FLOW APPARATUS

Concentration in mg./cu.m.	Time in days 2–3	3–4	4–5	5–6	6–7	7–8	8–9	9–10	10–11	11–12	12–13	13–14	14–15	15–16	16–17	17–18	18–19	19–20	20–21	21–22	22–23	23–24	24–25	25–26	Total mortality
A 67								2		2			2			1				1					8/30
B 93						5		1		2			3	2				1						1	15/30
C 116					1	1		1	1	1	4		3			2		2	1			1			18/30
D 160			1		1	3	5		2	2		1		1			1								17/30
E 260		4	2	2	2	2	6	7		1	3									1					30/30
F 300			5	8	2	4	2	3		2						1	3								30/30
G 375		5	9	7	4	3	2																		30/30
H 505	3	11	10	6																					30/30

Analysis of results

(i) As a preliminary analysis the median time for each group was plotted against the dosage; in general, it was clear that median survival time decreased with increase of dosage, but not as a linear function; in fact, a smooth curve drawn through the points resembled a rectangular hyperbola. A second very noticeable feature of the data was the increase in spread of the observations in the groups with larger average survival times.

(ii) The original data were converted to logarithms and, in the groups where all the animals died, the means and standard deviations were computed in the usual way. In the groups in which some animals survived, estimates of the means and standard deviations were obtained by plotting the logarithms of the individual observations in each group on a probit scale and estimating the mean from the intersections of the best "eye-fitted" line with the ordinate at 5 probits, and the standard deviation from the reciprocal of the slope of the line (Bliss, 1936, and Gaddum, 1945). It seemed doubtful whether very accurate estimates could be obtained by this method, since there appeared to be a general tendency for the lines to be convex upwards (i.e., for the distributions to be positively skew), and the assumption of normal distribution, on which this method is based, therefore seemed to be invalid. A second very noticeable feature about the data was that in spite of the log transformation the means and standard deviations for the groups were still highly correlated (the product moment correlation coefficient for the unweighted data was about 0.93, which is highly significant: $P = < 0.001$).

When these estimates of the mean log survival times were plotted against the logarithms of the dosage there appeared to be a linear relationship. A straight line drawn through the points had a slope of about −1, suggesting a reciprocal relationship between time and dosage.

(iii) The original data were therefore converted to reciprocals, and means and standard deviations of the reciprocal survival times for the groups were found as before. It was noticeable that when the results for each group were plotted on a probability scale there seemed to be no tendency for the points to be curvilinear, which suggested that after this transformation the data were more nearly normally distributed. A second feature of the use of this transformation was the apparent stability of the variance. (The product moment correlation coefficient for the unweighted data was about −0.26, which is quite non-significant: $P \simeq 0.5$.)

The values for the estimates of means and standard deviations for the groups using the logarithmic and the reciprocal transformations are given in Table II.

TABLE II

VALUES FOR THE ESTIMATED MEANS AND STANDARD DEVIATIONS OF TRANSFORMED SURVIVAL TIMES (DAYS)

Group	Mustard dosage (*Ct*) mg. min./cu.m.	Log transformation (log days)		Reciprocal transformation ($100 \times$ days^{-1})	
		Mean	Standard deviations	Mean	Standard deviations
A	670	1·43	0·31	1·3	6·8
B	930	1·30	0·35	3·9	7·8
C	1160	1·27	0·25	5·2	4·3
D	1600	1·10	0·21	7·1	6·0
E	2600	0·88	0·19	14·6	7·0
F	3000	0·87	0·20	14·6	5·4
G	3750	0·71	0·12	20·1	5·1
H	5050	0·60	0·10	26·0	6·4
		Correlation coeff. between mean and S.D. = 0·93 $P \simeq 0·001$		Correlation coeff. between mean and S.D. = −0·26 $P \simeq 0·5$	

When the estimates of the mean reciprocal survival times were plotted against the dosage the points appeared to fall near a straight line.

It is clear that the estimates of mean reciprocal survival time were not of equal reliability, since some were estimated from complete data in the ordinary way, whilst others had to be obtained by graphical methods. Some system of weighting should therefore be used to allow for this. Bliss (1936) and Stevens (1936) have worked out a method for an analogous case in which by successive approximation the maximum likelihood estimates of the means and standard deviations of the truncated normal curve can be calculated to any required degree of accuracy. They also give tables of a factor *E* which enables one to calculate the variance of the mean of the truncated curve. This factor E may be used to furnish a weighting factor N/E.

When the data are complete (i.e., when all the animals react), E is equal to unity and the weighting factor becomes equal to N, the number of animals in the group. When the exposure curve is truncated owing to some animals failing to react, N/E will give the equivalent number of animals which would have given as accurate a result if they had all reacted. With the reciprocal transformation, all the points seemed to fit the dosage–probit line and there was no difficulty in deciding at what point truncation* had occurred ; it was in every case determined simply by the proportions of animals dying.

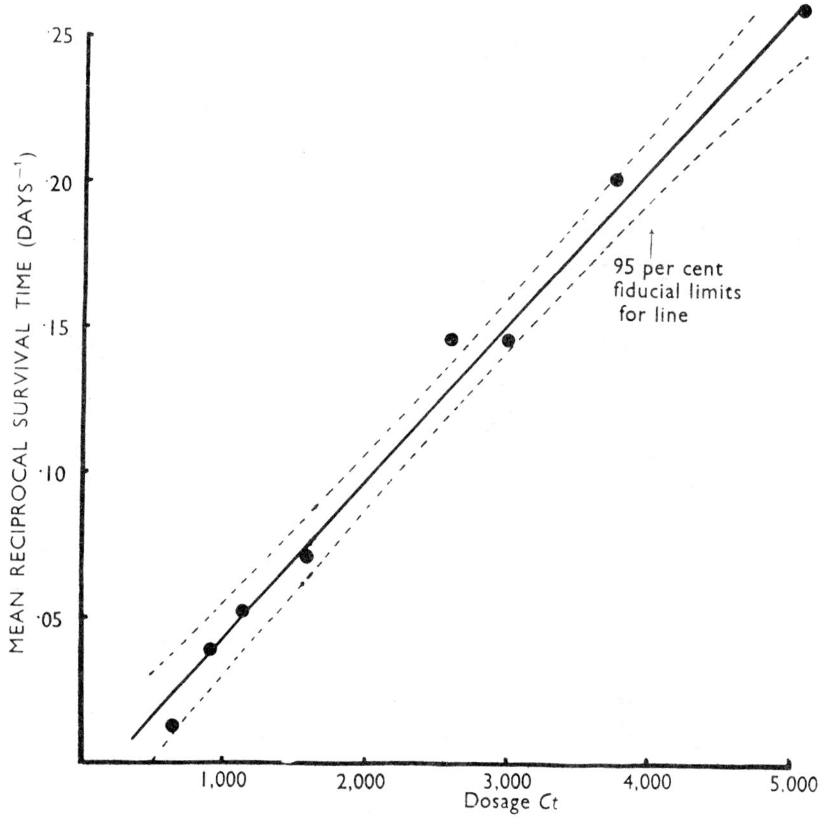

FIG. 1.—*Mustard*. The mean reciprocal survival time for groups of mice exposed to various dosages of mustard gas vapour, showing the line of best fit with its 95 per cent fiducial limits.

*Bliss (1936) dealt with a similar problem (reaction time of organisms when immersed in a toxic solution or gas). His data usually showed a tendency for the points corresponding to the last few organisms to react to fall away from the line. He regarded these organisms as atypical and found the point of truncation by noting where the line appeared to change in slope.

To test the hypothesis that the reciprocal of the survival time was linearly related to the dosage a regression line was fitted between the means of the reciprocal survival times and the dosages. This line, together with the approximate 95 per cent fiducial limits, is shown in Fig. 1.

An alternative hypothesis is that the survival time is proportional to the dosage (i.e., that the line passes through the origin); a second calculation was performed, therefore, and the best-fitting line obtained with the restriction that it should pass through the origin. The "goodness of fit" of the two lines may now be tested, since we may compare the sum of weighted squares of residuals in each case with the pooled "within groups" variance; the ratio of these two quantities will be distributed approximately as χ^2 (since the numbers of observations involved in the between groups variance is large). We obtain:

TABLE III

Hypothesis (1) DOSE AND RECIPROCAL OF SURVIVAL TIME LINEARLY RELATED			*Hypothesis* (2) DOSE AND RECIPROCAL OF SURVIVAL TIME PROPORTIONAL		
χ^2	Degrees of freedom	P (approx.)	χ^2	Degrees of freedom	P (approx.)
3·43	6	0·75	5·03	7	0·5

It is clear that either of the above hypotheses is acceptable.

A distinction is here drawn between the two possibilities, since it seems likely that for some toxic substances a definite "threshold" effect may exist when, although the relationship might be linear, the line would not pass through the origin. It seems useful to keep the more general case in mind, even though in our work we have found no significant departure from proportionality.

TABLE IV

DISTRIBUTION OF SURVIVAL TIMES IN GROUPS OF RATS EXPOSED TO VARIOUS CONCENTRATIONS OF PHOSGENE GAS FOR 10 MIN. IN A CONSTANT-FLOW APPARATUS

Concentration in mg./cu.m.	Time in hours: 3	3¼	3½	3¾	4	4¼	4½	4¾	5	5½	6	6½	7	8	9	10	11	12	14	16	20	24	30	40	60	Total mortality
T 151															1			1	1	1	1	1		1		7/12
U 156													1			1	2	1	1		1	1	4	2		14/20
V 300								1			1				1		1		1		1					6/12
W 411										1		1	2			1	2		1	1	1					10/12
X 589							1		2	1	1		1	2		1			1	1		1				12/12
Y 702				1		1		1	1	2	2		1	2			1									12/12
Z 923	1		1	2	3	1		1	4	1	1	1	2	2												20/20

Rats exposed to phosgene gas

The apparatus used for the phosgene exposure is described fully elsewhere (Box and Cullumbine, 1947). A number of groups of albino rats (weight 115–125 g.) were exposed all on the same day to various concentrations of phosgene; as with mice, each rat was exposed in a separate compartment. The exposure time was 10 minutes in each case. The rats were inspected at the times shown and the mortalities recorded. The results are given in Table IV.

Analysis of results

The results were analysed exactly as before, the features of the distribution of survival time being remarkably similar. Again we found a general decrease in survival time with increase in dosage, which assumed a linear form when plotted on a log–log scale with slope approximately equal to -1. Again the spread of observations was correlated with the mean value and this was still the case on transforming to logarithms ($r = 0.87$, $P \simeq 0.01$), but not the case with reciprocal transformation ($r = -0.28$, $P \simeq 0.5$). As before, plots of probit mortality against log dose suggested positively skew distribution with the log transformation, but normal distribution with the reciprocal transformation.

The values for the estimates of means and standard deviations for the individual groups are given in Table V.

TABLE V

VALUES FOR THE ESTIMATED MEANS AND STANDARD DEVIATIONS OF TRANSFORMED SURVIVAL TIMES (HOURS)

Group	Phosgene dosage (*Ct*) mg. min./cu.m.	Log transformation (log hours)		Reciprocal transformation ($100 \times$ hours^{-1})	
		Mean	Standard deviations	Mean	Standard deviations
T	1510	1·45	0·34	2·8	5·0
U	1560	1·43	0·33	3·3	5·2
V	3000	1·34	0·47	4·0	10·9
W	4110	1·07	0·27	8·9	6·0
X	5890	0·94	0·24	13·0	5·7
Y	7020	0·80	0·14	16·7	5·8
Z	9230	0·72	0·13	20·1	5·7
		Correlation coeff. between mean and S.D. = 0·87 $P \simeq 0{\cdot}01$		Correlation coeff. between mean and S.D. = −0·28 $P \simeq 0{\cdot}5$	

Regression lines were calculated exactly as before and the line obtained on the hypothesis of simple linearity is shown in Fig. 2.

Again the alternative hypothesis of proportionality between mean reciprocal time and dosage was tested, and the values obtained for χ^2 in the test of goodness of fit are given in Table VI.

TABLE VI

Hypothesis (1) DOSE AND RECIPROCAL OF SURVIVAL TIME LINEARLY RELATED			*Hypothesis* (2) DOSE AND RECIPROCAL OF SURVIVAL TIME PROPORTIONAL		
χ^2	Degrees of freedom	P (approx.)	χ^2	Degrees of freedom	P (approx.)
1·81	5	0·9	2·15	6	0·9

As before, neither of these tests demonstrates significant departure from the respective hypotheses.

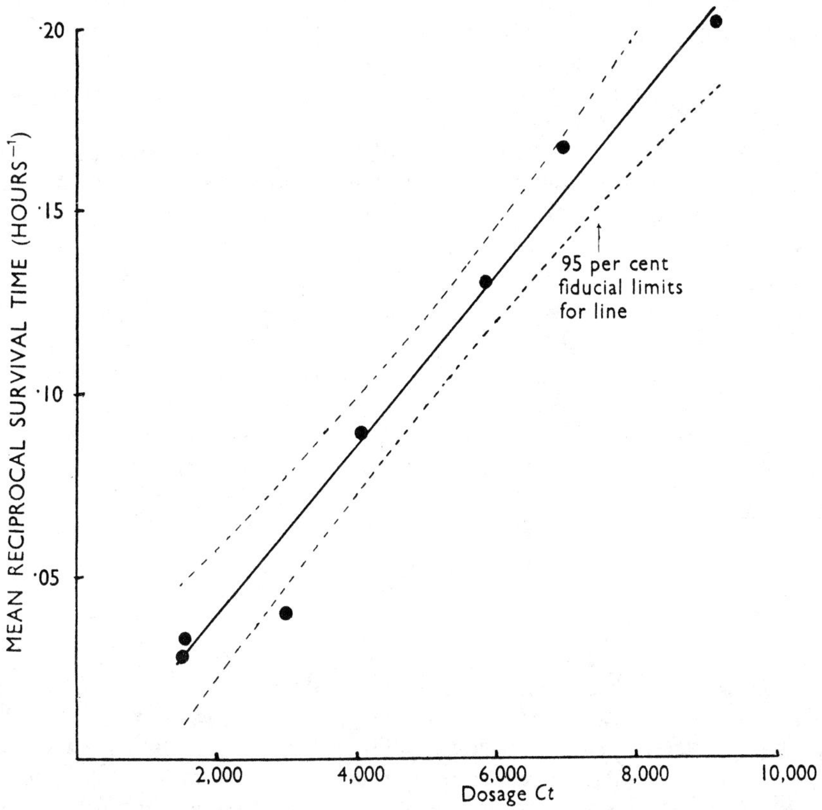

FIG. 2.—*Phosgene.* The mean reciprocal survival time for groups of rats exposed to various dosages of phosgene gas, showing the line of best fit with its 95 per cent fiducial limits.

DISCUSSION

Interpretation and use of the relationship

The use of survival time in pharmacological experiments has been largely confined to the standardizing of cortical extracts (Bulbring, 1937, *et al.*). We believe that where substantial correlation can be shown between dosage and a function of survival time, its use may be considerably extended. With the two substances and species of animals investigated our experiments suggest: (i) Rate of dying (i.e., the reciprocal of the survival time) is directly proportional to the dosage ; (ii) rate of dying tends to be homoscedastic (i.e., the variance is roughly constant for different survival times) ; (iii) rate of dying is probably more nearly normally distributed than survival time or log survival time. These facts were used in the design and analysis of experiments, including many to assess the value of suggested therapies for chemical warfare agents. In this type of experiment it is a great advantage to be able to test a large number of factors at once in accordance with the principles of experimental design described by Professor R. A. Fisher and his followers. For example, the injection of a certain substance may be suggested as a possible treatment for a certain type of poisoning, but normally there will be many uncertainties, such as the correct dose, the best route of administration, whether it should be given in single or divided doses, and so on. We could, of course, try to guess what would be the most promising combination of these factors, expose a control and treated group to (say) an LD90 of the poison, and observe whether any significant reduction of mortality in the treated group occurred. If it did, then attempts to enhance it by varying other factors could be carried out later. The disadvantages of this type of experimentation are (i) if we guess wrongly we may miss the treatment altogether, (ii) it is very uneconomical in the use of animals, and (iii) it cannot detect interaction between factors.

By means of a continuous variate, such as transformed survival time, many of the factors can be introduced into the first experiment *without increasing its size or losing efficiency in testing significance* (Fisher, 1942). Difficulties occur, however, when attempts are made to use factorial schemes in which percentage mortality is the variate. The most important of these is the lack of range of this type of experiment. Usually we try to get our stock of animals as uniform as possible, but the more uniform we make them the smaller will be the range over which we shall get a response between 0 and 100 per cent mortality. When a number of factors are introduced to be tested simultaneously, the range of the effects will usually be increased ; this will often result in a number of groups with 0 or 100 per cent mortality which have little weight and cannot be compared one with the other. We can widen the range of the experiment by carrying it out at more than one dosage level ; but this type of experiment will almost certainly have a low overall efficiency owing to the number of groups with little weight. Further, if we try to introduce a number of factors and dosage levels, we shall soon have to increase the overall size of the experiment, otherwise the individual group will become too small. In this way when a quantal response is used the advantage of the factorial design tends to be lost. Provided that the groups are large enough, this type of experiment can be analysed by the use of the transformation $x = \sin^{-1}\sqrt{p}$ (where p is the proportion dying).

This transformation overcomes the difficulty of inequality of variance in the groups ; its disadvantage, however, is that the transformed variate will not be a linear function of dosage. However, if we use a transformation which *is* a linear function of dosage (i.e., the probit transformation), the weights will depend upon the expected values and be different for different groups. Finney (1943) has given the solution in this case, but his analysis is by multiple regression and becomes laborious when there are many factors, especially if some of the interaction effects have to be included in the analysis.

In view of difficulties of this kind arising out of the use of a quantal response (percentage mortality), wherever it could be shown that some function of survival time was closely correlated with dosage, factorial experiments have been carried out with this function as the variate. The procedure was to give very large doses (usually 3–5 times the LD50), which it was known would probably result in 100 per cent mortality even if the treatments were fairly effective ; the therapies were then judged by the increase in survival time. For example, with mustard gas and phosgene, the use of the reciprocal transformation allows the ordinary methods of analysis of variance to be correctly applied. As an additional safeguard the most promising of the treatments may be compared by using percentage mortality and (if the experiment is carried out at more than one dosage level) by calculating the index $I_{LD50} = \dfrac{\text{LD50 for treated group}}{\text{LD50 for control group}}$; this index will give a true measure of the effectiveness of the treatment and is a similar measure to M (the log of the ratio of the potencies) used by Gaddum (1933). It is interesting to note that I is a function of the percentage mortality and b (the slope of the probit–dose line), so that unless b is known a test at one dosage level using percentage mortality as the variate does not provide us with any information about the *absolute effectiveness* of the treatment ; e.g., it is possible for two equally good therapies to give quite different reductions in mortality if the values of b in the two tests are different. However, in the case of survival time, the ratios of the means of the transformed survival times in control and treated groups will give a measure of effectiveness which is independent of the standard deviations of the groups. For instance, we found that, where the reciprocal transformation is appropriate, $I_T = \dfrac{\text{Median survival time of treated group}}{\text{Median survival time of control group}}$ gives a measure of effectiveness, which is often a good approximation to I_{LD50}.

Even if a relationship can be established between transformed survival time and dosage for a particular case, it will *not* follow that the method outlined here will necessarily be appropriate for the testing of therapies. Examples may occur where a therapy will increase the survival time without reducing mortality. Usually, however, a knowledge of the probable mechanism of a therapy will enable one to judge whether this method of experimentation will be appropriate. In the opinion of the authors the technique can often be used profitably, side by side with more orthodox methods, and they have conducted many experiments on these lines, some of which appear elsewhere. (Cullumbine and Box, 1946; Box and Cullumbine, 1947.)

Analysis of experiments in which some groups are incomplete

Occasionally when using factorial experiments of the type outlined above, a group or groups will partly, or wholly, survive. When all or most of the animals survive in a group or groups in marked contrast to other groups, usually no analysis will be needed to establish the superiority of the corresponding treatments. Experiments in which there are a few survivors in one or two groups only are not so easily dealt with.

(i) Maximum likelihood estimates of the means in the incomplete groups may be obtained by the method of successive approximation given by Bliss (1936) and Stevens (1936). The solution will then be given by regression analysis using the appropriate weighting coefficients; however, the labour involved in the calculations is scarcely justified.

(ii) Provided at least 50 per cent have died in all groups, the medians of the transformed values may be used and the ordinary analysis of variance procedure adopted. A certain amount of information is lost when this method is used, but with small groups the loss is not large. Pearson and Adyanthaya (1928) give the following values:

Sample size using median	2	3	4	5	7	10
Equivalent sample size using mean	2	3	4	4	5	8

(iii) The means may be estimated in the incomplete groups by graphic analysis and the experiments analysed as though the ordinary analysis of variance technique were appropriate (i.e., as though the means of the groups were all known with equal accuracy). This method is probably sufficiently accurate provided that the number of animals failing to respond is not too great.

(iv) Where the reciprocal transformation is appropriate, the " rate of dying " of an animal which survives is $\frac{1}{\infty}=0$, so that with this transformation we could score these animals as zero and carry out the analysis in the ordinary way. In our opinion this approximation will be more inaccurate than (iii). If the " within groups " variance is used in the analysis the truncated groups should be omitted from its calculation, since the values will be too small. In general it is better to use high order interactions between group means as the " error " estimate. We find that (iii) gives a satisfactory approximation and utilizes all the information without complicating the analysis.

SUMMARY

1. Experiments are described in which mice were exposed to mustard vapour and rats to phosgene gas. The properties of the survival time, its logarithm and reciprocal were investigated. Using the reciprocal transformation, the transformed

variate (rate of dying) was proportional to the dosage, had stable variance, and appeared to be more nearly normally distributed than time or log time.

2. The use in appropriate circumstances of relationships of this nature is discussed and the advantages of factorial experiments using transformed survival time are indicated.

3. Methods are suggested for dealing with the data when in some groups the animals do not all die.

We should like to express our indebtedness to Prof. R. A. Fisher for discussing this problem with one of us and for suggesting the use of the reciprocal transformation.

Our thanks are due to the Director-General of Scientific Research (Defence) for permission to publish these results, and to our colleagues at Porton for their advice and criticism.

References

Bliss, C. I. (1936). *Ann. appl. Biol.,* **24**, 815.
Box, G. E. P., and Cullumbine, H. (1947). *Brit. J. Pharmacol.,* **2**, 38.
Bulbring, E. (1937). *J. Physiol.,* **89**, 64.
Cullumbine, H., and Box, G. E. P. (1946). *Brit. med. J.,* **1**, 607.
Finney, D. J. (1943). *Ann. appl. Biol.,* **30**, 71.
Fisher, R. A. (1942). *The Design of Experiments,* 3rd Edition, page 90. London and Edinburgh: Oliver and Boyd.
Gaddum, J. H. (1933). *Med. Res. Council Spec. Rept.,* No. 183.
Gaddum, J. H. (1945). *Nature,* **156**, 463.
Pearson, E. S., and Adyanthaya, N. K. (1928). *Biometrika,* **20A**, 358.
Stevens, W. L. (1936). *Ann. appl. Biol.,* **24**, 847 (Appendix).

5.3
Problems in the Analysis of Growth and Wear Curves

G. E. P. BOX
Imperial Chemical Industries Ltd.

Reprinted from
BIOMETRICS
AMERICAN STATISTICAL ASSOCIATION, December 1950, Vol. 6, No. 4

1. INTRODUCTION

IN A NUMBER of different fields of application, the problem arises of comparing sets of growth and wear curves. The object of this paper is to describe methods of analysis which have been found useful, to point out what assumptions are being made, and show how these assumptions can themselves be put to the test.

1.1 Examples of wear and growth curves.

Much research is carried on by technologists with the object of improving the abrasion resistance of materials, and various machines have been devised to test this property. In most of these, specimens of materials to be compared are rubbed against a standard abrasive under standard conditions and the loss in weight or decrease in thickness of each of the specimens is noted at suitable intervals. For example, Fig. I shows the weight loss curves of 24 specimens of coated fabrics tested in the Martindale wear tester. The experiment was arranged in the form of a $2 \times 2 \times 3$ factorial design replicated twice, two different fillers F_1 and F_2 being tried in three different proportions Q_1, Q_2, and Q_3 with and without a surface treatment T, and weight losses were recorded after 1000, 2000, and 3000 revolutions of the machine.

Further examples of this type of test arose in road trial assessments of materials for tire treads (see for example Buist et al., 1950). In one of these investigations, for instance, a test vehicle was run with each of its tires constructed in 3 segments: 4 compounds A, B, C, and D were tested on the four tires of the car in a balanced incomplete block design as follows:

Tire 1	Tire 2	Tire 3	Tire 4
$B\ C\ D$	$A\ C\ D$	$A\ B\ D$	$A\ B\ C$

The Collected Works of George E. P. Box, 1984, Wadsworth, Inc., Belmont, CA 94002.
Originally published in *Biometrics,* vol. 6, no. 4 (1950), pp. 362–389.

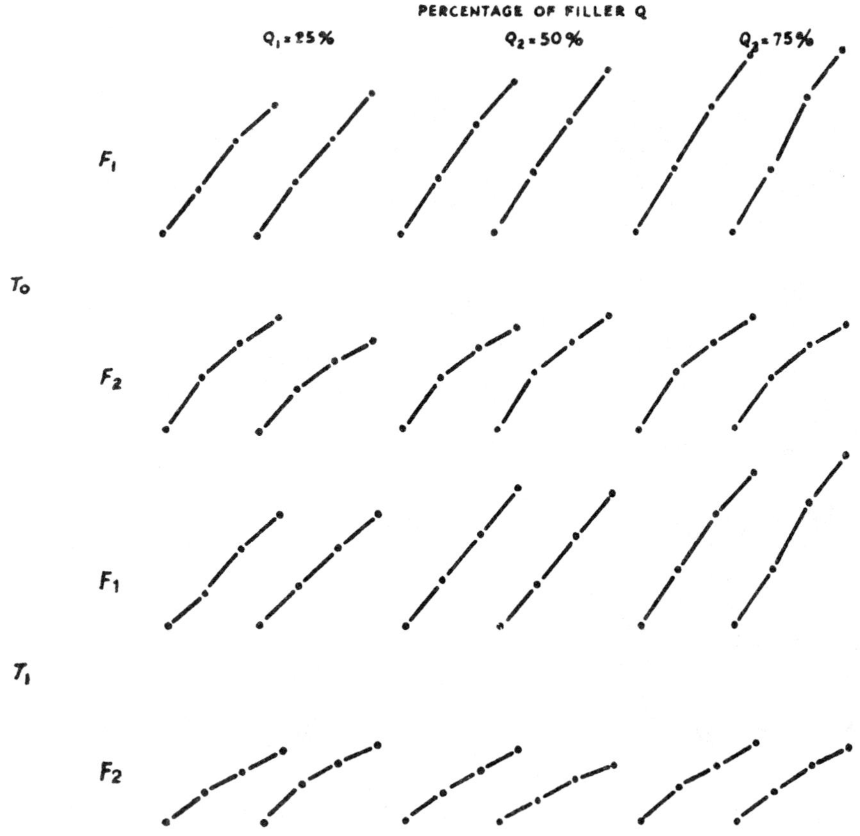

FIGURE I. WEIGHT LOSS CURVES FOR COATED FABRICS

The wear of the compounds was measured by the decrease in tread depth which was observed for each segment of each tire at a number of mileages. Thus for each compound in each tire, there again resulted, not a single result, but a set of results from each of which a wear curve could be plotted.

In biological investigations the growth of an animal or part of an animal is often the subject of study; Fig. II, for example, shows growth curves for 27 rats kept in separate cages. The rats were divided at random into 3 groups containing 10, 7, and 10 rats respectively, (the second group contains fewer rats than the other two, due to an accident at the beginning of the experiment). The first group were kept as a control, the second group had thyroxin, and the third group thiouracil added to their drinking water.

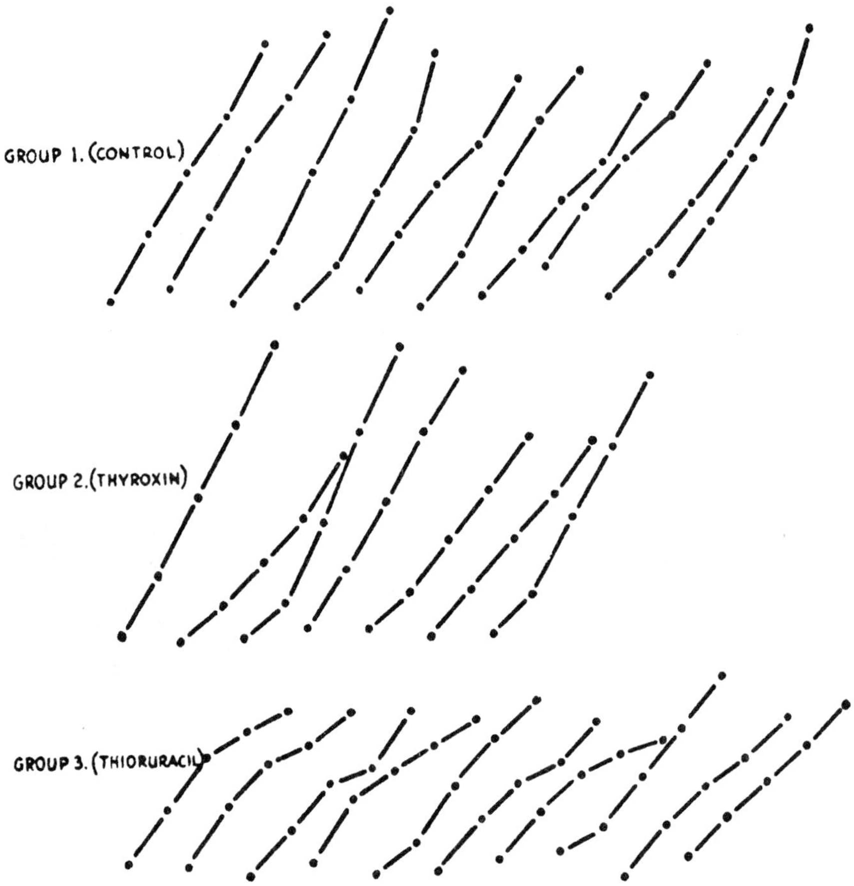

FIGURE II. GROWTH CURVES FOR 27 RATS

In all the examples, it will be seen that we are concerned with experiments which may be of a simple or complex character; each observation, however, consists not of a single value, but of a set of values recorded at intervals of time, from which curves can be plotted.

2. A SIMPLE ANALYSIS

We begin by considering the wear data plotted in Fig. I for coated fabrics prepared in a number of different ways. With data of this kind it has been found to be of value to consider not the wear, after say 1000, 2000, and 3000 revolutions of the machine, but the wear occurring *during* the first thousand, second thousand, and third thousand revolutions, that is to say to consider the first differences of the original data.

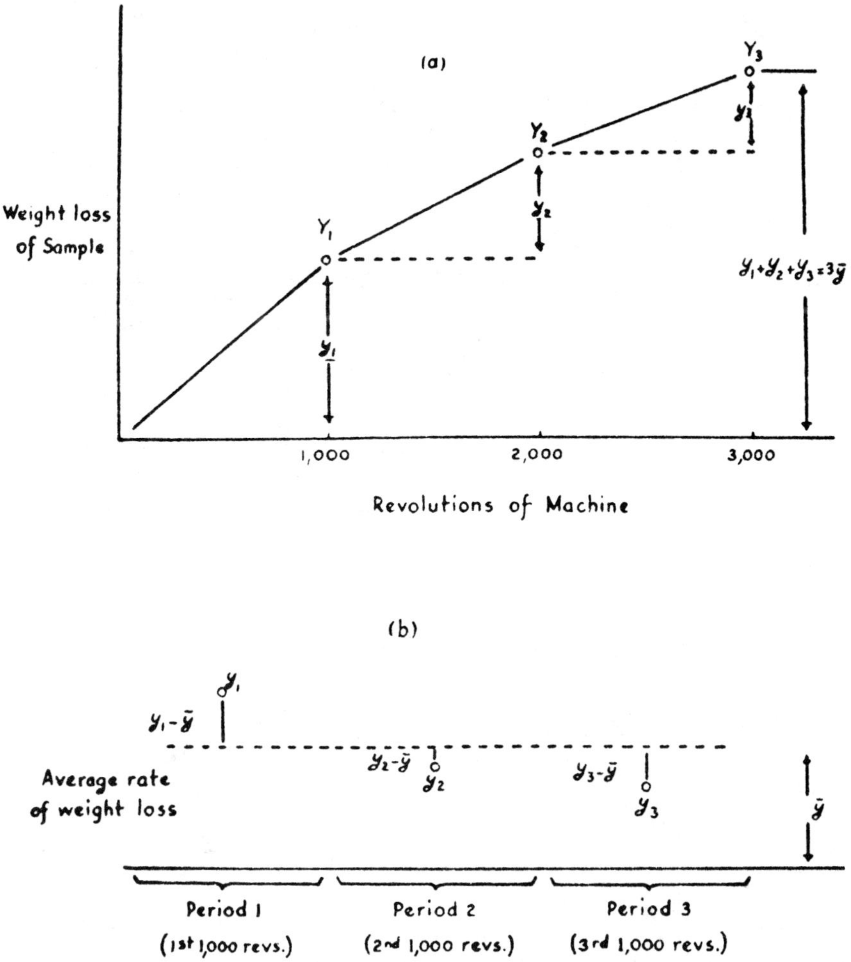

FIGURE III. ANALYSING FIRST DIFFERENCES OF WEAR DATA

In Fig. IIIa, Y_1, Y_2 and Y_3 are the total weight losses for a particular specimen at 1000, 2000, and 3000 revolutions of the machine. We consider $y_1 = Y_1$, $y_2 = Y_2 - Y_1$ and $y_3 = Y_3 - Y_2$; y_1, y_2 and y_3 can be regarded as measuring the average *rates* of wear in milligrams per 1000 revolutions during the three periods, and $\bar{y}$, the mean as measuring the *overall* average rate. Since $3\bar{y}$ is equal to Y_3, $\bar{y}$ is proportional to the total wear during the experiment. The three periods of wear considered can then formally be regarded as a further factor, "periods", and the curve obtained by plotting the differences y_1, y_2, y_3 (Fig. IIIb) indicates the approximate shape of the wear *rate* curve. Now whatever other information is required from the data it will usually be the case that the overall effects of

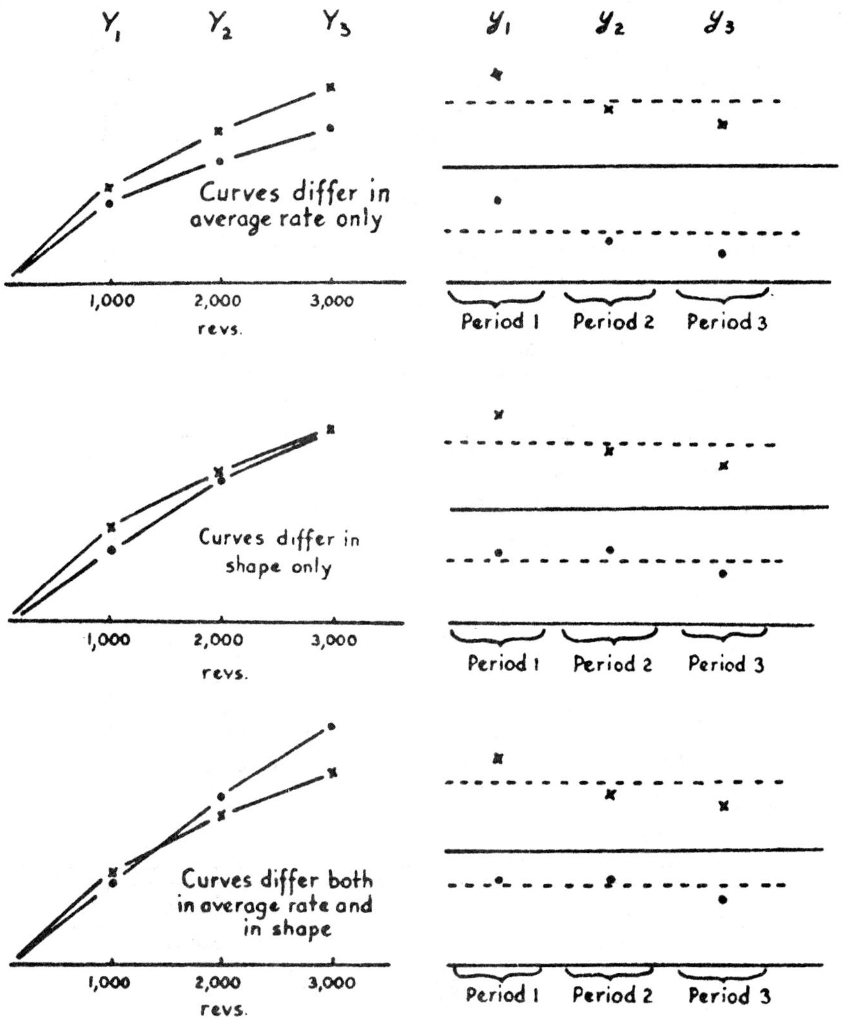

FIGURE IV. COMPARISON OF CURVES USING AVERAGE RATE AND "SHAPE"

treatments will be of major interest and these effects can be elucidated by analyzing the variate $\bar{y}$; it may be, however, that the treatments have effected not only the average rates, but have also influenced the configuration of the individual period rates about their averages, and such effects as this can be regarded as interactions between the factor concerned and "periods".

If (Fig. IV) a factor affects only the average wear rate $\bar{y}$, i.e. alters y_1, y_2, and y_3, equally, (and there is consequently no interaction with "periods"), we shall say that the wear rate curve has been altered only in level, but if the deviations $y_1 - \bar{y}$, $y_2 - \bar{y}$, $y_3 - \bar{y}$ are affected, we shall say that the wear rate curve is also altered in *shape*. This use of the average level $\bar{y}$ and the deviations $y_1 - \bar{y}$, $y_2 - \bar{y}$, $y_3 - \bar{y}$, to describe the curve is analogous to the size and shape analysis used in discrimination problems by Penrose (1947) and Smith (1947). On this definition, then, two curves are said to have different shapes when the wear rates differ by different amounts at different stages of wear.

The data for Fig. I, after differences have been taken, are set out in Table A. The figures in the table corresponding with periods 1, 2, and 3 refer to the wear between 0–1000, 1000–2000, 2000–3000 revolutions of the machine respectively.

TABLE A

WEAR OF COATED FABRICS IN MILLIGRAMS

Surface Treatment	Filler	Proportion of Filler								
		Q_1 (25%)			Q_2 (50%)			Q_3 (75%)		
		Period 1	2	3	Period 1	2	3	Period 1	2	3
T_0	F_1	194	192	141	233	217	171	265	252	207
		208	188	165	241	222	201	269	283	191
	F_2	239	127	90	224	123	79	243	117	100
		187	105	85	243	123	110	226	125	75
T_1	F_1	155	169	151	198	187	176	235	225	166
		173	152	141	177	196	167	229	270	183
	F_2	137	82	77	129	94	78	155	76	91
		160	82	83	98	89	48	132	105	67

2.1 A mathematical model for the experiment.

The error variances for the data are as follows:

Revolutions	Error Variance	Revolutions	Error Variance
0–1000	269	0–1000	269
0–2000	456	1000–2000	200
0–3000	935	2000–3000	222

A characteristic of the cumulative data is seen to be the increase of the error variance as the test proceeds, the variance of the differences however, remains stable from period to period. This is to be expected, for much of the variation probably arises in the operation of the testing machine itself and the variation occurring in equal periods of running might be expected to be the same. Also, in the original data the errors would be expected to be correlated from one observation to the next, since for example abnormally low wear occurring in the first 1000 revolutions would be reflected in subsequent values. So far as the variation of the machine was concerned however, correlation between wear in successive *periods* might be expected to be much smaller and possibly negligible.

For the moment then, we will assume that the departure of y_{ti} (the loss in weight during the ith period of wear of a sample having the tth factor combination) from its mean value η_{ti} can be represented by two independent random variables ϵ_t and δ_i each being normally distributed about zero, the first with variance σ_1^2 allows for *overall* variation in mean rate in duplicate specimens and the second with variance σ_0^2 allows for variations associated with individual periods.

$$y_{ti} - \eta_{ti} = \epsilon_t + \delta_i \tag{1}$$

If this is true, then the analysis of variance for the data in Table A, will be analogous to that for a split-plot agricultural experiment, the "plots" being the samples of material and the three values for wear in successive periods corresponding to the splitting of the plots. As with split plot experiments, the analysis will consist of two parts, each part having its own error estimate. This analysis for the wear data is set out in Table B. The entries in the table are the mean squares corresponding to the effects; the figures in brackets refer to the degrees of freedom available for the comparisons.

The mean squares in the left-hand column are obtained after averaging over "periods" and they enable hypotheses concerning the average wear rate (or what is equivalent, the overall wear) to be tested; the correction for the mean, denoted by I, is included for completeness. Significance is judged by comparison with the mean square of 312 at the foot of this column; this error term has 12 degrees of freedom and is an estimate of the "between samples" variance $3\sigma_1^2 + \sigma_0^2$ as is each of the mean squares in this column if the treatments are without effect. The asterisks indicate significance at the 5, 1 and 0.1% points respectively.

The mean squares in the right-hand column correspond to interactions with "periods" and significant interactions imply that a factor has

TABLE B
ANALYSIS OF VARIANCE FOR THE DATA OF TABLE A

Source		Averaged over "periods" (level of rate curve)		Interactions with "periods" (Shape of rate curve)	
Mean	(I)	(1)	1,866,956	(2)	30,479***
Main effects					
% Filler	(Q)	(2)	6,785***	(4)	440
Type of Filler	(F)	(1)	107,803***	(2)	9,144***
Surface Treatment	(T)	(1)	24,494***	(2)	4,124***
Interactions					
$Q \times F$		(2)	4,942***	(4)	354
$Q \times T$		(2)	397	(4)	172
$F \times T$		(1)	1,682*	(2)	1,164***
$Q \times F \times T$		(2)	150	(4)	116
Error		(12)	312	(24)	190

affected the quantities $y_1 - \bar{y}$, $y_2 - \bar{y}$, $y_3 - \bar{y}$, that is, that it has altered the "shape" of the rate curve. The interaction of I with the periods factor P is $P \times I = P$ the main effect for periods, which indicates whether the mean *rate* of growth has remained constant from period to period i.e. whether the mean growth curve is a straight line. The error mean square of 190 appropriate for testing these effects has 24 degrees of freedom and is an estimate of σ_0^2 the "within samples" variance, as is each of the mean squares in this column on the hypothesis that the treatments are without effect. The entries in this column can be most easily calculated by first carrying through an analysis of variance for each period separately; and then using the identity;

$$\left\{\begin{matrix}\text{Sum of squares}\\ \text{for interaction}\\ \text{with periods}\end{matrix}\right\} = \left\{\begin{matrix}\text{total of sums}\\ \text{of squares for}\\ \text{individual periods}\end{matrix}\right\} - \left\{\begin{matrix}\text{sums of squares}\\ \text{for average}\\ \text{effects.}\end{matrix}\right\}$$

For example, for the surface treatment T the sums of squares for the three periods of wear were found to be 26,268, 5,017 and 1,457 respectively; the sum of squares for the average effect was 24,494, and the sum of squares for interaction with periods was therefore 26,268 + 5,017 + 1,457 − 24,494 = 8,248 and the mean square, 4,124 as shown in the table; all the entries in this column including, of course, the error term, are calculated in this way. This procedure can be followed whether the remaining effects are orthogonal or not. For example, in the balanced incomplete block design used for the assessment of tire treads, the usual (nonorthogonal) analysis for incomplete block designs is first carried through for the means, averaging over periods; this analysis is then repeated for each period separately and the above identity used to determine the interactions with periods.

2.2 Interpretation of the analysis.

Considering first the left-hand column of table B, we see that all the main effects are highly significant and that interactions exist between Q and F, that is, between the proportion of filler and the type of filler used and between F and T, the type of filler and the surface treatment. The appropriate tables of mean values showing the nature of these interactions follow:

% FILLER (Q)

		25	50	75
Type of Filler	F_1	169	199	231
	F_2	121	120	126

SURFACE TREATMENT

		T_0	T_1
Type of Filler	F_1	214	145
	F_2	186	99

From the first table, we see that the average rate of wear is less with the second filler, and this average rate was virtually unaffected by the increase in the percentage of filler, whereas with the first filler, the material wears more, and the wear increases as the percentage is increased. From the second table of means we see that the beneficial effect of the surface treatment is even more pronounced with the second than with the first filler.

So far, we have considered the effect of factors only on the average rate of wear, we now need to consider whether these effects change at different stages of wear; that is to say whether the factors affect the shape of the rate curves. To do this, we consider the mean square for the interaction of each of the effects with periods, shown in the right-hand column of the table; the significance of these items is judged by comparison with the error term at the foot of this column. The % filler (Q) and % filler × type of filler (QF) effects have no significant interactions with periods, so we can regard our conclusions for these effects as probably true at all stages of wear; however both F (the type of filler) and T (the surface treatment) show strong interactions. The tables of mean values are as follows:

		Period 1	Period 2	Period 3
Type of Filler	F_1	215	213	172
	F_2	181	104	82
		Period 1	Period 2	Period 3
Surface Treatment	T_0	231	173	134
	T_1	165	144	119

From the first table we conclude that not only do the samples having filler (1) wear at a greater rate than those having filler (2), but also the fall off in wear is greater with filler (2). From the second table we conclude that although the surface treatment has a favorable effect at all stages of wear this is (as would be expected) most marked initially. Finally, we see that the interaction $F \times T$, between type of filler and surface treatment, interacts with periods and, on consulting a table of mean values, this is seen to be due to the fact that the interaction between filler and surface treatment found before, is confined to the first period of wear alone.

The conclusions from the experiment can therefore be set out as follows:

1. Filler (2) results in less wear than filler (1), the rate of fall off of wear is greater and the protective effect of surface treatment is more marked with (2) than with (1).
2. Whereas increasing the % of filler (1) results in increased rate of wear, the % of filler (2) can be increased, at least to 75%, without increasing wear.
3. The surface treatment markedly reduces wear especially during the early stages and when filler (2) is used.

It is clear that the two fillers are behaving differently and in a full analysis the data would be split into two, and an analysis made for each filler separately. We shall not however elaborate the analysis further here, since our purpose is merely to show that when the set-up given in equation (1) can be regarded as valid, an exceedingly simple and informative analysis can be made. We now proceed to show how it is possible to test whether significant departure from the simple set-up occurs or not.

3. A TEST FOR DEPARTURE FROM THE SIMPLE MODEL

Instead of equation (1) let us write

$$y_{ti} - \eta_{ti} = z_{ti} \tag{2}$$

Then if the simple set-up is valid

$$z_{ti} = \epsilon_t + \delta_i \tag{3}$$

and z_{ti} would be distributed in a three-variate multinormal distribution with each variance equal to $\sigma_1^2 + \sigma_0^2$ and each covariance equal to σ_1^2. (For simplicity the test is illustrated for three variates; it can of course be immediately generalized to the p-variate case.) To check the validity of our analysis therefore we must test the hypothesis, that all the variances are equal and all the covariances are equal; that is, that V^*, the matrix of variances and covariances is of the form:

$$V_0 = \begin{bmatrix} a & d & d \\ d & a & d \\ d & d & a \end{bmatrix} \tag{4}$$

against the alternative that the variances are not all the same and the covariances not all the same; that is, that the variance covariance matrix is of the form:

$$V_1 = \begin{bmatrix} a & d & e \\ d & b & f \\ e & f & c \end{bmatrix} \tag{5}$$

where a, b, c are not all equal and d, e, f, are not all equal.

The null hypothesis of (4) is a little more general than that implied by equation (3) for negative values of the covariance d are possible with (4) (although since V_0 must be positive definite d cannot be less than $-a/(p - 1)$), whereas correlation arising from a common component ϵ_t in (1) must clearly be positive. However, an analysis of the type given in Table B is valid for any positive definite matrix of form (4) so that this extension is appropriate.

3.1 The Test Criterion

A criterion for testing a statistical hypothesis of this form has been

*We tacitly assume here that the variance covariance matrix V remains constant from group to group, that is to say is itself unaffected by the treatments; we consider this assumption later in § 7.

obtained by Wilks (1946) using the likelihood ratio method of Neyman and Pearson (1928). Let $c_{ij} = c_{ji}$ denote the error sums of squares and products calculated from the sample for the ith and j^{th} variates, then the criterion is

$$\Lambda = \frac{\Delta_1}{\Delta_0} = \frac{\begin{vmatrix} c_{11} & c_{12} & c_{13} \\ c_{21} & c_{22} & c_{23} \\ c_{31} & c_{32} & c_{33} \end{vmatrix}}{\begin{vmatrix} \bar{c}_{ii} & \bar{c}_{ij} & \bar{c}_{ij} \\ \bar{c}_{ij} & \bar{c}_{ii} & \bar{c}_{ij} \\ \bar{c}_{ij} & \bar{c}_{ij} & \bar{c}_{ii} \end{vmatrix}} \tag{6}$$

where $\bar{c}_{ii}$ is the average sum of squares $(c_{11} + c_{22} + c_{33})/3$ or in general $(\sum_i^p c_{ii})/p$ and $\bar{c}_{ij}$ is the average sum of products $(c_{12} + c_{13} + c_{23})/3$, or in general

$$\left\{\sum_{i=1}^{p-1} \sum_{j=i+1}^{p} c_{ij}\right\} \Big/ \left\{\frac{1}{2} p(p-1)\right\}$$

and p is the number of variates (3 in this case). The value of the determinant Δ_0 in the denominator can be easily shown to be equal to $[\bar{c}_{ii} + (p-1)\bar{c}_{ij}][\bar{c}_{ii} - \bar{c}_{ij}]^{p-1}$ which simplifies the calculations.

For the wear test example the sums of squares and products for error can be most easily calculated from the differences between duplicate observations in table A. These differences are set out in table C, for example, 14 = 208 − 194, − 4 = 188 − 192, etc.

TABLE C

DIFFERENCES BETWEEN DUPLICATES

Period	T_0						T_1					
	F_1			F_2			F_1			F_2		
	Q_1	Q_2	Q_3	Q_1	Q_2	Q_3	Q_1	Q_2	Q_3	Q_1	Q_2	Q_3
1	14	8	4	−52	19	−17	18	−21	−6	23	−31	−23
2	−4	5	31	−22	0	8	−17	9	45	0	−5	29
3	24	30	−16	−5	31	−25	−10	−9	17	6	−30	−24
Total	34	43	19	−79	50	−34	−9	−21	56	29	−66	−18

The sums of squares and products are then obtained as follows:

		Period			
		1	2	3	S
Period	1	3225.0	−80.5	1656.5	4801.0
	2	−80.5	2405.5	−112.0	2213.0
	3	1656.5	−112.0	2662.5	4207.0

For example $3225.0 = [14^2 + 8^2 + 4^2 + \cdots + (-23)^2]/2$

$$-80.5 = [14 \times (-4) + 8 \times 5 + \cdots + (-23) \times 29]/2.$$

The divisor 2 is employed since the values concerned are differences between two observations. The check column S shows the sum of products between each of the variates and the column totals of table C; this supplies an independent check on the working, for example, for the second row of the matrix

$$-80.5 + 2405.5 - 112.0 = 2213.0.$$

The value of the determinant of this matrix which is the numerator Δ_1 of the criterion is found to be $14{,}026 \times 10^6$. To calculate the denominator Δ_0 of the criterion, we find:

$$\bar{c}_{ii} = 2764.3, \qquad \bar{c}_{ij} = 488.0,$$

$$\bar{c}_{ii} + 2\bar{c}_{ij} = 3740.3 \qquad \bar{c}_{ii} - \bar{c}_{ij} = 2276.3$$

and $\Delta_0 = 3740.3 \times (2276.3)^2 = 19{,}381 \times 10^6$, whence $\Lambda = 0.7237$

3.2 The test of significance

We now wish to test whether or not this value of Λ is exceptionally small. The exact distribution is not known in the general case; however, an expression for the moments of Λ has been given by Wilks (1946) and for these, sufficiently accurate approximations can be calculated. The present author (Box 1949) has given a general distribution theory for a very wide class of what may be called "Λ" statistics, whose moments can be written in the form

$$E(\Lambda^a)^h = \text{constant} \cdot \left[\frac{\prod_{j=1}^{k} (y_j^{y_j})}{\prod_{i=1}^{m} (x_i^{x_i})}\right]^h \frac{\prod_{i=1}^{m} [\Gamma\{x_i(1 + h) + \xi_i\}]}{\prod_{j=1}^{k} [\Gamma\{y_j(1 + h) + \eta_j\}]} \tag{7}$$

Given the value of k, m, and the x_i, ξ_i, y_i and η_i, general formulae are provided from which taking $M = 2a \log_e \Lambda^{-1}$ as a working statistic, an accurate χ^2 series solution can be obtained, and simple χ^2 and F approximations. For the statistic here considered, Wilks' expression for the moments can be written

$$E(\Lambda^{\nu/2})^h = \text{constant} \cdot (p-1)^{\frac{1}{2}(p-1)\nu h} \frac{\prod_{i=1}^{p-1}\left[\Gamma\left\{\frac{\nu}{2}(1+h)-\frac{i}{2}\right\}\right]}{\Gamma\left\{\frac{\nu}{2}(p-1)(1+h)\right\}} \tag{8}$$

ν being the degrees of freedom of the sums of squares and products tested. This is seen to be of the same form as (7) so the general theory can be applied.

Making the substitutions we find we should take for our working statistic

$$M = \nu \log_e \Lambda^{-1}$$

The more the data depart from the simple set-up, the smaller will be Λ and the greater the value of M. To test whether M is large enough to indicate a "significant" departure from the simple set-up we calculate

$$f_1 = (p^2 + p - 4)/2, \qquad A_1 = \frac{\{p(p+1)^2(2p-3)\}}{\{6\nu(p-1)(p^2+p-4)\}}$$

and refer $(1 - A_1) M$ to tables of χ^2 with f_1 degrees of freedom. An approximation which is rather more precise especially when p is large and/or ν is small is supplied by calculating

$$A_2 = \frac{(p-1)p(p+1)(p+2)}{6\nu^2(p^2+p-4)}, \quad f_2 = \frac{f_1+2}{A_2 - A_1^2}, \quad b = \frac{f_1}{1 - A_1 - f_1/f_2}$$

and referring M/b to tables of the F distribution with f_1 and f_2 degrees of freedom.

In the present example $\nu = 12$ and $p = 3$. Whence $f_1 = 4$, $A_1 = 0.125$, $A_2 = 0.0174$, $M = 12 \log_e (1/0.7237) = 3.880$. $(1 - A_1) M = 3.395$ and this quantity is to be referred to tables of χ^2 with four degrees of freedom, from which we can conclude at once that on this data there is no reason to question the simple set-up. The actual probability given by this approximation for the chance occurrence of a value of M as great or greater than this is slightly less than 0.5. Using the F approximation a very similar result would have been obtained, we find

$$f_1 = 4, \qquad f_2 = 3456, \qquad b = 4.578.$$

Thus $M/b = 0.8476$ is to be referred to tables of F with 4 and 3456 degrees of freedom, and consulting the table of Thompson and Merrington (1943) we again find a value for the probability of chance occurrence slightly below 0.5. In this particular case both approximations are quite accurate, and the values they give are almost identical. When p is larger and ν is smaller the χ^2 approximation is less accurate and the F approximation differs more markedly from it, f_2 being no longer very large as it is in this example.

4. AN ALTERNATIVE SET-UP

We have seen how, by the device of taking differences, the problem of interpreting wear data was facilitated and a simple analysis was possible if it was reasonable to assume a particular set-up. We have shown that for a particular example the hypothesis that the set-up was of this simple form was not contradicted by the data.

The simple set-up would be expected to represent wear data satisfactorily if most of the variation arose from the operation of the machine itself or if the variation within replicate specimens of material were mainly confined to changes in *average* abrasion resistance and not to changes in "shape" of the wear curves which might give rise to serial correlation between differences. Observational errors would also tend to cause departures from the set-up, for an apparent increase in wear rate during one period would be compensated by an apparent decrease in the next, and this would lead to negative correlation between successive differences. In some cases therefore we would not expect the variances and covariances, even after differencing of the data, to be capable of adequate representation by the simple pattern of equation (4) and the more general set-up typified by equation (5) would have to be adopted.

When a research program is being carried out in which the results will appear in the form of wear curves or growth curves, it is worthwhile paying particular attention, in the preliminary experiments, to the form of the variance co-variance matrices, so that decisions may be reached concerning a set-up, and method of analysis, suitable for use in this *particular* investigation; that is to say with this particular type of material and interval between observations. Consider now the rat growth data shown in detail in Table D at end of article, which was plotted in Figure II. For the reasons given above we should not expect the weight gains for these animals, even after differencing, to be uncorrelated from one period to another. In fact, if we denote the five variates recording initial weight and weight after one, two, three, and four weeks by Y_0, Y_1, Y_2, Y_3, Y_4 and the differences, $Y_1 - Y_0$, $Y_2 - Y_1$, $Y_3 - Y_2$, $Y_4 - Y_3$ by y_1, y_2, y_3, y_4, the matrix of sums of squares and products for the 24 error degrees of freedom for y_1, y_2, y_3 and y_4, is found to be

$$\begin{bmatrix} 582.3 & 42.5 & -55.5 & -74.6 \\ 42.5 & 609.0 & 626.5 & 344.5 \\ -55.5 & 626.5 & 1046.7 & 459.0 \\ -74.6 & 344.5 & 459.0 & 853.0 \end{bmatrix} \tag{9}$$

From which proceeding as before we find

$$\Lambda = 0.3534, \qquad M = 25.0, \qquad 1 - A_1 = 0.9277, \qquad f = 8.$$

Referring $(1 - A_1)M = 23.2$ to tables of χ^2 with 8 degrees of freedom we conclude that the probability of a chance deviation from the simple set-up as great as this is less than 0.01; a similar result is given by the F approximation and we are therefore led to reject this set-up.

Although differencing is not completely successful in transforming the data into a form in which the variances are equal and the covariances are equal, it is worth noting that the differences do show a great deal more uniformity in this respect than the data before differencing. The corresponding error matrix for sums of squares and products for the overall weight gains $Y_1 - Y_0$, $Y_2 - Y_0$, $Y_3 - Y_0$, and $Y_4 - Y_0$ is

$$\begin{bmatrix} 582.3 & 624.8 & 569.3 & 494.7 \\ 624.8 & 1276.3 & 1847.3 & 2117.2 \\ 569.3 & 1847.3 & 3465.0 & 4193.9 \\ 494.7 & 2117.2 & 4193.9 & 5775.8 \end{bmatrix} \tag{10}$$

and the value of χ^2 obtained on applying the test to these values is 109.1; the transformation has thus gone a good deal of the way towards bringing the data to the simpler form.

In making tests of an actual set-up, it should be borne in mind that the important consideration is how far departures from the assumptions made will affect the *tests based on these assumptions*. This problem has been investigated in a number of cases by the present author and it is hoped to publish the results elsewhere in the near future. It appears that minor departures of the data from independence and homoscedasticity of the type considered here will not seriously influence subsequent tests of significance; a similar conclusion was reached by Daniels (1938). Thus, in the wear curve example we found no significant departure from the simple set-up, and although this did not mean that no such departure could have occurred, but possibly, merely that the data were not sufficiently extensive for it to be detected, it is probable that little error was made by adopting the simple set-up in this case. In the growth curve example however a marked departure is apparent and it would be safer to adopt the less restrictive assumptions of the alternative hypothesis of equation (5).

We shall assume that the variates Y_0, Y_1, Y_2, Y_3 and Y_4 are distributed about their mean values in a 5-variate normal distribution, the same for each rat, and this of course implies that any set of variates derived from these, by linear transformation, will also be distributed multinormally; in particular the differences will be distributed in that form.

4.1 The Multivariate Test

Now if a single variate only, say the overall increase in weight, were being analyzed, the hypothesis concerning the significance of treatment differences could be tested by means of the analysis of variance, that is to say the criterion

$$\frac{\text{mean square for treatments}}{\text{mean square for error}}$$

would be referred to tables of the F distribution with the appropriate numbers of degrees of freedom. Alternatively (see for example Kolodziejczyk 1935) the criterion*

$$\Lambda = \frac{\text{sum of squares for error}}{\text{sum of squares for error} + \text{sum of squares for treatment}}$$

could be employed, and the test carried out by referring to tables of the incomplete B-function (Karl Pearson; 1934, Thomson; 1941), the result of course would be precisely the same.

For our present purpose the latter criterion is of more interest, because it can be directly generalized (Wilks; 1932, Pearson and Wilks; 1933, Bartlett; 1934, 1938) to the case where the observations are not single variates but multivariates, whereas the former criterion cannot. Thus the hypothesis that the mean value for each of the variates is the same from group to group; (for example, that the gains in weight during the first week are all equal, *and* the gains during the second week are all equal, etc.) can be tested by calculating the criterion;

*Λ is used in the paper to denote a criterion of the form associated with the likelihood ratio method of Neyman and Pearson. M is used to denote a logarithmic statistic derived from Λ. The likelihood statistic Λ and the derived quantity M referred to in this section are of course different from the criterion discussed in § 3.

$$\Lambda = \frac{|\text{sums of squares and products for error}|}{\left|\begin{array}{c}\text{sums of squares and products for error} \\ + \text{ sums of squares and products for treatment}\end{array}\right|}$$

where *determinants* whose elements are sums of squares and products replace the single sums of squares of the univariate criterion.

Thus, had we desired to use the more general set-up in the wear test example, an *overall* test for each of the main effects and interactions could have been applied by calculating the 3 × 3 matrix of sums of squares and products first for error and then for the particular main effect or interaction concerned, and hence calculating the criterion Λ. This test would not have distinguished between changes in average and changes in shape but would have been an overall test including both.

The exact distribution of Λ is known only for certain special cases, however, this is another of the general class of statistics whose moments can be written in the form of equation (7), and simple approximations which are perfectly general and which are usually sufficient for most practical purposes can be obtained. To preserve generality, even when the exact distribution is available, these approximations will be used in all the tests that follow. If n is the number of degrees of freedom for *treatments plus error*, q the number of degrees of freedom for treatments and p the number of variates then Bartlett's (1938) χ^2 approximation is obtained by calculating

$$M = n \log_e \Lambda^{-1} \qquad A_1 = (p + q + 1)/2n \qquad f_1 = pq$$

and referring $(1 - A_1)M$ to tables of χ^2 with f_1 degrees of freedom. This approximation tends to break down if n is small or p and q are large and in these cases it is worthwhile calculating the more accurate F approximation (Box 1949). For this we calculate in addition

$$f_2 = \frac{12n^2(pq + 2)}{p^2 + q^2 - 5}, \qquad b = \frac{pq}{1 - A_1 - f_1/f_2},$$

and refer M/b to tables of F with f_1 and f_2 degrees of freedom.

In the growth curve example if we consider the variates y_1, y_2, y_3, y_4 (that is, the first differences of the weight gains) the matrix for sums of squares and products for treatments is found to be

$$\begin{bmatrix} 81.7 & 37.2 & 11.5 & 112.9 \\ 37.2 & 476.9 & 782.7 & 787.4 \\ 11.5 & 782.7 & 1315.9 & 1260.1 \\ 112.9 & 787.4 & 1260.1 & 1334.0 \end{bmatrix} \qquad (11)$$

The corresponding matrix for sums of squares and products for error has already been given (9). The error plus treatment matrix is obtained by adding each element in the error matrix to the corresponding element in the treatment matrix. The ratio Λ of the error determinant to the error plus treatment determinant is then found to be 0.2661, and $n = 26$, $p = 4$, $q = 2$. For such large values of n and comparatively small values of p and q Bartlett's χ^2 approximation will be adequate and we find

$$M = 34.4, \qquad A_1 = 7/52, \qquad f_1 = 8$$

and referring $(1 - A_1)M = 29.8$ to tables of χ^2 with 8 degrees of freedom we conclude that the mean values for y_1, y_2, y_3, and y_4 representing the growth during the first, second, third, and fourth weeks differ very significantly from group to group ($P < 0.001$).

4.2 Special Properties of the Criterion.

Before proceeding further we note certain important properties of this criterion used in the multivariate extension of the analysis of variance.

1. The criterion is invariant under non-singular linear transformation of the variates. Thus, in the example above, if we had analyzed the total gains in weight instead of the differences, or had applied any other linear transformation of this sort to the data, the value of the overall criterion would have been unchanged.

2. The sums of squares and products matrix for any *new* set of variates obtained by linear transformation can be found directly by applying the transformation to the rows and columns of the matrix of sums of squares and products of the *old* set of variates. For example if the matrix (10) for total weight gains were known, (9) the corresponding matrix after differencing the data, could be obtained by applying the differencing process to the rows and columns of (10) itself; it is not necessary to make the transformation to the original data and recalculate.

3. In the calculation of determinants, the method of pivotal condensation (see for example Aitken, 1948) provides a rapid practical procedure; *this device also provides a useful method for the elimination of variables.* As an example consider a determinant Δ of sums of squares and products for, say, three variates y_1, y_2, y_3,

$$\Delta = \begin{vmatrix} c_{11} & c_{12} & c_{13} \\ c_{21} & c_{22} & c_{23} \\ c_{31} & c_{32} & c_{33} \end{vmatrix} \qquad \text{where } c_{ij} = c_{ji}$$

dividing through the first row by c_{11} we obtain a new first row

$$1 \quad \frac{c_{12}}{c_{11}} \quad \frac{c_{13}}{c_{11}}$$

if this is subtracted c_{12} times from the second row and c_{13} times from the third, we have

$$\Delta = c_{11} \begin{vmatrix} 1 & \frac{c_{12}}{c_{11}} & \frac{c_{13}}{c_{11}} \\ 0 & c_{22} - \frac{c_{12}^2}{c_{11}} & c_{23} - \frac{c_{12}c_{13}}{c_{11}} \\ 0 & c_{23} - \frac{c_{12}c_{13}}{c_{11}} & c_{33} - \frac{c_{13}^2}{c_{11}} \end{vmatrix}$$

and writing $c_{22} - c_{12}^2/c_{11}$ as $c_{22.1}$, $c_{23} - c_{12}c_{13}/c_{11}$, as $c_{23.1}$, etc. and expanding the determinant along the first column, we have

$$\Delta = c_{11} \begin{vmatrix} c_{22.1} & c_{23.1} \\ c_{23.1} & c_{33.1} \end{vmatrix}$$

that is

$$\Delta = \Delta_{123} = c_{11}\,\Delta_{23.1}\,.$$

As is well known, to compute the value of any $p \times p$ determinant, the process may be repeated $p - 1$ times till the determinant is reduced to the product of p known quantities, and this process is the basis of the Gauss-Doolittle method for the solution of the linear equations, and can be still further simplified (see for example Dwyer; 1942). What is interesting for our purpose is the fact that the elements of $\Delta_{23.1}$ are the sums of squares and products for y_2 and y_3 after eliminating the variable y_1 , that is to say they are the sums of squares and products of deviations from the regressions of y_2 and y_3 on y_1 . Now if condensation of this sort is applied simultaneously to numerator and denominator of Λ we have

$$\Lambda = \frac{c_{11}\text{ (error)}}{c_{11}\text{ (error + treatments)}} \quad \frac{\Delta_{23.1}\text{ (error)}}{\Delta_{23.1}\text{ (error + treatments)}}$$

$$= \Lambda_1\,\Lambda_{23.1}$$

In the above equation Λ_1 corresponds to a univariate analysis of variance for the variable y_1 and the second component to a multivariate analysis of covariance for the remaining variables with the first variate y_1 eliminated. Any number of variables can be eliminated in this way, the number of degrees of freedom for error being correspondingly reduced after each elimination. A successive elimination of variates of this sort was applied by Bartlett to the linear quadratic and cubic components fitted to the growth curves of pigs by Wishart (1939).

4.3 Further Analysis of the Data.

Now even though the taking of differences does not result in a simplification of the set-up, it allows the changes in the wear curves to be more easily appreciated and may still be employed in the interpretation of the overall criterion. We shall therefore again consider the mean growth rate $\bar{y}$ and the deviations from the mean $y_1 - \bar{y}$, etc. Only three of the four deviations from the mean are linearly independent and all the information concerning departures from the mean is contained in any three of them, we therefore consider the variates $\bar{y}$, $y_1 - \bar{y}$, $y_2 - \bar{y}$ and $y_3 - \bar{y}$; $y_4 - \bar{y}$ is omitted from the analysis, (exactly the same result will be obtained whichever of the deviations is omitted). Using the second property noted above, the entries for sums of squares and products for the new variates are obtained by direct transformation. For example we obtain the error matrix for the new variates from the error matrix (9) for y_1, y_2, y_3, and y_4 as follows; first applying the transformation to the rows, corresponding to the operation of taking the mean $\bar{y}$ we replace the elements of the first row of (9) by their column means, the second row is then obtained by subtracting these values from the elements of the first row of (9) corresponding to the operation of taking $y_1 - \bar{y}$; the third and fourth rows are found similarly, and the whole set of operations is then carried out on the columns. A partial check is supplied by the symmetry of the final transformed matrix and a complete check can be made by calculating the sums for each row and column of the final matrix and confirming that these totals agree with the values found by operating on the sums of rows and columns of the original matrix. From the error matrix (9) we obtain

	$\bar{y}$	$y_1 - \bar{y}$	$y_2 - \bar{y}$	$y_3 - \bar{y}$
$\bar{y}$	361.0	−237.3	44.6	158.2
$y_1 - \bar{y}$	−237.3	695.9	−125.8	−337.4
$y_2 - \bar{y}$	44.6	−125.8	158.7	62.7
$y_3 - \bar{y}$	158.2	−337.4	62.7	369.3

and applying the same procedure to the error plus treatment matrix we have

	$\bar{y}$	$y_1 - \bar{y}$	$y_2 - \bar{y}$	$y_3 - \bar{y}$
$\bar{y}$	935.5	−751.0	−8.8	426.2
$y_1 - \bar{y}$	−751.0	1230.5	−96.0	−654.7
$y_2 - \bar{y}$	−5.8	−96.0	168.0	56.3
$y_3 - \bar{y}$	426.2	−654.7	56.3	574.6

The Λ criterion for means alone is therefore:

$$\Delta\,(\bar{y}) = \frac{360.9}{935.4} = 0.3859$$

and for reasons already given we shall employ Bartlett's approximation to make the test of significance. We find $(1 - A_1)M = 22.9$, should be referred to χ^2 tables with two degrees of freedom whence we deduce that the mean growth rates differ very significantly ($P < .001$) from group to group. To test the deviations from the means, that is to test whether the "shape" of the growth curve varies from group to group we calculate the ratio of the 3 x 3 determinant for error to that for error + treatments for the three variates $y_1 - \bar{y}$, $y_2 - \bar{y}$, $y_3 - \bar{y}$. We find

$$\Lambda(y_1 - \bar{y},\ y_2 - \bar{y},\ y_3 - \bar{y}) = 0.4366$$

and $(1 - A_1)M = 19.0$ is referred to tables of χ^2 with 6 degrees of freedom. This value is significant ($P < .01$) and we therefore conclude that, not only the mean level, but also the shape of the curve is changing from group to group. A table of mean values indicates the nature of the differences.

MEAN GAINS IN WEIGHT (GRAMS)

Period	Group		
	1	2	3
1st week	24.5	20.3	21.6
2nd week	27.5	29.0	19.5
3rd week	24.1	29.3	12.4
4th week	30.5	30.1	15.8
Mean rate (grams/week)	26.7	27.2	17.3

Further tests show that no significant differences occur between groups 1 and 2, i.e., that the treatment of group 2 is without significant effect, however group 3 differs from the other groups both in average level and in shape. In the first two groups we find a fairly steady rate, which if anything, is tending to increase, whereas a fall in growth rate is found in the third group.

So far the average effects and "shape" effects have been treated separately, but it may be relevant to inquire whether the effects found in the two parts of the analysis can be regarded as separate entities, or whether they are really manifestations of the same thing. In Fig. II it is noticeable that the growth curves of group 3 not only show a low average rate, but also tend to be convex upwards whereas the growth curves of groups 1 and 2 have a higher average and are if anything concave. Now there may also be a tendency *within* the groups, for these curves with low average rates of growth to be also those which are most convex; we may therefore wish to test whether *given the change in mean growth rate*, any differences in "shape" occur, other than would be expected from the internal evidence of the groups concerning the relation between "shape" and mean value. To make the test we use the third property of the Λ criterion mentioned above; the criterion for variables $y_1 - \bar{y}$, $y_2 - \bar{y}$, $y_3 - \bar{y}$ *given* $\bar{y}$ is calculated by dividing the overall criterion for the 4 variables by that for the single variable $\bar{y}$.

$$\Lambda(y_1 - \bar{y}, y_2 - \bar{y}, y_3 - \bar{y} : \bar{y}) = \Lambda(y_1 - \bar{y}, y_2 - \bar{y}, y_3 - \bar{y}, \bar{y})/\Lambda(\bar{y})$$

$$= \Lambda(y_1, y_2, y_3, y_4)/\Lambda(\bar{y})$$

$$= \frac{0.2661}{0.3859} = 0.6896$$

Since one variable $\bar{y}$ has been eliminated we have $n = 25$, $q = 2$, $p = 3$, and $(1 - A_1)M = 8.18$ is referred to tables of χ^2 with 6 degrees of freedom. The probability of chance occurrence of such a value is about 0.25. We see, therefore, that there is no evidence of differences in shape other than would be expected from the *internal* relation between average and shape.

5. REDUCTION OF THE DATA

In some experiments, weighings are made at very short intervals of time, and the number of points p' for each growth curve is large. Usually however the salient features of the curves will be described by employing fewer than p' constants. Thus to reduce his data Wishart (1938) fitted orthogonal polynomials up to the third degree to the overall weight gain for each of a number of pigs receiving different rations, and analyzed the regression constants. The essential idea is to reduce the data without sacrificing the extra precision given by the larger number of points available. When the method of analysis given here is to be used, this can be done by first applying some process of graduation, to produce a smooth curve through each set of points corresponding to each animal, and analyzing the smoothed values read off from the curves, at a number of equal intervals sufficient to give an adequate description of the curves. This graduation of the data can sometimes be accomplished quite satisfactorily by fitting the curves by eye, but some may prefer a method which is more objective. Since any polynomial of degree p is uniquely determined by specifying $p + 1$ points through which it passes, the division of the smoothed curve into p periods, specified by $p + 1$ points, is equivalent to the description of the curve by a polynomial of degree p. In the growth experiment described here, two weighings were made per week, and the values actually plotted in Fig. II and analyzed, are the means of these pairs of values.

6. ELIMINATION OF THE INITIAL WEIGHT

In the analysis of growth curves the increases in weight of the animals may be correlated with their initial weights. In this case greater precision may be obtained if the analysis is made after elimination of the initial weight by covariance analysis (see for example Fisher 1941). The elimination of y_0, the initial weight, can be accomplished in a precisely similar manner to that used for the elimination of $\bar{y}$ from the criterion for "shape" analysis of 4.3, that is to say we have for the *overall* criterion after the elimination of y_0

$$\Lambda_{1234.0} = \Lambda_{01234}/\Lambda_0$$

This criterion serves to assess the differences in growth rates during the four periods, after the regression of each of these variates on the initial weight has been allowed for, and its significance is assessed as before, the degrees of freedom for error being one less than for the corresponding criterion Λ_{1234}. We shall normally wish to analyze $\Lambda_{1234.0}$ further, however, and to do this we shall need the corresponding matrices of sums of squares and products. These can be found in the manner described in §4.2. The matrices for error and for error plus treatments for the five variates y_0, y_1, y_2, y_3 and y_4 are reduced by a single pivotal condensation based on the element corresponding to the sum of squares for y_0. This gives the desired matrices for the numerator and denominator of $\Lambda_{1234.0}$. Further analysis of the data into differences in mean growth rate and differences in "shape" can be accomplished by operating on the determinants of $\Lambda_{1234.0}$ in precisely the same manner as has already been described for Λ_{1234}.

The two components $\Lambda(\bar{y} : y_0)$ and $\Lambda(y_1 - \bar{y}, y_2 - \bar{y}, y_3 - \bar{y} : y_0)$ assess respectively, the change in overall growth rate when change in initial weight is allowed for, and the change in shape when change in initial weight is allowed for; again apart from the loss of a degree of freedom the tests are the same.

Finally $\Lambda(y_1 - \bar{y}, y_2 - \bar{y}, y_3 - \bar{y} : y_0 , \bar{y})$ can be calculated by eliminating $\bar{y}$ in a precisely similar way as before and this criterion will assess whether group differences occur other than can be explained by the relation within the groups between the "shape" and the initial weight and average growth rate.

7. TESTING AN ASSUMPTION IN THE MULTIVARIATE ANALYSIS

Just as in a single-variate analysis of variance the assumption is usually made that the observations are normally distributed about their population mean values with *constant variance* so an analogous assumption that the variates are multinormally distributed about their mean values with constant variance-covariance matrix is made in the multivariate analysis of variance of §4.1. If the variance-covariance pattern changes markedly from group to group, this test may be invalidated. Also in the test of §3.1 concerning the form of the variance-covariance matrix, the sums of squares and products are pooled on the tacit assumption that the variances and covariances do not change from one treatment group to the next. If this were not so an averaging effect might occur so that even though individual groups showed departure from the simple set-up, the overall criterion computed from the pooled error sums of squares and products for all the groups, might show no such departure.

To test the assumption that the matrix of variances and covariances remains constant from one treatment group to the next, the present author (Box 1949) has employed the multivariate analogue of Bartlett's (1937) criterion which is used to test for constancy of variance in the univariate case.

We take as our criterion

$$M = N \log_e | s_{ij} | - \sum_l (\nu_l \log_e | s_{ijl} |)$$

where s_{ijl} is the usual unbiased estimate of variance or covariance between the i-th and j-th variable in the l-th sample based on sums of squares and products having ν_l degrees of freedom and there are k such samples, s_{ij} is the *average* variance or covariance $(\sum_l \nu_l s_{ijl})/N$ and $N = \sum_l \nu_l$ the total of the degrees of freedom. It will be noted that, as usual, the *determinants* of variances and covariances replace the single variances of the univariate criterion. It is perhaps worth noting that again this criterion is invariant under linear transformation of the data, that is to say data which show lack of constancy in variance and covariance cannot be made more homogeneous by linear transformation.

The test will be illustrated for the growth data of rats set out in Table D for the variates y_1, y_2, y_3 and y_4, recording increases in weight in four successive weekly periods.

The individual matrices for sums of squares and products are as follows:

GROUP I

$$\begin{bmatrix} 210.5 & 13.5 & -7.5 & -13.5 \\ 13.5 & 202.5 & 224.5 & 110.5 \\ -7.5 & 224.5 & 310.9 & 117.5 \\ -13.5 & 110.5 & 117.5 & 258.5 \end{bmatrix} \quad \nu_1 = 9$$

GROUP II

$$\begin{bmatrix} 111.4 & 83.0 & 78.4 & 39.7 \\ 83.0 & 246.0 & 292.0 & 157.0 \\ 78.4 & 292.0 & 473.4 & 264.7 \\ 39.7 & 157.0 & 264.7 & 174.9 \end{bmatrix} \quad \nu_2 = 6$$

GROUP III

$$\begin{bmatrix} 260.4 & -54.0 & -126.4 & -100.8 \\ -54.0 & 160.5 & 110.0 & 77.0 \\ -126.4 & 110.0 & 262.4 & 76.8 \\ -100.8 & 77.0 & 76.8 & 419.6 \end{bmatrix} \quad \nu_3 = 9$$

Since $|s_{ijl}| = |c_{ijl}|/\nu_l^p$ the determinants of the variance-covariance matrices can be obtained directly from the determinants for the sums of squares and products and we find

$$\log_e |s_{ij1}| = 11.2700, \ \log_e |s_{ij2}| = 10.8357, \ \log_e |s_{ij3}| = 12.7008$$

the determinant for the average variances and covariances is found in a similar way from the total sums of squares and products matrix (9);

$$\log_e |s_{ij}| = 12.4473$$

and $M = 24 \times 12.4473 - 9 \times 11.2700 - 6 \times 10.8357 - 9 \times 12.7008 = 17.9838$. This logarithmic statistic M is a further example of the class discussed in §3.2, and approximations have been derived using the general theory referred to before.

For the χ^2 approximation the following quantities are calculated

$$A_1 = \frac{2p^2 + 3p - 1}{6(p + 1)(k - 1)} \left(\sum_i \frac{1}{\nu_i} - \frac{1}{N} \right) \qquad f_1 = \frac{1}{2}(k - 1)p(p + 1)$$

and $(1 - A_1)M$ is distributed as χ^2 with f_1 degrees of freedom. As before, a more precise approximation, which is useful when some of the degrees of freedom ν_i are small or p and/or k are large can be obtained using tables of F. We calculate

$$A_2 = \frac{(p - 1)(p + 2)}{6(k - 1)} \left(\sum_i \frac{1}{\nu_i^2} - \frac{1}{N^2} \right)$$

and refer M/b to tables of F with f_1 and f_2 degrees of freedom where

$$f_2 = \frac{f_1 + 2}{A_2 - A_1^2} \qquad \text{and} \qquad b = \frac{f_1}{1 - A_1 - f_1/f_2}$$

In this particular example $A_1 = 0.2408$, $f_1 = 20$, $(1 - A_1)M = 13.5$ is therefore referred to tables of χ^2 with 20 degrees of freedom. The probability for the occurrence of a value as great or greater than this, when the variances and covariances are in fact constant from one group to the next, is thus about 0.85, and there is therefore no reason to doubt the homogeneity of the data in this respect.

This paper originated partly as the result of a note published by O. L. Davies (1947) criticising a method of analysis for growth curves proposed by W. S. Weil (1947).

I am indebted to Dr. Davies for proposing this problem, and to those of my colleagues who were responsible for the investigations which are mentioned. In conclusion I wish to warmly acknowledge the help and guidance I have received from Dr. H. O. Hartley in this work.

SUMMARY

In the analysis of growth and wear curves, the effects can often be simply interpreted by differencing the original data; these differences correspond to the average growth rates during successive periods. If the successive periods are treated as the level of a further factor "periods", the effect of treatments on mean rate is measured by the variation in the period averages and on the "shape" of the rate curve by the interaction of these treatments with "periods". The taking of differences sometimes results, at least approximately, in a very simple covariance pattern for the errors, and the analysis can then be made by a simple application of the technique of the analysis of variance. A test is given which makes it possible to decide whether this simple set-up is contradicted by the data. When the simple set-up is not appropriate, a multivariate extension of the analysis of variance is used to make the tests. Certain simple properties of the criterion are discussed which facilitate the analysis and the elimination of variables such as initial weight. Finally, it is shown how an important assumption made in the multivariate analysis may be tested.

REFERENCES

Aitken, A. C. *Proc. Roy. Soc. Edin.*, *55*, 42, 1935.
Aitken, A. C. *Determinants and Matrices*, 5th ed., Edinburgh: Oliver and Boyd, 1948.
Bartlett, M. S. *Proc. Camb. Phil. Soc.*, *30*, 327, 1934.
Bartlett, M. S. *Proc. Roy. Soc. A.*, *160*, 268, 1937.
Bartlett, M. S. *Proc. Camb. Phil. Soc.*, *34*, 33, 1938
Box, G. E. P. *Biometrika*, *36*, 317, 1949.
Buist, J. M., Newton, R. G., and Thornley, E. R. *Trans. I. R. I.* (In the press).
Daniels, H. E. *Proc. Camb. Phil. Soc.*, *34*, 321, 1938
Dwyer, P. S. *J. Am. Statist. Ass.*, *37*, 441, 1942.
Fisher, R. A. *Statistical Methods for Research Workers*, 8th. Ed., 1941. Edinburgh: Oliver and Boyd.
Kolodziejczyk, S. *Biometrika*, *27*, 161, 1935.
Neyman, J. and Pearson, E. S. *Biometrika*, *20A*, 175 and 263, 1928.
Pearson, E. S. and Wilks, S. S. *Biometrika*, *25*, 353, 1933.
Pearson, K. (Editor) Tables of the Incomplete Beta Function. London: Biometrika 1934.
Penrose, L. S. *Ann. Eugen., London*, *13*, 228, 1947.
Smith, C. A. B. *Ann. Eugen., London*, *13*, 272, 1947.
Thompson, C. M. *Biometrika*, *32*, 168, 1941.
Thompson, C. M. and Merrington, M. *Biometrika*, *33*, 74, 1943.
Wilks, S. S. *Biometrika*, *24*, 471, 1932.
Wilks, S. S. *Ann. Math. Statist.*, *17*, 257, 1946.
Wishart, J. *Biometrika*, *30*, 16, 1938.
Wishart, J. *J. R. Statist. Soc. Suppl.*, *6*, 1, 1939.

TABLE D

INITIAL WEIGHT AND WEEKLY GAINS IN WEIGHT FOR 27 RATS

Group 1. Control						Group 2. Thyroxin						Group 3. Thiouracil					
Rat	y_0	y_1	y_2	y_3	y_4	Rat	y_0	y_1	y_2	y_3	y_4	Rat	y_0	y_1	y_2	y_3	y_4
1	57	29	28	25	33	11	59	26	36	35	35	18	61	25	23	11	9
2	60	33	30	23	-31	12	54	17	19	20	28	19	59	21	21	10	11
3	52	25	34	33	41	13	56	19	33	43	38	20	53	26	21	6	27
4	49	18	33	29	35	14	59	26	31	32	29	21	59	29	12	11	11
5	56	2[illegible]	23	17	30	15	57	15	25	23	24	22	51	24	26	22	17
6	46	24	32	29	22	16	52	21	24	19	24	23	51	24	17	8	19
7	51	20	23	16	31	17	52	18	35	33	33	24	56	22	17	8	5
8	63	28	21	18	24							25	58	11	24	21	24
9	49	18	23	22	28							26	46	15	17	12	17
10	57	25	28	29	30							27	53	19	17	15	18

y_0 represents initial weight of rat
y_1 gain in 1st week
y_2 gain in 2nd week
y_3 gain in 3rd week
y_4 gain in 4th week

5.4
Pigment Strength Testing with the Automatic Muller*

G. E. P. BOX, M. T. HOBBS, and F. NORTH

Summary

An investigation of an existing palette method of testing pigments for strength has shown that the variation in the amount of grinding applied represents a potentially large source of error. The recent production in Great Britain of an automatic muller by which a constant level of grinding may be achieved therefore offers possibilities of improving the reproducibility for this test.

Statistically designed comparisons, using a range of pigments, of the reproducibilities of the automatic muller and the palette methods, have shown that the muller in fact offers no general advantage in reproducibility over highly skilled operators. It is not, however, subject to the systematic bias between operators which is liable to occur on the palette; it may also be expected to give superior reproducibility in respect of results given by operators of lower skill. The muller provides, in addition, a useful means of comparing the important property of the rate of strength development of pigments.

Certain factors, for example, heat developed during grinding, which do not affect the palette may in certain circumstances have a profound effect on the muller. Simple preliminary tests on the muller are therefore necessary to ensure the suitability of the particular method of test employed, before applying it to the testing of treatment samples.

INTRODUCTION

There can be few pigment users who have not at one time or another been goaded, by discrepancies in the results given by existing methods of testing strength of pigments, into attempting to devise a new method of test of their own. While much effort has been expended in the direction of improving the accepted methods of testing, the human operator represents a potentially large source of error and many attempts must have been made to design a machine to do this work. For this reason, the production in Great Britain of a pigment testing machine is a matter of considerable interest to pigment users, for although a machine of similar design has been available in the United States for a good many years, and a few models have been imported into this country, the cost factor and import difficulties have alone been sufficient to prevent its general use in Britain, quite apart from the question of whether its adoption would be an improvement on present methods. The stage has now been reached,

*Read before the Manchester Section, 28 November 1952, the London Section 17 February, and the Victorian Section 13 April, 1953.

The Collected Works of George E. P. Box, 1984, Wadsworth, Inc., Belmont, CA 94002. Originally published in *Journal of the Oil and Colour Chemists' Assoc.,* vol. XXXVI, no. 396 (1953), pp. 283–304.

however, when the muller is available for the British pigment industry, if the industry wants it, and the questions present themselves—is the automatic muller generally superior to our present accepted testing methods, and if so, is its superiority sufficiently marked to counterbalance the trouble involved in changing testing methods?

The present paper describes briefly an investigation of the behaviour of an automatic muller and the conclusions to be drawn therefrom. It should be emphasised that the investigation (which is still proceeding) is on a strictly practical basis; its object is to answer the questions—does the muller provide a more reproducible method of testing pigments than an existing palette method, and if so, how much more reproducible? The main criterion is thus reproducibility, and as this can more simply be measured in terms of tinctorial strength, rather than of shade, the investigation is concentrated on the former property. It should be mentioned that due to the British model not yet being freely available, the main work has been carried out on an American machine—the Hoover Automatic Muller. The two machines, while differing in appearance, and to some extent in design, are fundamentally similar, and such direct comparison as has been possible indicates that they are closely similar in behaviour, although the British machine is superior both in appearance and in ease of operation.

MEASUREMENT OF VARIABILITY

Since a direct comparison of the relative reproducibilities of the muller and palette methods is the main object of the investigation, it is first desirable to comment on these existing methods, and the reproducibility to be expected from them.

We are of course dealing with a field whose measurements can never be exactly reproducible; in order to arrive at trustworthy conclusions it is essential therefore to design and analyse the experiments on statistical principles. Excellent accounts of some of the more useful statistical techniques in their application to oil and colour chemistry are to be found in a recent issue of this *Journal* [1, 2, 3] where also are useful lists of further references. The whole of the present work has been treated in this way, the complete programme of work including many separate sets of experiments designed on statistical principles to answer the various questions which arose as the work proceeded. Some examples of the designs employed will be found in what follows.

One essential in any discussion of the accuracy of rival methods of testing is a suitable yard-stick to measure variability of the results. The measure of variability universally adopted in work of this sort is the "variance," usually denoted by σ^2. The variance is the mean square deviation—that is, the average of the squared deviations from the true mean. By taking the square root of the variance, we obtain a measure of variability called the "standard deviation," and denoted by σ, which has the same units as the original observations. If, as may safely be assumed with the present data, the distribution of deviations follows approximately the Gaussian or Normal Law, a deviation from the true value greater than the standard deviation would be expected about once in three times, whilst in the remaining two out of three times, the deviation would be less. Similarly, a deviation greater than twice the standard deviation would be expected once in twenty times, and a deviation smaller than this the remaining nineteen out of twenty times.

It will be appreciated that a precise value for the standard deviation can only be obtained from a large number of results; however an estimate of the standard deviation calculated from a moderate number of observations will often have sufficient accuracy for the purposes of the experimenter. Furthermore any two apparently different estimates (for example, for a standard and a modified method of test) may be compared by a statistical method which takes account of their accuracies, and it is a simple matter to determine whether such apparent differences are probably real, or could be easily due to chance.

Differences unlikely to be due to chance are said to be "statistically significant" or where the context is unambiguous simply "significant." Statistical significance is judged by calculating the probability of chance occurrence of differences as large or larger than that observed. If this probability is greater than 10 per cent. the effect is said to be "not significant" (denoted in Table II by a dot), if the probability is between 10 per cent. and 5 per cent. the effect should be regarded as suspiciously large, but not of itself large enough compared with the size of the error to reach a definite conclusion or "possibly significant" (denoted by ?). When the probability is between 5 per cent. and 1 per cent. the effect is conventionally regarded as being so large compared with the size of the error as to be unlikely to be due to chance and is said to be "significant" (denoted by an asterisk). When the probability is less than 1 per cent. the unlikeliness of chance errors is even greater and the effect is "highly significant" (two asterisks).

The Palette Method of Pigment Strength Testing

The present British Standard testing method for pigments in reduced shade and strength specifies the rubbing of given weights of the pigment sample in oil for a specified time, with subsequent reduction of the paste with given weights of zinc oxide and oil. Opportunities for the accuracy of this method to be reliably investigated rarely occur in normal laboratory working and perhaps for this reason there is room for differences of opinion regarding its accuracy. But in general it is regarded as capable of giving reproducible results, such discrepancies as are apparent normally being assigned to faulty working by the operator. It is accepted, however, that with a few pigments discrepancies are more liable to occur than with the majority.

It is instructive, therefore, to investigate the accuracy of the palette method over a range of pigments, by carrying out replicate tests on a statistical design, and subjecting the results to statistical analysis. Using a range of nine pigments, selected to cover the main chemical types, the overall reproducibility of the British Standard method has been estimated, together with the effect of certain variable factors—namely the operator rubbing, and the number of rubs given to the colour alone. (The measure of grinding given to the colour was selected as the number of rubs, rather than the time of rubbing as the former was considered to be more easily reproducible, but apart from this the B.S. method was strictly adhered to.)

TABLE I

FACTORIAL DESIGN OF PALETTE INVESTIGATION

	100	200	300	400
A	12	16	10	3
B	5	9	14	8
C	6	1	7	4
D	2	11	15	13

Number of Rubs	100, 200, 300, 400
Four Operators	A, B, C, D
Order of Rubbings	1, 2, 3, ..., 16

The factorial design of the experiment is illustrated in Table I. In this experiment four operators, A, B, C and D, each carried out four grindings on each pigment, of 100, 200, 300 and 400 rubs respectively. The sixteen grindings were performed in random order, as indicated, and the paints obtained from each grinding were assessed against an arbitrary standard paint. It should be noted that, in this respect, the experiment is not in accordance with normal practice, where the object is more often a relative strength assessment of one sample against another, given identical treatment; the effects which are given in Table II should be observed with this fact in mind.

TABLE II

RESULTS OF PALETTE INVESTIGATION

Source of Variation	Pigment Type								
	Monolite Fast Red 2GS	Monolite Red 4RHS	Toluidine Red	Rubine Toner	Lemon Chrome	Orange Chrome	Prussian Blue	Phthalocyanine Blue B	Turkey Red Oxide
Number of rubs 100, 200, 300, 400	**	?	*	.	.	**	.	?	.
Operator A, B, C, D	?	?	**	.	.	?	.	?	.
Estimated standard deviation	4.8	2.1	2.4	4.1	3.2	3.6	2.9	3.8	5.4

Levels of Significance	Indicated By
Not significant	.
Possibly significant . .	?
Significant	*
Highly significant . .	**

The main conclusions are first that there is wide variation in the effects of the rubbing factor and the operator factor from one pigment to another. Second, that in those pigments where the rubbing factor is significant, the operator factor also tends to be significant; where variation in the number of rubs given is not significant, however, as with the Rubine Toner and Lemon Chrome for example, the operator factor is also not significant.

The design allows not only the effects of the factors (operators and number of rubs) to be assessed, but also an estimate to be made of the residual variation that remains after the mean effects of the factors has been allowed for. The residual variation will be accounted for partly by experimental error, and partly by interaction effects between the factors. Interaction between factors would imply that the effect of increasing the number of rubs was not the same for different operators. The quantity "estimated Standard Deviation of method" shown in Table II is calculated for this residual variation, and therefore provides in reality an upper limit for the standard deviation: any interaction effects will have tended to inflate it.

Effect of Comparative Strength Development on a Strength Assessment

Consideration of the comparative strength development of two samples of a pigment indicates that interactions of the sort mentioned above may well occur.

It is generally recognised that the strength of one pigment sample relative to another is not an absolute figure, but may depend upon the "level of grind" at which the tests are performed. Usually the strength of a pigment increases with increased grinding, up to a limiting value, and it is possible, in a rough and ready way, to plot this relationship to give development curves of the type illustrated in Fig. 1. It is clear that, taking Sample A as standard, the relative strength of Sample B will be the same, within wide limits of grinding applied. The relative strength of Sample C, however, will vary appreciably, being apparently weaker if the test is carried out at the lower level of grinding represented by X, although apparently equal if tested at the higher level represented by Y. Further, small variations in the grinding applied will, in the case of both samples, be reflected in greater variations of the individual strength when tested at X then at Y.

Considering the many potential variations affecting the level of grinding, which exist in palette testing, one would anticipate that strength assessments carried out in one laboratory might differ systematically from those carried out in another laboratory, which would use different operators and probably a somewhat different method of test. In practice, however, a fair measure of agreement is found. This can probably be ascribed to the fact that the majority of pigments on the market today possess sufficiently good dispersibility in oil for their rates of strength development to have reached their limiting values at low levels of grinding, so that variations in grinding applied exert a relatively small effect on the strength. Possibly the measure of agreement even in these cases would not bear very close examination—in general there is a tendency to over-rate the reproducibility of existing testing methods—but broadly speaking the results are satisfactory.

There is, however, an appreciable minority of pigments which do not possess the desirable property of good dispersibility; although the products of different manufactures will vary, some showing a quicker strength development than

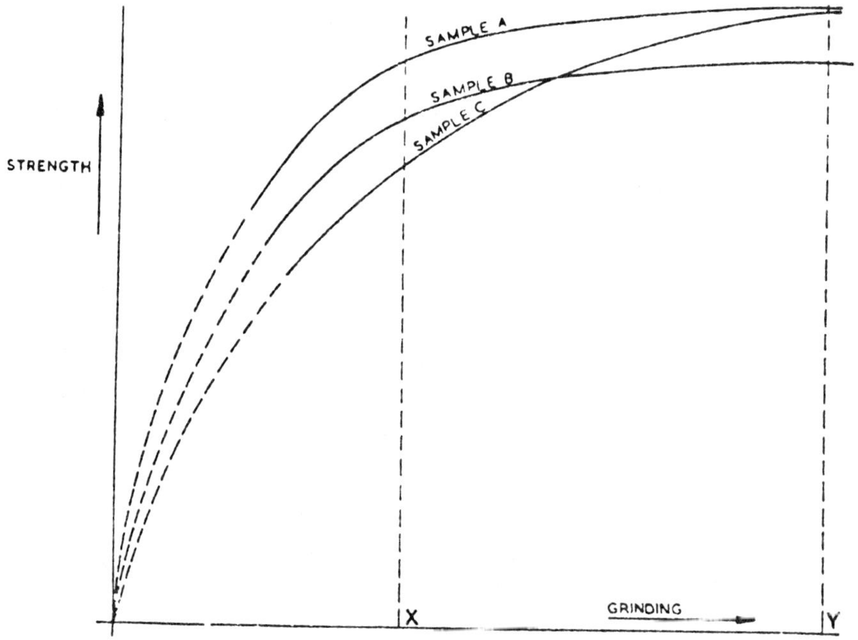

FIG. 1
CHARACTERISTIC COLOUR STRENGTH DEVELOPMENT CURVES

others, a distinct class of pigments having poor dispersibility can be broadly distinguished. Examples of these are the Chrome Greens, and Phthalocyanine Blues. In such cases variations in the level of grinding at which two samples were tested would be expected to reflect the variations of strength assessments made by different operators, and particularly between different laboratories, using different methods of test. It would clearly be desirable in such cases to specify the level of grinding at which the test is carried out, if this were practicable.

With a palette testing method, it would not appear to be possible to specify the level of grinding, however closely the conditions of test—that is to say, quantities taken, number of rubs, etc.—are specified, since fundamentally the amount of grinding applied depends on the operator. The automatic muller, on the other hand, has at first sight great advantages in that its use should eliminate the error found in the palette method arising from differences in the grinding applied in different laboratories by different operators.

THE AUTOMATIC MULLER

The automatic muller consists essentially of two steel plates, fitted with detachable glass grinding surfaces, of which the upper is fixed in a horizontal plane, and the lower rotated by a constant speed electric motor controlled by a tripswitch. The upper plate may be swung upwards through approximately 120°

being opened for charging the machine and closed by a spring-catch during grinding. A varying weight of approximately 150, 100, 50 or zero lb. can be applied to the upper plate by detachable weights working through a lever system, the weights being applied, and the plates brought into close contact, by an operating lever. A revolution counter is fitted to the machine which may be set to 100, 75, 50 or 25 rev.: provision is also made for continuous running if desired.

The general method of grinding is thus to place the measured amounts of pigment and medium on the lower plate, and lightly mix with a palette knife to wet out the pigment. The weights and revolutions are set as desired, the upper plate closed and locked, and the operating lever raised. On switching on the motor, grinding is carried on for the pre-set number of revolutions, when the motor is automatically switched off. The operating lever is lowered, to take off the weight and to separate the plates. After raising the upper plate, the mix may be picked up and collected on a palette knife.

It should be noted that the grinding applied is not constant over the area of these plates, being theoretically zero at the centre, and increasing to a maximum at the periphery. For this reason it is advisable to carry out a grinding in at least two stages, with a thorough mixing of the paint between each stage. For example, in a grinding of 400 revolutions an efficient grinding is obtained by giving four stages of 100 revolutions each, picking up the paint and returning to the centre of the plate between each stage. It is also conceivable that the position of the bulk of the mix on the plate prior to grinding might be a source of error, due to the inconstant grinding force applied over the plates. While no specific information is available on this point, it is a precaution to spread the mix in an approximate 4-in. circle round the centre of the plates, at the beginning of each stage of the grind.

It will be seen that the muller appears to be capable of grinding to a reproducible level, irrespective of the operator. Further, and this is a point to which reference will be made later, the grinding applied may be varied over wide limits to give any practical level desired.

The Effect of Heat Developed During Grinding

In the course of the initial investigation of the behaviour of the automatic muller, it became clear that the machine is, as might be expected, affected by certain factors which do not apparently affect the palette. While the nature and practical effect of such factors are so far by no means fully understood, one example, namely the heat developed during grinding, will be considered.

It is customary in comparing the strengths of one or more pigment samples with that of a standard to rub out the standard and samples in succession, the time interval between each rubbing being confined to that necessary for cleaning the grinding surface, or weighing out fresh quantities of colour and reducing white, etc. In what follows, the complete operation of grinding one or more samples and the standard is referred to as a "run," it being understood that the separate rubbings within each run are carried out successively with the minimum time interval between them. In carrying out such a run on the muller, there is a gradual rise in temperature of the glass plates: in a run of four successive grindings, for example, the initial plate temperature of the fourth grinding may be of the order of 10 or 15° C. above that of the first grinding. If the colour strength of the mix is affected by temperature differences of this order, it is clear that the apparent strength of a sample will depend on its order of grinding relative to the standard.

The susceptibility of pigments in this respect depends to an extent on the conditions of test used, and the majority of pigments tested in the present investigation are not affected under the testing conditions used. But an example of a pigment which may in certain circumstances be affected in this way is illustrated in Fig. 2, which shows a correlation between the initial plate tem-

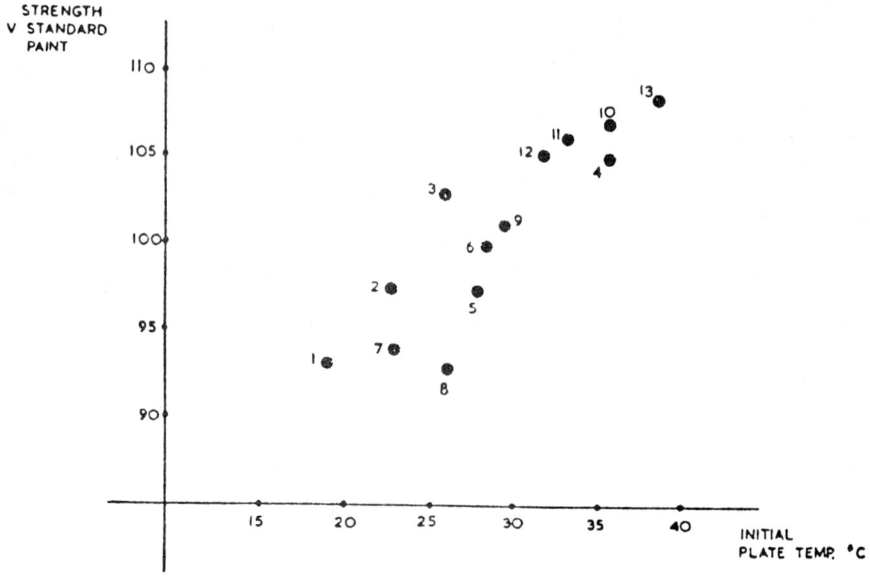

FIG. 2

TOLUIDINE RED: RELATION OF STRENGTH TO PLATE TEMPERATURE

perature and the strengths of 13 lithopone/linseed oil reductions of a sample of Toluidine Red. The assessments of the paints were made in this experiment against an arbitrary standard, and expressed in terms of percentages—so many parts of the pigment sample being equivalent to 100 parts of the standard, so that weakness of the sample is expressed by a figure greater than 100, and greater strength by a figure less than 100. In this experiment, the normal rise in temperature of the plates was artificially altered in certain cases by a current of hot or cold air, being increased between grindings 3 and 4, 9 and 10, and 12 and 13, and decreased between grindings 6 and 7. The correlation apparent from inspection of Fig. 2 is supported by mathematical treatment, although it is not suggested that the variation in strength of the paints is necessarily due solely to the variations in the plate temperature.

Whatever the ultimate cause, it will be seen that the practical effect in a run of say four grindings of samples of this pigment under these conditions of test is to produce a successive weakness superimposed on the inherent strengths of the samples, and reproducible results would not be expected (unless the initial plate temperatures and the order of grinding of the samples were specified). It is advisable therefore, before applying a particular method of muller test to a new type of pigment, to carry out a preliminary test to determine whether the pigment is susceptible to this or other factors. A test which has provided reliable results consists of making a number of successive grindings of the same sample of pigment, the paint produced from the first grinding being used as a standard of comparison for those of the remainder. Fig. 3 illustrates

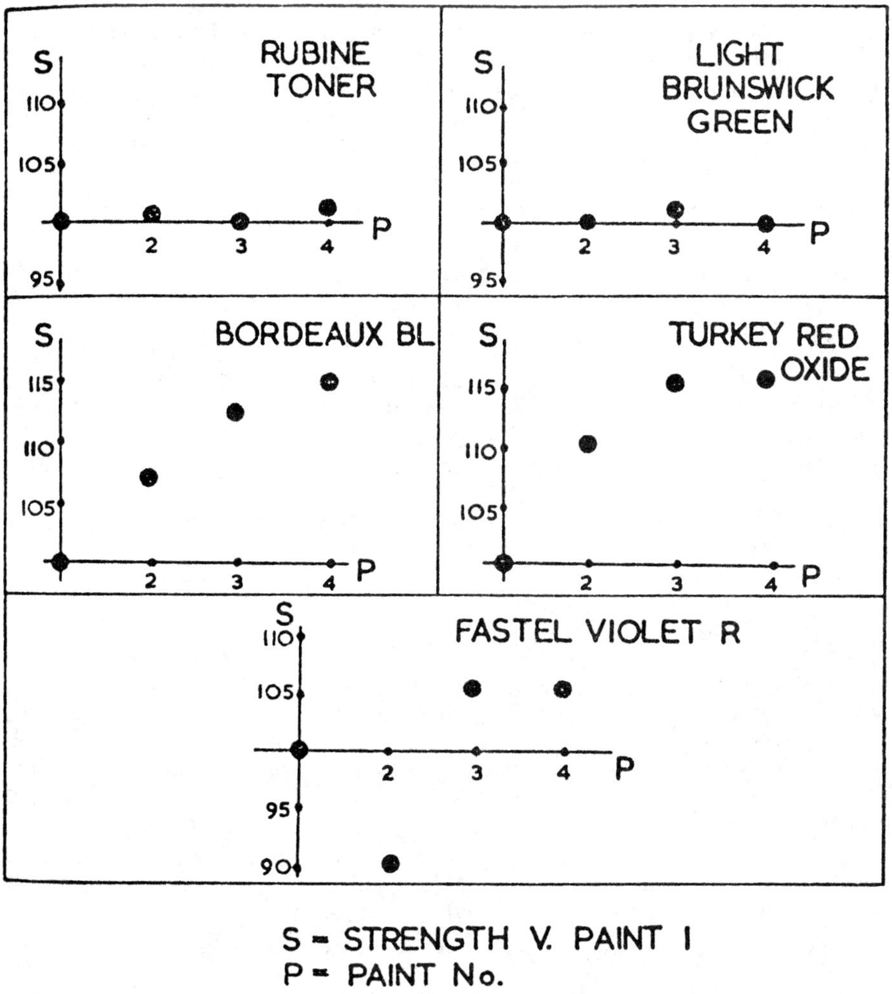

FIG. 3
TYPICAL RESULTS OF THE PRELIMINARY INVESTIGATION

the type of results obtained from lithopone linseed oil reductions. In the majority of cases, for example the Rubine Toner and Light Brunswick Green, a series of paints of approximately equal strength are obtained; in certain cases, however, such as Bordeaux BL and the Turkey Red Oxide, a successive decrease in strength is found, and occasionally, as with the Fastel type, a collection of points indicating large but random variation is found. In neither of the latter two cases are reproducible results to be expected on the muller, using this particular method of test—although reproducible results can be obtained with other methods, for example by using zinc oxide in place of lithopone.

It should be noted that this test does not attempt to assign a cause to any variation found; for example a change of strength on standing before assessment would exert an effect indistinguishable from one caused by susceptibility to changes in plate temperature. Also the test is not in itself a measure of reproducibility under normal conditions, since it involves a comparison of a pigment sample with itself; under normal conditions, a comparison is made of one sample with another.

Side-by-Side Comparisons of Reproducibility

As the primary object of the work described is to determine the relative reproducibilities under working conditions of the automatic muller and the palette, side-by-side comparisons have been made of the two methods by carrying out a number of replicate sample v. standard tests with each. In view of the wide variation in the reproducibility of the palette method between pigment and pigment, these side-by-side comparisons were carried out on a range of fifteen pigments covering the main chemical types.

The method of test used for the muller is as follows: given weights of colour and lithopone (for pure organic pigments 0.04 g. and 1.60 g. respectively) are lightly mixed on the lower plate with a given volume of oil. The paste is spread to a 4-in. circle round the centre of the plate, and ground under 150 lb. pressure for that number of revolutions necessary to obtain maximum strength development of the standard. The paste is mixed and re-spread at the halfway stage of the grinding.

The above method is the result of a programme of experiments which for lack of space cannot be described in detail here. Considerations which influenced the choice of this suitable method were briefly: First, ease of operation—which led to the adoption of a method in which pigment and white are ground together, in preference to grinding a full shade paint for subsequent reduction. Second, the level of grinding, which is affected by the type of white pigment and medium used, and the pigment-medium ratio, as well as the weight and number of revolutions applied.

The method used for the palette was the British Standard method, except that a specified number of rubs (200) was used to measure the amount of grinding applied, instead of a specified time of rubbing.

The design of the experiment as applied to each pigment is given in Table III. It will be noted that each run consisted of grindings of a "standard" and three samples in random order, the samples being in every case of unstandardised batch material; to minimise errors of assessment, however, samples showing

TABLE III

DESIGN FOR PALETTE/MULLER COMPARISON

Run No.	Operator A		Operator B	
	Muller	Palette	Muller	Palette
1	S, T_1, T_2, T_3			T_1, S, T_3, T_2
2	T_2, T_3, S, T_1			T_3, T_2, T_1, S
3		T_1, S, T_3, T_2	S, T_1, T_2, T_3	
4		T_3, T_2, S, T_1	T_2, T_3, S, T_1	
5	T_2, S, T_3, T_1			T_3, T_1, T_2, S
6		S, T_2, T_1, T_3	T_1, T_3, S, T_2	

S = Standard: T_1, T_2, T_3 = Samples 1, 2, and 3

large strength difference (that is over 10 per cent.) were not tested, and for this reason, in certain cases two samples were tested instead of three, the number of runs being increased to four. Furthermore, while one operator carried out a run on the muller, the other simultaneously made a similar run on the machine. The paints produced by each grinding were applied to glass slides with a bevel applicator at the conclusion of each run, and each sample was assessed for strength against the "standard" by three independent observers, the mean of these three assessments being taken as the strength.

It is obviously not possible to illustrate the detailed results of each of the fifteen pigments, but a typical set of results—namely of a Rubine Toner—is given in Table IV to illustrate the nature of the results obtained. The assessments of the samples against the "standard" were made as before in terms of percentages; in this case, however, for the sake of clarity and for ease of computation, the assessments recorded represent the true assessments minus 90. To

TABLE IV

TYPICAL RESULTS OF COMPARISON BETWEEN PALETTE AND MULLER

Rubine Toner	Operator A						Operator B					
	Machine			Palette			Machine			Palette		
	T_1	T_2	T_3	T_1	T_2	T_3	T_1	T_2	T_3	T_1	T_2	T_3
	5.8	5.0	11.7	7.5	5.8	8.3	5.0	4.2	8.3	5.0	4.2	7.5
	5.8	0	10.0	5.8	5.8	8.3	6.7	2.5	10.0	6.7	5.8	12.5
	3.3	4.2	10.0	9.2	7.5	7.5	9.2	7.5	12.5	1.7	1.7	5.0
Mean	5.0	3.1	10.6	7.5	6.4	8.0	7.0	4.7	10.3	4.5	3.9	8.3

Results recorded above are assessed strength minus 90.

obtain reliable estimates a statistical analysis is required, and since each sample is tested against the same "standard" in each group, it is necessary to allow for the intra-class correlation between results.

The results of such analyses are given for all the pigments tested in Table V, where the estimated standard deviations of each method applied to each pigment are given. It will be noted that these estimates are based on 12 degrees of freedom, and a ratio of S.D.s of 1.64 is required to establish that one method is significantly more accurate than the other. In the case of smaller ratios, as are found in the first group, small differences in the standard deviations of the two methods have no significance. The estimates are based in each case on results obtained *within* operators; there is no evidence indicating a consistent superiority of one operator over another. " Within operator error " refers to the variation in the results of the same operator performing a number of repeat tests; the values thus calculated, averaged over the two operators, are given in the table.

TABLE V

COMPARATIVE REPRODUCIBILITY OF PALETTE AND MULLER

Pigment Type	Estimated Standard Deviation	
	Palette	Muller
Monolite Red 4RHS	1.7	1.6
Rubine Toner	1.3	1.5
Lemon Chrome	1.5	1.1
Middle Chrome	1.4	1.0
Scarlet Chrome	2.4	2.0
Light Brunswick Green	1.4	2.1
Prussian Blue	2.5	1.9
Monastral Fast Blue LBS	2.3	1.7
Monastral Fast Green BS (Lake)	2.9	1.9
Monastral Fast Green GS	2.5	3.1
Pigment Green B	1.2	1.6
Monastral Fast Blue BS	2.7	1.0
Yellow Oxide 5GS	1.2	0.7
Monastral Fast Blue GS	1.3	4.1
Monastral Fast Blue LBS (Lake)	2.1	4.8

Degrees of Freedom = 12

Ratio of Standard Deviations required for significance at 10 per cent. level = 1.64

It will be seen that there is a wide variation in the reproducibility of each method, according to the pigment under test; and in general the pigments may be grouped as follows:—

(1) Eleven pigments—for example all the chromes, the Prussian Blue, the Monolite Red 4RHS and others, where there is no significant difference in reproducibility.

(2) Two pigments—Monastral Fast Blue BS Pdr. and a hydrated Iron Oxide where the muller is significantly more reproducible than the palette.

(3) Two pigments—Monastral Fast Blue GS, and a reduced lake of Monastral Fast Blue LBS where the muller is significantly less reproducible than the palette.

It is, however, pertinent to consider the question of operators, since although the experimental plan was designed to reproduce practical conditions of working, the operators concerned were more highly skilled than others might be. While the degree of skill of an operator is so individual a factor as to make an estimate of its practical effect a matter for speculation, it is to be expected that it will be a more critical factor on the palette than on the muller. All the sources of operator error (weighing, pre-mixing, application) that exist on the muller are common to the palette, but the palette has the additional source of error in variation of the grinding. It is justifiable to assume, therefore, that any difference in reproducibility of the two methods using skilled operators will be magnified, in favour of the machine, when less skilled operators are used. An indication of this is given in Table VI illustrating the result of a comparison between a skilled and an unskilled operator, although since this relates to one pigment only, and a poorly dispersible one at that, it is not considered to be conclusive.

TABLE VI

REPRODUCIBILITY OF SKILLED AND UNSKILLED OPERATORS

	Machine	Palette
Skilled Operator	2.2	2.0
Unskilled Operator	2.5	5.5

Each standard deviation based on 6 degrees of freedom:
Ratio required for significance at 5 per cent. level = 2.41

A second point to be borne in mind is that the estimates of reproducibility were based in each case on the results obtained *within* operators. For the same reasons as above, even though operators be comparing these results on a test material with a standard prepared by themselves, extension of the experiment to cover results obtained between operators would be expected to magnify the differences in reproducibility obtained, in favour of the machine. This point is important, since although reproducibility within operators is necessary, it is not an end in itself, since a satisfactory testing method must also be reproducible both between operators (including operators from different laboratories), and also between the laboratory and the works scale.

Analysis of the results obtained from this point of view is therefore of interest. Of the fifteen experiments in which the palette and muller were compared, ten were carried out by the same pair of operators, and by taking the difference between the mean results obtained by Operator A and Operator B, for each of the ten pigments, the systematic operator differences of the muller and palette may be compared. Fig. 4 shows these differences for each of the ten pigments. Further analysis shows first that the spread of results is much the same in both cases—Standard Deviation for the machine and palette being 1.40 and 1.20 respectively. Second, that while in the case of the muller the bias between operators is 0.35, which is not significantly different from zero, that on the palette is 1.26 which is highly significant. This indicates that on the palette Operator B tends to obtain a systematically lower value than Operator A over a wide range of pigments even though his sample is compared against a standard prepared by himself.

In general, therefore, it is concluded that while the automatic muller offers no advantage in reproducibility over a palette method operated by a single highly-skilled operator, it does offer a significant advantage when less highly-skilled operators are employed. Furthermore even with highly-skilled operators the palette method is significantly more subject to bias between operators than the machine.

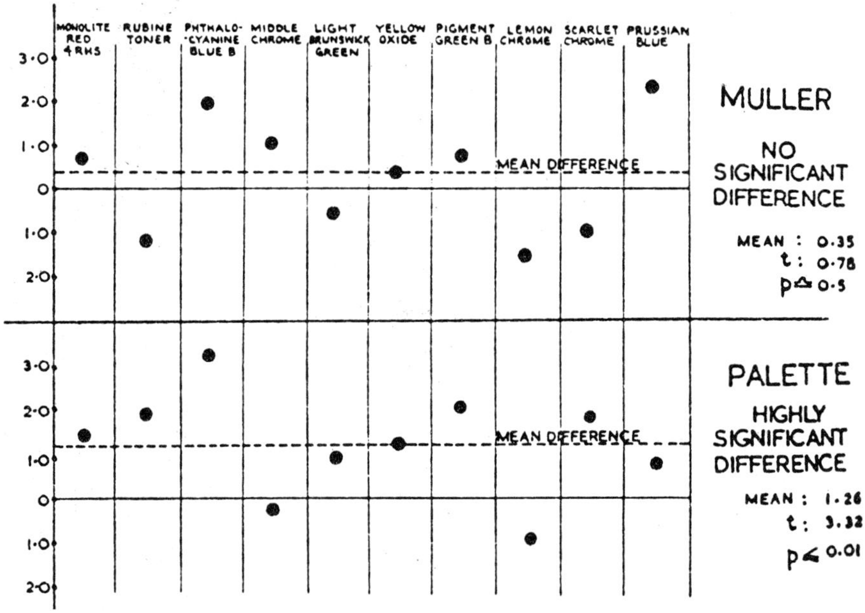

FIG. 4

DIFFERENCES BETWEEN OPERATORS (A—B)

Comparison of Pigment Strength Developments

Our remarks have hitherto been confined to the testing of pigments in reduced shade and strength. Before concluding it would not be out of place to mention briefly one other test for which the muller would also appear to offer possibilities —namely dispersibility in oil.

We have previously mentioned the close connection between the strength assessment of one sample against another of the same pigment type, and the shape of the development curve for the two samples. It has been pointed out that where the shapes of the curves differ, the strength assessment depends to a large degree on the level of grinding given in the test. For this reason, it is desirable when carrying out a straightforward test for strength to have some idea of the development curves. The determination of the characteristics of the curve is of considerable importance for in general a pigment that attains maximum strength quickly is more desirable than one which requires additional grinding to obtain the same strength from it. Indeed where, in extreme cases, prolonged grinding is required to obtain maximum strength, it is often more economical to forgo a certain amount of strength and save the additional grinding costs.

A palette test for strength is apt to give misleading results in this connection, because while the level of grinding may be artificially controlled—for example by altering the pigment-oil ratio—the operator can only grind to his own maximum strength, and not to any definable lower level; the effect of this is that the level of grinding tends to be high. A high level of grinding in the test may be suitable if the level of the user's works process is equally high; but it must not be forgotten that what is wanted in a pigment is strength at a low level of grinding—as given by good dispersibility, and a test that judges a pigment on its strength on prolonged grinding alone is setting up a false criterion. For example, referring to Fig. 2, a test carried out at a high level of grinding would indicate that Sample A was no better than Sample C in respect of strength; in fact, other things being equal, A would be the more desirable pigment, since the same strength as C could be obtained with considerably less grinding. This is particularly noteworthy by pigment manufacturers, who are concerned with the development of improved quality pigments. A reliable test for strength development in oil is therefore highly desirable, and though tests for this property are in use their value is limited. Tests exist of two main types:—

(1) in which pigments are stirred into white pastes—any form of grinding being carefully avoided,

(2) in which pigments are ground at various levels of grinding.

The objection to the first method is that, while it may act as a guide to comparative dispersibilities, it is an unreal test, in so far as, in practice, pigments are seldom dispersed by stirring, some form of grinding almost invariably taking place. The second method is therefore more generally favoured, and some form of test may be applied on the palette. The reliability of such a test is open to question, however, particularly when it is realised that the most interesting case—that is when there *is* a difference in the developments of the two samples —provides just those conditions under which the palette method is found to be least reliable. It would not be expected, therefore, that the reproducibility of such a method between operators would be very high.

The muller on the other hand would appear to be much more promising in this direction, and a suitable method of test might involve the grinding of a standard mix in stages; say 25 rev. at 50 lb., a further 25 rev. at 50 lb., and so on through two or three further stages at 100 and 150 lb., taking a sample of the paint at each stage. Reliable information concerning the reproducibility of such comparisons is not at present available, although there are indications that it may form the basis of a useful test; it certainly forms a quick and simple indication whether wide discrepancies between strength assessments carried out between different operators are due to differing strength developments of two samples or to some other cause.

One obvious source of error exists in such a method, however, in that while it is true to say that the grinding applied by the muller is independent of the operator, it must be remembered that prior to grinding, the pigment must be wetted out by the operator, and a certain variable force is applied to the pigment in this operation. This is normally small and negligible compared to the total force exerted in a full grind; but if grinding is done at a very low level, the work done in wetting out may form a disproportionate amount of the whole. In devising a method of test for comparing strength developments, therefore, any tendency to assess at too low a level of grinding must be resisted.

CONCLUSIONS

The conclusions to be drawn from the present investigation on the automatic muller are as follows:—

(1) The reproducibility of the muller depends, as does the palette, largely on the type of pigment tested; the results of the investigation show that the muller is in general of equivalent accuracy to the palette over a range of typical pigment types.

(2) It is emphasised that this conclusion refers to the palette method used by a single skilled operator. For unskilled operators using a standard method of test it would be expected that the reproducibility of the muller would be less adversely affected than that of the palette, since all the known variable factors (e.g. weighing) which affect the muller, also affect the palette, but the main variable factor affecting the palette—i.e. grinding—does not affect the muller. The same considerations apply in regard to reproducibility between operators. Experimental evidence supports these expectations.

(3) The muller appears to be a promising means of comparing development of strength in oil, which is difficult to compare reliably on the palette.

(4) Although there is no great difference in the times taken by the two methods, the muller causes less fatigue than does the palette, which may possibly affect the accuracy of routine testing where large numbers of samples must be tested.

(5) It must be appreciated, however, that there is a possibility of certain factors existing which do not affect the palette to the same extent as they affect the muller; an example of this is the heat factor, which is liable to give misleading results with certain types of pigment. For this

reason it is most desirable to carry out certain preliminary tests before accepting the results given by the muller as absolutely reliable.

It should not be supposed that an extensive programme of preliminary work is necessary before the muller may be usefully employed for testing a particular range of pigments. However, the small effort required by a short preliminary investigation will be more than justified by the greater confidence that can be placed in the subsequent testing. In short, it is considered that the automatic muller, if treated with the caution that all new methods deserve, is a useful testing instrument, even at the present stage; its usefulness may well be extended even more in the future, as more experience is gained with it.

References

1. Lord, A., *J.O.C.C.A.*, 1952, 35, 437.
2. Saunders, B., *J.O.C.C.A.*, 1952, 35, 448.
3. Bainbridge, J., *J.O.C.C.A.*, 1952, 35, 459.

Imperial Chemical Industries Ltd.,
Dyehouse Laboratories,
Hexagon House,
Manchester, 9.

Discussions

Manchester Section

Dr. M. E. D. Jarrett, opening the discussion, said it was commonly held that there was a parallelism between tinting strength and hiding power. While this might often be true, would it be correct to say that it was not a fundamental and invariable relationship? Taking just two physical properties of pigments, he would expect tinting strength and hiding power to vary similarly with light absorption, but in opposite directions with changes in refractive index.

Mr. North said that such a question was most difficult to answer with any degree of certainty. However, his own opinion was that there was no fundamental relationship between tinting strength and hiding power. The fundamental properties of pigments which controlled these technical properties were imperfectly understood, and few quantitative relationships existed.

Mr. H. Gosling commented that, in one of the tables, Turkey Red Oxide indicated no significant variation between the number of rubs or between different operators, yet showed the greater figure of deviation. Elsewhere Phthalocyanine Blue B and Phthalocyanine Blue B Lake gave no significant difference between muller and palette and yet showed considerable variation. What was the explanation?

Dr. Box explained that the significance of the variable factors was assessed by comparison with the residual variance, the latter being a measure of the experimental error of the method. In the case of the Turkey Red Oxide, the effects of the two variable factors might well be of the same order as those for the other pigment types; they were not significant, however, compared with the relatively large experimental error, which arose from factors other than the two variables discussed.

Mr. Hobbs said that two types of Phthalocyanine Blue B had been tested, namely Monastral Fast Blue BS Powder, and Monastral Fast Blue LBS Powder. The particular data produced did not distinguish between these two types, but in fact a significant difference was shown by Monastral Fast Blue BS Powder and a lake of Monastral Fast Blue LBS Powder, and no significant difference by Monastral Fast Blue LBS Powder and a lake of Monastral Fast Blue BS Powder.

MR. THOMAS enquired whether the difference in strength was measured visually or by mechanical means.

MR. HOBBS stated that in every case strength assessments were made visually. As had been mentioned, samples were selected for test which did not exceed a 10 per cent. difference in strength from the "standard," and under these conditions the assessment error of the experienced observers used was not statistically significant.

DR. GILLAN referred to the relative behaviour of zinc oxide and lithopone on the muller, and said this applied also when the palette method was used. For example, if Sample B were weaker than Sample A when tested with zinc oxide in oil, and had poorer dispersibility than the latter, then testing with lithopone would often return the result that Sample B was equal in strength to Sample A. The explanation seemed to lie in the fact that lithopone was a more abrasive pigment than zinc oxide, and developed more strength than zinc oxide in pigments which possessed moderate to poor dispersibility.

MR. HOBBS agreed that the type of white pigment used could, in certain cases, have a considerable effect on the strength developed especially where the dispersibilities of the samples under test differed, and that this was particularly noticeable in the case of zinc oxide compared with lithopone.

MR. C. G. BELL asked whether the automatic muller reduced the incidence of the reversal effect which was often met when viewing rubbings from different angles, and if the speakers had any comments to make on the use of the ordinary laboratory ball mill for pigment strength testing.

MR. HOBBS said that he had not personally experienced the effect described on examination either of palette or muller rubbings. He imagined, however, that such an effect would depend on the particular paint system used, and that if the same system were used, the method of rubbing would in itself have little effect.

MR. HOBBS said that the authors were primarily concerned with the development of a routine testing method, and he did not think a ball-mill would be suitable for this. In general he did not think that ball-mills were less likely to be subject to error than palette or muller testing.

DR. JARRETT, commenting on the use by the authors of lithopone as the white pigment, said that when Phthalocyanine Blue was rubbed out in oil with this pigment, the strength developed depended very greatly on the degree to which the mixture was worked. With zinc oxide or white lead, no comparable progressive development of strength was observed. He asked if similar observations had been made with the automatic muller.

MR. HOBBS agreed that Phthalocyanine Blue belonged to a class of pigments difficult to grind, and that the strength developed depended to a large extent on the amount of rubbing given. This was more noticeable when the pigment was rubbed with lithopone than with zinc oxide, presumably, as Dr. Gillan had said, because lithopone itself exerted a grinding action on the pigment: but he had found that a similar, though smaller, increase of strength also occurred on continued rubbing with zinc oxide. This applied both on the muller and on the palette. He had not tried white lead.

MR. STOREY stated that where comparison of a delivery of a pigment against the standard was being made, grinding out the reduction test to the limit did not necessarily give the information required, because, for instance, in the case of red oxide, standard and delivery might well be made from the same crude material and would, therefore, show the same strength when ground until full staining power has been developed.

Comparing the colours at earlier stages of the grinding with the muller, as carried out by Mr. North, would give a more valuable indication of the comparative fineness or dispersibility of the two samples.

MR. HOBBS agreed; the real difficulty, however, was to decide at just what stage in the grinding to carry out the test. Grinding methods employed by different pigment users varied so widely in efficiency that it was not possible to select one level which would suit all pigment users. It had been the aim of the authors, however, to devise a test, the level of grinding of which would give results comparable to those obtained by efficient triple-roll milling.

London Section

MR. J. GREY said he had come to the conclusion that variation in test results was due not so much to the effect of the temperature on the pigment as to its effect on the viscosity of the oil and varnish used. With that in mind he asked if there were any reason why metal plates, which would conduct the heat away better, should not be used in the muller instead of glass plates. Glass plates tended to wear and slip, which also affected reproducibility. He suggested that stainless steel plates would keep clean better and probably wear less than glass; it would also be an advantage to use an oil with a very small coefficient of viscosity change with temperature.

The only fault he had found in the handling of the automatic muller was that the brass bezels tended to jam tight, making it very difficult to remove the plates for cleaning and sometimes difficult to bed down the glass plates properly.

MR. NORTH said he thought that the change of viscosity of the oil with temperature was not the whole answer. Although there seemed to be a heat effect, it was not so much on the pigment itself, but on the system as a whole. The authors had found that the easiest way out of the difficulty was to use as much material as possible on the muller, at the highest possible consistency. Difficulties did not then arise, but he agreed that the use of an oil having a very low coefficient of viscosity change with temperature would be advantageous.

The authors had tried using metal plates on the muller, but had not had much success with them. It was difficult to obtain very hard steel plates with perfectly flat surfaces; the plates also became dirty and were extremely difficult to clean.

The authors had investigated the degree of reproducibility with varying condition of the plates. They had used well-matched brand new plates, plates which were specially smoothed, others that did not fit so well, and so on, but they had not been able to find any significant differences as a result. They too had had difficulty with the bezel binding. They cured it by having the threads re-cut and fitting a paper gasket under the glass plate.

MR. J. A. L. HAWKEY said that, in testing the strength of a pigment, it was important to use the medium which would be employed on the manufacturing scale. The use of linseed oil could lead to inaccuracy where other media were used in the manufacturing processes. He thought that small deviations in strength of the order of 5 per cent. were insignificant and counselled against securing strength at the sacrifice of more important properties. He particularly condemned the grinding in linseed oil of a mixture of small amounts of coloured pigment with large amounts of white pigment for he believed that, in comparing two pigment samples, it was possible to get one result by this method and find that the relationship was reversed in another system.

MR. NORTH agreed, and said that linseed oil was used because it was convenient. But he pleaded for a practical outlook on the colour strength assessment of pigments. In their work the authors had been striving to provide a basis for an agreed testing method. It was unlikely that it could ever cope with all the variations that existed in practice, but it would nevertheless be an advantage to set down a method which was reproducible within itself on a machine of the kind discussed in the paper.

Mr. J. D. Cohen asked Mr. North to give more details concerning the type of persons, described as "unskilled" or "semi-skilled," employed on the work and the extent of their experience. Inasmuch as they had to weigh quantities of pigment as small as 0.04 g., they must have acquired sufficient skill to be able to use an analytical balance, and presumably they had acquired corresponding skill on the plate.

Mr. North said that in the work he had described it was not so much a matter of skill in the use of a palette knife, but a matter of brawn. At his works about fifty people were continuously employed on rubbing pigments in oil and although in the main the results were reasonably good, from time to time they varied very much. For example, after 4 p.m. there might be a sharp decline in both reproducibility and output and one could not wonder at it, for the operatives had been working hard all day. The skilled man was the sort who had started there as a boy with a certain amount of technical qualification, knew the pigments, and had considerably above the normal level of intelligence for that sort of job.

The qualities required were not so much skill and experience, but rather inherent reliability, a willingness to do a good job all day and every day, and the strength of character to be able to avoid being upset by circumstances.

Dr. R. H. Leach said that he doubted whether the scheme put forward by Mr. North would result in colour agreement between the raw materials supplier and the user and he disagreed with Mr. North when he suggested that it was not necessary to grind pigments to a fine dispersion. Mr. North had pointed out that the muller did not produce fine dispersions, but claimed that it gave reproducible results, although admitting that the degree of grinding obtained was materially affected by manipulation with the palette knife. In order to achieve uniform pigment dispersion, maximum colour strength, working properties and protective properties from the raw materials, ink makers were using machinery of 20-30 h.p. to drive the mills, and therefore one could not expect to achieve that result with the palette knife.

On a few occasions he had found that his own strength results and those of raw materials suppliers had differed by 10-15 per cent. In one case he had received a material which was stated to have the tinctorial strength of the material he had used previously, a claim which he had disputed. It was then discovered that the raw materials supplier used the muller for testing where his organisation was using a three-roll mill. He believed that it was not possible to achieve the dispersion necessary in the printing ink industry and, he believed, in the paint industry by the methods discussed in the paper. He asked whether the muller applied shearing forces as great as those which were used in practice.

Secondly, he believed certain coloured pigments could not be adequately ground in the presence of much extender. He asked if the authors had tried the experiment of grinding the pigment in the oil on the muller and then mixing with a ground white paint and comparing the tinctorial value. Colours could not be compared in a half-ground state, but only in the condition in which they would come off the factory mills.

Mr. North disagreed entirely. The palette knife and the slab of glass, he said, was the most efficient method known for the dispersion of pigments. He would guarantee to take a pigment from any three-roll mill, put it on a palette and get some more strength out of it. In fact, one thought which had influenced him a very great deal was that, when a pigment supplier tested a pigment using an efficient palette method, he was testing at a very unreal level of grinding, a level which Mr. North believed was not achieved in practice. He believed that the muller gave results which were in every way comparable with those obtained when using conventional apparatus such as the three-roll mill.

The authors' work had mainly been concerned with paint pigments, but they had looked into some aspects of printing ink manufacture, and had found that very good dispersions indeed were obtained on the automatic muller. They claimed no great experience of the production capabilities of three-roll mills, but they believed that the levels of grinding which he had been discussing did represent commercial grinding practice. He agreed that with certain pigments different strength results were obtained according to whether they were rubbed together with a white or by rubbing them alone with medium and then bleaching with a white paste; that was one essential difference between the palette method and the muller, but if one wished to do it that way on the muller, it could be done.

Once more he stressed that what the authors were after in their investigation was the simplest, quickest and most reproducible method possible, and they believed they had gone some part of the way towards it. They had a machine on which could be built a generally agreed method or methods, and he considered that both pigment users and suppliers would be very foolish if they did not take every advantage they possibly could of the opportunity offered.

Mr. R. L. Frost agreed with Dr. Leach that the colours obtained when using mills in the works were, for example, in the simple case of white and blue, stronger or deeper than those obtained when the same mixtures were prepared with the manual muller. That could cause trouble in the works if mixtures of pigments to provide a specific colour were worked out on the manual muller in the laboratory.

It seemed to him that the power operated muller might be very useful for this purpose and he asked whether Mr. North had had experience of comparing the results obtained when the same pigment mixtures were prepared on the power operated muller and on paint grinding machinery in the works.

Mr. North said that if the laboratory did not achieve the same strength by hand mulling a given pigment-oil-white composition as was achieved in the works, he could only suggest that the people in the laboratory were not doing their job properly. The British Standard method laid down at the moment was capable of wide interpretation, but the method used in the investigation presented in the paper was the method which definitely rubbed the pigment to its maximum strength.

Dr. R. F. Bowles took the opportunity to place on record some experiments with the original American automatic muller, which he had conducted some years ago in an endeavour to compare results obtained on the muller with identical materials ground on a large three-roll mill. Using bronze blues and Brunswick greens in stand oils, it was found that the three-roll mill developed a higher colour strength from the pigments than did the muller, and that as the pigments were ground more and more, a maximum colour development was approached, but only asymptotically. On the other hand, with the Hoover muller, the maximum colour strength though less than on the mill was obtained very rapidly. It was concluded that the two methods of test were not comparable, but that the automatic muller was capable of giving reproducible experimental results under defined conditions much more satisfactorily than a hand operated muller. Possibly an investigation of the rate of shear/time factors of the muller and the mill might provide some explanation for the difference.

Mr. North said he presumed that Dr. Bowles had used a composition which had been designed in terms of consistency to work well on a three-roll mill, and had used that composition on the muller.

Dr. Bowles agreed.

Mr. North suggested, without knowing the details, that that might give misleading results because the rheological properties of the mix that one put into the muller to obtain optimum grinding were not necessarily those of a mix which would be put on a three-roll mill.

Mr. H. A. Idle confirmed that if one adjusted the consistency of the mix one could obtain practically the same colour strength from the muller as from a three-roll mill. Indeed, under certain conditions the muller might give slightly higher strength.

He was interested in the statistical method of approach adopted by the authors and as one who had always fought shy of tackling the problem in that way, he asked at what stage of the work it was decided to adopt that method.

MR. NORTH replied that his experience with a number of other investigations had confirmed his opinion that, where there were a large number of variables, the only method to adopt was the statistical method. It was not possible to derive the greatest advantage from the statistical approach if one did the experiments first. It was absolutely necessary to put the questions he was required to answer to the statistician before starting the experimental work and, although he might know nothing about the subject being investigated, he would indicate how the work should be done to obtain the answers. The factorial design the authors had shown could contain a very large number of variables, trace the effects of the variables and make an estimate of them, the work involved being about a quarter of that necessary when using the classical (one variable at a time) method and very much greater precision being achieved.

The Chairman said that skilled operators could exercise far greater pressure by hand than could be produced by any mechanical apparatus. The advantage of the automatic muller was that it gave reproducible results and it enabled comparisons to be made by unskilled operators. To prepare pigment systems by hand for test must be a tedious job, as well as being one which led to no great development of technique. He complimented the producers of the automatic muller for their courage and vision in making it available.

Victorian Section

MR. WINTER asked whether the strength of colour was determined by an optical instrument or visually.

MR. WHITE said it was doubtful whether satisfactory absolute measurements were possible. The present investigations were statistically designed so that any significant interaction due to variations in the determination would be detected.

MR. JOHNSON said that using a triple roll mill for evaluation of phthalocyamine green, good correlation had been found with full scale production of printing inks. For general testing, the roller mill technique might not be economically feasible, but in the case of this particular green, the trials had been warranted.

MR. WOINARSKI enquired whether in deriving the graph whereby number of rubs is plotted against strength, an end-point was agreed upon as being the most desirable stage to grind to.

MR. WHITE replied that no special end-point was selected, an arbitrary number of rubs thought to be suitable for the particular pigments in question being applied. Several attempts had been made in the past to analyse pigment strength by optical means. Optical machines give good results for transmission techniques, but on reflecting surfaces results were less favourable, particularly on dark shades.

MR. GOSS remarked that all the measurements described applied to pigments reduced in white paste and asked if comparisons could be made of mass tones.

MR. WHITE answered that no work had been done in that respect.

MR. PETZOLD asked whether the use of wetting agents or other vehicles such as alkyds had been investigated.

MR. WHITE said that wetting agents had taken no part in these investigations, but experiments in another direction had shown that wetting agents did not improve the rub-outs on bad pigments. For easily dispersed pigments some advantage might be gained, but from general experience it appeared that, using alkyds in place of linseed oil, similar rub-outs were obtained.

5.5
Mathematical Statistics and Rubber Technology

G. E. P. BOX
Imperial Chemical Industries Ltd.

Introduction

The connexion between mathematical statistics and rubber technology may not be an obvious one. First of all, what is meant by mathematical statistics? To some of you the word 'statistics' probably conjures up ideas of dull figures concerning such things as exports and imports and fat stock prices. These are statistics, it is true, but the subject we are to discuss is not concerned with these things. Mathematical statistics is the study of variation of natural phenomena. The operative word is variation and with this definition the connexion with rubber becomes clearer; for as all of you will know, the one thing of which the rubber technologist can be absolutely certain is that his material will vary.

Mathematical statistics makes use of probability theory and had its origin in the eighteenth century with writers such as De Moivre and the Bernoullis, who studied games of chance. Later, in the nineteenth century, astronomers who were making measurements that were subject to error studied the problem of combining information from such measurements. Here the great mathematicians Gauss and Laplace made notable contributions.

In this country later in the nineteenth century, the studies of Darwin inspired Sir Francis Galton, Weldon, and Professor Karl Pearson at University College, London, to apply probability theory to problems of biological evolution. Professor Pearson was the originator of mathematical statistics as we know it today. From 1895 onwards courses have been given at University College and since 1915 it has been possible to take an honours degree in Mathematical Statistics at the University of London. More recently similar departments have appeared in universities throughout the world, although this country has continued to take a leading part in new developments.

In the 1920's R. A. Fisher went to Rothamsted Experimental Station as statistician. There, and later as a professor at London and Cambridge Universities, he made fundamental contributions to scientific thought, for as well as solving many complicated and previously intractable mathematical problems in statistics that made the subject of much greater practical value, Fisher showed the basic relation between statistical theory and the planning of experiments, and laid the foundation for modern ideas on this subject. He pointed out that when experiments are conducted on variable material it is not sufficient merely to subject the results to statistical analysis. The statistical considerations must be brought in at the very beginning when the experiment is being planned. Only in this way is it possible to ensure unambiguous conclusions and to gain the maximum possible amount of information from a given amount of effort.

With the time at my disposal all I can hope to do this afternoon is to give you, by means of very simple examples, some preliminary notion of the way in which statistical methods are employed in experimental work.

The Collected Works of George E. P. Box, 1984, Wadsworth, Inc., Belmont, CA 94002.
Originally published in *Imperial Chemical Industries Technical Publication* (1955).

Need for Application of Statistical Methods

The old-fashioned attitude towards experimentation was that in a well-designed experiment, all the factors except the one to be studied should be held constant, then this factor should be varied and the effect noted. When the experiment was repeated the same measured result should be observable each time; some slight experimental error would be tolerated, but this should be small. If the variation were not small, then this was due to somebody's carelessness and the experiment was a bad one. Problems to which these principles could not be applied were presumably out of the scope of scientific investigation. If this line of thought is logically followed, we must exclude from numerical investigation all problems where the experimental error is large compared with the differences in which the experimenter is interested. This would preclude nearly all biological problems and, so far as industry is concerned, most of the chemical and physical ones.

A Rubber Tensile Strength Test

The following are the results obtained when tensile strength tests on apparently identical samples of rubber were carried out:

261 kg./cm.^{2}
235
223
257
248
237

These six tests were performed under identical conditions and the experimental technique was probably as near perfect as possible. None the less the results show considerable variation. The fact is simply that rubber test pieces *do* give variable tensile strengths owing to a multitude of imponderables concerned with the structure of the rubber, etc.

In circumstances such as this, what are we to do ? The attitude adopted to this embarrassing variation differs from one experimenter to another. One school of thought believes in pretending that the variation is not there, taking the average of the results and hoping for the best. It is only fair to warn you that with variation as large as that encountered in most experiments on rubber this attitude is unlikely to be very successful. The intelligent attitude is to accept the fact that in the real world every operation both in the laboratory and out of it is subject to inherent variation, but that a science exists which is specifically concerned with methods whereby experiments may be designed and valid conclusions drawn, even though this variation occurs. As soon as it is recognized that this is so, the realm in which numerical science can be properly applied is enormously extended.

Distributions

For the purpose of checking the machines, tensile strength tests on the same standard rubber mix are done every few days and records of the results from these tests are available over a long period. To show you the sort of results one may get when trying to do the same thing again and again, we have collected together 480 of these results. Now, if I read a list of these values, I should get very tired and you would be convinced that your worst fears concerning statisticians were justified. However, it is possible to summarize the information contained in these 480 separate results in a very simple way. The results vary between 216 and 285 kg./cm.^{2}. Suppose we create 14 classes, viz. 216–220, 221–225, 226–230, etc. We can then go through the results and group them into appropriate classes. In this case we find 9 values between 216 and 220, 18 values between 221 and 225, etc. This is shown diagrammatically in Figure 1(*a*).

FIG. 1

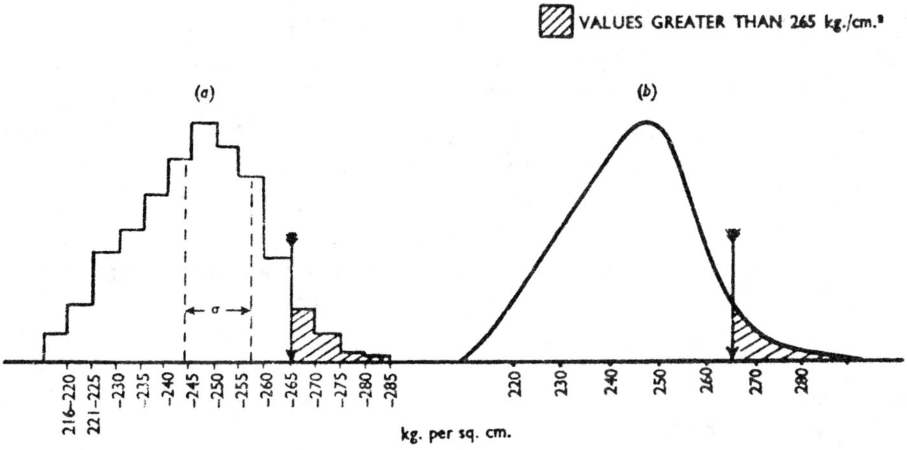

(*a*) (*b*)

Frequency distribution for tensile strength tests

The tensile strength is plotted from left to right and the frequency of occurrence in each class is denoted by the vertical height in each group. The resulting diagram is called a *frequency distribution*. You will see that the diagram has a good deal of regularity and you can imagine that if more and more results were taken with narrower and narrower classes, a smooth curve would eventually be obtained like Figure 1(*b*).

The frequency distribution allows us to see what is the probability that a tensile strength as great as or greater than any particular value will be obtained when the standard mix is tested. For example, 31 out of 480 results are greater than 265. The probability of obtaining a value greater than 265 with the standard mix is therefore about $31/480 \times 100 = 6{\cdot}5\%$. Even this collection of 480 results only represents a small and particular sample of all the results we would get if we went on testing this standard mix indefinitely. If we did this our frequency distribution would approach the smooth curve in Figure 1(*b*), and the ratio of the shaded area on the right of the arrow line to the whole enclosed area would give the exact probability of obtaining a value greater than 265.

A Measure of Centre—The Mean

We have succeeded in summarizing the complete list of 480 results in the frequency distribution of figure 1(a). To summarize the results still further we need constants that will describe this distribution. In the first place, we need a measure of the centre of the distribution. A number of different measures exist and one purpose of statistics is to decide which measure is the most appropriate with different types of distributions. The measure that we shall consider here is the sample average or arithmetic mean obtained by summing all the results and dividing by their number. In this case the average tensile strength is 246·5.

A Measure of Spread—The Variance

On its own, the average can be very misleading. If we told a man from Mars that Englishmen had an average height of 5 ft. 7½ in. he might imagine them to be all exactly that height, or he might imagine them to vary from, say 6 in. to 13 ft. To describe a set of data which vary we need some measure of spread. Perhaps the most obvious measure of spread is the range (i.e. the distance between the largest and smallest of the results). This measure, for reasons that I cannot go into, is not always a satisfactory one. A more useful measure is the mean square deviation, obtained by averaging the squares of the deviations from the mean, and this measure is called the *variance*. Its square root (the root mean square deviation) is called the *standard deviation*, usually denoted by the Greek letter σ (sigma). In this case the variance σ^2 is 167·3 and the standard deviation σ is 12·9.

Analysis of Variance

One reason why the variance σ^2 is so important is because it is additive. Suppose, for example, that we determined the tensile strengths of rubber samples which had originated from many different mixers. The variance of the individual tests would be the testing variance plus the variance between the mixers. If then the testing variance could be calculated separately, by simple subtraction the variance between mixers could be found. This fact is often utilized for finding out which of a series of operations chiefly contributes to variation in a finished product.

Table I

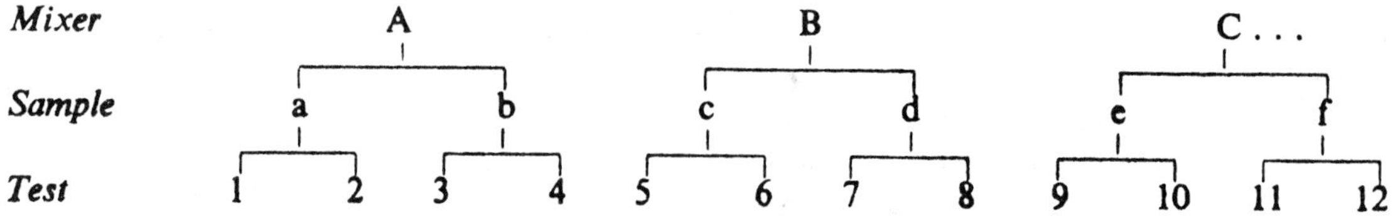

An example of this kind is shown in Table I. In such an experiment a number of samples would be taken from each mixer and a number of tests made on each sample. By considering the variation within samples, that is to say the differences between tests 1 and 2, 3 and 4, 5 and 6, etc., we obtain an estimate of the testing variance alone. We next calculate the sample means, that is to say the means of tests 1 and 2, 3 and 4, 5 and 6, etc. The discrepancies between the means for *samples* a and b, c and d, e and f, etc. are caused partly by sampling variation and partly by testing variation. A separate estimate for the testing variance has already been found; consequently we can find out what the variance due to sampling alone would be. Finally, by comparing mixer means, an estimate of the variance of the mixers is found. In this way the total variance is split up into its component parts and we can see how much of it is due to differences between the mixers, how much to differences between samples (inhomogeneity of the mix) and how much to testing. The same principles can be applied for any number of components. Once we know where the main variations are, we can set about trying to reduce them. For example, if we find variation due to inhomogeneity in the mix we would look to the method of mixing. Apart from indicating at what point mechanical attention is needed, knowledge of the components of variance enables the experimenter to use the most efficient sampling and testing procedures. If, for instance, it were known that large but unavoidable variation occurred between samples, it would be necessary to improve the accuracy by averaging a number of results. In these circumstances, to obtain a representative mean it would be better to take n samples and test each once than take one sample and test it n times. For, in the former method, when we take an average of the n tests the variations due to both sampling and testing are reduced, while in the latter only the testing variation is reduced.

The procedure utilized above is a particular example of an important statistical method having wide application. It is called the Analysis of Variance, which we shall refer to again later. Its name means (as can be seen from the example) exactly what it says. Just as the chemist breaks down a mixture of chemicals into component parts by chemical analysis, so the statistician breaks down a mixture of variations into component parts by statistical analysis.

Reduction in Variation by working with the Mean

A second property of the variance (which arises out of the fact that it is additive) is that the mean of n results has variance one nth of the original variance. Since the standard deviation, which measures the spread of a distribution (see Figure 1), is the square root of the variance, we can see the effect of averaging sets of results. If we average n results the spread is reduced by a factor $\sqrt{n}$. Thus if we take four results and average them, the spread of the average will be one-half of the spread of the individual results. To reduce it to one-third of the original spread we should need to average nine results, and so on. It is worth while to bear this simple rule in mind when averages are used, because it seems to be a common fallacy to suppose that variation can be reduced by averaging more than is, in fact, possible.

FIG. 2

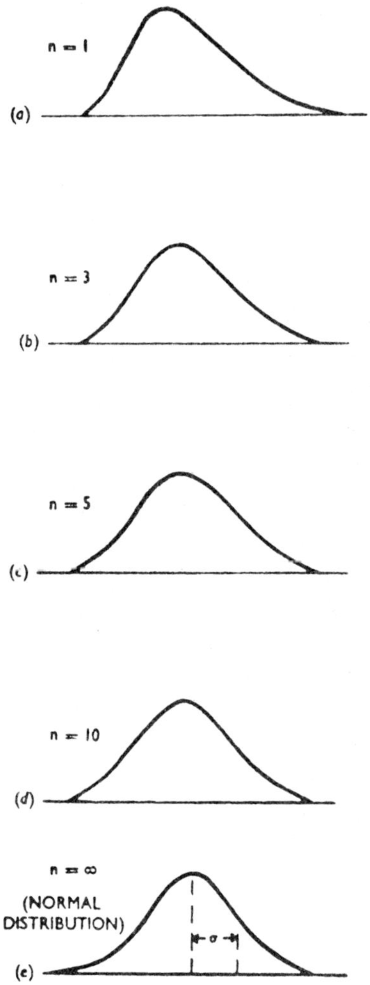

Skew distribution

FIG. 3

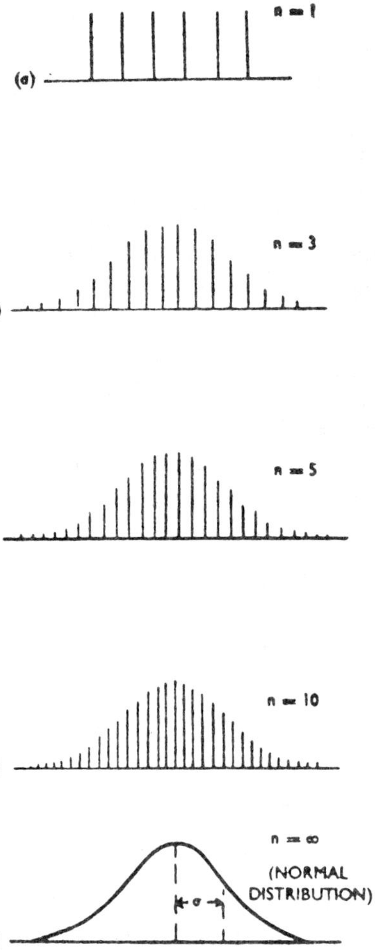

Rectangular distribution

Change in shape of a distribution curve due to averaging of results

Tendency for the Mean to be distributed in a Particular Form

Reduction of the variation is not the only effect of averaging results. Consider Figure 2. Here you see in Figure 2(*a*) a distribution that is rather skew, that is to say a distribution in which there is not an equal number of positive and negative deviations from the mean. You can imagine if you wish that this distribution is a summary of thousands of results for an experiment repeated under standard conditions, just as Figure 1(*a*) summarized the results of the tensile tests. Suppose all of those thousands of results were written on separate tickets and all these were placed in a bag, then Figure 2(*a*) summarizes the contents of the bag. Now suppose we took out of the bag a set of three tickets, say, calculated the mean of the numbers on the tickets and wrote it down. Suppose we then took another three tickets out of the bag and made the same calculation, and so on. The means of the sets of the results that we had written down would form another distribution. This you see in Figure 2(*b*). We already know that the spread of this second distribution would only be $1/\sqrt{3}$ times the spread of the first; in order to make it easier to compare their shapes, however, all the distributions have been drawn with the same standard deviation. You will notice that the second distribution is more symmetrical than the first. If means of five and ten results were taken the distributions would be like those in Figures 2(*c*) and (*d*). These distributions are more symmetrical still, and in fact when $n = 10$ it is difficult to notice any lack of symmetry at all.

Figure 3(*a*) shows the frequency distribution that would occur for throws of a single six-sided die. If the die were unbiased you would get an equal frequency of scores of 1, 2, 3, 4, 5 and 6. Suppose we threw three dice and each time took the average score, then the frequency distribution would be that shown in Figure 3(*b*). (Again, the spread of the curves has been adjusted so that the standard deviations are equal.) Figures 3(*c*) and (*d*) show the distributions for five and ten dice respectively.

We notice that, in both the examples, although we have started from distributions that were in one case rather skew and in the other rectangular in shape, the distribution of the mean has in each case rapidly assumed a characteristic symmetrical shape. This is a particular aspect of a more general and very important fact, i.e. that, as the number of results from which a mean is taken is increased, so the distribution of the mean approaches closer and closer to a particular form that can be represented by a comparatively simple function. This is called the Normal distribution and its shape is shown in Figure 2(*e*) and Figure 3(*e*). You can see that in both cases there is little to distinguish between the distribution of the means of sets of ten results and the Normal distribution. Now, the exact frequency of occurrence of any given deviation from the mean is known for the Normal distribution and is conveniently summarized in tables, given in most textbooks. For example, approximately 2/3 of the results lie within plus or minus one standard deviation from the mean. That is, about 1/6 of the results exceed the mean plus one standard deviation, and about 1/6 fall short of the mean minus one standard deviation. Similarly about 19/20 or 95% of the results lie within plus or minus two standard deviations from the mean, or in other words about 1/40 of the results exceed the mean plus two standard deviations, and about 1/40 fall short of the mean minus two standard deviations.

Significance Tests

How can we apply this knowledge to help in research on rubber? Well, suppose we are trying to increase the tensile strength in a particular way and we know from past results that the normal tensile has average 250 and standard deviation 10. Suppose nine test pieces have been prepared by a new technique which it is hoped will give a greater tensile strength and the mean of these results is 261. Can we conclude that this represents a real improvement or is it the sort of discrepancy we might expect owing to inherent variation in the test? Now we know that the standard deviation of the mean of nine results will be one-third ($1/\sqrt{9}$) of the standard deviation of the individual results, that is, it will be $10/3 = 3{\cdot}33$. Suppose there had been no real improvement, then a deviation of $t = (261 - 250)/3{\cdot}33 = 3{\cdot}30$ standard deviations has occurred. We have seen that within reason, whatever the distribution of the individual results, the means of nine results will follow closely the Normal distribution for which we have tables. Consulting the tables it will be found that a departure from the mean as great as or greater than 3·30 standard deviations would occur less than once in about 2,000 times if the true mean tensile were really 250. We should therefore feel confident that the new technique had almost certainly produced a real improvement, and we should say that the result was highly significant. On the other hand, had we found the probability of this deviation to be say 1 in 3 instead of 1 in 2,000, we should have concluded that this was just the sort of result to expect if the new technique had had no effect at all, and we should remain sceptical of its value. Of course we could not always expect to know in advance what the standard deviation of our results would be. Fortunately we could still make a significance test by estimating the standard deviation from the variation of the nine results themselves. This estimate of the standard deviation will not be so reliable as one which is calculated from a large quantity of data, but exact allowance for this uncertainty can be made by referring the ratio t not to tables of the Normal distribution, but to tables of what is called 'Student's' t-distribution (named after its discoverer W. S. Gosset, who used the pseudonym 'Student'). In these tables exact allowance for the uncertainty in the estimate of the standard deviation is made. Thus, if the standard deviation were calculated from the nine results, the ratio t would have to be 5·04 for a probability of 1 in 2,000 instead of 3·30, which is the value appropriate when the standard deviation is accurately known.

Deciding on the Number of Observations

The argument we have employed here may be used the other way round to answer a question of great importance to the experimenter. That is, how many experiments should he make in given circumstances to be reasonably sure of his conclusions? When the experimenter is carrying out development work on experimental material that is variable, he runs two risks. In the first place, owing to experimental error he may be led to believe that a particular modification is valuable when it is not, and on the other hand he may fail to detect the effect of modification when it is really of value.

Suppose, in the experiment just described, it were decided that to be of real value a modification would have to increase the tensile strength by at least 10 units, from 250 to 260. Suppose, also, that the experimenter wanted an experiment that was on a scale sufficient for a real improvement to 260 to have a 97·5% probability of detection, while if there were no improvement the probability would be 97·5% that no improvement would be found. As will be shown below, the number of experiments that it would be necessary to carry out to fulfil these conditions would be 16, and the decision as to whether or not a real improvement had occurred would depend on whether or not the mean of the 16 results was above or below 255. The argument upon which this is based will be clear from Figure 4.

FIG. 4

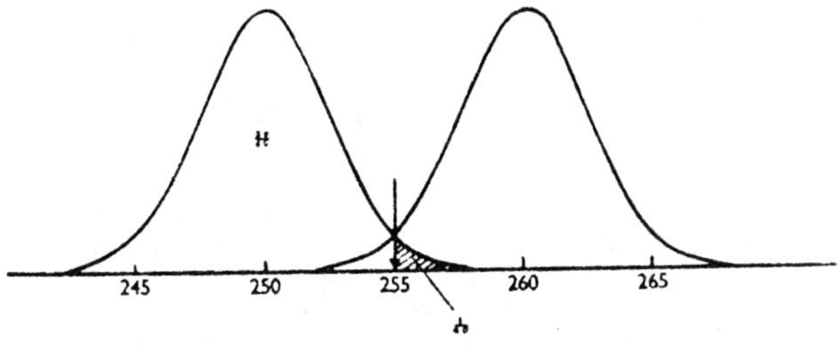

Distributions of means of 16 *results when true tensile strength is* 250 *and* 260

With standard deviation 10 for an individual observation, if the true mean is 250 and we take n observations then the mean of these n results will be distributed about 250 with standard deviation $10/\sqrt{n}$. Similarly, if the true mean is 260 then the mean of the n results will be distributed about 260 with standard deviation $10/\sqrt{n}$. We must take a value C somewhere between these extremes and say that if our mean is greater than C we shall conclude that we have an improvement, while if the mean is less than C we shall conclude that there is no improvement. Now, when the true mean is 250 the probability must be only 1 in 40 that the mean of the n results exceeds the value C, and from what we know of the Normal distribution we conclude that $C - 250$ must equal twice the standard deviation of the mean. Also, if the true mean is 260 the probability must be only 1 in 40 that the mean of the n results is less than C, that is $260 - C$ must also equal twice the standard deviation of the mean. Thus $250 - 260$ must be equal to four times the standard deviation of the mean, i.e. the standard deviation of the mean must be 2·5. Then $10/\sqrt{n} = 2{\cdot}5$, $\sqrt{n} = 4$, $n = 16$, and it follows that $C = 255$. The experimenter would make 16 observations, therefore. If the mean tensile strength were less than 255 he would say that the modification had probably produced no real improvement of importance, while if it were more than 255 he would say that a real improvement had probably occurred. The risk he would take of being wrong in either of the two ways would not exceed 1 in 40.

In ways such as these it is possible for the experimenter to utilize statistical ideas to plan his experiments so that even though his data are subject to variation his conclusions are nearly always correct.

The significance test with which I have dealt is appropriate for the comparison of two mean values. Many other sorts of problem occur; the experimenter may be interested in the *proportion* of times a certain thing happens (for example, a rubber band breaks when stretched a given amount). Alternatively, the point of interest may be the variability: Is method A less *variable* than method B ? Again, the point of interest may be whether there is really any significant *correlation* between one property and another (for example, between the abrasion resistance of rubber and its tensile strength). Statistical tests are available for these and the many other queries that arise in practice. The same kind of argument is used in all these situations, although the details of some of these tests are more complicated.

Sequential Tests

Before I leave the subject of making tests of significance I shall mention what is called a sequential test.

In the last section we saw how the experimenter could decide on the number of observations necessary in a given situation. The factors that affected the number were the size of the difference which it was important to detect and the risks the experimenter was prepared to take, either of finding a difference when none existed or of failing to find one which did. The distinctive feature of the technique used was that the number of experiments to be carried out was fixed in advance. It is frequently necessary or convenient to proceed in this way. On the other hand, situations often occur when the experimenter, left to himself, would proceed sequentially. That is to say, he would carry out one experiment or perhaps a small group of experiments and then consider the results before carrying out any more. It might be that so large a difference would be revealed that even on the few experiments he had carried out the results would be conclusive and no further experiments would be necessary. On the other hand, it might be that the results were inconclusive, in which case he would carry out further experiments and then reconsider all the results. He would proceed in this way until sufficient information had accumulated for him to be sufficiently sure of his conclusions. In situations where such a procedure is not inconvenient this method is preferable, since it is clearly more economical than the method of planning a fixed number of observations. Sequential significance tests designed for this type of situation have become available in recent years. They are, in essence, exceedingly simple and I propose to demonstrate such a test to you.

As Mr. Buist explained in his lecture last week, the resilience of rubber balls may be determined by dropping them on to a steel plate and measuring the height of fall (h_1) and the rebound height (h_2); the percentage resilience is then $100h_2/h_1$. Suppose we wish to devise a scheme that will control the quality of rubber balls as measured by their resilience. For example, the average resilience for a standard type of rubber ball is 67%, so that, on the average, if such a ball is dropped from a height of 100 cm. it will rebound to a height of 67 cm. Of course the balls vary from one to another, so that the rebound height will have a distribution and in this case the standard deviation of the distribution has been found to be just over 1%. Suppose that it is decided that if the average resilience of a batch of balls falls to 66% we should want to reject it as being below standard with 97·5% certainty, while if the batch were really up to standard we should desire with the same certainty that we should not reject it. You will see that, applying the argument we discussed previously, we could easily obtain a non-sequential test procedure that would satisfy these requirements. We should have to measure the resilience of a given number of balls, calculate the mean and then accept the batch if this value exceeded some value C and reject the batch if it fell below C. Now it is not particularly easy to measure the height of rebound, so that alternatively we could count the number that reached or surpassed a given height. Knowing that the standard deviation of the distribution is about 1 and that the distribution is approximately Normal, a simple calculation tells us that if we drop the balls from a height of 100 cm., about 70% of balls having mean resilience 67% will rebound above a string stretched at a height of 66·5 cm., while if the mean resilience of the batch we were testing were as low as 66%, only about 30% of them would rebound above it. It can then easily be shown that, if we adopt the fixed-number method, 20 balls should be tested, the batch being accepted if more than 10 balls rebounded above the string and rejected if less than this number did so. Such a test would ensure, as was required, that the probability of rightly accepting a good batch (mean resilience 67%) and of rejecting a bad batch (mean resilience 66%) was in each case at least 97·5%.

Alternatively we can use a sequential test, that is to say one in which the number of tests we make depends on the results we get, and we can arrange that this test gives exactly the same probability of arriving at the correct conclusions as the non-sequential test described above. The test takes the form of a sort of game, and for this particular case is as follows: We start with a score of 13 points, then if a ball rebounds above the string we add 3 points, whereas if it falls below the string we deduct 3 points. As we proceed with the testing, if the score reaches zero we stop and reject the batch as a bad one, but if the score reaches 26 we stop and accept the batch as a good one, the testing being carried on until one of these events happens. I cannot discuss here how these numbers are decided upon, but the calculations involved are not difficult.

Advantages and Disadvantages of Sequential and Non-sequential Tests

The advantage of the sequential scheme over the non-sequential test is that, on the average, it requires only about half as many observations. In situations (e.g. weathering trials) where the observations do not come to hand in sequence it cannot of course be applied and it will only have a real advantage compared with the non-sequential scheme when the labour involved in the experiment is mainly confined to the test itself. For example, if for a particular experiment special test pieces were required then it would probably be more appropriate to use a non-sequential test, for we should have to decide in advance how many test pieces to prepare. When sequential tests are first considered the objection is usually raised that in a particular case they might not terminate. Although it is theoretically possible that a sequential test could go on indefinitely, the probability that, for example, twice as many observations as are required by the non-sequential test would be necessary is very remote indeed. In fact the test would usually require considerably fewer observations than the non-sequential test (about half as many on the average).

The problem of making the decision as to whether one batch is better than another by sequential or non-sequential tests is analogous to that of deciding between one player and another in a game of skill. In the game of darts the scheme is non-sequential, for the players play for a fixed score, but in tennis and similar games a rule is introduced so that if the players have reached an equal score at a given point (40 all or deuce in tennis) then a sequential scheme is introduced, the game now continuing until one player has a two-point lead over the other. Just as in a sequential scheme, so in tennis is it possible for the game to continue indefinitely, but we know in practice that the probability of this happening is so remote that we can dismiss it altogether.

FIG. 5

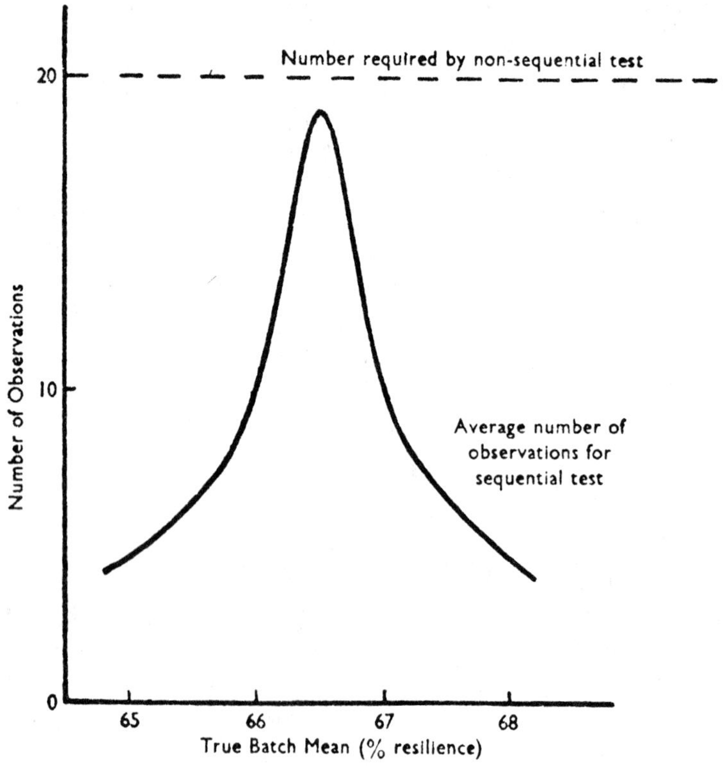

Average number of observations required to reach a decision in a sequential scheme

Before a sequential scheme is instituted we need to know at least the average number of tests that will be required. The number must of course depend on what the true value of the mean happens to be. In the example we have considered, if a very bad batch of balls came along none of the balls tested might rebound higher than the string, and we should reject the batch very quickly (in five observations, in fact). On the other hand, if the batch were very good, all the balls tested might rebound higher than the string and we should accept the batch very quickly (again in five observations). For intermediate values the number of observations would clearly be more. In Figure 5 is shown the average sample numbers in the scheme I have just discussed. The dotted line indicates the number necessary with the non-sequential scheme. We see that even under the most unfavourable conditions the sequential test is better on the average than the non-sequential test. For most of the situations in which non-sequential tests can be used, sequential tests have been or are now being developed. These tests were discovered independently by Professor G. A. Barnard in England and Professor Abraham Wald in America about 1943, and were used extensively during the war for the rapid checking of batches of war equipment (varying from light ammunition to tents). The tests were considered so valuable that the work was kept secret until after the end of the war, in 1946.

The Design of Experiments

So far we have seen how mathematical statistics can help us to decide which effects are probably real and which not, and how we can decide on a suitable scale to prepare an experiment. In order to focus attention on the principles involved we have assumed so far that the actual experiments with which we were concerned were of the simplest possible type. We can now consider a more elaborate type of experiment.

In order to make the conclusion from our experiment as precise as possible we are naturally careful to carry it out so that the uncontrollable variation will be as small as possible. In many of the situations that we encounter, even when this is done, the variation would still be large. The precision of an experiment can often be greatly increased by careful design.

Reduction of Variation by Experimental Design

The way in which experimental design can help to make experimental conclusions more precise is best illustrated by an example. The Martindale Wear Tester is a machine in which the abrasion resistance of four pieces of material can be tested simultaneously. Suppose we have four types of rubber A, B, C and D, referred to subsequently as the 'materials' A, B, C and D, whose abrasion resistances we desire to compare, and suppose it is decided to make four tests on each material. Then we shall make four runs on the machine. Table II shows three possible ways in which these runs could be made.

In design 1, the four samples of material A were tested in the first run, then the four samples of material B were tested in the second run, etc. This design is bad, because any systematic difference between runs would be confused with the differences between materials.

Table II

Positions in Machine	*Design 1 Runs*				*Design 2 Runs*				*Design 3 Runs*			
	1	2	3	4	1	2	3	4	1	2	3	4
1	A	B	C	D	A	A	A	A	A	B	C	D
2	A	B	C	D	B	B	B	B	D	A	B	C
3	A	B	C	D	C	C	C	C	C	D	A	B
4	A	B	C	D	D	D	D	D	B	C	D	A

This difficulty is overcome in design 2, but this is also open to objection, because differences between materials A, B, C and D would now be confused with systematic differences that might occur from position to position in the machine. In design 3, however, both these difficulties are overcome, since each material is tested once and once only in each run and in each position in the machine. This is the design which, in fact, we always use with this machine. By statistical analysis of past results it is found that large run differences do actually occur and that there are also systematic differences between positions on the machine. Using design 3, which is called a Latin Square, the variation due to both these causes is eliminated. It has been shown that in this particular problem design 3 can be as much as six times as efficient as design 1. That is to say, the means of four tests using design 3 (in which effects due to runs and to positions are eliminated) are as reliable as the means of twenty-four tests in which the effects of runs and positions are not eliminated in this way.

It often happens that we wish to compare more than four types of rubber in the wear tester, which can only compare four samples simultaneously. This may be done so that effects of runs and positions are still eliminated, by using what are called Youden Squares. A square for comparing seven materials in seven runs is shown below in Table III.

Table III

Youden Square to test Seven Materials in Seven Runs

Materials	*Runs:* R_1	R_2	R_3	R_4	R_5	R_6	R_7
M_1	.	P_2	.	P_1	.	P_4	P_3
M_2	P_1	.	.	P_2	P_3	.	P_4
M_3	.	P_1	P_3	.	P_4	.	P_2
M_4	P_3	.	P_4	.	.	P_2	P_1
M_5	.	.	P_2	P_4	P_1	P_3	.
M_6	P_4	P_3	.	.	P_2	P_1	.
M_7	P_2	P_4	P_1	P_3	.	.	.

Note: P_1, P_2, P_3 and P_4 refer to positions in the machine.

It will be seen that in each run four of the materials are tested, one in each of the four positions on the machine, and on completion of the experiment each material has been tested four times, once in each of the four positions on the machine. The symmetry of the design enables the means of the materials corrected for run and position differences to be calculated very simply.

This and other types of designs which I have not time to discuss now are of immense value in making comparisons when only a limited amount of homogeneous material is available. For example, if we were carrying out experiments on tyres, by running a number of test cars we could use the Latin Square to eliminate variation due to differences between cars and between the four positions (wheels) on the car. If we have more than four types of tyres to compare we could, by using designs like Youden Squares, still effectively confine our comparison to within cars and not introduce car-to-car and position-to-position variation.

The same principles hold, for example, in footwear trials. We have just completed a trial in which each of twelve subjects had four shoes, each identical in appearance but made with four different types of leather. By using a special type of statistical design, even though *four* different types of leather were compared, all the results were based on direct comparison between the comfort of left and right shoes worn at the same time. The precision of the trial was surprisingly high and was largely due to the special design used.

Factorial Experiments

Before the problem of experimental design was given close study it used to be accepted as a basic axiom that only one experimental variable should be changed at a time. It was assumed that if more than one factor were varied it would not be possible to separate the effects of individual factors. R. A. Fisher has shown that this is a misconception. In fact, great advantages can be obtained by varying a number of factors in what are called factorial designs. In these designs every level of each factor is tested in combination with every level of each other factor. The idea is best illustrated by an example.

The experiment I have chosen was designed to test the wearing properties of rubber-coated fabrics prepared in different ways. Two different fillers, F_1 and F_2, were tried at three different percentages, $Q_1 = 25\%$, $Q_2 = 50\%$ and $Q_3 = 75\%$, the samples being prepared with and without a surface treatment, T_0 and T_1. If we test all combinations of the factor levels there will be twelve different ways of preparing the coated fabrics. Duplicate preparations of these twelve types of coated fabric were made and their wear curves are shown in Figure 6. The first curve shows the weight losses observed after 1,000, 2,000 and 3,000 revolutions of the wear machine on a sample for which 25% of filler F_1 was used with no surface treatment. The second curve shows the corresponding losses of a sample from a duplicate preparation, and so on. The great advantage of the factorial experiment is that it enables us to determine not only the primary effects of the factors but also their interactions. If we had tried the effect of increasing the filler content with filler F_1 only, then our conclusions would only be true for this filler. The factorial experiment shows us whether or not the effect of increasing the content is independent of which filler we use, or whether we get one effect with filler F_1 and a different effect with filler F_2. In the latter case the factors are said to interact.

FIG. 6

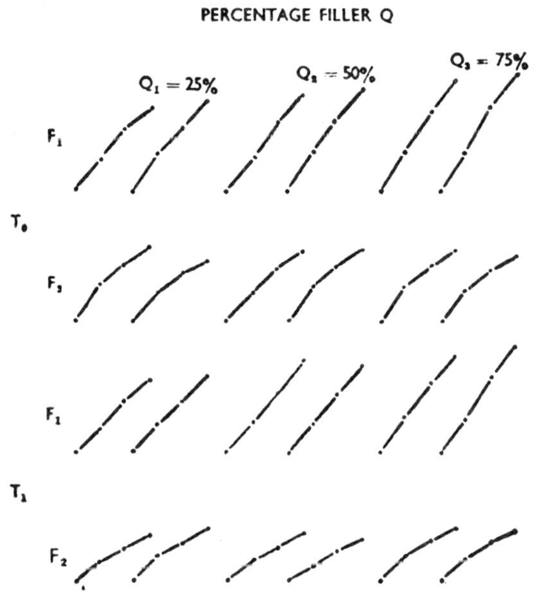

Weight loss curves for coated fabrics

The statistical analysis is performed by the Analysis of Variance technique mentioned earlier. The Analysis of Variance for these data is shown in Table IV.

Table IV

Analysis of Variance for Wear Data

	Average Rate of Wear	*Shape*
Filler content (Q)	(2) 6,785***	(4) 440
Type of filler (F)	(1) 107,803***	(2) 9,144***
Surface treatment (T) ..	(1) 24,494***	(2) 4,124***
Interactions		
Q × F	(2) 4,942***	(4) 354
Q × T	(2) 397	(4) 172
F × T	(1) 1,682*	(2) 1,164***
Q × F × T	(2) 150	(4) 116
Error	(12) 312	(24) 190

From the short talk I have given today you will not be able to understand how the table is derived, but I want you just to accept that such a table can be constructed from the data, and I shall use it to illustrate how valuable the design and its corresponding analysis can be in breaking down the problem. In this particular case it is possible to see broad differences by studying the curves in Figure 6, but the position is made much clearer by the analysis.

In Table IV the total variation is split up into parts corresponding to the various factors and their interactions. In this particular analysis the effect of the factors on *average* wear is shown in column 2, while in column 3 the effect of the factors on the *shape* of the curves is shown. Consider the column for average wear rate. The numbers are the variances measuring the variation due to the factor plus experimental error. These are compared with the estimate of error variance (312) at the foot of the column which estimates experimental error alone. If the apparent changes were chance effects, then the variance corresponding to a factor would be of the same order as the error variance. For filler content we find an estimate of variance of 6,785, which is 20 times as large as the error variance. As would be expected, on consulting the appropriate significance tables we find this value is highly significant* and we conclude that effect due to the filler content is much greater than can be attributed to chance. In this particular case all the factors varied are found to have highly significant effects. We now consider the interactions, and we find significant interactions between the quantity and type of filler and between the filler and the surface treatment but none between quantity of filler and surface treatment. In order to interpret these interacting effects we construct tables of mean values (see Table V). From the first part of the table we see that the average rate of

Table V

Type of Filler	*Filler Content*		
	$Q_1 = 25\%$	$Q_2 = 50\%$	$Q_3 = 75\%$
F_1	169	199	231
F_2	121	120	126

Type of Filler	*Surface Treatment*	
	T_0	T_1
F_1	214	145
F_2	186	99

* Significance of the items is denoted by asterisks. Three asterisks denote that the probability of chance occurrence is less than one in a thousand, two asterisks less than one in a hundred, one asterisk less than one in 20. A result with no asterisks is not significant, that is to say it could well have arisen owing to an experimental error.

wear is less with the second filler, and this average rate is virtually unaffected by the increase in the filler content, whereas with the first filler the material wears more and the wear increases as the content is increased. From the second half of the table of means we see that the beneficial effect of the surface treatment is even more pronounced with the second than with the first filler.

So far we have considered the effect of factors only on the average rate of wear; we now need to consider whether these effects change at different stages of wear; that is to say whether the factors affect the shape of the curves. To do this we consider the variances shown in the right-hand column of the table; the significance of these items is judged by comparison with the error term (190) at the foot of this column. The filler content (Q) and filler content × type of filler (QF) effects are not associated with significant changes in the shape of the wear curves; however, both F (the type of filler) and T (the surface treatment) markedly affect the shape of the curves. Again, tables of mean values show the nature of these effects (see Table VI).

Table VI

		Wear in first 1,000 *revs.*	*Wear in second* 1,000 *revs.*	*Wear in third* 1,000 *revs.*
Type of Filler	F_1	215	213	172
	F_2	181	104	82
Surface Treatment . .	T_0	231	173	134
	T_1	165	144	119

From the first portion of Table VI we conclude that not only do the samples having filler F_1 wear at a greater rate than those having filler F_2, but the fall-off in wear is greater with filler F_2. From the second portion we conclude that although the surface treatment has a favourable effect at all stages of wear this is (as would be expected) most marked initially. Finally, we see that the interaction F × T, between type of filler and surface treatment, affects the shape of the curves. On consulting a table of mean values this is seen to be due to the fact that the interaction between filler and surface treatment, found before, is confined to the first period of wear alone.

The conclusions from the experiment can therefore be set out as follows:

(1) Filler F_2 results in less wear than filler F_1; the rate of fall-off of wear is greater and the protective effect of surface treatment is more marked with F_2 than with F_1.

(2) Whereas increasing the content of filler F_1 results in increased rate of wear, the content of filler F_2 can be increased, at least to 75%, without increasing wear.

(3) The surface treatment markedly reduces wear, especially during the early stages and when filler F_2 is used.

The advantages of the factorial design and the accompanying analysis will be apparent from the example just given. The factorial design has ensured that not only the principal effects of the factors but also the effects of their interactions can be determined. Each observation has been made to contribute knowledge about all these effects, thus giving a greater amount of information per experiment than would be possible had this design not been used. The accompanying statistical analysis has made it possible to determine which apparent effects were real and which were probably due to experimental error and to analyse the situation clearly and without ambiguity.

Conclusion

In the time at my disposal I have been able to deal with only a very few of the statistical techniques that are of use in rubber technology and, moreover, in all scientific fields where variation occurs. I hope, nevertheless, that I have said enough to convince you of their value and of the necessity for all workers in this field to have at least some idea of the potentialities of statistical techniques.

The two following books, both of which were written to meet the needs of students without any previous knowledge of statistical methods, should be consulted by those desirous of gaining a fuller knowledge of the subject:

Statistical Methods in Research and Production, edited by O. L. Davies (Oliver and Boyd, Edinburgh, 1949).

The Design and Analysis of Industrial Experiments, edited by O. L. Davies (Oliver and Boyd, Edinburgh, 1954).

Lists of reference for more advanced reading will be found in these two books.

5.6
Evolutionary Operation:* A Method for Increasing Industrial Productivity

G. E. P. BOX†
Imperial Chemical Industries Ltd.

The rate at which industrial processes are improved is limited by the present shortage of technical personnel. Dr Box describes a method of process improvement which supplements the more orthodox studies and is run in the normal course of production by plant personnel themselves. The basic philosophy is introduced that industrial processes should be run so as to generate not only product, but also information on how the product can be improved.

Introduction

Much scientific effort in industry is directed on the one hand to the discovery of new products and processes and on the other to their development and improvement. This paper is concerned with a particular aspect of the problem of *improving* industrial processes.

Industrial organisations usually have specialist groups of scientific workers in research, development, and experimental departments, permanently occupied with improving manufacturing processes, who employ a wide variety of techniques, ranging from the fundamental study of reaction mechanisms to the purely empirical assessment of the effects of changes in variables. Associated experimentation may be conducted in the laboratory, on the pilot plant, and on the full scale process, and in particular may involve the use of statistical techniques having a fairly high degree of sophistication.[1–7] As a result of the application of this variety of specialised effort a steady rate of increase in productivity is usually attained.

Ultimately the rate of improvement is limited by the shortage of technical personnel. This shortage can be expected to become more rather than less severe, and in searching for further ways of attaining greater process efficiency one must look for methods which are sparing in their use of scientific manpower. The object of this paper is to outline one such device, which has been applied with considerable success over the past few years.

This is called 'Evolutionary Operation'. It is a method of process operation which has a 'built-in' procedure to increase productivity. It uses some simple statistical ideas and is run during normal routine

* Based on a paper given at the International Conference on Statistical Quality Control organised by the European Productivity Agency of OEEC in Paris in July 1955.

† Now Director of the Statistical Techniques Group at Princeton University.

The Collected Works of George E. P. Box, 1984, Wadsworth, Inc., Belmont, CA 94002.
Originally published in *Applied Statistics,* vol. VI, no. 2 (1957), pp. 3–23.

production largely by plant personnel themselves. Its basic philosophy is that it is nearly always inefficient to run an industrial process to produce product alone. A process should be run so as to generate product *plus information on how to improve the product.*

The technique is in no sense a substitute for the more fundamental investigations referred to above. On the contrary, the effects discovered by the application of evolutionary operation, particularly those which are of an unexpected kind, help to indicate new areas where fundamental research might be rewarding. Although the method has been specifically developed as a production technique for the chemical industry, it is believed that it has more general applicability.

Plant-scale and Small-scale Experiments

In the chemical industry the plant process will usually have been arrived at after considerable experimentation on the small scale. Now the optimum conditions of operation on the small scale usually provide no more than a good first approximation to the full-scale optimum. Because of this it is commonly found that considerable modification of the conditions arrived at from small-scale work is necessary before a comparable result can be obtained on the plant itself.

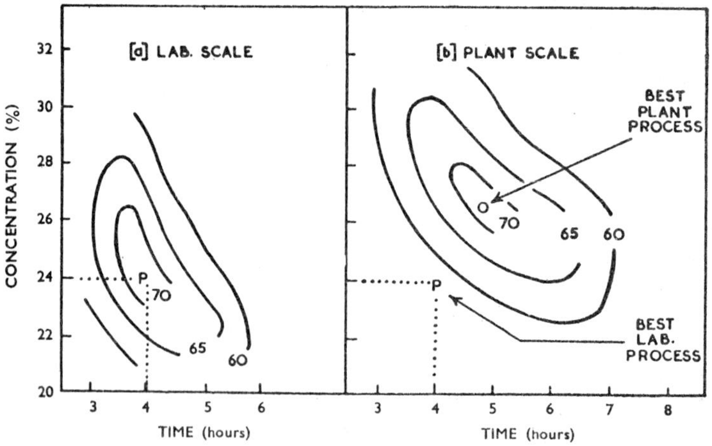

FIG. 1. Possible appearance of yield surfaces, showing contours of percentage yield, for a process conducted (*a*) on the laboratory scale and (*b*) on the plant scale.

Fig. 1 shows how, in translating the process from the laboratory to the plant, the yield surface can become distorted and displaced due to scale-up effects. It will be seen that when this happens one might expect the best laboratory conditions to give disappointingly low yields on the full scale. The effort to move from the laboratory maximum-yield conditions at P to the plant maximum-yield conditions at O must evidently be exerted *on the plant itself* and not in the laboratory—investigation in the laboratory can only lead back to P.

For those unfamiliar with the representation adopted in Fig. 1 it should be explained that the relationship between the response ('yield') and the process variables ('time of reaction' and 'concentration of one of the reactants') is imagined to be represented by a solid graph or 'response surface'. In the neighbourhood of maximum yield such a surface may have the appearance of a mound. The height of the mound at any particular point represents the yield at some set of reaction conditions. To allow representation in two dimensions the yield is shown by contours in the same way that the height of land is represented by contours on a map. Although for simplicity the above discussion is conducted in terms of yield, it should be understood that conditions giving highest yield would often not represent the optimum process. There would, for example, be no advantage in obtaining a higher yield if to do so involved the use of a disproportionate amount of some expensive starting material. The principal response usually considered, therefore, is 'the cost of producing unit quantity of product under the specified manufacturing conditions', or some other measure of productivity which takes account of the cost of running the process.

In addition to improvements made possible by adjustment of process conditions already studied on the small scale, further progress is usually possible by the introduction of new modifications not considered—and often not capable of being studied—at the small-scale stage of development.

Adjustments are made when the plant is first installed, but these seldom result in the location of the ultimate plant optimum, and as a result of special experimental campaigns, chance discoveries, and new ideas, improvement usually continues over many years. The object of evolutionary operation is to speed up this process.

Analogy with Evolutionary Process

The method used to speed the improvement is illustrated by the following analogy. Living things advance by means of two mechanisms:

(i) Genetic variability due to various agencies such as mutation.

(ii) Natural selection.

Chemical processes advance in a similar manner. Discovery of a new route for manufacture corresponds to a mutation. Adjustment of the process variables to their best levels, once the route is agreed, involves a process of natural selection in which unpromising combinations of the levels of process variables are neglected in favour of promising ones.

Fig. 2 illustrates diagrammatically the possible evolution of a species of lobster. It is supposed that a particular mutation produces a type of lobster with 'length of claws' and 'pressure attainable between claws' corresponding to the point P on the diagram and that in a given environment the contours of 'percentage surviving long enough to reproduce' are like those shown in the figure.

The dots around P indicate offspring produced by the initial type of lobster. Since those in the direction of the arrow have the greatest chance of survival, over a period of time the scatter of points representing succeeding generations of lobsters will automatically move up the survival surface. This automatic process of natural selection ensures, without any special effort on the part of the lobsters, that optimum-type lobsters exist. It also ensures that if the environment changes so that the survival surface is altered, the lobsters will change correspondingly to the new point of maximum survival.

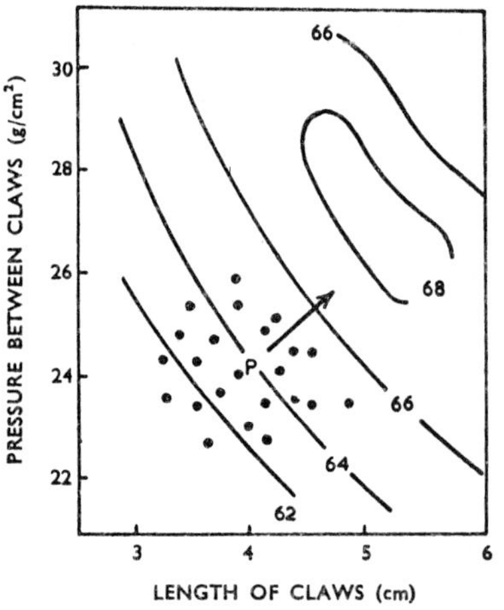

Fig. 2. Evolution of a species of lobster. Contours show the percentage surviving long enough to reproduce in a given environment.

What we have to do is to imitate this process. That is to say, we have to institute a set of rules for *normal plant operation* so that (without serious danger of loss through manufacture of unsatisfactory material) an evolutionary force is at work which steadily and automatically moves the process towards its optimum conditions if it is not operating there already. Such a technique will gradually nudge the operating procedure into the form that is ideally suited to the particular piece of equipment which happens to be available. The two essential features of the evolutionary process are:

(i) *Variation.*

(ii) *Selection* of 'favourable' variants.

Static and Evolutionary Operation

Routine production is normally conducted by running the plant at rigidly defined operating conditions called the 'works process'. The works process embodies the best conditions of operation known at the time. The manufacturing procedure, in which the plant operator aims always to reproduce exactly this same set of conditions, will be called the method of *Static Operation.* Although this method of operation, if strictly adhered to, clearly precludes the possibility of evolutionary development, yet the *objectives* which it sets out to achieve are nevertheless essential to successful manufacture, for in practice we are interested not only in the productivity of the process, but also in the physical properties of the product which is manufactured. These physical properties might fall outside specification limits if arbitrary deviations from the works process were allowed. Our modified method of operation must therefore include safeguards which will ensure that the risk of producing appreciable amounts of material of unsatisfactory quality is acceptably small.

In the method of Evolutionary Operation a carefully planned cycle of minor variants on the works process is agreed. The routine of plant operation then consists of running each of the variants in turn and continually repeating the cycle. The cycle of variants follows a simple pattern, the persistent repetition of which allows evidence concerning the yield and physical properties of the product in the immediate vicinity of the works process to accumulate during routine manufacture. In this way we use routine manufacture to generate not only the product we require but also the information we need to improve it.

Controlled variation having thus been introduced into the manufacture, the effect of selection is introduced by arranging that the results are continuously presented to the plant manager in a way which is easily comprehended. This allows him to see what changes ought to be made to improve manufacture. The stream of information concerning the products from the various manufacturing conditions is summarised on an *Information Board* prominently displayed in the plant manager's office. This is continuously brought up to date by a clerk to whom the duty is specifically assigned. The information is set out in such a way that the plant manager can at any time see what weight of evidence exists for moving the centre of the scheme of variants to some new point, what types of change are undesirable from the standpoint of producing material of inferior quality, how much the scheme is costing to run, and so on.

In making a permanent change in the routine of plant operation the situation is very different from that which we meet in running specialised experiments on the plant. The latter will last a limited time, during which special facilities can be made available. Furthermore some manufacture of substandard material is to be expected and will be budgeted for. Evolutionary operation, however, is virtually a *permanent* method of running the plant and cannot therefore demand special facilities and concessions. For this reason only small changes in the levels of the variables can be permitted, and only techniques simple enough to be run continuously by works personnel themselves under actual conditions of manufacture can be employed.

The effects of the deliberate changes in the variables will usually be masked by large errors customarily found on the full scale. *However, since production will continue anyway, a cycle of variants which does not significantly effect production can be run almost indefinitely, and because of constant repetition the effect of small changes can be detected.*

An Example

To illustrate the procedure we consider one phase of evolutionary operation for a particular batch process. At this stage of development two process factors—the percentage concentration of one of the reactants, and the temperature at which the reaction was conducted—

were being studied following the scheme of variants shown in **Fig. 3.** The works process is labelled (1) and the four variants are labelled (2), (3), (4), and (5). One batch of product was made at each set of conditions, which were run successively in the order 1, 2, 3, 4, 5; 1, 2, 3, 4, 5; and so on. Three responses were recorded:

(i) The cost of manufacturing unit weight of product. This was obtained by dividing 'the cost of running at the specified conditions' by 'the observed weight yield at those conditions'. It was desired to bring this cost to the smallest value possible, subject to certain restrictions listed in (ii) and (iii) below.

(ii) The percentage of a certain impurity. It was desired that this should not exceed 0·5.

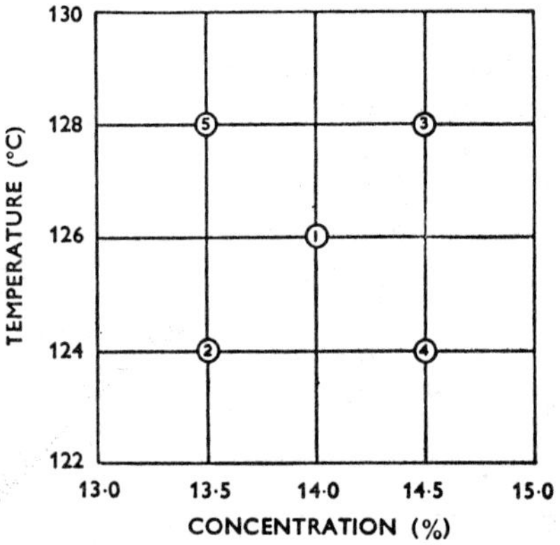

Fig. 3. A cycle of variants about the works process.

(iii) A measure of fluidity. It was preferred that this should lie between the limits 55 and 80.

The information coming from the experiment was recorded by writing in chalk on an ordinary blackboard. Alternatively, wax pencil on a white plastic board or magnetic letters and numbers on a steel board could have been used. The essential thing is that it should be a simple matter to erase or remove one number and replace it by another. The scheme set out in Fig. 4 is not the only one which could have been adopted, but is intended to show the sort of calculations and layout of the results which have been found useful.

The phase number at the top left-hand corner of the board indicates that two previous phases of evolutionary operation have already been completed on this process. In general these might have involved other variables or the same variables at other levels. In order that the new results may be considered in proper relation to those obtained previously, the final average values recorded in previous phases should also be available (for example on sheets of paper pinned to the board). The cycle number in the top right-hand corner indicates that 16 cycles of this third phase of operation have been completed. There follows a plan of the cycle of variants being run.

The table below this summarises the current situation. First are shown the requirements which it is desired to satisfy. These are followed by the running (i.e. up-to-date) averages at the various manufacturing conditions set out so as to follow the plan of the cycle of variants. This arrangement makes it easy to appreciate the general implications of the results. A measure of the reliability of the individual running averages is supplied by the '95% error limits'. These are simply the quantities

$\pm ts/\sqrt{n}$ appropriate to the calculation of fixed sample-size confidence limits for individual means. In this expression s is the standard deviation, n the number of cycles, and t the appropriate significance point of Student's t distribution. The scheme can be run until these limits for the means of the principal response have been reduced to an acceptably small preassigned width.[8]

PHASE 3 LAST CYCLE COMPLETED 16

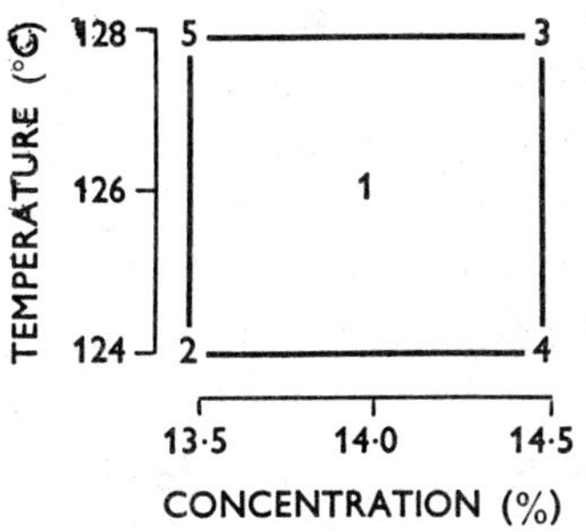

		Cost	Impurity (%)	Fluidity
Requirement		Minimum	Less than 0·50	Between 55 and 80
Running Averages		32·6 33·9 32·8 32·3 33·4	0·29 0·35 0·27 0·17 0·19	73·2 76·2 71·3 60·2 67·6
95% Error Limits		±0·7	±0·03	±1·1
Effects with 95% Error Limits	Conc.	1·2 ± 0·7	0·04 ± 0·03	5·2 ± 1·1
	Temp.	0·4 ± 0·7	0·14 ± 0·03	10·8 ± 1·1
	C × T	0·1 ± 0·7	0·02 ± 0·03	−2·2 ± 1·1
	Change in Mean	0·2 ± 0·6	−0·02 ± 0·03	−1·6 ± 1·0
Standard Deviation		1·44	0·059	2·12
95% Error Limits		1·22 1·76	0·050 0·072	1·80 2·59
Prior Estimate		2·71	0·054	3·22

FIG. 4. Appearance of information board at the end of cycle 16.

Below this are shown the 'effects' of the variables and their 95% error limits. The concentration and temperature effects are each calculated in the usual way as the difference in the average values of the response at the higher and the lower level of the variable. The value at the centre conditions does not enter the calculations except in computing the 'change in mean' effect. If y_1, y_2, y_3, y_4, and y_5 are the running averages of one of the responses after n cycles of operation, the effects and their limits of error for this particular example are as follows:

Effect		*Value*	*Limits of Error*
Concentration	..	$(y_3 + y_4 - y_2 - y_5)/2$	$\pm ts/\sqrt{n}$
Temperature	..	$(y_3 + y_5 - y_2 - y_4)/2$	$\pm ts/\sqrt{n}$
Interaction	..	$(y_2 + y_3 - y_4 - y_5)/2$	$\pm ts/\sqrt{n}$
Change in mean	..	$(y_2 + y_3 + y_4 + y_5 - 4y_1)/5$	$\pm 2ts/\sqrt{(5n)}$

The effect referred to as the 'change in mean' is simply the grand mean for all the runs $(y_1 + y_2 + y_3 + y_4 + y_5)/5$ less the mean for the 'standard' conditions y_1. It is therefore an estimate of the difference in average response resulting from the use of the evolutionary scheme. In the present example the effect of introducing the evolutionary scheme is thus:

(i) To increase the cost per batch by $0{\cdot}2 \pm 0{\cdot}6$ units.

(ii) To reduce the average impurity by $0{\cdot}02 \pm 0{\cdot}03\%$.

(iii) To reduce the average fluidity by $1{\cdot}6 \pm 1{\cdot}0$ units.

(It will be seen that if the effect of blending batches of slightly different qualities was to average the physical properties, the 'change in mean' would measure the difference between the product of the works process and a blend of the products from evolutionary operation. In many actual examples partial or complete blending of the products does naturally occur at the later stages of manufacture, so that where physical properties behave approximately additively there may in fact be remarkably little overall change in the manufactured product due to evolutionary operation. In some cases, especially if the effect of introducing small variations in the levels of the variables was unexpectedly large, blending could be deliberately introduced to produce an acceptable product.)

If the variants cover a region of the response surface which is a sloping plane, then it is not difficult to see that the true 'change in mean' will be zero. For a convex surface such as that near a maximum it will be negative, whereas for a concave surface such as that near a minimum it will be positive. It can be shown that, on certain plausible assumptions, the 'change in mean' is proportional to the sum of the quadratic constants which measure curvature in the directions of the variables. When the interaction effect and the change in mean effect are not small compared with the single effect of the variables, this indicates that a maximum or minimum is being approached and for exact location and exploration a technique fully set out elsewhere[4–6] is adopted.

For the cost response the change in mean supplies a continuous measure of the cost of obtaining information by the process of evolutionary operation. In the present example this is estimated at $0{\cdot}2 \pm 0{\cdot}6$ units of cost per batch. If the cost surface were locally planar, evolutionary operation would cost nothing. In practice, concavity of the cost surface is to be expected, since a minimum is usually being approached, so that there will usually be a small cost associated with running the evolutionary process. Except when the process has been brought very close to its ultimate optimum this cost will be redeemed many times over by the value of the permanent process improvements that occur from time to time as a result of the information generated by the evolutionary scheme.

After the calculated effects, the experimental error standard deviations calculated from the observations themselves are shown. Except in the initial stages of the scheme these are used in computing the limits of error for the running means and for the 'effects' of the variables. The normal theory fixed sample-size 95% confidence limits for these estimates of the standard deviations are also shown. The final items are the estimates obtained from prior data used in initiating the scheme.

In general, by inspecting the results set out on the information board in the light of the requirements it is desired to satisfy and from expert knowledge of other factors which affect plant operation, the plant manager decides at any particular stage whether

(*a*) to wait for further information;

(*b*) to modify operation.

Under (*b*) some of the alternatives open to him are:

(i) To adopt one of the variants as the new 'works process' and to recommence the cycle about this new centre point.

(ii) To explore an indicated favourable direction of advance and recommence the cycle around the best conditions found. (This exploration may be done, for example, by making a series of tentative advances in the indicated direction, at each stage running the new conditions and the previous best conditions alternately, until sufficient evidence has been gathered.)

(iii) To change the pattern of variants to one in which the levels are more widely spaced.

(iv) To substitute new variables for one or more of the old variables.

In the example actually discussed it is seen that a decrease in concentration would be expected to result in reduced cost, reduced impurity, and reduced fluidity. The effect on cost of a decrease in temperature is uncertain but is more likely to be favourable than not, and would almost certainly result in marked reductions in impurity and fluidity. These facts have to be considered, bearing in mind that a fluidity of less than 55 is undesirable and that although a further large reduction in impurity is welcome it is not necessary in order to meet the specification. It was decided in the event to explore the effects of reducing concentration alone. Phase 3 was terminated, and in the next phase the three processes (13%, 126°), (13·5%, 126°), (14%, 126°) were compared. The first of these gave a mean cost of 32·1 with an impurity of 0·25 and a fluidity of 60·7, and was adopted as a base for further development.

A geometrical display of the results on the information board may also be used when three variables are jointly considered. In this case perspective drawings of the sort illustrated in Fig. 5 are used. The

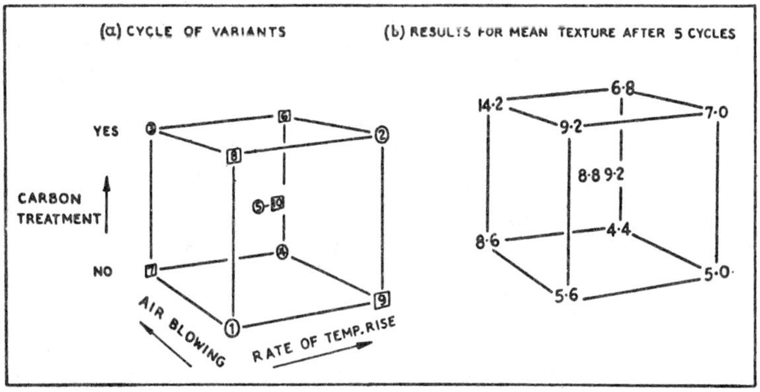

Fig. 5. Display of pattern of variants and results for three variables.

cycle of variants for a three-variable scheme is shown in Fig. 5(*a*), while 5(*b*) shows the results from one of the responses after five cycles of operation. This method may, of course, be applied to responses which cannot be measured directly. In the case illustrated in Fig. 5(*b*) an important property of the product was a somewhat esoteric quality referred to as 'texture'. A set of artificial standard samples were prepared which were judged by experts to have a range of 'textures' in approximately uniform steps, and these were arbitrarily scored. The texture of a sample of each manufactured material was then matched against the standards and an appropriate score given to it. In a similar way a scheme to evolve conditions which would give a product less inclined to 'cake' has been run, in which caking was judged by visual inspection and scored by comparison with a verbally defined scale.

Selection of the Variants

The technique outlined differs from the natural evolutionary process in one vital respect. In nature the variants occur spontaneously, but in our artificial evolutionary process we have to introduce them. Variants involving the levels of temperature, concentration, pressure, etc., are natural choices, but there are usually an almost unlimited number of less obvious ways in which manufacturing procedure can be tentatively modified. Frequent instances of marked improvement due to some innovation never previously considered in a process which has been running for many years testify to the existence of valuable modifications waiting to be thought of.

To make our artificial evolutionary process really effective, therefore, one more circumstance is needed—we must set up a situation in which useful ideas are continually forthcoming. An atmosphere for the generation of such ideas is perhaps best induced by bringing together at suitable intervals a group of people with special, but different, technical backgrounds. In addition to plant personnel themselves, obvious candidates for such a group are, for example, a research man with an intimate knowledge of the chemistry of the similar reactions and a chemical engineer with special knowledge of the type of plant in question. The intention should be to have complementary rather than common disciplines represented.

These people should form the nucleus of a small evolutionary operation committee, meeting perhaps once a month, whose duty it is to help and advise the plant manager in the performance of evolutionary operation. The major task of such a group is to discuss the implications of current results and make suggestions for future phases of operation. Their deliberations will frequently lead to the formulation of theories which in turn suggest new modifications that can be tried with profit.

Since questions of modification of certain physical properties of the manufactured product may arise, a representative of the department responsible for the quality of manufacture should also be on the evolutionary operation committee. Rather more may be got from the results and more ambitious techniques adopted if a statistician is also present at the meetings.

With the establishment of this committee all the requirements for an efficient evolutionary method of production are satisfied and the 'closed loop' illustrated in Fig. 6 is obtained. We are thus provided with a practical method of 'automatic optimisation' which requires no special equipment and which can be applied to almost any manufacturing process, whether the plant concerned is simple or elaborate.

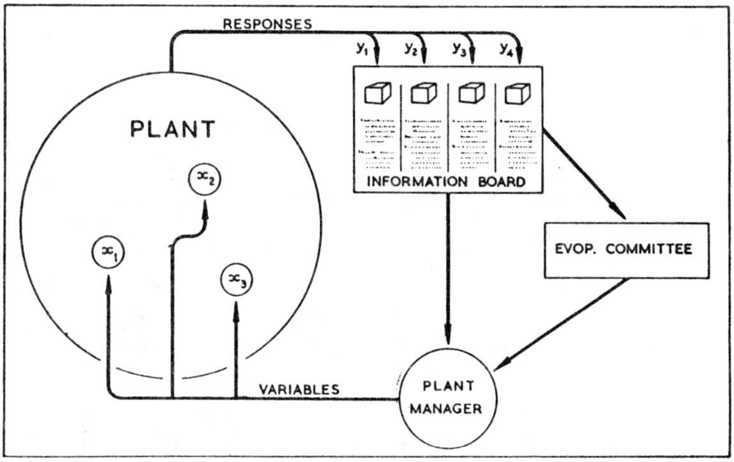

FIG. 6. Diagrammatic representation of the 'closed loop' provided by Evolutionary Operation.

At the beginning of this article I spoke of evolutionary operation as being run largely by plant personnel themselves rather than by specialists. The use of some specialists as advisers on the evolutionary operation committee does not seriously vitiate this principle. In practice, the time spent by the specialists is perhaps one afternoon a month, and the ultimate responsibility for running the scheme still rests with the plant manager and not with the specialists.

When not to Stop

With an alert team of workers new ideas should be continually forthcoming and the evolutionary method becomes virtually a permanent mode of operation and should be so regarded. Only if it seemed that more would be lost than gained from the evolutionary procedure would the reintroduction of static operation be justified. In practice it is found that even very small gains will justify the continual operation of the evolutionary method. The situation at any given time can be appraised by the use of a pictorial log like that shown in Fig. 7.

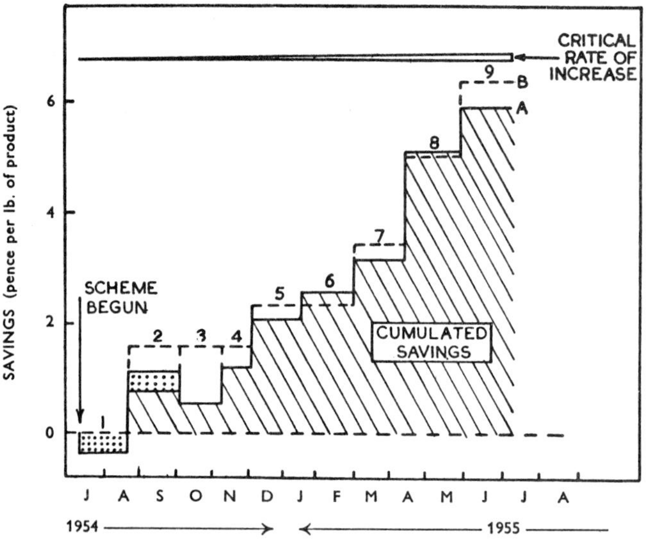

FIG. 7. Progress of an evolutionary scheme, showing critical rate of increase.

The full line A in this diagram shows the savings in pence per lb. which have been achieved in the various phases of operation. The dotted line B shows the savings that would have resulted if the centre or 'works process conditions', appropriate for each particular phase, had been run. On the assumption of constant output the shaded area is proportional to the accumulated savings resulting from the scheme, while each of the rectangular areas between the dotted and full lines shows the accumulated 'expenses' of running the scheme during that phase. The speckled area in phase 1 represents the cumulated expenses for the scheme during this phase and is debited from the cumulated savings in phase 2.

Whereas each phase of evolutionary operation, and consequently the expenses associated with it, lasts for only a limited time, any improvements which result go on for as long as the process is used. Suppose it is assumed that process improvements will go on earning money for p years after they are discovered and that the running of the evolutionary scheme adds c pence per lb. to the cost of the product. Then the question of whether or not, at any instant of time, evolutionary operation should be continued may be resolved by comparing the rate of improvement r which it is expected may be produced by the evolutionary process (measured in pence per lb. per year) with the critical rate of improvement r_0 given by $r_0 = c/p$. For if it is expected that the evolutionary scheme will need to be run for time t years to produce an improvement at the end of that period of rt pence per lb., and if k lb. of product is made per year, then the total saving during the p years for which the discovery is used will be $rtkp$ pence. During this time, kt lb. of product will be made and the loss due to running the evolutionary scheme will be ckt pence. Thus the scheme will pay off if $rtkp$ is greater than ckt, that is if r is greater than $r_0 = c/p$.

As an example consider the situation in Fig. 7. Should the scheme there shown be continued or not? Let us suppose (very conservatively) that improvements on this process are expected to go on earning money for 3 years after their discovery, so that p is put equal to 3. Suppose also that c is taken to be the average of the values experienced in the 9 previous phases. This gives the value $c = 0{\cdot}3$ pence per lb. We then find for the critical rate $r_0 = 0{\cdot}3/3 = 0{\cdot}1$ pence per lb. per year. Thus so long as the rate of improvement due to the evolutionary process is expected to be at least as great as 0·1 pence per lb. per year the evolutionary scheme should be continued. This critical rate of increase is shown diagrammatically at the top of Fig. 7. It will be seen that the actual rate of improvement which had been experienced over the previous year was about 6 pence per lb. per year (about 60 times the critical rate), and there is no evidence as yet of any flagging in this rate of improvement. There is therefore no doubt whatever that this scheme should be continued.

The example given is by no means atypical, which explains my insistence that the evolutionary method should be regarded as virtually a permanent mode of operation. It is psychologically wrong to talk of production under such a scheme as 'experimental manufacture', since an experiment is something which is done for a limited period and is not part of the normal run of things.

Some Questions and Tentative Answers

Like statistical quality control, evolutionary operation is designed for application in the factory itself. Its aims are different and more ambitious than those of quality control, however, since it is directed to ensure not a more uniform product but a cheaper and better product. I believe that this technique, if applied sufficiently widely, can have a marked affect in achieving greater productivity in industry by ensuring that the plant that is available, whether old or new, is operated in the best possible manner. The outline above is intentionally general because it is the general attitude and philosophy that is important here, and not the particular details of application. A full account is nevertheless being prepared in which a number of technical questions are discussed. In the present paper I shall do no more than indicate some of these questions and some tentative replies, which it is hoped to amplify and to justify in the later discussion.

Q. How many variables should be included at one time in an evolutionary scheme?

A. Usually two or three variables can be handled satisfactorily under the normal production conditions with which I am familiar. It should perhaps be emphasised once again that what is being discussed is the normal production situation in which evolutionary operation is applied. For specialist short-period investigations (which, as has been explained in the introduction, are *not* the subject of this account but which nevertheless play an extremely important part in the general scheme of process development to which evolutionary operation is a supplement) the situation is entirely different. In these specialist investigations where, *for a limited period*, it is permissible to interfere with production, and where special supervision and other facilities are made available, the object should be to saturate (or following Satterthwaite[9] possibly 'super-saturate') the experiment with as many factors as possible.

By studying the variables in groups of three or so at a time we forgo the possibility of detecting dependence (interaction) between variables not in the same group. The effect of this limitation should be minimised as much as possible by examining in the same group sets of variables which are expected to be interrelated. Periodically variables whose effects have been found to be important in different phases of operation should be tested together.

Q. What patterns of variants are of most value?

A. A variety of patterns of variants are useful for particular purposes. Among the most valuable for initial use are those based on two-level factorial designs with one or more added points at the centre conditions. These are the arrangements shown for two and three variables in Figs. 3 and 5. They have the advantages that:

(i) They are simple to comprehend, perform and analyse.

(ii) The added centre points allow continual reference to the 'standard' process and permit the 'cost' of the evolutionary scheme to be assessed.

(iii) Complexity of the surface is easily detected by considering the relative magnitudes of the simple 'main effects' on the one hand and the 'change in mean' and interaction effects on the other.

(iv) They can be made the nucleus of more elaborate designs (in particular of composite second-order rotatable designs[10]) by which complexity of the surface may be elucidated.

(v) They lend themselves conveniently to 'blocking arrangements' whereby extraneous disturbances due to such uncontrolled factors as time trends may be reduced. In Fig. 5(*a*), for example, the circles and squares indicate two sub-cycles into which the complete cycle may be divided. A general change in mean occurring between sub-cycles will not bias the estimation of effects.

Q. How should past plant-records be used in planning the evolutionary scheme?

A. When, as is often the case, past plant-records covering long periods of normal operation are available, the planning of an evolutionary scheme should always begin with a careful study of these records. In particular they may be used to determine the approximate magnitude and nature of the uncontrolled variation in the various responses, and consequently the number of repetitions of the cycle likely to be needed to detect effects of a given size.

From these records we can find out whether the errors in the principal response can be regarded as effectively independent and, if not, we can determine the nature of the dependence that exists. This is of some importance in choosing the period for which each variant should be run before changing to the next variant. In practice, of course, this period depends partly on convenience of operation and, for continuous processes, on the time it takes for the plant to settle down after a change in reaction conditions. The time for running each variant which gives the maximum amount of information for a given total period of production can be shown to depend on the nature of the dependence between the observations, and may be determined by a fairly simple use of the correlogram along the lines considered by Jowett[11] or, equivalently, by considering the spectral properties of the record.

Q. Should the variants be run in random order?

A. Faced with the possibility of serial correlation between successive observations, the statistician would normally wish to perform the variants in random order within each cycle, thus guaranteeing the validity of the simple type of analysis used in the example of Fig. 4. However, in some cases, particularly where the time for running each cycle is short, it is much simpler to run a systematic routine of variants on the plant than a random one. In these circumstances randomisation is usually abandoned (as was in fact done in the example considered). The observations after n cycles of k variants can be written in a table having n rows and k columns, and we are only concerned with comparison of *column* means. Now the major part of the dependence occurs *within* rows, and in this situation, as was pointed out by R. A. Fisher,[14,15] the simple analysis of the type we have considered above will usually not be seriously invalidated. Where the correlogram is available it is possible to determine how far dependence between observations will affect the simple analysis and what remedial measures, if any, need be taken, although this refinement would seldom be worth while.

Q. How should multiple responses be considered?

A. Although it is theoretically possible to equate all responses to a single criterion such as profitability, this usually presents great practical difficulties. As a general rule it is best to represent the problem as one of improving a *principal* response (for example the cost per lb. of product) subject to satisfying certain conditions on a number of *auxiliary* responses. These auxiliary responses usually measure the quality and important physical properties of the product.

Very careful thought in the selection of the principal response is essential. The vital question to ask is: 'If this response is improved will it mean *necessarily* that the process is improved?'

In the example of Fig. 4 the reconciliation of the requirements for the various responses in the light of the experimental results was done intuitively and led to the decision to reduce concentration alone. This intuitive approach has the virtue of simplicity and allows background information not emanating from the experimental results themselves to be taken into account. It is fairly satisfactory in the situation specifically dealt with here when there are only two or three variables to consider. However, as has been pointed out elsewhere,[12] the problem is really one of programming, in the sense of linear programming, with the added complications that the problems are not always approximately linear and that the restraints are not known exactly but must be estimated. In the fuller account we show how certain calculations can help with the more difficult cases.

Q. How best can the stream of information coming from the plant during the evolutionary process be presented to those responsible for deciding what to do?

A. Two things are necessary: first, to show how much weight ought to be attached to the results, and second, to present them in such a way that their interpretation is facilitated as much as possible.

To convey a sense of the degree of reliability which the plant manager should associate with the results, a number of ideas have been tried. In the original schemes various types of sequential charts and significance tests were used. It is now felt, however, that the problem is not one of significance testing and that what is needed is a presentation of the information contained in the data unweighted by external features subsequently injected into the situation. For example, the particular choice of the risks α and β and of the hypotheses 'which it is desired to test' (subtleties not readily comprehended by plant managers) can completely alter the apparent implications of a set of data when these are plotted on a sequential chart. If the observations are roughly normally distributed, are independent, and have constant variance, then all the information they contain is included in the mean, the standard deviation, and the number of observations. These statistics seem best comprehended in the form of a mean with its 95% confidence limits. In appraising the results prior information about the importance of different sorts of effect must be used. It seems best, however, to separate this from the presentation of the results, which then refer to information supplied by the observations and to nothing else. This problem is regarded as being far from solved. It involves many intricacies which cannot be discussed in the present account and is probably best considered in terms of stochastic approximation[16] and servo-mechanism theory. All that would be claimed for the present method is that it does allow a satisfactory evolutionary process to go on.

To show the implications of the mean results once it becomes apparent that these are determined sufficiently accurately, there is no doubt that for two or three variables geometrical representation, such as that shown in Figs. 4 and 5, is ideal. It allows the general trend in the responses and their relationship to each other to be appreciated in a manner not possible in any other way.

The plant manager should run evolutionary operation in much the same way as he would play a card game. The information board shows him his 'hand' at any given time, and depending on that hand there are a number of actions he can take (including drawing a further card and deciding what to do then).

Where one or more of the variants is clearly better than the works process or where clear-cut trends in the results exist, the plant manager will have no difficulty in following the indications of the information board. Where the results indicate that complexity exists he will be able to obtain the help of the statistician on the evolutionary operation committee in elucidating the results and, where necessary, in augmenting and modifying[4-6] the cycle of variants in the next phase of operation so as to resolve the complexity.

A duplicate information board may be kept on the plant itself and its significance explained to process workers. This provides added interest and is an incentive to accurate operation, which itself can result in general improvement in productivity.

Q. In what way can the results from small-scale experimental studies be used in planning the evolutionary scheme, and how should this affect the way in which these small studies are conducted?

A. The complexity mentioned above arises principally because the variables studied fail to behave independently in their effects on the response; that is, they interact. The plant process will usually have been arrived at as the result of a small-scale investigation of at least some of the variables. This investigation should have culminated in a study of the local 'geography' of the response surfaces in the neighbourhood of the proposed operating conditions. The principal features of the laboratory response surfaces will normally be preserved on the plant scale even though some distortion occurs. If the characteristics of the laboratory surfaces have been determined in the manner mentioned above, it is frequently possible to discover transforms of the variables originally considered which act approximately independently, at least for the principal response. By working in terms of these new variables, difficulties due to complexity of the surface can be greatly reduced in the plant-scale investigation.

Q. In practice it is impossible to attain truly static operation. Small variations in the process conditions are bound to occur from one run to the next. In cases where these changes are recorded, why should one bother to carry out a special pattern of variants? Why not use the 'pattern of variants' supplied by the natural variation of the process to supply information on which evolutionary improvement can be based?

A. It is true, of course, that for data generated by natural variation the simple type of analysis of the results which has been used above would no longer be applicable. However, this itself is no reason why the natural pattern of variants should not be used. Suppose that the level of response is denoted by y and that there are k variables whose levels are denoted by $x_1, x_2, \ldots, x_k$; suppose that the works process is defined by the particular set of conditions $x_{10}, x_{20}, \ldots, x_{k0}$ and that, owing to imperfect control, fluctuations about those levels occur and are recorded. Then we can, for example, assume a local relationship of the form

$$y = b_0 + b_1x_1 + b_2x_2 + \ldots + b_kx_k$$

and estimate the coefficients continuously by the method of least squares (multiple regression). If our assumptions were correct, these coefficients would measure the individual effects of the variables. The calculations required to fit the equations are laborious but, as has been suggested by Professor Goodman,[13] could in principle be done mechanically (e.g. by an electronic computer).

At first sight such a method appears attractive, for here we seem to have an evolutionary scheme in which we do not need to bother about introducing variants deliberately. On closer examination, however, its value seems much more doubtful. Many investigations have been made by statisticians over the years in which plant records have been analysed by multiple regression in an attempt to determine the 'effects' of the variables and so to improve the process. In my experience the results of such investigations are nearly always disappointing. The reasons are not far to seek:

1. Many of the factors that may vitally affect the efficiency of the process are not in the normal course of events altered at all.

2. Those factors which vary naturally do so, not over the ranges we should like, but over ranges dictated by the degree of control which happens to exist. The more control is improved, the less information we get.

3. The fluctuations that naturally occur in the variables are often heavily correlated. This results in poor precision of the estimates when we try to disentangle the effects of the variables one from the other.

4. Accidental modifications often tend to happen in 'phases' and so become spuriously correlated with causally unrelated time-trends in response. Such effects can lead to completely wrong conclusions. Attempts to eliminate time-trends computationally usually eliminate the effects of the factors at the same time.

What this all amounts to is that a naturally occurring scheme of variants is not very likely to provide a good, or even passable, 'design' and consequently that the amount of information generated by natural variation may be scarcely worth salvaging.

Q. Can evolutionary operation be made automatic?

A. The procedure of evolutionary operation so far described is a 'manual' one. It requires no special facilities and can be immediately applied in one form or another to a very large proportion of industrial processes. This is so whether the available plant is of the crudest kind or whether it includes such refinements as automatic controllers and recorders. The plant manager is himself a part of the 'closed loop', thus ensuring that sensible action will be taken even in unforeseen circumstances.

With a sufficiently instrumented plant the evolutionary procedure is, of course, capable of being made completely automatic. Thus variables whose levels are regulated by a controller can be automatically changed at regular intervals so as to follow a cycle of variants, and a response such as cost per lb. can be automatically computed from the readings of instruments which measure the properties of the product. The cumulated differences in response at the various process conditions can be used to trigger off adjustments in the location of the pattern of variants, so completing the evolutionary process.

In continuous processes (where there is a continuous input of starting materials and a continuous output of product) the 'pattern of variants', instead of being a discrete set of points as in Fig. 3, can consist of a continuous locus. The problem of detecting the effects of the variables is then precisely that which arises in communication theory, of detecting a signal of known form in a noisy channel.

The introduction of automatic evolutionary operation would usually be worth while only if the response surface itself was changing in some way and it was desirable to attempt to follow that change. For many chemical processes the response surfaces are reasonably stable. In some, however, unpredictable but steady changes can occur owing to slow changes in raw material (such as crude oil) or in catalyst activity. Here unpredictable differences in the position of the optimum conditions may occur between batches of catalyst and also within the life of a single catalyst batch. In these cases automatic evolutionary operation may be effective in keeping the plant operating near its best performance, but only if the rate at which information is generated is sufficiently large compared with the rate at which the optimum conditions are changing. This is essentially a problem in the theory of servomechanisms.

Discussion

The device I have described is of course only a more powerful and concentrated form of the naturally occurring evolutionary process which goes on during all manufacture. In the ordinary course of events once the favourable effect of a deliberate or accidental modification is *recognised*, that modification is included in the works process. Unfortunately, because of a high level of variation, which usually obscures all but very large effects, favourable modifications frequently go unrecognised unless they are forced to reveal themselves by the device used here of constant repetition and consequent averaging-out of errors.

That many of the problems touched on in the later part of this paper are still the subject of active investigation should not obscure the fact that evolutionary operation, as set out in earlier sections, is a practical and immediately available method which ought to be more widely applied.

Both practical experience and theoretical consideration show that very little can be lost and a great deal can be gained by application of the technique, and for this reason evolutionary operation should be adopted as a *normal production method.* Static operation should be tolerated only if good reasons for not using the evolutionary procedure can be advanced.

REFERENCES

[1] Youden, W. J. (1954 onwards). Bimonthly articles on 'Statistical Design', *Industrial and Engineering Chemistry.*

[2] Daniel, C. and Riblett, E. W. (1954). 'A multifactor experiment', *Industrial and Engineering Chemistry*, **46**, 1465.

[3] Vaurio, V. W. and Daniel, C. (1954). 'Evaluation of several sets of constants and several sources of variability', *Chemical Engineering Progress*, **50**, 81.

[4] Box, G. E. P., Connor, L. R., Cousins, W. R., Davies, O. L. (Editor), Himsworth, F. R., and Sillito, G. P. (1954). *The Design and Analysis of Industrial Experiments.* Oliver and Boyd, Edinburgh and London.

[5] Box, G. E. P. and Wilson, K. B. (1951). 'On the experimental attainment of optimum conditions', *J. R. Statist. Soc.*, B, **13**, 1.

[6] Box, G. E. P. (1954). 'The exploration and exploitation of response surfaces: some general considerations and examples', *Biometrics*, **10**, 16.

[7] Box, G. E. P. and Youle, P. V. (1955). 'The exploration and exploitation of response surfaces: an example of the link between the fitted surface and the basic mechanism of the system', *Biometrics*, **11**, 287.

[8] Anscombe, F. J. (1954). 'Fixed-sample-size analysis of sequential observations', *Biometrics*, **10**, 89.

[9] Satterthwaite, F. E. (1956). (Unpublished communication.)

[10] Box, G. E. P. and Hunter, J. S. (1956). 'Multifactor experimental designs for exploring response surfaces', *Ann. Math. Statist.* (In the press.)

[11] Jowett, G. H. (1955). 'The comparison of means of industrial time series', *Applied Statistics*, **4**, 32.

[12] Box, G. E. P. (1955). Discussion at Symposium on Linear Programming, *J. R. Statist. Soc.*, B, **17**, 198.

[13] Wilkes, M. V. (1956). Discussion of paper 'Application of digital computers in the exploration of functional relationships' by G. E. P. Box and G. A. Coutie at the Convention on Digital Computer Techniques. *Proceedings of the Institution of Electrical Engineers*, **103**, Part B, Supplement No. 1, 108.

[14] Fisher, R. A. (1941). *Statistical Methods for Research Workers*, 8th edition, p. 226. Oliver and Boyd, Edinburgh and London.

[15] Box, G. E. P. (1954). 'Some theorems on quadratic forms applied in the study of analysis of variance problems. II: Effects of inequality of variance and of correlation between errors in the two-way classification'. *Ann. Math. Statist.*, **25**, 484.

[16] Robbins, H. and Monro, S. (1951). 'A stochastic approximation method', *Ann. Math. Statist.*, **22**, 400.

5.7

The Challenge of Statistical Computation

Keynote Address

G. E. P. BOX
University of Wisconsin

"When you can measure what you are speaking about and express it in numbers, you know something about it, but when you cannot measure it, when you cannot express it in numbers, your knowledge is of a meagre and unsatisfactory kind." This famous remark of Lord Kelvin reminds us how very important to scientific progress is the proper handling of numbers. And so it's not surprising that we should find this gathering today with statisticians and computer scientists and various hybrids meeting together. We are here to discuss the business of efficient scientific investigation particularly as it involves, on the one hand, data gathering and generation (as exemplified by the design of experiments and sample surveys), and on the other, data analysis. But more than that, we must consider the iterative interplay of data generation and analysis on the efficient production of information and how this may be facilitated by the discerning use of computers.

I think sometimes people wonder why statistics, with its emphasis on probability and the theory of errors, is so important in science. In fact, statisticians sometimes are criticized on the grounds that they are so busy looking at the part of the observations that is due to error that they fail to pay enough attention to the other part which contains the essential information. One answer is that the only way to know that we have an adequate model of a given system is to study the errors. It's rather like a chemist who is doing a filtration -- he can discover whether his filtration is fully effective by testing the filtrate and seeing if it is pure water.

The Collected Works of George E. P. Box, 1984, Wadsworth, Inc., Belmont, CA 94002.
Originally published in *Statistical Computation* (New York: Academic Press, 1969), pp. 3–10.

And that's the sort of thing we do. An adequate statistical model is a transformation of the data that provides random noise -- random noise that is uncorrelated with any possible input variable that we can think of. To know we have a model which fully accounts for some physical phenomenon we must be sure that what is left, after the effect of the model is allowed for, is informationless; and information must be discussed in terms of probability.

The business of model building is an interesting iterative process. It seems to consist of three stages, used in alternation, which may be called

Model Identification,
Model Fitting, and
Diagnostic Checking.

Model identification is an informal technique which statisticians have been regretably loath to own up to, and to discuss. Here one is trying to get some idea of what model or class of models (or of hypotheses or of conjectures) is worthy to be tentatively entertained. This will obviously include such questions as what variables should be considered. We cannot, of course, use efficient statistical methods at this stage because we don't know yet what the model is.

Model fitting or estimation is a much more popular field of study because at first sight at least it seems to be associated with the purely mathematical question "If A were true would B follow?" which is a sensible mathematical question even if A is patently false.

Diagnostic checking is partly involved with what have been called tests of goodness of fit. However, merely testing fit is not enough. It is insufficient to say to the experimenter "It doesn't fit, good afternoon." He wants to know how it doesn't fit and when he knows this he can begin to consider how he should modify the model and thus to commence a second iterative cycle.

All of these procedures can benefit enormously from the use of the computer and, in particular, from imaginative choice of the form and display of output in such a way as is likely to interact with the human mind and so allow the creative process to proceed.

In some problems we may be dealing with very simple models but the sheer amount of the data may make the computer invaluable. In other problems the data may not be numerous, but the power of the computer is essential in coping with the complexity of the models that the scientist and statistician may originate or be led to.

Now of course human beings are supposed to be differentiated from other animals largely because they discovered how to use tools. Also, it is clear that there is enormous interaction between the things that humans do, and the tools they have -- the one producing development of the other. In particular, the nature and direction of enquiries which humans have undertaken have often been functions of the development of suitable tools and vice versa. Major quantum effects in the sciences have followed the development of suitable tools -- the theory and practice of astronomy made little progress before the development of adequate telescopes, and giant strides have been made since the introduction of radio telescopes.

And so it is with the computer. The existence of the new tool has created a revolution not only in the kinds of things that scientists do, but in their thinking, in their theorizing and in their demands for new tools to elaborate these thoughts.

This same revolution is also influencing the kinds of things that statisticians do. However, there is less of a revolution here than there should be, perhaps because there just aren't enough doers among the statisticians. This may even apply to some computer scientists -- but I get ahead of myself.

We are fortunate indeed to be living in a time when exciting developments can take place in the theory of efficient data generation and data analysis. For example, one class of problems in data generation and analysis, which the computer has made it possible to tackle and which has interested us here for some time, arises in the building of mechanistic models. Thus in the study of a chemical reaction we may wish to choose experimental conditions which will best discriminate between a group of possible physical models defined in terms of sets of differential equations, each set of equations being appropriate on some specific view of the nature and ordering of the component molecular combinations. Another problem of this kind occurs when we believe we have the right model and may wish to plan experiments which will estimate its parameters with greatest precision. Although the necessary numerical and statistical theory was certainly available, such problems were not considered until recently, chiefly, one supposes, because the computational aspects were too daunting. Similar developments have recently been taking place in the solution of nonlinear problems arising in the analysis of multiple time series, dynamic systems and control schemes. In more standard data analysis we have seen many advances, particularly in the relaxing of distribution assumptions, in the introduction of wider classes of models, in the analysis of residuals and in the introduction of new computer-oriented algorithms of analysis such as Wilkinson and Nelder have discussed.

The world is presently faced with tremendous problems which can only be solved by efficient use of scientific effort. The statistician and the computer scientist interested in efficient data acquisition and analysis have the satisfaction of working in a period when great contributions are needed from them and can be put into effect immediately. But as seems to be true with almost all innovations, on the one hand they give reason for hope of progress and on the other for alarm because of possible regression (and I use the last word advisedly).

There was a time when people who shouldn't do regression analysis or factor analysis didn't because it would have taken them too much time to find out how to do the calculations and to carry them out. Only the more determined spirits went ahead, and these were perhaps also willing to spend time to find out something about the assumptions and the pitfalls of the calculations they were making.

But now it's really too easy; you can go to the computer and with practically no knowledge of what you are doing, you can produce sense or nonsense at a truly astonishing rate. At this University, for instance, there are about ten thousand statistical jobs run on the computer per year. It is fairly certain that many of the people using these programs don't know very much about the methods they are employing and one suspects that some of the applications are nonsensical. The task of setting this right is rather like that of Hercules in cleaning up the Augean stables. Hercules was, of course, rather short handed, and that is the other difficulty.

Not only are we faced with this problem, but also the people who are concerned about it and can usefully lend a hand are extremely thin on the ground. Part of this is due to the something very peculiar that has been happening to statistics over the last 25 years or so. I spoke earlier about the iterative interplay of data generation and

STATISTICAL COMPUTATION

analysis. Of course, an iterative process of this kind in which practice and theory continually interact is also necessary to the progress of statistical methods for scientific investigation (and I don't know what else statistical methods are for).

Unfortunately, statistics has got divided up. There is a U group called Mathematical Statisticians and a Non-U group called Applied Statisticians. The effect of all the U-manship has, not surprisingly, been to produce a U-shaped distribution of talents with these two groups of people either ignoring each other or else eyeing each other distrustfully and getting further and further apart. The result is that instead of having a productive iteration between theory and practice, which history and common sense both show is the key to progress, we have theoreticians with less and less acquaintance with the real world, and we have work being done by (and advice being given by) applied people having less and less acquaintance with important theoretical ideas. Among this group there are many specialists in cookbook application who are insufficiently equipped to consider scientific problems on their merits and who have cut themselves off from the mainstream development of the subject of statistics.

Both of these extremes are dangerous and undesirable. Happily, individuals differ, but surely statisticians ought to have a bell-shaped rather than a U-shaped distribution of talent. The typical statistician should not be half a man -- whether that half be applied or mathematical.

We might need to coin a new term to define the golden mean, and indeed in recent years the term "data analyst" has been mooted. But surely a data analyst is another half man cut a different way. We all know that of even greater importance than data analysis is the gathering and generating of the right data to begin with. We surely need not be reminded that this job can easily be fumbled so that no analysis can help us. We ought not to choose a name such as may encourage those who would cast the statistician in the impotent role of trying to make something of someone else's sloppy data. I suppose we could speak of an Applied Mathematical Statistician, but then it might turn out that this was rated lower than a Mathematical Applied Statistician, so perhaps our ideal man ought just to be called a Statistician (who it should go without saying is trained in both the theory and the application of statistics).

We ought to learn to talk of a person as a Good Statistician in very much the same way that we would talk of a bridge builder as a Good Engineer if he had the necessary knowledge of theory and the necessary wisdom and practical experience to build bridges that didn't fall down (and hopefully were also aesthetically pleasing and inexpensive).

The existence of this U-shaped distribution of practical and theoretical talent becomes painfully clear to anyone who, for example, attempts to find a group of speakers to discuss topics such as the design of experiments or data analysis before an audience of scientists. One is looking for a set of people who will have something useful to say on topics which are surely the heart of statistics and who will not disgrace the profession because of one or another kind of naïveté. You count off a few familiar names, and you rapidly reach an end.

We have to try to cope with this problem. If we don't, we are really missing a good bet. There are all kinds of new and important theoretical problems which would quickly come to light if more competently trained statisticians became involved in real scientific investigations.

So I would like to suggest two things. First, that we ought to get people to see that to be half a statistician (whichever half one is, upper or lower, left or right) is not a desirable goal. We ought to talk this idea up at every possible opportunity and to try to get agreement that some kind of whole statistician ought to be our objective. Our teaching ought to be directed to this end and everything possible should be done to encourage inhabitants of the two ends of our distribution to educate themselves toward a better balance and to encourage once more the dialogue between theory and practice.

Part of the difficulty of communication rests on an elementary misunderstanding. Pure mathematics need be only concerned with the absolute truth or otherwise of a conditional statement. If A is true then B is true. Such statements are of mathematical interest even when there is no case in which A is true. Statistics, on the other hand, is a branch of applied mathematics used in the generation and analysis of scientific knowledge and must be concerned with the possible truth of A.

A first attempt at Wisconsin to alleviate the problems that I have mentioned employs the twin idea of a Statistical Consulting Lab and the Statistician in Residence

STATISTICAL COMPUTATION

program.

Graham Wilkinson is this year's Statistician in Residence. Each year we try to fill this visiting appointment with a candidate experienced in consulting and with wide theoretical and practical knowledge of statistics. He and Don Watts, the director of the Statistical Consulting Lab, spend about half their time discussing problems brought to them by experimenters from all over the University, with our graduate students attending the sessions so that they can learn how to consult. The Consulting Lab, of course, helps to ensure that better and more efficiently conducted research is performed in the University and that better use is made of statistics and of computing. Perhaps even more important, it provides an essential (and compulsory) part of the training of our graduate students. Consulting counts as a course which all must take for credit. Actual involvement of the student in the investigative process and the writing up of satisfactory reports are necessary for the student to pass the course. A grant from the General Electric Company has helped us to initiate the Statistician in Residence Program. The Consulting Lab is supported by the Computing Center and the Graduate School as well as by the College of Letters and Science. We are hoping that we can find a further source of funds so that other interested faculty can simultaneously contribute to the research efforts of their colleagues in other departments and to their own expertise as statisticians.

Another plan might be to try to do something with statistical computer programs along the following lines. Suppose we have a regression program -- a good program when used intelligently -- then perhaps the program should have on it a red label which says "If you haven't listened to tape 73, please do so." And tape 73 is a "Dutch Uncle" talk about regression and it's given by somebody who is a wise and experienced man in this area. Now, once a tape has been made, it can be reproduced so that it doesn't necessarily represent a large amount of labor. So we could perhaps go to some of the best people and ask them to provide appropriate talks. We could make a good start, for example, if we could persuade Frank Anscombe to talk about the analysis of residuals, Bill Cochran to tell us about the wise use of analysis of variance, and Cuthbert Daniel to discuss the criticism of experimental data. There might also be written material listing other references, but, I think, a tape too -- a tape with a sensible person talking about the philosophy of the thing, about what can be done, what cannot be done, what is being assumed, what the experimenter should be particularly concerned about, and so on. We are fortunate to live in such challenging times.

5.8
Statistics and the Environment

G. E. P. BOX[1]
University of Wisconsin

The Problem

It seems only a little time ago that we were concerned with matters which now seem comparatively trivial. We had for some time lived with the knowledge that our survival was threatened by nuclear attack by a foreign enemy, but it seems only recently that we have noticed a more insidious threat of our own making. Most of us now recognize that we are well on the way to destroying ourselves by overpopulation, pollution, the frittering away of our raw materials, and the poisoning of our food by inadequately tested chemicals.

Opinions differ as to how long it will take before various predictable crises occur and how much each problem will complicate the solution of the others, but it is very clear that we will be hard pressed and we will be lucky to escape by the skin of our teeth. The truth is that although we are called on to meet very difficult problems of great urgency we know pathetically little of the facts. So we must learn fast.

Now it is precisely this ability to learn fast that has got us into our present difficulties. It was only a few hundred years ago that men's minds seriously turned to the question of how the, normally very slow, process of learning by chance experience might be accelerated. Scientific method, the secret of learning fast, has altered the normal birth and death process, yielding perhaps a more comfortable world but at the cost of world overpopulation. Scientific method has provided us with motor cars and factories producing convenient products, but the by-products of both are threatening the air we breathe and the water we drink. Furthermore, their insatiable appetite for raw materials is stripping the earth of its irreplaceable treasure. Scientific method has provided us with conveniently packaged foods with chemical additives which make them taste good, look good, and last a long time on the shelves of the supermarket, but pharmacologists will tell you that it is almost impossible to keep up with the flood of these new substances which we ingest, and to be sure what are their long term effects on human beings.

So we are hopeful that the same scientific method which has in a period of a few hundred years got us to where we are now, can in a few decades get us to where we would like to be.

I believe it can, but with two provisos:

First, we must release, by public education, the will to make it happen.

Second, because with so little time we cannot afford inefficient investigation, we must catalyze the learning process still further. The catalyst is the proper use of Statistical Methods.

Science and Statistics

It was Lord Kelvin who said, "When you can measure what you are speaking about and express it in numbers, you know something about it; but when you cannot measure it, when you cannot express it in numbers, your knowledge is of a meagre and unsatisfactory kind: it may be the beginning of knowledge, but you have scarcely, in your thoughts, advanced to the stage of science." But, in case that should seem too much an encouragement to those who believe that mere unthinking accumulation of numbers is synonymous with good science

[1] Supported by the United States Office of Army Research under grant number DA-ARU-D-31-124-72-G162.

The Collected Works of George E. P. Box, 1984, Wadsworth, Inc., Belmont, CA 94002. Originally published in *J. Wash. Acad. Sci.,* vol. 64, no. 2 (1974), pp. 52–59.

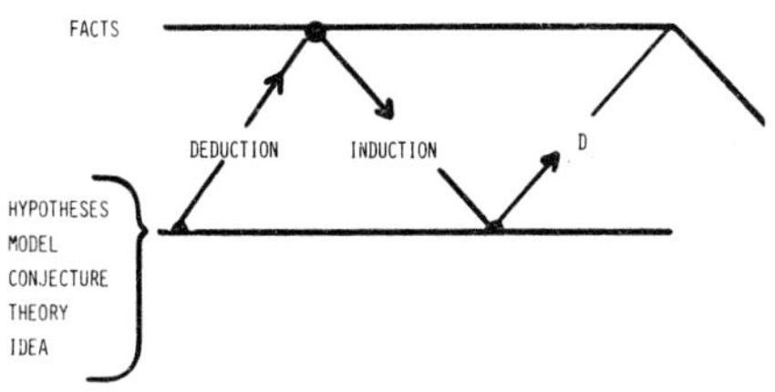

Fig. 1. The iterative learning process.

and will of itself solve the problem, I hasten to add the well known words attributed, among others, to Mark Twain, a contemporary of the noble lord's: "There are three kinds of lies—lies, damn lies, and Statistics."

What then is scientific method and what part does Statistics play within it?

Scientific method is a process of controlled learning. The object of statistical method is to make that learning process as efficient as possible.

Learning is an iterative process, illustrated in Fig. 1, in which a hypothesis (or theory or model or conjecture) leads by a process of deduction to certain consequences which may be compared with known facts. Usually the consequences and the facts fail to agree, leading by a process called induction to modification of the hypothesis. Thus a second iteration is initiated, the consequences of the modified hypothesis are worked out and again compared with facts (old or newly acquired) which, with luck, leads to further modification and to further gaining of knowledge.

This process of learning can be thought of in terms of the feedback loop shown in Fig. 2, where discrepancy between the facts and the consequences of the initial hypothesis H leads to the modified hypothesis H′. This view makes it clear why there is no place in science for the man who wants to demonstrate that he has always been right. For it is by arranging matters so that there is maximum opportunity to find out where he may be wrong, that most progress is made.

Suppose at a certain stage in an investigation the situation is that shown in the bottom half of Fig. 3. A hypothesis H concerning the state of nature has been formulated, leading to certain consequences that have been compared with the facts deduced from analysis of the available data. Discrepancies have suggested a modification from H to H′. Consequences of H′ may now be in accord with the data analysis or may still be discordant. When it is not clear what modification should be made to an unsatisfactory hypothesis or, alternatively, when confirmation of an apparently satisfactory hypothesis is needed, further data must be sought. Depending on the context, the further data may come from a designed experiment, a sample survey, or already existing results. Whatever the source of the data, careful attention must be given to its selection or design. As illustrated in Fig. 3, the direction of the effort at data getting will inevitably depend on our latest view of the state of nature and the hopes and fears which surround that view.

While, at a particular stage, the conjectured state of nature may be false or at least inexact, the data themselves are generated by the true state of nature. It is because of this that the comparison of

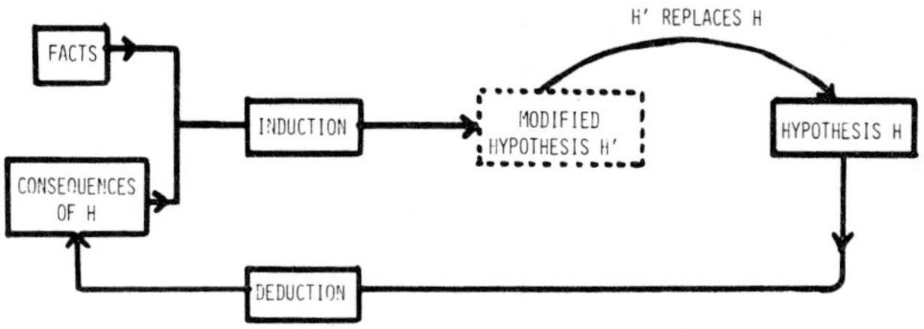

Fig. 2. The learning process as a feedback loop.

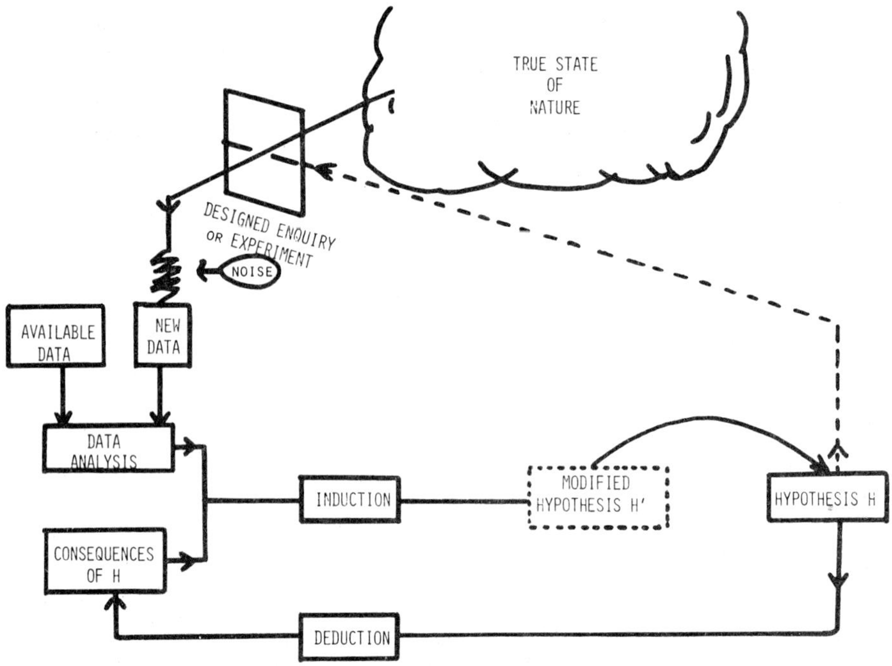

Fig. 3. Data analysis and data getting in the process of scientific investigation.

successively conjectured states of nature with actual data can lead to convergence on the truth. Even if we could see such data free of experimental error, however, the task of discovery would usually not be easy because of the complexity of the systems that need to be studied. So in practice in addition to complexity, we have to cope with an added difficulty—that the data contains experimental error (or noise), which tends to mislead.

Scientific investigation, then, is not easy, and obviously the process we have described depends crucially on the scientific wit and subject matter knowledge of the investigator. The statistician's job is to advise and assist the investigator in two crucial tasks, so as to allow the investigator to employ his talents most efficiently. These tasks are:

(1) deciding what would be appropriate data to get at each stage of the investigation. Broadly we can call this the *design* problem.

(2) deciding what the data entitles us to believe at each stage of the investigation. We can call this the *analysis* problem.

Of the two, *design*—the decision as to what are the appropriate data to get—is of paramount importance. This is equally true whether by actual design of an appropriate experiment, the planning of a suitable sample survey, or the proper choice of a data base. No amount of skill in data analysis can extract information which is not there to begin with. The second task of the statistician, although not so vital as the first, is still very important. Inappropriate analysis of data can produce unjustifiable conclusions or fail to discover justifiable ones. *Worse,* it can fail to unearth those hints of, perhaps unexpected, phenomena which often catalyze the investigator's progress to a solution. In any case, inappropriate analysis of data will greatly hamper convergence of the scientific iteration.

In summary then, we learn through numbers. But what numbers or data should we try to get and what do they mean when we have them? These are the questions that good statisticians are trained to answer. It is very easy to acquire useless or irrelevant data. It is very easy to be misled by data once they are acquired. The design of *each stage* of an enquiry so as to produce useful data with the minimum of time and expense, and the analysis of data of each stage so as to produce, not only valid conclusions, but also valuable hints on how the investigation ought to proceed, these are the two critical tasks in which the statistician plays a key role.

Part of the statistician's job is also, I think, to encourage and accompany the scientist in the slightly schizophrenic role that he has perforce to play.

Having entertained a tentative model (hypothesis, etc.) it is up to the statistician to see that fully efficient means are used to investigate the consequences of that model. That is the inference step in Fig. 4. However, having then produced the best analysis possible, supposing the model to be accurate, he must now change his stance from that of a sponsor to a critic. He becomes a doubting Thomas prepared to find fault by inspecting residuals for suspicious features, etc. This criticism can lead to modification of the model, either at once or at some time after more data has been taken.

Switching alternately from sponsor to critic and back again is a painful business but one which we must steel ourselves to pursue. The Pygmalions who have fallen in love with their models somewhere along the way are a nuisance and a hindrance to progress.

Another part of the statistician's job is to make sure that Statistics and Computers do not separate the investigator from his data but, on the contrary, help him to see his data from many different angles. We must remember that the best induction machine so far devised is the human mind, and if modern methods of dealing with data result in separating the investigator from his data, they are almost certainly doing more harm than good.

Going now into a little more detail, what then are some of the difficulties that appropriate use of statistical methods can alleviate or avoid?

Coping with Natural Variation

We live in a world which is universally variable. How much air a man breathes depends on the particular man, his temporary physiological state, the atmosphere he is presently in, and so forth. And yet, until quite recently, attempts were made to study variable phenomena in an entirely deterministic manner. Variation was frowned upon, as if disapproval

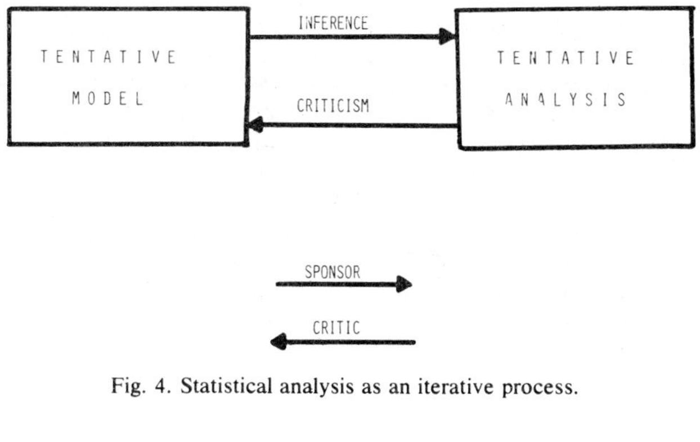

Fig. 4. Statistical analysis as an iterative process.

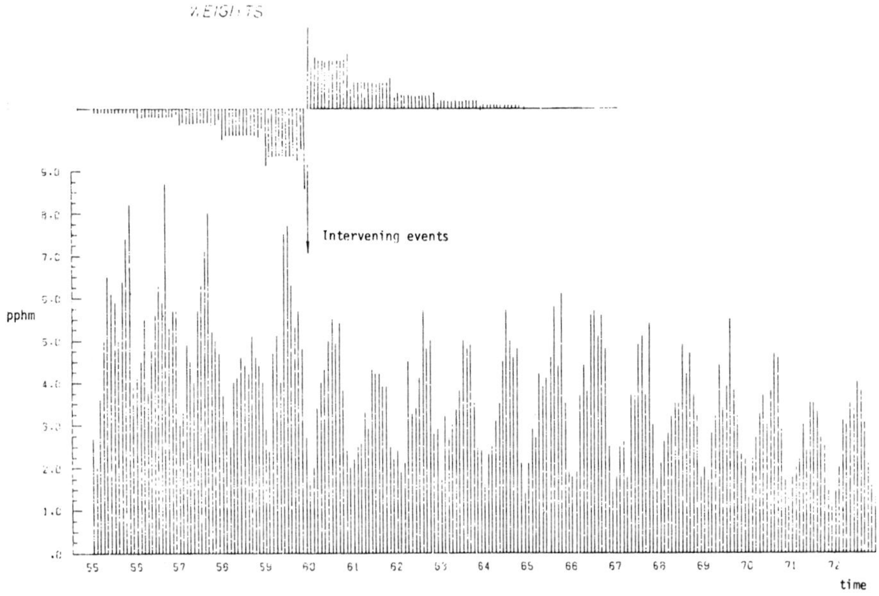

Fig. 5. Monthly average of hourly readings of O_3 (pphm) at downtown Los Angeles (1955–1972), with the weight function for estimating the effect of intervening events in 1960.

would make it go away, and probability statements were treated as in some way unsatisfactory. There was little readiness to admit that everything varies and, except perhaps from God himself, every statement, if exactly made, would have to be a probability statement.

Increasing Accuracy by Exploiting the Variational Structure

Environmental data are usually highly variable. It is by facing this fact, rather than running away from it, that we can solve some of our problems. Indeed, it is a fascinating fact, that it is the *structure* of the variation or noise, which determines how we can extract the information which the data contain. As an illustration, Fig. 5 shows monthly averages of oxidant (O_3) levels observed in downtown Los Angeles from 1955 to 1972. These data are highly seasonal and variable. About the beginning of 1960 two events occurred which might have been expected to change these levels. These events were the diversion of traffic by the opening of the Golden State Freeway and the coming into effect of a new law (Rule 63), which reduced the allowable proportion of reactive hydrocarbons in the gasoline sold locally. By a study of the structure of the variation it is possible to obtain (Box and Jenkins, 1970; Box and Tiao, 1965, 1973; Tiao et al., 1973) a valid and most sensitive test of the possibility that the events in January 1960 changed the oxidant level and to estimate the change. For this data the estimate turns out to be -1.10 ± 0.10 p.p.h.m. The function shown at the top of the diagram displays the manner in which the data are weighted in the optimal difference estimate. As common sense might expect (i) most weight is given to values obtained immediately before and after the events and remote data are suitably discounted, (ii) the weighting is automatically chosen so that seasonal effects are eliminated.

I believe that the use of "Intervention

Analysis'' such as the above, in which difference equation models are used to represent dynamic and stochastic systems, has much to contribute in uncovering possible effects of public policy changes. For example, it could show the effect of the opening of a nuclear power station on the ecology of the river from which cooling water is drawn and returned. It is clear that studies of this kind are vital to the intelligent framing of new laws.

We owe to Sir Ronald Fisher the concept that we can exploit the patterns of natural variation in data to *design* enquiries and experiments so that errors are minimized. For example, randomized block designs and stratified sampling plans can eliminate major sources of disturbance and ensure that important comparisons are made within the least variable material.

Another tool which should, I believe, find much application is the use of components of variance to improve tests of environmental quality. The analysis of variance table used in the analysis of data from the randomized block designs I mentioned above may also be employed in conjunction with suitable designs to estimate components of variation, for example, in tests of environmental quality. Suppose we take a sample from a stream and perform an analysis. How accurate is the result we get? What do we mean by that question? Certainly not how closely repeated chemical analyses of that same sample would agree with one another. What we want to know is how nearly does our analysis give a picture of the quality of that stream at that time and place.

An appropriate study of components of variance—how much variation is associated with chemical analysis, how much with the sampling method, how much with change of location in the river, together with knowledge of how much it will cost to take a sample and perform a chemical analysis—enables us to devise a testing scheme which can be dramatically more accurate and economical than one naively chosen.

Causation and Correlation

Many years ago when I studied statistics at University College London there was a plot of some data which none who saw could easily forget. On the x axis was the number of storks' nests observed each year in a certain town; on the y axis was the corresponding human birth rate for that year. The data showed an almost perfect straight line relationship. It is perhaps superfluous to explain that the correlation arose because, over the period of years in which the data were taken, the stork population was increasing and so was the human population. It is also unnecessary to point out that our over-population problem will not be solved by shooting storks.

In case these remarks should seem frivolous we must remember that it was precisely this kind of question which was debated in some of the early discussions on smoking and lung cancer and which bedevil much data analysis in other fields.

Again it was Fisher who showed how in planned experimentation the introduction of randomization could break the purely correlative chain and enable causation to be distinguished. In cases where planned experiments are not possible the situation is always very tricky, and very careful analysis is needed to decide in any given case precisely what the data allow us to conclude.

Complexity in the Face of High Noise Levels

Many of the phenomena we face in considering the environment are complex. To cope with problems which are complex, as well as being obscured by experimental error, we would be wise to welcome whatever help we can get. Even though the complexity of problems is admitted, the idea that variables should only be studied one at a time dies hard. The one variable at a time method would, of course, only be a satisfactory mode of study if nature were so obliging as to have its variables affect the environment independently. Again, it was Fisher who pointed out that by the use of suitable design the effect of experimental error could be averaged out at the same time that provision was made for the estimation of complex effects. Designs of this kind may be used, not only for empirical descriptions of phenomena, but also for testing mechanisms. This is done by treating as data the estimated constants of the system. If the model is correct, these should remain constant when extraneous conditions are varied. When, as is usually the case initially, the model is not wholly correct, analysis of the changes in the "constants" provides a valuable diagnostic tool for model testing, pointing to where the model needs attention. Endelman (1973) has recently used these methods at Wisconsin to study nitrogen changes in the soil and soil water. In many ways this study was a model one, in which the Departments of Soil Science, Chemical Engineering, and Statistics all cooperated.

While on the subject of complexity a word should be said about the models needed to represent complex phenomena. In any given investigation it seems to me we can err in two ways. We can have too simple a model or too elaborate a model. My recent experience has been that investigators have often erred in building models that are too elaborate. There is a tendency to try to model each step that the investigator can imagine, whether there is strong evidence that that step really occurs in the system or not, whether the step affects the solution or not, and whether the data could possibly supply any information about that step or not. Even if he had a 50% chance of being right about any given step, the investigator need only introduce a few such steps into a system and the chance of error becomes overwhelming. My experience is that we must borrow William of Occam's razor and use it rather ruthlessly to remove deadwood. Usually, models are best built up from simple beginnings, elaboration being introduced only as it is shown to be necessary by actual comparison with data, as in Fig. 3.

The Peril of the Open Loop

Perhaps of all the problems that face us, whether personal, professional, scientific or statistical, the most menacing of all is the danger of the open loop.

I have spoken of the process of scientific learning in terms of a feedback loop. If the loop is open, learning stops, of course. The idea applies more generally. As an earlier speaker has so ably pointed out, feedback is essential between scientists and legislators; otherwise, even when the scientists know what to do, it cannot get done.

As another example, I recently attended a seminar where the speaker was building a pollution model for a city. The method he used was to calculate by dead reckoning the amount of every substance going into the atmosphere over each small area of the city. For example, he could calculate over, say, a given hundred yards square area, how much rubber was worn off the tires of automobiles passing through that area and hence presumably going into the atmosphere. There was nothing wrong with that, but I was surprised to hear him explain, as he commenced his seminar, that there were two kinds of modelling—his kind based on dead reckoning and statistical modelling based on data. Learning happens surely only when the loop is closed and what can be calculated from dead reckoning is compared with what the data actually say.

The Supply of Competent Statisticians

Perhaps finally I should say something about the supply of statisticians. A little while ago I saw a report prepared by a distinguished panel of mathematicians on the current need for graduate training in mathematics and mathematically related subjects. One conclusion was that since a principal outlet for Ph.D. mathematicians was as university teachers and since the great expansion of the universities had now ceased, we must plan for a major cut-back in the production of Ph.D.'s or face the possibility of producing a glut of unemployed mathematicians. I was alarmed because the "mathematically related subjects" which the report claimed to cover included statistics!

Now whatever may be true about the future need for pure mathematicians, the fact is that we face a scarcity of trained statisticians competent to deal with real problems. Furthermore as, one by one, the various environmental crises become more obviously imminent and the need for hard facts on which to take sensible action becomes inescapable, the demand for such people will markedly increase. It takes many years to produce a properly trained statistician. It cannot be over-emphasized that steps must be taken now not to restrict but to expand the educational facilities available for the training of competent statisticians.

How do we get competent statisticians? Neither surely by producing mere theorem provers nor mere users of a cook book. A proper balance of theory and practice is needed and, most important, statisticians must learn how to be good scientists, a talent which, I think, has to be learned by example. At Wisconsin, we have taken a number of steps to help this along:

- To obtain any graduate degree, a student must have spent a period of time in the Statistical Consulting Lab working with the statistician in residence and other faculty to deal with clients' problems. This counts as a course for credit, and no student can graduate without passing this course.
- The Masters Degree, which all students are encouraged to take, whether or not they proceed to a Ph.D., is not a "failed Ph.D." degree but is awarded on their demonstrated competence to becoming a practicing statistician.
- A Monday night beer session is held in the basement of my house where research problems are discussed on an ongoing basis.
- The department is deliberately diversified with joint appointments and research interests in engineering, business, medicine and agriculture.
- Students act as research assistants in projects such as the Analysis of the Los Angeles Air Pollution data, the improvement of operating methods for the local sewage works, etc.

When we look at the history of the subject of statistics itself, there is no doubt that it develops most rapidly when there is active feedback, with practical problems initiating new theory and new theory in turn showing new ways to handle real situations. I believe we are moving now into a period of great statistical activity where, because of the service it will render to the community, our science will come into its own. In doing so, it will inevitably undergo new and exciting development.

Literature References

Box, G. E. P., and Jenkins, G. M. 1970. *Time Series Analysis: Forecasting and Control.* Holden-Day.

Box, G. E. P., and Tiao, G. C. 1965. A change in level of a non-stationary time series. *Biometrika,* Vol. 52.

Box, G. E. P., and Tiao, G. C. 1973. "Intervention Analysis with Applications to Environmental Problems", Technical Report No. 335, Department of Statistics, University of Wisconsin, Madison.

Endelman, F. 1973. "Systems Studying of the Transport and Transformations of Soil Nitrogen", Ph.D. Thesis, University of Wisconsin, Madison.

Tiao, G. C., Box, G. E. P., and Hamming, W. J. 1973. "Analysis of Los Angeles Photochemical Smog Data: A Statistical Overview", Technical Report No. 331, Department of Statistics, University of Wisconsin, April.

5.9 Analysis of Los Angeles Photochemical Smog Data: A Statistical Overview

G. C. TIAO and G. E. P. BOX
University of Wisconsin

W. J. HAMMING
Los Angeles County Air Pollution Control District

A research project has been under way to investigate air pollution problems in Los Angeles County with the help of the data supplied by the Los Angeles County Air Pollution Control District. These data consist of measurements of primary pollutants such as nitric oxide, hydrocarbons, carbon monoxide, sulfur dioxide and particulates, and secondary pollutants such as ozone and nitrogen dioxide, recorded hourly at a number of different stations in Los Angeles County over the past seventeen years. This present discussion deals in a preliminary way with a particular aspect of this analysis, namely, the occurrence of photochemical smog in Los Angeles. The paper is divided into two main sections. The first is intended to provide a brief survey of the problem of photochemical smog in Los Angeles as presently understood in relation to the available field data and also in relation to chamber experiments which have been run in various laboratories. The second part of the paper discusses a class of intervention problems that arise in studying the data. It is noted that parallel problems occur in the study of other ecological material and elsewhere. Statistical methods for dealing with this class of problems are illustrated with some of the Los Angeles data.

Dr. Tiao is Professor and Chairman of the Department of Statistics and Dr. Box is R. A. Fisher Professor of Statistics, University of Wisconsin—Madison, 1210 West Dayton Street, Madison, WI 53706. Mr. Hamming has recently retired from the Los Angeles County Air Pollution Control District. This is the revised version of Paper No. 73-79, presented at the 66th Annual Meeting of APCA at Chicago in 1973.

A research project has been underway to perform statistical analysis of aerometric data from January 1955 to December 1972 assembled by the Los Angeles Air Pollution Control District. The principal pollutant source data are hourly readings on such primary contaminants as nitric oxides, hydrocarbons, carbon monoxide, sulfur dioxide and particulates, and secondary pollutants such as ozone* and nitrogen dioxide, recorded at seven appropriately distributed locations in the Los Angeles Basin. They are: Downtown Los Angeles, West Los Angeles, Burbank, Pasadena, Azusa, Long Beach, and Lennox. Atmospheric variables such as mixing height, wind speed and direction, and base inversion height are considered. Additional exogenous factors, such as the chronology of control measures, vehicle population, and traffic patterns will also be included in the analysis.

The eventual objective of this study is to utilize information from the data to build mathematical models which will adequately represent the chronological and spatial movements of the pollutants. It will then be possible to produce efficient forecasts of current and future trends of the pollutants and to assess the effectiveness of control measures.

This paper is the first of a series reporting on our research findings. It deals in a preliminary way with only one particular aspect of the problem, namely, the movement of photochemical smog in Log Angeles. The paper is divided into two main sections. The first is intended to provide a brief survey of the problem of photochemical smog as presently understood. Using O_3 as the basic indicator of smog, various graphical devices are employed to illustrate the effect of atmospheric variables and the historical development of the pollutant over the seven locations. In particular, it will be seen that considerable differences exist in the

* Strictly, other oxidants as well as O_3 are measured by the chemical test.

The Collected Works of George E. P. Box, 1984, Wadsworth, Inc., Belmont, CA 94002.
Originally published in *APCA Journal,* vol. 25, no. 3 (1975), pp. 260–268.

historical variations of O_3 over these locations. In places such as Downtown Los Angeles and West Los Angeles, the level of concentration was lower in the 1960's compared with the 1950's and there seems to have been some progressive improvement over the last several years. Corresponding changes are, however, not evident in the more easterly regions such as Pasadena and Azusa.

The second part of the paper discusses a number of statistical problems that arise in studying the data. As pollutant data are expected to be correlated through time, appropriate time series models which can adequately represent the data and allow for projection of current and future trends are considered. A new and widely applicable technique, called "intervention analysis," has been further developed and is here illustrated in assessing the effect of exogenous factors such as control legislation, opening of new freeways and other environmental changes.

The Photochemical Smog Problem in Los Angeles

Los Angeles has what may be called *general* and *special* air pollution problems. The nature of these problems has been reasonably well understood for some time.[1–3] The general pollution problem, which is currently found to some extent in any large city, is attributable directly to the properties of the primary contaminants themselves. For example, NO_x, SO_2, and particulates are of themselves undesirable and can cause lung damage if inhaled for sufficiently long periods in high concentrations.

The Product of Photochemical Smog

The special pollution problem of Los Angeles, and the most serious one, comes not from the primary pollutants themselves. It comes from substances produced by *chemical reactions* among certain of these primary pollutants. The products of these chemical reactions are responsible for the famous Los Angeles Smog. They include eye and lung irritants, and other agents, which in high enough concentrations can produce damage to living things both animal and vegetable. In addition, these products are also associated with visibility-reducing haze and high levels of ozone. The principal contaminants taking part in the reaction are: (1) oxides of nitrogen, denoted by NO_x, whose main constituent is nitric oxide, denoted by NO, and (2) reactive hydrocarbons, denoted by HC. A measured product which is indicative of the degree of photochemical pollution is ozone, denoted by O_3.

The chemistry is very complicated and is still not completely understood, but very roughly we can say that under certain very special conditions NO_x and HC take part in a photochemical reaction which eventually forms sufficient O_3 and other associated substances to produce undesirable effects. Thus, in several steps,

$$O_2 + NO + HC \xrightarrow{\text{sunlight}} O_3; \text{ and other products}$$

The very special conditions needed are: (1) that the main reactants (NO and HC) be present; and (2) that sunlight be present of sufficient intensity and duration to initiate and sustain the reaction in the air for several hours. The reaction is rather slow and it is typically three to five hours after its initiation before significant quantities of O_3 build up.

Daily Movement of Ozone

The behavior of O_3 during the day is illustrated in Figure 1. The chart is based on a two way table of O_3 measured in Downtown Los Angeles(DOLA). The entries are the monthly averages (in pphm, parts per hundred million) of each hour of the day from Jan. 1955 to Dec. 1972. Each column represents the average behavior of O_3 during the day for a particular month, and each row indicates the chronological variation of the pollutant concentration for a particular hour (Pacific Standard Time) of the day. To facilitate comparison, the entries are classified into four ranges by shading as labelled. The solid line in the figure indicates the time of occurrence of the peak level of O_3.

By reading Figure 1 vertically, we see that in Downtown Los Angeles the level of O_3 begins to rise about 7 A.M., peaks around the noon hour and dissipates in the afternoon. Most of the accumulation of O_3 during the daytime hours is the result of photochemical reaction of contaminants from automobile emissions and other sources.

The Seasonal Effect

From June to October which we shall loosely call "the summer," Los Angeles County provides an ideal natural reactor in which these chemical changes can take place. During this period occur (i) strong and persistent night and daytime inversions; (ii) weak and, for several hours each day, stagnant winds; (iii) bright sunshine.

Because of (i) and (ii) the primary pollutants from heavy morning automobile traffic and other sources are not dispersed and occur in rather high concentrations while (iii) allows the photochemical reaction to proceed. During the remainder of the year November–May, which we shall call "the winter," the inversion, although initially lower, usually rises higher by noon each day than in the summer. Also, the sunlight is of shorter duration and is less intense. In the winter, therefore, there is much less time and less available energy for the chemical reactions to take place and the problem is less serious.

The seasonal difference between the summer and the winter can be seen by studying Figure 1 horizontally which shows that the daytime concentration of O_3 is much lower in the winter time than in the summer. Also, the peak hour in DOLA shifts from 10 to 11 A.M. in the summer to 1–2 P.M. in the winter.

The figure also seems to suggest that the level of concentration in Downtown Los Angeles was lower in the 1960's compared with the 1950's and that there has been some progressive improvement over the last few years. This will be discussed in further detail later in the paper.

Effect of Inversion Mixing Height and Wind Speed

It is of interest to know to what extent changes in atmospheric conditions within a season may affect the concentration level of the pollutant. The maximum mixing height affects the volume available for diffusion of the pollutants, while the wind speed affects their dilution and transport.

This is illustrated in Table I showing means of daily averages of O_3 for various wind speeds and for various maximum mixing heights in Downtown Los Angeles during the four summers (1967–1970). As one would expect, on the days on which the mixing height and wind speed are lower, the concentration of O_3 is increased.

Table I. Means of daily averages of O_3 (pphm) for various wind speeds and maximum mixing heights in Downtown Los Angeles: June–October 1967–1970.

Wind Speed (MPH) 6-12 AM	Maximum Mixing Ht. (feet)			
	0-2000	2100-2500	2600-3500	3600 and up
0 – 4	5.7	5.6	4.3	3.8
4.1 – 5	4.8	4.5	4.1	3.1
5.1 & up	4.3	3.9	3.0	2.2

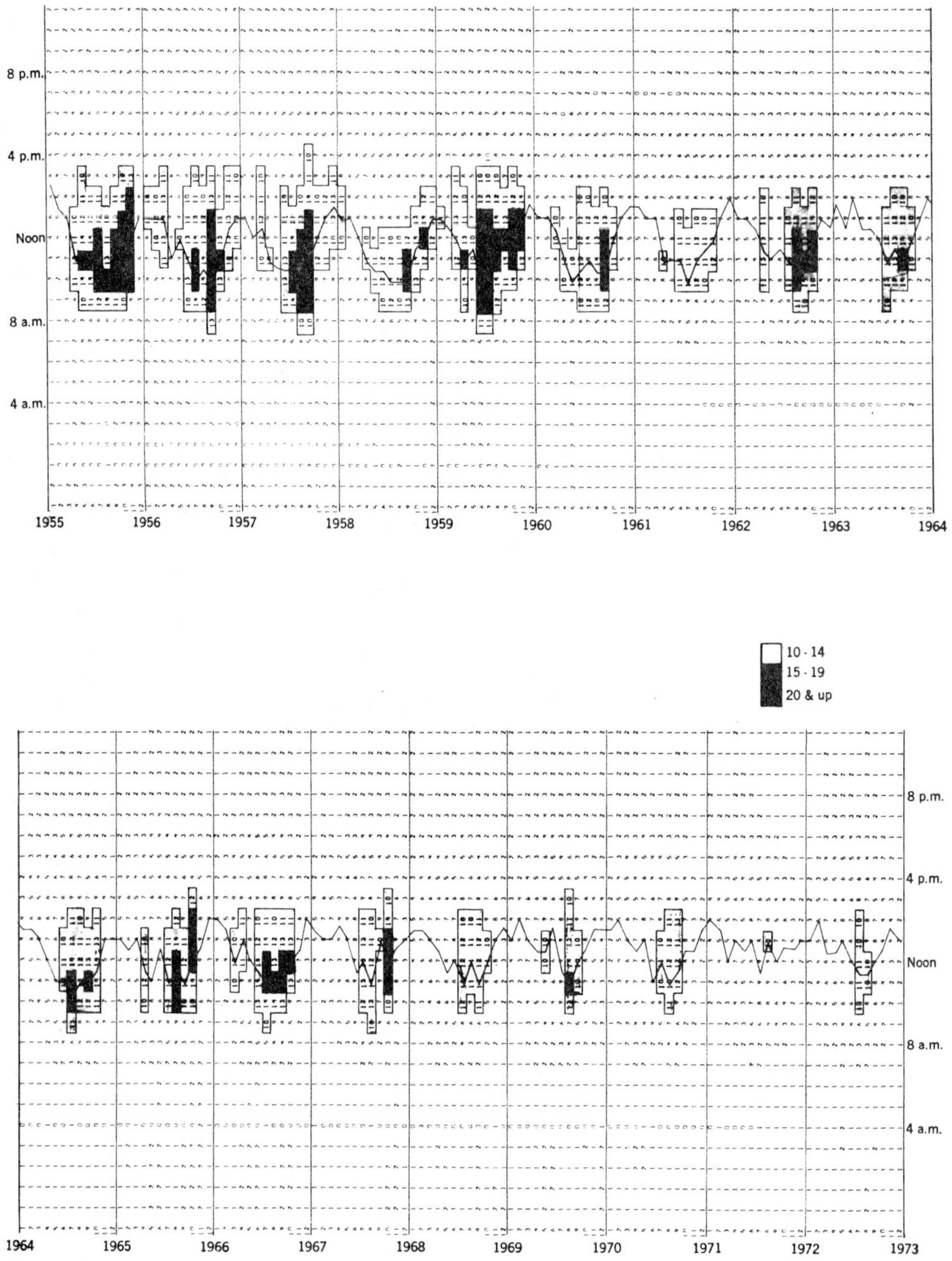

Figure 1. Monthly average of hourly readings of O_3 (pphm) at DOLA 1955–1972.

Geographical Location

In the above, we have confined our discussion to one location—Downtown Los Angeles. Charts and tables similar to Figure 1 and Table I have been developed for the other six locations.[4]

Preliminary study indicates that appreciable differences exist in the level of O_3 concentration over these stations. Some clue as to why this is so may be found in Figure 2 which shows the geographical location of the seven recording stations. At each location, the first two numbers are, respectively, the gross seventeen-year (1956–1972) summer averages for O_3 and the primary pollutant NO, and the third number is the average vehicle miles traveled per acre within a three-mile radius of the air monitoring station, estimated from actual counts collected over the last three or four years.

The arrows in the figure roughly indicate the direction of the prevailing winds and show the "pipe reactor effect." Thus, although the heaviest concentration of traffic is in the region around Downtown Los Angeles and southwest of it, the highest concentration of O_3 is at Azusa at the end of the "reactor pipe line." This is so even though Azusa itself has a traffic density which is much lower than that in Downtown Los Angeles. Similarly the concentration of NO is in general reduced as we move through the pipe line. This may be partly because it is used up in the reaction and partly because the local emissions are in general less in the Pasadena and Azusa region.

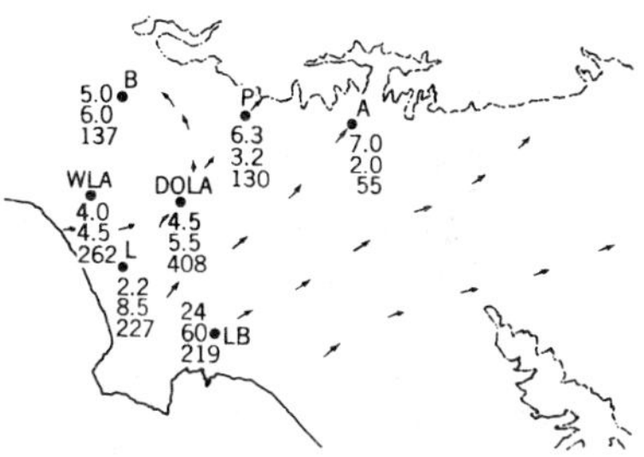

Figure 2. Geographical location of recording stations, with the gross seventeen-year (1956–1972) summer averages for O_3 and NO, and the estimated vehicle miles traveled per acre. Azusa (A), Burbank (B), Downtown Los Angeles (DOLA), Lennox (L), Long Beach (LB), Pasadena (P), West Los Angeles (WLA).

An Overall Look at the Past and Present

A natural question to ask is "Are things getting better, worse, or staying about the same?" Some preliminary answer may be found (Figure 3) in the relative frequency diagrams of daily maximum hourly concentrations of O_3 in four ranges for the seven locations over the last eighteen years. The range 8 pphm or below is of particular importance since it is the Federal Air Quality Standard for oxidant.

The diagrams illustrate how the situation has developed differently at different locations. Thus, from 1966 to 1972 the proportional frequency with which the Federal Air Quality Standard was met, increased in Downtown Los Angeles from 0.35 to 0.62 and in West Los Angeles from 0.51 to 0.82. However, in Pasadena and Azusa only minor changes occurred from 1966 to 1970, although an apparent improvement of about 0.10 did occur in 1971 for both locations. It should be noted that the sudden change in frequency occurring at Long Beach in 1963 is not necessarily an indication of improvement, since there was a change in the location of the monitoring station in that year.

Some Statistical Problems

What are some of the problems that must be considered in the analysis of data of this kind? Of course, a major problem is that of coping with so many numbers. There are in these records collected over 18 years more than twenty million separate pieces of data. Considerable effort has had to be expended in the organization of the data and numerous problems have been encountered, which we will not discuss here. At a more technical level analysis of these data has raised some interesting problems in statistical methodology. In the remaining part of this paper, we discuss one such problem which has resulted in new procedures for time series intervention analysis.

Intervention Problems

A class of problems likely to be important in many different kinds of ecological study concerns the question "Did a change of a certain kind made on a certain date make any difference?" More specifically "Did it produce the kind of difference to be expected?"

For example, in 1960 the Golden State Freeway which may have had an effect on the traffic pattern in Downtown Los Angeles was opened. Also, at about this time Rule 63 was introduced which reduced the proportion of reactive hydrocarbons in the gasoline sold in Los Angeles County. Thus about 1960 two events occurred either or both of which, could have, produced a change in level of ozone concentration in Downtown Los Angeles. Is there evidence that such a change did occur at this time? Figures 1 and 3 seem to indicate that it did. However, appearances can be deceiving and more sophisticated statistical methods are needed in considering questions of this kind.

Available procedures, such as Student's t test for estimating and testing for a change in level, have played an important part in Statistics for a very long time. Unfortunately, however, the employment of most such procedures requires assumptions that are invalid in the present context. The ordinary t test for comparison of two means (i) would be valid if the observations before and after the event of interest varied about means μ_1 and μ_2, with errors normally and *independently* distributed with constant variance, or (ii) would be approximately valid if randomization could be introduced into the conduct of the data gathering process. In fact, as we see for the monthly averages of O_3 in Downtown Los Angeles in Figure 4 (obtained by averaging the entries of each of the columns of Figure 1 from January 1955 to December 1965), our data are time series in which not only are successive observations autocorrelated but also strong seasonal effects occur. Thus the ordinary statistical procedures (parametric or non-parametric) which rely on independence or special symmetry in the distribution function are not available to us nor are the blessings endowed by randomization.

One way to proceed which we initiated earlier[5] is to build a stochastic model for the data which takes cognizance of a possible change of the form hypothesized.

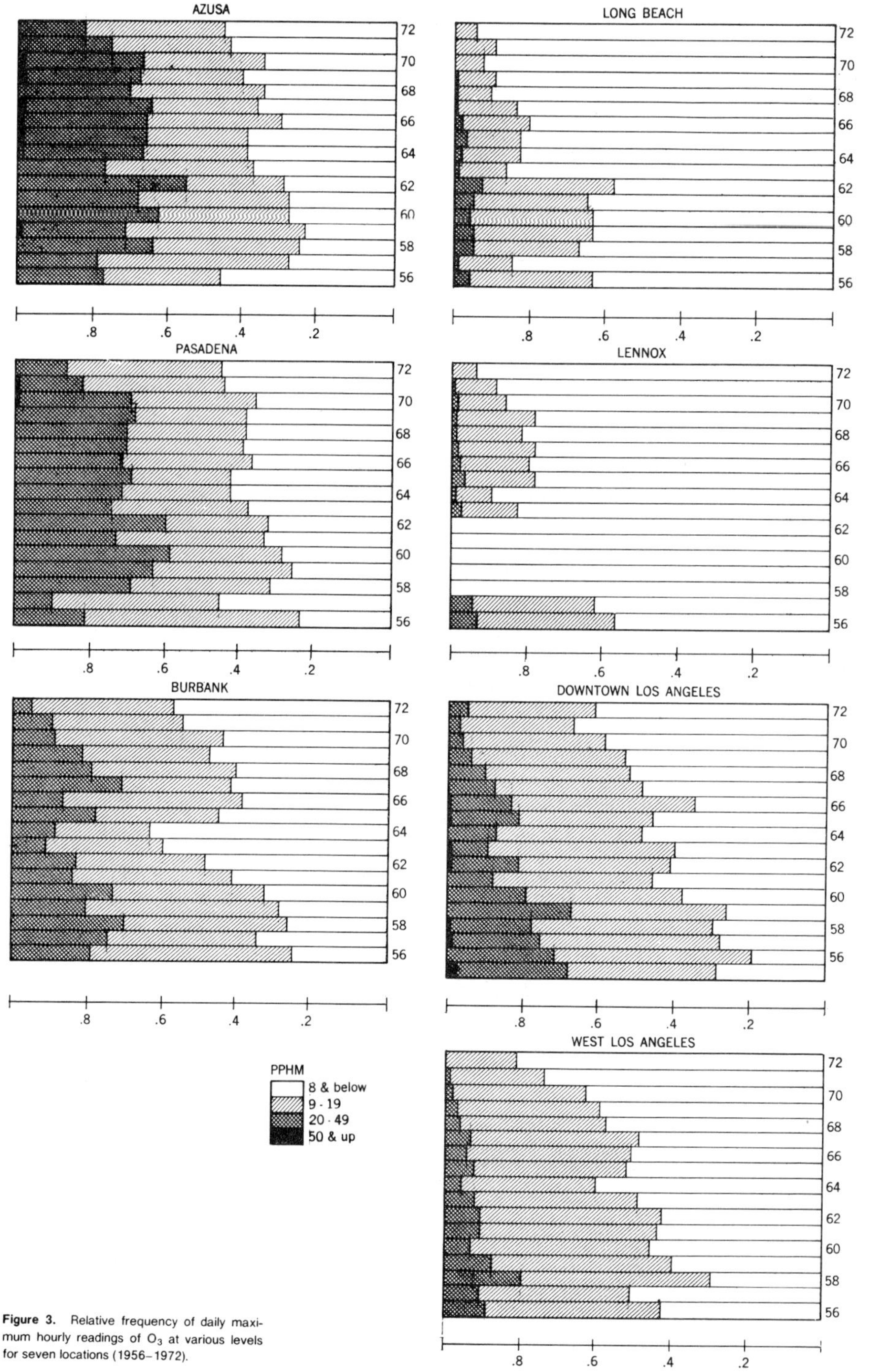

Figure 3. Relative frequency of daily maximum hourly readings of O_3 at various levels for seven locations (1956–1972).

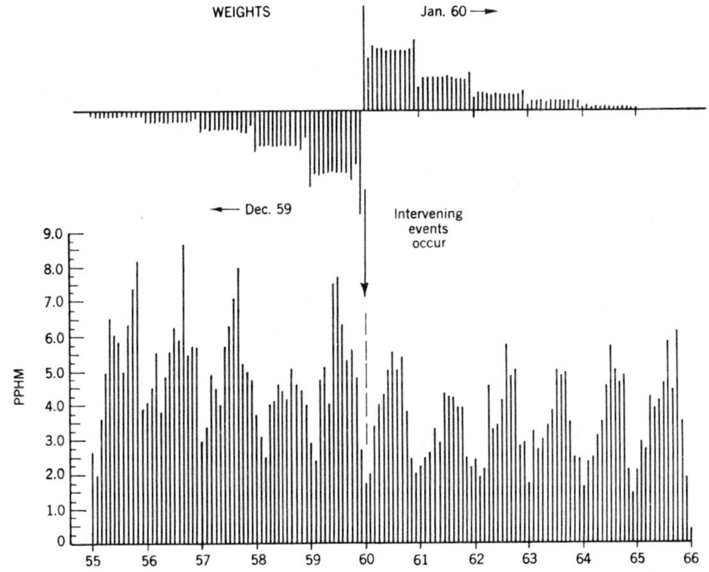

Figure 4. Monthly averages of O_3 in Downtown Los Angeles (January 1955 to December 1965) and weight function for determining the change in 1960.

Time Series Models

Let us denote the time series obtained at equal intervals of time by $\ldots y_{t-1}, y_t, y_{t+1}, \ldots$ and use B for the back shift operator such that $Bz_t = z_{t-1}$. Also let $\ldots a_{t-1}, a_t, a_{t+1}, \ldots$ be a sequence of independently distributed normal variables having mean zero and variance σ_a^2 which for brevity we refer to as "white" noise. Then a useful model[6] for such a series may be written

$$Y_t = f(\kappa, t) + \frac{\theta(B)}{\varphi(B)} a_t \quad (1)$$

where $Y_t = F(y_t)$ is some appropriate transformation of y_t (say $\log y_t$ or $y_t^{1/2}$ or perhaps y_t itself), $f(\kappa, t)$ is some function of time containing parameters κ,

$$\theta(B) = 1 - \theta_1 B - \theta_2 B^2 \ldots - \theta_q B^q,$$

$$\varphi(B) = 1 - \varphi_1 B - \varphi_2 B^2 \ldots - \varphi_p B^p$$

are "moving average" and "autoregressive" polynomials in B, and we shall require that the zeroes of $\theta(B)$ lie outside and those of $\varphi(B)$ lie on or outside the unit circle. Then if $z_t = Y_t - f(\kappa, t)$ the model may be written

$$\varphi(B) z_t = \theta(B) a_t. \quad (2)$$

For the representation of certain kinds of stable non-stationary series the operator $\varphi(B)$ can be factored so that

$$\varphi(B) = (1 - B)^d \phi(B) \quad (3)$$

where the zeroes of $\phi(B)$ all lie outside the unit circle. This corresponds to the use of a stationary model in the dth difference. Thus if we let

$$(1 - B)^d z_t = w_t \quad (4)$$

the model may be written

$$\phi(B) w_t = \theta(B) a_t, \quad (5)$$

a stationary autoregressive moving average process.

In representing seasonal models with period s (for monthly data $s = 12$) it is often convenient to write $\varphi(B) = \varphi_1(B)\,\varphi_2(B^s)$ and $\theta(B) = \theta_1(B)\,\theta_2(B^s)$. Also, non-stationarity can frequently be eliminated by seasonal differencing so that we come finally to seasonal models of the form

$$\phi_1(B)\phi_2(B^s)(1 - B)^d(1 - B^s)^D z_t = \theta_1(B)\theta_2(B^s) a_t \quad (6)$$

where the polynomials $\phi_1(B)$, $\phi_2(B^s)$, $\theta_1(B)$, $\theta_2(B^s)$ are of degrees p_1, p_2, q_1, q_2, respectively.

The process of model building is necessarily iterative and, as discussed for example in Ref. 6, entails the successive use of Identification, Fitting, and Diagnostic Checking in establishing the model. Applied in the present context our strategy is as follows: (i) first frame a model for change, which describes what is expected to occur given knowledge of the actions which are known to have been taken; (ii) work out the appropriate data analysis based on that model; (iii) if diagnostic checks show no inadequacy in the model form, conclusions as to the effects of the change may be drawn directly from the analysis; but (iv) if the model appears to be inadequate, then one tries to learn how and why and to check up on new possibilities which suggest themselves to those having understanding of the problem.

Thus we have a strategy for learning which we shall illustrate with a few examples.

A Model for Ozone

For monthly ozone averages denoted below by z_t, preliminary identification studies show that, apart from a possible deterministic component to be discussed later, a model of the simple form

$$(1 - B^{12})z_t = (1 - \theta_1 B)(1 - \theta_2 B^{12})a_t \qquad (7)$$

seems adequate to describe the stochastic time-dependent relationship of the data for all the seven locations. In particular, the factors $1\text{-}B^{12}$ and $1\text{-}\theta_2 B^{12}$ jointly take account of the pronounced but somewhat drifting seasonal pattern shown for example in Figure 4 for Downtown Los Angeles. The nature of the two factors can best be understood as follows. Suppose θ_1 were zero, then the forecast of, say, next January would be a weighted average of the data on all past Januarys, with weights $(1\text{-}\theta_2)$, $\theta_2(1\text{-}\theta_2)$, $\theta_2^2(1\text{-}\theta_2)$, ... decreasing exponentially into the remote past. The factor $1\text{-}\theta_1 B$ represents the "non-seasonal" relationship between observations in successive months. Using the model, forecasts follow a stable seasonal pattern, which is appropriately updated as new data become available. The model in Eq. (7) is of a similar nature to those obtained in Ref. 7 where, however, biweekly averages of daily maximum hourly concentrations were considered.

A General Model for Intervention Analysis

In the above, we have argued that variation in, say the monthly result can be represented by a time series model which is a stochastic difference equation. That is, a difference equation $\varphi(B)z_t = \theta(B)a_t$ driven by "white" noise a_t. The effects of environmental changes are often dynamic in character with transfer functions which may also be represented by a suitable difference equation model.

Let ξ_t be a variable (which, for example, could take the value zero before 1960 and unity after that time thus representing a step change) and consider the model

$$z_t = \theta_0 + \frac{\omega(B)}{\delta(B)}\xi_t + \frac{\theta(B)}{\varphi(B)}a_t \qquad (8)$$

In general, $\delta(B) = 1\text{-}\delta_1 B\text{-} \ldots -\delta_r B^r$ and $\omega(B) = \omega_0 - \omega_1 B\text{-} \ldots \text{-}\omega_s B^s$ are polynomials in B of degrees r and s, respectively. We shall normally require that $\omega(B)$ has zeroes outside and $\delta(B)$ zeroes outside or on the unit circle. Figure 5 shows the output response to a step change for various simple transfer functions.

For the Freeway and/or Rule 63 change of 1960, an appropriate model will then be of the form of (8) with

$$\text{the indicator variable } \xi_t = \begin{cases} 0, \text{prior to 1960} \\ 1, \text{after 1960} \end{cases}$$

and

$$\text{the transfer function } \frac{\omega(B)}{\delta(B)} = \omega_0 \qquad (9)$$

as illustrated in Figure 5(a).

Some events might not be expected to produce an immediate response but rather a "first order" dynamic response like that in Figure 5(b) for which the transfer function is $\omega_0/(1\text{-}\delta_1 B)$. With this model it is readily shown that the time constant of the system is estimated by $T = \{-\log_e \delta_1\}^{-1}$ and the steady state gain is $\omega_0/(1\text{-}\delta_1)$.

It is to be noted that the model for Eq. (8) has great flexibility, since ξ need not merely be an indicator variable. It could be any exogenous variable such as atmospheric conditions, traffic patterns, vehicle population, etc. whose effect on the response needed to be modelled dynamically. Given this model the likelihood function associated with any given set of data can be written down. This can be used to produce maximum likelihood estimates of the parameters of both the stochastic and dynamic parts of the model and the approximate standard errors. Alternatively we may adopt a Bayesian viewpoint and analyze the estimation results in terms of posterior distributions of the parameters on the basis of noninformative prior distributions.[8]

Some Examples of Time Series and Intervention Analysis Applied to the Pollution Data.

We now apply the above time series methods and intervention analysis techniques to three selected but important cases.

Possible Change of O_3 Level in Downtown Los Angeles in 1960

For this study we employ the series of monthly averages of O_3 in Downtown Los Angeles from January 1955 to December 1965, denoted by z_t, shown earlier in Figure 4. Combining the model in Eq. (7) together with the transfer function in Eq. (9), we have

$$z_t = \omega_0 \xi_t + \frac{(1 - \theta_1 B)(1 - \theta_2 B^{12})}{1 - B^{12}} a_t \qquad (10)$$

where

$$\xi_t = \begin{cases} 0, \text{ prior to January, 1960} \\ 1, \text{ beginning January 1960.} \end{cases}$$

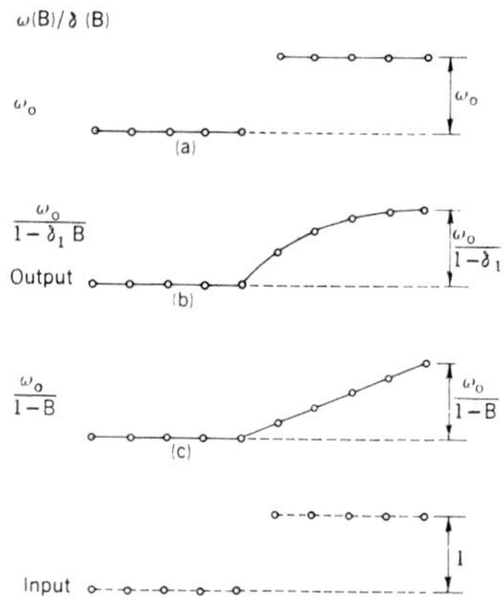

Figure 5. Response to a step change for various transfer functions $\omega(B)/\delta(B)$.

The maximum likelihood estimates (MLE) of the parameters together with their estimated standard errors are given below.

	MLE	*S.D.*
ω_0	−1.10	0.10
θ_1	−0.24	0.07
θ_2	0.54	0.06

Examination of residuals fails to show any obvious inadequacies in the model so we proceed to discuss the findings.

From a Bayesian viewpoint, we may interpret the result concerning ω_0 as saying that, on the basis that little is known initially about these parameters, the posterior distribution of ω_0 is nearly a normal distribution centered at $\hat{\omega}_0 = -1.10$ with standard deviation 0.10. The evidence is thus very strong that a reduction in the level of O_3 in Downtown Los Angeles occurred in 1960, the best estimate of the reduction being 1.10 pphm.

It is illuminating to consider the form of the estimate $\hat{\omega}_0$. Let the month January 1960, when the effect took place be denoted by $T+1$. Suppose for the moment that the parameters θ_1 and θ_2 are known. Then it is shown[9] that $\hat{\omega}_0$ can be written as

$$\hat{\omega}_0 = f_1(z_{T+1}, z_{T+2}, \ldots | \theta_1, \theta_2) - f_2(z_T, z_{T-1}, z_{T-2}, \ldots | \theta_1, \theta_2)$$

where f_1 is a weighted average of the observations $z_{T+1}, z_{T+2}, \ldots$ after the event occurred and f_2 is a weighted average of the observations $z_T, z_{T-1}, \ldots$ prior to the event. In practice, θ_1 and θ_2 would be unknown, but for large samples, as in the present example, they can be approximated by the corresponding maximum likelihood estimates. Using the values $\hat{\theta}_1 = -0.24$ and $\hat{\theta}_2 = 0.54$, the weights applied to the observations z_t in forming $\hat{\omega}_0$ are shown above the series in Figure 4.

The weight function gives large weight to observations in the years immediately before and after the event took place and less and less weight to observations remote from it. This is an intuitively pleasing result, as one would certainly expect that, for data of this kind, observations farther removed from the time the event took place should be less and less relevant in determining its effect.

A Test of the Progressive Reduction of O_3 at Downtown Los Angeles since 1966

This second example illustrates a situation where the model for change must take a more elaborate form. From 1966 to 1970 regulations for the reduction of hydrocarbon emissions resulted in engine design changes in new cars which had the effect of decreasing HC but sometimes at the expense of increasing NO_x.[3] Also, legislation since 1970 progressively reduced the emissions of both HC and NO_x in the new cars. It has been suggested that these changes could slow down or reduce the photochemical reactions and hence could lead to a reduction in the level of O_3 near a primary pollutant center such as Downtown Los Angeles. Visual inspection of Figures 1 and 3 indicate this may very well be the case. In this example the change in O_3 should show, not as a step function, but as a trend reflecting an increasing proportion of new design vehicles in the car population. Furthermore, because of the summer-winter inversion differential, the net effect would be different in the winter from that in the summer. To examine these possibilities the following model was postulated for all the available monthly averages of O_3 in Downtown Los Angeles from January 1955 to December 1972,

$$z_t = \omega_0 \xi_t + \lambda_1 \frac{\xi_t'}{1 - B^{12}} + \lambda_2 \frac{\xi_t''}{1 - B^{12}} + \frac{(1 - \theta_1 B)(1 - \theta_2 B^{12})}{1 - B^{12}} a_t \quad (11)$$

where

$$\xi_t' = \begin{cases} 1, & \text{“summer” months June-October beginning 1966} \\ 0, & \text{otherwise} \end{cases}$$

$$\xi_t'' = \begin{cases} 1, & \text{“winter” months November-May beginning 1966} \\ 0, & \text{otherwise} \end{cases}$$

The parameters λ_1 and λ_2 represent, respectively, for the summer and winter months, progressive changes in the levels of the observations from one year to the next beginning in 1966 and are, therefore, slopes of two deterministic yearly trends for these two "seasons." As before, the parameter ω_0 is introduced to allow for the change in 1960.

Estimation results are as follows:

	MLE	*S.D.*
λ_1	0.25	.07
λ_2	−0.07	.06
ω_0	−1.09	.13
θ_1	−0.24	.03
θ_2	0.55	.04

The estimate of λ_1 provides very strong evidence that there has been a steady improvement for the summer seasons, the yearly reduction being estimated at 0.25 pphm. On the other hand, the improvement, if any, in the winter is very slight. In particular, the parameter value $\lambda_2 = 0$, representing no improvement, is well inside the approximate 95% highest posterior density interval -0.07 ± 0.12, and hence is not contradicted by the data.

A Useful General Indicator of Change

A device which is sometimes useful, at least as a preliminary identification tool, is to compare the forecast made from some point at which change is known to have been instituted with what has actually occurred. This can be done graphically. But it is easy to deceive the eye and more formal statistical procedures are needed. Suppose we have n observations $(z_1, \ldots, z_n)$ and it is suspected that changes might have occurred in the pattern level after time $T (T < n)$. The nature of the changes may be studied by comparing forecasts $\hat{z}_T(1), \hat{z}_T(2), \ldots, \hat{z}_T(n-T)$ of $z_{T+1}, \ldots, z_n$ made at time T with the actual observations themselves, from which $a_{T+1}, \ldots, a_n$ can be readily calculated.† If in fact there were no change, the a_t's would be "white noise" and hence

$$\sum_{t=T+1}^{n} a_t^2$$

† Notice that, for example, a_{T+l} is not the forecast error $z_{T+l} - \hat{z}_T(l)$ but may be computed from the complete set of forecast errors.

would be distributed as $\sigma_a^2\chi^2$ with $(n\text{-}T)$ degrees of freedom. This χ^2 test could thus provide a useful general indicator of various changes in pattern* from that expected. For illustration we consider the recent ozone results for Azusa.

Has there been improvement in Azusa since 1971? We now employ monthly averages of O_3 in Azusa to investigate whether changes have actually occurred since 1971, as a result of new auto emissions standards introduced at that time. First, the data from January 1956 through December 1970 was used to estimate the parameters in the basic model (7). The results are

	MLE	*S.D.*
θ_1	−0.15	0.07
θ_2	0.91	0.04
σ_a	1.00	

Based on the fitted model, forecasts of the 24 months of 1971 and 1972 made at time *T*, December 1970, are shown in Figure 6(a) together with the actual observations. The errors a_t's for these months are plotted against time in Figure 6(b). The sum of squares value

$$(\sigma_a^2)^{-1} \sum_{t=T+1}^{n} a_t^2 \doteq 35$$

is larger than the χ^2 value with 24 degrees of freedom at the 10% level, suggesting that the hypothesis of no-change is untenable. Indeed, Figures 6(a) and 6(b) suggest that (i) there has been improvements in these two years, (ii) the improvement is much greater in 1972, and (iii) the magnitude of the reduction is appreciably larger in the summer months. If desired, one can proceed to introduce dynamic models of the type in (11) to estimate the magnitude of the reductions as well as to project the level of O_3 for future years.

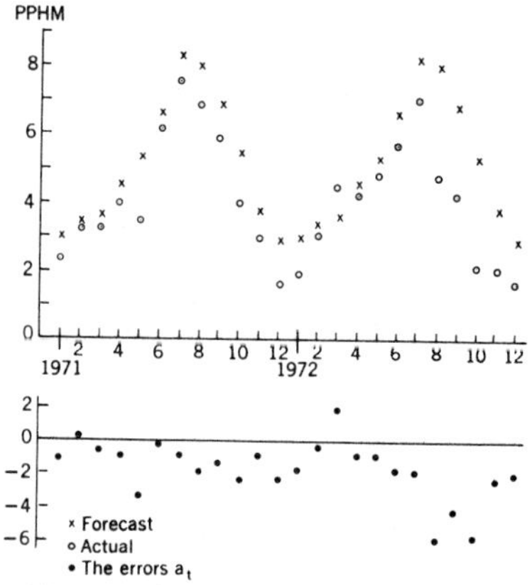

Figure 6. Comparison of forecasts with actual observations: monthly averages of O_3 in Azusa January 1971–December 1972.

Conclusions and Further Work

In the preceding sections, a preliminary statistical analysis has been presented for the ozone data over the seven locations in the Los Angeles Basin. The level of ozone concentration is subject to strong seasonal fluctuations and, within the summer season, to the influence of changes in atmospheric conditions. In Downtown Los Angeles and West Los Angeles, the level was lower in the 1960's compared with that in the 1950's and there have been progressive improvements beginning around 1966. Corresponding improvements were, however, not evident at the end of the pipe line in Pasadena and Azusa until 1971–1972.

Various graphical devices have been presented to bring out important features of the data. We believe that such presentations provided a powerful means to promote discussion about the nature of the pollution problem and thus play an indispensable role in the preliminary stage of model building. At the same time, analysis of the data gives rise to the need for new statistical methods. In particular, the intervention analysis technique has been developed and is illustrated using the ozone data.

Using similar statistical tools, much preliminary work has already been done on other pollutants including nitric oxide, carbon monoxide, sulphur dioxide and particulates, and the findings will be reported in the near future. Work has begun to explore (i) the interdependence between the primary and the secondary pollutants, (ii) the phase relations between the locations (iii) the influence of exogenous atmospheric variables and other factors such as vehicle population and traffic density and (iv) the effect of control measures. New statistical tools in modelling multivariate time series which are necessary to achieve these goals are currently being investigated.

* There is evidence[10,11] that examination of the a_t's is not sensitive to all kinds of model change. Especially it tends to be insensitive to changes in the memory parameters of $\theta(B)$.

Acknowledgments

This research has been carried out under a joint project between the Department of Statistics, the University of Wisconsin, Madison and the Los Angeles County Air Pollution Control District, supported by a grant from the American Petroleum Institute. The authors wish to thank W. S. Wei, J. Schaap, and S. Grover for computing assistance and for preparing the figures in this paper.

References

1. B. Dimitriades, "Effect of hydrocarbon and nitrogen oxides on photo-chemical smog formation, "*Environ. Sci. Technol.*, **6:** 253 (1972).
2. A. J. Haagen-Smit, "Chemistry and physiology of Los Angeles smog," *Ind. Eng Chem*, **44:** 1342 (1952).
3. W. J. Hamming and J. E. Dickinson, "Contrl of photochemical smog by alteration of initial reactant ratio," *J. Air Poll. Control Assoc.*, **16:** 317 (1966).
4. G. C. Tiao, G. E. P. Box, M. Grupe, S. T. Liu, S. Hillmer, W. S. Wei, and W. J. Hamming, "Los Angeles Aerometric Ozone Data 1955–1972." Technical Report #346, Department of Statistics, University of Wisconsin, Madison, Oct. 1973.
5. G. E. P. Box and G. C. Tiao, "A change in level of a non-stationary time series," *Biometrika*, **52:** 181 (1965).
6. G. E. P. Box and G. M. Jenkins, *Time Series Analysis, Forecasting and Control,* Holden-Day, San Francisco 1970.
7. P. H. Merz, L. M. Painter, and P. R. Ryason, "Aerometric data analysis—time series analysis and forecast and an atmospheric smog diagram," *Atmos. Environ.*, **6:** 319 (1972).
8. G. E. P. Box. and G. C. Tiao, *Bayesian Inference in Statistical Analysis,* Addison Wesley, Reading 1973.
9. G. E. P. Box and G. C. Tiao, "Intervention Analysis with Applications to Economic and Environmental Problems," Technical Report #335, Department of Statistics, University of Wisconsin, Madison (June 1973). (To appear in the *Journal of American Statistical Association.)*
10. G. E. P. Box and G. M. Jenkins, "Models for Prediction and Control VI. Diagnostic Checking," Technical Report #99, Department of Statistics, University of Wisconsin, Madison Dec., 1966.
11. R. A. Johnson and M. Bagshaw, Private communication.

5.10
Some Empirical Models for the Los Angeles Photochemical Smog Data

G. C. TIAO, M. S. PHADKE, and G. E. P. BOX
University of Wisconsin

This paper presents (i) an empirico-mechanistic model which describes the dependence of CO, NO, NO_2, and O_3 on total hydrocarbons, traffic, wind speed, inversion base height, and solar radiation as well as the photochemical reactions associated with these pollutants; (ii) a detailed study of weather conditions when the instantaneous daily maximum O_3 exceeds the L.A. County alert level of 50 pphm; and (iii) regression models for the prediction of daily maximum O_3 values.

This is a further paper in which we report on our analysis of the aerometric data assembled by the Los Angeles County Air Pollution Control District. In two earlier papers,[1,2] some preliminary analyses of the oxidant and the carbon monoxide data were given. Also, a series of four reports[3–6] have been issued in which various useful data summaries on O_3, CO, NO_x, and SO_2 and particulates are presented. The present paper is concerned mainly with the modeling of O_3 and is divided into four sections as follows:

i) An empirico-mechanistic model is presented which describes the diurnal variation of CO, NO, NO_2, and O_3 in relation to hydrocarbons (HC),* traffic flow, wind speed, solar radiation, and inversion base height.

ii) A study is made of conditions occurring during the 46 days in the period 1960–1972 when the daily instantaneous (5 min) maximum O_3 level exceeds the L.A. County "alert" level of 50 pphm.

iii) An empirical model is obtained relating daily maximum hourly O_3 levels to weather variables. In particular, consideration is given to how accurate a forecast could be using such a model and employing only data available in the early morning.

iv) A general discussion of our results is presented and goals indicated for our continuing work.

A Photochemical Smog Model

The observed average concentration levels on Wednesdays during the summer months (July–Sept) of 1972 at Downtown Los Angeles for the primary and secondary pollutants CO, NO, NO_2, and O_3 are shown in Figure 1. Also shown in the figure are the values calculated from a fitted empirico-mechanistic model which takes account simultaneously of traffic flow, wind speed, solar radiation, inversion base height, and total hydrocarbons. Table I gives the diurnal variations of these five exogenous variables. The wind speed, total HC, and solar radiation values are averages over the same period

* HC here means total hydrocarbons.

The Collected Works of George E. P. Box, 1984, Wadsworth, Inc., Belmont, CA 94002.
Originally published in *APCA Journal,* vol. 26, no. 5 (1976), pp. 485–490.

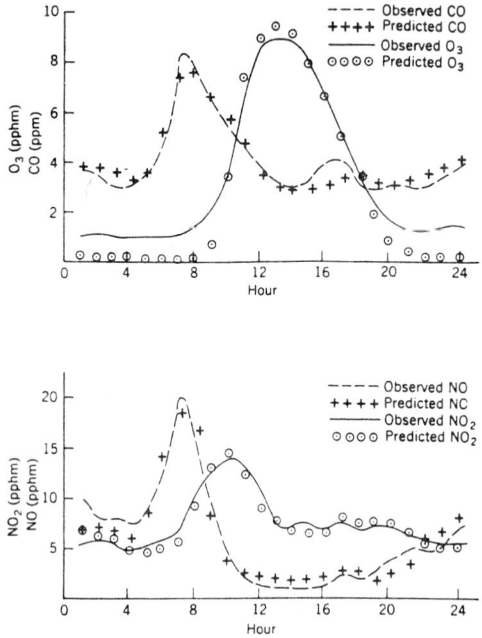

Figure 1. Diurnal variations of various pollutants.

as the pollutants. Due to lack of corresponding hourly data on the remaining two variables, the values shown are (i) hourly cordon count of vehicles in Downtown Los Angeles on a Wednesday in May, 1972, and (ii) averages of hourly inversion base height at El Monte for the four days July 26–29, 1974.*

Although the model is still being developed and tested further on other sets of data, the agreement so far obtained is extremely close and we now describe its deviation.

Modeling Philosophy

The system producing pollutants in Los Angeles County involving pollutant sources, meteorological variables, and chemical reactions is obviously extremely complex. Also, the measurements actually available are quite limited. It is, therefore, both unrealistic and unnecessary to attempt to include in a mechanistic model every feature that might occur in the real system. We can, however, realistically seek a simplified model which accounts for the principal physical and chemical phenomena relating the variables on which data are available. Unknown constants may then be estimated by standard statistical techniques and the model checked for representational adequacy. We refer to such a model as an empirico-mechanistic model.

Such a model is derived for the prediction of CO, NO, NO_2, and O_3 in Downtown Los Angeles. We suppose that it is a rainless summer weekday with sunshine and with a typical

* The data were supplied by Dr. Steven Reynolds of Systems Applications, Inc.

elevated temperature inversion whose height may, however, change during the day. The diurnal variations of CO, NO, NO_2, and O_3 depend on

1. the photochemical reactions among HC, NO, NO_2, and O_3;
2. the emission sources. For Downtown Los Angeles, the automobile traffic constitutes the major source of emission for the primary pollutants CO, NO, and HC[7]; and
3. the variations in the meteorological variables, the wind speed, and the inversion base height being the predominant factors.

Photochemical Reactions

The following set of "lumped" parameter reactions are proposed for the quantitative description of the O_3 formation in the atmosphere:

$$NO_2 + SR + O_2 \xrightarrow{k_1'} NO + O_3$$

$$NO + O_3 \xrightarrow{k_2} NO_2 + O_2$$

$$HC + O_3 + \beta_1 NO \xrightarrow{k_3} \beta_1 NO_2 + \text{other products}$$

$$2NO + O_2 \xrightarrow{k_4'} 2NO_2 \quad (1)$$

where SR refers to the level of solar radiation, β_1 is a stoichiometric constant, and k_1', k_2, k_3, k_4' are rate constants.

The first reaction is the O_3 formation reaction. The second reaction is the back reaction between O_3 and NO. The third reaction is the branching reaction where one molecule of O_3 together with one molecule of HC oxidizes β_1 ($\beta_1 > 1$) molecules of NO into NO_2. Such a branching reaction is essential to describe the buildup of NO_2 and O_3 in the atmosphere. The fourth reaction, generally called the dark phase reaction, permits the slow oxidation of NO into NO_2 without the presence of HC, O_3, or the sunlight.

Reaction mechanisms with a much larger number of steps have been proposed in the literature for the generation of O_3 in the atmosphere.[8–10] It is believed that the lumped parameter reaction system proposed here can reflect the principal changes which are taking place while not involving chemical species on which no measurements are available.

On the assumption of a simple kinetic mechanism the rates of change of NO, NO_2, and O_3 associated with chemical reactions may be written as:†

$$\left(\frac{d[NO]}{dt}\right)_{\text{chem.}} = k_1[NO_2][SR] - k_2[NO][O_3] - \beta_1 k_3[HC][O_3][NO]^{\beta_1} - 2k_4[NO]^2 \quad (2)$$

$$\left(\frac{d[NO_2]}{dt}\right)_{\text{chem.}} = -k_1[NO_2][SR] + k_2[NO][O_3] + \beta_1 k_3[HC][O_3][NO]^{\beta_1} + 2k_4[NO]^2 \quad (3)$$

$$\left(\frac{d[O_3]}{dt}\right)_{\text{chem.}} = k_1[NO_2][SR] - k_2[NO][O_3] - k_3[HC][O_3][NO]^{\beta_1} \quad (4)$$

where $k_1 = k_1'[O_2]$ and $k_4 = k_4'[O_2]$. Since the concentration of O_2 in the atmosphere remains practically constant, k_1 and k_4 may be treated as modified rate constants.

† A similar equation may be obtained for the rate of change of HC. At this stage of development of our model, HC is tentatively regarded as an exogenous factor, and its prediction will not be considered.

Effects of Traffic, Wind Speed, and Inversion Base Height

We now propose a model which relates the concentrations of CO, NO, NO_2, and O_3 to the three exogenous variables traffic (T), wind speed (W), and inversion base height (h).

Using the single box representation for Downtown Los Angeles, it is shown in the Appendix that the model can be written as

$$\frac{d[CO]}{dt} = \frac{k_T T}{h^\alpha} - k_w W \times [CO] - I_h \frac{d\log h^\alpha}{dt}[CO] \quad (5)$$

$$\frac{d[NO]}{dt} = \frac{k_{TNO} k_T T}{h^\alpha} - k_w W \times [NO] - I_h \frac{d\log h^\alpha}{dt}[NO] + \left(\frac{d[NO]}{dt}\right)_{chem.} \quad (6)$$

$$\frac{d[NO_2]}{dt} = -k_w W \times [NO_2] - I_h \frac{d\log h^\alpha}{dt}[NO_2] + \left(\frac{d[NO_2]}{dt}\right)_{chem.} \quad (7)$$

$$\frac{d[O_3]}{dt} = -k_w W \times [O_3] - I_h \frac{d\log h^\alpha}{dt}[O_3] + \left(\frac{d[O_3]}{dt}\right)_{chem.} \quad (8)$$

where

$$I_h = \begin{cases} 1 \text{ if } dh/dt > 0 \\ 0 \text{ otherwise} \end{cases},$$

$$\left(\frac{d[\cdot]}{dt}\right)_{chem.}$$

describes the changes in chemical reactions, and k_T, k_{TNO}, k_w and α are constants. Physically, k_T is a measure of the CO emissions, k_{TNO} the ratio of NO to CO emissions, k_w the wind transport parameter, and α the effective mixing height constant.

Table I. Diurnal variations of weather variables, traffic, and total hydrocarbons.

Hour of the day	Wind speed (mph)	Inversion base height (1000 ft)	Solar radiation (cal/cm²/hr)	Traffic (10^5 vehicles)	Total hydrocarbons (pphm)
1	4.0	0.31	00.0	0.92	331.0
2	3.3	0.34	00.0	0.39	338.0
3	3.0	0.32	00.0	0.35	346.0
4	3.0	0.34	00.0	0.50	362.0
5	3.0	0.37	00.0	1.15	385.0
6	3.0	0.35	01.0	2.50	400.0
7	3.0	0.49	09.0	5.61	438.0
8	5.0	0.71	21.0	5.17	431.0
9	4.0	1.01	35.0	3.80	408.0
10	4.3	1.24	48.0	3.76	377.0
11	5.3	1.58	62.0	4.00	354.0
12	7.0	2.02	68.0	4.10	331.0
13	8.3	2.12	69.0	4.17	300.0
14	9.0	1.95	65.0	4.20	283.0
15	10.0	1.79	56.0	4.78	283.0
16	9.3	1.73	43.0	6.84	285.0
17	9.0	1.66	28.0	5.77	285.0
18	8.3	1.39	12.0	3.03	277.0
19	6.7	1.16	02.0	2.11	285.0
20	5.7	0.86	00.0	1.66	292.0
21	5.0	0.53	00.0	1.20	308.0
22	4.3	0.46	00.0	0.95	315.0
23	4.0	0.43	00.0	0.95	308.0
24	3.0	0.38	00.0	0.93	323.0

Table II. Parameter estimates for the photochemical smog model.

	1970	1971	1972
k_T	0.637	0.647	0.545
k_W	0.051	0.050	0.052
α	0.96	0.90	1.00
k_{TNO}	3.18	3.07	3.36
k_1	1.43×10^{-2}	0.97×10^{-2}	1.62×10^{-2}
k_3	1.17×10^{-4}	0.27×10^{-4}	0.79×10^{-4}
β	2.34	3.46	2.75
k_4	0.75×10^{-2}	1.15×10^{-2}	1.29×10^{-2}
Standard deviations of model predictions			
σ_{CO}(ppm)	0.55	0.44	0.46
σ_{NO}(pphm)	1.64	1.54	1.31
σ_{NO_2}(pphm)	0.95	0.98	0.86
σ_{O_3}(pphm)	1.00	1.00	0.96

It will be noted that Eq. (5) does not contain a chemical component since CO is essentially inert. Also, Eq. (7) and (8) do not contain traffic components since NO_2 and O_3 are regarded as wholly secondary pollutants.

Upon substituting Eq. (2), (3), and (4), respectively in Eq. (6), (7), and (8) we obtain a set of four first-order, coupled, nonlinear, ordinary differential equations which describe the diurnal variations of the pollutants CO, NO, NO_2, and O_3 in terms of the exogenous variables traffic, wind speed, inversion base height, solar radiation, and total hydrocarbons.

The Fitting Results

Given the initial concentrations of CO, NO, NO_2, and O_3 at midnight, and the diurnal curves of HC, T, W, h, and SR, then for any set of trial values of the constants

$$k_1, k_2, k_3, k_4, \beta_1, k_T, k_{TNO}, k_w, \text{ and } \alpha$$

Eq. (5)–(8) may be solved numerically and the resulting hourly averages compared with the observed values. Using a process of iterative nonlinear least squares, best values may now be found for these constants. For the 1972 data given in Figure 1 and also for similar data in 1970 and 1971 (not shown) the fitted constants are given in Table II.* The bottom part of the table gives the estimated standard deviations of the predicted values for each of the four pollutants.

From this part of the study we can draw the following conclusions. (i) The fitted model yields predicted values which are in very close agreement with those actually observed. The average standard deviation of the prediction errors of CO is 0.48 ppm and those for NO, NO_2, and O_3 are, respectively, 1.50, 0.93, and 0.99 pphm. The results graphed for 1972 in Figure 1 are typical of those obtained in all three years. (ii) The estimated constants remain satisfactorily stable from year to year. (iii) As we shall discuss at greater length later, we intend to test and further develop this model using data on individual days and from different locations and seasons.

* It will be noted that the fitted value of k_2 is not shown. In all of the three years considered it turns out to be approximately zero. This, of course, does not imply that the 2nd chemical reaction in (1) is unimportant. It simply means that the 1st, 3rd, and 4th reactions with parameters suitably adjusted can adequately describe the observed relationships among the pollutants.

Meteorological Conditions when Daily Instantaneous Peak Level of O_3 Exceeds 50 pphm

For short term prediction purposes, it is particularly important to study the relationship between high O_3 levels and meteorological variables occurring at that time. One important measure of the severity of O_3 pollution is the daily instantaneous (5 min) maximum (DIM) reading.

There are 46 days in the period 1962–1972 on which the 50 pphm "alert" level for DIM ozone was reached or exceeded at one or more of the following five locations: Downtown Los Angeles, West Los Angeles, Burbank, Pasadena, and Azusa. A table is presented in an earlier publication[11] which gives the date, location, time of occurrence, 8–12 A.M. average wind speed (WS), inversion base height at 4 A.M. (IBH), maximum mixing height (MMH), maximum temperature (TMAX), and total 7–12 A.M. solar radiation (TSR).†

The temporal and spatial distributions as well as the weather conditions for these 46 days can be summarized as follows:

(i) The frequencies of exceedance of the alert level and the average time of occurrence for each of the locations are

	Burbank	West Los Angeles	Downtown L.A.	Pasadena	Azusa
Frequency	2	1	9	18	24
Hour (PST)	12.5	10	11.1	12.0	12.6

Exceedances occurred most frequently at Pasadena and Azusa and the average time of occurrence is about 1 to 1.5 hr later than that at Downtown Los Angeles. This is to be expected because the prevailing daytime wind is from the southwest direction and these two locations are at the end of the reactor pipeline.[1]

(ii) The occurrences tabulated by day of the week are:

Sun.	Mon.	Tues.	Wed.	Thur.	Fri.	Sat.
0	8	7	7	10	12	2

These are roughly evenly distributed over the weekdays, but there were distinctively fewer exceedances on the weekends. This is so even though on the average, O_3 concentrations are about the same between weekdays and weekends.[12] The matter will be clarified by considering the entire frequency distributions of daily instantaneous maximum and daily maximum hourly readings which will be given in a later report.

(iii) The occurrences tabulated by month are:

Month	J	F	M	A	M	J	J	A	S	O	N	D
	0	0	1	1	3	2	4	9	16	10	0	0

As expected, the occurrences were concentrated mostly in the summer months, especially August, September, and October.

(iv) Figure 2 summarizes results for the four months July–Oct. and compares them with corresponding overall averages. The most striking differences are seen for the 4 A.M. inversion base heights. The exceedances are also seen to be associated with somewhat higher wind speeds, lower maximum mixing heights, higher maximum temperatures, and higher levels of solar radiation. The fall off in the solar radiation as the summer advances is clearly seen.

† IBH was measured at the Los Angeles International Airport; WS, TMAX, and TSR are the values measured in Downtown Los Angeles; and MMH is the calculated maximum mixing height for Downtown Los Angeles.

Regression Models for the Daily Maximum Hourly Readings of O_3

Another important index of O_3 is the daily maximum hourly reading. We present in this section some findings on empirical relationships between this index and meteorological and other explanatory variables. For data of this kind, which in some cases cover wide ranges, better approximations are usually obtained by working with the logarithms of all the variables. After taking logarithms, we consider a linear model of the form

$$O_{3t} = \omega_0 + \omega_1 X_{1t} + \ldots + \omega_k X_{kt} + a_t \qquad (9)$$

where the subscript t denotes the date, $X_1, \ldots, X_k$ are explanatory variables, and the a_t are supposed to be errors independently and normally distributed with zero means and common variance σ^2.

The daily data for Downtown Los Angeles during the four months July–Oct. 1970, were employed to fit a number of such regression models and the results are summarized in Table III. Apart from the constant term, the explanatory variables are:

1. $O_{3(t-1)}$: daily maximum hourly reading of O_3 for the $(t-1)$th day.
2. Aug_t, $Sept_t$, Oct_t: indicator variables taking the values 1 or 0 to indicate whether or not t falls in a particular one of these months.
3. NO_{2t}: the 4 A.M. reading of NO_2 for the tth day.
4. IBH_t: the 4 A.M. reading of the height of the inversion base for the tth day.
5. IBH_t^2: square of (logarithm of) IBH_t.
6. $\bar{T}_t$: the difference between the inversion breaking temperature and the surface temperature at 4 A.M. for the tth day. This variable can be constructed from the vertical temperature profile available at 4 A.M. and it provides a measure of the strength of temperature inversion. In the original metric, the value 1 is added when the difference is zero.
7. $WS1_t$: average 1–4 A.M. wind speed for the tth day.
8. $WS2_t$: average 8–12 A.M. wind speed for the tth day.

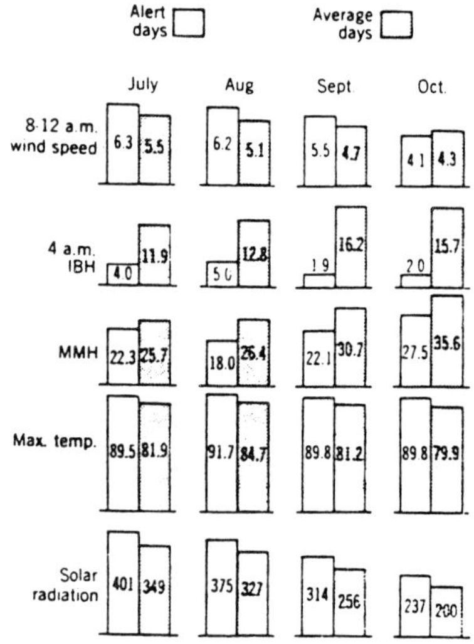

Figure 2. Comparison of the average meteorological conditions on the "alert" days with overall average conditions. The periods for the overall averages are: 1960–72 for the WS, IBH, and MMH; 1965–71 for Max temp.; and 1968–71 for solar radiation.

9. TSR_t: total 7–12 A.M. solar radiation for the tth day.
10. MMH_t: maximum mixing height for the tth day.
11. $TMAX_t$: maximum temperature recorded for the tth day.

The regression coefficients and their estimated standard errors corresponding to 4 different models are given in columns (1)–(4). Also given are the estimated variance $\hat{\sigma}^2$ and estimated standard deviation $\hat{\sigma}$ of the error term a_t. The variance $\hat{\sigma}^2$ indicates the expected accuracy in forecasting future values.

In column (1) only the constant term is used. Thus 2.40 and 0.248 are, respectively, the sample mean and variance of O_{3t}. It should be remembered that all values refer to logged data.

In column (2), $O_{3(t-1)}$ and the three indicator variables Aug_t, $Sept_t$, Oct_t were added. The reason for the introduction of these indicator variables is to take into account the possible "seasonal" changes. The large coefficient associated with $O_{3(t-1)}$ indicates the necessity to allow for a fairly strong positive autocorrelation in the data. The results also show that the mean level in October is lower than that of the other months. The introduction of these 4 variables achieves a 45% reduction in the error variance, from 0.248 to 0.138.

Column (3) shows that a further substantial reduction in the error variance occurred when the additional variables NO_{2t}, IBH_t, IBH_t^2, $\bar{T}_t$, and $WS1_t$ were added. Except for IBH, the effects are linear (in the logarithmic metric) such that increases in NO_{2t}, $\bar{T}_t$, and a decrease in $WS1_t$ will cause O_{3t} to increase. On the other hand, the effect of IBH is quadratic. In particular, holding other variables constant, O_{3t} will tend to a peak at IBH = 0.94 (about 250 ft in the original metric). Figure 3 shows a scatter plot of O_{3t} vs. IBH_t where the quadratic effect is clearly seen.

Finally, in column (4) we added further the four variables $WS2_t$, TSR_t, MMH_t, and $TMAX_t$. The effects of the first two variables are appreciable and are in the expected directions. On the other hand, given all the other variables in the model, MMH_t and $TMAX_t$ appear to provide little, if any, additional explanatory power.

Table III. Results for the regression of the daily maximum hourly reading of O_{3t} on various explanatory variables. July 2–Oct. 31, 1970[ab]

Explanatory Variables	(1)	(2)	(3)	(4)
Const.	2.40	1.28 ± 0.19	1.48 ± 0.29	−2.89 ± 2.46
$O_{3(t-1)}$		0.49 ± 0.07	0.21 ± 0.07	0.14 ± 0.06
Aug_t		0.04 ± 0.10	0.09 ± 0.07	0.13 ± 0.06
$Sept_t$		0.04 ± 0.09	0.24 ± 0.08	0.36 ± 0.08
Oct_t		−0.28 ± 0.10	−0.01 ± 0.01	0.18 ± 0.11
NO_{2t}			0.09 ± 0.05	0.06 ± 0.04
IBH_t			0.17 ± 0.09	0.16 ± 0.09
IBH_t^2			−0.08 ± 0.03	−0.06 ± 0.03
$\bar{T}_t$			0.21 ± 0.05	0.18 ± 0.04
$WS1_t$			−0.34 ± 0.09	−0.27 ± 0.08
$WS2_t$				−0.64 ± 0.14
TSR_t				0.42 ± 0.11
MMH_t				−0.08 ± 0.13
$TMAX_t$				0.74 ± 0.60
$\hat{\sigma}^2$	0.248	0.138	0.081	0.057
$\hat{\sigma}$	0.50	0.37	0.28	0.24

Available midnight [(1)–(2)]

Available 4 A.M. [(1)–(3)]

[a] Data for July 29 which caused an extremely large residual is excluded.
[b] Except for the three indicator variables Aug_t, $Sept_t$, and Oct_t, all variables entering in the regression models refer to logged data.

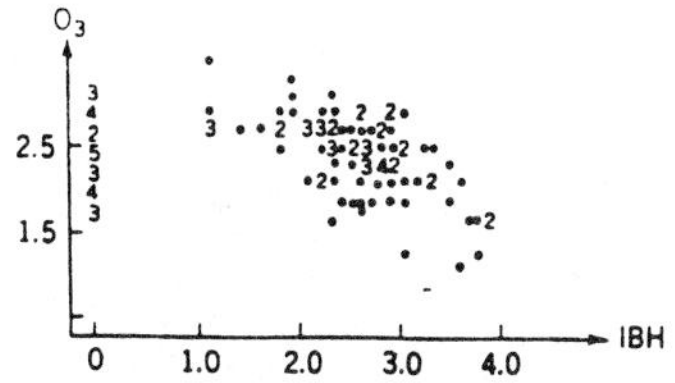

Figure 3. Daily maximum hourly readings of O_3 vs. inversion base heights, July–October, 1970 (logged data). The numbers in the diagram give the frequencies at those points.

The results corresponding to Columns (2) and (3) can be used for prediction purpose. Specifically, the model with only $O_{3(t-1)}$ and the indicator variables as given in column (2) can be employed to predict O_3 one day ahead. All the variables in column (3) will become available shortly after 4 A.M. on the tth day and hence the corresponding model can be used for early morning prediction of O_3 for the same day.

Discussion

A thorough understanding of the principal meteorological and chemical mechanisms in the formation of atmospheric O_3 is essential: (i) for assessing the effects of past control measure, (ii) for making wise choices for future control strategies, and (iii) for deriving the best means for making short term forecasts of hazardous conditions. The results presented in the preceding sections on a photochemical smog model, on the analysis of weather conditions for the "alert" days and on the empirical relationship between daily maximum hourly O_3 and exogenous variables have, we believe, contributed to such understanding. Our attack is many sided, but information from one approach has frequently proved helpful to another and will eventually lead to a unified model.

An extensive effort is being made to extend the analysis and test the models, especially the empirico-mechanistic model, on a wide range of data. In an investigation of this kind, it is not possible to develop and test models by running desired experiments. However, naturally occurring changes would make it possible to test models under different conditions. In particular, a detailed examination[11] of the diurnal variations of CO, NO, NO_2, and O_3 on summer weekdays, summer weekends, winter weekdays, and winter weekends at each of the three locations—Lennox, Downtown Los Angeles, and Pasadena shows that:

(i) weekday vs. weekend—the primary pollutants CO and NO are markedly different, but O_3 is not;
(ii) summer vs. winter—the primary pollutants CO and NO are much higher in winter, while the converse is true for O_3; and
(iii) locations—even though Lennox and Downtown Los Angeles have much larger input of CO and NO, O_3 is highest at Pasadena.

A satisfactory model should reproduce the changes from weekdays to weekends when the traffic patterns are different, and from summer to winter when the meteorological variables markedly differ. Also, knowing the geographical locations of Lennox, Downtown Los Angeles, and Pasadena and the speed and direction of the wind, one would need to allow for transport of the pollutants from one site to another. Finally, we would wish to test our models not merely on averages over a certain period of time, but on individual days with markedly different exogenous conditions.

Acknowledgment

This research has been carried out under a joint project between the Department of Statistics, the University of Wisconsin, Madison, and the Los Angeles County Air Pollution Control District, supported by a grant from the American Petroleum Institute. The authors wish to thank W. J. Hamming for useful discussions; and to thank A. Krug, S. T. Liu, S. C. Wu, and M. Grupe for computing assistance.

References

1. G. C. Tiao, G. E. P. Box, and W. J. Hamming, "Analysis of the Los Angeles photochemical smog data: a statistical overview," *J. Air Poll. Control Assoc.*, 25: 260, 1975.
2. G. C. Tiao, G. E. P. Box, and W. J. Hamming, "A statistical analysis of the Los Angeles ambient carbon monoxide data," *J. Air Poll. Control Assoc.* 25: 1137 (1975).
3. M. S. Phadke *et al.*, "Los Angeles Aerometric Data on Oxides of Nitrogen 1957–1972," Tech. Rept. No. 395, Department of Statistics, Univ. Wisconsin, 1974.
4. M. S. Phadke, G. C. Tiao, M. Grupe, S. C. Wu, A. Krug, and S. T. Liu, "Los Angeles Aerometric Data on Sulphur Dioxide, Particulate Matter and Sulphate 1955–1972." Technical Report No. 410, Department of Statistics, the University of Wisconsin, Madison, 1975.
5. G. C. Tiao, G. E. P. Box, M. Grupe, S. T. Liu, S. Hillmer, W. S. Wei, and W. J. Hamming, "Los Angeles Aerometric Ozone Data." Technical Report No. 346, Department of Statistics, the University of Wisconsin, Madison.
6. G. C. Tiao, M. S. Phadke, M. Grupe, S. Hillmer, S. T. Liu, and W. Fortney, "Los Angeles Aerometric Carbon Monoxide Data." Technical Report No. 377, Department of Statistics, the University of Wisconsin, Madison, 1974.
7. *Profile of Air Pollution.* Air Pollution Control District, County of Los Angeles, 1971.
8. A. Q. Eschenroeder and J. R. Martinez, "Concepts and applications of photochemical smog models." *Advances in Chemistry*, 113: 101, (1972).
9. S. D. Reynolds, P. M. Roth and J. H. Seinfeld, "Mathematical modeling of photochemical air pollution I. Formulation of the model," *Atmos. Environ.* 7: 1033 (1973).
10. T. E. Graedel, L. A. Farrow, and T. A. Weber, "The influence of aerosols on the chemistry of the troposphere." *Intern. J. Chemical Kinetics* (in press 1975).
11. G. C. Tiao, M. S. Phadke, and G. E. P. Box, "Some Empiric Models for the Los Angeles Photochemical Smog Data." Technical Report No. 412, Department of Statistics, University of Wisconsin-Madison, 1975.
12. B. Elkus and K. Wilson, "Air Basin Pollution Response Function: The Weekend Effect." Technical Report, Department of Chemistry, University of California, San Diego, 1974.

Appendix

Derivation of a Single Box Photochemical Smog Model

In what follows we use notations already defined earlier. Consider a cylindrical box of constant diameter D and variable height H, equal to the effective mixing height at that time. Let the base of the box be centered on the monitoring station at which data are collected. It is assumed that (i) the mixing is instantaneous (ii) the air is incompressible, (iii) there is no diffusion across the boundaries of the box, (iv) the air above the inversion base is clean, and (v) the effective mixing height is given by $H = h^{\alpha}$ where h is the inversion base height and α is a positive constant.

We consider first the rate of change of CO, which is essentially chemically inert, associated with changes in traffic density T, wind speed W and inversion base height h.

Traffic T

For zero W and constant h, the rate of change of CO in the box is proportional to T and inversely proportional to H. Thus

$$\frac{d[CO]}{dt} = k_T T h^{-\alpha} \tag{A.1}$$

where k_T is a constant measuring the CO emission.

Wind Speed W

With zero T and constant h, the rate of change of CO in the box is proportional to W and also to the difference in concentration of CO in the outgoing and incoming air. If we assume, to a first approximation, that the concentration of CO in the incoming air is a constant multiple of that in the outgoing air, then

$$\frac{d[CO]}{dt} = -k_w W[CO] \tag{A.2}$$

where k_w is the wind transport constant.

Inversion Base Height h

Suppose $T = 0$ and $W = 0$. (i) If the inversion base lifts due to heating of the air, then the pollutant is diluted by the clean air such that

$$[CO] \times H = \text{constant}$$

Upon differentiating, we get

$$\frac{d[CO]}{dt} = -h^{-\alpha}\frac{dh^{\alpha}}{dt}[CO] = -\frac{d\log h^{\alpha}}{dt}[CO].$$

(ii) On the other hand, lowering of the inversion base has no influence on CO concentration inside the box.

Thus, both cases can be covered by

$$\frac{d[CO]}{dt} = -I_h\frac{d\log h^{\alpha}}{dt}[CO] \tag{A.3}$$

where I_h is an indicator function defined earlier.

To first-order approximation, the effects of traffic, wind speed, and inversion height can be obtained by linearly combining Eq. (A.1), (A.2), and (A.3) yielding Eq. (5). Eq. (6), (7), and (8) for NO, NO_2, and O_3 are similarly obtained except that we need to account for the facts that (i) these pollutants are chemically reactive, (ii) automobile emission of NO_2 can be ignored, and (iii) O_3 is not emitted by automobiles.

Dr. Tiao is Professor and Chairman of the Department of Statistics, Dr. Phadke and Dr. Box are Professors in the Department of Statistics, University of Wisconsin, Madison, 1210 West Dayton St., Madison, WI 53706. This is a revised version of Paper No. 75-51.5, presented at the 68th Annual Meeting of APCA at Boston in June 1975.

Books and Articles Written by Box

Books

1. *Statistical Methods in Research and Production* (with W. R. Cousins, O. L. Davies, F. R. Himsworth, H. Kenney, M. Milbourn, W. Spendley, and W. L. Stevens). Edinburgh: Oliver and Boyd, 1963.
2. *Design and Analysis of Industrial Experiments* (with L. R. Connor, W. R. Cousins, O. L. Davies, F. R. Himsworth, and G. P. Sillitto). Edinburgh: Oliver and Boyd, 1963.
3. *Evolutionary Operation—A Statistical Method for Process Improvement* (with N. R. Draper). New York: John Wiley & Sons, 1969.
4. *Time Series Analysis Forecasting and Control,* 2nd ed. (with G. M. Jenkins). Oakland, Calif: Holden-Day, 1970.
5. *Bayesian Inference in Statistical Analysis* (with G. C. Tiao). Reading, Mass.: Addison-Wesley, 1973.
6. *Statistics for Experimenters* (with W. G. Hunter and J. S. Hunter). New York: John Wiley & Sons, 1977.

Articles

1. "The effect of exposure to sub-lethal doses of phosgene on the subsequent L(*ct*)50 for rats and mice" (with H. Cullumbine). *British Journal of Pharmacology and Chemotherapy,* vol. 2, no. 1 (1947), pp. 38–55.
2. "The relationship between survival time and dosage with certain toxic agents" (with H. Cullumbine). *British Journal of Pharmacology and Chemotherapy,* vol. 2, no. 1 (1947), pp. 27–37.
3. "A general distribution theory for a class of likelihood criteria." *Biometrika,* vol. XXXVI, parts IXX and IV (1949), pp. 317–346.
4. "Problems in the analysis of growth and wear curves." *Biometrics,* vol. 6, no. 4. (1950), pp. 362–389.
5. "On the experimental attainment of optimum conditions" (with K. B. Wilson). *J. Roy. Stat. Soc.,* Series B, vol. XIII, no. 1 (1951), pp. 1–45.
6. "Multifactorial designs of first order." *Biometrika,* vol. 39 (1952), pp. 49–57.
7. "Plan statistique dans l'etude des methodes de l'analyse chimique." *The Analyst,* vol. 77 (1952), pp. 879–891; Proceedings of the International Congress of Analytical Chemistry (1952), pp. 323–355.
8. "Non-normality and tests on variances." *Biometrika,* vol. 40 (1953), pp. 318–335.
9. "Pigment strength testing with the automatic muller" (with M. T. Hobbs and P. North). *Journal of the Oil and Colour Chemists' Assoc.,* vol. XXXVI, no. 396 (1953), pp. 283–299.

10. "A note on regions for tests of kurtosis." *Biometrika,* vol. 40, parts 3 and 4 (1953), pp. 465–466.
11. "A statistical design for the efficient removal of trends occurring in a comparative experiment with an application in biological assay" (with W. A. Hay). *Biometrics,* vol. 9, no. 3 (1953), pp. 304–319.
12. "The exploration and exploitation of response surfaces: Some general considerations and examples." *Biometrics,* vol. 10, no. 1 (1954), pp.16–60.
13. "A confidence region for the solutions of a set of simultaneous equations with an application to experimental design" (with J. S. Hunter). *Biometrika,* vol. 41, parts 1 and 2 (1954), pp. 190–198.
14. "Some theorems on quadratic forms applied in the study of analysis of variance problems: I. Effect on inequality of variance in the one way classification." *Ann. Math. Stat.,* vol. 25, no. 2 (1954), pp. 290–302.
15. "Some theorems on quadratic forms applied in the study of analysis of variance problems: II. Effects on inequality of variance and of correlation between errors in the two way classification." *Ann. Math. Stat.,* vol. 25, no. 3 (1954), pp. 484–498.
16. "Mathematical statistics and rubber technology." *Imperial Chemical Industries Technical Publication* (1955).
17. "Permutation theory in the derivation of robust criteria and the study of departures from assumptions" (with S. L. Andersen). *J. Roy. Stat. Soc.,* Series B, vol. XVII, part 1 (1955), pp. 1–34.
18. "The exploration and exploitation of response surfaces: An example of the link between the fitted surface and the basic mechanism of the system" (with P. V. Youle). *Biometrics,* vol. 11, no. 3 (1955), pp. 287–323.
19. "Application of digital computers in the exploration of functional relationships" (with G. A. Coutie). *Proc. Inst. Elec. Engrs.,* vol. 103, part B, supplement no. 1 (1956), pp. 100–107.
20. "Evolutionary operation: A method for increasing industrial productivity." *Applied Statistics,* vol. VI, no. 2 (1957), pp. 3–23.
21. "Multifactor experimental designs for exploring response surfaces" (with J. S. Hunter). *Ann. Math. Stat.,* vol. 28, no. 1 (1957), pp. 195–241.
22. "Integration of techniques in process development." *Trans. 11th Annual Convention of American Society for Quality Control* (1957), pp. 687–702.
23. "Use of statistical methods in the elucidation of basic mechanisms." *Bull. Int. Inst. of Stat.,* Stockholm (1957), pp. 215–225.
24. "A note on the generation of random normal deviates" (with M. E. Muller). *Ann. Math. Stat.,* vol. 29, no. 2 (1958), pp. 610–613.
25. "Experimental designs for the exploration and exploitation of response surfaces" (with J. S. Hunter). In *Experimental Designs in Industry,* V. Chew (ed.). New York: John Wiley & Sons (1958), pp. 610–613.
26. "Discussion of the papers of Messrs. Satterthwaite and Budne." *Technometrics,* vol. 1, no. 2 (1959), pp. 174–18C.
27. "Design of experiments in non-linear situations" (with H. L. Lucas). *Biometrika,* vol. 46, parts 1 and 2 (1959), pp. 77–90.
28. "A basis for the selection of a response surface design" (with N. R. Draper). *J. Am. Stat. Assoc.,* vol. 54 (1959), pp. 622–654.
29. "Condensed calculations for evolutionary operation programs" (with J. S. Hunter). *Technometrics,* vol. 1, no. 1 (1959), pp. 77–95.
30. "Some general considerations in process optimization." *J. Basic Engineering* (1960), pp. 113–119.

31. "Simplex-sum designs: A class of second order rotatable designs derivable from those of first order" (with D. W. Behnken). *Ann. Math. Stat.,* vol. 31, no. 4 (1960), pp. 838–864.
32. "Some new three level designs for the study of quantitative variables" (with D. W. Behnken). *Technometrics,* vol. 2, no. 4 (1960), pp. 455–475.
33. "Fitting empirical data." *Ann. N. Y. Acad. of Sciences,* vol. 86, no. 3 (1960), pp. 792–816.
34. "The effects of errors in the factor levels and experimental design." *Bull. International Stat. Inst.,* vol. XXXVII, part III (1961). Reprinted in *Technometrics,* vol. 5, no. 2 (1963), pp. 247–262.
35. "The theory of errors." *Encyclopedia Britannica* (1961).
36. "The 2^{k-p} fractional factorial designs, part I" (with J. S. Hunter). *Technometrics,* vol. 3, no. 3 (1961), pp. 311–351.
37. "The 2^{k-p} fractional factorial designs, part II" (with J. S. Hunter). *Technometrics,* vol. 3, no. 4 (1961), pp. 449–458.
38. "Adaptive optimization of continuous processes" (with J. Chanmugam). *I and EC Fundamentals,* vol. 1 (1962), pp. 2–16.
39. "A useful method for model-building" (with W. G. Hunter). *Technometrics,* vol. 4, no. 3 (1962), pp. 301–318.
40. "Robustness to non-normality regression tests" (with G. S. Watson). *Biometrika,* vol. 49, parts 1 and 2 (1962), pp. 93–106.
41. "Some statistical aspects of adaptive optimization and control" (with G. M. Jenkins). *J. Roy. Stat. Soc.,* Series B, vol. 24, no. 2 (1962), pp. 297–343.
42. "A further look at robustness via Bayes' Theorem" (with G. C. Tiao). *Biometrika,* vol. 49, parts 3 and 4 (1962), pp. 419–432.
43. "Transformation of the independent variables" (with P. W. Tidwell) *Technometrics,* vol. 4, no. 4 (1962), pp. 531–550.
44. "Further contributions to adaptive quality control: Simultaneous estimation of dynamics: Non-zero costs" (with G. M. Jenkins). *Proceedings of the International Statistical Institute* (1963), pp. 943–974.
45. "The choice of a second order rotatable design" (with N. R. Draper). *Biometrika,* vol. 50, parts 3 and 4 (1963), pp. 335–352.
46. "An analysis of transformations" (with D. R. Cox). *J. Roy. Stat. Soc.,* Series B, vol. 26, no. 2 (1964), pp. 211–252.
47. "A Bayesian approach to the importance of assumptions applied to the comparison of variances" (with G. C. Tiao). *Biometrika,* vol. 51, parts 1 and 2 (1964), pp. 153–167.
48. "A note on criterion robustness and inference robustness" (with G. C. Tiao). *Biometrika,* vol. 51, parts 1 and 2 (1964), pp. 169–173.
49. "Sequential design of experiments for non-linear models" (with W. G. Hunter). *Proceedings of IBM Scientific Computing Symposium on Statistics* (1963), pp. 113–137.
50. "A simple system of evolutionary operation subject to empirical feedback." *Technometrics,* vol. 8, no. 1 (1966), pp. 19–26.
51. "The experimental study of physical mechanisms" (with W. G. Hunter). *Technometrics,* vol. 7, no. 1 (1965), pp. 23–42.
52. "A change in level of a non-stationary time series" (with G. C. Tiao). *Biometrika,* vol. 52, parts 1 and 2 (1965), pp. 181–192.
53. "Mathematical models for adaptive control and optimization" (with G. M. Jenkins). *A. I. Ch. E. Joint Meetings* (1965), pp. 4:61–4:68.

54. "The Bayesian estimation of common parameters from several responses" (with N. R. Draper). *Biometrika,* vol. 52, parts 3 and 4 (1965), pp. 355–365.
55. "Multi-parameter problems from a Bayesian point of view" (with G. C. Tiao). *Ann. Math. Stat.,* vol. 36, no. 5 (1965), pp. 1468–1482.
56. "A note on augmented designs." *Technometrics,* vol. 8, no. 1 (1966), pp. 184–188.
57. "A discrete predictor controller applied to sinusoidal perturbation adaptive optimization" (with R. J. Altpeter and K. D. Kotnour). *Instrument Society of America Transactions,* vol. 5, no. 3 (1966), pp. 255–262.
58. "Use and abuse of regression." *Technometrics,* vol. 8, no. 4 (1966), pp. 625–629.
59. "Some aspects of randomization" (with I. Guttman). *J. Roy. Stat. Soc.,* Series B, vol. 28, no. 3 (1966), pp. 543–558.
60. "Discrimination among mechanistic models" (with W. J. Hill). *Technometrics,* vol. 9, no. 1 (1967), pp. 57–71.
61. "Models for forecasting seasonal and non-seasonal time series" (with G. M. Jenkins and D. W. Bacon). In *Spectral Analysis of Time Series,* B. Harris (ed.). New York: John Wiley & Sons (1967), pp. 271–311.
62. "Bayesian analysis of a three-component hierarchical design model" (with G. C. Tiao). *Biometrika,* vol. 54, parts 1 and 2 (1967), pp. 109–125.
63. "Experimental strategy." *Proceedings of 6th International Biometric Conference,* Sydney, Australia (1967).
64. "A Bayesian approach to some outlier problems" (with G. C. Tiao). *Biometrika,* vol. 55, no. 1 (1968), pp. 119–130.
65. "Bayesian estimation of means for the random effect model" (with G. C. Tiao). *J. Amer. Stat. Assoc.,* vol. 63 (1968), pp. 174–181.
66. "Bayesian approaches to some bothersome problems in data analysis." *Proceedings of Seventh Annual Phi Delta Kappa Symposium on Education Research* (1968), pp. 61–101.
67. "Discrete models for forecasting and control." *Encyclopedia of Linguistics, Information, and Control.* Oxford: Pergamon Press (1968), pp. 1–6.
68. "Experimental design: Response surfaces." *International Encyclopedia of the Social Sciences.* New York: Macmillan (1968), pp. 254–259.
69. "The future of department of statistics." Panel discussion chaired by J. W. Tukey in *The Future of Statistics,* D. G. Watts (ed.). New York: Academic Press (1968), pp. 103–137.
70. "Discrete models for feedback and feedforward control" (with G. M. Jenkins). In *The Future Of Statistics,* D. G. Watts (ed.). New York: Academic Press (1968), pp. 201–240.
71. "Some recent advances in forecasting and control" (with G. M. Jenkins). *Applied Statistics,* vol. 17, no. 2 (1968), pp. 91–109.
72. "Isn't my process too variable for EVOP?" (with N. R. Draper). *Technometrics,* vol. 10 (1969), pp. 439–444.
73. "The challenge of statistical computation." In *Statistical Computation* (1969). New York: Academic Press, pp. 3–10.
74. "Distributions of residual autocorrelations in autoregressive-integrated moving average time series models" (with D. A. Pierce). *J. Amer. Stat. Assoc.,* vol. 65, no. 332 (1970), pp. 1509–1526.
75. "Some comments on a paper of Coen, Gomme, and Kendall" (with P. Newbold). *J. Roy. Stat. Soc.,* Series A, vol. 134, no. 2 (1971), pp. 229–240.

76. "Statistical techniques for mechanistic modelling" (with W. G. Hunter). *Proceedings of Second International Symposium on Chemical Reaction Engineering,* Amsterdam (1972), pp. B 4.9–4.19.
77. "Partial autocorrelations from a Bayesian viewpoint and orthogonal parameterization" (with G. M. Jenkins and I. Guttman). *METRON,* vol. XXX — N.1–4, 31-XII (1972), pp. 87–112.
78. "Some problems associated with the analysis of multiresponse models" (with W. G. Hunter, J. Erjavec, and J. F. MacGregor). *Technometrics,* vol. 15, no. 1 (1973), pp. 33–51.
79. "Some comments on Bayes' estimators" (with G. C. Tiao). *The American Statistician,* vol. 27, no. 1 (1973), pp. 12–14.
80. "Some comments on a paper by Chatfield and Prothero and on a review by Kendall" (with G. M. Jenkins). *J. Roy. Stat. Soc.,* Series A, vol. 136, part 3 (1973), pp. 337–352.
81. "Some recent advances in forecasting and control, Part II" (with G. M. Jenkins and J. F. MacGregor). *Applied Statistics,* vol. 23, no. 2 (1974), pp. 158–179.
82. "Statistics and the environment." *J. Wash. Acad. Sci.,* vol. 64, no. 2 (1974), pp. 52–59.
83. "Correcting inhomogeneity of variance with power transformation weighting" (with W. J. Hill). *Technometrics,* vol. 16, no. 3 (1974), pp. 385–389.
84. "The analysis of closed-loop dynamic-stochastic systems" (with J. F. MacGregor). *Technometrics,* vol. 16, no. 3 (1974), pp. 391–398.
85. "Analysis of Los Angeles photochemical smog data: A statistical overview" (with G. C. Tiao and W. J. Hamming). *APCA Journal,* vol. 25, no. 3 (1975), pp. 260–268.
86. "Intervention analysis with applications to economic and environmental problems" (with G. C. Tiao). *J. Amer. Stat. Assoc.,* vol. 70, no. 349 (1975), pp. 70–79.
87. "Parameter estimation for dynamic-stochastic models using closed-loop operating data" (with J. F. MacGregor). Invited address, *Proceedings of Sixth Triennial World Congress of International Federation of Automatic Control,* Boston, Mass., August (1975). Reprinted *Technometrics,* vol. 18, no. 4 (1976), pp. 371–380.
88. "Robust designs" (with N. R. Draper). *Biometrika,* vol. 62, no. 2 (1975), pp. 347–352.
89. "A statistical analysis of the Los Angeles ambient carbon monoxide data 1955–1972" (with G. C. Tiao and W. J. Hamming). *APCA Journal,* vol. 25, no. 11 (1975), pp. 1129–1136.
90. "Some empirical models for the Los Angeles photochemical smog data" (with G. C. Tiao and M. S. Phadke). *APCA Journal,* vol. 26, no. 5 (1976), pp. 485–490.
91. "Comparison of forecasts and actuality" (with G. C. Tiao). *Applied Statistics,* vol. 25, no. 3 (1975), pp. 195–200.
92. "Science and statistics." *J. Amer. Stat. Assoc.,* vol. 71, no. 356 (1976), pp. 791–799.
93. "Comment on 'Strong inconsistency from uniform priors' by M. Stone" (with G. C. Tiao). *J. Amer. Stat. Assoc.,* vol. 71, no. 353 (1976), p. 122.
94. "Identification of dynamic regression (distributed lag) models connecting two time series" (with L. D. Haugh). *J. Amer. Stat. Assoc.,* vol. 72, no. 357 (1977), pp. 121–130.

95. "Empirical-mechanistic modeling of air pollution" (with M. S. Phadke and G. C. Tiao). *Proceedings of 4th Symposium on Statistics and the Environment* (1976), pp. 91-101.

96. "Analysis and modeling of seasonal time series" (with S. C. Hillmer and G. C. Tiao). *Proceedings of Conference on Seasonal Analysis of Economic Time Series* (1976), pp. 309-344.

97. "A canonical analysis of multiple time series" (with G. C. Tiao). *Biometrika,* vol. 64, no. 2 (1977), pp. 355-365.

98. "On a measure of lack of fit in time series models" (with G. Ljung). *Biometrika,* vol. 65, no. 2 (1978), pp. 297-303.

99. "Deterministic and forecast-adaptive time-dependent models" (with B. Abraham). *Applied Statistics,* vol. 27, no. 3 (1978), pp. 120-130.

100. "Applications of time series analysis" (with G. C. Tiao). *Contributions to Survey Sampling and Applied Statistics,* papers in honor of H. O. Hartley, H. A. David (ed.). New York: Academic Press (1978), pp. 203-219.

101. "Conditions for the optimality of exponential smoothing forecasts procedures" (with J. Ledolter). *Metrika,* vol. 25 (1978), pp. 77-93.

102. "Linear models and spurious observations" (with B. Abraham). *Applied Statistics,* vol. 27, no. 2 (1978), pp. 131-138.

103. "Sampling interval and feedback control" (with B. Abraham). *Technometrics,* vol. 21, no. 1 (1979), pp. 1-8.

104. "The likelihood function of stationary autoregressive-moving average models" (with G. Ljung). *Biometrika,* vol. 66, no. 2 (1979), pp. 265-270.

105. "Some problems of statistics and everyday life." *J. Amer. Stat. Assoc.,* vol. 74, no. 365 (1979), pp. 1-4.

106. "Robustness in the strategy of scientific model building." In *Robustness in Statistics.* New York: Academic Press (1979), pp. 229-236.

107. "Bayesian analysis of some outlier problems in time series" (with B. Abraham). *Biometrika,* vol. 66, no. 2 (1979), pp. 229-236.

108. "The variance function of the difference between two estimated responses" (with N. Draper). *J. Roy. Stat. Soc.,* Series B, vol. 42, no. 1 (1980), pp. 79-82.

109. "Sampling and Bayes' inference in scientific modelling and robustness." *J. Roy. Stat. Soc.,* Series A, vol. 143, part 4 (1980), pp. 383-430.

110. "Analysis of variance and autocorrelated errors" (with G. Ljung). *Scandinavian Journal of Statistics,* vol. 7, part 4 (1980), pp. 172-180.

111. "Sampling inference, Bayes' inference, and robustness in the advancement of learning." In *Bayesian Statistics,* J. M. Bernardo, M. H. DeGroot, D. V. Lindley, and A. F. M. Smith (eds.) (1980), pp. 366-381.

112. "Modelling multiple time series with applications" (with G. C. Tiao). *J. Amer. Stat. Assoc.,* vol. 76, no. 376 (1981), pp. 802-816.

113. "Measures of lack of fit for response surface designs and predictor variable transformations" (with N. R. Draper). *Technometrics,* vol. 24, no. 1 (1982), pp. 1-8.

114. "An analysis of transformations revisited, rebutted" (with D. R. Cox). *J. Amer. Stat. Assoc.,* vol. 77, no. 377 (1982), pp. 209-210.

115. "Choice of response surface design and alphabetic optimality." *Utilitas Mathematica,* vol. 21B (1982), pp. 11-55.

116. "An apology for ecumenism in statistics." In *Scientific Inference, Data Analysis, and Robustness,* G. E. P. Box, Tom Leonard, Chien-Fu Wu (eds). New York: Academic Press (1983), pp. 51-84.

117. "The importance of practice in the development of statistics." To appear in *Technometrics,* Feb. 1984.

118. "Anatomy of time series models." To appear in *Statistics: An Appreciation,* H. A. David and H. T. David (eds.). Ames, Iowa: Iowa State University Press.

119. "Constrained nonlinear least squares" (with H. Kanemasu). To appear in *Contributions to Experimental Design Linear Models, and Genetic Statistics.* Essays in honor of Oscar Kempthorne. New York: Marcel Dekker (1984).

120. "Gwilym Jenkins, experimental design and time series." (1983). Invited address, *First Catalan Symposium on Statistics,* with special time series sessions dedicated to the memory of G. M. Jenkins (1983).

Acknowledgments (continued from page iv)

STATISTICS. Articles 5.1, 5.2 are reprinted with permission from MACMILLAN PRESS LIMITED. Article 4.6 is reprinted with permission from the NEW YORK ACADEMY OF SCIENCES. Articles 3.4, 3.8 from *Applied Statistics* and 3.1, 3.6, 4.9 from the *Journal of the Royal Statistical Society* are reprinted with permission from the ROYAL STATISTICAL SOCIETY. Article 3.3 is reprinted with the permission of JOHN WILEY AND SONS, INCORPORATED.